HIV/AIDS AND THE NERVOUS SYSTEM

HANDBOOK OF CLINICAL NEUROLOGY

Series Editors

MICHAEL J. AMINOFF, FRANÇOIS BOLLER AND DICK F. SWAAB

VOLUME 85

ELSEVIER

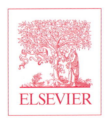

EDINBURGH LONDON NEW YORK OXFORD PHILADELPHIA
ST LOUIS SYDNEY TORONTO 2007

HIV/AIDS AND THE NERVOUS SYSTEM

Series Editors

MICHAEL J. AMINOFF, FRANÇOIS BOLLER AND DICK F. SWAAB

Volume Editors

PETER PORTEGIES AND JOSEPH R. BERGER

VOLUME 85

3rd Series

ELSEVIER

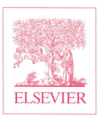

EDINBURGH LONDON NEW YORK OXFORD PHILADELPHIA
ST LOUIS SYDNEY TORONTO 2007

ELSEVIER B.V.
Radarweg 29, 1043 NX, Amsterdam, The Netherlands

First published 2007

ISBN: 978 0 444 52010 4

British Library Cataloguing in Publication Data
A catalogue record for this book is available from the British Library

Library of Congress Cataloging in Publication Data
A catalog record for this book is available from the Library of Congress

Notice

Knowledge and best practice in this field are constantly changing. As new research and experience broaden our knowledge, changes in practice, treatment and drug therapy may become necessary or appropriate. Readers are advised to check the most current information provided (i) on procedures featured or (ii) by the manufacturer of each product to be administered, to verify the recommended dose or formula, the method and duration of administration, and contraindications. It is the responsibility of the practitioner, relying on their own experience and knowledge of the patient, to make diagnoses, to determine dosages and the best treatment for each individual patient, and to take all appropriate safety precautions. To the fullest extent of the law, neither the Publisher nor the Editors and Authors assume any liability for any injury and/or damage to persons or property arising out or related to any use of the material contained in this book.

The Publisher

Printed in China

For Elsevier:
Commissioning Editor: Michael Parkinson
Development Editor: Lynn Watt
Project Manager: Anne Dickie
Design Direction: George Ajayi

The publisher's policy is to use **paper manufactured from sustainable forests**

Handbook of Clinical Neurology 3rd Series

Available titles

The human hypothalamus: basic and clinical aspects, Part I, vol. 79, D.F. Swaab ISBN 0444513574
The human hypothalamus: basic and clinical aspects, Part II, vol. 80, D.F. Swaab ISBN 0444514902
Pain, vol. 81, F. Cervero and T.S. Jensen, eds ISBN 0444519017
Motor neuron disorders and related diseases, vol. 82, A.A. Eisen and P.J. Shaw, eds, ISBN 0444518940
Parkinson's disease and related disorders, Part I, vol. 83, W.C. Koller and E. Melamed, eds, ISBN 0444519009
Parkinson's disease and related disorders, Part II, vol. 84, W.C. Koller and E. Melamed, eds, ISBN 0444528938
Myopathies, vol. 86, F. Mastaglia and D. Hilton-Jones, eds, ISBN 0444518991

Forthcoming titles

Malformations of the nervous system, vol. 87, H. B. Sarnat and P. Curatolo, eds, ISBN 0444518967

Foreword

We are proud at the publication of this seventh volume in the third series of the Handbook of Clinical Neurology. The Handbook was originally conceived by Pierre Vinken and George Bruyn, and the first volume was published in 1968. This new — third — series started in 2004 and covers advances in traditional areas of clinical neurology and the neurosciences in subjects embraced by earlier volumes, as well as new contemporary topics that were not previously considered. The neurobiological aspects of the nervous system in health and disease are included in order to clarify physiological and pathogenic mechanisms and to provide new therapeutic strategies for neurological disorders. Data related to epidemiology, imaging, genetics, and therapy are also emphasized in the new series.

The present volume, edited by Peter Portegies and Joseph R. Berger, is the first Handbook volume dedicated to HIV infection and neuro-AIDS and has all the characteristics we wanted in the series. AIDS was first described in 1981 and was soon found to be accompanied by serious complications, such as opportunistic infections, central nervous system tumors, spinal cord disorders, peripheral neuropathies, myopathies, and a progressive encephalopathy. In response to advances in molecular biology and the advent of other investigative techniques, understanding of the basic mechanisms of the disease and the development of therapeutic strategies for the disorder grew very rapidly, and now, almost 25 years later, the outlook for HIV-infected patients has totally changed in Western countries: patients on antiretroviral therapy may have prolonged survival with few or no complications. However, due to financial and political constraints, AIDS is still a deadly disease in other parts of the world, such as Africa.

Portegies and Berger are to be congratulated on bringing together such an excellent group of experts in this truly multidisciplinary volume, which includes chapters on neuroepidemiology, third-world characteristics of the disease, neuropharmacology, clinical neurology, neuropsychology, neuroradiology, neuropathology, neurophysiology, neurovirology, neuropathogenesis, and animal models of HIV-related neurological diseases. As series editors we are also very grateful to the authors of the chapters and to the editorial staff of Elsevier BV, and particularly to Lynn Watt and Michael Parkinson in Edinburgh, who have all provided invaluable assistance.

Michael J. Aminoff
François Boller
Dick F. Swaab

List of contributors

R. J. Baumann
Department of Neurology, University of Kentucky, Lexington, KY, USA

J. R. Berger
Department of Neurology, University of Kentucky, Lexington, KY, USA

G. L. Birbeck
Departments of Neurology, Epidemiology & African Studies, Michigan State University, East Lansing, MI, USA and Chikankata Health Services, Mazabuka, Zambia

B. J. Brew
Department of Neurology, St Vincent's Hospital, University of New South Wales, Sydney, Australia

M. C. Brouwer
Department of Neurology, Academic Medical Center, University of Amsterdam, The Netherlands

J. L. Buescher
The Center for Neurovirology and Neurodegenerative Disorders, Department of Pharmacology, Nebraska Medical Center, Omaha, NE, USA

A. T-A. Chan
The Mount Sinai Medical Center, New York, NY, USA

L. M. Chirch
Stony Brook University School of Medicine, Stony Brook, NY, USA

P. Cinque
Clinic of Infectious Diseases, San Raffaele Scientific Institute, Milan, Italy

B. A. Cohen
Feinberg School of Medicine, Northwestern University, Chicago, IL, USA

S. Dawes
HIV Neurobehavioral Research Center, University of California, San Diego, CA, USA

A. Di Rocco
New York University School of Medicine, New York, NY, USA

P. S. Espinosa
Department of Neurology, University of Kentucky, Lexington, KY, USA

L. B. Estanislao
The Mount Sinai Medical Center, New York, NY, USA

B. B. Gelman
Texas NeuroAIDS Research Center, University of Texas Medical Branch, Galveston, TX, USA

H. E. Gendelman
The Center for Neurovirology and Neurodegenerative Disorders, Department of Pharmacology, Nebraska Medical Center, Omaha, NE, USA

M. Gisslen
Department of Infectious Diseases, Göteborg University, Sahlgrenska University Hospital, Göteborg, Sweden

C. A. Given II
Department of Radiology, University of Kentucky Chandler Medical Center, Lexington, KY, USA

I. Grant
HIV Neurobehavioral Research Center, University of California, San Diego, CA, USA

S. Gross
The Center for Neurovirology and Neurodegenerative Disorders, Department of Pharmacology, Nebraska Medical Center, Omaha, NE, USA

L. Hagberg
Department of Infectious Diseases, Göteborg University, Sahlgrenska University Hospital, Göteborg, Sweden

S. A. Houff
Department of Neurology, University of Kentucky, Lexington, KY, USA

T. Ikezu
The Center for Neurovirology and Neurodegenerative Disorders, Department of Pharmacology, Nebraska Medical Center, Omaha, NE, USA

A. Jackson
University College Dublin, Mater University Hospital, Dublin, Ireland

C. Kirton
The Mount Sinai Medical Center, New York, NY, USA

B. J. Luft
Stony Brook University School of Medicine, Stony Brook, NY, USA

E. O. Major
Laboratory of Molecular Medicine and Neuroscience, NINDS, NIH, Bethesda, MD, USA

P. Portegies
Department of Neurology, OLVG Hospital, Amsterdam, The Netherlands

W. G. Powderly
University College Dublin, Mater University Hospital, Dublin, Ireland

R. W. Price
Department of Neurology, University of California, San Francisco General Hospital, San Francisco, CA, USA

E. Richard
Department of Neurology, Academic Medical Center, University of Amsterdam, Amsterdam, The Netherlands

D. M. Simpson
The Mount Sinai Medical Center, New York, NY, USA

S. Verma
The Mount Sinai Medical Center, New York, NY, USA

Contents

Handbook of Clinical Neurology, Vol. 85 (3rd series)
HIV/AIDS and the Nervous System
P. Portegies, J. R. Berger, Editors

Chapter 1

Introduction to HIV infection and neuro-AIDS

PETER PORTEGIES* AND JOSEPH R. BERGER

OLVG Hospital, Amsterdam, The Netherlands and University of Kentucky, College of Medicine, Lexington, KY, USA

The acquired immune deficiency syndrome (AIDS), caused by infection with the retrovirus known as human immunodeficiency virus (HIV), was first described in 1981. Not long afterwards, reports began to appear of neurological complications, including an unusual encephalopathy, in affected patients. In the ensuing years, the nature and spectrum of the neurological complications occurring in these immunosuppressed patients became better defined. These disorders included opportunistic infections rarely seen in immunologically healthy individuals, CNS tumors, spinal cord disorders, peripheral neuropathies, myopathies and a progressive encephalopathy. Some of the opportunistic infections could be treated well; others had a very poor prognosis, partly due to the fact that the underlying immunodeficiency could not be treated. The face of AIDS changed when antiretroviral drugs were developed. Zidovudine was introduced in 1986, and combination therapy of two or three drugs became available in the early 1990s. In patients on highly active antiretroviral therapy (HAART) morbidity and mortality dropped dramatically. The incidence of many of the opportunistic infections, including the neurological infections, decreased. For some of the neurological complications that were thought to be more directly related to HIV itself, HAART was beneficial (dementia); for others, such as AIDS-related peripheral neuropathy and myelopathy, the influence of treatment was limited, at best.

Almost 25 years later HIV infection has evolved into a different disease. Patients on HAART live normal lives, and have little or no complications of their infection. Untreated patients are still at risk for developing a severe immunodeficiency with all the well-known complications, including the neurological ones. The approach to these neurological complications did not change; the treatment for many of them is still the same as in the early 1980s. However, for clinical neurologists the HIV-infected patient with neurological disease is a diagnostic and therapeutic enigma. The challenges arise because these patients often present in an atypical fashion, exhibit unpredictable reactions to HAART and specific therapies, and often have more than one neurological complication developing simultaneously. Furthermore, the neurological disease may be entirely unrelated to the underlying HIV infection. Therefore, the problems in these patients remain complex and need neurologists with knowledge of and expertise in HIV infection.

1.1. The epidemic in 2006

The global infection is not under control. The global burden of the disease remains enormous. More than 40 million people worldwide are infected today. Still there are almost 5 million new infections each year and in 2005 the estimated death toll was 3.1 million people [AIDS Epidemic Update: December 2005, available online at http://data.unaids.org/pub/EpiReport/2006/2006_EpiUpdate_en.pdf]. Only a small fraction of patients has access to antiretroviral treatment. Despite the efforts of many (non-governmental organizations, state governments, WHO, UNAIDS), the hurdles we face in providing adequate care to those infected continue to be extraordinarily challenging. This is particularly true in areas of the world where the pandemic has been especially devastating, such as sub-Saharan Africa. Epidemiological studies regarding the burden of AIDS-related neurological disease from large parts of the world are hardly available and neurological care, in general, and neuro-AIDS expertise, specifically, is very limited in many countries where HIV infection has reached epidemic proportions. An up-to-date report of the

*Correspondence to: Peter Portegies, MD, PhD, Department of Neurology, OLVG Hospital, PO Box 95500, 1090 HM Amsterdam, The Netherlands. E-mail: p.portegies@olvg.nl, Tel: +31-20-599-3044, Fax: +31-20-599-3845.

epidemiology of HIV infection focused on the neurological complications of AIDS and the situation in underdeveloped and developing countries is discussed in depth in this book.

1.2. Neuro-AIDS in 1985 and 2005

Since the beginning of the HIV epidemic approximately 10% of patients have presented with a neurological complication as the first manifestation of their HIV infection, 40–60% develop neurological disease in the course of their disease and most patients (80%) have neuropathological abnormalities at autopsy. As similar studies have not been performed in the recent past, particularly since the introduction of HAART, it is unclear whether these percentages still pertain. Clearly, in regions of the world where people have access to HAART, the incidence of many of the common neurological complications of AIDS has decreased. For instance, as with the non-neurological opportunistic infections, the CNS infections of cerebral toxoplasmosis and cryptococcal meningitis have become rare in patients with normal CD^4-cell counts. HAART may also prevent the development of primary central nervous system lymphoma (PCNSL) and progressive multifocal leukoencephalopathy (PML) in those individuals responding to the medications. Similarly, the classic picture of HIV-dementia has declined dramatically, although cognitive abnormalities still develop even in the face of normal immune function. Likewise, despite HAART, peripheral neuropathy is still a frequent complication.

The treatment of patients who present with a neurological disease is not uncomplicated. Specific treatment is available for some, particularly opportunistic infections, such as toxoplasmosis and cryptococcal meningitis; however, for other conditions there is no treatment of demonstrated value, such as PML. The recovery of immunological impairment that attends the initiation of HAART is generally beneficial, but sometimes the immune reconstitution inflammatory syndrome (IRIS) develops resulting in temporary worsening of the patient's condition or, in some instances, contributing to their death. When no specific treatment is available, HAART with its associated immune recovery is enough to stabilize or reverse a neurological disorder as may be seen in some patients with AIDS-related PML. An explanation for the success of such treatment in some patients and the failure in others remains unclear. The impact of HAART on the various neurological complications of AIDS is widely discussed in the following chapters.

1.3. What have we learned from HIV and its interaction with the nervous system?

Our understanding of the nature and pathogenesis of the neurological complications of HIV, and, for that matter, of viral illness of the nervous system, has increased dramatically in the past 25 years. We know that neurological disease in HIV-infected persons is common, though increasingly less so in the face of treatment with HAART. These illnesses are a cause of significant morbidity and mortality. No part of the neuraxis is invulnerable to the HIV-related disease; the brain, spinal cord, peripheral nerves and muscles may all be affected. Many of the neurological diseases that have been described since the recognition of AIDS are unique. Often, the clinical picture is confusing; patients may present with two or more neurological disorders that have developed simultaneously and expected findings, such as headache and neck stiffness with cryptococcal meningitis or focal deficits with large CNS mass lesions from toxoplasmosis or PCNSL, are absent. Therefore, we have approached these patients with greater diagnostic aggressiveness and are more inclined to perform neuroimaging studies and cerebrospinal fluid analysis than we might otherwise have done. We have also increasingly used new techniques, like polymerase chain reaction, in establishing diagnoses. We have also learned that the CNS may serve as a reservoir for HIV; one that may be relatively protected from otherwise effective antiretroviral therapy and one from which the virus may reseed the body.

We have learnt a great deal of the interaction of HIV and the nervous system and of the pathogenesis of CNS viral infection, in general, such as the role of microglia cells, cytokines and chemokines in the defense and damage of the CNS in viral infection have been studied extensively.

This volume provides an in-depth overview of the neurology of HIV infection as at 2006. We have attempted to make this volume as comprehensive as possible with all aspects of neurology of AIDS covered. Among the chapters are those dealing with clinical neurology, neuroradiology, neuropathology, cerebrospinal fluid, neurophysiology, neurovirology, neuropathogenesis and animal models of HIV-related neurological disease.

Handbook of Clinical Neurology, Vol. 85 (3rd series)
HIV/AIDS and the Nervous System
P. Portegies, J. R. Berger, Editors

Chapter 2

Neuroepidemiology of HIV/AIDS

ROBERT J. BAUMANN* AND PATRICIO S. ESPINOSA

Department of Neurology, University of Kentucky, Lexington, KY, USA

2.1. Introduction

Human immunodeficiency virus (HIV) infection and acquired immunodeficiency syndrome (AIDS) have developed as one of our most important public health problems. Current estimates are that 38 million people are living with HIV worldwide. In 2003 annual mortality was estimated at 2.9 million with over 20 million deaths having occurred since the first cases of AIDS were identified in 1981 (UNAIDS, 2004). Also in 2003, an estimated 4.8 million people became newly infected with HIV (UNAIDS, 2004). According to the Centers for Disease Control and Prevention (CDC), in 2003 an estimated 1,039,000 to 1,185,000 persons in the USA were living with HIV/AIDS. The CDC estimates that approximately 40,000 persons become infected with HIV each year (CDC, 2005).

2.1.1. The origin of HIV

It is believed that HIV was present in rural Africa for decades before the USA outbreak in the 1980s. Serum from residents of rural Zaire obtained in the 1970s provides evidence of the existence of this infection before its identification as a specific disorder (Nemeth et al., 1986; Nzilambi et al., 1988). In June 1981 the CDC reported five homosexual men in Los Angeles, California, who developed *Pneumocystis carinii* and later in the same month 20 homosexual men in New York were diagnosed with Kaposi sarcoma, marking the beginning of the world epidemic (CDC, 1981).

2.1.2. Genetic diversity of HIV

HIV, a retrovirus, infects humans causing impairment of the ability of the immune system to battle and prevent disease (UNAIDS, 2004). Two viral types have been described in humans, HIV-1 and HIV-2, both of which originated in Africa. HIV-1 has now become dispersed around the world while HIV-2 is primarily concentrated in Africa. HIV-1 virus has 3 subgroups: M, O and N; the M subgroup has several subtypes: A, B, C, D, E, F, G, H, J and K. One epidemiological feature of these subtypes is that dual infection has been reported between types and subtypes; however, clinically, they are impossible to differentiate.

2.1.3. Case definition

The case definition for HIV infection, updated in December 1999 by the CDC, applies to both HIV-1 and HIV-2 infections (Table 2.1) (CDC, 1999). It is the basis of the USA HIV case surveillance effort to categorize, track and control this disease, though elsewhere in the world case definitions can vary.

2.1.4. Mode and trend of transmission of HIV

HIV can be transmitted by sexual contact, parenteral inoculation and perinatally from mother to child. Other categories of transmission such as insect bites, saliva, tears and casual contact (hand shake, sharing clothes, etc.) have not been proven (Lifson, 1988).

Primary transmission is by sexual contact with epidemiological evidence suggesting anal intercourse as the most efficient mode of transmission (Biggar & Rosenberg, 1993; Folch et al., 2005). Since the onset of the epidemic the group with highest risk of HIV infection have been men who have sex with men (CDC, 1981, 2003). Surveillance reports show that most reported cases of HIV infection in the USA occur among men who have sex with men, followed by het-

*Correspondence to: Robert J. Baumann, MD, Professor of Neurology and Pediatrics, Director, Child Neurology Program, Department of Neurology, University of Kentucky, Kentucky Clinic L 409, Speed Sort 0284, Lexington, KY 40536-0284, USA. E-mail: Baumann@uky.edu, Tel: +1-859-323-6702. Ext 242.

Table 2.1

Case definition for HIV infection

I. In adults, adolescents, or children aged greater than or equal to 18 months,[a] a reportable case of HIV infection must meet at least one of the following criteria:

Laboratory Criteria:

a) Positive result on a screening test for HIV antibody (e.g., repeatedly reactive enzyme immunoassay), followed by a positive result on a confirmatory (sensitive and more specific) test for HIV antibody (e.g., Western blot or immunofluorescence antibody test)

b) Positive result or report of a detectable quantity on any of the following HIV virologic (non-antibody) tests:
- HIV nucleic acid (DNA or RNA) detection (e.g., DNA polymerase chain reaction [PCR] or plasma HIV-1 RNA)[b]
- HIV p24 antigen test, including neutralization assay
- HIV isolation (viral culture)

Clinical or Other Criteria (if the above laboratory criteria are not met)

a) Diagnosis of HIV infection, based on the laboratory criteria above, that is documented in a medical record by a physician.

b) Conditions that meet criteria included in the case definition for AIDS

II. In a child aged less than 18 months, a reportable case of HIV infection must meet at least one of the following criteria:

Laboratory Criteria:

Definitive:

a) Positive results on two separate specimens (excluding cord blood) using one or more of the following HIV virologic (non-antibody) tests:
- HIV nucleic acid (DNA or RNA) detection
- HIV p24 antigen test, including neutralization assay, in a child greater than or equal to 1 month of age
- HIV isolation (viral culture)

Presumptive: A child who does not meet the criteria for definitive HIV infection but who has:

b) Positive results on only one specimen (excluding cord blood) using the above HIV virologic tests and no subsequent negative HIV virologic or negative HIV antibody tests.

Clinical or Other Criteria (if the above laboratory criteria are not met)

a) Diagnosis of HIV infection, based on the laboratory criteria above, that is documented in a medical record by a physician.

b) Conditions that meet criteria included in the 1987 pediatric surveillance case definition for AIDS

III. A child aged less than 18 months born to an HIV-infected mother will be categorized for surveillance purposes as "not infected with HIV" if the child does not meet the criteria for HIV infection but meets the following criteria:

Laboratory Criteria:

Definitive:

a) At least two negative HIV antibody tests from separate specimens obtained at greater than or equal to 6 months of age

or

b) At least two negative HIV virologic tests[c] from separate specimens, both of which were performed at greater than or equal to 1 month of age and one of which was performed at greater than or equal to 4 months of age

AND

c) No other laboratory or clinical evidence of HIV infection (i.e., has not had any positive virologic tests, if performed, and has not had an AIDS-defining condition)

Presumptive: A child who does not meet the above criteria for definitive "not infected" status but who has:

a) One negative EIA HIV antibody test performed at greater than or equal to 6 months of age and NO positive HIV virologic tests, if performed

or

b) One negative HIV virologic test[c] performed at greater than or equal to 4 months of age and NO positive HIV virologic tests, if performed

or

c) One positive HIV virologic test with at least two subsequent negative virologic tests,[d] at least one of which is at greater than or equal to 4 months of age; or negative HIV antibody test results, at least one of which is at greater than or equal to 6 months of age

AND

d) No other laboratory or clinical evidence of HIV infection (i.e., has not had any positive virologic tests, if performed, and has not had an AIDS-defining condition).

Clinical or Other Criteria (if the above laboratory criteria are not met)

a) Determined by a physician to be "not infected", and a physician has noted the results of the preceding HIV diagnostic tests in the medical record

AND

b) No other laboratory or clinical evidence of HIV infection (i.e., has not had any positive virologic tests, if performed, and has not had an AIDS-defining condition).

IV. A child aged less than 18 months born to an HIV-infected mother will be categorized as having perinatal exposure to HIV infection if the child does not meet the criteria for HIV infection (II) or the criteria for "not infected with HIV" (III).

Adapted from CDC, 1999, December 10. Report No.: MMWR 48(RR13);29–31.

Revised Surveillance Case Definition for HIV Infection available online: http://www.cdc.gov/mmwr/preview/mmwrhtml/rr4813a2.htm

[a]Children aged greater than or equal to 18 months but less than 13 years are categorized as "not infected with HIV" if they meet the criteria in III.

[b]In adults, adolescents, and children infected by other than perinatal exposure, plasma viral RNA nucleic acid tests should NOT be used in lieu of licensed HIV screening tests (e.g., repeatedly reactive enzyme immunoassay). In addition, a negative (i.e., undetectable) plasma HIV-1 RNA test result does not rule out the diagnosis of HIV infection.

[c]Draft revised surveillance criteria for HIV infection were approved and recommended by the membership of the Council of State and Territorial Epidemiologists (CSTE) at the 1998 annual meeting (11).

[d]HIV nucleic acid (DNA or RNA) detection tests are the virologic methods of choice to exclude infection in children aged less than 18 months. Although HIV culture can be used for this purpose, it is more complex and expensive to perform and is less well standardized than nucleic acid detection tests. The use of p24 antigen testing to exclude infection in children aged less than 18 months is not recommended because of its lack of sensitivity.

erosexual contact and injection drug use (IDU). In the pediatric population most HIV infection results from mother to infant transmission in the perinatal period (CDC, 2003) (Table 2.2). In Africa, the predominant mode of transmission of HIV is through heterosexual contact; 57% of infected adults are women in sub-Saharan Africa. This trend can be explained in part by the tendency of young African women to have male partners who are older than themselves. These male partners are more likely to be infected with HIV (UNAIDS, 2004). Another factor is gender inequality. Some African women are the victims of domestic and sexual violence which can damage the genital mucosa increasing the risk of HIV transmission. In a South African study high levels of male control in a woman's relationship was associated with an increased risk of HIV seropositivity with a 95% CI: 1.52 (1.13–2.04) (Dunkle et al., 2004).

2.1.5. Natural history of HIV infection

The HIV virus can be detected as early as two weeks after infection and HIV antibodies can be detected within four to six weeks (Allain et al., 1986). In a group of 12 men who have sex with men in whom seroconversion to HIV infection was recorded, 11 subjects developed an acute infectious-mononucleosis-like illness. The illness was of sudden onset, lasted from 3 to 14 days and was associated with fevers, sweats, malaise, lethargy, anorexia, nausea, myalgia, arthralgia, headaches, sore throat, diarrhea, generalized lymphadenopathy, a macular erythematous rash and thrombocytopenia (Cooper et al., 1985). The seroconversion period can vary. While one subject seroconverted 6 days after presumed exposure to HIV, 3 other subjects seroconverted 19, 32 and 56 days after onset (Cooper et al., 1985).

The duration of the incubation period can be influenced by factors such as age and comorbid condition. In a cohort study among persons with hemophilia, children developed AIDS more slowly than adults ($P = 0.02$) and hemophilic adults developed AIDS more slowly than homosexual men ($P < 0.05$). Progression time to AIDS was significantly faster in older hemophilic men, but not in older homosexual men. The authors suggest that these differences may be related to different risks of exposure to potential opportunistic pathogens (Biggar, 1990). In a cohort of 84 men who have sex with men and bisexual men a model for the incubation period for AIDS and for the proportion likely to develop AIDS showed a maximum likelihood estimate for the mean incubation period for AIDS in homosexual men of 7.8 years (90% CI: 4.2 to 15.0 years). This study also showed that a maximum likelihood estimate for the proportion of HIV-infected homosexual men developing AIDS was 0.99 (90% CI: 0.38 to 1) (Lui et al., 1988).

Table 2.2

Reported cases of HIV infection (not AIDS), by age category, transmission category, and sex, cumulative through 2003 – 41 areas with confidential name-based HIV infection reporting

	Cumulative through 2003[a]	
Category of transmission	No.	%
Adult or adolescent		
Male-to-male sexual contact	72745	34
Injection drug use	31133	14
Male-to-male sexual contact and injection drug use	8623	4
Hemophilia/coagulation disorder	584	0
Heterosexual contact:	41152	19
Sex with injection drug user	8173	4
Sex with bisexual male	1757	1
Sex with person with hemophilia	199	0
Sex with HIV-infected transfusion recipient	291	0
Sex with HIV-infected person, risk factor not specified	30732	14
Receipt of blood transfusion, blood components, or tissue	1016	0
Other/risk factor not reported or identified	61233	28
Child (<13 years at diagnosis)		
Hemophilia/coagulation disorder	108	2
Mother with the following risk factor for, or documented, HIV infection:	3898	85
Injection drug use	1034	23
Sex with injection drug user	393	9
Sex with bisexual male	44	1
Sex with person with hemophilia	9	0
Sex with HIV-infected transfusion recipient	11	0
Sex with HIV-infected person, risk factor not specified	905	20
Receipt of blood transfusion, blood components, or tissue	33	1
Has HIV infection, risk factor not specified	1469	32
Receipt of blood transfusion, blood components, or tissue	51	1
Other/risk factor not reported or identified	522	11

Adapted from CDC Surveillance Report 2003, 2003.
[a]Includes persons with a diagnosis of HIV infection (not AIDS), reported from the beginning of the epidemic through December 2003. Cumulative total includes seven persons of unknown sex.

2.1.6. The HAART era

Since the outbreak of the epidemic in the early 1980s the numbers of HIV-infected persons has steadily increased. In contrast, the mortality from HIV-related diseases has decreased (CDC, 2003). All deaths among persons with HIV are not due to HIV disease. They can be secondary to causes not related to HIV infection, such as myocardial infarction or motor vehicle accident. With improved treatment, survival after a diagnosis of AIDS has become longer, allowing a greater proportion (about 25%) of deaths of persons with AIDS to result from other causes (CDC, 2003).

The estimated number of adults and adolescents living with AIDS in the USA increased from 1993 through 2003. This increase was due primarily to the widespread use of highly active antiretroviral therapy (HAART), introduced in 1996, which has delayed the progression of AIDS to death (Albrecht et al., 1998; Antinori et al., 2001; CDC, 2003; de Gaetano Donati et al., 2003; Sanchez et al., 2003; Casalino et al., 2004). Before the HAART era it was estimated that the survival following the AIDS diagnosis was 2 years (CDC, 2001a). Current estimates are that survival after the diagnosis of AIDS can exceed a decade (CDC, 2003).

In the USA the age-adjusted rate of death due to HIV decreased 28% from 1995 to 1996, 45% from 1996 to 1997, 18% from 1997 to 1998 and 3% or less in each of the next 4 years (CDC, 2003). The 1996 and 1997 decrease was largely due to HAART, prophylactic medications for opportunistic infections and national campaigns for prevention of HIV infection (CDC, 2003). The same trend has been observed in Europe (d'Arminio Monforte et al., 2004). This increased survivorship accounts for an increased prevalence of the neurological manifestations of AIDS such as dementia and neuropathy since these and other neurological disorders develop over time (McArthur et al., 2003).

2.2. Epidemiology of neurological complications of HIV infection

As described above, the acute manifestations of HIV infection seldom involve the nervous system; nervous system infection and manifestations occur later in the course of the illness. HIV can injure the nervous system directly (as with HIV dementia) or indirectly by making it susceptible to other infections (such as toxoplasmosis, tuberculosis and cytomegalovirus (CMV)) or to neoplasms (such as CNS lymphoma). These manifestations of HIV usually occur after profound immunosuppression is present (see Table 2.3).

Table 2.3

Neurological complications of HIV

HIV related:
HIV dementia
HIV progressive encephalopathy of childhood
HIV cerebrovascular disease
HIV myelopathy
HIV motor neuron disease
HIV neuropathy
HIV myopathy
Neoplasms:
Primary CNS lymphoma
Metastatic systemic lymphoma
Metastatic Kaposi sarcoma
Opportunistic infections in CNS:
BACTERIA:
Mycobacterium tuberculosis and atypical mycobacteria
Syphilis
Bartonella
Listeria
Nocardia
FUNGI:
Cryptococus
Coccidiomycosis
Histoplasmosis
Blastomycosis
Aspergillus
Candida albicans
Mucormycosis
Sporotrichosis
Cladosporosis
VIRUSES:
Herpes viruses
Cytomegalovirus
Herpes simplex virus
Varicella zoster virus
JC virus
PARASITES:
CNS toxoplasmosis
Acanthamoeba
Tripanosoma cruzi (Chagas' disease)
Strongyloides stercolaris
Cisticercosis (*Taenia solium*)

2.2.1. HIV dementia

The essential feature of dementia is a loss of intellectual ability sufficiently severe as to interfere with social and occupational functioning (APA, 2000). In recognition of the variations in terminology and diagnostic criteria then prevalent in both clinical practice and research protocols, the American Academy of Neurology (AAN) assembled a group of American and European professionals experienced in the care of patients with AIDS, as well as representatives of a number of American and international professional organizations. They developed consensus definitions of the primary CNS illnesses associated with HIV-1 infections (AAN, 1991). All of these 1991 criteria require that patients have laboratory evidence of systemic HIV-1 infection. Since the neurological complications of HIV-2 had not been defined (and though they may be identical to those caused by HIV-1) the group confined themselves to HIV-1. The report points out that the essential feature of HIV-1-associated dementia is disabling cognitive impairment usually accompanied by motor dysfunction, behavioral change or both (Table 2.4). To allow for subgroups of patients who have cognitive impairment but no behavioral abnormality or who have cognitive impairment but no motor dysfunction, the classification allows patients to be coded by major features. A patient with all three features, cognitive impairment, motor dysfunction and behavioral change, has *HIV-1 associated dementia*. With cognitive and motor impairment but no behavioral change the patient is classified as *HIV-1 associated dementia complex (motor)*; with cognitive impairment and behavioral change the patient is classified as *HIV-1 associated dementia complex (behavior)*. This allows for research studies examining differences in patient manifestations while acknowledging that cognitive, motor and behavioral changes in these patients are likely to have a common etiology.

Dementia associated with HIV-1 is divided into two categories: HIV-1 associated minor cognitive/motor impairment (MCMD) and HIV-1 associated dementia (HIV-D) (AAN, 1991). The pivotal difference between the two is the degree of impairment in activities of

Table 2.4

Criteria for clinical diagnosis of HIV-1 associated dementia complex, modified from AAN. (AAN, 1991)

Requires laboratory evidence for systemic HIV-1 infections.
Probable (must have *each* of the following):
1. Acquired abnormality in at least *two* of the following cognitive abilities (present for at least 1 month):
   A. attention/concentration, speed of processing of information, abstraction/reasoning, visuospatial skills, memory/learning and speech/language.
   B. The decline should be verified by reliable history and mental status examination.
   C. In all cases, when possible, history should be obtained from an informant and examination should be supplemented by neuropsychological testing.
   The cognitive dysfunction should cause impairment of work activities or of daily living (objectively verifiable or by report of key informant). This impairment should not be attributable solely to severe systemic illness.
2. At least *one* of the following:
   A. Acquired abnormality in motor function or performance verified by clinical examination (e.g., slowed rapid movements, abnormal gait, limb incoordination, hyperreflexia, hypertonia or weakness), neuropsychological tests (e.g., fine motor speed, manual dexterity, perceptual motor skills), or both.
   B. Decline in motivation or emotional control or change in social behavior. This may be characterized by any of the following: change in personality with apathy, inertia, irritability, emotional liability, or new onset of impaired judgment characterized by socially inappropriate behavior or disinhibition
3. Absence of clouding of consciousness during a period long enough to establish the presence of 1.
4. Evidence of another etiology, including active CNS opportunistic infection or malignancy, psychiatric disorder (e.g., depressive disorder), active alcohol or substance use, or acute or chronic substance withdrawal, *must be sought* from history, psychical and psychiatric examination and appropriate laboratory and radiologic investigation (e.g., lumbar puncture, neuroimaging). *If another potential etiology is present, HIV-1 associated dementia complex is not the cause of the above cognitive, motor or behavioral symptoms and signs.*

Possible (must have *one* of the following):
1. Other potential etiology is present (must have *each* of the following):
   a) As above (see *Probable*) 1, 2 and 3.
   b) Other potential etiology is present but the cause of 1 above is uncertain.
2. Incomplete clinical evaluation (must have *each* of the following):
   a) As above (see *Probable*) 1, 2 and 3.
   b) Etiology can not by determined (appropriate laboratory or radiologic investigations not performed).

Adapted from AAN (1991).

daily living. In order to meet the criteria for HIV-D a combination of acquired abnormality in cognitive abilities (e.g. attention, concentration, memory, learning, speech, reasoning), acquired abnormality in motor function (e.g. abnormal gait, ataxia, hyperreflexia, hypertonia, weakness) and decline in motivational and emotional control (e.g. apathy, inertia, irritability, emotional liability, new onset impaired judgment, disinhibition, inappropriate behavior) must be sufficiently severe as to impair activities of daily living (especially handling money) and the ability to function at work (AAN, 1991). Persons who only evidence impairment when attempting the most demanding activities of daily living are classified as MCMD.

Clinically patients with HIV-D have "difficulty with concentration and forgetfulness. Mental slowness and speech alterations (such as hypophonia and slowness) are also common. Language defects are uncommon and, if present, reflect additional focal dysfunction. Frank confusion can occur with either spatial or temporal disorientation" (AAN, 1991). Patients can also exhibit "apathy, lethargy, loss of sexual drive and diminished emotional responsiveness. Withdrawal from social and business contacts may occur, and others may notice irritability or increasing inflexibility to change. Early motor symptoms include unsteady gait, leg weakness and tremor. Late manifestations include severe global neurologic dysfunction" (AAN, 1991).

2.2.2. Occurrence of HIV-D

The exact burden of HIV-D is difficult to determine. The available studies have been performed in relatively small groups of selected patients and it is uncertain to what extent they are representative of the larger populations. This is in large part because the diagnosis of

HIV-D requires the integration of clinical and laboratory data, including neurological examination and neuropsychological testing. It is labor intensive, expensive and requires personnel with high levels of skill who are scarce in many care settings. A number of disorders including opportunistic CNS infections and CNS tumors can cause diagnostic confusion as can fatigue or depression, common problems in persons with AIDS. Multiple large-scale studies would be difficult to accomplish — especially in lower income countries.

In a pre-HAART era group of patients (n = 1580) who were carefully evaluated (including neuropsychological testing) the prevalence of HIV-D was 0.4% during the asymptomatic phase of infection. Once patients became symptomatic the rate of abnormality in neuropsychologic testing (especially measures of memory and timed motor functions) rose to 16% (Miller et al., 1990a). Subsequently 21% of patients in this clinic developed HIV-D (Sacktor et al., 2001). Data from this prospectively followed patient group were analyzed to determine the changes in the incidence of HIV-D in the period 1990–1998 covering the introduction of HAART. There was a statistically significant decrease in incidence. The rate was 21% in 1990–1992 (patients predominantly receiving monotherapy), 18% in 1993–1995 (predominantly dual therapy) and 10% in 1996–1998 (predominantly HAART) (Sacktor et al., 2001). In this interesting group of homosexual, urban men, 79% were Caucasian and 50% had at least a college degree. Median age at study entry was 37 years. From 1990 to 1992, 70% of HIV-D cases in this group occurred with advanced immunosuppression (CD^4 count <200 cells/μL). After the introduction of HAART, from 1996 to 1998, the cases of HIV dementia were distributed among CD^4 count ranges as follows: CD^4 200, 26%; 201–350, 13%; and >350, 6% (HIV-D cases).

The San Francisco, California, Department of Public Health found through its AIDS case surveillance program that the annual incidence of AIDS dementia declined from 3.71 per 100 persons living with AIDS in 1992 to 0.34 per 100 persons living with AIDS in 2002. There was a further decline to 0.24 per 100 in persons living with AIDS in 2003 (Dilley et al., 2005). After the introduction of HAART the incidence of HIV-D declined. Presumably HAART decreases the incidence of HIV-D and related diseases by decreasing the systemic HIV viral load and consequently the risk of seeding HIV in the brain (Sacktor et al., 2001; d'Arminio Monforte et al., 2004).

The Multicenter AIDS Cohort Study (MACS) also investigated the pre-HAART era risk factors for HIV-D in the Baltimore subset of this group. During the first 2 years after AIDS, HIV-D developed at an annual rate of 7% and overall 15% of the cohort developed dementia after four years. In this cohort the median survival after dementia was 6 months. Anemia, low weight, constitutional symptoms and older age at the onset of AIDS were statistically significant risk factors for the onset of HIV-D. Ethnicity, gender, AIDS-defining illness, zidovudine use before AIDS or CD^4 lymphocyte counts were not associated with increased risk for HIV-D (McArthur et al., 1993).

Age as a risk factor for developing HIV-D has been examined in other groups with conflicting results. In the Hawaii Aging with HIV-1 Cohort and in the Multicenter AIDS cohort study older age was associated with an increased risk for HIV-D with an odds ratio of 3.26 (95% CI, 1.32–8.07) in older individuals in comparison with the younger group (McArthur et al., 1993; Valcour et al., 2004). In a HAART era study that compared the relationship between age and cognitive function in HIV-infected men, a comparison between 66 seronegative men and 188 HIV-seropositive men found that older seropositive individuals were not at an increased risk for HIV-related cognitive impairment when adjustments were made for expected age-related cognitive changes and for education. There was no interaction between cognitive impairment and HIV status ($F = 0.273$, p-value: 0.844) (Kissel et al., 2005).

With the marked increase of HIV-infected women there has been concern about gender as a risk factor. A prospective, longitudinal study with standardized neurological, neuropsychological and laboratory investigations compared 42 HIV-negative women, 52 HIV-positive women and 52 HIV-positive males. Though the HIV positive patients (male and female) had poorer neurologic functioning than the control group there was no difference between HIV-positive women and HIV-positive men (Robertson et al., 2004).

2.2.3. Progressive encephalopathy of childhood

In pediatric AIDS multiple events including direct infection, secondary infections and central nervous system neoplasms can all injure the developing brain. Early in the AIDS epidemic it was apparent that infants and young children with AIDS had high rates of neurological and developmental abnormality. Determining the etiologies of these problems was complicated by the children's multiple medical and social problems.

In the 1980s several centers observed that HIV-infected infants and young children had patterns of neurological disability that could not be adequately explained by the occurrence of central nervous system opportunistic infections or by the occurrence of central nervous system neoplasms (Belman et al., 1985;

Epstein et al., 1985; Ultmann et al., 1985). It was soon reported that the HIV virus could be identified in the brains of some of these patients (Shaw et al., 1985). These infants and children exhibit striking developmental lag with some children having a progressive neurological decline, presumably a primary effect of HIV-1 infection. This progressive disorder is characterized by reduced brain growth producing acquired microcephaly (Belman et al., 1985; Epstein et al., 1985; Dickson et al., 1993; Kozlowski et al., 1993), failure to achieve age appropriate milestones and progressive motor dysfunction (Rosenfeldt et al., 2000; Chiriboga et al., 2005). The most common motor abnormality is spasticity but patients have been reported with cerebellar signs (ataxia and tremor) and other abnormalities of movement (rigidity, opisthotonia, dystonia, tremor) in addition to spasticity (Belman, 1993).

Initially there were varying terms used to describe these patients as well as variations in case definitions. Under the auspices of the American Academy of Neurology a group of American and European professionals experienced in adult and pediatric AIDS as well as representatives of a number of American and international professional organizations gathered to define the CNS illnesses associated with HIV-1 infections. One of the definitions was for HIV-1 associated progressive encephalopathy of childhood (Table 2.5) (Collaborative, 1990; AAN, 1991). These diagnostic criteria were subsequently incorporated into the International Classification of Diseases as well as the Diagnostic and Statistical Manual of Mental Disorders (DSM-IV) of the American Psychiatric Association.

Before the widespread use of HAART (introduced in 1996), HIV-1-associated progressive encephalopathy of childhood manifests itself in a variety of clinical patterns with varying rates of progression. Longitudinal studies found that some progressive encephalopathy of childhood-affected patients would progress and then plateau, others would show progression without any remission or stabilization and still others would resume a deteriorating course after apparently being stable (Belman, 1993). Many of the children demonstrated attentional and other behavioral problems though it was unclear to what extent this was a manifestation of the encephalopathy and to what extent it reflected possible in utero drug exposures, psychosocial problems or manifestations of other disorders or treatments.

In the European Collaborative Study 5 of 16 children with AIDS had serious neurologic manifestations (31%). In two of the children these manifestations were not ascribed to AIDS yielding a rate of 19% for AIDS-related neurological problems. In a comparison group of children where the mothers were HIV-positive but the children HIV-negative the rate of neurologic manifestations was 2 of 164 children or 1% (Collaborative, 1990). The Mothers and Infants Cohort Study also provided reassuring data for HIV-1-negative children born to HIV-positive mothers. At age 2 years these children had identical scores on a structured neurological examination when compared to control children of similar social and economic status whose mothers were not HIV infected (Belman et al., 1996).

In a multi-clinic study of 1811 HIV-infected children from the USA, 178 of 766 children (23%) with perinatally acquired AIDS were diagnosed with HIV encephalopathy (Lobato et al., 1995). The median age at diagnosis was 19 months and the risk of an infected child having HIV encephalopathy by age 12 months was 4.0% (95% CI, 2.6–6.0%). Progressive encephalopathy of childhood is associated with severe morbidity. The children with HIV encephalopathy had more hospitalizations and lower CD^4 T-lymphocyte counts in the first year of life than children with AIDS without encephalopathy. Their median survival after diagnosis was 22 months, similar to that for children with another major complication, *Pneumocystis carinii* pneumonia. A similar rate of progressive encephalopathy (21%, 27 out of 128 children) and of mortality (mean survival after diagnosis of 14 months) was observed in the Women and Infants Transmission Study cohort of vertically (perinatally) infected infants (Cooper et al., 1998).

Sanchez-Ramon et al. found progressive HIV encephalopathy in 58 out of 189 vertically infected children (31%). Comparing CD^8 T-lymphocyte counts obtained in the first months of life and before the onset of symptoms of progressive encephalopathy between infants who subsequently developed progressive encephalopathy and those who did not showed that CD^8 T-lymphocyte counts <25% indicated a 4-fold higher risk of developing progressive encephalopathy (95% CI, 1.2–13.9). This risk factor was significant after adjustment for differing treatments (Sanchez-Ramon et al., 2003). The authors suggest that a CD^8 T-lymphocyte determination could be used in drug trials and for selecting subjects requiring treatment for CNS infection.

With the advent of antiviral therapies there has been a profound change in the incidence, clinical characteristics and course of progressive encephalopathy of childhood. In 1988 Pizzo and colleagues demonstrated that intravenous zidovudine (AZT) increased IQ scores for 13 neurologically abnormal children infected with HIV-1 (Pizzo et al., 1988). Subsequently with the introduction of HAART there have been even more impressive changes. In a cohort of perinatally infected children from Harlem Hospital Center (NYC) whose rate of progressive encephalopathy was 31% in 1992 the rate was

Table 2.5

Criteria for clinical diagnosis of HIV-1 associated progressive encephalopathy of childhood. (AAN, 1991)

Probable (must have *each* of the following):
1. Evidence for systemic HIV-1 infection:
   a. Infants and children < 15 months
      i. Virus in blood or tissues, or
      ii. Presence of HIV-1 antibody and evidence of cellular and humoral immune deficiency or other conditions meeting CDC case definition for AIDS
   b. Children >15 months
      i. Antibody or virus in blood or tissues
2. At least one of the following progressive findings present at least 2 months:
   a. Failure to attain or loss of developmental milestones or loss of intellectual ability, verified by standard developmental scale or neuropsychological tests.
   b. Impaired brain growth (acquired microcephaly or brain atrophy demonstrated on serial CT or MRI).
   c. Acquired symmetric motor deficits manifested by *two or more* of the following: paresis, abnormal tone, pathologic reflexes, ataxia, or gait disturbance.
3. Evidence of another etiology, including active CNS opportunistic infection or malignancy, must be sought from history, physical examination, and appropriate laboratory and radiologic investigation (e.g. lumbar puncture, neuroimaging). If another potential etiology is present, it is *not* thought to be the cause of the above cognitive/motor/behavioral/developmental symptoms and signs.

Possible (must have *one* of the following):
1. Other potential etiology present (must have *each* of the following):
   a. As above (see *Probable*) 1 and 2
   b. Other potential etiology is present but the cause of 2 is uncertain
2. Incomplete clinical evaluation (must have *each* of the following):
   a. As above (see *Probable*) 1 and 2
   b. Etiology cannot be determined (appropriate laboratory or radiologic investigations not performed).

Adapted form AAN (1991).

reduced to 1.6% (2 out of 126 children) by the year 2000 (Chiriboga et al., 2005). For 13 children (10%) progression of the encephalopathy was arrested and their neurologic abnormalities became static. Comparing viral load, CD^4 and CD^8 T-lymphocyte percentages between patients with and without progressive encephalopathy of childhood near the time of diagnosis of the encephalopathy, they found only viral load to be significantly different between cases (those who developed progressive encephalopathy of childhood) and controls. The rate of developmental delay in this population was high (17% or 22 out of 126 children) but did not show an association with HIV infection or viral load.

Among 146 perinatally HIV-infected children studied in a special clinic in Philadelphia between June 1990 and May 2003 the prevalence of progressive encephalopathy of childhood was 29.6% in 113 children born between 1990 and 1996 (pre-HAART era) and 12% in 33 children born after 1996, a statistically significant difference ($p = 0.049$). There was an association between viral load ($p = 0.06$) and CD^4 T-lymphocyte percentages ($p < 0.001$) and the occurrence of progressive encephalopathy. A meta-analysis of five studies from the Pediatric AIDS Clinical Trials Group (USA) also indicated that among children on anti-viral therapy, virologic markers but not CD^4 cell count were indicators of increased risk for encephalopathy (Lindsey et al., 2000).

It is widely observed that children born to HIV-positive mothers have a high rate of behavioral problems and there has been concern that this is a manifestation of HIV-encephalopathy (Misdrahi et al., 2004). The Women and Infant Transmission Study Group (USA) followed 307 children longitudinally and obtained caretaker completed behavioral rating scales beginning when the children were 3 years old. Multivariate analyses comparing the HIV-infected children with perinatally exposed but uninfected children from similar backgrounds failed to find an association between HIV status (or prenatal drug exposure) and poor behavioral outcomes (Lui et al., 1988; Mellins et al., 2003).

The rate of viral suppression at delivery is predictive of the HIV infection risk to the infant. Initially with zidovudine therapy (Connor et al., 1994) and subsequently with multiple preventative measures including prenatal testing, antiretroviral prophylaxis, cesarean delivery and avoidance of breast-feeding the rate of perinatal HIV infection has become <2% (Mofenson, 2000; Cooper et al., 2002).

2.2.4. HIV-associated central nervous system lymphoma

HIV-infected patients are at increased risk of developing CNS lymphoma (B cell non-Hodgkin's lymphoma), an association that has been observed since the beginning of the AIDS epidemic (Ziegler et al., 1984). Based on data from nine local health department cancer registries and including pre-HAART HIV cases there was a 9-fold increase in the incident cases of CNS lymphoma from 1980 to 1989 with a reported incidence of 4.7 cases per 1000 person years. This study also estimated that the occurrence of HIV-related lymphomas was 3600 times higher than seen in the general population (Cote et al., 1996). Since the introduction of HAART the incidence of CNS lymphoma has decreased. In the MACS study it decreased from 2.8 per 1000 person years (1990–1992, pre-HAART era) to 4.3 per 1000 person years (1993–1995) and to 0.4 per 1000 person years (1996–1998, post-HAART era) (p value $=$ 0.005, between these periods) (Sacktor et al., 2001). The association between Epstein–Barr virus (EBV) and HIV-related CNS lymphoma has been reported in several studies. EBV was identified in 51 of 65 evaluated tumors and in 14 of 16 cases of HIV-related CNS lymphoma and in another study in 32 out 50 cases diagnosed with systemic lymphoma (11 had CNS lymphoma) (Cingolani et al., 2000; Hansen et al., 2000).

The clinical presentation is characterized by confusion, headache, memory loss, personality changes, seizures, cranial nerve palsies and varying focal neurological symptoms depending on the location of the lesions (Rosenblum et al., 1988; Corti et al., 2004). CSF EBV DNA testing has a sensitivity of 90% (95% CI, 54.1–99.5) with a specificity of 100% (95% CI, 62.9–100) (Cingolani et al., 2000).

Brain CT imaging can show an enhancing isodense or hyperdense lesion as well as mass effect and edema. After radiotherapy lesions can decrease in diameter, become hypodense and lose their enhancement with contrast media (Goldstein et al., 1991). T1 weighted magnetic resonance imaging (MRI) shows isointense or low signal intensity lesions which are hyperintense on T2 and FLAIR sequences. Gadolinium can demonstrate ring and peripheral enhancement without homogenous enhancement of the lesion. Most lesions are supratentorial (Thurnher et al., 2001; Corti et al., 2004). With customary treatment (radiation) survival is poor and on average is less than 3 months (Remick et al., 1990; Ling et al., 1994; Khoo et al., 2000).

2.2.5. HIV and stroke risk

Cerebrovascular disease has been noted in association with HIV infection (Berger et al., 1987a; Engstrom et al., 1989; Evers et al., 2003). Early in the AIDS epidemic, in a case-control study from a large urban hospital, Berger et al. compared the prevalence of cerebrovascular disease in 154 adult autopsied patients with and without AIDS who were between 20 and 50 years old. Cerebrovascular disease was confirmed by neuropathological criteria, in 13 (8%) of 154 patients with AIDS compared with 23% of controls. Among the causes of stroke were: thrombosis (6 cases), emboli (4 cases) and hemorrhage (3 cases). The spectrum of cerebrovascular diseases was similar in patients with and without AIDS with the exception of cerebral vasculitis which was only observed in those with AIDS (Berger et al., 1990). Other less common causes of stroke are: disseminated intravascular coagulation, hyperviscosity syndrome, lupus anticoagulant thrombocytopenia causing intracerebral hemorrhage, ruptured mycotic aneurism, vasculitis and intracranial malignancy (Gluck et al., 1990). A population based study of young adults hospitalized with a stroke in 1988 or 1991 found the AIDS-associated incidence of both ischemic stroke and intracerebral hemorrhage to be 0.2% per year in individuals with AIDS (Cole et al., 2004). The incidence of ischemic stroke was 0.14% per year and intracerebral hemorrhage was 0.11% per year. The relative risks for both ischemic stroke and intracerebral hemorrhage were both significantly increased at 9.1 (95% CI, 3.4–24.6) for ischemic stroke and 12.7 (95% CI, 4.0–40.0) for intracerebral hemorrhage, with a combined adjusted relative risk of 10.4.

2.2.6. HIV myelopathy

The myelopathy associated with HIV infection usually occurs in the setting of advanced HIV disease. Patients can complain of numbness, tingling, leg weakness or unsteadiness, as well as incontinence of bladder and bowel (Helweg-Larsen et al., 1988). In an autopsy series 23 out of 70 autopsied patients with AIDS had a vacuolar myelopathy. Myelopathy was generally more common and more severe in those patients with advanced brain pathology (Navia et al., 1986). In a case series 3 out of 128 neurologically symptomatic patients had a viral myelitis (Levy et al., 1985).

The diagnosis is suggested by MRI. In a case series of 21 patients with clinically diagnosed HIV myelopathy the MRI scans were abnormal in 18 (86%) with spinal cord atrophy as the most common finding ($n = 15$). There was no correlation between the extent of cord involvement on MR images and the clinical severity of myelopathy as independently assessed by a neurologist (Chong et al., 1999). T2 weighted MRI of the cervical cord can reveal a focal, symmetrical, well-defined area of high signal intensity along the posterior columns with predominance in the gracile tracts (Shimojima et al., 2005). Cerebrospinal fluid (CSF) examination and serology can help to rule out opportunistic infections.

2.2.7. Motor neuron disease associated with HIV infection

Motor neuron disease associated with HIV infection was first reported in 1985 (Hoffman et al., 1985), four years after the initial description of AIDS. Clinical patterns of HIV-associated motor neuron disease may imitate those of amyotrophic lateral sclerosis (ALS), progressive spinal muscular atrophy (PSMA) and brachial amyotrophic diplegia (MacGowan et al., 2001; Moulignier et al., 2001). Brachial amyotrophic diplegia has been recently reported in association with HIV infection without signs of other opportunistic infections. Brachial amyotrophic diplegia results in severe lower motor neuron weakness with atrophy of the upper extremities in the absence of bulbar or lower extremity involvement, pyramidal features, bowel and bladder incontinence or sensory loss (Berger et al., 2005).

2.2.8. HIV neuropathy

Peripheral neuropathy is the most frequent of the neurological complications associated with adult HIV infection (Barohn et al., 1993) though it is seldom seen in children (Raphael et al., 1991). Neuropathy seldom appears until there is moderate to severe immunosuppression from the HIV infection (Schifitto et al., 2002). Distal symmetrical polyneuropathy (DSP) is the most common of the HIV-1-associated neuropathies. It generally has an indolent and protracted clinical course with the primary manifestations in symptomatic patients being distal pain, paresthesias and numbness. Some patients with DSP remain asymptomatic (Verma, 2001). In a pre-HAART cohort of HIV-positive, predominantly male (75%) patients who were followed prospectively for 30 months with a structured neurological examination, DSP was diagnosed if the subjects had (a) decreased or absent ankle jerks, (b) decreased or absent vibratory perception at the toes or (c) decreased pinprick or temperature in a stocking distribution. At initiation of the study 55% of patients had DSP and of these 20% were asymptomatic. Interestingly, gender, use of dideoxynucleosides and presence of asymptomatic DSP were not risk factors for developing symptomatic DSP (Schifitto et al., 2002).

A similar examination was used for a New York City cohort in the HAART era. DSP was diagnosed by the presence of two or more abnormalities: ankle reflexes, vibratory sensation or pinprick perception. 99 of 187 patients (53%) had DSP, similar to the rate seen in the pre-HAART cohort. Patients with neuropathy were older (45.3 years, SD 0.7, versus 41.2 years, SD 0.8 years) and were more likely to be men, 83 out of 99 (83%), than women, 16 out of 99 (16%). No other risk factors, not CD^4 cell counts, plasma viral load nor the use of neurotoxic antiretroviral therapy, were associated with DSP. Asymptomatic DSP was correlated with histories of opiate and sedative abuse. Symptomatic DSP correlated with ethanol and hallucinogen syndromes but sural nerve morphometric findings (performed in a subset of subjects at autopsy) did not distinguish between patients with substance use syndromes and those without (Morgello et al., 2004).

Patients with HIV-associated neuropathies can have axonal, demyelinating or mixed axonal and demyelinating pathological features (Brinley et al., 2001). The most common type of peripheral neuropathy, DSP, is axonal. Though DSP occurs primarily in patients with advanced immunosuppression it can also be secondary to the neurotoxicity of HAART (especially the combination of zidovudine and zalcitabine (Wulff et al., 2000)) as well as to diabetes, alcoholism, nutritional deficiencies and other disorders commonly associated with AIDS (Bailey et al., 1988; Moore et al., 2000).

Additionally, acute inflammatory demyelinating polyneuropathy (AIDP) and chronic inflammatory demyelinating polyneuropathy (CIDP) can be associated with HIV infection (Brinley et al., 2001) but are reported to be uncommon (Wulff et al., 2000). AIDP can occur at the time of primary HIV infection or seroconversion (Thornton et al., 1991; Verma, 2001) and is generally a rapidly progressive ascending weakness associated with generalized areflexia (Wulff et al., 2000). Examination of the CSF can be helpful in the diagnosis since it may show a high cell count, predominately lymphocytes. This abnormality is not always present (Cornblath et al., 1987) but when seen it helps to differentiate HIV-1-associated AIDP from AIDP in non-HIV-infected persons (Thornton et al., 1991). CIDP evidences a slower progression and a tendency to improve and relapse (Wulff et al., 2000). In some patients it may be due to CMV infection (Morgello & Simpson, 1994).

Progressive polyradiculopathy (PP) is another type of neuropathy associated with HIV infection, that can be present as a herald sign of HIV infection (Mahieux et al., 1989). It is most commonly seen in advanced immunosuppression as an expression of CMV infection (Eidelberg et al., 1986). Rapidly progressive flaccid paraparesis, radiating pain and paresthesias, areflexia and sphincter dysfunction are the cardinal clinical features (Wulff et al., 2000). Rapid diagnosis and anti-CMV treatment with ganciclovir has been shown to prevent irreversible injury and to promote functional recovery (Miller et al., 1990b; Brinley et al., 2001).

Mononeuropathy multiplex has been described with HIV infection (Lipkin et al., 1985; Stricker et al., 1992) though it occurs infrequently. Similar to AIDP it can occur early in the illness when it is considered to have an autoimmune etiology or it can be present in advanced HIV infection when it is likely to be caused by CMV infection (Said et al., 1991; Roullet et al., 1994; Wulff et al., 2000). This disorder is recognized by the acute onset of sensory deficits in the distribution of two or more peripheral or cranial nerves (Roullet et al., 1994; Wulff et al., 2000).

Diffuse infiltrative lymphocytosis syndrome neuropathy represents a separate and uncommon entity among HIV-associated neuropathies (Gherardi et al., 1998). These patients develop persistent CD^8 hyperlymphocytosis and a Sjogren's syndrome-like syndrome associated with multivisceral CD^8 T cell infiltration. They present with a painful, acute or subacute, symmetrical neuropathy, sicca syndrome and multi-visceral involvement (Moulignier et al., 1997). Antiretroviral therapy and steroids may be effective treatments (Moulignier et al., 1997).

2.2.9. HIV-related skeletal muscle disorders

Skeletal muscle disorders in AIDS can be divided into three major groups: (1) HIV-associated myopathies and related conditions; (2) muscle complications of antiretroviral therapy; and (3) opportunistic infections and infiltrating tumors (François-Jérôme Authier, 2005).

The HIV-related myopathies, which can occur in any stage of HIV infection, are characterized by proximal and often symmetric muscle weakness that develops subacutely, over weeks to months. The weakness tends to be proximal (affecting arms and legs) and is associated with serum creatine phosphokinase (CK) elevations and electromyographic abnormalities (AAN, 1991). HIV-associated myopathies (and related conditions) include HIV polymyositis, inclusion-body myositis, nemaline myopathy, diffuse infiltrative lymphocytosis syndrome, HIV-wasting syndrome, vasculitic processes, myasthenic syndromes and possibly chronic fatigue. The prevalence of these disorders is unclear since they are relatively uncommon but their occurrence in clinical practice appears unchanged with the advent of HAART (Maschke et al., 2000; François-Jérôme Authier, 2005). An evaluation of 64 HIV patients referred to an urban clinic because of elevated CK levels or muscle weakness determined that 13 subjects had biopsy proven polymyositis. Duration of HIV infection ranged from 0 to 11 years. With treatment these patients had a generally favorable prognosis: 5 of 8 patients treated with prednisone achieved a complete remission while three others had normalization of strength and CK level on azathioprine or methotrexate combined with intravenous immunoglobulin (Johnson et al., 2003).

Muscle complications of antiretroviral therapy are primarily a mitochondrial myopathy related to zidovudine, toxicity to other nucleoside-analogue reverse-transcriptase inhibitors (NRTIs), HIV-associated lipodystrophy syndrome and immune restoration syndrome related to HAART (François-Jérôme Authier, 2005). Mitochondrial toxicity is generally gradual in onset. It may start early in therapy but the prevalence and severity seem to increase with prolonged therapy with a reported overall rate of 17% (Carr & Cooper, 2000). In a short-term, placebo-controlled trial of zidovudine that was terminated after 21 months (mean duration of follow-up was 39.4 weeks) 5 subjects, all on zidovudine (5 of 360, 1.4% of zidovudine-treated subjects), met the criteria for myopathy which required both limb weakness and CK elevation (Simpson et al., 1997).

2.3. Opportunistic infections of the nervous system in AIDS

Since the introduction of HAART there has been a decrease in both overall mortality rates and in the incidence of neurological opportunistic infections (Sacktor, 2002). A review of autopsied AIDS patients found the brain to be the second most frequently involved organ in the HAART era (Jellinger et al., 2000). We describe the most common neurological infections seen in patients with AIDS by etiological agent.

2.3.1. Bacteria

2.3.1.1. *Mycobacterium tuberculosis*

Tuberculosis (TB) remains a global burden. Using both surveillance and survey data the World Health Organization (WHO) estimated that there were 8.8 million new cases of TB in 2003 with 674,000 (8%) of these cases in persons thought to be infected with HIV. In the same year there were 15.4 million prevalent cases of TB with an estimated mortality of 1.7 million (11%). Among these fatal cases there were an estimated 229,000 (13%) persons co-infected with HIV (Dye et al., 1999; WHO, 2005). In published series HIV is the strongest risk factor for TB infection and the risk of death among HIV-infected individuals co-infected with TB is three times higher than among those without HIV infection (Perriens et al., 1991; Bucher et al., 1999). There has been a decrease in TB incidence, morbidity and mortality since the introduction of HAART (Kirk et al., 2000; Dheda et al., 2004). In a prospective, observational cohort study, the pre-HAART era estimated incidence for TB was 0.8 per 100 person-years of follow-up (PYF) (1994) compared to 0.3 cases/100 PYF in the HAART era (1997) (Kirk et al., 2000). A retrospective study that compared patient outcomes after starting TB treatment during the pre-HAART era (before 1996; $n = 36$) with outcomes after starting treatment during the HAART era (during or after 1996; $n = 60$) found that those in the HAART group had a lower risk of death (cumulative at 4 years, 43% vs. 22%; $P = 0.01$) and of having an AIDS event (69% vs. 43%; $P = 0.02$) (Dheda et al., 2004).

2.3.1.2. Tuberculous meningitis

Patients infected with HIV and TB are at increased risk for meningitis. A cross sectional study that compared the occurrence of TB meningitis in patients with and without HIV infection found that only 2% of TB patients without HIV developed meningitis versus 10% of HIV-infected patients. Interestingly, HIV infection did not change the clinical manifestations or the outcome of tuberculous meningitis in this pre-HAART era study (Berenguer et al., 1992). Patients with TB meningitis typically present with seizures, meningismus, altered mental status and fever (Bishburg et al., 1986). Lumbar puncture can show a high opening pressure, and CSF studies usually reveal pleocytosis, hypoglycorrhachia and an elevated or normal protein (Kent et al., 1993). The diagnosis is made by a *M. tuberculosis* (M-TB)-positive CSF culture, ELISA or polymerase chain reaction (PCR) (Kaneko et al., 1990). MRI or computed tomography (CT) imaging of the head may show basilar meningeal enhancement, periventricular edema, areas of infarction or hydrocephalus. Less commonly a mass lesion due to tuberculoma or TB abscess may be imaged (Villoria et al., 1992). Recommended treatment calls for a 6-month regimen of isoniazid, rifabutin, pyrazinamide and ethambutol (CDC, 1998).

2.3.1.3. Atypical mycobacteria

Mycobacterium avium complex (MAC) is the most common atypical mycobacteria found in patients with HIV infection and caused disseminated disease in up to 40% of patients with advanced HIV in the pre-HAART period (Chaisson et al., 1992). In the Johns Hopkins cohort with advanced HIV disease, the proportion developing MAC has fallen from 16% before 1996 to 4% after 1996 (HAART era) with a current rate of less than 1% per year (Karakousis et al., 2004). Infection with MAC resembles TB meningitis symptomatically with fever, chills, night sweats, meningismus, generalized weakness and weight loss (Young et al., 1986). The initial treatment of MAC infection consists of two antimycobacterial drugs to prevent or delay the emergence of resistance: clarithromycin and ethambutol have proven efficacy (CDC, 2004a). Patients with HIV infection and CD^4 counts less than 100 cells/μL are recommended to receive lifetime prophylaxis against MAC with clarithromycin or azithromycin (CDC, 1993).

2.3.1.4. Syphilis

The CDC reported that the rates for primary and secondary syphilis in the USA decreased during the 1990s and in 2000 was the lowest since reporting began in 1941 (CDC, 2001b). However, the number of cases of primary and secondary syphilis increased during 2000–2002 and continued to increase from 2002 (6862 cases) to 2003 (7177 cases) (CDC, 2004b). This 2000–2003 increase in cases was observed only among men who have sex with men in outbreaks characterized by high rates of HIV co-infection (CDC, 2004b). In a retrospective study of 767 HIV-infected patients, 238 were co-infected with syphilis. In this group the prevalence of neurosyphilis was 3% (7 cases) (Brandon et al., 1993). There were significant differences in the clinical presentation of neurosyphilis in persons with and without HIV infection. As a group HIV-infected patients were younger, and more frequently had features of secondary syphilis, such as rash, fever, lymph adenopathy, headache and meningismus (Katz et al., 1993). The diagnosis of neurosyphilis is based on the clinical presentation and CSF presence of reactive CSF-VDRL, high protein and pleocytosis (Marra, 1992). The CSF-VDRL is considered the standard test for neurosyphilis; the sensitivity ranges from 30 to 70% (Hart, 1986); in some cases the CSF-VDRL can be falsely negative. A recent study that investigated the CSF of patients with neurosyphilis found that 27% of patients with typical symptoms of neurosyphilis had the combination of a negative VDRL and a positive FTA (fluorescent treponema antibody test-absorption) in the CSF. The FTA-abs test in the CSF is a highly sensitive marker for the presence of neurosyphilis; however, due to its high sensitivity a high number of false positives may result. The main objective of obtaining a FTA test in CSF is to rule out the possibility of neurosyphilis when the clinical suspicion is high but the VDRL CSF is negative (Timmermans & Carr, 2004).

The CDC recommends that HIV-infected persons with early-stage (i.e. primary, secondary or early latent) syphilis should receive a single intramuscular (IM) injection of 2.4 million units of benzathine penicillin G (CDC, 2004b). HIV-infected patients with clinical or laboratory evidence of neurosyphilis should receive intravenous aqueous crystalline penicillin G, 18–24 million units daily, administered 3–4 million units IV every 4 hours or by continuous infusion for 10–14 days or procaine penicillin 2.4 million units IM once daily plus probenecid 500 mg orally four times a day for 10–14 days (CDC, 2004b).

2.3.1.5. Bartonella

Bacillary angiomatosis (BA), caused by bacteria of the genus *Bartonella* (most commonly *Bartonella henselae B. quintana*), is not seen with substantially increased frequency in HIV infection (Spach & Koehler, 1998). BA often presents with vascular skin lesions (that resemble Kaposi sarcoma), anemia and fever (Moore et al., 1995). Infection with *Bartonella* has been associated with traumatic exposure (usually scratches) to domestic cats (CDC, 2004a). BA occurs most often late in HIV infection in patients with CD^4 counts of <50 cells/μL. The neurological manifestations associated with *Bartonella* are cerebral and retinal bacillary angiomatosis, encephalitis, miliary cerebral arteritis and retinitis (Schwartzman et al., 1990). The diagnosis is suspected by the clinical presentation and a history of exposure to cats. The diagnosis can be confirmed by tissue biopsy that shows vascular proliferative histopathology, modified silver stain demonstrating numerous bacilli, or by a CSF PCR (available through the CDC) (CDC, 2004a). MRI of the brain can show white matter changes with hyperintense signal in T2 weighted sequences (Schwartzman et al., 1990). Case reports have shown erythromycin and doxycycline to be successful in treating BA (Berger et al., 1989; Schwartzman, 1996; CDC, 2004a).

2.3.1.6. Listeria

Listeria monocytogenes infection is uncommon in association with HIV infection, despite the severe impairment in cellular immunity. The association has been noted in a few case reports (Valencia Ortega et al., 2000). Listeriosis presents with acute meningitis, that can progress to chronic meningitis, and brain abscess. Ampicillin, the drug of choice for listeriosis, has proven efficacy in patients with HIV infection (Decker et al., 1991).

2.3.1.7. Nocardia

Nocardiosis is also an infrequent infection in patients with HIV. A recent retrospective study analyzed the clinical record of 27 HIV-infected patients with nocardiosis and found pulmonary involvement to be the most frequent manifestation followed by cutaneous and disseminated infection (Biscione et al., 2005). Cerebral abscess and meningitis are well known complications of nocardiosis (Ogg et al., 1997). The overall mortality was high (37%). Diagnosis is by blood culture and CSF analysis. Treatment with cotrimoxazole-amikacin has been shown to be effective (Biscione et al., 2005).

2.3.2. Fungi

2.3.2.1. Cryptococcus

Cryptococcus neoformans is a frequent, opportunistic CNS infection and the most common opportunistic infection in HIV patients (Berger et al., 1987a). A population-based surveillance study conducted during 1992–2000 in two major cities in the USA found a total of 1491 incident cases of cryptococcosis, of which 1322 (89%) occurred in HIV-infected persons (Mirza et al., 2003). In 1992 (pre-HAART era) the annual incidence of cryptococcosis was 66 per 1000 persons with HIV, which decreased to 7 per 1000 in 2000 (post-HAART era) (Mirza et al., 2003). In HIV infection, cryptococcosis usually presents as a subacute meningitis or meningoencephalitis with fever, malaise and headache (Powderly, 2000). In patients with meningitis the most common presenting symptoms were fever and headache with photophobia and neck stiffness occurring in one-third of the cases. Patients may also present with an acute encephalopathy (Clark et al., 1990; CDC, 2004a). Lumbar puncture can reveal an elevated opening pressure and may be inadvisable. CSF typically shows mild elevation in protein, hypoglycorrhachia and mild or absent pleocytosis. India ink stains can show the fungi and cryptococcus antigen titers are usually positive (Clark et al., 1990). MRI of the brain can demonstrate punctate hyperintensities in the white matter on T2 weighted images that correspond to pathologically dilated Virchow–Robin space crowded with cryptococci (Mathews et al., 1992; Bos et al., 2001). The recommended treatment by the CDC for acute disease is amphotericin B, usually combined with flucytosine, for a 2-week duration followed by fluconazole alone for an additional 8 weeks. Patients with HIV infection treated for cryptococcus require lifetime suppressive therapy (van der Horst et al., 1997; CDC, 2004a).

2.3.2.2. Coccidioidomycosis

Coccidioidomycosis is caused by *Coccidioides immitis*, a fungus endemic to the southwestern USA and Central and South America (Galgiani, 1993).

A retrospective study of 77 HIV patients with coccidioidomycosis found that 9 of them had meningitis (the majority had pulmonary disease) (Fish et al., 1990). The clinical presentation among patients with meningeal involvement consists of fever, headache, nausea, vomiting or confusion. Cavitary necrosis can result in abscess formation making lumbar puncture inadvisable (Jarvik et al., 1988). CSF analysis shows a lymphocyte pleocytosis and high protein. The diagnosis of coccidioidomycosis can be confirmed by culture of the organism in CSF or blood. The treatment of choice for meningitis is fluconazole, which has been reported to be successful in 79% of patients in an uncontrolled trial of 47 patients with coccidioidal meningitis. After the acute infection patients need lifelong suppressive therapy (Galgiani et al., 1993; CDC, 2004a).

2.3.2.3. Histoplasmosis

Histoplasma capsulatum, a dimorphic fungus with a worldwide distribution, has its highest prevalence in temperate and tropical environments. In the USA histoplasmosis is endemic in the north central and south central regions. It is the most common of the endemic mycoses in patients with AIDS with an estimated prevalence of 2 to 5% among patients with AIDS residing in endemic areas compared to 1% of those in other regions (Wheat et al., 1990a). Histoplasmosis can represent the first manifestation of HIV infection and usually occurs when CD^4 counts are <150 cells/μL (Johnson et al., 1986; Hajjeh et al., 2001). A case control study found that the most common symptoms included fever (88%), weight loss (71%), respiratory symptoms (55%), abdominal pain (33%) and altered mental status (16%). In this group the mortality was 12% within 3 months of diagnosis of histoplasmosis (Hajjeh et al., 2001). CNS meningeal involvement can be seen in as many as 20% of cases. It can present with headache, fever, seizures, cranial nerve palsies, mental status changes, lethargy and meningismus (Wheat et al., 1990a). The diagnosis of histoplasmosis can be made by detection of the antigen in blood or urine. With meningeal involvement the CSF may reveal a lymphocytic pleocytosis with protein elevation and hypoglycorrhachia though lumbar puncture may not be advisable (Wheat et al., 1990a; Williams et al., 1994). Cerebral imaging can show contrast enchasing mass lesions, meningeal enhancement and areas of acute infarction (Wheat et al., 1990a). The treatment for HIV patients with disseminated histoplasmosis is intravenous amphotericin, for 3 to 10 days (until clinical improvement), and then oral therapy with itraconazole for 12 weeks (Wheat et al., 2005). HIV-infected patients who complete the initial therapy for histoplasmosis require lifelong suppressive treatment with itraconazole 200 mg twice daily (Hecht et al., 1997).

2.3.2.4. Blastomycosis

Blastomyces dermatitidis is a dimorphic fungus endemic in the Southern and Midwestern USA and in Canada (Witzig et al., 1994; Lortholary et al., 1999). Cases have also been reported from Africa, India, the Middle East and Central and South America (Witzig et al., 1994; Cury et al., 2003). Blastomycosis is infrequently seen in association with HIV. In an autopsy case series of 92 HIV-infected patients who died of opportunistic diseases there was only one case of blastomycosis (Cury et al., 2003). Central nervous system involvement can range from 40% to 46% when blastomycosis infection occurs in persons with AIDS (Pappas et al., 1992; Witzig et al., 1994). The mortality rate for these patients is 54%, about 5 times the rate of blastomycosis deaths in the general population (Witzig et al., 1994). The neurological symptoms at presentation are non-specific and include headache, lethargy, confusion, seizures and focal signs if a mass lesion is present. The systemic symptoms are also non-specific consisting of fever, chills, night sweats, anorexia and weight loss (Witzig et al., 1994; Ludmerer & Kissane, 1996; Lortholary et al., 1999). CSF analysis can show hypoglycorrhachia with a high protein and a normal cell count though lumbar puncture may not be advisable. The fungus can be cultured from CSF (Ludmerer & Kissane, 1996). CT and MR imaging have shown epidural abscess, parenchymal abscess, ring enhancing lesions and leptomeningeal enhancement (Ward et al., 1995). Amphotericin B is the treatment of choice for life-threatening blastomycosis including acute respiratory distress syndrome and CNS infection (Pappas et al., 1992). There are no definitive studies indicating whether an antifungal agent is effective for secondary prophylaxis for AIDS patients who have been treated for blastomycosis; however, itraconazole has been used with apparent efficacy (Lortholary et al., 1999).

2.3.2.5. Aspergillus

Aspergillus fumigatus is a life-threatening infection in patients with HIV (Singh et al., 1991), although it is uncommon (Holding et al., 2000). Reported risk factors for aspergillosis include neutropenia and corticosteroid therapy (Minamoto et al., 1992). A retrospective study of 35,252 HIV-infected patients found 228 cases of aspergillosis with an estimated incidence of 3.5 cases per 1000 person-years (95% CI, 3.04 per 1000 person-years) and a higher incidence among people 35 years or older and those with CD^4 counts of ≤ 99 cells/mm^3.

In this group the median survival time after diagnosis of aspergillosis was 3 months and 26% survived for over a year. Patients with aspergillosis were unlikely to be receiving HAART and were more likely to have a history of other AIDS-defining opportunistic infections (Holding et al., 2000). The patients with CNS involvement present with headache, fever, meningismus and focal neurological signs secondary to infarction (Lortholary et al., 1993). A review of the CNS imaging studies for 8 immunosuppressed patients (none had AIDS) with CNS aspergillosis showed three distinct patterns. The first pattern was multiple areas of hypodensity on CT scans or hyperintensity on T2-weighted MR images involving the cortex and/or subcortical white matter consistent with embolic infarctions. The second pattern consisted of multiple intracerebral ring-enhancing lesions consistent with abscesses, while the third pattern was meningeal enhancement associated with enhancing lesions in the adjacent paranasal sinus or orbit (Ashdown et al., 1994). CSF analysis can show hypoglycorrhachia, increased protein, with lymphocytic or neutrophilic pleocytosis, though lumbar puncture may not be advisable (Palo et al., 1975; Carrazana et al., 1991). The definite diagnosis is made by histopathologic demonstration or culture of organisms obtained by biopsy (Minamoto et al., 1992). The recommended treatment for invasive aspergillosis is voriconazole with amphotericin B an alternative regimen (CDC, 2004a). There are no data available to recommend a prophylactic or suppressive therapy for patients who survive the acute infection (CDC, 2004a); nevertheless, for patients with HIV, it would be reasonable to start antifungals that have been proven useful in similar infections (CDC, 2004a).

2.3.2.6. Candida albicans

Oropharyngeal candidiasis is common in HIV-infected individuals (Klein et al., 1984) yet few cases of CNS or systemic candidiasis have been reported (Casado et al., 1998). In the reported cases candidiasis has presented with both meningeal inflammation and abscess formation (Casado et al., 1997). In a review of 14 HIV patients with meningeal candidiasis the most frequent symptoms were headache (13 of 14), fever (12 of 14), nuchal rigidity (7 of 14), third nerve palsy and diplopia (2 of 14) and altered mental status (4 of 14) (Casado et al., 1997). CT and MRI imaging showed meningeal enhancement, hydrocephalus and mass lesions secondary to abscess formation. All cases demonstrated a CSF pleocytosis with most patients also demonstrating hypoglycorrhachia (11 of 14) and elevated protein (though lumbar puncture may not be advisable) (Casado et al., 1997). There is no randomized clinical trial available for the antifungal treatment of CNS infection in these patients; however, amphotericin B and fluconazol appear to be effective (Casado et al., 1997; Rodriguez-Arrondo et al., 1998).

2.3.2.7. Mucormycosis

Mucormycosis, caused by zygomycetous fungi, has seldom been associated with HIV infection. In the few cases in the literature, the most common clinical presentations are indicative of cerebral, cutaneous and renal involvement (Cuadrado et al., 1988; Guardia et al., 2000). Patients can present with headache, fever, chills and lethargy. They can develop cranial nerve palsies and other focal deficits (Lee et al., 1996; Oliveira & Costa, 2001). CSF analysis can show lymphocytic pleocytosis with elevated protein (though lumbar puncture may not be advisable) and biopsy may be required for diagnosis (Cuadrado et al., 1988). CT and MRI imaging can show mass lesions, areas of infarction and hematomas secondary to the dissemination of the infection. Amphotericin has been used to treat this often fatal infection (Blazquez et al., 1996; Oliveira & Costa, 2001).

2.3.2.8. Sporotricosis

Sporothrix schenckii is a dimorphic fungus that results in chronic granulomatous mycosis; infection occurs with traumatic inoculation of the fungi through the skin (Scott et al., 1987). There have been five cases of HIV infection and CNS involvement reported in the literature (Silva-Vergara et al., 2005). The most common manifestations reported are headache, seizures, confusion and coma. The diagnosis can be made with fugal serum culture, CSF antibodies and culture, and skin biopsy (Scott et al., 1987). Treatment with amphotericin B and itraconazole has been reported to be effective (Bonifaz et al., 2001; Silva-Vergara et al., 2005).

2.3.2.9. Cladosporiosis

Cladophialophora bantiana is a dematiaceous fungus, which is a common cause of skin infections in humans, but a rare cause of systemic infection (Jayakeerthi et al., 2004). There have been 2 reported cases of CNS cladosporiosis in association with HIV infection, both in IV drug users. These patients presented with headache, fever and focal neurological signs (Colon et al., 1988; Jayakeerthi et al., 2004). The contrasted CT showed multiple ring-enhancing lesions in both patients. One patient was treated empirically for toxoplasmosis with no clinical response, the other was treated with amphotericin B and fluconazol with a survival of 5 weeks (Colon et al., 1988; Jayakeerthi et al., 2004).

2.3.3. Viruses

2.3.3.1. Cytomegalovirus

Cytomegalovirus (CMV) is a double-stranded DNA herpes virus that can reactivate and cause disseminated or localized end-organ disease among patients with HIV who have experienced a previous CMV infection (CDC, 2004a). Since the introduction of HAART the incidence, morbidity and mortality of CMV has declined (Palella et al., 1998; Detels et al., 2001). The estimated rate of CMV before the HAART era was 16 per 100 person years with a decline to 2–3 per 100 person years in the post-HAART era (Palella et al., 1998). "In a failure-rate model, increases in the intensity of antiretroviral therapy (classified as none, monotherapy, combination therapy without a protease inhibitor, and combination therapy with a protease inhibitor) were associated with stepwise reductions in morbidity and mortality. Combination antiretroviral therapy was associated with the most benefit; the inclusion of protease inhibitors in such regimens conferred additional benefit" (Palella et al., 1998). CMV can present as retinitis, encephalitis, esophagitis, colitis, gastritis and hepatitis (Gallant et al., 1992). CMV neurologic disease can cause dementia, ventriculitis, encephalitis, myelitis and polyradiculomyelopathy (Arribas et al., 1996). While the clinical picture is heavily dependent upon the location of the CNS infection, the most frequent symptoms are fever and headache. In CMV encephalitis and ventriculitis patients can present with confusion, seizures, coma, aphasia or dysphasia, and cranial nerve palsies (Arribas et al., 1996). In CMV-associated necrotizing myelitis patients can present with acute or progressive paraplegia and bowel and/or bladder incontinence (Said et al., 1991; Gungor et al., 1993; Berger & Sabet, 2002). In polyradiculomyelopathy patients can present initially with paresthesias, burning and painful sensation in lower extremities or with progressive paraparesis and absence of reflexes (Miller et al., 1990b). The CSF in CNS CMV infection can show a lymphocytic pleocytosis, hypoglycorrhachia and high protein (Kalayjian et al., 1993; Arribas et al., 1996). Brain imaging findings can range from normal to the demonstration of ependymal and meningeal enhancement, ring-enhancing lesions, hemorrhage and ventricle dilation (Ramsey & Geremia, 1988; Grafe et al., 1990). The diagnosis of CMV neurological disease is facilitated by the detection of the virus in blood or CSF using PCR. Other diagnostics tests such as antigen–antibody assays and viral blood culture have been shown to be less reliable (Wolf & Spector, 1992; Dodt et al., 1997; Quereda et al., 2000). While the treatment for CMV has been studied extensively in CMV retinitis, there are no studies to date with precise recommendation for CMV neurological disease (Group, 1994; Musch et al., 1997). In the setting of CNS neurological disease immediate treatment with foscarnet, valganciclovir or ganciclovir is recommended (Hengge et al., 1993; CDC, 2004a).

2.3.3.2. Herpes simplex virus

Human herpes simplex virus type 1 (HSV-1) and type 2 (HSV-2) infections are common and endemic throughout the world (Langenberg et al., 1999). HSV-1 infections result from contact with infectious salivary secretions and can result in both symptomatic mucocutaneous HSV-1 lesions in the mouth and asymptomatic infection of the trigeminal ganglion. This latent infection can subsequently become symptomatic producing a focal encephalitis (Sacks et al., 2004). HSV-2 infection is usually transmitted sexually and can cause recurrent, painful genital ulcers (Fleming et al., 1997). It is estimated that 60 to 80% of the general population have serum antibodies to HSV-1, HSV-2 or both (Gibson et al., 1990; Schacker et al., 1998), and as many as 95% of HIV-positive persons are seropositive for HSV-1, HSV-2 or both (Siegel et al., 1992; Schacker et al., 1998). HSV-1 CNS infection can cause encephalitis localized to the temporal lobes and HSV-2 can cause aseptic meningitis (Boivin, 2004). In spite of the high seropositive rates only a few cases of CNS HIV infection have been reported in patients with HIV — it seems to be an uncommon complication. In a small series, patients presented with confusion (3 of 9), fever and headache (2 of 9), anxiety and depression (2 of 9), slow mentation and memory impairment (1 of 9) and expressive aphasia (1 of 9). Five of these patients had previous AIDS-defining diagnoses and four of the five had previous cutaneous HSV infection (Schmutzhard, 2001; Grover et al., 2004). The CSF analysis can show lymphocytic pleocytosis, high protein, normal glucose or mild hypoglycorrhachia though lumbar puncture may not be advisable (Tyler, 2004). HSV CSF PCR has shown to be very sensitive in diagnosis HSV (Lakeman & Whitley, 1995). One of the nine patients in the series described above had a negative PCR on CSF and was diagnosed by biopsy. MRI is more sensitive than CT for identifying herpes simplex encephalitis and it is considered the brain imaging examination of choice. Images can show increased signal in T2 and FLAIR sequences in one or both temporal lobes and meningeal enhancement (Bakken et al., 1989; Demaerel et al., 1992). Electroencephalography (EEG) can show periodic lateralized epileptiform discharges (Watemberg & Morton, 1996). The treatment of choice for HSV encephalitis is intravenous acyclovir for 14 to 21 days (Tyler, 2004).

2.3.3.3. Varicella zoster virus

Varicella zoster virus (VZV) is a human herpes virus that causes chickenpox (varicella). It has the potential to become latent in cranial nerve and dorsal-root ganglia and upon reactivation produces shingles (zoster) and post-herpetic neuralgia. In immunocompetent elderly persons or immunocompromised patients, VZV can produce disease of the central nervous system (Gilden et al., 2000). A retrospective nested case-control analysis of patients enrolled in an urban HIV clinic between January 1, 1997 and December 31, 2001 investigated the incidence, risk factors, and clinical sequelae of herpes zoster (including ocular, visceral and neurological complications). The estimated incidence of VSV infection in the pre-HAART-era patients of the Amsterdam Cohort Study was 3.31 per 1000 person-years in HIV-1 VSV seronegative subjects and 51.51 per 1000 person-years in HIV-1-seropositive subjects. Recurrences only occurred in seropositive subjects (26%) though the cumulative incidences of first infections increased linearly with the duration of follow-up. Risk of infection increased inversely to the number of CD^4 cells (Veenstra et al., 1995). In a post-HAART retrospective cohort study from an urban clinic the incidence was 3.2 per 100 person-years of follow-up. A total of 28 patients (18%) developed post-herpetic neuralgia and 29 (18%) had other complications, including aseptic meningitis, transverse myelitis, trigeminal neuralgia and Ramsay–Hunt syndrome. The authors considered these rates as high when compared with the general population of similar age. They did not see evidence of a reduction in VZV incidence with HAART therapy (Gebo et al., 2005). VZV clinical presentation depends on the localization of the infection. Patients with VSV encephalitis can present with headache, fever, altered level of consciousness, seizures and focal neurological deficits. MRI imaging of the brain can show ischemic or hemorrhagic strokes and white matter changes consistent with demyelination. The CSF can show mild mononuclear pleocytosis, a normal or very high protein (Froin's syndrome) and a normal glucose (Kleinschmidt-DeMasters et al., 1998; Gilden et al., 2000). In VSV myelits patients present with progressive weakness, sensory impairment and bladder and bowel dysfunction. MRI imaging of the brain can show hyperintense lesions and CSF analyses have the same features described above (Gilden et al., 1994; de Silva et al., 1996). VZV neuropathy is characterized by severe sharp, lancinating, radicular pain and rash which is most common over the thorax and face. The diagnosis is suggested by the combination of the clinical presentation and the characteristic rash that can follow dermatomes (Gilden et al., 2000). The recommended treatment for localized dermatomal herpes zoster by the CDC is famciclovir or valacyclovir for 7–10 days; there are no evidence based recommendations for CNS infection but there have been case reports with successful treatment with acyclovir (Tattevin et al., 2001; Hernandez et al., 2002; Toledo et al., 2004).

2.3.3.4. JC virus

JC virus is a DNA papovavirus that causes progressive multifocal leucoencephalopathy (PML), a demyelinating disease of the CNS (Padgett et al., 1971) with the name JC being derived from the initials of the patient from whom this virus was first isolated (CDC, 2004a). HIV infection is now the major risk factor for PML (Berger et al., 1987b; Berger, 2003). Before the AIDS epidemic PML was infrequently diagnosed. Between 1979 and 1987 deaths related to PML increased fourfold (from 1.5 per 10,000,000 persons to 6.1 per 10,000,000 persons) (Holman et al., 1991). A retrospective hospital-based cohort study reported 205 cases of PML in HIV-infected persons during the years 1980 to 1994 with a 12-fold increase over this 14-year period. Seven cases were diagnosed from 1980 to 1984 while 71 cases were discovered from 1991 to 1994 (Berger et al., 1998). In as many as 45% of the cases PML is the initial presentation of HIV (Berger et al., 1987b; Albrecht et al., 1998). There has not been a documented change in the incidence of PML since the introduction of HAART. In the Multicenter AIDS Cohort Study ($n = 2734$) the incidence of PML in 1990–1992 was 2.0 per 1000 person-years (treatment predominantly monotherapy), in 1993–1995 it was 1.8 per 1000 person-years (treatment predominantly dual therapy) and in 1996–1998 it was 1.5 per 1000 person-years (treatment predominantly HAART). While there appears to be a downward trend, the rates were not statistically different ($p = 0.84$). It should be noted that all the cases of PML occurred in patients with a CD^4 count <200 (Sacktor et al., 2001). A retrospective multicenter cohort study of 35 HIV-infected patients with PML confirms that PML continues to occur even during treatment with HAART (11 out of 35 HIV-infected patients). There was a suggestion that patients who developed PML during HAART treatment tended to have a shorter median survival compared with patients not treated with HAART; however, this difference was not statistically significant ($p = 0.15$) (Wyen et al., 2004).

In 25 carefully studied patients with PML and HIV infection the most common symptoms were: arm or leg weakness, cognitive dysfunction (symptoms range from impaired concentration to stupor), visual loss, limb incoordination, headache and dysphasia (Berger et al., 1987b). The CSF can show normal to mild protein elevations, elevated myelin basic protein, high IGG index and mild pleocytosis (predominately mononuclear cells) (Berger et al., 1987b). CSF PCR for JC virus is usually diagnostic (sensitivity and specificity of 92%) (McGuire et al., 1995). EEG can show generalized to focal slowing with delta and theta waves in the areas of leucoencephalopathy (Berger et al., 1987b). The MRI changes associated with PML are high signal intensity with T2-weighted images and low signal intensity abnormalities on T1-weighted images. These lesions do not enhance with gadolinium, which helps to differentiate PVL from primary CNS lymphoma (Weiss & DeMarco, 1994). The lesions are generally localized to the frontal and parieto-occipital white matter and the cerebellum. Lesions elsewhere, such as in the medulla, have been reported (Power et al., 1997; Mathew & Murnane, 2004). The presence of JC virus in the brain can be demonstrated on biopsy or post mortem obtained tissue (Silver et al., 1995). There is no effective treatment for PML. Interestingly, HAART appears to significantly improve the prognosis in HIV patients with PML. In a retrospective cohort study there was a medial survival of 131 days after the diagnosis of PML in patients (n = 14) who never received or stopped taking antiviral therapy; a median survival of 127 days for patients (n = 10) who continued to take or were switched to antiviral nucleosides alone; and a median survival of over 500 days for those patients (n = 5) who were started on HAART (Albrecht et al., 1998).

2.3.4. Parasites

2.3.4.1. CNS toxoplasmosis

The protozoan *Toxoplasma gondii* is the most common cause of cerebral mass lesions in patients with HIV. If untreated CNS toxoplasmosis can be fatal (Luft & Remington, 1992; Antinori et al., 2004). In the USA, CNS toxoplasmosis can develop in 3–10% of patients with HIV while in Latin America, where the overall seroprevalence of toxoplasma is higher, the estimated rate of CNS toxoplasmosis is 14% (Luft & Remington, 1992; Vidal et al., 2005). This makes toxoplasmosis among the most common opportunistic CNS infections in patients with HIV (Kure et al., 1991; Renold et al., 1992) and in some cohorts it is the most frequent (Bouckenooghe & Shandera, 2002; Antinori et al., 2004). CNS toxoplasmosis can be a primary infection but it usually represents reactivation of a latent infection (Zangerle et al., 1991). The incidence of CNS toxoplasmosis has declined since the introduction of HAART. In the French Hospital Database study there were 19,598 pre-HAART era patients with HIV of whom 1259 had CNS toxoplasmosis (incidence = 3.87 cases per 100 person-years, 95% CI = 3.664.09). After the initiation of HAART therapy 17,016 patients were studied with 319 cases observed (incidence = 0.97 cases per 100 person-years, 95% CI = 0.861.08) (Abgrall et al., 2001). The same trend was apparent in the US-based Multicenter AIDS Cohort Study where the CNS toxoplasmosis incidence declined from 5.4 per 1000 person-years (1990–1992), to 3.8 per 1000 person-years (1993–1995) and to 2.2 per 1000 person-years (1996–1998) (p-value = 0.08) (Sacktor et al., 2001). A multivariate analysis of baseline characteristics that significantly correlated with a final diagnosis of toxoplasmic encephalitis was performed using a prospective cohort of 186 consecutive HIV-positive inpatients undergoing empiric anti-toxoplasma therapy for a first episode of presumed CNS toxoplasmosis. The data showed the following risk factors expressed in hazard ratios (HR): toxoplasma serology, HR 15.0 (p-value < 0.001); CT/MRI suggestive of toxoplasmosis, HR 18 (p-value < 0.001); headache, HR 2.5 (p-value 0.004); seizures, HR 2.3 (p-value 0.05); fever, HR 2.2 (p-value 0.01); and CD^4 count, HR 2.3 (p-value 0.001). Prophylaxis for toxoplasmosis had a protective HR of 0.2 (p-value < 0.001) (Raffi et al., 1997). Among 115 patients at San Francisco General Hospital (1981–1990) the most common presenting symptoms were headache (55%), confusion (52%) and fever (47%). A total of 69% had focal neurological deficits on examination (Porter & Sande, 1992). Lumbar puncture may not be advisable and CSF analysis may not be helpful. CSF findings include a mild protein elevation, mild lymphocytic pleocytosis and, occasionally, hypoglycorrhachia (Renold et al., 1992). MRI can show lesions that have increased signal on T2-weighted scans with a hypointense center and hyperintense rings. On T1-weighted images the lesions are predominantly hypointense. Gadolinium-enhanced scans can show typical ring-enhancing lesions (the target/pea-pod appearance) with peripheral edema. Diffusion weighted images also show the hyperintense ring target appearance (Batra et al., 2004). The definite diagnosis is by demonstration of the parasite on brain biopsy. In clinical practice patients are treated empirically and brain biopsy is reserved as a last resource for those patients whose imaging lesions do not resolve with treatment (Skiest, 2002). The treatment of choice is a combination of pyrimethamine and sulfadiazine which has been shown to be highly efficacious. Life-long therapy is needed to prevent relapse in HIV-infected individuals (Leport et al., 1988).

2.3.4.2. *Trypanosoma cruzi* (Chagas' disease)

Chagas' disease is caused by the parasite *Trypanosoma cruzi*, a flagellated protozoan transmitted to humans by the blood-sucking insect triatomine as well as by blood transfusion. Endemic in Central and South America there are an estimated 16–18 million persons who are infected and 100 million in this region who are at risk of developing Chagas' disease (WHO, 1997; Antunes et al., 2002; Moncayo, 2003). Reactivation of Chagas' disease has been reported in 15 patients with HIV infection with brain involvement being the most common manifestation (Ferreira et al., 1997). There are no data on the impact of HAART on HIV-related Chagas' disease. Chagas' disease with HIV co-infection has a high mortality rate with survival after the onset of neurological symptoms ranging from a few days to a few weeks (Ferreira et al., 1997; Antunes et al., 2002). The most common clinical presentation is meningoencephalitis with fever, headache, seizures and focal neurological signs in the presence of mass lesions (Rosemberg et al., 1992; Rocha et al., 1994; Ferreira et al., 1997). The diagnosis can be made based upon the clinical presentation, exposure in an endemic area and positive serological studies for *T. cruzi* (Rosemberg et al., 1992; Ferreira et al., 1997; Yoo et al., 2004). Lumbar puncture may be inadvisable but when performed the CSF analysis can show an elevated opening pressure, high protein, hypoglycorrhachia and lymphocytic pleocytosis. Patients have also shown normal CSF findings (Ferreira et al., 1997; Yoo et al., 2004). CNS imaging can show ring-enhancing lesions and areas of acute infarction (Yoo et al., 2004). In some patients treatment with benznidazole and nifurtinox has been effective (Antunes et al., 2002; CDC, 2004a).

2.3.4.3. Acanthamoeba

The parasite *Acanthamoeba* is found in the soil of tropical and subtropical areas, with infection the result of inhalation or cutaneous contact (Ma et al., 1990). Three genera of free-living amoebae, *Naegleria*, *Acanthamoeba* and *Balamuthia*, can cause a brain infection (Martinez et al., 1994; Seijo Martinez et al., 2000) which results in meningoencephalitis and granulomatous amoebic encephalitis (GAE). Prognosis is poor and death within a few days to a few weeks from the onset of symptoms is common (Zagardo et al., 1997). In the scattered reports of patients with both HIV and amoebic meningoencephalitis presenting findings included headache, fever, confusion, seizures, cranial nerve palsies and focal neurological symptoms which varied depending upon the location of granulomas (Ma et al., 1990; Gardner et al., 1991; Di Gregorio et al., 1992; Gordon et al., 1992; Martinez et al., 1994; Murakawa et al., 1995; Zagardo et al., 1997; Seijo Martinez et al., 2000). Arriving at an accurate clinical diagnosis can be challenging in patients with HIV. A final diagnosis has often required brain biopsy which can show the parasite as well as characteristic granulomatous changes with multinucleated giant cells, histiocytes, necrotizing vasculitis and an inflammatory perivascular infiltrate (Gordon et al., 1992; Zagardo et al., 1997; Seijo Martinez et al., 2000). Lumbar puncture may be inadvisable and the organism has not been isolated from the CSF. CSF findings are nonspecific with high protein, low glucose and lymphocytic pleocytosis (Zagardo et al., 1997). CT and MRI imaging can show one or multiple ring-enhancing mass lesions that can be located in the cerebral hemispheres and brain stem (Zagardo et al., 1997; Seijo Martinez et al., 2000). There is currently no accepted treatment for amoebic encephalitis. One patient is reported to have improved after surgical excision of the granuloma and treatment with fluconazol (Seijo Martinez et al., 2000). Another patient treated with miconazole, rifampin and pentamidine died within 3 weeks of onset of symptoms (Zagardo et al., 1997).

2.3.4.4. Strongyloides stercolaris

The intestinal parasite *Strongyloides stercolaris* (SS) is found in tropical and subtropical climates. Acquired by direct skin contact SS can remain asymptomatic in the intestines for decades (Walker & Zunt, 2005). HIV patients have occasionally presented with meningitis and encephalitis due to SS (Dutcher et al., 1990; Harcourt-Webster et al., 1991). Usually gastrointestinal symptoms preceded the neurological complaints which included confusion, headache, meningismus and paresis (Morgello et al., 1993; Walker & Zunt, 2005). The parasite has been isolated from the stool and has also been found in CSF (Dutcher et al., 1990; Harcourt-Webster et al., 1991). Thiabendazole has been reported to be effective in the treatment of SS; nevertheless when the infection is disseminated the mortality is high (Dutcher et al., 1990; Harcourt-Webster et al., 1991; Jain et al., 1994).

2.3.4.5. Cysticercosis (Taenia solium)

Taenia solium neurocysticercosis is a common infection in many lower income countries due to poor sanitation and is a major cause of epilepsy in the world (Cruz et al., 1999). Neurocysticercosis has been reported in case series in association with HIV infection (Thornton et al., 1992) but there is no evidence that it occurs in higher rates among HIV-infected populations. An important limitation is that AIDS populations from lower income countries (where cisicercoisis is prevalent) are underrepresented in the literature. Neurocysticercosis is suspected when the CT scan shows a characteristic pattern of multiple cystic lesions in the parenchyma of the brain, especially at the gray white junction. Treatment with praziquantel has been shown to be effective in non-HIV-infected populations (Cruz et al., 1999).

2.4. Data limitations

Given the worldwide nature of the HIV/AIDS epidemic, our knowledge of the epidemiology of HIV-D and of other neurological complications rests on a very narrow base of data. The vast majority of the studies cited in this review come from specialized clinics in the USA and Europe, yet there has been a rapid increase in the number of persons with HIV/AIDS in lower income countries (The Antiretroviral Therapy in Lower Income Countries Study, 2005). Because of cost many of these persons will not receive specialized care or effective therapy. Among those with access to specialized clinics there is marked variation in the availability of drugs and various combinations are in use. In the ART-LINC centers only 61% used one of the four WHO recommended first-line HAART regimens. Other clinical circumstances also vary such as intensity and frequency of follow-up, level of laboratory support, and the availability of professionals with scarce medical skills such as neurologists (The Antiretroviral Therapy in Lower Income Countries Study, 2005). Most of the available data for adults involve men who have sex with men, yet heterosexual men and women are represented in rapidly increasing numbers in the HIV/AIDS populations. Variations in social economic factors such as early malnutrition, differences in genetic makeup, availability of therapy for co-morbid conditions and other local factors could also limit the applicability of the currently available data.

Most of the data about neurological complications come from therapeutic trials and specialized clinics. A variety of factors including varying treatment protocols, patient attrition, poor adherence to regimen and co-morbid conditions tend to render therapies less effective in general practice than in these specialized settings (Green, 2005). This is especially important since the prevalence and manifestations of neurological complications, especially dementia, vary depending upon overall efficacy of therapy. The patients in these specialized settings may be atypical of affected patients in the general population due to selection factors unique to the clinics. The tendency to conduct such clinics in high incidence areas may mean that these data are not strictly applicable to other geographic locations (Sacktor et al., 2001; Eltom et al., 2002).

References

AAN (1991). Nomenclature and research case definitions for neurologic manifestations of human immunodeficiency virus-type 1 (HIV-1) infection. Report of a Working Group of the American Academy of Neurology AIDS Task Force. Neurology 41(6): 778–785.

Abgrall S, Rabaud C, Costagliola D (2001). Incidence and risk factors for toxoplasmic encephalitis in human immunodeficiency virus-infected patients before and during the highly active antiretroviral therapy era. Clin Infect Dis 33(10): 1747–1755.

Albrecht H, Hoffmann C, Degen O, et al (1998). Highly active antiretroviral therapy significantly improves the prognosis of patients with HIV-associated progressive multifocal leukoencephalopathy. AIDS 12(10): 1149–1154.

Allain JP, Laurian Y, Paul DA, et al (1986). Serological markers in early stages of human immunodeficiency virus infection in haemophiliacs. Lancet 2(8518): 1233–1236.

Antinori A, Cingolani A, Alba L, et al (2001). Better response to chemotherapy and prolonged survival in AIDS-related lymphomas responding to highly active antiretroviral therapy. AIDS 15(12): 1483–1491.

Antinori A, Larussa D, Cingolani A, et al (2004). Prevalence, associated factors, and prognostic determinants of AIDS-related toxoplasmic encephalitis in the era of advanced highly active antiretroviral therapy. Clin Infect Dis 39 (11): 1681–1691.

Antunes AC, Cecchini FM, Bolli FB, et al (2002). Cerebral trypanosomiasis and AIDS. Arq Neuropsiquiatr 60(3-B): 730–733.

APA (2000). Diagnostic and Statistical Manual of Mental Disorders DSM-IV-TR, American Psychiatric Association, Washington, DC.

Arribas JR, Storch GA, Clifford DB, et al (1996). Cytomegalovirus encephalitis. Ann Intern Med 125(7): 577–587.

Ashdown BC, Tien RD, Felsberg GJ (1994). Aspergillosis of the brain and paranasal sinuses in immunocompromised patients: CT and MR imaging findings. AJR Am J Roentgenol 162(1): 155–159.

Bailey RO, Baltch AL, Venkatesh R, et al (1988). Sensory motor neuropathy associated with AIDS. Neurology 38 (6): 886–891.

Bakken JS, Camenga DL, Glazier MC, et al (1989). [Herpes simplex 1 virus encephalitis. Early diagnosis with magnetic tomography (MT)]. Tidsskr Nor Laegeforen 109(24): 2430–2432.

Barohn RJ, Gronseth GS, LeForce BR, et al (1993). Peripheral nervous system involvement in a large cohort of human immunodeficiency virus-infected individuals. Arch Neurol 50(2): 167–171.

Batra A, Tripathi RP, Gorthi SP (2004). Magnetic resonance evaluation of cerebral toxoplasmosis in patients with the acquired immunodeficiency syndrome. Acta Radiol 45(2): 212–221.

Belman A (1993). Neurologic syndromes. Ann N Y Acad Sci 693: 107–122.

Belman AL, Ultmann MH, Horoupian D, et al (1985). Neurological complications in infants and children with acquired immune deficiency syndrome. Ann Neurol 18(5): 560–566.
Belman AL, Muenz LR, Marcus JC, et al (1996). Neurologic status of human immunodeficiency virus 1-infected infants and their controls: a prospective study from birth to 2 years. Mothers and Infants Cohort Study. Pediatrics 98(6 Pt 1): 1109–1118.
Berenguer J, Moreno S, Laguna F (1992). Tuberculous meningitis in patients infected with the human immunodeficiency virus. N Engl J Med 326(10): 668–672.
Berger JR (2003). Progressive multifocal leukoencephalopathy in acquired immunodeficiency syndrome: explaining the high incidence and disproportionate frequency of the illness relative to other immunosuppressive conditions. J Neurovirol 9(Suppl 1): 38–41.
Berger JR, Sabet A (2002). Infectious myelopathies. Semin Neurol 22(2): 133–142.
Berger JR, Moskowitz L, Fischl M, et al (1987a). Neurologic disease as the presenting manifestation of acquired immunodeficiency syndrome. South Med J 80(6): 683–686.
Berger JR, Kaszovitz B, Post MJ, et al (1987b). Progressive multifocal leukoencephalopathy associated with human immunodeficiency virus infection. A review of the literature with a report of sixteen cases. Ann Intern Med 107 (1): 78–87.
Berger JR, Harris JO, Gregorios J, et al (1990). Cerebrovascular disease in AIDS: a case-control study. AIDS 4(3): 239–244.
Berger JR, Pall L, Lanska D, et al (1998). Progressive multifocal leukoencephalopathy in patients with HIV infection. J Neurovirol 4(1): 59–68.
Berger JR, Espinosa PS, Kissel J (2005). Brachial amyotrophic diplegia in a patient with human immunodeficiency virus infection: widening the spectrum of motor neuron diseases occurring with the human immunodeficiency virus. Arch Neurol 62(5): 817–823.
Berger TG, Tappero JW, Kaymen A, et al (1989). Bacillary (epithelioid) angiomatosis and concurrent Kaposi's sarcoma in acquired immunodeficiency syndrome. Arch Dermatol 125(11): 1543–1547.
Biggar RJ (1990). AIDS incubation in 1891 HIV seroconverters from different exposure groups. International Registry of Seroconverters. AIDS 4(11): 1059–1066.
Biggar RJ, Rosenberg PS (1993). HIV infection/AIDS in the United States during the 1990s. Clin Infect Dis 17(Suppl 1): S219–S223.
Biscione F, Cecchini D, Ambrosioni J, et al (2005). [Nocardiosis in patients with human immunodeficiency virus infection]. Enferm Infecc Microbiol Clin 23(7): 419–423.
Bishburg E, Sunderam G, Reichman LB, et al (1986). Central nervous system tuberculosis with the acquired immunodeficiency syndrome and its related complex. Ann Intern Med 105(2): 210–213.
Blazquez R, Pinedo A, Cosin J, et al (1996). Nonsurgical cure of isolated cerebral mucormycosis in an intravenous drug user. Eur J Clin Microbiol Infect Dis 15(7): 598–599.
Boivin G (2004). Diagnosis of herpesvirus infections of the central nervous system. Herpes 11(Suppl 2): 48A–56A.
Bonifaz A, Peniche A, Mercadillo P, et al (2001). Successful treatment of AIDS-related disseminated cutaneous sporotrichosis with itraconazole. AIDS Patient Care and STDs 15(12): 603–606.
Bos HM, Hofman PAM, Schreij G, et al (2001). Overwhelming CNS cryptococcus in AIDS. Neurology 57(9): 1560.
Bouckenooghe AR, Shandera WX (2002). The epidemiology of HIV and AIDS among Central American, South American, and Caribbean immigrants to Houston, Texas. J Immigr Health 4(2): 81–86.
Brandon WR, Boulos LM, Morse A (1993). Determining the prevalence of neurosyphilis in a cohort co-infected with HIV. Int J STD AIDS 4(2): 99–101.
Brinley FJ, Jr, Pardo CA, Verma A (2001). Human Immunodeficiency Virus and the Peripheral Nervous System Workshop. Arch Neurol 58(10): 1561–1566.
Bucher HC, Griffith LE, Guyatt GH, et al (1999). Isoniazid prophylaxis for tuberculosis in HIV infection: a meta-analysis of randomized controlled trials. AIDS 13(4): 501–507.
Carr A, Cooper DA (2000). Adverse effects of antiretroviral therapy. Lancet 356(9239): 1423–1430.
Carrazana EJ, Rossitch E, Jr, Morris J (1991). Isolated central nervous system aspergillosis in the acquired immunodeficiency syndrome. Clin Neurol Neurosurg 93(3): 227–230.
Casado JL, Quereda C, Oliva J, et al (1997). Candidal meningitis in HIV-infected patients: analysis of 14 cases. Clin Infect Dis 25(3): 673–676.
Casado JL, Quereda C, Corral I (1998). Candidal meningitis in HIV-infected patients. AIDS Patient Care STDs 12(9): 681–686.
Casalino E, Wolff M, Ravaud P, et al (2004). Impact of HAART advent on admission patterns and survival in HIV-infected patients admitted to an intensive care unit. AIDS 18(10): 1429–1433.
CDC (1981). Kaposi's Sarcoma and Pneumocycstis Pneumonia Among Homosexual Men: New York City and California. CDC, pp. 305–308.
CDC (1993). Recommendations on Prophylaxis and Therapy for Disseminated *Mycobacterium Avium* Complex for Adults and Adolescents Infected with Human Immunodeficiency Virus 42(RR-9). CDC, pp. 14–20.
CDC (1998). Prevention and treatment of tuberculosis among patients infected with human immunodeficiency virus: principles of therapy and revised recommendations. MMWR 47(RR20): 1–51.
CDC (1999). Appendix: Revised Surveillance Case Definition for HIV Infection. CDC, pp. 29–31.
CDC (2001a). Sexually Transmitted Disease Surveillance 2000. USDoHaH Services.
CDC (2001b). Surveillance Report.
CDC (2003). HIV/AIDS Surveillance Report 2003.
CDC (2004a). Sexually Transmitted Disease Surveillance 2003 Supplement: Syphilis Surveillance Report. Division of STD Prevention, pp. 1–18.

CDC (2004b). Treating opportunistic infections among HIV-infected adults and adolescents. Recommendations from CDC, the National Institutes of Health, and the HIV Medicine Association/Infectious Diseases Society of America. MMWR 53(RR15): 1–112.

CDC (2005). At a glance the HV/AIDS Epidemic. CDC, pp. 1–2.

Chaisson RE, Moore RD, Richman DD, et al (1992). Incidence and natural history of *Mycobacterium avium*-complex infections in patients with advanced human immunodeficiency virus disease treated with zidovudine. The Zidovudine Epidemiology Study Group. Am Rev Respir Dis 146(2): 285–289.

Chiriboga CA, Fleishman S, Champion S, et al (2005). Incidence and prevalence of HIV encephalopathy in children with HIV infection receiving highly active anti-retroviral therapy (HAART). J Pediatr 146(3): 402–407.

Chong J, Rocco AD, Tagliati M, et al (1999). MR findings in AIDS-associated myelopathy. AJNR Am J Neuroradiol 20 (8): 1412–1416.

Cingolani A, Gastaldi R, Fassone L, et al (2000). Epstein-Barr virus infection is predictive of CNS involvement in systemic AIDS-related non-Hodgkin's lymphomas. J Clin Oncol 18(19): 3325–3330.

Clark RA, Greer D, Atkinson W, et al (1990). Spectrum of Cryptococcus neoformans infection in 68 patients infected with human immunodeficiency virus. Rev Infect Dis 12 (5): 768–777.

Cole JW, Pinto AN, Hebel JR, et al (2004). Acquired immunodeficiency syndrome and the risk of stroke. Stroke 35 (1): 51–56.

Collaborative SE (1990). Neurologic signs in young children with human immunodeficiency virus infection. The European Collaborative Study. Pediatr Infect Dis J 9: 402–406.

Colon L, Lasala G, Kanzer M, et al (1988). Cerebral cladosporiosis in AIDS. J Neuropathol Exp Neurol. 47: 387.

Connor EM, Sperling RS, Gelber R, et al (1994). Reduction of maternal–infant transmission of human immunodeficiency virus type 1 with zidovudine treatment. Pediatric AIDS Clinical Trials Group Protocol 076 Study Group. N Engl J Med 331(18): 1173–1180.

Cooper DA, Gold J, Maclean P, et al (1985). Acute AIDS retrovirus infection. Definition of a clinical illness associated with seroconversion. Lancet 1(8428): 537–540.

Cooper ER, Hanson C, Diaz C, et al (1998). Encephalopathy and progression of human immunodeficiency virus disease in a cohort of children with perinatally acquired human immunodeficiency virus infection. Women and Infants Transmission Study Group. J Pediatr 132(5): 808–812.

Cooper ER, Charurat M, Mofenson L, et al (2002). Combination antiretroviral strategies for the treatment of pregnant HIV-1-infected women and prevention of perinatal HIV-1 transmission. J Acquir Immune Defic Syndr 29(5): 484–494.

Cornblath DR, McArthur JC, Kennedy PG, et al (1987). Inflammatory demyelinating peripheral neuropathies associated with human T-cell lymphotropic virus type III infection. Ann Neurol 21(1): 32–40.

Corti M, Villafane F, Trione N, et al (2004). [Primary central nervous system lymphomas in AIDS patients]. Enferm Infecc Microbiol Clin 22(6): 332–336.

Cote TR, Manns A, Hardy CR, et al (1996). Epidemiology of brain lymphoma among people with or without acquired immunodeficiency syndrome. AIDS/Cancer Study Group. J Natl Cancer Inst 88(10): 675–679.

Cruz ME, Schantz PM, Cruz I, et al (1999). Epilepsy and neurocysticercosis in an Andean community. Int J Epidemiol 28(4): 799–803.

Cuadrado LM, Guerrero A, Garcia Asenjo JA, et al (1988). Cerebral mucormycosis in two cases of acquired immunodeficiency syndrome. Arch Neurol 45(1): 109–111.

Cury PM, Pulido CF, Furtado VM, et al (2003). Autopsy findings in AIDS patients from a reference hospital in Brazil: analysis of 92 cases. Pathol Res Pract 199(12): 811–814.

d'Arminio Monforte A, Cinque P, Mocroft A, et al (2004). Changing incidence of central nervous system diseases in the EuroSIDA cohort. Ann Neurol 55(3): 320–328.

de Gaetano Donati K, Tumbarello M, Tacconelli E, et al (2003). Impact of highly active antiretroviral therapy (HAART) on the incidence of bacterial infections in HIV-infected subjects. J Chemother 15(1): 60–65.

de Silva SM, Mark AS, Gilden DH, et al (1996). Zoster myelitis: improvement with antiviral therapy in two cases. Neurology 47(4): 929–931.

Decker CF, Simon GL, DiGioia RA, et al (1991). Listeria monocytogenes infections in patients with AIDS: report of five cases and review. Rev Infect Dis 13(3): 413–417.

Demaerel P, Wilms G, Robberecht W, et al (1992). MRI of herpes simplex encephalitis. Neuroradiology 34(6): 490–493.

Detels R, Tarwater P, Phair JP, et al (2001). Effectiveness of potent antiretroviral therapies on the incidence of opportunistic infections before and after AIDS diagnosis. AIDS 15 (3): 347–355.

Dheda K, Lampe FC, Johnson MA, et al (2004). Outcome of HIV-associated tuberculosis in the era of highly active antiretroviral therapy. J Infect Dis 190(9): 1670–1676.

Di Gregorio C, Rivasi F, Mongiardo N, et al (1992). Acanthamoeba meningoencephalitis in a patient with acquired immunodeficiency syndrome. Arch Pathol Lab Med 116 (12): 1363–1365.

Dickson DW, Llena JF, Nelson SJ, et al (1993). Central nervous system pathology in pediatric AIDS. Ann N Y Acad Sci 693: 93–106.

Dilley JW, Schwarcz S, Loeb L, et al (2005). The decline of incident cases of HIV-associated neurological disorders in San Francisco, 1991–2003. AIDS 19(6): 634–635.

Dodt KK, Jacobsen PH, Hofmann B, et al (1997). Development of cytomegalovirus (CMV) disease may be predicted in HIV-infected patients by CMV polymerase chain reaction and the antigenemia test. AIDS 11(3): F21–F28.

Dunkle KL, Jewkes RK, Brown HC, et al (2004). Gender-based violence, relationship power, and risk of HIV infection in women attending antenatal clinics in South Africa. Lancet 363(9419): 1415–1421.

Dutcher JP, Marcus SL, Tanowitz HB, et al (1990). Disseminated strongyloidiasis with central nervous system involvement diagnosed antemortem in a patient with acquired immunodeficiency syndrome and Burkitts lymphoma. Cancer 66(11): 2417–2420.

Dye C, Scheele S, Dolin P, et al (1999). Global burden of tuberculosis: estimated incidence, prevalence, and mortality by country. JAMA 282(7): 677–686.

Eidelberg D, Sotrel A, Vogel H, et al (1986). Progressive polyradiculopathy in acquired immune deficiency syndrome. Neurology 36(7): 912–916.

Eltom MA, Jemal A, Mbulaiteye SM, et al (2002). Trends in Kaposi's sarcoma and non-Hodgkin's lymphoma incidence in the United States from 1973 through 1998. J Natl Cancer Inst 94(16): 1204–1210.

Engstrom JW, Lowenstein DH, Bredesen DE (1989). Cerebral infarctions and transient neurologic deficits associated with acquired immunodeficiency syndrome. Am J Med 86 (5): 528–532.

Epstein LG, Sharer LR, Joshi VV, et al (1985). Progressive encephalopathy in children with acquired immune deficiency syndrome. Ann Neurol 17(5): 488–496.

Evers S, Nabavi D, Rahmann A, et al (2003). Ischaemic cerebrovascular events in HIV infection: a cohort study. Cerebrovasc Dis 15(3): 199–205.

Ferreira MS, Nishioka Sde S, Silvestre MT, et al (1997). Reactivation of Chagas' disease in patients with AIDS: report of three new cases and review of the literature. Clin Infect Dis 25(6): 1397–1400.

Fish DG, Ampel NM, Galgiani JN, et al (1990). Coccidioidomycosis during human immunodeficiency virus infection. A review of 77 patients. Medicine (Baltimore) 69(6): 384–391.

Fleming DT, McQuillan GM, Johnson RE, et al (1997). Herpes simplex virus type 2 in the United States, 1976 to 1994. N Engl J Med 337(16): 1105–1111.

Folch C, Casabona J, Munoz R, et al (2005). [Trends in the prevalence of HIV infection and risk behaviors in homo- and bisexual men]. Gac Sanit 19(4): 294–301.

François-Jérôme Authier PC RKG (2005). Skeletal muscle involvement in human immunodeficiency virus (HIV)-infected patients in the era of highly active antiretroviral therapy (HAART). Muscle Nerve 32(3): 247–260.

Galgiani JN (1993). Coccidioidomycosis. West J Med 159 (2): 153–171.

Galgiani JN, Catanzaro A, Cloud GA, et al (1993). Fluconazole therapy for coccidioidal meningitis. The NIAID-Mycoses Study Group. Ann Intern Med 119(1): 28–35.

Gallant JE, Moore RD, Richman DD, et al (1992). Incidence and natural history of cytomegalovirus disease in patients with advanced human immunodeficiency virus disease treated with zidovudine. The Zidovudine Epidemiology Study Group. J Infect Dis 166(6): 1223–1227.

Gardner HA, Martinez AJ, Visvesvara GS, et al (1991). Granulomatous amebic encephalitis in an AIDS patient. Neurology 41(12): 1993–1995.

Gebo KA, Kalyani R, Moore RD, et al (2005). The incidence of, risk factors for, and sequelae of herpes zoster among HIV patients in the highly active antiretroviral therapy era. J Acquir Immune Defic Syndr 40(2): 169–174.

Gherardi RK, Chretien F, Delfau-Larue MH, et al (1998). Neuropathy in diffuse infiltrative lymphocytosis syndrome: an HIV neuropathy, not a lymphoma. Neurology 50(4): 1041–1044.

Gibson JJ, Hornung CA, Alexander GR, et al (1990). A cross-sectional study of herpes simplex virus types 1 and 2 in college students: occurrence and determinants of infection. J Infect Dis 162(2): 306–312.

Gilden DH, Beinlich BR, Rubinstien EM, et al (1994). Varicella-zoster virus myelitis: an expanding spectrum. Neurology 44(10): 1818–1823.

Gilden DH, Kleinschmidt-DeMasters BK, LaGuardia JJ, et al (2000). Neurologic complications of the reactivation of varicella-zoster virus. N Engl J Med 342(9): 635–645.

Gluck D, Kubanek B, Gaus W, et al (1990). [Current data on the prevalence and epidemiology of HIV from the HIV study by the German Red Cross of West Germany]. Infusionstherapie 17(3): 160–162.

Goldstein JD, Zeifer B, Chao C, et al (1991). CT appearance of primary CNS lymphoma in patients with acquired immunodeficiency syndrome. J Comput Assist Tomogr 15(1): 39–44.

Gordon SM, Steinberg JP, DuPuis MH, et al (1992). Culture isolation of Acanthamoeba species and leptomyxid amebas from patients with amebic meningoencephalitis, including two patients with AIDS. Clin Infect Dis 15(6): 1024–1030.

Grafe MR, Press GA, Berthoty DP, et al (1990). Abnormalities of the brain in AIDS patients: correlation of postmortem MR findings with neuropathology. AJNR Am J Neuroradiol 11(5): 905–911; discussion 912–913.

Green L (2005). Benefits of early invasive treatment for acute coronary syndromes: lost in translation? BMJ 330 (7500): E351–E352.

Group RG with ACT (1994). Foscarnet-Ganciclovir Cytomegalovirus Retinitis Trial. 4. Visual outcomes. Studies of Ocular Complications of AIDS Research Group in collaboration with the AIDS Clinical Trials Group. Ophthalmology 101(7): 1250–1261.

Grover D, Newsholme W, Brink N, et al (2004). Herpes simplex virus infection of the central nervous system in human immunodeficiency virus-type 1-infected patients. Int J STD AIDS 15(9): 597–600.

Guardia JA, Bourgoignie J, Diego J (2000). Renal mucormycosis in the HIV patient. Am J Kidney Dis 35(5): E24.

Gungor T, Funk M, Linde R, et al (1993). Cytomegalovirus myelitis in perinatally acquired HIV. Arch Dis Child 68 (3): 399–401.

Hajjeh RA, Pappas PG, Henderson H, et al (2001). Multicenter case-control study of risk factors for histoplasmosis in human immunodeficiency virus-infected persons. Clin Infect Dis 32(8): 1215–1220.

Hansen PB, Penkowa M, Kirk O, et al (2000). Human immunodeficiency virus-associated malignant lymphoma in eastern Denmark diagnosed from 1990–1996: clinical features, histopathology, and association with Epstein-Barr virus and human herpesvirus-8. Eur J Haematol 64(6): 368–375.

Harcourt-Webster JN, Scaravilli F, Darwish AH (1991). Strongyloides stercoralis hyperinfection in an HIV positive patient. J Clin Pathol 44(4): 346–348.

Hart G (1986). Syphilis tests in diagnostic and therapeutic decision making. Ann Intern Med 104(3): 368–376.

Hecht FM, Wheat J, Korzun AH, et al (1997). Itraconazole maintenance treatment for histoplasmosis in AIDS: a prospective, multicenter trial. J Acquir Immune Defic Syndr Hum Retrovirol 16(2): 100–107.

Helweg-Larsen S, Jakobsen J, Boesen F, et al (1988). Myelopathy in AIDS. A clinical and electrophysiological study of 23 Danish patients. Acta Neurol Scand 77(1): 64–73.

Hengge UR, Brockmeyer NH, Malessa R, et al (1993). Foscarnet penetrates the blood-brain barrier: rationale for therapy of cytomegalovirus encephalitis. Antimicrob Agents Chemother 37(5): 1010–1014.

Hernandez N, del Castillo F, Garcia-Miguel MJ, et al (2002). [Encephalitis as the first manifestation of herpes zoster]. Enferm Infecc Microbiol Clin 20(8): 415.

Hoffman PM, Festoff BW, Giron, Jr, LT, et al (1985). Isolation of LAV/HTLV-III from a patient with amyotrophic lateral sclerosis. N Engl J Med 313(5): 324–325.

Holding KJ, Dworkin MS, Wan PC, et al (2000). Aspergillosis among people infected with human immunodeficiency virus: incidence and survival. Adult and Adolescent Spectrum of HIV Disease Project. Clin Infect Dis 31(5): 1253–1257.

Holman RC, Janssen RS, Buehler JW, et al (1991). Epidemiology of progressive multifocal leukoencephalopathy in the United States: analysis of national mortality and AIDS surveillance data. Neurology 41(11): 1733–1736.

Jain AK, Agarwal K, el-Sadr W (1994). Streptococcus bovis bacteremia and meningitis associated with Strongyloides stercoralis colitis in a patient infected with human immunodeficiency virus. Clin Infect Dis 18(2): 253–254.

Jarvik JG, Hesselink JR, Wiley C, et al (1988). Coccidioidomycotic brain abscess in an HIV-infected man. West J Med 149(1): 83–86.

Jayakeerthi SR DM, Nagarathna S, Anandh B, et al (2004). Brain abscess due to cladophialophora bantiana. Indian J Med Microbiol 2004; 22: 193–195.

Jellinger KA, Setinek U, Drlicek M, et al (2000). Neuropathology and general autopsy findings in AIDS during the last 15 years. Acta Neuropathol (Berl) 100(2): 213–220.

Johnson PC, Sarosi GA, Septimus EJ, et al (1986). Progressive disseminated histoplasmosis in patients with the acquired immune deficiency syndrome: a report of 12 cases and a literature review. Semin Respir Infect 1(1): 1–8.

Johnson RW, Williams FM, Kazi S, et al (2003). Human immunodeficiency virus-associated polymyositis: a longitudinal study of outcome. Arthritis Rheum 49(2): 172–178.

Kalayjian RC, Cohen ML, Bonomo RA, et al (1993). Cytomegalovirus ventriculoencephalitis in AIDS. A syndrome with distinct clinical and pathologic features. Medicine (Baltimore) 72(2): 67–77.

Kaneko K, Onodera O, Miyatake T, et al (1990). Rapid diagnosis of tuberculous meningitis by polymerase chain reaction (PCR). Neurology 40(10): 1617–1618.

Karakousis PC, Moore RD, Chaisson RE (2004). *Mycobacterium avium* complex in patients with HIV infection in the era of highly active antiretroviral therapy. Lancet Infect Dis 4(9): 557–565.

Katz DA, Berger JR, Duncan RC (1993). Neurosyphilis. A comparative study of the effects of infection with human immunodeficiency virus. Arch Neurol 50(3): 243–249.

Kent SJ, Crowe SM, Yung A, et al (1993). Tuberculous meningitis: a 30-year review. Clin Infect Dis 17(6): 987–994.

Khoo VS, Wilson PC, Sexton MJ, et al (2000). Acquired immunodeficiency syndrome-related primary cerebral lymphoma: response to irradiation. Australas Radiol 44(2): 178–184.

Kirk OLE, Gatell JM, Mocroft A, et al (2000). Infections with *Mycobacterium tuberculosis* and *Mycobacterium avium* among HIV-infected patients after the introduction of highly active antiretroviral therapy. Am J Respir Crit Care Med 162(3): 865–872.

Kissel EC, Pukay-Martin ND, Bornstein RA (2005). The relationship between age and cognitive function in HIV-infected men. J Neuropsychiatry Clin Neurosci 17(2): 180–184.

Klein RS, Harris CA, Small CB, et al (1984). Oral candidiasis in high-risk patients as the initial manifestation of the acquired immunodeficiency syndrome. N Engl J Med 311(6): 354–358.

Kleinschmidt-DeMasters BK, Mahalingam R, Shimek C, et al (1998). Profound cerebrospinal fluid pleocytosis and Froin's syndrome secondary to widespread necrotizing vasculitis in an HIV-positive patient with varicella zoster virus encephalomyelitis. J Neurol Sci 159(2): 213–218.

Kozlowski PB, Sher JH, Rao C, et al (1993). Central nervous system in pediatric AIDS. Results from Neuropathologic Pediatric AIDS Registry. Ann N Y Acad Sci 693: 295–296.

Kure K, Llena JF, Lyman WD, et al (1991). Human immunodeficiency virus-1 infection of the nervous system: an autopsy study of 268 adult, pediatric, and fetal brains. Hum Pathol 22(7): 700–710.

Lakeman FD, Whitley RJ (1995). Diagnosis of herpes simplex encephalitis: application of polymerase chain reaction to cerebrospinal fluid from brain-biopsied patients and correlation with disease. National Institute of Allergy and Infectious Diseases Collaborative Antiviral Study Group. J Infect Dis 171(4): 857–863.

Langenberg AGM, Corey L, Ashley RL, et al (1999). A prospective study of new infections with herpes simplex virus type 1 and type 2. N Engl J Med 341(19): 1432–1438.

Lee BL, Holland GN, Glasgow BJ (1996). Chiasmal infarction and sudden blindness caused by mucormycosis in AIDS and diabetes mellitus. Am J Ophthalmol 122(6): 895–896.

Leport C, Raffi F, Matheron S, et al (1988). Treatment of central nervous system toxoplasmosis with pyrimethamine/sulfadiazine combination in 35 patients with the acquired immunodeficiency syndrome. Efficacy of long-term continuous therapy. Am J Med 84(1): 94–100.

Levy RM, Bredesen DE, Rosenblum ML (1985). Neurological manifestations of the acquired immunodeficiency syndrome (AIDS): experience at UCSF and review of the literature. J Neurosurg 62: 475–495.

Lifson AR (1988). Do alternate modes for transmission of human immunodeficiency virus exist? A review. JAMA 259(9): 1353–1356.

Lindsey JC, Hughes MD, McKinney RE, et al (2000). Treatment-mediated changes in human immunodeficiency virus (HIV) type 1 RNA and CD4 cell counts as predictors of weight growth failure, cognitive decline, and survival in HIV-infected children. J Infect Dis 182(5): 1385–1393.

Ling SM, Roach 3rd M, Larson DA, et al (1994). Radiotherapy of primary central nervous system lymphoma in patients with and without human immunodeficiency virus. Ten years of treatment experience at the University of California San Francisco. Cancer 73(10): 2570–2582.

Lipkin WI, Parry G, Kiprov D, et al (1985). Inflammatory neuropathy in homosexual men with lymphadenopathy. Neurology 35(10): 1479–1483.

Lobato MN, Caldwell MB, Ng P, et al (1995). Encephalopathy in children with perinatally acquired human immunodeficiency virus infection. Pediatric Spectrum of Disease Clinical Consortium. J Pediatr 126(5 Pt 1): 710–715.

Lortholary O, Meyohas MC, Dupont B, et al (1993). Invasive aspergillosis in patients with acquired immunodeficiency syndrome: report of 33 cases. French Cooperative Study Group on Aspergillosis in AIDS. Am J Med 95(2): 177–187.

Lortholary O, Denning DW, Dupont B (1999). Endemic mycoses: a treatment update. J Antimicrob Chemother 43(3): 321–331.

Ludmerer KM, Kissane JM (1996). Deterioration and death in a 30-year-old male with AIDS. Am J Med 100(5): 571–578.

Luft BJ, Remington JS (1992). Toxoplasmic encephalitis in AIDS. Clin Infect Dis 15(2): 211–222.

Lui KJ, Darrow WW, Rutherford 3rd GW (1988). A model-based estimate of the mean incubation period for AIDS in homosexual men. Science 240(4857): 1333–1335.

Ma P, Visvesvara GS, Martinez AJ, et al (1990). Naegleria and Acanthamoeba infections: review. Rev Infect Dis 12 (3): 490–513.

MacGowan DJ, Scelsa SN, Waldron M (2001). An ALS-like syndrome with new HIV infection and complete response to antiretroviral therapy. Neurology 57(6): 1094–1097.

Mahieux F, Gray F, Fenelon G, et al (1989). Acute myeloradiculitis due to cytomegalovirus as the initial manifestation of AIDS. J Neurol Neurosurg Psychiatry 52(2): 270–274.

Marra CM (1992). Syphilis and human immunodeficiency virus infection. Semin Neurol 12(1): 43–50.

Martinez AJ, Guerra AE, Garcia-Tamayo J, et al (1994). Granulomatous amebic encephalitis: a review and report of a spontaneous case from Venezuela. Acta Neuropathol (Berl) 87(4): 430–434.

Maschke M, Kastrup O, Esser S, et al (2000). Incidence and prevalence of neurological disorders associated with HIV since the introduction of highly active antiretroviral therapy (HAART). J Neurol Neurosurg Psychiatry 69(3): 376–380.

Mathew RM, Murnane M (2004). MRI in PML: bilateral medullary lesions. Neurology 63(12): 2380.

Mathews VP, Alo PL, Glass JD, et al (1992). AIDS-related CNS cryptococcosis: radiologic-pathologic correlation. AJNR Am J Neuroradiol 13(5): 1477–1486.

McArthur JC, Hoover DR, Bacellar H, et al (1993). Dementia in AIDS patients: incidence and risk factors. Multicenter AIDS Cohort Study. Neurology 43(11): 2245–2252.

McArthur JC, Haughey N, Gartner S, et al (2003). Human immunodeficiency virus-associated dementia: an evolving disease. J Neurovirol 9(2): 205–221.

McGuire D, Barhite S, Hollander H, et al (1995). JC virus DNA in cerebrospinal fluid of human immunodeficiency virus-infected patients: predictive value for progressive multifocal leukoencephalopathy. Ann Neurol 37(3): 395–399.

Mellins CA, Smith R, O'Driscoll P, et al (2003). High rates of behavioral problems in perinatally HIV-infected children are not linked to HIV disease. Pediatrics 111(2): 384–393.

Miller EN, Selnes OA, McArthur JC, et al (1990a). Neuropsychological performance in HIV-1-infected homosexual men: The Multicenter AIDS Cohort Study (MACS). Neurology 40(2): 197–203.

Miller RG, Storey JR, Greco CM (1990b). Ganciclovir in the treatment of progressive AIDS-related polyradiculopathy. Neurology 40(4): 569–574.

Minamoto GY, Barlam TF, Vander Els NJ (1992). Invasive aspergillosis in patients with AIDS. Clin Infect Dis 14 (1): 66–74.

Mirza SA, Phelan M, Rimland D, et al (2003). The changing epidemiology of cryptococcosis: an update from population-based active surveillance in 2 large metropolitan areas, 1992–2000. Clin Infect Dis 36(6): 789–794.

Misdrahi D, Vila G, Funk-Brentano I, et al (2004). DSM-IV mental disorders and neurological complications in children and adolescents with human immunodeficiency virus type 1 infection (HIV-1). Eur Psychiatry 19(3): 182–184.

Mofenson LM (2000). Technical report: perinatal human immunodeficiency virus testing and prevention of transmission. Committee on Pediatric AIDS. Pediatrics 106(6): E88.

Moncayo A (2003). Chagas disease: current epidemiological trends after the interruption of vectorial and transfusional transmission in the Southern Cone countries. Mem Inst Oswaldo Cruz 98(5): 577–591.

Moore EH, Russell LA, Klein JS, et al (1995). Bacillary angiomatosis in patients with AIDS: multiorgan imaging findings. Radiology 197(1): 67–72.

Moore RD, Wong WM, Keruly JC, et al (2000). Incidence of neuropathy in HIV-infected patients on monotherapy versus those on combination therapy with didanosine, stavudine and hydroxyurea. AIDS 14(3): 273–278.

Morgello S, Simpson DM (1994). Multifocal cytomegalovirus demyelinative polyneuropathy associated with AIDS. Muscle Nerve 17(2): 176–182.

Morgello S, Soifer FM, Lin CS, et al (1993). Central nervous system *Strongyloides stercoralis* in acquired immunodeficiency syndrome: a report of two cases and review of the literature. Acta Neuropathol (Berl) 86(3): 285–288.

Morgello S, Estanislao L, Simpson D, et al (2004). HIV-associated distal sensory polyneuropathy in the era of highly active antiretroviral therapy: the Manhattan HIV Brain Bank. Arch Neurol 61(4): 546–551.

Moulignier A, Authier FJ, Baudrimont M, et al (1997). Peripheral neuropathy in human immunodeficiency virus-infected patients with the diffuse infiltrative lymphocytosis syndrome. Ann Neurol 41(4): 438–445.

Moulignier A, Moulonguet A, Pialoux G, et al (2001). Reversible ALS-like disorder in HIV infection. Neurology 57 (6): 995–1001.

Murakawa GJ, McCalmont T, Altman J, et al (1995). Disseminated acanthamebiasis in patients with AIDS. A report of five cases and a review of the literature. Arch Dermatol 131(11): 1291–1296.

Musch DC, Martin DF, Gordon JF, et al (1997). Treatment of cytomegalovirus retinitis with a sustained-release ganciclovir implant. The Ganciclovir Implant Study Group. N Engl J Med 337(2): 83–90.

Navia BA, Cho ES, Petito CK, et al (1986). The AIDS dementia complex: II. Neuropathology. Ann Neurol 19 (6): 525–535.

Nemeth A, Bygdeman S, Sandstrom E, et al (1986). Early case of acquired immunodeficiency syndrome in a child from Zaire. Sex Transm Dis 13(2): 111–113.

Nzilambi N, De Cock KM, Forthal DN, et al (1988). The prevalence of infection with human immunodeficiency virus over a 10-year period in rural Zaire. N Engl J Med 318(5): 276–279.

Ogg G, Lynn WA, Peters M, et al (1997). Cerebral nocardia abscesses in a patient with AIDS: correlation of magnetic resonance and white cell scanning images with neuropathological findings. J Infect 35(3): 311–313.

Oliveira V, Costa A (2001). [Cerebral hematoma caused by mucormycosis]. Rev Neurol 33(10): 951–953.

Padgett BL, Walker DL, ZuRhein GM, et al (1971). Cultivation of papova-like virus from human brain with progressive multifocal leucoencephalopathy. Lancet 1(7712): 1257–1260.

Palella FJ, Jr, Delaney KM, Moorman AC, et al (1998). Declining morbidity and mortality among patients with advanced human immunodeficiency virus infection. HIV Outpatient Study Investigators. N Engl J Med 338(13): 853–860.

Palo J, Haltia M, Uutela T (1975). Cerebral aspergillosis with special reference to cerebrospinal fluid findings. Eur Neurol 13(3): 224–231.

Pappas PG, Pottage JC, Powderly WG, et al (1992). Blastomycosis in patients with the acquired immunodeficiency syndrome. Ann Intern Med 116(10): 847–853.

Perriens JH, Colebunders RL, Karahunga C, et al (1991). Increased mortality and tuberculosis treatment failure rate among human immunodeficiency virus (HIV) seropositive compared with HIV seronegative patients with pulmonary tuberculosis treated with standard chemotherapy in Kinshasa, Zaire. Am Rev Respir Dis 144(4): 750–755.

Pizzo PA, Eddy J, Falloon J, et al (1988). Effect of continuous intravenous infusion of zidovudine (AZT) in children with symptomatic HIV infection. N Engl J Med 319(14): 889–896.

Porter SB, Sande MA (1992). Toxoplasmosis of the central nervous system in the acquired immunodeficiency syndrome. N Engl J Med 327(23): 1643–1648.

Powderly WG (2000). Cryptococcal meningitis in HIV-infected patients. Curr Infect Dis Rep 2(4): 352–357.

Power C, Nath A, Aoki FY, et al (1997). Remission of progressive multifocal leukoencephalopathy following splenectomy and antiretroviral therapy in a patient with HIV infection. N Engl J Med 336(9): 661–662.

Quereda C, Corral I, Laguna F, et al (2000). Diagnostic utility of a multiplex herpesvirus PCR assay performed with cerebrospinal fluid from human immunodeficiency virus-infected patients with neurological disorders. J Clin Microbiol 38(8): 3061–3067.

Raffi F, Aboulker JP, Michelet C, et al (1997). A prospective study of criteria for the diagnosis of toxoplasmic encephalitis in 186 AIDS patients. The BIOTOXO Study Group. AIDS 11(2): 177–184.

Ramsey RG, Geremia GK (1988). CNS complications of AIDS: CT and MR findings. AJR Am J Roentgenol 151 (3): 449–454.

Raphael SA, Price ML, Lischner HW, et al (1991). Inflammatory demyelinating polyneuropathy in a child with symptomatic human immunodeficiency virus infection. J Pediatr 118(2): 242–245.

Remick SC, Diamond C, Migliozzi JA, et al (1990). Primary central nervous system lymphoma in patients with and without the acquired immune deficiency syndrome. A retrospective analysis and review of the literature. Medicine (Baltimore) 69(6): 345–360.

Renold C, Sugar A, Chave JP, et al (1992). Toxoplasma encephalitis in patients with the acquired immunodeficiency syndrome. Medicine (Baltimore) 71(4): 224–239.

Robertson KR, Kapoor C, Robertson WT, et al (2004). No gender differences in the progression of nervous system disease in HIV infection. J Acquir Immune Defic Syndr 36(3): 817–822.

Rocha A, de Meneses AC, da Silva AM, et al (1994). Pathology of patients with Chagas' disease and acquired immunodeficiency syndrome. Am J Trop Med Hyg 50(3): 261–268.

Rodriguez-Arrondo F, Aguirrebengoa K, De Arce A, et al (1998). Candidal meningitis in HIV-infected patients: treatment with fluconazole. Scand J Infect Dis 30(4): 417–418.

Rosemberg S, Chaves CJ, Higuchi ML, et al (1992). Fatal meningoencephalitis caused by reactivation of Trypanosoma cruzi infection in a patient with AIDS. Neurology 42(3 Pt 1): 640–642.

Rosenblum ML, Levy RM, Bredesen DE, et al (1988). Primary central nervous system lymphomas in patients with AIDS. Ann Neurol 23(Suppl): S13–S16.

Rosenfeldt V, Valerius NH, Paerregaard A (2000). Regression of HIV-associated progressive encephalopathy of childhood during HAART. Scand J Infect Dis 32(5): 571–574.

Roullet E, Assuerus V, Gozlan J, et al (1994). Cytomegalovirus multifocal neuropathy in AIDS: analysis of 15 consecutive cases. Neurology 44(11): 2174–2182.

Sacks SL, Griffiths PD, Corey L, et al (2004). HSV shedding. Antiviral Res 63(Suppl 1): S19–S26.

Sacktor N (2002). The epidemiology of human immunodeficiency virus-associated neurological disease in the era of highly active antiretroviral therapy. J Neurovirol 8 (Suppl 2): 115–121.

Sacktor N, Lyles RH, Skolasky R, et al (2001). HIV-associated neurologic disease incidence changes: Multicenter AIDS Cohort Study, 1990–1998. Neurology 56(2): 257–260.

Said G, Lacroix C, Chemouilli P, et al (1991). Cytomegalovirus neuropathy in acquired immunodeficiency syndrome: a clinical and pathological study. Ann Neurol 29 (2): 139–146.

Sanchez JM, Ramos Amador JT, Fernandez de Miguel S, et al (2003). Impact of highly active antiretroviral therapy on the morbidity and mortality in Spanish human immunodeficiency virus-infected children. Pediatr Infect Dis J 22(10): 863–867.

Sanchez-Ramon S, Bellon JM, Resino S, et al (2003). Low blood CD8+ T-lymphocytes and high circulating monocytes are predictors of HIV-1-associated progressive encephalopathy in children. Pediatrics 111(2): E168–E175.

Schacker T, Hu HL, Koelle DM, et al (1998). Famciclovir for the suppression of symptomatic and asymptomatic herpes simplex virus reactivation in HIV-infected persons: a double-blind, placebo-controlled trial. Ann Intern Med 128(1): 21–28.

Schifitto G, McDermott MP, McArthur JC, et al (2002). Incidence of and risk factors for HIV-associated distal sensory polyneuropathy. Neurology 58(12): 1764–1768.

Schmutzhard E (2001). Viral infections of the CNS with special emphasis on herpes simplex infections. J Neurol 248 (6): 469–477.

Schwartzman W (1996). Bartonella (Rochalimaea) infections: beyond cat scratch. Annu Rev Med 47: 355–364.

Schwartzman WA, Marchevsky A, Meyer RD (1990). Epithelioid angiomatosis or cat scratch disease with splenic and hepatic abnormalities in AIDS: case report and review of the literature. Scand J Infect Dis 22(2): 121–133.

Scott EN, Kaufman L, Brown AC, Muchmore HG (1987). Serologic studies in the diagnosis and management of meningitis due to *Sporothrix schenckii*. N Engl J Med 317(15): 935–940.

Seijo Martinez M, Gonzalez-Mediero G, Santiago P, et al (2000). Granulomatous amebic encephalitis in a patient with AIDS: isolation of acanthamoeba sp. Group II from brain tissue and successful treatment with sulfadiazine and fluconazole. J Clin Microbiol 38(10): 3892–3895.

Shaw GM, Harper ME, Hahn BH, et al (1985). HTLV-III infection in brains of children and adults with AIDS encephalopathy. Science 227(4683): 177–182.

Shimojima Y, Yazaki M, Kaneko K, et al (2005). Characteristic spinal MRI findings of HIV-associated myelopathy in an AIDS patient. Intern Med 44(7): 763–764.

Siegel D, Golden E, Washington AE, et al (1992). Prevalence and correlates of herpes simplex infections. The population-based AIDS in Multiethnic Neighborhoods Study. JAMA 268(13): 1702–1708.

Silva-Vergara ML, Maneira FR, De Oliveira RM, et al (2005). Multifocal sporotrichosis with meningeal involvement in a patient with AIDS. Med Mycol 43(2): 187–190.

Silver SA, Arthur RR, Erozan YS, et al (1995). Diagnosis of progressive multifocal leukoencephalopathy by stereotactic brain biopsy utilizing immunohistochemistry and the polymerase chain reaction. Acta Cytol 39(1): 35–44.

Simpson DM, Slasor P, Dafni U, et al (1997). Analysis of myopathy in a placebo-controlled zidovudine trial. Muscle Nerve 20(3): 382–385.

Singh N, Yu VL, Rihs JD (1991). Invasive aspergillosis in AIDS. South Med J 84(7): 822–827.

Skiest DJ (2002). Focal neurological disease in patients with acquired immunodeficiency syndrome. Clin Infect Dis 34 (1): 103–115.

Spach DH, Koehler JE (1998). Bartonella-associated infections. Infect Dis Clin North Am 12(1): 137–155.

Stricker RB, Sanders KA, Owen WF, et al (1992). Mononeuritis multiplex associated with cryoglobulinemia in HIV infection. Neurology 42(11): 2103–2105.

Tattevin P, Schortgen F, de Broucker T, et al (2001). Varicella-zoster virus limbic encephalitis in an immunocompromised patient. Scand J Infect Dis 33(10): 786–788.

The Antiretroviral Therapy in Lower Income Countries Study G (2005). Cohort Profile: Antiretroviral Therapy in Lower Income Countries (ART-LINC): international collaboration of treatment cohorts. Int J Epidemiol 34(5): 979–986.

Thornton CA, Latif AS, Emmanuel JC (1991). Guillain-Barre syndrome associated with human immunodeficiency virus infection in Zimbabwe. Neurology 41(6): 812–815.

Thornton CA, Houston S, Latif AS (1992). Neurocysticercosis and human immunodeficiency virus infection. A possible association. Arch Neurol 49(9): 963–965.

Thurnher MM, Rieger A, Kleibl-Popov C, et al (2001). Primary central nervous system lymphoma in AIDS: a wider spectrum of CT and MRI findings. Neuroradiology 43(1): 29–35.

Timmermans M, Carr J (2004). Neurosyphilis in the modern era. J Neurol Neurosurg Psychiatry 75(12): 1727–1730.

Toledo PV, Pellegrino LN, Cunha CA (2004). Varicella-zoster virus encephalitis in an AIDS patient. Braz J Infect Dis 8 (3): 255–258.

Tyler KL (2004). Herpes simplex virus infections of the central nervous system: encephalitis and meningitis, including Mollaret's. Herpes 11(Suppl 2): 57A–64A.

Ultmann MH, Belman AL, Ruff HA, et al (1985). Developmental abnormalities in infants and children with acquired immune deficiency syndrome (AIDS) and AIDS-related complex. Dev Med Child Neurol 27(5): 563–571.

UNAIDS (2004). 2004 report on the global HIV/AIDS epidemic: 4th global report.

Valcour V, Shikuma C, Shiramizu B, et al (2004). Higher frequency of dementia in older HIV-1 individuals: the Hawaii Aging with HIV-1 Cohort. Neurology 63(5): 822–827.

Valencia Ortega ME, Enriquez Crego A, Laguna Cuesta F, et al (2000). [Listeriosis: an infrequent infection in patients with HIV]. An Med Interna 17(12): 649–651.

van der Horst CM, Saag MS, Cloud GA, et al (1997). Treatment of cryptococcal meningitis associated with the acquired immunodeficiency syndrome. National Institute of Allergy and Infectious Diseases Mycoses Study Group and AIDS Clinical Trials Group. N Engl J Med 337(1): 15–21.

Veenstra J, Krol A, van Praag RM, et al (1995). Herpes zoster, immunological deterioration and disease progression in HIV-1 infection. AIDS 9(10): 1153–1158.

Verma A (2001). Epidemiology and clinical features of HIV-1 associated neuropathies. J Peripher Nerv Syst 6(1): 8–13.

Vidal JE, Dauar RF, Melhem MS, et al (2005). Cerebral aspergillosis due to *Aspergillus fumigatus* in AIDS patient: first culture-proven case reported in Brazil. Rev Inst Med Trop Sao Paulo 47(3): 161–165.

Villoria MF, de la Torre J, Fortea F, et al (1992). Intracranial tuberculosis in AIDS: CT and MRI findings. Neuroradiology 34(1): 11–14.

Walker M, Zunt JR (2005). Parasitic central nervous system infections in immunocompromised hosts. Clin Infect Dis 40(7): 1005–1015.

Ward BA, Parent AD, Raila F (1995). Indications for the surgical management of central nervous system blastomycosis. Surg Neurol 43(4): 379–388.

Watemberg N, Morton LD (1996). Images in clinical medicine. Periodic lateralized epileptiform discharges. N Engl J Med 334(10): 634.

Weiss PJ, DeMarco JK (1994). Images in clinical medicine. Progressive multifocal leukoencephalopathy. N Engl J Med 330(17): 1197.

Wheat LJ, Connolly-Stringfield PA, Baker RL, et al (1990). Disseminated histoplasmosis in the acquired immune deficiency syndrome: clinical findings, diagnosis and treatment, and review of the literature. Medicine (Baltimore) 69(6): 361–374.

Wheat LJ, Musial CE, Jenny-Avital E (2005). Diagnosis and management of central nervous system histoplasmosis. Clin Infect Dis 40(6): 844–852.

WHO (1997). Control of Chagas' Disease: Report of the WHO Expert Committee, World Health Organization, Geneva.

WHO (2005). Global Tuberculosis Control Report, World Health Organization, Geneva.

Williams B, Fojtasek M, Connolly-Stringfield P, et al (1994). Diagnosis of histoplasmosis by antigen detection during an outbreak in Indianapolis, Ind. Arch Pathol Lab Med 118(12): 1205–1208.

Witzig RS, Hoadley DJ, Greer DL, et al (1994). Blastomycosis and human immunodeficiency virus: three new cases and review. South Med J 87(7): 715–719.

Wolf DG, Spector SA (1992). Diagnosis of human cytomegalovirus central nervous system disease in AIDS patients by DNA amplification from cerebrospinal fluid. J Infect Dis 166(6): 1412–1415.

Wulff EA, Wang AK, Simpson DM (2000). HIV-associated peripheral neuropathy: epidemiology, pathophysiology and treatment. Drugs 59(6): 1251–1260.

Wyen C, Hoffmann C, Schmeisser N, et al (2004). Progressive multifocal leukencephalopathy in patients on highly active antiretroviral therapy: survival and risk factors of death. J Acquir Immune Defic Syndr 37(2): 1263–1268.

Yoo TW, Mlikotic A, Cornford ME, et al (2004). Concurrent cerebral American trypanosomiasis and toxoplasmosis in a patient with AIDS. Clin Infect Dis 39(4): e30–e34.

Young LS, Inderlied CB, Berlin OG, et al (1986). Mycobacterial infections in AIDS patients, with an emphasis on the *Mycobacterium avium* complex. Rev Infect Dis 8(6): 1024–1033.

Zagardo MT, Castellani RJ, Zoarski GH, et al (1997). Granulomatous amebic encephalitis caused by leptomyxid amebae in an HIV-infected patient. AJNR Am J Neuroradiol 18 (5): 903–908.

Zangerle R, Allerberger F, Pohl P, et al (1991). High risk of developing toxoplasmic encephalitis in AIDS patients seropositive to *Toxoplasma gondii*. Med Microbiol Immunol (Berl) 180(2): 59–66.

Ziegler JL, Beckstead JA, Volberding PA, et al (1984). Non-Hodgkin's lymphoma in 90 homosexual men. Relation to generalized lymphadenopathy and the acquired immunodeficiency syndrome. N Engl J Med 311(9): 565–570.

Handbook of Clinical Neurology, Vol. 85 (3rd series)
HIV/AIDS and the Nervous System
P. Portegies, J. R. Berger, Editors

Chapter 3

HIV neurology in the developing world

GRETCHEN L. BIRBECK*

Michigan State University, East Lansing, MI, USA and Chikankata Health Services, Mazabuka, Zambia

3.1. HIV neurology in the developing world

Until recently HIV neurology in the developing world generally represented the natural history of HIV untouched by therapies or interventions and to a large extent this is still the case. In regions of the developing world HIV prevalence rates remain >20% and antiretroviral agents (ARVs) are not yet routinely available although they are becoming more so. Furthermore, prophylactic therapies to prevent opportunistic infections are frequently not available or used routinely among most people living with AIDS (PLWAs). The definition of "developing" or low-income country is generally determined by the World Bank based upon gross national income (GNI) per capita (http://www.worldbank.org/data/aboutdata/errata03/Class.htm). Today, lower income countries are defined as those with a GNI of less than US$735. Sub-Saharan Africa has the largest number of countries in this category (48) with Latin America following at 30, Central Europe and Asia (27), East Asia/Pacific Islands (24), the Middle East/North Africa (15) and Southern Asia (8). Several countries that have recently risen to "middle" income status were still developing countries when the AIDS epidemic began and data from those regions are also presented here.

What we know of the epidemiology of HIV neurology in developing regions is limited by the dearth of neurologic expertise in most of these regions compounded by little diagnostic capacity in imaging or neurophysiology. No population-based study of the epidemiology of HIV neurology in a developing country has been completed. Although some authors of hospital-based clinical and autopsy studies have claimed their reports were "population-based", they failed to recognize that in developing countries patients presenting for care at a tertiary care center are almost certainly not representative of the general population. In the countries hardest hit by the AIDS epidemic, most people dying of AIDS do so in the community without accessing care at secondary or tertiary care centers (MPowell et al., unpublished data).

HIV neurology in this environment clearly presents challenges as dual- and poly-central nervous system (CNS) infections are probably even more common where ARVs are not available, prophylaxis for opportunistic infections is not utilized and infectious conditions are rife independent of the HIV epidemic.

3.2. Diagnostic limitations in developing countries

Less developed countries throughout the world suffer from a severe lack of healthcare providers with neurologic expertise. A 2002 survey conducted by the World Federation of Neurology found that in all of the sub-Saharan, excluding South Africa, only 11 neurologists were present (Bergen, 2002). Even in those countries with some neurologic expertise, the population served per neurologist indicates the lack of human resources available to systematically study HIV neurology or provide care. On average in the USA, there is one neurologist for every 26,200 people. Compare this with India at 2,180,000, the Philippines at 556,000 or Guatemala at 444,000. Moreover, this situation is unlikely to change in the next several years. Among the poorest 69 countries in the world, only four have neurologic training programs. Unfortunately, physician emigration from developing countries continues to be problematic. Neurologists from developing countries who complete part of their studies abroad are especially likely to relocate to more developed regions (Bergen, 2005).

Of course, good neurologic care might be delivered in the absence of neurologists — but studies of the

*Correspondence to: Gretchen L. Birbeck, MD, MPH, Associate Professor & Director, Michigan State University International Neurologic & Psychiatric Epidemiology Program, #138 Service Road, A217, East Lansing, MI 48824-1313, USA. E-mail: gretchen.birbeck@ht.msu.edu or gbirbeck@zamnet.zm, Tel: +1-517-353-8122 or +260-(0)32-30765, Fax: 1-517-432-9414 ot.

primary healthcare providers in Zambia indicate that less than a third of the non-physician healthcare providers in primary care centers have received training in the diagnosis and treatment of common neurologic conditions in their region. Unfortunately, access to physician-level services is limited by social, financial and geographic barriers to care and referral (Birbeck and Munsat, 2002; Birbeck and Kalichi, 2004). Even when care is sought and neurologic expertise is available, advanced neuroimaging serologic studies, basic metabolic assessments and neurophysiologic studies often cannot be acquired.

3.3. Autopsy versus clinical studies of HIV neurology in developing regions

Given the limited diagnostic capacity available in these settings, most autopsy studies have found that clinical studies probably under-diagnose CNS problems, especially cytomegalovirus (CMV), primary CNS lymphoma (PCNSL) and non-Hodgkin's lymphoma (NHL) (Mohar et al., 1992). Therefore, data from developing countries based purely upon clinical diagnoses without autopsy evidence should be viewed with some skepticism.

A retrospective review of 10 HIV-positive, criminal autopsy cases from Tanzania revealed that 80% had evidence of CNS pathology including TB meningitis and abscess, bacterial meningitis and lymphocytic meningitis. None of these cases had CNS disease suspected as the cause of death, supporting the fact that CNS pathology is almost certainly under-diagnosed clinically in developing world settings (Kibayashi et al., 1999).

3.3.1. Clinical studies

See Table 3.1 for an overview of clinical studies detailing neurologic problems found among AIDS patients in developing countries. Inconsistent case definitions, varying diagnostic capacity and different case selection methods make comparisons across populations difficult. Nonetheless, the conditions described will be familiar to neurologists caring for patients in a more developed setting.

As these reports indicate, neurologic disorders are common among patients in developing regions presenting with AIDS-related conditions. Bane et al. completed a clinical assessment of 237 patients suspected to have

Table 3.1

Clinical neurologic diagnoses or conditions in developing country case series

Country	No. of cases	Source	Case selection	Diagnosis
Central African Republic	51	Di Costanzo et al., 1989	Hospitalized patients with HIV	Hemiparesis 18% Cryptococcal meningitis 16% Bacterial meningitis 16% Encephalopathy 16% Zoster 14% Facial palsy 8% Lymphocytic meningitis 4%
Tanzania	200	Howlett et al., 1989	Among inpatients suspected to have HIV who were positive and met WHO clinical criteria for AIDS	Tremor 76% Frontal release sign 72% HIV dementia 54% Retinopathy 23% Areflexia 21% Pyramidal tract signs 19% Mass lesion 11%
Zaire	104	Perriens et al., 1992	Hospital-based cross-sectional study of consecutive patients admitted to internal medicine	Cryptococcal meningitis 6% Metabolic encephalopathy 6% CNS mass lesion 3% Stroke 4% Myelopathy 1% Delirium 3%
Camaroon	51	Atangana et al., 2003	Consecutive HIV positive patients admitted to ICU over 2-year period	Subacute encephalopathy 42% Aseptic meningitis 20% Abscess 16% Toxoplasmosis 12% Tumor 8% Bacterial meningitis 4%

Ethiopia	237	Bane et al., 2003	Identified people "suspected" to have AIDS who tested positive	CNS mass lesion 31% TB meningitis 8% HIV dementia 6% Cryptococcal meningitis 6% Peripheral neuropathy 5% Myelopathy 5% Tuberculoma 3% Spinal TB 2%
South Africa	32	Modi and Modi, 2004	Series of HIV-positive patients with focal brain lesions assessed by imaging, serologies and biopsy	CNS TB 53% Neurocysticercosis (NCC) 19% TB and NCC 6% Multiple CVAs 6% TB + cryptococcal meningitis 3% Toxoplasmosis 3% PML 3% PCNSL 3%
Bangalore, India	100	Satishchandra et al., 2000	Retrospective chart review of known HIV-positive patients	Cryptococcal meningitis 80% CNS TB 24% Toxoplasmosis 9% Cryptococcal meningitis + TBM 6% Toxoplasmosis + TBM 4% Pyomeningitis 1% CNS syphilis 1%
Mumbai, India	85	Satishchandra et al., 2000	Retrospective clinical data from autopsy series for patients with HIV referred to neurology	Delirium 20% Seizures 17% Motor problems 17% Headache 17% Abnormal behavior 12% Cognitive problems 7%
South India	594	Kumarasamy et al., 2003	594 attendees of HIV clinic who had CD^4 counts available and attended more than once	Cryptococcal meningitis 5% Toxoplasmosis 4%
Mexico	40	Trujillo et al., 1995	Sample of patients hospitalized with AIDS	HIV encephalopathy 33% HIV dementia 33% Cryptococcal meningitis 17% Peripheral neuropathy 15% CNS TB 10% Toxoplasmosis 8% Aseptic meningitis 8% Post-herpetic neuralgia 5% PML 3% PCNSL 3% Myelopathy 3%

HIV in Addis Ababa, Ethiopia, in 2001. A total of 92% of cases were in WHO stage 4 AIDS and over half of the patients who died had a CNS mass lesion as the cause of death with TB or cryptococcal meningitis being the commonest cause (Bane et al., 2003). Similarly in Tanzania neurologic examinations of people with AIDS revealed significant abnormalities in more than half the individuals examined and neurologic abnormalities were more common among patients in the later stages of HIV (Howlett et al., 1989). In a series of Central African Republic patients admitted to internal medicine, simply having a neurologic complaint was strongly predictive of being HIV positive with a positive predictive value of 66% (Di Costanzo et al., 1989).

3.3.2. Autopsy data

Table 3.2 describes the neuropathologic findings from biopsy and/or autopsy case series in developing countries. Again, case selection methods make cross-country comparisons difficult. Premorbid under-diagnosis of CNS disease is evident. For example, in the study from Mexico of 160 brain autopsies the researchers found that clinically CMV, toxoplasmosis and cryptococcal meningitis were under-diagnosed and CNS TB was over-diagnosed (presumptive diagnoses being false). Furthermore, of the CNS malignancies identified only 50% were recognized prior to the autopsy (Mohar et al., 1992).

3.4. Specific conditions in the developing world also common in developed countries

3.4.1. Cryptococcal meningitis

Early cases of cryptococcal meningitis reported in young patients from the Congo River Basin (now the Democratic Republic of Congo) in the 1950s and 1960s have been proposed as evidence that HIV originated in

Table 3.2

Neurologic findings in biopsy/autopsy series from developing regions

Country	Source	Number of cases	Pathology
Ivory Coast	Bell et al., 1997	70 children	HIV encephalitis 6% Toxoplasmosis 4% CNS CMV 3%
Bangalore, India	Santosh et al., 1995	10 adults	Cryptococcal meningitis 50% Toxoplasmosis 20% Acanthomeba 10% HIV leukoencephalopathy 10%
South Korea	Oh et al., 1999[1]	173 adults	HIV leukoencephalopathy 5% Cryptococcal meningitis 2% PML 1% PCNSL 0.6%
India (Mumbai)	Satishchandra et al., 2000	85 adults	Lymphocytic meningitis 25% Toxoplasmosis 13% CNS TB 12% Cryptococcal meningitis 8% CMV 7% AIDP 6% Brainstem lesion NOS 3% Mononeuritis multiplex 2% PML 1% Other findings of primary HIV in brain 25%[2]
Mexico	Mohar et al., 1992	160 adults	Toxoplasmosis 19% Cryptococcal meningitis 10% CNS TB 7% CNS CMV 7% NHL 5% PCNSL 1%
Brazil	Chimelli et al., 1992	252	Toxoplasmosis 34% Cryptococcal meningitis 14% CNS CMV 8% Nodular encephalopathy 7% PCNSL 4% HIV leukoencephalopathy 11% Chagas' encephalitis (1 case)

[1]ARVs used by most patients
[2]Including microglial nodules, angiocentric pallor, choroids plexitis, and calcifications

this region (Molez, 1998). *C. neoformans* is common in wood used in hut-building and the digestive tracts of cockroaches in humid African climates (Molez, 1998). In Burundi, *C. neoformans* has been found in common house dust and the pigeon coops of persons infected with cryptococcal meningitis (Swinne et al., 1991). With an environment so rich in this fungus, it is not surprising that cryptococcal meningitis appears so often among AIDS patients in sub-Saharan Africa.

A review of 31 Malawian patients with cryptococcal meningitis seen in 1991–1993 revealed that the median duration of symptoms was 2 weeks, with a range of 1 day to 5 months. Some 97% presented with headache, 74% nuchal rigidity, 61% fever and 58% with altered consciousness. Since most could not afford treatment, death ensued within ~4 days after diagnosis. Among those who could afford treatment, the median survival was 4–9 months (Maher and Mwandumba, 1994). Malawi is not unique. Until recently, the public medical systems of many sub-Saharan African countries were unable to provide treatment for cryptococcal meningitis. Moreover, the iatrogenic risks of amphotericin without precise delivery systems or access to blood chemistries, drug levels or renal function testing was of concern even if the drug could be purchased by the patient. Symptomatic treatment with narcotics and serial LPs might be all a patient could be offered. Recently free fluconazole donated by the pharmaceutical industry to many regions has offered life-sustaining therapy.

In India, cryptococcal infections are also common. A study of AIDS defining illness and cause of death completed in southern India found cryptococcal meningitis was the initial AIDS defining illness among 4.7% of people with AIDS with a median survival with treatment of 22 months (Kumarasamy et al., 2003). Studies in Mexico suggest that cryptococcal meningitis may be more common in Mexico than in bordering US states (Trujillo et al., 1995).

Dual CNS pathology with cryptococcal meningitis plus another opportunistic infection is especially problematic in developing countries with limited diagnostic options. In a series of 38 South African miners who underwent lumbar puncture, cryptococcal meningitis was found in seven with three cases demonstrating additional infections — two with TB and one with syphilis (Silber et al., 1998). In an Indian study, cryptococcal meningitis was associated with low-grade fevers and headaches. These symptoms were difficult to distinguish from the symptoms seen with confirmed TB meningitis. Head CTs are normal in the majority of cases. Furthermore, cell count and protein levels from the CSF in these patients were minimally helpful in distinguishing cryptococcal meningitis from TB meningitis (Satishchandra et al., 2000). TB was more likely to present with fever and seizures were more common in the cryptococcal meningitis group. India ink staining, the CSF diagnostic test most likely to be available in developing regions, has a sensitivity of 70–75% compared to 90–100% for antigen and cultures (John and Coovadia, 1998), making under-diagnosis of dual pathology particularly problematic. Some have advocated combination therapy awaiting cultures. However, this is not practical in clinics and hospitals with insufficient drug supplies. Furthermore, the routine practice of giving polytherapy to patients regardless of LP results may encourage physicians to proceed directly to treatment forgoing LP altogether.

3.4.2. Toxoplasmosis

Whether toxoplasmosis is seen in a developing country appears to be primarily determined by the primary *Toxoplasma gondii* infection rates within the general population. In the USA, 10–40% of the population has evidence on serologic testing of previous *T. gondii* infection. In Africa, serologies suggest that rates of primary infection are extremely high and toxoplasmosis is responsible for a great deal of morbidity and mortality among AIDS patients. Within South Africa, toxoplasmosis serologies are positive in 29% of blacks. In Kwa-Zulu Natal, rates are 46% (Modi and Modi, 2004). A total of 53% of AIDS-related deaths in an autopsy series in the Ivory Coast were due to toxoplasmosis (Lucas et al., 1993).

Within Latin America, toxoplasmosis may be more common than in bordering US states (Trujillo et al., 1995). Toxoplasmosis serology prevalence in the Brazilian population is 50–80% and toxoplasmosis is the third most common AIDS-associated opportunistic infection and the commonest cause of CNS lesions in this population (Vidal et al., 2004). Even within the same country, positive toxoplasmosis serologies may vary substantially by SES. In Brazil rates are 84% for lower SES, 62% among the middle class and 23% in the higher income individuals, with exposure among the lower and middle classes likely related to drinking unfiltered water (Bahia-Oliveira et al., 2003).

Within India, serologies are estimated to range from 3.3 to 18.9% (Lanjewar et al., 1998) with rates as high as 57% in the Uttar Pradesh region (Singh and Nautiyal, 1991). In the first autopsy series from India to confirm toxoplasmosis, out of 49 consecutive autopsies in HIV-positive individuals, 10 had toxoplasmosis and the condition had been misdiagnosed in all 10 cases, with eight attributed to CNS TB, one case diagnosed as a stroke and another as HIV encephalitis (Lanjewar et al., 1998). Toxoplasmosis was common in Mumbai autopsies, but was also under-diagnosed premorbidly (Mathew and Chandy, 1999; Satishchandra et al., 2000).

In contrast, toxoplasmosis is very rare in South Korea, where *T. gondii* serologies are positive in only 4% of South Koreans. These low rates of infection in Korea are attributed to the fact that Koreans have little exposure to cats and dogs as pets (Oh et al., 1999).

Treatment of toxoplasmosis in developing countries using regimens from developed countries is often not possible. However, cotrimoxazole (CTX; sulfamethoxazole plus trimethoprim) is routinely available in many developing countries and open label studies have shown improvement among patients with toxoplasmosis who receive cotrimoxazole (regimen 960 mg QID for 2 days, TID for 2 weeks, BID to CT scan resolution, then QD maintenance; Smadja et al., 1998). Sulfadoxine–pyramethamine antimalarial combinations, where available, may also be effective treatment for toxoplasmosis. In US studies, the risk of Stevens Johnson syndrome is higher in PLWAs compared with uninfected populations, and this has certainly been my experience in Africa. The use of CTX for treatment of toxoplasmosis may be limited by insufficient drug supplies where CTX is purchased for acute treatment of infections unrelated to AIDS that only require a brief course of treatment. Given the doses required for toxoplasmosis and the chronic nature of treatment, a handful of toxoplasmosis patients could consume all of the CTX supplied for large hospital catchment areas. De facto triage is forced upon physicians who must decide the best use of available drugs. UNAIDS recommends CTX for prophylaxis among PLWAs, which of course would require an even greater supply of medication than CTX for treatment (UNAIDS: http://www.who.int/3by5/mediacentre/en/Cotrimstatement.pdf).

3.4.3. CNS TB

Multiple studies confirm that CNS TB involvement is more likely among PLWAs in developing regions (Trujillo et al., 1995) relative to those in developed countries. And multiple studies have confirmed CNS TB as a common cause of death in developing regions, especially Africa and India where the TB epidemic rages (Gray, 1997; Satishchandra et al., 2000).

TB may accelerate the progression from HIV to AIDS and HIV certainly increases the likelihood of active TB. In sub-Saharan Africa, 20–67% of people with TB are co-infected with HIV (De Cock et al., 1992). Hence, the two conditions behave almost symbiotically. Among persons with both CNS TB and AIDS, a fulminant TB course is more likely to occur and often involves hydrocephalus and dissemination (Gray, 1997).

3.4.4. Bacterial infections

Although not typically thought of as an opportunistic infection, purulent bacterial meningitis, especially streptococcal pneumonia, does occur more often among PLWAs in developing regions. In fact, in West Africa bacterial meningitis is the cause of death in up to 5% of HIV-related deaths (Gray, 1997). Tertiary syphilis, or rather the quaternary syphilis described among PLWAs, is more likely to occur in developing regions where syphilis remains a common sexually transmitted disease and antibiotics are less likely to be prescribed for incidental infections. One might question whether the "strokes" described in many autopsy series result from syphilitic vasculitis.

3.4.5. Less common infections (CMV, PCNSL, NHL, progressive multifocal leucoencephalopathy)

These conditions are difficult to diagnose in developing countries and larger autopsy series are probably needed to accurately estimate their prevalence. One autopsy series in India showed rates of CMV to be comparable to USA rates pre-ARVs (Satishchandra et al., 2000). This autopsy series did not identify PCNSL among the 85 cases (Satishchandra et al., 2000) and PCNSL appears to be less common in Mexico than bordering US states (Trujillo et al., 1995). In Zimbabwe, three cases of PCNSL have been reported and all presented with typical symptoms and imaging consistent with PCNSL (Levy et al., 1997). An autopsy series of AIDS-related deaths in the Ivory Coast identified 1.2% with PCNSL (Lucas et al., 1994).

Few cases of progressive multifocal leucoencephalopathy (PML) have been reported in Africa. This is somewhat surprising since the JC virus exists there as a geographically based phenotype, likely secondary to co-evolution with the human species. It is possible that JC subtypes in Africa have less virulence (Chima et al., 1999). Competing causes of death and poor diagnostic capacity may also be the reason for the lack of reported PML cases.

3.4.6. HTLV-1

The impact of co-infection with HTLV-1 and HIV has been minimally studied. However, one prospective, nested case-control study in Brazil found that patients co-infected with HIV and HTLV-1 are at higher risk of myelopathy and neuropathy than patients with either infection alone (Harrison et al., 1997). Whether the co-infection results in a multiplicative or additive effect is unclear.

3.4.7. Neuropathy

Similar to developed regions, acute inflammatory distal polyneuropathy (AIDP) is more common among PLWAs in developing countries. In Burkina Faso among 32 consecutive AIDP cases seen in a teaching hospital over 5 years, 27 were associated with HIV (this in a population with an HIV prevalence of 20.1%). These AIDP cases included individuals with both HIV-1 and HIV-2. Autonomic problems were evident in 30% and over half of those with CD^4 counts assessed had >200/μl (Millogo et al., 2004).

Clinical anecdote indicates that during the routine care of AIDS patients in developing regions, peripheral neuropathies (PNs) are extremely common and debilitating. Yet formal study of this has been limited. One survey of patients in an HIV clinic in Uganda showed that 46% had evidence of a peripheral neuropathy (Frankish and Butcher, 2004). Further characterization of the neuropathies was limited. In Kenya, among 200 consecutive AIDS patients admitted to a medical ward, attempts were made to assess them for PNs. Among the 200, 150 were "too sick for evaluation", and 10 had non-HIV-related etiologies for a PN and were excluded. Among the remaining 40, 32 had evidence of PN on examination with 18 asymptomatic and 14 suffering from painful PNs. Vibratory loss was evident in 60% and parasthesias on the palms and soles in 35%, hyperpathia in 15%, areflexia in 22.5%, loss of proprioception in 12.5% and loss of strength in 10%. All had stage 4 AIDS and 40% had abnormal EMG/NCV tests showing distal sensory PN, polyneuropathy or mononeuritis multiplex. In Zimbabwe, a study of people with HIV with EMG/NCV studies completed showed that subclinical neuropathies were evident at all stages of AIDS, but were more common in the later stages (Parry et al., 1997).

3.4.8. Myelopathy

Since most autopsy studies of AIDS in developing regions have only included pathologic assessments of the brain, our knowledge of spinal cord pathology is especially limited. A hospital-based study of 33 AIDS patients with myelopathy in Natal, South Africa, reported 36% with HTLV-1, 18% spinal TB, 9% zoster myelitis, 6% syphilitic disease, 6% bilharzias, 3% vacuolar myelopathy and 1% attributed to the seroconversion response (Bhigjee et al., 1993).

3.4.9. HIV dementia

Diagnosing HIV dementia in developing countries can be particularly challenging. Ecologically valid measures of cognition appropriate for use in illiterate populations are not available for most regions, so study of HIV dementia requires development of these instruments or adaptation and validation of existing instruments. In 1992, based upon unspecified diagnostic criteria, researchers in Zaire concluded that HIV dementia was rare in their population of 104 consecutive AIDS patients admitted to internal medicine. However, they also noted that among patients with stage 3 and 4 HIV, 42% reported concentration problems, 33% complained of memory problems, 40% noted slowed thinking and 21% complained of word finding. Behavioral problems and motor complaints were also common (Perriens et al., 1992). The WHO Neuropsychiatric AIDS study attempted to examine more subtle cognitive deficits associated with HIV by taking advantage of the cross-cultural (i.e. culture neutral) findings associated with a typically subcortical dementia (Maj et al., 1994). More recent African studies using standardized and validated measures have demonstrated that these complaints are highly correlated with HIV dementia (Alcorn, 2004). Ned Sacktor and his colleagues in Uganda found that in an infectious disease clinic among PLWAs, 11% had some degree of dementia. This figure rose to 50% among patients with a CD^4 count < 200. These individuals had evidence of impaired verbal memory, motor function and functional performance compared to HIV-negative Ugandans seen in the clinic. In Lusaka among hospice patients dying of AIDS, 69% had clinical evidence of HIV dementia, although diagnostic limitations made it difficult to rule out other opportunistic infections as a cause for the cognitive impairment (M Powell et al., unpublished data).

3.5. Pediatric HIV neurology

Our understanding of the effects of HIV upon the immature nervous system in developing regions is hampered by a poor understanding of the baseline neurologic status of children from these regions. Clearly, children who acquire HIV through vertical transmission have poorer general physical development, poorer motor development and lower global mental processing scores than children born to mothers with HIV who do not become infected. Among 218 children in Rwanda born to HIV-positive mothers, infected children were at significant risk of developmental delay and poor gross motor scores: rates were 12.5% at 6 months, 16% at 12 months, 20% at 18 months and 9% at 24 months with increased rates of developmental delay noted in the later stages of AIDS (Msellati et al., 1993). A hospital-based, retrospective study of 185 Zimbabwean children with symptomatic HIV found that 38% had failure-to-thrive and developmental delay

(Nkrumah et al., 1990). The physical and emotional deprivation associated with having a chronically ill mother in the developing world cannot be ignored. Uninfected children born to HIV-positive mothers perform well below their peers with uninfected mothers, particularly in measures of verbal skills (Boivin et al., 1995).

Specific CNS opportunistic infections among children in developing regions have been poorly studied and described. An autopsy study of 70 children who died of AIDS in the Ivory Coast compared their pediatric autopsy data to US studies. Notably, toxoplasmosis was more common among this African population, but CMV, PCNSL and HIV encephalitis were less likely. Cryptococcal meningitis has been reported in Malawian children (Maher and Mwandumba, 1994). Clinical studies of children in developing regions indicate severe failure to thrive is an early finding in HIV. Given the many competing causes of death and high under-fives mortality rates in these regions, it is possible that children with AIDS in developing countries have too short a survival time to develop high rates of CNS opportunistic infections (Bell et al., 1997).

3.6. Special issues in tropical settings

3.6.1. Impact of HIV-2 compared to HIV-1

Little is known about how HIV-2 relative to HIV-1 affects the nervous system. HIV-2 remains relatively uncommon in developed countries where systematic study is more likely to be undertaken. Long-term follow-up comparing 175 HIV-1-infected to 294 HIV-2-infected individuals in the Gambia found HIV-2 is associated with longer survival (Whittle et al., 1994). In an autopsy series in the Ivory Coast, HIV-2 infection appeared to be associated with higher rates of CMV and HIV encephalitis supporting the proposition that HIV-2 is associated with a more prolonged terminal course (Lucas et al., 1993).

3.6.2. Trypanosomiasis

Among tropical infections that can manifest uniquely in the setting of HIV, Chagas' disease must be considered. This infection occurs in Latin America (Fig. 3.1) due to *Trypanosoma cruzi* transmitted to humans via

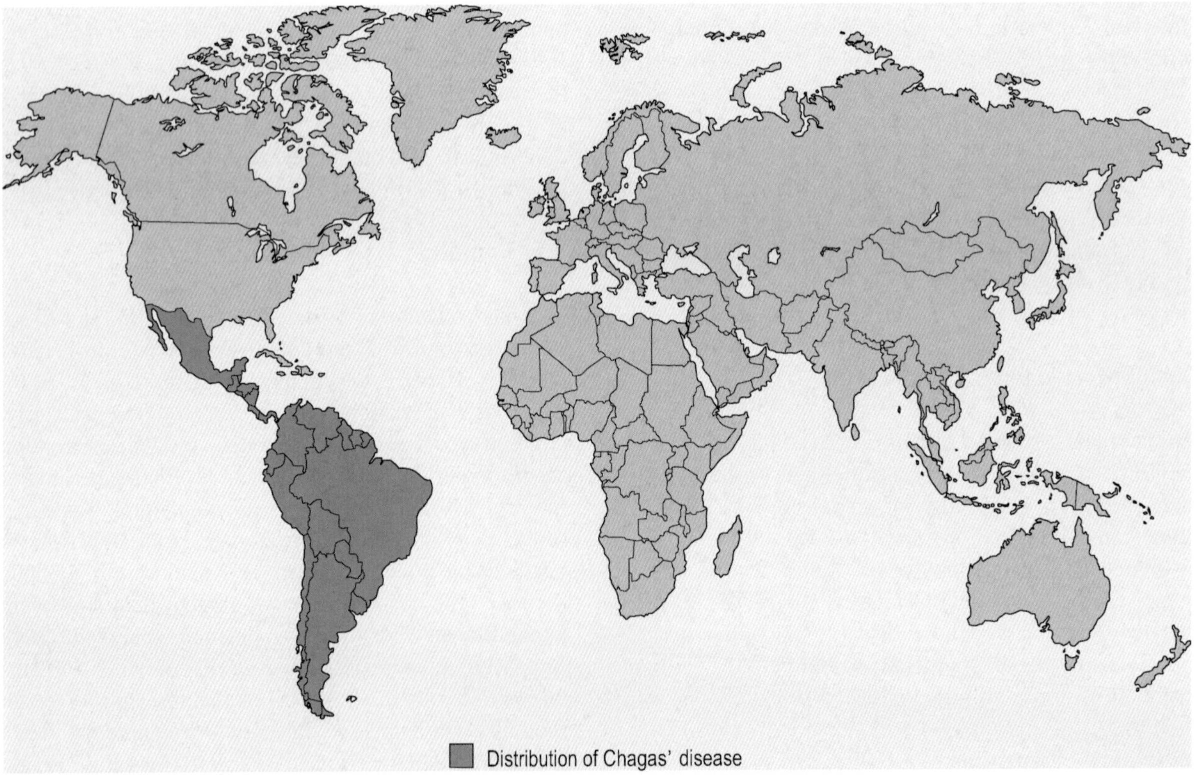

Fig. 3.1. Global distribution of Chagas' disease: a protozoal infection that occurs in Latin America due to *Trypanosoma cruzi* transmitted to humans via blood-sucking triatomes. Adapted from the World Health Organization: http://www.who.int/tdr/dw/chagas2003.htm.

blood-sucking triatomes. Typical manifestations of chronic infection are cardiac in nature. Numerous reports have established that among PLWAs, both reactivated previously latent Chagas' disease and acute infections may present with CNS mass lesions resembling toxoplasmosis or necrotizing encephalitis with angiitis (Ferreira et al., 1991; Gluckstein et al., 1992; Oddo et al., 1992; Rosemberg et al., 1992; Villanueva, 1993). Cases present with a meningoencephalitis and mass lesions that fail to respond to toxoplasmosis therapy. Diagnosis can be made by lesion biopsy with amastigotes identified. Unfortunately, diagnosis is often made only at autopsy. For AIDS patients with potential exposure to Chagas' disease through residents, travel or blood transfusion, CNS Chagas' disease needs to be included in the differential diagnosis of CNS mass lesion. Biopsy should be considered early since mortality rates appear to be high without treatment. The impact of HIV on African trypanosomiasis has not been described to date.

3.6.3. Malaria

Multiple studies have shown that HIV and malaria interact to impact the maintenance of acquired immunity to malaria in pregnant women residing in malaria endemic regions by altering the typical gravid-specific patterns of malaria risk. Among HIV-negative women with acquired immunity, only primiparous women experience a substantial decrease in their malarial immunity and this does not recur in future pregnancies. In the setting of HIV infection, the risk of malaria during pregnancy remains high during subsequent pregnancies (Ayisi et al., 2003). One would assume that the risk of cerebral malaria is also increased.

Some data have shown HIV-1 infection to be associated with an increased frequency of clinical malaria and parasitemia and this association is more pronounced in advanced HIV (Whitworth et al., 2000; French et al., 2001). This finding may be somewhat confounded by the fact that an otherwise asymptomatic parasitemia in someone with an AIDS-related fever would invariably be labeled "symptomatic malaria". Whether progression to cerebral malaria is more likely remains unclear. Children with gross failure-to-thrive and malnutrition seem to be protected somewhat from cerebral malaria. Whether this holds true for HIV-positive children is unknown.

A retrospective review of pediatric autopsy cases found cerebral malaria significantly less likely to be the cause of death among children with HIV compared to HIV-negative children (Bell et al., 1997). However, analysis of 30-years monthly databases in Kwa-Zulu Natal revealed a substantial increase in malaria rates over the same period that HIV rates increased from 1.6 to 36.2%. After controlling for other non-climatic factors, HIV still appeared to be associated with the increased malaria rates (Craig et al., 2004). If loss of acquired immunity occurs in HIV-positive individuals residing in malaria endemic regions, then malaria recovery rates may be diminished and could increase the estimated number of cases. HIV infection seems to increase parasite load among infected individuals and this might lead to increasing infectiousness of malaria to feeding vectors, also resulting in higher transmission rates even among the non-HIV-infected population.

3.6.4. Leprosy

Several studies have failed to find evidence that AIDS changes the risk of leprosy or the clinical disease subtype or progression (Leonard et al., 1990; Jayasheela et al., 1994). Studies assessing leprosy incidence pre- and post-AIDS in Ethiopia have noted no change in the number of new cases presenting for care (Frommel et al., 1994). Consider, however, that as a highly stigmatized condition, leprosy is still identified primarily by active case-finding (i.e. patients do not present for care but rather are found in their communities by leprosy control officers routinely surveying known regions of infections and known contacts). One of the consequences of the AIDS epidemic has been deterioration in the existing medical infrastructure of poor countries, and resources previously directed to leprosy care have been redirected toward HIV/AIDS care (personal communication, Charles Mang'ombe and Ellie Kalichi, leprosy control officers for the Zambian Southern Province). The static rates of leprosy in some countries could represent decreased active case-finding rather than stable infection and transmission rates. Neglect of this ancient scourge due to the competing needs for HIV care and prevention might ultimately "link" the two conditions epidemiologically, even if a biological association is absent.

3.7. Potential impact of ARVs

Much of what we need to know about HIV and AIDS in the nervous system is limited by poor understanding of the neurologic system in the "average" person residing in developing regions. Protein deprivation, recurrent infections, environmental toxins, chronic anemia, episodic famine, micronutrient deficiency — all might be expected to have some impact on nervous system function and resiliency.

Bringing ARVs into developing regions must happen and many efforts are being made to facilitate this process. However, careful longitudinal assessment of individuals receiving these agents is needed. How individuals with a life course of nervous system challenges will respond to the toxicity of ARVs is unclear.

References

Alcorn K (2004). Dementia and neuropathy common in Ugandan HIV-positive patients. Aidsmap News, April 28, 2004.

Atangana R, Bahebeck J, Mboudou ET, et al (2003). Troubles neurologiques chez les porteurs du virus d'immunodeficience humaine a Yaounde. Sante 13: 155–158.

Ayisi JG, Branch OH, Rafi-Janajreh A, et al (2003). Does infection with human immunodeficiency virus affect the antibody responses to Plasmodium falciparum antigenic determinants in asymptomatic pregnant women? J Infect 46: 164–172.

Bahia-Oliveira LM, Jones JL, Azevedo-Silva J, et al (2003). Highly endemic, waterborne toxoplasmosis in north Rio de Janeiro state, Brazil. Emerg Infect Dis 9, 55–62.

Bane A, Yohannes AG, Fekade D (2003). Morbidity and mortality of adult patients with HIV/AIDS at Tikur Anbessa Teaching Hospital, Addis Ababa, Ethiopia. Ethiopian Med J 41: 131–140.

Bell JE, Lowrie S, Koffi K, et al (1997). The neuropathology of HIV-infected African children in Abidjan, Cote d'Ivoire. J Neuropathol Exp Neurol 56: 686–692.

Bergen DC, Good D (2006). Neurology training programs worldwide: a World Federation of Neurology survey. J Neurol Sci 246: 59–64.

Bergen DC (2002). Training and distribution of neurologists worldwide. J Neurol Sci 198: 3–7.

Bhigjee AI, Vinsen C, Windsor IM, et al (1993). Prevalence and transmission of HTLV-I infection in Natal/KwaZulu. S Afr Med J 83: 665–667.

Birbeck GL, Kalichi EM (2004). Primary healthcare workers' perceptions about barriers to health services in Zambia. Trop Doct 34: 84–86.

Birbeck GL, Munsat T (2002). Neurologic services in sub-Saharan Africa: a case study among Zambian primary healthcare workers. J Neurol Sci 200: 75–78.

Boivin MJ, Green SD, Davies AG, et al (1995). A preliminary evaluation of the cognitive and motor effects of pediatric HIV infection in Zairian children. Health Psychol 14: 13–21.

Chima SC, Agostini HT, Ryschkewitsch CF, et al (1999). Progressive multifocal leukoencephalopathy and JC virus genotypes in West African patients with acquired immunodeficiency syndrome: a pathologic and DNA sequence analysis of 4 cases. Arch Pathol Lab Med 123: 395–403.

Chimelli L, Rosemberg S, Hahn MD, et al (1992). Pathology of the central nervous system in patients infected with the human immunodeficiency virus (HIV): a report of 252 autopsy cases from Brazil. Neuropathol Appl Neurobiol 18: 478–488.

Craig MH, Kleinschmidt I, Le Sueur D, et al (2004). Exploring 30 years of malaria case data in KwaZulu-Natal, South Africa: II. The impact of non-climatic factors. Trop Med Int Health 9(12): 1258–1266.

De Cock KM, Soro B, Coulibaly IM, et al (1992). Tuberculosis and HIV infection in sub-Saharan Africa. JAMA 268: 1581–1587.

Di Costanzo B, Belec L, Georges AJ, et al (1989). High predictivity of neurological manifestations for HIV infection in Africa. Lancet 2: 270.

Ferreira MS, Nishioka Sde A, Rocha A, et al (1991). Acute fatal Trypanosoma cruzi meningoencephalitis in a human immunodeficiency virus-positive hemophiliac patient. Am J Trop Med Hyg 45: 723–727.

Frankish H, Butcher J (2004). Newsdesk: HIV in sub-Saharan Africa. Lancet Neurol 3: 324.

French N, Nakiyingi J, Lugada E, et al (2001). Increasing rates of malarial fever with deteriorating immune status in HIV-1-infected Ugandan adults. AIDS 15: 899–906.

Frommel D, Tekle-Haimanot R, Verdier M, et al (1994). HIV infection and leprosy: a four-year survey in Ethiopia. Lancet 344: 165–166.

Gluckstein D, Ciferri F, Ruskin J (1992). Chagas' disease: another cause of cerebral mass in the acquired immunodeficiency syndrome. Am J Med 92: 429–432.

Gray F (1997). Bacterial infections. Brain Pathol 7: 629–647.

Harrison LH, Vaz B, Taveira DM, et al (1997). Myelopathy among Brazilians coinfected with human T-cell lymphotropic virus type I and HIV. Neurology 48: 13–18.

Howlett WP, Nkya WM, Mmuni KA, et al (1989). Neurological disorders in AIDS and HIV disease in the northern zone of Tanzania. AIDS 3: 289–296.

Jayasheela M, Sharma RN, Sekar B, et al (1994). HIV infection amongst leprosy patients in south India. Indian J Lepr 66: 429–433.

John MA, Coovadia Y (1998). Meningitis due to a combined infection with Cryptococcus neoformans and Streptococcus pneumoniae in an AIDS patient. J Infect 36: 231–232.

Kibayashi K, Ng'walali PM, Mbonde MP, et al (1999). Neuropathology of human immunodeficiency virus 1 infection. Significance of studying in forensic autopsy cases at Dar es Salaam, Tanzania. Arch Pathol Lab Med 123: 519–523.

Kumarasamy N, Solomon S, Flanigan TP, et al (2003). Natural history of human immunodeficiency virus disease in southern India. Clin Infect Dis 36: 79–85.

Lanjewar DN, Jain PP, Shetty CR (1998). Profile of central nervous system pathology in patients with AIDS: an autopsy study from India. AIDS 12: 309–313.

Leonard G, Sangare A, Verdier M, et al (1990). Prevalence of HIV infection among patients with leprosy in African countries and Yemen. J Acquir Immune Defic Syndr 3: 1109–1113.

Levy LM, Coutts AM, Abayomi EA, et al (1997). Primary cerebral lymphoma in Zimbabwe: a report of three patients. Central Afr J Med 43: 328–331.

Lucas SB, Hounnou A, Peacock C, et al (1993). The mortality and pathology of HIV infection in a west African city. [See comment.] AIDS 7: 1569–1579.

Lucas SB, Diomande M, Hounnou A, et al (1994). HIV-associated lymphoma in Africa: an autopsy study in Cote d'Ivoire. Int J Cancer 59: 20–24.

Maher D, Mwandumba H (1994). Cryptococcal meningitis in Lilongwe and Blantyre, Malawi. J Infect 28: 59–64.

Maj M, Satz P, Janssen R, et al (1994). WHO Neuropsychiatric AIDS study, cross-sectional phase II. Neuropsychological and neurological findings. Arch Gen Psychiatry 51: 51–61.

Mathew MJ, Chandy MJ (1999). Central nervous system toxoplasmosis in acquired immunodeficiency syndrome: an emerging disease in India. Neurol India 47: 182–187.

Millogo A, Sawadogo A, Lankoande D, et al (2004). Syndrome de Guillain-Barre chez les patients infectes par le VIH a Bobo-Dioulasso (Burkina Faso). Revue Neurologique 160: 559–562.

Modi MMA, Modi G (2004). Management of HIV-associated focal brain lesions in developing countries. QJM 97: 413–421.

Mohar A, Romo J, Salido F, et al (1992). The spectrum of clinical and pathological manifestations of AIDS in a consecutive series of autopsied patients in Mexico. AIDS 6: 467–473.

Molez JF (1998). The historical question of acquired immunodeficiency syndrome in the 1960s in the Congo River basin area in relation to cryptococcal meningitis. Am J Trop Med Hyg 58: 273–276.

Msellati P, Lepage P, Hitimana DG, et al (1993). Neurodevelopmental testing of children born to human immunodeficiency virus type 1 seropositive and seronegative mothers: a prospective cohort study in Kigali, Rwanda. Pediatrics 92: 843–848.

Nkrumah FK, Choto RG, Emmanuel J, et al (1990). Clinical presentation of symptomatic human immuno-deficiency virus in children. Central Afr J Med 36: 116–120.

Oddo D, Casanova M, Acuna G, et al (1992). Acute Chagas' disease (Trypanosomiasis americana) in acquired immunodeficiency syndrome: report of two cases. Hum Pathol 23: 41–44.

Oh MD, Park SW, Kim HB, et al (1999). Spectrum of opportunistic infections and malignancies in patients with human immunodeficiency virus infection in South Korea. Clin Infect Dis 29: 1524–1528.

Parry O, Mielke J, Latif AS, et al (1997). Peripheral neuropathy in individuals with HIV infection in Zimbabwe. Acta Neurol Scand 96: 218–222.

Perriens JH, Mussa M, Luabeya MK, et al (1992). Neurological complications of HIV-1-seropositive internal medicine inpatients in Kinshasa, Zaire. J Acq Immune Defic Syndr 5: 333–340.

Rosemberg S, Chaves CJ, Higuchi ML, et al (1992). Fatal meningoencephalitis caused by reactivation of Trypanosoma cruzi infection in a patient with AIDS. Neurology 42: 640–642.

Santosh V, Shankar SK, Das S, et al (1995). Pathological lesions in HIV positive patients. Indian J Med Res 101: 134–141.

Satishchandra P, Nalini A, Gourie-Devi M, et al (2000). Profile of neurologic disorders associated with HIV/AIDS from Bangalore, south India (1989–96). Indian J Med Res 111: 14–23.

Silber E, Sonnenberg P, Koornhof HJ, et al (1998). Dual infective pathology in patients with cryptococcal meningitis. Neurology 51: 1213–1215.

Singh S, Nautiyal BL (1991). Seroprevalence of toxoplasmosis in Kumaon region of India. Indian J Med Res 93: 247–249.

Smadja D, Fournerie P, Cabre P, et al (1998). Efficacite et bonne tolerance du cotrimoxazole comme traitement de la toxoplasmose cerebrale au cours du SIDA. Presse Medicale 27: 1315–1320.

Swinne D, Deppner M, Maniratunga S, et al (1991). AIDS-associated cryptococcosis in Bujumbura, Burundi: an epidemiological study. J Med Vet Mycol 29: 25–30.

Trujillo JR, Garcia-Ramos G, Novak IS, et al (1995). Neurologic manifestations of AIDS: a comparative study of two populations from Mexico and the United States. J Acq Immune Defic Syndr Hum Retrovirol 8: 23–29.

Vidal JE CF, Penalva de Oliveira AC, Focaccia R, et al (2004). PCR assay using cerebrospinal fluid for diagnosis of cerebral toxoplasmosis in Brazilian AIDS patients. J Clin Microbiol 42: 4765–4768.

Villanueva MS (1993). Trypanosomiasis of the central nervous system. Semin Neurol 13: 209–218.

Whittle H, Morris J, Todd J, et al (1994). HIV-2-infected patients survive longer than HIV-1-infected patients. AIDS 8: 1617–1620.

Whitworth J, Morgan D, Quigley M, et al (2000). Effect of HIV-1 and increasing immunosuppression on malaria parasitaemia and clinical episodes in adults in rural Uganda: a cohort study. Lancet 356: 1051–1056.

Handbook of Clinical Neurology, Vol. 85 (3rd series)
HIV/AIDS and the Nervous System
P. Portegies, J. R. Berger, Editors

Chapter 4

The neuropathogenesis of HIV-1 infection

JAMES L. BUESCHER, SARA GROSS, HOWARD E. GENDELMAN, AND TSUNEYA IKEZU*

Center for Neurovirology and Neurodegenerative Disorders, Department of Pharmacology, University of Nebraska Medical Center, Omaha, NE, USA

4.1. Introduction

Significant neurological complications associated with HIV-1 infection occur years after the acute viral seroconversion reaction and is commonly coincident with progressive immunosuppression and high viral loads. Disease processes start soon after initial viral infection, initiated through exchange of infected body fluids by blood transfusion, accidental needle sticks, sexual intercourse and maternal–fetal transmission (Royce et al., 1997; Lackritz, 1998; Newell, 1998; Beltrami et al., 2000). Infection is "highly" restricted for varying time periods but usually measured in years (Krishnakumar et al., 2005). Evasion of the innate and acquired immune surveillance mechanisms leads to dissemination and high-level replication to regional lymphatics and HIV-1 target tissues, including the bone marrow, lung and brain. Importantly, HIV-1 enters the central nervous system (CNS) early in the course of disease (Davis et al., 1992; An et al., 1999). Virus is carried into the brain principally through CD^4 T lymphocytes (Haase, 1999) and mononuclear phagocytes (MPs), dendritic cells, monocytes and macrophages (Tardieu and Boutet, 2002).

Following acute infection the host engages the virus in a commensal relationship. This allows a subclinical stage in which HIV-1 persists in the infected human host but at low levels. Virus persists in CD^4 T lymphocytes and cell death occurs as a result of active replication. The destroyed CD^4 T lymphocytes are replaced by new bone marrow-derived cells. Inevitably, as disease progresses, the birth of new cells cannot replace those destroyed by virus and immune suppression ensues with uncontrolled viral growth and the breakdown of adaptive antiretroviral immune surveillance mechanisms (Ho et al., 1995; Krishnakumar et al., 2005). CD^4 T lymphocyte depletion may also occur as a consequence of HIV-1-induced deficits in hematopoiesis by preventing production of new lymphocytes. Here, viral glycoproteins and inflammatory cytokines can impair bone marrow cell repopulation functions (Geissler et al., 1991; Maciejewski et al., 1994; Douek et al., 1998). The lymphopenia that occurs affects the host's ability to combat a broad spectrum of opportunistic infections and neoplasms and HIV-1-associated tissue damage ensues.

The most severe form of HIV-1-associated tissue damage occurs in the CNS. The lack of adequate innate viral control mechanisms and adaptive immunity allows active viral replication to occur in brain. Here, a metabolic encephalopathy results as a consequence of continued and uncontrolled brain MP (perivascular macrophage and microglia) infection and immune activation (Anderson et al., 2002). The length of time prior to the development of CNS disease and its associated profound immune suppression has increased following the institution of highly active antiretroviral therapy (HAART) (d'Arminio Monforte et al., 2000; Yang, 2004). However, in patients without therapy or those that respond poorly to it, the immune system is damaged and disease occurs (Ho et al., 1995; Wei et al., 1995; Krishnakumar et al., 2005). When the numbers of CD^4 T lymphocytes drop below 200/μl patients not only become susceptible to opportunistic infections but also to virus induced tissue injury including, most notably, HIV-associated dementia (HAD).

*Correspondence to: Tsuneya Ikezu, MD, PhD, Center for Neurovirology and Neurodegenerative Disorders, Department of Pharmacology, 985880 Nebraska Medical Center, Omaha, NE 68198-5880, USA. E-mail: tikezu@unmc.edu, Tel: +1-402-559-4035.

4.2. Epidemiology of HAD in the era of HAART

Incidence, prevalence and severity of HIV-1-associated neurocognitive disorders have changed remarkably in the past decade following the advent of HAART. In the early 1990s the prevalence of HAD was between 20 and 30% of individuals with advanced disease with CD^4 T lymphocyte counts of less than 200/μl (Navia et al., 1986; McArthur et al., 2003). The introduction of HAART increased life expectancy and decreased severity and incidence of HAD to 10% in HIV-infected people (Dore et al., 1997; Ferrando et al., 1998; Sacktor et al., 2001). Nonetheless, despite such improvements in disease outcomes, HIV-associated neurocognitive disorders (dementia, motor and behavioral impairments) continue to affect many infected individuals. Notably, minor cognitive/motor disorder (MCMD) is currently on the rise as is neurological disease prevalence (McArthur et al., 2003; McArthur, 2004). Such changes in the epidemiological patterns of neurocognitive deficits in the HAART era may occur through poor penetration of the brain by some anti-retroviral drugs, back mutation of virus from brain to blood, persistent low level infection in astrocytes and other neural cells, untoward side effects of drug administrations and poor compliance (Enting et al., 1998; Dore et al., 1999; Major et al., 2000). Longer life expectancy also permits CNS infection to independently evolve over time in a "seemingly" protected brain reservoir and viral sanctuary (Gendelman et al., 2004). In fact, distinct viral drug resistance patterns in plasma and cerebrospinal fluid (CSF) compartments are not uncommon (Cunningham et al., 2000; Letendre et al., 2004). Consequently, the prevalence of HIV dementia continues to rise with the increase in life expectancy of individuals with HIV infection. In the past half decade the incidence of HAD as an acquired immune deficiency syndrome (AIDS)-defining illness has increased (Lipton, 1997; Dore et al., 1999; Clifford, 2000). Furthermore, the proportion of new cases of HAD with CD^4 T lymphocyte counts greater than 200/μl is growing (Sacktor et al., 2001). MCMD, a more subtle form of HIV-associated CNS dysfunction, has become more common as HAD has diminished (McArthur et al., 2003). MCMD is diagnosed in at least 30% of symptomatic HIV-1-seropositive adults (Janssen et al., 1989; Sacktor et al., 2002) and is one indication of a worsening prognosis in disease (Sacktor et al., 1996). These observations suggest that neurocognitive dysfunctions will continue to be a significant complication in advanced HIV-1 disease (McArthur et al., 1999; Carpenter et al., 2000; Krebs et al., 2000) and remain an independent risk factor for death (Ellis et al., 1997).

4.3. Viral infection and entry into the brain

4.3.1. Viral receptors and infection

Cellular susceptibility to HIV-1 infection is regulated by CD^4, CXCR4 and CCR5 receptor expression on virus target cells. HIV-1 enters CD^4 T lymphocytes by fusion at the plasma membrane after interactions with CD^4 and a co-receptor. The identity of the co-receptor determines HIV-1 macrophage or T cell tropism (M- or T-tropic). The genetic polymorphism in the HIV-1 envelope glycoprotein gp120 underlies co-receptor usage and the host cell type infected. The hypervariable region at V3 loop of HIV-1gp120 is a principal determinant for cell tropism. The V3 functional domain is distinct from the gp120–CD^4 interaction and affects post-CD^4 binding events such as proteolytic cleavage and fusion in the viral life cycle.

M-tropic viruses predominate the brain in infected individuals (Brew et al., 1990, 1996b, c; Brew and Miller, 1996). Such isolates replicate well in cultured human microglia and monocyte-derived macrophages (MDM) and are associated with CCR5 utilization (Strizki et al., 1996; Ghorpade et al., 1998). These viruses affect neurotoxic responses after infection of macrophages or following macrophage–astrocyte cell interactions (Genis et al., 1992). T-tropic strains are typically associated with the α-chemokine receptor CXCR4; however, T-cells may be infected by either M- or T-tropic viral strains. For MDM and microglia the pathways for viral infection are similar, if not identical. Microglia express CD^4 as well as CCR3, CCR5, and CXCR4 (Lavi et al., 1997; Vallat et al., 1998); however, CCR5 is most commonly used as the viral co-receptor (Shieh et al., 1998; Albright et al., 1999). Because HIV-1 relies on chemokine receptors for entry, chemokine receptor expression by brain macrophages and microglia may influence viral evolution in brain leading to neurovirulence.

CCR5 is critical for HIV-1 transmission, underscored by the observation that individuals who are repeatedly exposed to HIV-1, but remained uninfected, show CCR5 polymorphisms (Liu et al., 1996). CD^4 T lymphocytes from such individuals are resistant to in vitro infection of M-tropic HIV-1, but were readily infected with viruses adapted to grow in T cell lines (Liu et al., 1996). These individuals are homozygous for a defective CCR5 allele that contains an internal 32-base pair deletion (CCR5 $\Delta 32$). The truncated protein encoded by this gene is not expressed at the cell surface. CCR5 $\Delta 32$ homozygous individuals comprise 1% of the Caucasian population, and heterozygous individuals comprise 20%. Those that are heterozygous for this deletion progress more slowly to AIDS than wild-type homozygous individuals (Liu et al., 1996). Additional mutations associated with HIV-1 resistance include the chemokine receptor CCR2b, and the ligand, stromal cell-derived factor (SDF)-1/CXCL12 (Smith and Hale, 1997; Winkler et al., 1998).

4.3.2. Blood–brain barrier: structure and biochemical considerations

The blood–brain barrier (BBB) separates the CNS from the peripheral blood supply such that HIV-1 must penetrate in order to infect the CNS. Entry into the CNS can occur as cell-free virus or within macrophages and lymphocytes. During the systemic immune activation that occurs during an initial viral infection, leukocytes egress across the BBB and can transmit virus to primary glial elements including, most notably, the perivascular macrophage (Gendelman et al., 2004). However, once inside the brain, virus is acted on by both innate and adaptive immune surveillance mechanisms and remains highly restricted for years, often without significant demonstrable effects on cognitive, behavior or motor functions.

BBB structure consists of a monolayer of brain microvascular endothelial cells (BMVECs) that is made impermeable through a combination of tight junctions and lack of transcellular pores (Pardridge, 1983; Banks, 2005). The BBB is also composed of a capillary basement membrane on the abluminal side and astrocytes. The end-feet of the astrocytes are in close proximity to the BMVECs and play a part in the integrity of the BBB. The integrity of tight junctions relies on a high concentration of integral membrane proteins (such as occludin and claudin-5) and the intercellular signaling between the adjacent astrocytes and the BMVEC determine the extent of phosphorylation of the junctional proteins (Janzer and Raff, 1987; Staunton et al., 1988; Hession et al., 1990; Takeshima et al., 1994; Kim et al., 1995; Rubin and Staddon, 1999; Ballabh et al., 2004). This unique structure of the BBB and its high electrical resistance (Pardridge, 1983) appear to be responsible for its function as a barrier between the CNS and the periphery, and the selective transport of factors across the BBB necessary for the homeostasis of the CNS. Normally, in an intact BBB, leukocyte brain infiltration is limited (Nottet and Dhawan, 1997; Carson and Sutcliffe, 1999), and the movements of ions, proteins and polar molecules across the BBB are restricted (Risau and Wolburg, 1990).

4.3.3. Viral entry into the CNS

Early entry of HIV-1 into the CNS during the acute seroconversion reaction is commonly marked by a meningoencephalitis (Davis et al., 1992; An et al., 1999; Newton et al., 2002). Aseptic meningitis is associated with HIV infection of meningeal macrophage. This is distinct from the perivascular macrophage and microglial infections seen in the later stages of disease (Gendelman et al., 2004). If encephalopathy occurs after initial virus exposure, it is typically self-limited.

The process for HIV-1 entry into the CNS revolves around secretory products from immune-activated and virus-infected MP (Fig. 4.1) that affect BBB function, expression of cell adhesion molecules and chemokines, and leads to disruption of brain microvessel integrity (Nottet et al., 1996; Persidsky et al., 1997). For example, tumor necrosis factor (TNF)-α from infiltrating lymphocytes and macrophages, as well as some HIV-1 proteins such as TAT, increases BBB permeability (Abraham et al., 1996; Avraham et al., 2004). BBB endothelial cells undergo apoptosis following exposure to HIV-1gp120, Tat, Vpr and Nef and open the barrier to large-scale infiltration (Huang and Bond, 2000; Avraham et al., 2004). In the infected human host, the BBB remains relatively intact until late in the course of disease when functional disruption occurs allowing easy access of virus and virus-infected cells to the CNS (Persidsky et al., 2000). Four possible mechanisms that are supportive of viral entry into the CNS and currently under investigation include: the surreptitious transmission of virus in infected macrophages (the Trojan horse model); direct infection of the BBB by HIV; transcytosis of HIV; and BBB disruption (Eugenin and Berman, 2005).

Chemokines are key regulators of monocyte recruitment into the CNS. Endothelial cells (EC), microglia and astrocytes are major cellular sources of β-chemokines [for example, monocyte chemotactic protein (MCP)-1/CCL2, macrophage inflammatory protein (MIP)-1α/CCL3 and MIP-1β/CCL4] whose production during HIV-1 encephalitis (HIVE) sets up an inflammatory chemoattractant gradient (Nottet, 2004) drawing cells into the nervous system (Boven et al., 2000; Anderson et al., 2002; Cho and Miller, 2002; Williams and Hickey, 2002). Egress of virus into the CNS is facilitated through virus and immune-mediated compromise of the BBB structure. The recruitment of infected macrophages into the brain during HIVE is termed the Trojan horse model of viral dissemination. Such a model for viral dissemination into brain and other target tissues was described first for ruminant lentiviral systems and later for HIV-1 (Peluso et al., 1985; Gendelman et al., 1986; Orenstein et al., 1988;

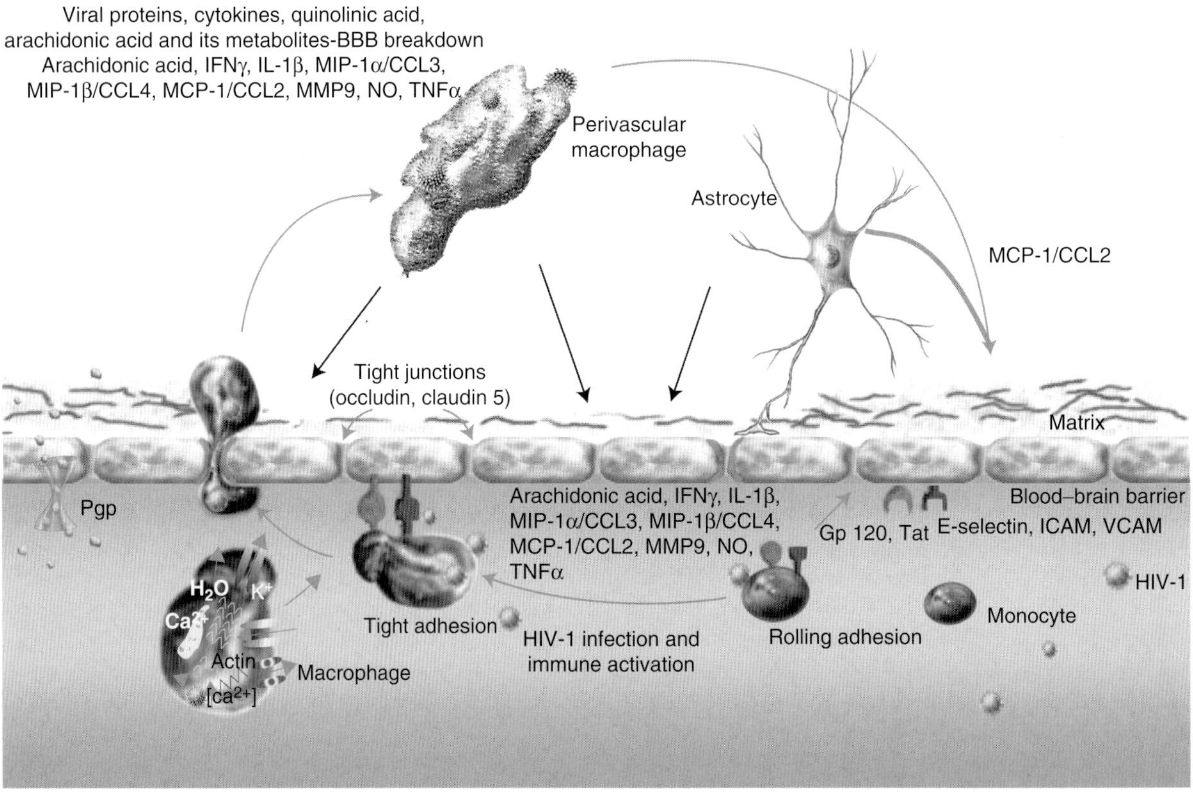

Fig. 4.1. For full color figure, see plate section. BBB and HIV-1 entry into an inflamed CNS. The structure of the BBB consists of matrix and BMVEC connected by tight junctions consisting of claudin-5 and occludin proteins among others. Astrocyte end-feet are in close proximity to the matrix and the BMVEC and contribute to the integrity of the BBB. Pgp is an ATP-dependent transporter located in the BMVEC that negatively impacts drug delivery into the CNS through its removal of antiretroviral drugs from the CNS. Entry of virus into the brain and across the BBB involves monocyte–macrophage maturation, ion channel expression, viral infection and immune activation leading to the production and release of viral and cellular toxins that affect the integrity and function of the BBB. The factors released by MPs that damage the BBB include, but are not limited to, viral proteins (Tat and gp120), pro-inflammatory cytokines (including TNF-α and IL-1β), MMP9 and free radicals such as NO. Chemokines such as MCP-1/CCL2 and MIP-1α/β (CCL3/CCL4) are released from astrocytes, microglia, endothelial cells and neurons and attract more macrophages into the brain. Virus-infected and immune-activated macrophages and microglia MPs alter the BBB by increasing the number of cell adhesion molecules present on the BMVEC, promote rolling and tight adhesion of macrophages with BMVEC through upregulation of cell adhesion molecules and changes in cell shape and volume mediated by alterations in ion channels in macrophages. This leads to the transendothelial migration of macrophages and serving as an increased nidus for continued inflammatory activities and perpetuating the entry of virus and macrophages into the brain.

Gendelman and Meltzer, 1989). This is the most favored model for the entry of virus into the brain as virus can accumulate into intracytoplasmic vesicles and thereby escape immune surveillance. Once inside the CNS, the infected macrophage in conjunction with microglia and astrocytes secrete chemokines and other inflammatory factors causing functional BBB disruption and recruitment of additional macrophages. This process serves to perpetuate brain inflammation and ongoing viral replication (Fig. 4.1 and reviewed by Gendelman et al., 2004). A body of laboratory evidence supports this model as a mechanism of viral entry into brain during disease. For example, elevated numbers of CD^{16} and CD^{69} circulating monocytes emerge during HIV-1 infection, and these cells have been shown to transmigrate through the BBB (Pulliam et al., 1997; Gartner, 2000; Williams et al., 2001).

4.3.4. Inflammation and BBB integrity

The ability of activated peripheral macrophages to disrupt the BBB during HIV-1 infection is largely due to their production of inflammatory toxins (Persidsky et al., 2000). As the infection progresses, pro-inflammatory cytokines like TNF-α and interleukin (IL)-1β are secreted by activated and infected macrophages.

These cytokines have multiple biological effects. Proinflammatory cytokines activate brain EC. Following inflammation EC undergo a number of morphological and functional alterations that lead to alterations in BBB permeability, cell hypertrophy and proliferation, accumulation of intracellular organelles, expression of major histocompatibility complex antigens and expression of adhesion molecules: intercellular adhesion molecule (ICAM)-1, vascular cell adhesion molecule (VCAM)-1 and E-selectin among others (Staunton et al., 1988; Bevilacqua et al., 1989; Cavender et al., 1989; Osborn et al., 1989; Hession et al., 1990; Carlos et al., 1991; Hurwitz et al., 1992). The upregulation of EC adhesion molecules by infected and uninfected but activated monocytes increases monocyte rolling and tight adhesion to the BBB endothelium. This upregulation of cell adhesion molecules is further enhanced by HIV-1 gp120-induced changes in astrocyte ICAM-1 (Shrikant et al., 1996). Furthermore, HIV-1 infected macrophages themselves upregulate the expression of E-selectin, a molecule that increases adhesion of cells to the endothelium (Nottet, 2004). MCP-1/CCL2 injection into rodent brains serves as a monocyte chemoattractant (Bell et al., 1996). Moreover, transgenic mice overexpressing TNF-α driven by the GFAP promoter develop meningoencephalitis (Stalder et al., 1998). These mice show increased expression of cell adhesion molecules and expression of α- and β-chemokines precede inflammatory cell infiltration into the brain.

Another mechanism that promotes entry of macrophages into the brain is through matrix metalloproteinases (MMPs). These are proteolytic enzymes that under normal circumstances affect remodeling of the extracellular matrix (Leppert et al., 1995). In disease, they may disrupt the integrity of the BBB and increase its permeability. HIV-1-infected cells secrete MMP-9 promoting the weakening of the basement membrane matrix of the EC layer of the BBB, which can be inhibited through pretreatment with an MMP inhibitor (Sporer et al., 2000). In addition, antagonizing MMP-9 are tissue inhibitors of metalloproteinases (TIMP)-1 and TIMP-2 (Gardner and Ghorpade, 2003), which are shown to diminish the permeability of the BBB (Nottet, 2004).

Other mediators of BBB disruption include nitric oxide (NO). NO is secreted from HIV-1-infected macrophages, and may affect macrophage movement across the BBB by causing the cells to move slowly along the endothelium thereby assisting in adhesion (Bukrinsky et al., 1995; Boven et al., 1999; Nottet, 2004). Furthermore, tight junction proteins are degraded by NO and interfere with kinase phosphorylation, preventing the normal assembly of the tight junction proteins (Stuart and Nigam, 1995). NO from activated macrophages may also lead to apoptosis of brain EC (Huang and Bond, 2000).

4.3.5. Direct entry of virus into the CNS

There are a number of other models for entry of HIV-1 into the CNS. In the transcytosis model, HIV-1 is internalized by either EC or astrocyte foot processes via endosomes or through macropinocytosis followed by virus transfer into the CNS. Prior electron microscopic studies indicate that the uptake of virions into the endothelial cells is pH-dependent and that viral particles accumulate in endosome-like structures (Banks et al., 1998; Banks et al., 2001). Other groups demonstrated HIV-1 uptake by macropinocytosis mechanisms (Marechal et al., 2001), which are dependent on intact lipid rafts and mitogen-activated protein kinase signaling pathways (Liu et al., 2002a). Another model for viral entry into the CNS is via direct infection of BBB cells. Although astrocytes and endothelial cells can be infected in vitro (Bagasra et al., 1996; An et al., 1999), there is limited evidence that continued infection occurs in vivo. HIV-1 infection of astrocytes, for example, appears to be severely restricted. Endothelial infection is controversial. Virus infection of astrocytes, nonetheless, may have multiple effects on BBB and neural function by contributing to the release of chemokines and cytokines and affecting the structural integrity of cells and tissues. HIV-1 may also gain entry into the brain through the choroid plexus. The choroid plexus is located in the ventricular cavities of the brain and is covered with a single layer of epithelial cells connected by tight junctions through which molecules must pass to get from the choroid plexus to the CSF. For virus to go from the CSF into the brain it must then pass through a second barrier of epithelial cells connected by tight junctions. Virus can freely diffuse from the blood into the interior of the choroid plexus since the blood vessel epithelial layer does not have tight junctions. The mechanisms by which the virus passes from the stroma to the CSF and then from the CSF into the brain through cell barriers connected by tight junctions may explain how virus can bypass the BBB (Burkala et al., 2004).

4.3.6. BBB and drug delivery

Not only does the BBB play a key role in viral entry into the brain it also has a major impact on the antiretroviral delivery of drugs to the CNS. Cerebral capillary endothelium expresses a number of efflux transporters, which actively remove a broad range of drug molecules before they cross into the brain parenchyma (Cordon-Cardo et al., 1989; Sugawara et al., 1990; Zhang et al.,

2000). P-glycoprotein (Pgp), the most investigated brain efflux transport protein, has a broad affinity for dissimilar lipophilic and amphiphilic substrates (Cordon-Cardo et al., 1989; Sugawara et al., 1990). The ability of Pgp to restrict brain entry of drugs has been shown in numerous studies, including a murine model deficient for *mdr-1*, the gene coding Pgp (Kawahara et al., 1999). Accumulation of Pgp substrates (molecules that Pgp transports across the BBB) in the brain was about 100-fold higher in the knockout mice as compared to the control wild-type animals (Schinkel et al., 1994). Pgp hampers penetration of anti-HIV-1 protease inhibitors (ritonavir, nelfinavir, indinavir and saquinavir) across the BBB because they are substrates for Pgp (Kim et al., 1998; Lee et al., 1998; Choo et al., 2000). Thus, despite successful suppression of the viral infection in the blood or lymphoid tissues, active replication of the virus continues in the CNS (Persidsky and Gendelman, 2003). Effective levels of antiviral agents may not be achieved in the CNS within the dose limits of clinical toxicity (Groothuis and Levy, 1997). Pgp inhibition provides a key strategy aimed at enhancing the uptake of drugs into the brain.

An emerging strategy for enhanced BBB penetration of drugs is the co-administration of competitive or non-competitive inhibitors of the efflux transporter together with the desired CNS drug. First-generation low molecular weight Pgp inhibitors (cyclosporine A, verapamil, among others) are substrates of the drug efflux transporter, which compete for the active site with the therapeutic agent (Lu et al., 2001). Second-generation inhibitors (LY335979, XR9576 and GF120918) are non-competitive inhibitors, which allosterically bind to Pgp, inactivating it and increasing drug transport to the brain (Martin et al., 2000). Despite their high efficiency in cell culture models, the small therapeutic range of these inhibitors, high in vivo toxicity and fast clearance are the main obstacles for their therapeutic application.

Recently, a new class of inhibitors (high molecular weight nonionic surfactants) was identified as promising agents for drug formulations. These compounds are two- or three-block copolymers arranged in a linear ABA or AB structure. The A block is a hydrophilic poly(ethylene oxide) chain and the B block a hydrophobic lipid- or a poly(propylene oxide) (in copolymers such as Pluronics) chain. Selected polymer nonionic surfactants with optimal structure, Pluronic block copolymers (PBC), are potent biological modulators of the Pgp drug efflux transporter. PBC reduce the resistance against drugs in Pgp-overexpressing cancer cells, thereby increasing the therapeutic efficacy of a drug by 2 to 3 orders of magnitude (Venne et al., 1996). There is significant evidence that PBC can inhibit the Pgp efflux system in various cells, including the BMVEC (Batrakova et al., 2001a, b). One of the mechanisms of PBC action is ATP depletion, which is necessary for effective Pgp efflux, after exposure of BMVEC to PBC (Batrakova et al., 2003). Furthermore, in vivo studies suggested that PBC significantly enhances brain penetration of Pgp substrates in mice from the peripheral blood through the BBB (Batrakova et al., 2001a). PBC could be clinically applicable for more efficient transfer of antiretroviral drugs.

4.4. Viral load in the brain: disease progression and immune suppression

HIV-1 infection induces effective cell-mediated and humoral immune responses, which, in the earliest stages of disease, curbs HIV-1 infection (Krishnakumar et al., 2005). Indeed, HIV-1 antibodies and the virus itself can be detected in the CSF during primary HIV-1 infection (Ho et al., 1985; Goudsmit et al., 1986). The initial protective response to HIV-1 infection comes from CD^8 cytotoxic T cells. CD^4 T helper cells affect B cell activation/differentiation and the production of antibodies as well as a variety of cellular and effector immune responses including the release of a variety of cytokines in response to activation by antigen presenting cells. The humoral response occurs after the cell-mediated response, when cytokines released from CD^4 T lymphocytes stimulate B cell antibody production (Koup et al., 1994; Krishnakumar et al., 2005). Nonetheless, as the immune system becomes ravaged by progressive viral infection a concomitant increase in viral load occurs in the cerebrospinal fluid and brain. Such increases in viral burden often correlate with the later development of neurocognitive defects and monitoring of CSF HIV-1 RNA levels during therapy has been suggested as a means to follow patients who could be at risk for HAD (Ellis et al., 2002).

4.5. Neuropathogenesis of HIV infection: cell biology and disease mechanisms

4.5.1. Neurotoxic output of brain MP

MPs are a significant source of neurotoxins during disease (Gabuzda et al., 1986; Koenig et al., 1986; Genis et al., 1992). Microglia represent up to 10% of the parenchymal brain cell population in some regions. Neighboring elliptical microglia contact each other in series and in parallel, forming a network. Recent studies, by several groups, support the idea that perivascular macrophages are preferentially infected and pose the greatest threat as sources of neurotoxic activities (Pulliam et al., 1997; Rappaport et al., 2001; Williams et al., 2001; Williams and Hickey, 2002). MPs are the primary reservoir for virus in the brain. MP-secreted toxins (viral and cellular) influence neuronal injury and lead to HAD (Fig. 4.2) (Milligan et al., 1991; Kaul et al., 2001; Luo et al., 2003). Under normal conditions MPs eliminate foreign material and secrete trophic factors important in maintaining CNS homeostasis (Elkabes et al., 1996; Lazarov-Spiegler et al., 1996; Rapalino et al., 1998; Zheng et al., 1999; Gras et al., 2003). A partial list of trophic factors include: brain-derived neurotrophic factor (BDNF) (Miwa et al., 1997), β-nerve growth factor (βNGF) (Garaci et al., 1999), transforming growth factor beta-β (Chao et al., 1995), neurotrophin (NT)-3 (Kullander et al., 1997) and glial derived neurotrophic factor (GDNF) (Batchelor et al., 1999). Cytokines released by MPs are also important regulators of neurotrophic factors signaling pathways. During HAD the regulation of the neurotrophic factors becomes impaired, and this disruption works together with the production of neurotoxins leading to HAD.

Infected MPs secrete multiple neurotoxic factors that contribute to HAD (Fig. 4.2). These factors include: cytokines [interferon (IFN)-γ, TNF-α and platelet activating factor (PAF)]; chemokines (MIP-1α/CCL3, MIP-1β/CCL4, RANTES/CCL5, SDF-1/CXCL12 and fractalkine/CX3CL1); eicosanoids (prostaglandins, thromboxane); excitatory amino acids (EAA); reactive oxygen species (ROS); reactive nitrogen species; TNF-related apoptosis-inducing ligand (TRAIL); and viral proteins (gp120, gp41, Tat, Nef, Vpr) (Gendelman et al., 1998; Tardieu, 2004). In fact, the severity of dementia has been found to correlate more strongly with the levels of activated macrophages and microglia than actual viral count in the CNS (Adle-Biassette et al., 1999). This evidence indicates that neuronal damage is a result of factors related to macrophage activation rather than direct attack

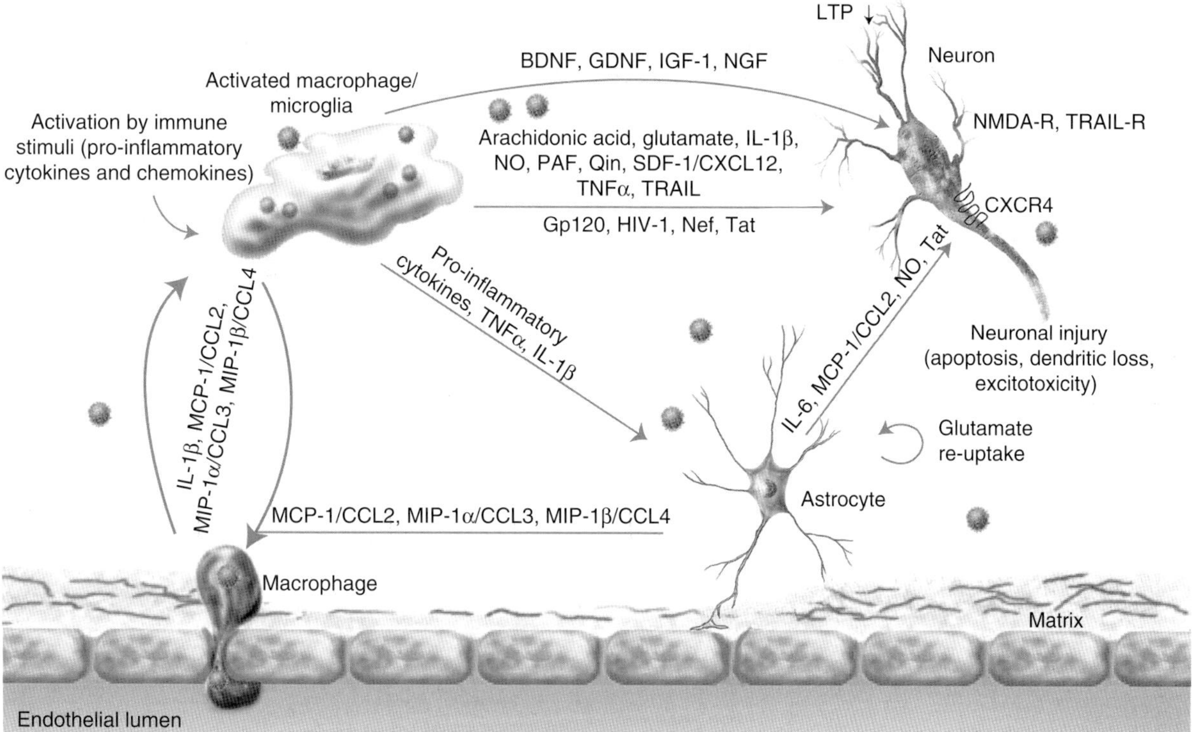

Fig. 4.2. For full color figure, see plate section. Mechanisms for HIV-1 neuropathogenesis. MP secretory products affect a cascade of immunomodulatory activities that engage neurons and astrocyte effector functions. Activated perivascular macrophages and microglia secrete viral and cellular neurotoxins which include, but are not limited to, viral proteins (Nef, Tat and gp120), pro-inflammatory cytokines (for example, TNF-α and IL-1β), PAF, free radicals such as NO, glutamate-like agonists, and quinolate, arachidonic acid and its metabolites and amines. Chemokines such as MCP-1/CCL2 and MIP-1α/β (CCL3/CCL4) are also produced from glial cells attracting more macrophages into an inflamed brain. The cumulative effects of MP neurotoxic factors include LTP inhibition and neuronal injury through excitotoxicity, apoptosis and dendritic process loss. Astrocyte activation also plays a critical regulatory role in neuronal injury. Innate astrocyte immunity includes the secretion of chemokines, cytokines and ROS (such as MCP-1/CCL2, IL-6 and NO).

by HIV-1. How MPs switch from a neurotrophic phenotype to a neurotoxic one remains unknown. The term "metabolic encephalopathy" has been used to describe the neurotoxic effects of HIV-1 replication in the brain, mediated by inflammatory factors (Gendelman et al., 1998).

A role for EAA in HAD has also been elucidated. EAAs include glutamate (Jiang et al., 2001), and quinolinic acid (Kerr et al., 1997). Glutamate is thought to cause neurotoxicity via over-stimulation of neurons, leading to Ca^2-mediated cell death (Di Stefano, 2004). Quinolinic acid, a metabolite of tryptophan, is another mediator of glutamate-induced apoptosis that is found in the CSF of patients with HAD (Tardieu, 2004). Macrophages appear to be the primary source of quinolinic acid in the CNS, damage cells by activating N-methyl-d-aspartate (NMDA) receptors, causing overexcitability of neurons and neuronal loss. Quinolinic acid levels correspond with the HAD patient's CSF HIV-1 RNA levels and progressive cerebral atrophy, indicating neuronal loss (Heyes et al., 2001).

Multiple inflammatory mediators are secreted by MPs (Fig. 4.2). Cytokines, PAF (Gelbard et al., 1994), and arachidonic acid and its metabolites (Nottet et al., 1995) promote inflammation through interactions with other leukocytes and astrocytes. Macrophages secrete a number of cytokines, including: IFN-α/β, IL-1α/β, IL-6 and TNF-α/β. IL-1β is a potent astrocyte stimulator while TNF-α is a potent activator of macrophages. These cytokines can lead to astrocyte and macrophage amplification of immune activation leading to the downward spiral of increasing numbers of activated macrophages promoting increasingly higher levels of inflammation.

The pro-inflammatory cytokine TNF-α plays a prominent role in the neuropathogenesis of HAD. Activated macrophages that have been infected with HIV-1 release both TNF-α and PAF (Genis et al., 1992; Nottet et al., 1995). TNF-α is released from CD^8 cytotoxic T cells, which are mainly responsible for direct killing of HIV-1-infected cells, but also appear to participate in cytokine-induced neurotoxicity as well (Jassoy et al., 1993). Macrophages in patients with HAD express higher levels of TNF-α and PAF mRNA. Furthermore, the PAF level in the CSF is increased (Wesselingh et al., 1993; Gelbard et al., 1994). Both TNF-α and PAF have been found to cause dose-dependent neuronal apoptosis (Gelbard et al., 1994; Talley et al., 1995). TNF-α is thought to cause neuronal damage through several mechanisms: direct damage to oligodendrocytes and myelin (Nottet et al., 1995), stimulating production of other inflammatory factors (Janabi et al., 1996, 1998) and inhibiting glutamate uptake by astrocytes and promoting glutamate receptor-mediated toxicity (Nottet et al., 1995).

A number of other inflammatory mediators figure significantly into HAD. Prostaglandins, which are detected in the CSF of patients with HAD, are believed to inhibit glutamate uptake in astrocytes thereby increasing the amount of free glutamate. This leads to stimulation of NMDA receptors in the brain and subsequently neuronal cell death (Tardieu, 2004). HIV-1 gp120, the viral coat protein, begins a cascade of effects starting with the upregulation of IL-1β, an inflammatory cytokine. IL-1β in turn regulates cyclooxygenase type 2, an inflammatory mediator involved in conversion of arachidonic acid to prostaglandin E2 (PGE2), leading to the potentiation of glutamate release and toxicity (Corosaniti, 2000).

Monocytes may cause disease through changes in levels of cell surface molecules, such as TRAIL (Green, 2003). TRAIL is an integral membrane protein that is upregulated in HIV-1-infected macrophages by both TNF-α and IFN-γ. $TRAIL^+$ macrophages are associated with neuronal apoptosis, possibly by stimulating death receptors on neurons, or by influencing association between infected macrophages and uninfected macrophages and astrocytes, causing inflammation and subsequent neuronal damage (Ryan et al., 2004).

4.5.2. Virotoxins

Extensive research has gone into characterizing the direct effect HIV-1 and HIV-1 proteins have on neuronal death. As previously mentioned the primary reservoir for HIV-1 are the MPs in the brain which release infectious virus and viral proteins (gp120, gp41 and Tat; Figure 4.2), which were shown to be neurotoxins, while astrocytes have a restricted infection but still express the early viral proteins: Tat, Rev and Nef. Cells in the CNS are exposed to viral proteins in a number of ways: release of viral proteins from cytopathic infection, formation of viral proteins during restricted viral infection, formation of defective viral particles, active release of viral proteins, shedding of viral coat and interaction with viral proteins through cell to cell contact. The effect of other viral proteins, including Nef and Vpr, in the CNS has also been examined.

HIV-1 gp120 is an external viral glycoprotein that together with gp41 binds to CD^4 and chemokine receptors in order to infect cells. It is not only located in the virion, but is also found in the plasma membrane of infected cells. Multiple studies have shown that gp120 acts as a neurotoxin in vitro on primary cortical neurons (Dawson et al., 1993), human CNS cultures (Iskander et al., 2004), neural cell lines (Corasaniti et al., 2001a) and rat brain in vivo (Corasaniti et al., 2001b). Different neuronal populations apparently have variable susceptibility to gp120 neurotoxicity. Calbindin-containing neurons are resistant both in vitro (Diop et al., 1995) and in HIVE brains (Masliah et al., 1995) while dopaminergic neurons are susceptible to gp120 neurotoxicity (Bennett et al., 1995; Barks et al., 1997). The mechanisms by which gp120 affects neuronal cells are incompletely understood. Picomolar amounts of gp120 injure cultured neurons in a dose dependent manner, which can be blocked by anti-gp120 but not anti-CD^4 antibodies, indicating that CD^4 receptor binding is not required for HIV-1 neurotoxicity (Kaiser et al., 1990). Furthermore, treatment in vitro with calcium channel antagonists also prevents gp120 neurotoxicity (Dreyer et al., 1990).

Gp120 interaction with astrocytes and microglia modifies the secretion of cytokines, chemokines and other soluble factors. These secreted factors can then act in an autocrine or paracrine fashion leading to modification in intracellular signaling, gene expression, protein synthesis, transport mechanisms and cell death. They include, but are not limited to, IL-1, IL-6, and TNF-α (Merrill et al., 1992; Koka et al., 1995), inducible nitric oxide synthase (iNOS) and NO release (Koka et al., 1995; Hori et al., 1999), which could further contribute to neuropathological changes. In addition, gp120 interaction with astrocytes alone or in conjunction with cytokines leads to multiple alterations in astrocyte function, such as reduced expression of excitatory amino acid transporter (EAAT) 2 (see Section 4.5.3).

The presence of gp41 leads to an elevation of iNOS (Adamson et al., 1996). Specifically the N-terminal region of gp41 induces iNOS leading to the production of NO and increased oxidative stress, an important mechanism for neuronal damage (Adamson et al., 1999). Furthermore IL-10 and chemokine receptor expression by monocytes, astrocytes and neurons in vitro are altered by exposure to gp41 (Speth et al., 2000).

Vpr has been shown to have a direct neurotoxic effect in vitro. Vpr is an accessory protein that modestly increases HIV-1 transcription, increases viral production in MP and may be involved in nuclear translocation of the viral preintegration complex. Vpr has also been implicated in HIV-1 neuropathogenesis (for review see Pomerantz, 2004). Vpr has been shown to induce apoptosis through mitochondria disruption by both caspase-dependent and independent mechanisms (Muthumani et al., 2003). Vpr induced apoptosis has been shown to occur in human fibroblasts, T cell lines and in primary cultured cells. Furthermore, Vpr induced apoptosis has been shown in human neuronal cell line NT2 and in cultured rat hippocampal neurons in vitro (Piller et al., 1998). In addition, Vpr could contribute to neuropathogenesis through its ability to form cation-permeable channels (Piller et al., 1998).

Tat, another viral protein, helps increase HIV-1 RNA transcription and processing, and its transcripts have been located in the CNS of HAD patients (Di Stefano, 2004). Significant quantities of Tat are found in the CSF of demented patients (Cheng et al., 1998). Tat is believed to be involved in neuronal injury both indirectly through macrophage released soluble factors (Lipton and Gendelman, 1995) and by direct neurotoxicity through interfering with neural electrophysiological activity (Xiong et al., 1999a, b, 2000). It is also involved in neuronal excitotoxicity (Liu et al., 2000; Nath et al., 2000). Gp120 and Tat synergistically activate NMDA receptors leading to neuronal overexcitability. Tat stimulates production of MMP-2 and -7, thereby increasing BBB permeability (Di Stefano, 2004). It has also been shown that Tat can induce iNOS expression in human astroglia (Liu et al., 2002b). Chemokine and chemokine-receptor expression is also altered upon exposure to Tat (McManus et al., 2000).

The HIV-1 accessory protein Nef has also been implicated in HIV-1 neuropathogenesis. Nef is strongly expressed during HIV-1 infection in infected MPs and astrocytes. Expression of Nef seems to be essential for maintaining a high rate of replication (Kestler et al., 1991). Nef has been shown in vitro to be toxic to neural cell lines, primary human neurons and glial cell cultures (Trillo-Pazos et al., 2000). Furthermore, Nef is able to induce histopathological and cognitive defects in vivo (Mordelet et al., 2004). Nef may also contribute to HAD by acting as a chemotactic factor recruiting macrophages into the CNS (Koedel et al., 1999). Furthermore, Nef interaction with CXCR4 induces apoptosis (Huang et al., 2004; James et al., 2004).

4.5.3. Astrocytes

During HIV-1 CNS infections astrocytes can proliferate and undergo apoptosis (Toggas et al., 1994; Petito and Roberts, 1995; Wyss-Coray et al., 1996). Astrocytes can be infected by both M- and T-tropic virus (Messam and Major, 2000). HIV-1 infection of astrocytes appears to be severely restricted. Astrocyte infection leads to expression of only HIV regulatory genes (the early HIV-1 proteins Tat, Rev and Nef). Structural proteins, such as HIV-1 gp120, are rarely seen. Astrocyte infection by HIV-1 is CD^4-independent (Sabri et al., 1999). CCR3, CCR5 and CXCR4 receptors are, however, found on astrocytes (Boutet et al., 2001). Nonetheless, astrocyte cell lines can be infected independent of both CD[4] and CXCR4 (Schweighardt et al., 2001). The HIV-1 infection may be maintained for years and therefore astrocytes may serve as a viral reservoir (Messam and Major, 2000). Although the percent of astrocytes infected is believed to be small, the actual extent of astrocyte infection by HIV-1 remains unknown. Astrocytes may contribute to HIV-1 pathogenesis by different mechanisms. *First*, astrocytes contribute to neuronal excitotoxicity. In HIV-1 infection, the normal astrocyte function of regulating extracellular glutamate levels is disrupted in a number of ways. Astrocyte-gp120 interaction (Patton et al., 2000) and infection (Messam and Major, 2000) both lead to disruption of astrocyte re-uptake of glutamate via EAAT (Wang et al., 2003). Astrocyte glutamate release is induced by activated macrophages (Genis et al., 1992; Fine et al., 1996). This astrocyte-released glutamate could stimulate astrocyte receptors contributing to even higher levels of glutamate (Bezzi et al., 1998). *Second*, astrocytes can amplify neurotoxic signals through multiple mechanisms. HIV-1 Tat induces astrocyte expression of MCP-1/CCL2, which attract macrophages (Conant et al., 1998), and IL-8 and IP-10, which attract multiple types of leukocytes (Kutsch et al., 2000). Cytokines and viral proteins induce iNOS in astrocytes (Adamson et al., 1996). NO production has been found to be overproduced in astrocytes after HIV-1 infection due to upregulation of iNOS, and can cause neuronal damage via NMDA stimulation, leading to neuronal overexcitability (Hori et al., 1999). *Third*, Fas ligand (FasL), present in the CSF of HIV-1 patients (Sabri et al., 2001), is also upregulated by astrocytes after exposure to IL-1β, IL-6, IFN-γ and TNF-α (Choi et al., 1999; Ghorpade et al., 2003). Since Fas, the receptor for FasL, is expressed in neurons (Beer et al., 2000; Medana et al., 2000), it is likely that astrocyte expression of FasL contributes to neuronal injury in HAD.

4.5.4. Mechanisms of neuronal dysfunction and death

Neuronal injury is the ultimate end of the pathological processes occurring during progressive HIV-1 infection of the nervous system. Histopathology demonstrates clear changes in neuronal morphology in HIVE. Such changes include decreases in synaptic density and vacuolation of dendritic spines. Apoptosis appears to be the major means of neuronal death during HAD and is found early on in the disease (Adle-Biassette et al., 1995; An et al., 1996). Neural injury and death by apoptosis appears to be the cumulative result of multiple neuronal insults (Gray et al., 2000) from activated MPs, activated astrocytes and virotoxins. TNF-α, for example, can directly induce apoptosis by itself (Pulliam et al., 1998) or can act synergistically with Tat to induce both apoptosis and oxidative stress (Shi et al., 1998). TNF-α can also act with HIV-1 gp120 to promote apoptosis. The trigger for neuronal apoptosis is calcium influx resulting from either NMDA receptor dysfunction, or some other mechanism. Increased calcium levels induce chromatin condensation through p38 mitogen activated protein kinase and cause mitochondria to release cytochrome-C (cytC) and ROS. CytC also activates caspases, leading to chromatin condensation. ROS and free radicals produced in this process lead to lipid peroxidation, chromatin condensation and eventual neuronal death.

Pro-inflammatory cytokines released from activated or infected MPs mediate neurotoxicity but may also cause more subtle alterations in the functions of neurons. For example, TNF-α and IL-1β are prominent in neuronal injury. IL-1β inhibits long-term potentiation (LTP) in rat hippocampus (Xiong et al., 1999a). This inhibition of LTP by IL-1β may act through activation of nuclear factor κB (NFκB) and stimulation of stress-activated protein kinase p38, possibly leading to the alteration of glutamate release by cells in the CNS (Kelly et al., 2003). A second possible mechanism for this LTP inhibition by IL-1β is through the increased activity of superoxide dismutase and increased ROS (Vereker et al., 2001). IL-8, a chemokine secreted by infected MPs, has also been shown to inhibit LTP (Xiong et al., 2003a).

4.6. Bioimaging for monitoring inflammation and neurodegeneration

The HIV-1-associated brain inflammation can be visualized through the use of magnetic resonance imaging (MRI) and magnetic resonance spectroscopy (MRS) (Boska et al., 2004), two very promising techniques for examining the effect of HAD on brain metabolites and functioning. MRS can detect and measure multiple metabolites in the brain, including: N-acetyl (NA)-containing compounds (predominantly N-acetyl aspartate), choline-containing compounds, glutamate and glutamine, and myoinositol (MI). The metabolite levels in HIV-1-positive and HAD patients have been examined and are usually expressed in ratios. Most studies have observed decreases in NA/Cr or NA/Cho ratios in HIV-1 patients with cognitive impairment but not in HIV-1 patients without impairment and these changes are believed to be an indication of neuronal loss (Wilkinson et al., 1997; Pavlakis et al., 1998). Some studies have shown these decreases to be reversed upon antiretroviral treatment (Wilkinson et al., 1997). MI, a putative glial marker, has been examined as a possible marker for inflammation that occurs during HAD. Unfortunately, the few studies on MI and HAD have not borne this out, although there are difficulties in determining the metabolite concentrations (Chang et al., 2000; Suwanwelaa et al., 2000; von Giesen et al., 2001). Studies seem to be moving from MRS to functional MRI (fMRI), which can measure brain activation through neuronal activity and glycolysis. There have been two reports using fMRI to determine the correlation of neural activities to cognitive deficits in HIV-1 patients (Chang et al., 2001; Ernst et al., 2002). In these studies HIV-1 patients show a larger increase in activation of the lateral prefrontal cortex as compared with the activation observed in matched noninfected individuals when performing memory tasks. This increased activation precedes any clinical signs or observed deficits on cognitive tests. Increased activation may result from early neuronal injury leading to an increased usage of the brain reserve in order to compensate. The fMRI appears to be more sensitive than clinical signs or cognitive tests by being able to pick up neural injury at an earlier stage. If so, increased brain activation of the lateral prefrontal cortex may be a direct neural correlate to memory deficits.

4.7. Adjunctive therapies

The goal of adjunctive therapies is the treatment and prevention of HAD and MCMD (Dou et al., 2004) through minimizing the cell damage that occurs with HIV-1 infection of the brain, such as lithum chloride and valproic acid (Dou et al., 2003, 2004). An understanding of how cells of the CNS are damaged and how such damage can lead to CNS dysfunction upon HIV-1 infection has identified new therapeutic targets for adjunctive therapies. Enhancing the protective action of neurotrophins (Bachis et al., 2003) is one way to accomplish this. Neurotrophins include BDNF, GDNF, NT-3, nerve growth factor (NGF), fibroblast growth factor (FGF) and insulin-like growth factor (IGF)-1 (Connor and Dragunow, 1998; Hayashi et al., 2000; Namiki et al., 2000; Everall et al., 2001; Paula-Barbosa et al., 2003). Expression of neurotrophins may also be associated with neurotoxicity during the early stages of neurodegenerative processes (Finklestein, 1996; Speliotes et al., 1996; Fiedorowicz et al., 2001; Ganat et al., 2002). For example, dysregulation of neurotrophic factors are known to affect HAD pathogenesis (Chauhan et al., 2001; Felderhoff-Mueser et al., 2002; Johnson and Sharma, 2003). The neuroprotective effects of neurotrophins, upon binding their cognate receptors, act to limit neurotoxin- and lesion-induced neuropathologic damage and can affect individual neuronal populations, dendritic length, spine density, synaptic transmission, anti-apoptotic signaling or signaling to limit oxidative stress (Meucci and Miller, 1996; Connor and Dragunow, 1998; Ramirez et al., 2001; Titanji et al., 2003). Neurotrophins can also confer protection by preventing apoptosis (Meucci and Miller, 1996; Macdonald et al., 1999; Ramirez et al., 2001).

Moreover, the effects of anti-inflammatory cytokines, such as IL-4, IL-10 and IFN-α, may confer protection against pro-inflammatory substances released from MPs. These anti-inflammatory cytokines are produced by and act on both neurons and glia and are upregulated in CNS degenerative disorders (Vitkovic et al., 2001; Koeberle et al., 2004). Several lines of evidence have demonstrated that some cytokines (IL-4 and IL-10) may be neuroprotective through anti-inflammatory effects (Spera et al., 1998; Dietrich et al., 1999; Sholl-Franco et al., 2002; Abraham et al., 2004).

Another approach is through the inhibition of glycogen synthase kinase (GSK)-3β. GSK-3β was isolated from skeletal muscle, but is widely expressed in all tissues, and particularly abundant in brain (Yao et al., 2002; Schaffer et al., 2003). Among the intriguing links between GSK-3β and cell survival are the findings that GSK-3β activation is directly linked to increased neuronal apoptosis and associated with the downregulation of transcription factors that protect neurons from toxic insults. Overexpression of catalytically active GSK-3β at levels that induce apoptosis has been linked with HIV-1 protein-mediated neurotoxicity (Maggirwar et al., 1999; Tong et al., 2001). The importance of GSK-3β in many apoptotic conditions is further supported by evidence that selective small-molecule inhibitors of GSK-3β provide considerable protection from apoptotic cell death. Several inhibitors of enzymes have been found to be capable of mediating this modification following HIV-1-mediated neurotoxicity (Everall et al., 2002; Dou et al., 2003), such as lithium and sodium valproate (Chen et al., 1999; Linseman et al., 2003). Lithium reduces GSK-3β-mediated tau phosphorylation (Lee et al., 2003) and has been shown to promote neuronal survival (Hongisto et al., 2003). Taken together with present data, the regulation of GSK-3β activity, by enzyme inhibitors within the brain, suggests that modulation of GSK-3β in neurons may be an important strategy for neuroprotection.

NMDA antagonists are another promising approach for neuroprotection. Previous studies have revealed that the over-activation of NMDA receptors with its resultant excitotoxicity, disruption of the cellular calcium homeostasis and free radical formation are all key mechanisms involved in brain damage and neurodegenerative disease accompanied by deficits in cognition (Bi and Sze, 2002; Xiong et al., 2003b; Anderson and Xiong, 2004). Over-activation of NMDA receptors by glutamate or NMDA results in neuronal cell death (Lipsky et al., 2001; Jiang et al., 2003; Baptiste et al., 2004). NMDA neuroprotection can occur through neurotrophins including BDNF, NGF and NT-3, with the subsequent regulation of glutamate positively affecting neuronal survival (Rocha et al., 1999; Jiang et al., 2003; Marmigere et al., 2003).

Memantine, a non-competitive NMDA antagonist, has been clinically used in the treatment of vascular dementia and Alzheimer's disease (AD) in Germany since the 1980s (Bormann, 1989). The neurotoxicity caused by HIV-1 proteins Tat and gp120 can be blocked by memantine (Jain, 2000). Furthermore, memantine improved hippocampal synaptic transmission in the severe combined immune-deficient mouse model of HIV-1-associated neurologic disease (Anderson and Xiong, 2004). In clinical applications, memantine treatment has been directed primarily toward HAD, AD and other senile dementias (Jain, 2000; Anderson and Xiong, 2004; Tariot et al., 2004).

4.8. Prognostic biomarkers for HAD

4.8.1. Potential biomarkers in CSF and plasma for HAD

As previously stated, HIV-1 replication is tightly controlled in the brain for years following initial infection (Navia et al., 1986). The underlying molecular mechanisms of innate CNS immune responses regulating viral replication and influencing MP neurotoxic reactions are not completely understood. A critical question in the pathogenesis of HAD is what instigates high levels of viral replication late in the course of disease. Productive viral infection of brain MP is necessary but not sufficient to induce HAD, since absolute levels of virus in the brain do not always correlate with the degree of cognitive impairment (Kure et al., 1990; Dickson et al., 1991; Wiley, 1994). A complex interplay of both viral and host cellular factors regulate neurological disease. Thus, quantification of viral concentration in CSF is not adequate as a prognostic marker for HAD (Stankoff et al., 1999). As a result, identification of prognostic markers for HAD has been an important area of research (Syndulko et al., 1994).

Unfortunately, the search for HAD prognostic markers remains limited. Multiple factors involved in HIV-1 neuropathogenesis have been examined but lack sensitivity and specificity as disease biomarkers. Some reported markers include neopterin (Fuchs et al., 1989), quinolinic and kynurenic acid ratios (Heyes et al., 1991), β2-microglobulin (McArthur et al., 1992), soluble vascular cell adhesion molecule-1 (for simian immunodeficiency virus encephalitis) (Sasseville et al., 1992), quinolinic acid (Brouwers et al., 1993), HIV-1 p24 antigen (Royal et al., 1994), soluble intracellular adhesion molecule-1 (Heidenreich et al., 1994), TNF-α (Calvo Manuel et al., 1995), soluble Fas and FasL (Sabri et al., 2001) and S-100β (Pemberton and Brew, 2001). Tau protein and HIV-1 RNA levels remain controversial as markers for HAD. Combination of biomarkers such as CSF β2-microglobulin and neopterin levels in conjunction with blood CD^4 lymphocyte count shows some utility to predict disease (Brew et al., 1996a).

4.8.2. OTK18: a putative biomarker for HAD

Our group is now working on a promising candidate marker for HAD, OTK18. OTK18 has been recently shown by our group to be induced by and suppress HIV-1 infection (Carlson et al., 2004a). OTK18 gene expression is enhanced by HIV-1 infection in macrophages. However, its expression is tightly regulated in brain macrophages, and is only detectable in severe HIVE cases (Carlson et al., 2004b), making it an attractive HAD candidate marker to be used in combination with other reported markers. OTK18 is classified as a transcription factor because of its 13 C2H2-type zinc fingers (Saito et al., 1996). There is a mass of evidence that Krüppel-associated box (KRAB) and C2H2 zinc finger proteins can regulate HIV-1 replication (Reynolds et al., 2003). C2H2 zinc finger motifs are capable of binding to a wide range of DNA sequences, including the HIV-1 long terminal repeat (LTR) (Wu et al., 1995; Isalan et al., 2001). OTK18 is the first KRAB-containing C2H2 zinc finger protein endogenously expressed in macrophages and affects its antiviral activities.

MPs have multiple approaches to controlling HIV-1 production. One way is through cellular factors, such as NFκB and C/EBPβ, which are important in regulating HIV-1 expression in MPs (Griffin et al., 1989; Akira and Kishimoto, 1992; Tesmer et al., 1993; Henderson et al., 1995, 1996). There are also a number of cellular transcription factors that can repress virus, often through the viral LTR. These factors include: *c-myc* promoter binding protein (MBP-1), tumor suppressor p53, 16kD inhibitory CCAAT/enhancer binding protein-β (C/EBPβ), peroxisome proliferator-activated receptor-γ (PPARγ) and regulatory protein T lymphocyte-1 (Patarca et al., 1988; Subler et al., 1994; Ray and Srinivas, 1997; Weiden et al., 2000; Cicala et al., 2002; Hayes et al., 2002). Research in this area supports the idea that transcription factor malfunction plays an important role in permitting ongoing viral replication and dissemination in the infected human host. Taken together with the pattern of OTK18 gene expression in HIV-1 macrophages and its detection in only severe HIVE cases, this suggests that HIV-1-induced OTK18 expression is tightly regulated through some feedback mechanism, which is destroyed in HIVE brains.

OTK18 generates multiple protein fragments (mainly 75, 65 and 35 kD). We have found that full-length OTK18α (75 kD) and OTK18β (65 kD) containing a nuclear localization signal are localized to the nucleus. Whereas the processed N-terminal fragment (OTK18N, 35 kD) lacking the nuclear localization signal is localized to the cytoplasm. Cleavage of OTK18 appears to be important in the pathogenesis of HIV-1 (Fig. 4.3). During HIV-1 infection of MPs, OTK18 production is increased and suppresses HIV-1 replication through down-modulating viral LTR activity. During HIVE, OTK18 undergoes proteolytic cleavage and accumulates in the cytoplasm of brain macrophages. OTK18-suppressive activity is therefore lost, possibly contributing to the progression to HIVE and its clinical counterpart, HAD.

To examine the utility of OTK18 as a HAD marker we studied 26 cases of HIVE with distinct pathological endpoints and have found that OTK18 expression paralleled the severity of disease, including the degree of giant cell encephalitis, viral replication and microglial nodules (Carlson et al., 2004b). Interestingly, OTK18 is also detected in the cytoplasm of brain macrophages in moderate to severe HAD cases. There was a limited co-localization of OTK18 and HIV-1 p24 antigens in HIVE brain tissue. Expression of OTK18 paralleled the severity of HIVE and the degree of cognitive dysfunction in infected individuals. Thus OTK18 could be used in conjunction with other markers to diagnose and follow the course of dementia.

4.9. Future prospects for the neuropathogenesis of HIV-1 infection

The neuropathogenesis of HIV-1 infection revolves around inflammatory factors secreted from virus-infected and immune-competent brain MPs. This classic view of HAD has undergone some re-evaluation in the post-HAART era. Diagnostic tools for neurocognitive disorders will likely be improved through the use of advanced brain imaging techniques. In particular, understanding the molecular mechanism of MP activation by characterizing cell-specific molecules will aid in biomarker discoveries. Combinations of antiretroviral and adjunctive drugs will likely improve therapeutic outcomes. In closing, the elucidation of the mechanisms for how HIV-1-infected MPs affect neuronal injury will not only provide a path for new therapies and diagnostics for HAD but will likely have broad applicability in

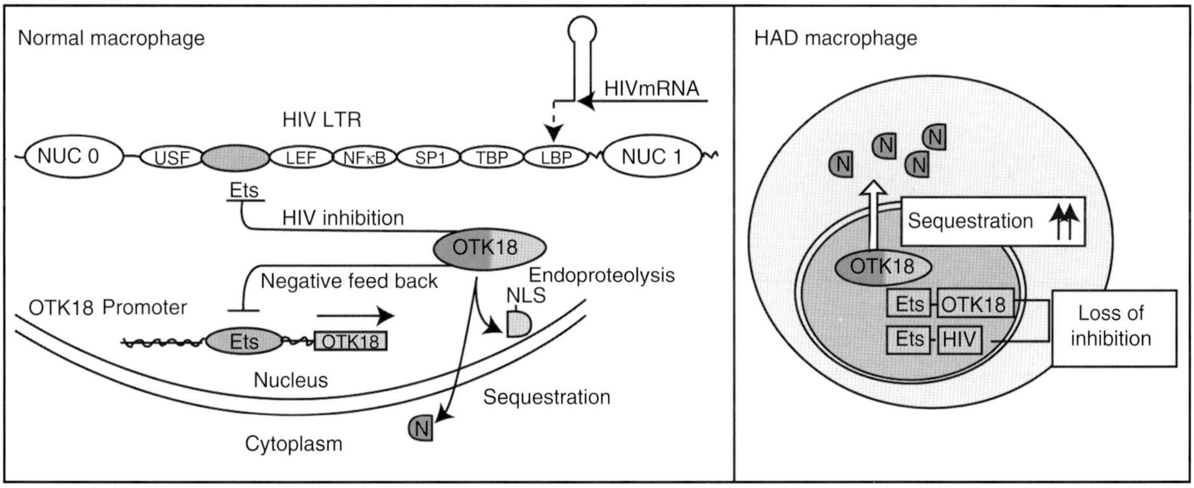

Fig. 4.3. For full color figure, see plate section. Regulation and antiretroviral activities of OTK18. In macrophages, HIV-1 infection induces OTK18 expression mediated through a proximal promoter region (Ets). Viral suppression results when OTK18 binds to the HIV-1 LTR Ets. Transient expression of OTK18 occurs as the result of a negative feedback where OTK18 binds to an Ets proximal to its own transcriptional start site. In HIVE, endoproteolysis of OTK18 is enhanced due to chronic viral infection and inflammation, leading to increased endoproteolysis and accumulation of N-terminal fragments in the cytosol. This results in failure of OTK18 suppression of the HIV-1 LTR and OTK18 promoter. As a result the cytoplasmic accumulation of OTK18 in macrophages may serve as a predictor of advanced HIVE and HAD.

many other neurodegenerative disorders where inflammatory responses underlie the processes of neuronal injury and neurologic impairments.

References

Abraham CS, Deli MA, Joo F, et al (1996). Intracarotid tumor necrosis factor-alpha administration increases the blood-brain barrier permeability in cerebral cortex of the newborn pig: quantitative aspects of double-labelling studies and confocal laser scanning analysis. Neurosci Lett 208: 85–88.

Abraham KE, McMillen D, Brewer KL (2004). The effects of endogenous interleukin-10 on gray matter damage and the development of pain behaviors following excitotoxic spinal cord injury in the mouse. Neuroscience 124: 945–952.

Adamson DC, Wildemann B, Sasaki M, et al (1996). Immunologic NO synthase: elevation in severe AIDS dementia and induction by HIV-1 gp41. Science 274: 1917–1921.

Adamson DC, Kopnisky KL, Dawson TM, et al (1999). Mechanisms and structural determinants of HIV-1 coat protein, gp41-induced neurotoxicity. J Neurosci 19: 64–71.

Adle-Biassette H, Levy Y, Colombel M, et al (1995). Neuronal apoptosis in HIV infection in adults. Neuropathol Appl Neurobiol 21: 218–227.

Adle-Biassette H, Chretien F, Wingertsmann L, et al (1999). Neuronal apoptosis does not correlate with dementia in HIV infection but is related to microglial activation and axonal damage. Neuropathol Appl Neurobiol 25: 123–133.

Akira S, Kishimoto T (1992). IL-6 and NF-IL6 in acute-phase response and viral infection. Immunol Rev 127: 25–50.

Albright AV, Shieh JT, Itoh T, et al (1999). Microglia express CCR5, CXCR4, and CCR3, but of these, CCR5 is the principal coreceptor for human immunodeficiency virus type 1 dementia isolates. J Virol 73: 205–213.

An SF, Giometto B, Scaravilli T, et al (1996). Programmed cell death in brains of HIV-1-positive AIDS and pre-AIDS patients. Acta Neuropathol (Berl) 91: 169–173.

An SF, Groves M, Gray F, et al (1999). Early entry and widespread cellular involvement of HIV-1 DNA in brains of HIV-1 positive asymptomatic individuals. J Neuropathol Exp Neurol 58: 1156–1162.

Anderson E, Zink W, Xiong H, et al (2002). HIV-1-associated dementia: a metabolic encephalopathy perpetrated by virus-infected and immune-competent mononuclear phagocytes. J Acquir Immune Defic Syndr 31(Suppl 2): S43–S54.

Anderson ERG, Xiong H (2004). Memantine protects hippocampal neuronal function in murine human immunodeficiency virus type 1 encephalitis. J Neurosci 24: 7194–7198.

Avraham HK, Jiang S, Lee TH, et al (2004). HIV-1 Tat-mediated effects on focal adhesion assembly and permeability in brain microvascular endothelial cells. J Immunol 173: 6228–6233.

Bachis A, Major EO, Mocchetti I (2003). Brain-derived neurotrophic factor inhibits human immunodeficiency virus-1/gp120-mediated cerebellar granule cell death by preventing gp120 internalization. J Neurosci 23: 5715–5722.

Bagasra O, Lavi E, Bobroski L, et al (1996). Cellular reservoirs of HIV-1 in the central nervous system of infected individuals: identification by the combination of in situ polymerase chain reaction and immunohistochemistry. AIDS 10: 573–585.

Ballabh P, Braun A, Nedergaard M (2004). The blood-brain barrier: an overview: structure, regulation, and clinical implications. Neurobiol Dis 16: 1–13.

Banks WA (2005). The blood–brain barrier: its structure and function. In: H Gendelman, IP Everall, S Lipton, S Swindells (Eds.), The Neurology of AIDS. Oxford University Press.

Banks WA, Akerstrom V, Kastin AJ (1998). Adsorptive endocytosis mediates the passage of HIV-1 across the blood-brain barrier: evidence for a post-internalization coreceptor. J Cell Sci 111(Pt 4): 533–540.

Banks WA, Freed EO, Wolf KM, et al (2001). Transport of human immunodeficiency virus type 1 pseudoviruses across the blood-brain barrier: role of envelope proteins and adsorptive endocytosis. J Virol 75: 4681–4691.

Baptiste DC, Hartwick AT, Jollimore CA, et al (2004). An investigation of the neuroprotective effects of tetracycline derivatives in experimental models of retinal cell death. Mol Pharmacol 66: 1113–1122.

Barks JD, Liu XH, Sun R, et al (1997). gp120, a human immunodeficiency virus-1 coat protein, augments excitotoxic hippocampal injury in perinatal rats. Neuroscience 76: 397–409.

Batchelor PE, Liberatore GT, Wong JY, et al (1999). Activated macrophages and microglia induce dopaminergic sprouting in the injured striatum and express brain-derived neurotrophic factor and glial cell line-derived neurotrophic factor. J Neurosci 19: 1708–1716.

Batrakova EV, Li S, Vinogradov SV, et al (2001a). Mechanism of pluronic effect on P-glycoprotein efflux system in blood-brain barrier: contributions of energy depletion and membrane fluidization. J Pharmacol Exp Ther 299: 483–493.

Batrakova EV, Miller DW, Li S, et al (2001b). Pluronic P85 enhances the delivery of digoxin to the brain: in vitro and in vivo studies. J Pharmacol Exp Ther 296: 551–557.

Batrakova EV, Li S, Alakhov VY, et al (2003). Sensitization of cells overexpressing multidrug-resistant proteins by pluronic P85. Pharm Res 20: 1581–1590.

Beer R, Franz G, Schopf M, et al (2000). Expression of Fas and Fas ligand after experimental traumatic brain injury in the rat. J Cereb Blood Flow Metab 20: 669–677.

Bell MD, Taub DD, Perry VH (1996). Overriding the brain's intrinsic resistance to leukocyte recruitment with intraparenchymal injections of recombinant chemokines. Neuroscience 74: 283–292.

Beltrami EM, Williams IT, Shapiro CN, et al (2000). Risk and management of blood-borne infections in health care workers. Clin Microbiol Rev 13: 385–407.

Bennett BA, Rusyniak DE, Hollingsworth CK (1995). HIV-1 gp120-induced neurotoxicity to midbrain dopamine cultures. Brain Res 705: 168–176.

Bevilacqua MP, Stengelin S, Gimbrone MA, Jr, et al (1989). Endothelial leukocyte adhesion molecule 1: an inducible receptor for neutrophils related to complement regulatory proteins and lectins. Science 243: 1160–1165.

Bezzi P, Carmignoto G, Pasti L, et al (1998). Prostaglandins stimulate calcium-dependent glutamate release in astrocytes. Nature 391: 281–285.

Bi H, Sze CI (2002). N-methyl-D-aspartate receptor subunit NR2A and NR2B messenger RNA levels are altered in the hippocampus and entorhinal cortex in Alzheimer's disease. J Neurol Sci 200: 11–18.

Bormann J (1989). Memantine is a potent blocker of N-methyl-D-aspartate (NMDA) receptor channels. Eur J Pharmacol 166: 591–592.

Boska MD, Mosley RL, Nawab M, et al (2004). Advances in neuroimaging for HIV-1 associated neurological dysfunction: clues to the diagnosis, pathogenesis and therapeutic monitoring. Curr HIV Res 2: 61–78.

Boutet A, Salim H, Leclerc P, et al (2001). Cellular expression of functional chemokine receptor CCR5 and CXCR4 in human embryonic neurons. Neurosci Lett 311: 105–108.

Boven LA, Gomes L, Hery C, et al (1999). Increased peroxynitrite activity in AIDS dementia complex: implications for the neuropathogenesis of HIV-1 infection. J Immunol 162: 4319–4327.

Boven LA, Middel J, Breij EC, et al (2000). Interactions between HIV-infected monocyte-derived macrophages and human brain microvascular endothelial cells result in increased expression of CC chemokines. J Neurovirol 6: 382–389.

Brew BJ, Miller J (1996). Human immunodeficiency virus type 1-related transient neurological deficits. Am J Med 101: 257–261.

Brew BJ, Bhalla RB, Paul M, et al (1990). Cerebrospinal fluid neopterin in human immunodeficiency virus type 1 infection. Ann Neurol 28: 556–560.

Brew BJ, Dunbar N, Pemberton L, et al (1996a). Predictive markers of AIDS dementia complex: CD4 cell count and cerebrospinal fluid concentrations of beta 2-microglobulin and neopterin. J Infect Dis 174: 294–298.

Brew BJ, Evans L, Byrne C, et al (1996b). The relationship between AIDS dementia complex and the presence of macrophage tropic and non-syncytium inducing isolates of human immunodeficiency virus type 1 in the cerebrospinal fluid. J Neurovirol 2: 152–157.

Brew BJ, Wesselingh SL, Gonzales M, et al (1996c). Managing HIV: 3. Mechanisms of diseases. 3.7 How HIV leads to neurological disease. Med J Aust 164: 233–234.

Brouwers P, Heyes MP, Moss HA, et al (1993). Quinolinic acid in the cerebrospinal fluid of children with symptomatic human immunodeficiency virus type 1 disease: relationships to clinical status and therapeutic response. J Infect Dis 168: 1380–1386.

Bukrinsky MI, Nottet HS, Schmidtmayerova H, et al (1995). Regulation of nitric oxide synthase activity in human immunodeficiency virus type 1 (HIV-1)-infected monocytes: implications for HIV-associated neurological disease. J Exp Med 181: 735–745.

Burkala E, West JT, Jun He, et al (2004). The choroid plexus and viral entry into the brain. In: HE Gendelman, IP Everall, SA Lipton, S Swindells (Eds.), The Neurology of AIDS. Oxford University Press.

Calvo Manuel E, Arranz Garcia F, Sanchez-Portocarrero J, et al (1995). [Alpha tumor necrosis factor in central nervous system disease associated with HIV infection]. An Med Interna 12: 263–266.

Carlos T, Kovach N, Schwartz B, et al (1991). Human monocytes bind to two cytokine-induced adhesive ligands on cultured human endothelial cells: endothelial-leukocyte adhesion molecule-1 and vascular cell adhesion molecule-1. Blood 77: 2266–2271.

Carlson K, Leisman G, Limoges J, et al (2004a). Molecular characterization of a putative anti-retroviral transcriptional factor, OTK18. J Immunol 172: 381–391.

Carlson KA, Limoges J, Pohlman GD, et al (2004b). OTK18 expression in brain mononuclear phagocytes parallels the severity of HIV-1 encephalitis. J Neuroimmunol 150: 186–198.

Carpenter CC, Cooper DA, Fischl MA, et al (2000). Antiretroviral therapy in adults: updated recommendations of the International AIDS Society-USA Panel. JAMA 283: 381–390.

Carson MJ, Sutcliffe JG (1999). Balancing function vs. self defense: the CNS as an active regulator of immune responses. J Neurosci Res 55: 1–8.

Cavender DE, Edelbaum D, Ziff M (1989). Endothelial cell activation induced by tumor necrosis factor and lymphotoxin. Am J Pathol 134: 551–560.

Chang L, Ernst T, Leonido-Yee M, et al (2000). Perfusion MRI detects rCBF abnormalities in early stages of HIV-cognitive motor complex. Neurology 54: 389–396.

Chang L, Speck O, Miller EN, et al (2001). Neural correlates of attention and working memory deficits in HIV patients. Neurology 57: 1001–1007.

Chao CC, Hu S, Sheng WS, et al (1995). Tumor necrosis factor-alpha production by human fetal microglial cells: regulation by other cytokines. Dev Neurosci 17: 97–105.

Chauhan NB, Siegel GJ, Lee JM (2001). Depletion of glial cell line-derived neurotrophic factor in substantia nigra neurons of Parkinson's disease brain. J Chem Neuroanat 21: 277–288.

Chen G, Zeng WZ, Yuan PX, et al (1999). The mood-stabilizing agents lithium and valproate robustly increase the levels of the neuroprotective protein bcl-2 in the CNS. J Neurochem 72: 879–882.

Cheng J, Nath A, Knudsen B, et al (1998). Neuronal excitatory properties of human immunodeficiency virus type 1 Tat protein. Neuroscience 82: 97–106.

Cho C, Miller RJ (2002). Chemokine receptors and neural function. J Neurovirol 8: 573–584.

Choi C, Park JY, Lee J, et al (1999). Fas ligand and Fas are expressed constitutively in human astrocytes and the expression increases with IL-1, IL-6, TNF-alpha, or IFN-gamma. J Immunol 162: 1889–1895.

Choo EF, Leake B, Wandel C, et al (2000). Pharmacological inhibition of P-glycoprotein transport enhances the distribution of HIV-1 protease inhibitors into brain and testes. Drug Metab Dispos 28: 655–660.

Cicala C, Arthos J, Selig SM, et al (2002). HIV envelope induces a cascade of cell signals in non-proliferating target cells that favor virus replication. Proc Natl Acad Sci USA 99: 9380–9385.

Clifford DB (2000). Human immunodeficiency virus-associated dementia. Arch Neurol 57: 321–324.

Conant K, Garzino-Demo A, Nath A, et al (1998). Induction of monocyte chemoattractant protein-1 in HIV-1 Tat-stimulated astrocytes and elevation in AIDS dementia. Proc Natl Acad Sci USA 95: 3117–3121.

Connor B, Dragunow M (1998). The role of neuronal growth factors in neurodegenerative disorders of the human brain. Brain Res Brain Res Rev 27: 1–39.

Corasaniti MT, Bilotta A, Strongoli MC, et al (2001a). HIV-1 coat protein gp120 stimulates interleukin-1beta secretion from human neuroblastoma cells: evidence for a role in the mechanism of cell death. Br J Pharmacol 134: 1344–1350.

Corasaniti MT, Piccirilli S, Paoletti A, et al (2001b). Evidence that the HIV-1 coat protein gp120 causes neuronal apoptosis in the neocortex of rat via a mechanism involving CXCR4 chemokine receptor. Neurosci Lett 312: 67–70.

Cordon-Cardo C, O'Brien JP, Casals D, et al (1989). Multidrug-resistance gene (P-glycoprotein) is expressed by endothelial cells at blood-brain barrier sites. Proc Natl Acad Sci USA 86: 695–698.

Corosaniti MTEA (2000). Apoptosis induced by gp120 in the neocortex of rat involves enhanced expression of cyclooxygenase type 2 (COX-2) and is prevented by NMDA receptor antagonists and by the 21-aminosteroid U-72389G. Biochem Biophys Res Commun 274: 664–669.

Cunningham PH, Smith DG, Satchell C, et al (2000). Evidence for independent development of resistance to HIV-1 reverse transcriptase inhibitors in the cerebrospinal fluid. AIDS 14: 1949–1954.

d'Arminio Monforte A, Duca PG, Vago L, et al (2000). Decreasing incidence of CNS AIDS-defining events associated with antiretroviral therapy. Neurology 54: 1856–1859.

Davis LE, Hjelle BL, Miller VE, et al (1992). Early viral brain invasion in iatrogenic human immunodeficiency virus infection. Neurology 42: 1736–1739.

Dawson VL, Dawson TM, Uhl GR, et al (1993). Human immunodeficiency virus type 1 coat protein neurotoxicity mediated by nitric oxide in primary cortical cultures. Proc Natl Acad Sci USA 90: 3256–3259.

Di Stefano M (2004). HIV-1 structural and regulatory proteins and neurotoxicity. In: HE Gendelman, I Grant, IP Everall, SA Lipton, S Swindells (Eds.), The Neurology of AIDS. Oxford University Press.

Dickson DW, Mattiace LA, Kure K, et al (1991). Microglia in human disease, with an emphasis on acquired immune deficiency syndrome. Lab Invest 64: 135–156.

Dietrich WD, Busto R, Bethea JR (1999). Postischemic hypothermia and IL-10 treatment provide long-lasting neuroprotection of CA1 hippocampus following transient global ischemia in rats. Exp Neurol 158: 444–450.

Diop AG, Lesort M, Esclaire F, et al (1995). Calbindin D28K-containing neurons, and not HSP70-expressing neurons, are more resistant to HIV-1 envelope (gp120) toxicity in cortical cell cultures. J Neurosci Res 42: 252–258.

Dore GJ, Hoy JF, Mallal SA, et al (1997). Trends in incidence of AIDS illnesses in Australia from 1983 to 1994: the Australian AIDS cohort. J Acquir Immune Defic Syndr Hum Retrovirol 16: 39–43.

Dore GJ, Correll PK, Li Y, et al (1999). Changes to AIDS dementia complex in the era of highly active antiretroviral therapy. AIDS 13: 1249–1253.

Dou H, Birusingh K, Faraci J, et al (2003). Neuroprotective activities of sodium valproate in a murine model of human immunodeficiency virus-1 encephalitis. J Neurosci 23: 9162–9170.

Dou H, Kingsley JD, Mosley RL, et al (2004). Neuroprotective strategies for HIV-1 associated dementia. Neurotox Res 6: 503–521.

Douek DC, McFarland RD, Keiser PH, et al (1998). Changes in thymic function with age and during the treatment of HIV infection. Nature 396: 690–695.

Dreyer EB, Kaiser PK, Offermann JT, et al (1990). HIV-1 coat protein neurotoxicity prevented by calcium channel antagonists. Science 248: 364–367.

Elkabes S, DiCicco-Bloom EM, Black IB (1996). Brain microglia/macrophages express neurotrophins that selectively regulate microglial proliferation and function. J Neurosci 16: 2508–2521.

Ellis RJ, Deutsch R, Heaton RK, et al (1997). Neurocognitive impairment is an independent risk factor for death in HIV infection. San Diego HIV Neurobehavioral Research Center Group. Arch Neurol 54: 416–424.

Ellis RJ, Moore DJ, Childers ME, et al (2002). Progression to neuropsychological impairment in human immunodeficiency virus infection predicted by elevated cerebrospinal fluid levels of human immunodeficiency virus RNA. Arch Neurol 59: 923–928.

Enting RH, Hoetelmans RM, Lange JM, et al (1998). Antiretroviral drugs and the central nervous system. AIDS 12: 1941–1955.

Ernst T, Chang L, Jovicich J, et al (2002). Abnormal brain activation on functional MRI in cognitively asymptomatic HIV patients. Neurology 59: 1343–1349.

Eugenin EA, Berman JW (2005). Mechanisms of viral entry through the blood-brain barrier. In: H Gendelman, L Epstein, S Swidells (Eds.), The Neurology of AIDS. Oxford University Press.

Everall IP, Trillo-Pazos G, Bell C, et al (2001). Amelioration of neurotoxic effects of HIV envelope protein gp120 by fibroblast growth factor: a strategy for neuroprotection. J Neuropathol Exp Neurol 60: 293–301.

Everall IP, Bell C, Mallory M, et al (2002). Lithium ameliorates HIV-gp120-mediated neurotoxicity. Mol Cell Neurosci 21: 493–501.

Felderhoff-Mueser U, Sifringer M, Pesditschek S, et al (2002). Pathways leading to apoptotic neurodegeneration following trauma to the developing rat brain. Neurobiol Dis 11: 231–245.

Ferrando S, van Gorp W, McElhiney M, et al (1998). Highly active antiretroviral treatment in HIV infection: benefits for neuropsychological function. AIDS 12: F65–70.

Fiedorowicz A, Figiel I, Kaminska B, et al (2001). Dentate granule neuron apoptosis and glia activation in murine hippocampus induced by trimethyltin exposure. Brain Res 912: 116–127.

Fine SM, Angel RA, Perry SW, et al (1996). Tumor necrosis factor alpha inhibits glutamate uptake by primary human astrocytes. Implications for pathogenesis of HIV-1 dementia. J Biol Chem 271: 15303–15306.

Finklestein SP (1996). The potential use of neurotrophic growth factors in the treatment of cerebral ischemia. Adv Neurol 71: 413–417; discussion 417–418.

Fuchs D, Chiodi F, Albert J, et al (1989). Neopterin concentrations in cerebrospinal fluid and serum of individuals infected with HIV-1. AIDS 3: 285–288.

Gabuzda DH, Ho DD, de la Monte SM, et al (1986). Immunohistochemical identification of HTLV-III antigen in brains of patients with AIDS. Ann Neurol 20: 289–295.

Ganat Y, Soni S, Chacon M, et al (2002). Chronic hypoxia up-regulates fibroblast growth factor ligands in the perinatal brain and induces fibroblast growth factor-responsive radial glial cells in the sub-ependymal zone. Neuroscience 112: 977–991.

Garaci E, Caroleo MC, Aloe L, et al (1999). Nerve growth factor is an autocrine factor essential for the survival of macrophages infected with HIV. Proc Natl Acad Sci USA 96: 14013–14018.

Gardner J, Ghorpade A (2003). Tissue inhibitor of metalloproteinase (TIMP)-1: the TIMPed balance of matrix metalloproteinases in the central nervous system. J Neurosci Res 74: 801–806.

Gartner S (2000). HIV infection and dementia. Science 287: 602–604.

Geissler RG, Ottmann OG, Eder M, et al (1991). Effect of recombinant human transforming growth factor beta and tumor necrosis factor alpha on bone marrow progenitor cells of HIV-infected persons. Ann Hematol 62: 151–155.

Gelbard HA, Nottet HS, Swindells S, et al (1994). Platelet-activating factor: a candidate human immunodeficiency virus type 1-induced neurotoxin. J Virol 68: 4628–4635.

Gendelman HE, Meltzer MS (1989). Mononuclear phagocytes and the human immunodeficiency virus. Curr Opin Immunol 2: 414–419.

Gendelman HE, Narayan O, Kennedy-Stoskopf S, et al (1986). Tropism of sheep lentiviruses for monocytes: susceptibility to infection and virus gene expression increase during maturation of monocytes to macrophages. J Virol 58: 67–74.

Gendelman HE, Zheng J, Coulter CL, et al (1998). Suppression of inflammatory neurotoxins by highly active antiretroviral therapy in human immunodeficiency virus-associated dementia. J Infect Dis 178: 1000–1007.

Gendelman HE, Diesing S, Gelbard H, et al (2004). The neuropathogenesis of HIV-1 infection. In: GP Wormser (Ed.), AIDS and Other Manifestations of HIV Infection Elsevier Press.

Genis P, Jett M, Bernton EW, et al (1992). Cytokines and arachidonic metabolites produced during human immunodeficiency virus (HIV)-infected macrophage-astroglia interactions: implications for the neuropathogenesis of HIV disease. J Exp Med 176: 1703–1718.

Ghorpade A, Nukuna A, Che M, et al (1998). Human immunodeficiency virus neurotropism: an analysis of viral replication and cytopathicity for divergent strains in monocytes and microglia. J Virol 72: 3340–3350.

Ghorpade A, Holter S, Borgmann K, et al (2003). HIV-1 and IL-1 beta regulate Fas ligand expression in human astrocytes through the NF-kappa B pathway. J Neuroimmunol 141: 141–149.

Goudsmit J, de Wolf F, Paul DA, et al (1986). Expression of human immunodeficiency virus antigen (HIV-Ag) in serum and cerebrospinal fluid during acute and chronic infection. Lancet 2: 177–180.

Gras G, Chretien F, Vallat-Decouvelaere AV, et al (2003). Regulated expression of sodium-dependent glutamate transporters and synthetase: a neuroprotective role for activated microglia and macrophages in HIV infection? Brain Pathol 13: 211–222.

Gray F, Adle-Biassette H, Brion F, et al (2000). Neuronal apoptosis in human immunodeficiency virus infection. J Neurovirol 6(Suppl 1): S38–S43.

Green DR (2003). The suicide in the thymus, a twisted trail. Nat Immunol 4: 207–208.

Griffin GE, Leung K, Folks TM, et al (1989). Activation of HIV gene expression during monocyte differentiation by induction of NF-kappa B. Nature 339: 70–73.

Groothuis DR, Levy RM (1997). The entry of antiviral and antiretroviral drugs into the central nervous system. J Neurovirol 3: 387–400.

Haase AT (1999). Population biology of HIV-1 infection: viral and CD4+ T cell demographics and dynamics in lymphatic tissues. Annu Rev Immunol 17: 625–656.

Hayashi M, Ueyama T, Nemoto K, et al (2000). Sequential mRNA expression for immediate early genes, cytokines, and neurotrophins in spinal cord injury. J Neurotrauma 17: 203–218.

Hayes MM, Lane BR, King SR, et al (2002). Peroxisome proliferator-activated receptor gamma agonists inhibit HIV-1 replication in macrophages by transcriptional and post-transcriptional effects. J Biol Chem 277: 16913–16919.

Heidenreich F, Arendt G, Jander S, et al (1994). Serum and cerebrospinal fluid levels of soluble intercellular adhesion molecule 1 (sICAM-1) in patients with HIV-1 associated neurological diseases. J Neuroimmunol 52: 117–126.

Henderson AJ, Zou X, Calame KL (1995). C/EBP proteins activate transcription from the human immunodeficiency virus type 1 long terminal repeat in macrophages/monocytes. J Virol 69: 5337–5344.

Henderson AJ, Connor RI, Calame KL (1996). C/EBP activators are required for HIV-1 replication and proviral induction in monocytic cell lines. Immunity 5: 91–101.

Hession C, Osborn L, Goff D, et al (1990). Endothelial leukocyte adhesion molecule 1: direct expression cloning and functional interactions. Proc Natl Acad Sci USA 87: 1673–1677.

Heyes MP, Brew BJ, Martin A, et al (1991). Quinolinic acid in cerebrospinal fluid and serum in HIV-1 infection: relationship to clinical and neurological status. Ann Neurol 29: 202–209.

Heyes MP, Ellis RJ, Ryan L, et al (2001). Elevated cerebrospinal fluid quinolinic acid levels are associated with region-specific cerebral volume loss in HIV infection. Brain 124: 1033–1042.

Ho DD, Rota TR, Schooley RT, et al (1985). Isolation of HTLV-III from cerebrospinal fluid and neural tissues of patients with neurologic syndromes related to the acquired immunodeficiency syndrome. N Engl J Med 313: 1493–1497.

Ho DD, Neumann AU, Perelson AS, et al (1995). Rapid turnover of plasma virions and CD4 lymphocytes in HIV-1 infection. Nature 373: 123–126.

Hongisto V, Smeds N, Brecht S, et al (2003). Lithium blocks the c-Jun stress response and protects neurons via its action on glycogen synthase kinase 3. Mol Cell Biol 23: 6027–6036.

Hori K, Burd PR, Furuke K, et al (1999). Human immunodeficiency virus-1-infected macrophages induce inducible nitric oxide synthase and nitric oxide (NO) production in astrocytes: astrocytic NO as a possible mediator of neural damage in acquired immunodeficiency syndrome. Blood 93: 1843–1850.

Huang MB, Bond VC (2000). Involvement of protein kinase C in HIV-1 gp120-induced apoptosis in primary endothelium. J Acquir Immune Defic Syndr 25: 375–389.

Huang MB, Jin LL, James CO, et al (2004). Characterization of Nef-CXCR4 interactions important for apoptosis induction. J Virol 78: 11084–11096.

Hurwitz AA, Lyman WD, Guida MP, et al (1992). Tumor necrosis factor alpha induces adhesion molecule expression on human fetal astrocytes. J Exp Med 176: 1631–1636.

Isalan M, Klug A, Choo Y (2001). A rapid, generally applicable method to engineer zinc fingers illustrated by targeting the HIV-1 promoter. Nat Biotechnol 19: 656–660.

Iskander S, Walsh KA, Hammond RR (2004). Human CNS cultures exposed to HIV-1 gp120 reproduce dendritic injuries of HIV-1-associated dementia. J Neuroinflammation 1: 7.

Jain KK (2000). Evaluation of memantine for neuroprotection in dementia. Expert Opin Investig Drugs 9: 1397–1406.

James CO, Huang MB, Khan M, et al (2004). Extracellular Nef protein targets CD4+ T cells for apoptosis by interacting with CXCR4 surface receptors. J Virol 78: 3099–3109.

Janabi N, Chabrier S, Tardieu M (1996). Endogenous nitric oxide activates prostaglandin F2 alpha production in human microglial cells but not in astrocytes: a study of interactions between eicosanoids, nitric oxide, and superoxide anion (O^{2-}) regulatory pathways. J Immunol 157: 2129–2135.

Janabi N, Di Stefano M, Wallon C, et al (1998). Induction of human immunodeficiency virus type 1 replication in human glial cells after proinflammatory cytokines stimulation: effect of IFNgamma, IL1beta, and TNFalpha on differentiation and chemokine production in glial cells. Glia 23: 304–315.

Janssen RS, Saykin AJ, Cannon L, et al (1989). Neurological and neuropsychological manifestations of HIV-1 infection: association with AIDS-related complex but not asymptomatic HIV-1 infection. Ann Neurol 26: 592–600.

Janzer RC, Raff MC (1987). Astrocytes induce blood-brain barrier properties in endothelial cells. Nature 325: 253–257.

Jassoy C, Harrer T, Rosenthal T, et al (1993). Human immunodeficiency virus type 1-specific cytotoxic T lymphocytes release gamma interferon, tumor necrosis factor alpha (TNF-alpha), and TNF-beta when they encounter their target antigens. J Virol 67: 2844–2852.

Jiang X, Zhu D, Okagaki P, et al (2003). N-methyl-D-aspartate and TrkB receptor activation in cerebellar granule cells: an in vitro model of preconditioning to stimulate intrinsic survival pathways in neurons. Ann N Y Acad Sci 993: 134–145; discussion 159–160.

Jiang ZG, Piggee C, Heyes MP, et al (2001). Glutamate is a mediator of neurotoxicity in secretions of activated HIV-1-infected macrophages. J Neuroimmunol 117: 97–107.

Johnson VJ, Sharma RP (2003). Aluminum disrupts the proinflammatory cytokine/neurotrophin balance in primary brain rotation-mediated aggregate cultures: possible role in neurodegeneration. Neurotoxicology 24: 261–268.

Kaiser PK, Offermann JT, Lipton SA (1990). Neuronal injury due to HIV-1 envelope protein is blocked by anti-gp120 antibodies but not by anti-CD4 antibodies. Neurology 40: 1757–1761.

Kaul M, Garden GA, Lipton SA (2001). Pathways to neuronal injury and apoptosis in HIV-associated dementia. Nature 410: 988–994.

Kawahara M, Sakata A, Miyashita T, et al (1999). Physiologically based pharmacokinetics of digoxin in mdr1a knockout mice. J Pharm Sci 88: 1281–1287.

Kelly A, Vereker E, Nolan Y, et al (2003). Activation of p38 plays a pivotal role in the inhibitory effect of lipopolysaccharide and interleukin-1 beta on long term potentiation in rat dentate gyrus. J Biol Chem 278: 19453–19462.

Kerr SJ, Armati PJ, Pemberton LA, et al (1997). Kynurenine pathway inhibition reduces neurotoxicity of HIV-1-infected macrophages. Neurology 49: 1671–1681.

Kestler HW, 3rd, Ringler DJ, Mori K, et al (1991). Importance of the nef gene for maintenance of high virus loads and for development of AIDS. Cell 65: 651–662.

Kim JS, Gautam SC, Chopp M, et al (1995). Expression of monocyte chemoattractant protein-1 and macrophage inflammatory protein-1 after focal cerebral ischemia in the rat. J Neuroimmunol 56: 127–134.

Kim RB, Fromm MF, Wandel C, et al (1998). The drug transporter P-glycoprotein limits oral absorption and brain entry of HIV-1 protease inhibitors. J Clin Invest 101: 289–294.

Koeberle PD, Gauldie J, Ball AK (2004). Effects of adenoviral-mediated gene transfer of interleukin-10, interleukin-4, and transforming growth factor-beta on the survival of axotomized retinal ganglion cells. Neuroscience 125: 903–920.

Koedel U, Kohleisen B, Sporer B, et al (1999). HIV type 1 Nef protein is a viral factor for leukocyte recruitment into the central nervous system. J Immunol 163: 1237–1245.
Koenig S, Gendelman HE, Orenstein JM, et al (1986). Detection of AIDS virus in macrophages in brain tissue from AIDS patients with encephalopathy. Science 233: 1089–1093.
Koka P, He K, Zack JA, et al (1995). Human immunodeficiency virus 1 envelope proteins induce interleukin 1, tumor necrosis factor alpha, and nitric oxide in glial cultures derived from fetal, neonatal, and adult human brain. J Exp Med 182: 941–951.
Koup RA, Safrit JT, Cao Y, et al (1994). Temporal association of cellular immune responses with the initial control of viremia in primary human immunodeficiency virus type 1 syndrome. J Virol 68: 4650–4655.
Krebs FC, Ross H, McAllister J, et al (2000). HIV-1-associated central nervous system dysfunction. Adv Pharmacol 49: 315–385.
Krishnakumar D, Lal RB, Dhawan S (2005). Immunology of HIV-1. In: HE Gendelman, I Grant, IP Everall, SA Lipton, S Swindells (Eds.), The Neurology of AIDS. Oxford University Press.
Kullander K, Kylberg A, Ebendal T (1997). Specificity of neurotrophin-3 determined by loss-of-function mutagenesis. J Neurosci Res 50: 496–503.
Kure K, Lyman WD, Weidenheim KM, et al (1990). Cellular localization of an HIV-1 antigen in subacute AIDS encephalitis using an improved double-labeling immunohistochemical method. Am J Pathol 136: 1085–1092.
Kutsch O, Oh J, Nath A, et al (2000). Induction of the chemokines interleukin-8 and IP-10 by human immunodeficiency virus type 1 tat in astrocytes. J Virol 74: 9214–9221.
Lackritz EM (1998). Prevention of HIV transmission by blood transfusion in the developing world: achievements and continuing challenges. AIDS 12(Suppl A): S81–S86.
Lavi E, Strizki JM, Ulrich AM, et al (1997). CXCR-4 (Fusin), a co-receptor for the type 1 human immunodeficiency virus (HIV-1), is expressed in the human brain in a variety of cell types, including microglia and neurons. Am J Pathol 151: 1035–1042.
Lazarov-Spiegler O, Solomon AS, Zeev-Brann AB, et al (1996). Transplantation of activated macrophages overcomes central nervous system regrowth failure. FASEB J 10: 1296–1302.
Lee CG, Gottesman MM, Cardarelli CO, et al (1998). HIV-1 protease inhibitors are substrates for the MDR1 multidrug transporter. Biochemistry 37: 3594–3601.
Lee CW, Lau KF, Miller CC, et al (2003). Glycogen synthase kinase-3 beta-mediated tau phosphorylation in cultured cell lines. Neuroreport 14: 257–260.
Leppert D, Waubant E, Galardy R, et al (1995). T cell gelatinases mediate basement membrane transmigration in vitro. J Immunol 154: 4379–4389.
Letendre SL, McCutchan JA, Childers ME, et al (2004). Enhancing antiretroviral therapy for human immunodeficiency virus cognitive disorders. Ann Neurol 56: 416–423.
Linseman DA, Cornejo BJ, Le SS, et al (2003). A myocyte enhancer factor 2D (MEF2D) kinase activated during neuronal apoptosis is a novel target inhibited by lithium. J Neurochem 85: 1488–1499.
Lipsky RH, Xu K, Zhu D, et al (2001). Nuclear factor kappaB is a critical determinant in N-methyl-D-aspartate receptor-mediated neuroprotection. J Neurochem 78: 254–264.
Lipton SA (1997). Treating AIDS dementia. Science 276: 1629–1630.
Lipton SA, Gendelman HE (1995). Seminars in medicine of the Beth Israel Hospital, Boston. Dementia associated with the acquired immunodeficiency syndrome. N Engl J Med 332: 934–940.
Liu NQ, Lossinsky AS, Popik W, et al (2002a). Human immunodeficiency virus type 1 enters brain microvascular endothelia by macropinocytosis dependent on lipid rafts and the mitogen-activated protein kinase signaling pathway. J Virol 76: 6689–6700.
Liu R, Paxton WA, Choe S, et al (1996). Homozygous defect in HIV-1 coreceptor accounts for resistance of some multiply-exposed individuals to HIV-1 infection. Cell 86: 367–377.
Liu X, Jana M, Dasgupta S, et al (2002b). Human immunodeficiency virus type 1 (HIV-1) tat induces nitric-oxide synthase in human astroglia. J Biol Chem 277: 39312–39319.
Liu Y, Jones M, Hingtgen CM, et al (2000). Uptake of HIV-1 tat protein mediated by low-density lipoprotein receptor-related protein disrupts the neuronal metabolic balance of the receptor ligands. Nat Med 6: 1380–1387.
Lu L, Leonessa F, Clarke R, et al (2001). Competitive and allosteric interactions in ligand binding to P-glycoprotein as observed on an immobilized P-glycoprotein liquid chromatographic stationary phase. Mol Pharmacol 59: 62–68.
Luo X, Carlson KA, Wojna V, et al (2003). Macrophage proteomic fingerprinting predicts HIV-1-associated cognitive impairment. Neurology 60: 1931–1937.
Macdonald NJ, Perez-Polo JR, Bennett AD, et al (1999). NGF-resistant PC12 cell death induced by arachidonic acid is accompanied by a decrease of active PKC zeta and nuclear factor kappa B. J Neurosci Res 57: 219–226.
Maciejewski JP, Weichold FF, Young NS (1994). HIV-1 suppression of hematopoiesis in vitro mediated by envelope glycoprotein and TNF-alpha. J Immunol 153: 4303–4310.
Maggirwar SB, Tong N, Ramirez S, et al (1999). HIV-1 Tat-mediated activation of glycogen synthase kinase-3beta contributes to Tat-mediated neurotoxicity. J Neurochem 73: 578–586.
Major EO, Rausch D, Marra C, et al (2000). HIV-associated dementia. Science 288: 440–442.
Marechal V, Prevost MC, Petit C, et al (2001). Human immunodeficiency virus type 1 entry into macrophages mediated by macropinocytosis. J Virol 75: 11166–11177.
Marmigere F, Rage F, Tapia-Arancibia L (2003). GABA-glutamate interaction in the control of BDNF expression in hypothalamic neurons. Neurochem Int 42: 353–358.
Martin C, Berridge G, Higgins CF, et al (2000). Communication between multiple drug binding sites on P-glycoprotein. Mol Pharmacol 58: 624–632.
Masliah E, Ge N, Achim CL, et al (1995). Differential vulnerability of calbindin-immunoreactive neurons in HIV encephalitis. J Neuropathol Exp Neurol 54: 350–357.

McArthur JC (2004). HIV dementia: an evolving disease. J Neuroimmunol 157: 3–10.

McArthur JC, Nance-Sproson TE, Griffin DE, et al (1992). The diagnostic utility of elevation in cerebrospinal fluid beta 2-microglobulin in HIV-1 dementia. Multicenter AIDS Cohort Study. Neurology 42: 1707–1712.

McArthur JC, Sacktor N, Selnes O (1999). Human immunodeficiency virus-associated dementia. Semin Neurol 19: 129–150.

McArthur JC, Haughey N, Gartner S, et al (2003). Human immunodeficiency virus-associated dementia: an evolving disease. J Neurovirol 9: 205–221.

McManus CM, Weidenheim K, Woodman SE, et al (2000). Chemokine and chemokine-receptor expression in human glial elements: induction by the HIV protein, Tat, and chemokine autoregulation. Am J Pathol 156: 1441–1453.

Medana IM, Gallimore A, Oxenius A, et al (2000). MHC class I-restricted killing of neurons by virus-specific CD8+ T lymphocytes is effected through the Fas/FasL, but not the perforin pathway. Eur J Immunol 30: 3623–3633.

Merrill JE, Koyanagi Y, Zack J, et al (1992). Induction of interleukin-1 and tumor necrosis factor alpha in brain cultures by human immunodeficiency virus type 1. J Virol 66: 2217–2225.

Messam CA, Major EO (2000). Stages of restricted HIV-1 infection in astrocyte cultures derived from human fetal brain tissue. J Neurovirol 6(Suppl 1): S90–S94.

Meucci O, Miller RJ (1996). gp120-induced neurotoxicity in hippocampal pyramidal neuron cultures: protective action of TGF-beta1. J Neurosci 16: 4080–4088.

Milligan CE, Cunningham TJ, Levitt P (1991). Differential immunochemical markers reveal the normal distribution of brain macrophages and microglia in the developing rat brain. J Comp Neurol 314: 125–135.

Miwa T, Furukawa S, Nakajima K, et al (1997). Lipopolysaccharide enhances synthesis of brain-derived neurotrophic factor in cultured rat microglia. J Neurosci Res 50: 1023–1029.

Mordelet E, Kissa K, Cressant A, et al (2004). Histopathological and cognitive defects induced by Nef in the brain. Faseb J 18: 1851–1861.

Muthumani K, Choo AY, Hwang DS, et al (2003). Mechanism of HIV-1 viral protein R-induced apoptosis. Biochem Biophys Res Commun 304: 583–592.

Namiki J, Kojima A, Tator CH (2000). Effect of brain-derived neurotrophic factor, nerve growth factor, and neurotrophin-3 on functional recovery and regeneration after spinal cord injury in adult rats. J Neurotrauma 17: 1219–1231.

Nath A, Haughey NJ, Jones M, et al (2000). Synergistic neurotoxicity by human immunodeficiency virus proteins Tat and gp120: protection by memantine. Ann Neurol 47: 186–194.

Navia BA, Jordan BD, Price RW (1986). The AIDS dementia complex: I. Clinical features. Ann Neurol 19: 517–524.

Newell ML (1998). Mechanisms and timing of mother-to-child transmission of HIV-1. AIDS 12: 831–837.

Newton PJ, Newsholme W, Brink NS, et al (2002). Acute meningoencephalitis and meningitis due to primary HIV infection. BMJ 325: 1225–1227.

Nottet HS (2004). The blood-brain barrier: monocyte and viral entry into the brain. In: HE Gendelman, I Grant, IP Everall, SA Lipton, S Swindells (Eds.), The Neurology of AIDS. Oxford University Press.

Nottet H, Dhawan S (1997). HIV-1 entry into the brain: mechanisms for the infiltration of HIV-1 infected macrophages across blood-brain barrier. In: H Gendelman, L Epstein, S Swidells (Eds.), The Neurology of AIDS. Chapman & Hall, New York.

Nottet HS, Jett M, Flanagan CR, et al (1995). A regulatory role for astrocytes in HIV-1 encephalitis. An overexpression of eicosanoids, platelet-activating factor, and tumor necrosis factor-alpha by activated HIV-1-infected monocytes is attenuated by primary human astrocytes. J Immunol 154: 3567–3581.

Nottet HS, Persidsky Y, Sasseville VG, et al (1996). Mechanisms for the transendothelial migration of HIV-1-infected monocytes into brain. J Immunol 156: 1284–1295.

Orenstein JM, Meltzer MS, Phipps T, et al (1988). Cytoplasmic assembly and accumulation of human immunodeficiency virus types 1 and 2 in recombinant human colony-stimulating factor-1-treated human monocytes: an ultrastructural study. J Virol 62: 2578–2586.

Osborn L, Hession C, Tizard R, et al (1989). Direct expression cloning of vascular cell adhesion molecule 1, a cytokine-induced endothelial protein that binds to lymphocytes. Cell 59: 1203–1211.

Pardridge WM (1983). Brain metabolism: a perspective from the blood-brain barrier. Physiol Rev 63: 1481–1535.

Patarca R, Freeman GJ, Schwartz J, et al (1988). rpt-1, an intracellular protein from helper/inducer T cells that regulates gene expression of interleukin 2 receptor and human immunodeficiency virus type 1. Proc Natl Acad Sci USA 85: 2733–2737.

Patton HK, Zhou ZH, Bubien JK, et al (2000). gp120-induced alterations of human astrocyte function: Na(+)/H(+) exchange, K(+) conductance, and glutamate flux. Am J Physiol Cell Physiol 279: C700–C708.

Paula-Barbosa MM, Pereira PA, Cadete-Leite A, et al (2003). NGF and NT-3 exert differential effects on the expression of neuropeptides in the suprachiasmatic nucleus of rats withdrawn from ethanol treatment. Brain Res 983: 64–73.

Pavlakis SG, Lu D, Frank Y, et al (1998). Brain lactate and N-acetylaspartate in pediatric AIDS encephalopathy. AJNR Am J Neuroradiol 19: 383–385.

Peluso R, Haase A, Stowring L, et al (1985). A Trojan horse mechanism for the spread of visna virus in monocytes. Virology 147: 231–236.

Pemberton LA, Brew BJ (2001). Cerebrospinal fluid S-100beta and its relationship with AIDS dementia complex. J Clin Virol 22: 249–253.

Persidsky Y, Gendelman HE (2003). Mononuclear phagocyte immunity and the neuropathogenesis of HIV-1 infection. J Leukoc Biol 74: 691–701.

Persidsky Y, Stins M, Way D, et al (1997). A model for monocyte migration through the blood-brain barrier during HIV-1 encephalitis. J Immunol 158: 3499–3510.

Persidsky Y, Zheng J, Miller D, et al (2000). Mononuclear phagocytes mediate blood-brain barrier compromise and neuronal injury during HIV-1-associated dementia. J Leukoc Biol 68: 413–422.

Petito CK, Roberts B (1995). Evidence of apoptotic cell death in HIV encephalitis. Am J Pathol 146: 1121–1130.

Piller SC, Jans P, Gage PW, et al (1998). Extracellular HIV-1 virus protein R causes a large inward current and cell death in cultured hippocampal neurons: implications for AIDS pathology. Proc Natl Acad Sci USA 95: 4595–4600.

Pomerantz RJ (2004). Effects of HIV-1 Vpr on neuroinvasion and neuropathogenesis. DNA Cell Biol 23: 227–238.

Pulliam L, Gascon R, Stubblebine M, et al (1997). Unique monocyte subset in patients with AIDS dementia. Lancet 349: 692–695.

Pulliam L, Zhou M, Stubblebine M, et al (1998). Differential modulation of cell death proteins in human brain cells by tumor necrosis factor alpha and platelet activating factor. J Neurosci Res 54: 530–538.

Ramirez SH, Sanchez JF, Dimitri CA, et al (2001). Neurotrophins prevent HIV Tat-induced neuronal apoptosis via a nuclear factor-kappaB (NF-kappaB)-dependent mechanism. J Neurochem 78: 874–889.

Rapalino O, Lazarov-Spiegler O, Agranov E, et al (1998). Implantation of stimulated homologous macrophages results in partial recovery of paraplegic rats. Nat Med 4: 814–821.

Rappaport A, Shaked M, Landau M, et al (2001). Sweet's syndrome in association with Crohn's disease: report of a case and review of the literature. Dis Colon Rectum 44: 1526–1529.

Ray RB, Srinivas RV (1997). Inhibition of human immunodeficiency virus type 1 replication by a cellular transcriptional factor MBP-1. J Cell Biochem 64: 565–572.

Reynolds L, Ullman C, Moore M, et al (2003). Repression of the HIV-1 5 LTR promoter and inhibition of HIV-1 replication by using engineered zinc-finger transcription factors. Proc Natl Acad Sci USA 100: 1615–1620.

Risau W, Wolburg H (1990). Development of the blood-brain barrier. Trends Neurosci 13: 174–178.

Rocha M, Martins RA, Linden R (1999). Activation of NMDA receptors protects against glutamate neurotoxicity in the retina: evidence for the involvement of neurotrophins. Brain Res 827: 79–92.

Royal W, 3rd, Selnes OA, Concha M, et al (1994). Cerebrospinal fluid human immunodeficiency virus type 1 (HIV-1) p24 antigen levels in HIV-1-related dementia. Ann Neurol 36: 32–39.

Royce RA, Sena A, Cates W, Jr, et al (1997). Sexual transmission of HIV. N Engl J Med 336: 1072–1078.

Rubin LL, Staddon JM (1999). The cell biology of the blood-brain barrier. Annu Rev Neurosci 22: 11–28.

Ryan LA, Peng H, Erichsen DA, et al (2004). TNF-related apoptosis-inducing ligand mediates human neuronal apoptosis: links to HIV-1-associated dementia. J Neuroimmunol 148: 127–139.

Sabri F, Tresoldi E, Di Stefano M, et al (1999). Nonproductive human immunodeficiency virus type 1 infection of human fetal astrocytes: independence from CD4 and major chemokine receptors. Virology 264: 370–384.

Sabri F, De Milito A, Pirskanen R, et al (2001). Elevated levels of soluble Fas and Fas ligand in cerebrospinal fluid of patients with AIDS dementia complex. J Neuroimmunol 114: 197–206.

Sacktor NC, Bacellar H, Hoover DR, et al (1996). Psychomotor slowing in HIV infection: a predictor of dementia, AIDS and death. J Neurovirol 2: 404–410.

Sacktor N, Lyles RH, Skolasky R, et al (2001). HIV-associated neurologic disease incidence changes: Multicenter AIDS Cohort Study, 1990–1998. Neurology 56: 257–260.

Sacktor N, McDermott MP, Marder K, et al (2002). HIV-associated cognitive impairment before and after the advent of combination therapy. J Neurovirol 8: 136–142.

Saito H, Fujiwara T, Takahashi EI, et al (1996). Isolation and mapping of a novel human gene encoding a protein containing zinc-finger structures. Genomics 31: 376–379.

Sasseville VG, Newman WA, Lackner AA, et al (1992). Elevated vascular cell adhesion molecule-1 in AIDS encephalitis induced by simian immunodeficiency virus. Am J Pathol 141: 1021–1030.

Schaffer B, Wiedau-Pazos M, Geschwind DH (2003). Gene structure and alternative splicing of glycogen synthase kinase 3 beta (GSK-3beta) in neural and non-neural tissues. Gene 302: 73–81.

Schinkel AH, Smit JJ, van Tellingen O, et al (1994). Disruption of the mouse mdr1a P-glycoprotein gene leads to a deficiency in the blood-brain barrier and to increased sensitivity to drugs. Cell 77: 491–502.

Schweighardt B, Shieh JT, Atwood WJ (2001). CD4/CXCR4-independent infection of human astrocytes by a T-tropic strain of HIV-1. J Neurovirol 7: 155–162.

Shi B, Raina J, Lorenzo A, et al (1998). Neuronal apoptosis induced by HIV-1 Tat protein and TNF-alpha: potentiation of neurotoxicity mediated by oxidative stress and implications for HIV-1 dementia. J Neurovirol 4: 281–290.

Shieh JT, Albright AV, Sharron M, et al (1998). Chemokine receptor utilization by human immunodeficiency virus type 1 isolates that replicate in microglia. J Virol 72: 4243–4249.

Sholl-Franco A, Marques PM, Ferreira CM, et al (2002). IL-4 increases GABAergic phenotype in rat retinal cell cultures: involvement of muscarinic receptors and protein kinase C. J Neuroimmunol 133: 20–29.

Shrikant P, Benos DJ, Tang LP, et al (1996). HIV glycoprotein 120 enhances intercellular adhesion molecule-1 gene expression in glial cells. Involvement of Janus kinase/signal transducer and activator of transcription and protein kinase C signaling pathways. J Immunol 156: 1307–1314.

Smith GM, Hale JH (1997). Macrophage/microglia regulation of astrocytic tenascin: synergistic action of transforming growth factor-beta and basic fibroblast growth factor. J Neurosci 17: 9624–9633.

Speliotes EK, Caday CG, Do T, et al (1996). Increased expression of basic fibroblast growth factor (bFGF) following focal cerebral infarction in the rat. Brain Res Mol Brain Res 39: 31–42.

Spera PA, Ellison JA, Feuerstein GZ, et al (1998). IL-10 reduces rat brain injury following focal stroke. Neurosci Lett 251: 189–192.

Speth C, Joebstl B, Barcova M, et al (2000). HIV-1 envelope protein gp41 modulates expression of interleukin-10 and chemokine receptors on monocytes, astrocytes and neurones. AIDS 14: 629–636.

Sporer B, Koedel U, Paul R, et al (2000). Human immunodeficiency virus type-1 Nef protein induces blood-brain barrier disruption in the rat: role of matrix metalloproteinase-9. J Neuroimmunol 102: 125–130.

Stalder AK, Carson MJ, Pagenstecher A, et al (1998). Late-onset chronic inflammatory encephalopathy in immune-competent and severe combined immune-deficient (SCID) mice with astrocyte-targeted expression of tumor necrosis factor. Am J Pathol 153: 767–783.

Stankoff B, Calvez V, Suarez S, et al (1999). Plasma and cerebrospinal fluid human immunodeficiency virus type-1 (HIV-1) RNA levels in HIV-related cognitive impairment. Eur J Neurol 6: 669–675.

Staunton DE, Marlin SD, Stratowa C, et al (1988). Primary structure of ICAM-1 demonstrates interaction between members of the immunoglobulin and integrin supergene families. Cell 52: 925–933.

Strizki JM, Albright AV, Sheng H, et al (1996). Infection of primary human microglia and monocyte-derived macrophages with human immunodeficiency virus type 1 isolates: evidence of differential tropism. J Virol 70: 7654–7662.

Stuart RO, Nigam SK (1995). Regulated assembly of tight junctions by protein kinase C. Proc Natl Acad Sci USA 92: 6072–6076.

Subler MA, Martin DW, Deb S (1994). Activation of the human immunodeficiency virus type 1 long terminal repeat by transforming mutants of human p53. J Virol 68: 103–110.

Sugawara I, Hamada H, Tsuruo T, et al (1990). Specialized localization of P-glycoprotein recognized by MRK 16 monoclonal antibody in endothelial cells of the brain and the spinal cord. Jpn J Cancer Res 81: 727–730.

Suwanwelaa N, Phanuphak P, Phanthumchinda K, et al (2000). Magnetic resonance spectroscopy of the brain in neurologically asymptomatic HIV-infected patients. Magn Reson Imaging 18: 859–865.

Syndulko K, Singer EJ, Nogales-Gaete J, et al (1994). Laboratory evaluations in HIV-1-associated cognitive/ motor complex. Psychiatr Clin North Am 17: 91–123.

Takeshima H, Kuratsu J, Takeya M, et al (1994). Expression and localization of messenger RNA and protein for monocyte chemoattractant protein-1 in human malignant glioma. J Neurosurg 80: 1056–1062.

Talley AK, Dewhurst S, Perry SW, et al (1995). Tumor necrosis factor alpha-induced apoptosis in human neuronal cells: protection by the antioxidant N-acetylcysteine and the genes bcl-2 and crmA. Mol Cell Biol 15: 2359–2366.

Tardieu M (2004). Mechanisms of macrophage-mediated neurotoxicity. In: HE Gendelman, I Grant, IP Everall, SA Lipton, S Swindells (Eds.), The Neurology of AIDS. Oxford University Press.

Tardieu M, Boutet A (2002). HIV-1 and the central nervous system. Curr Top Microbiol Immunol 265: 183–195.

Tariot PN, Farlow MR, Grossberg GT, et al (2004). Memantine treatment in patients with moderate to severe Alzheimer disease already receiving donepezil: a randomized controlled trial. JAMA 291: 317–324.

Tesmer VM, Rajadhyaksha A, Babin J, et al (1993). NF-IL6-mediated transcriptional activation of the long terminal repeat of the human immunodeficiency virus type 1. Proc Natl Acad Sci USA 90: 7298–7302.

Titanji K, Nilsson A, Morch C, et al (2003). Low frequency of plasma nerve-growth factor detection is associated with death of memory B lymphocytes in HIV-1 infection. Clin Exp Immunol 132: 297–303.

Toggas SM, Masliah E, Rockenstein EM, et al (1994). Central nervous system damage produced by expression of the HIV-1 coat protein gp120 in transgenic mice. Nature 367: 188–193.

Tong N, Sanchez JF, Maggirwar SB, et al (2001). Activation of glycogen synthase kinase 3 beta (GSK-3beta) by platelet activating factor mediates migration and cell death in cerebellar granule neurons. Eur J Neurosci 13: 1913–1922.

Trillo-Pazos G, McFarlane-Abdulla E, Campbell IC, et al (2000). Recombinant nef HIV-IIIB protein is toxic to human neurons in culture. Brain Res 864: 315–326.

Vallat AV, De Girolami U, He J, et al (1998). Localization of HIV-1 co-receptors CCR5 and CXCR4 in the brain of children with AIDS. Am J Pathol 152: 167–178.

Venne A, Li S, Mandeville R, et al (1996). Hypersensitizing effect of pluronic L61 on cytotoxic activity, transport, and subcellular distribution of doxorubicin in multiple drug-resistant cells. Cancer Res 56: 3626–3629.

Vereker E, O'Donnell E, Lynch A, et al (2001). Evidence that interleukin-1beta and reactive oxygen species production play a pivotal role in stress-induced impairment of LTP in the rat dentate gyrus. Eur J Neurosci 14: 1809–1819.

Vitkovic L, Maeda S, Sternberg E (2001). Anti-inflammatory cytokines: expression and action in the brain. Neuroimmunomodulation 9: 295–312.

von Giesen HJ, Wittsack HJ, Wenserski F, et al (2001). Basal ganglia metabolite abnormalities in minor motor disorders associated with human immunodeficiency virus type 1. Arch Neurol 58: 1281–1286.

Wang Z, Pekarskaya O, Bencheikh M, et al (2003). Reduced expression of glutamate transporter EAAT2 and impaired glutamate transport in human primary astrocytes exposed to HIV-1 or gp120. Virology 312: 60–73.

Wei X, Ghosh SK, Taylor ME, et al (1995). Viral dynamics in human immunodeficiency virus type 1 infection. Nature 373: 117–122.

Weiden M, Tanaka N, Qiao Y, et al (2000). Differentiation of monocytes to macrophages switches the Mycobacterium tuberculosis effect on HIV-1 replication from stimulation to inhibition: modulation of interferon response and CCAAT/enhancer binding protein beta expression. J Immunol 165: 2028–2039.

Wesselingh SL, Power C, Glass JD, et al (1993). Intracerebral cytokine messenger RNA expression in acquired immunodeficiency syndrome dementia. Ann Neurol 33: 576–582.
Wiley CA (1994). Pathology of neurologic disease in AIDS. Psychiatr Clin North Am 17: 1–15.
Wilkinson ID, Lunn S, Miszkiel KA, et al (1997). Proton MRS and quantitative MRI assessment of the short term neurological response to antiretroviral therapy in AIDS. J Neurol Neurosurg Psychiatry 63: 477–482.
Williams KC, Hickey WF (2002). Central nervous system damage, monocytes and macrophages, and neurological disorders in AIDS. Annu Rev Neurosci 25: 537–562.
Williams KC, Corey S, Westmoreland SV, et al (2001). Perivascular macrophages are the primary cell type productively infected by simian immunodeficiency virus in the brains of macaques: implications for the neuropathogenesis of AIDS. J Exp Med 193: 905–915.
Winkler C, Modi W, Smith MW, et al (1998). Genetic restriction of AIDS pathogenesis by an SDF-1 chemokine gene variant. ALIVE Study, Hemophilia Growth and Development Study (HGDS), Multicenter AIDS Cohort Study (MACS), Multicenter Hemophilia Cohort Study (MHCS), San Francisco City Cohort (SFCC). Science 279: 389–393.
Wu H, Yang WP, Barbas CF, 3rd (1995). Building zinc fingers by selection: toward a therapeutic application. Proc Natl Acad Sci USA 92: 344–348.
Wyss-Coray T, Masliah E, Toggas SM, et al (1996). Dysregulation of signal transduction pathways as a potential mechanism of nervous system alterations in HIV-1 gp120 transgenic mice and humans with HIV-1 encephalitis. J Clin Invest 97: 789–798.
Xiong H, Zeng YC, Zheng J, et al (1999a). Soluble HIV-1 infected macrophage secretory products mediate blockade of long-term potentiation: a mechanism for cognitive dysfunction in HIV-1-associated dementia. J Neurovirol 5: 519–528.
Xiong H, Zheng J, Thylin M, et al (1999b). Unraveling the mechanisms of neurotoxicity in HIV type 1-associated dementia: inhibition of neuronal synaptic transmission by macrophage secretory products. AIDS Res Hum Retroviruses 15: 57–63.
Xiong H, Zeng YC, Lewis T, et al (2000). HIV-1 infected mononuclear phagocyte secretory products affect neuronal physiology leading to cellular demise: relevance for HIV-1-associated dementia. J Neurovirol 6(Suppl 1): S14–S23.
Xiong H, Boyle J, Winkelbauer M, et al (2003a). Inhibition of long-term potentiation by interleukin-8: implications for human immunodeficiency virus-1-associated dementia. J Neurosci Res 71: 600–607.
Xiong H, McCabe L, Skifter D, et al (2003b). Activation of NR1a/NR2B receptors by monocyte-derived macrophage secretory products: implications for human immunodeficiency virus type one-associated dementia. Neurosci Lett 341: 246–250.
Yang QE (2004). Eradication of HIV in infected patients: some potential approaches. Med Sci Monit 10: RA155–165.
Yao HB, Shaw PC, Wong CC, et al (2002). Expression of glycogen synthase kinase-3 isoforms in mouse tissues and their transcription in the brain. J Chem Neuroanat 23: 291–297.
Zhang Y, Han H, Elmquist WF, et al (2000). Expression of various multidrug resistance-associated protein (MRP) homologues in brain microvessel endothelial cells. Brain Res 876: 148–153.
Zheng J, Thylin MR, Ghorpade A, et al (1999). Intracellular CXCR4 signaling, neuronal apoptosis and neuropathogenic mechanisms of HIV-1-associated dementia. J Neuroimmunol 98: 185–200.

Handbook of Clinical Neurology, Vol. 85 (3rd series)
HIV/AIDS and the Nervous System
P. Portegies, J. R. Berger, Editors

Chapter 5

Neurological sequelae of primary HIV infection

BRUCE J. BREW*

Departments of Neurology, HIV Medicine, Centre for Immunology, and National Centre in HIV Epidemiology and Clinical Research, St Vincent's Hospital, University of New South Wales, Sydney, Australia

5.1. Introduction

Primary HIV-1 infection (PHI) is defined as the period from initial infection with HIV to complete seroconversion (Taiwo and Hicks, 2002). Patients may or may not be symptomatic, but in those who are ill it may afford an early diagnosis of HIV infection and the opportunity for consideration of early antiretroviral therapy. Prior to discussion of these sequelae, the pertinent aspects of symptomatic PHI, also known as seroconversion illness, will be summarized. These are important for the neurologist to appreciate as the neurological presentations are almost always "grafted onto" the classic features of symptomatic PHI.

5.2. Epidemiology

Symptomatic PHI occurs in up to 90% of patients (Clark et al., 1991; Kahn and Walker, 1998). There is no obvious relationship to the mode of transmission, whether it be through mucous membranes, skin or intravenous routes.

5.3. Natural history

The clinical illness is typically two to four weeks (with a reported range from four days to just over eight weeks) from the time of HIV exposure (Gaines et al., 1988; Kinloch-de Loes et al., 1993; Schacker et al., 1996; Quinn, 1997; Apoola et al., 2002). The illness itself is acute in onset with a median duration of 18 days (Gaines et al., 1988; Quinn, 1997; Kaufman et al., 1998). In most patients there is rapid resolution followed by a period of asymptomatic infection lasting years. The precise factors determining the development of symptomatic PHI are yet to be defined but in broad terms include the tropism, virulence and inoculum of the infecting strain, as well as the nature of the individual host immune response. Importantly, transmitted viruses are almost always macrophage tropic (Zhu et al., 1996), possibly explaining the high potential for neurological complications during PHI.

Patients with symptoms of PHI, especially one that lasts longer than two months, progress to AIDS more rapidly: 68% developed AIDS after 56 months compared to 20% who had an asymptomatic seroconversion (Keet et al., 1993). This difference was also maintained after over seven years of follow-up (Lindbacl et al., 1994). Furthermore patients with symptoms of PHI progress more rapidly to cognitive impairment (Wallace et al., 2001). Moreover, patients who have neurological complications of PHI are more likely to rapidly progress to AIDS (Boufassa et al., 1995).

5.4. Clinical features

Combining the highest figures from two reviews (Clark et al., 1991; Kahn and Walker, 1998) reveals that the most common symptoms and signs are fever (97%), adenopathy (77%), pharyngitis (73%), headache (70%), rash (70%), myalgia or arthralgia (77%) and mucocutaneous ulceration (35%) (see Table 5.1). The average frequency (taken from the published cohorts to date) of each abnormality is also shown in Table 5.1. The average temperature is approximately 38.6°C while the most commonly affected lymph nodes are those in the cervical, axillary and inguinal regions (Apoola et al., 2002). The rash most often involves the upper parts of the body and is typically macular or maculopapular (Apoola et al., 2002).

*Correspondence to: Professor B. J. Brew, MBBS, MD, FRACP, Department of Neurology, Level 4 Xavier Building, St Vincent's Hospital, Victoria St Darlinghurst, Sydney 2010, Australia. Email: B.Brew@unsw.edu.au, Tel: +61-2-8382-4100, Fax: +61-2-8382-4101.

Table 5.1

Characteristic features of primary HIV-1 infection

Clinical findings	Laboratory findings
Fever 97% (Brew, 2001a) 75% (Kaufman et al., 1998)	Thrombocytopenia (weeks 1–3) 74% (Kinloch-de Loes et al., 1993)
Myalgia/arthralgia 77% (Brew, 2001a) 47% (Kaufman et al., 1998)	Leukopenia 49% (Kinloch-de Loes et al., 1993)
	Neutropenia 35% (Kinloch-de Loes et al., 1993)
Adenopathy 77% (Brew, 2001a) 42% (Kaufman et al., 1998)	Lymphocytopenia (weeks 1–2) 30% (Kinloch-de Loes et al., 1993)
Pharyngitis 73% (Brew, 2001a) 47% (Kaufman et al., 1998)	Anemia 26% (Kinloch-de Loes et al., 1993)
	Abnormal liver function tests (weeks 1–2)
Rash (usually truncal) 70% (Brew, 2001a) 45% (Kaufman et al., 1998) (days 2–5)	Detectable HIV viral RNA in the plasma 87% (Schacker et al., 1996)
Headache 70% (Brew, 2001a)	Positive serum HIV-1 p24 antigen 6–79% (Kinloch-de Loes et al., 1993; Schacker et al., 1996)
Mucocutaneous ulceration 35% (Brew, 2001a) 33% (Kaufman et al., 1998)	Negative/indeterminate HIV-1 antibody (ELISA/Western blot)
Aseptic meningitis (fever, headache, photophobia and neck stiffness) 24% (Brew, 2001a) 45% (Kaufman et al., 1998)	Lymphocytosis (weeks 2–4)
	CD^4 cell count (Schacker et al., 1996): >600 cells/mm^3: 41% (Brew, 2001a) 400–599 cells/mm^3: 37% (Brew, 2001a) 300–399 cells/mm^3: 15% (Brew, 2001a) <300 cells/mm^3: 7% (Brew, 2001a) 458 median (178–1082) (Kinloch-de Loes et al., 1993)

PHI illness is often associated with some degree of immunodeficiency, occasionally moderate to severe, as assessed by depressed CD^4 and CD^8 cell counts (Table 5.1) (Gaines et al., 1990). Consequently, patients may develop complications usually only seen in advanced HIV-1 disease, such as cytomegalovirus pneumonitis (Tindall et al., 1989). Over the subsequent few months, the immunodeficiency significantly improves, but does not revert entirely to normal (Gaines et al., 1990).

5.5. Differential diagnosis

The differential diagnosis of the clinical picture is wide but as shown in Tables 5.2 and 5.3 there are particular diseases and several clinical "clues" that deserve consideration.

5.6. Investigations

The diagnosis of PHI illness is contingent on a high index of clinical suspicion in order to request the appropriate tests: HIV-1 serology and Western blot analysis, p24 antigen (the core antigen of HIV-1) and plasma HIV-1 RNA concentrations.

Suspicion of PHI can be heightened by the results of other laboratory investigations that are routinely performed in patients who are unwell (Table 5.1). The full blood count is almost always abnormal. A transient lymphopenia during the first week followed

Table 5.2

Major differential diagnoses for PHI

Cytomegalovirus
Epstein–Barr
Toxoplasmosis
Viral hepatitis
Rubella
Secondary syphilis
Disseminated gonorrhea
Herpes simplex

Table 5.3

Clinical clues to PHI

Rash (specificity 82%; Hecht et al., 2002) especially when there is involvement of palms and soles
Mucocutaneous ulcerations especially when diffuse (oral ulcers: specificity 85%; Hecht et al., 2002)
Recent weight loss exceeding 2.5 kg (specificity 86%; Hecht et al., 2002)
Absence of jaundice (Apoola et al., 2002)

by a lymphocytosis (mainly consisting of CD^8 lymphocytes) in the second to fourth weeks of the illness is very frequent (Gaines et al., 1990). Transient, mild to moderate thrombocytopenia is also common in the first two weeks of illness (Gaines et al., 1990). Elevated serum levels of hepatic transaminases (AST and alkaline phosphatase) are also relatively common returning to normal after several weeks (Boag et al., 1992).

In relation to the more definitive tests, HIV-1 serology and Western blot analysis are often negative or indeterminate (Henrard et al., 1994; Busch et al., 1995; Busch and Satten, 1997) in the context of PHI illness and for the subsequent few months. IgM antibodies against the envelope or core antigens of HIV appear approximately two weeks after acquisition (Kaufman et al., 1998) and then become undetectable after approximately three months (Kaufman et al., 1998). IgG antibodies are positive at between two and six weeks after symptoms of PHI (not after acquisition) although false positive results may occur. In this regard IgG_3 against p24 may be useful as there are no known false positive results (Wilson et al., 2004). It is detectable approximately 34 days after HIV acquisition and persists for approximately four months. The Western blot becomes positive at approximately two weeks after acquisition but again at this early stage the results are most often indeterminate. Well-known false positive Western blots include the p24 band which can occur in patients with systemic lupus erythematosus and Sjogren's syndrome (Wilson et al., 2004). The commonly accepted criteria for a positive Western blot are the presence of at least one reactive band in each of the gag, pol and env domains (Kinloch-de Loes et al., 1993). A positive Western blot was found in only 15% of patients with symptomatic PHI (Kinloch-de Loes et al., 1993). Serum p24 antigen concentrations are often detectable (Table 5.1). Indeed, p24 antigen is found on average 7 days before any evidence of HIV-1 antibody response (Kaufman et al., 1998). Serum p24 antigen becomes progressively less reliably positive in the subsequent weeks due to the development of p24 antibodies which form complexes with p24 antigen. Plasma HIV RNA levels, however, are more sensitive with detection of HIV infection some 3–5 days before the p24 test is positive, and almost three weeks before standard serological tests begin to become abnormal (Kahn and Walker, 1998). It is positive as early as 10 days after acquisition and peaks between days 20 and 30 (Taiwo and Hicks, 2002). Rarely, there can be a false positive plasma HIV RNA result. There are two "clues" to this: low levels (<2000 cpml) and a normal CD^4 to CD^8 ratio (Taiwo and Hicks, 2002).

Thus in practical terms symptomatic PHI is associated with detectable HIV RNA and p24 levels while HIV antibodies by ELISA kits and Western blot are negative or indeterminate. Hecht et al. (2002) found that tests of plasma HIV RNA were 100% sensitive with false positive rates of 5% for the Chiron/Bayer b-DNA assay, 3% for the Roche RT-PCR assay and 2% for the Gen-Probe transcription-mediated amplification test. These data contrasted with p24 antigen test with a sensitivity of 77% and specificity of 99.5%. As mentioned previously, p24 antigen is less likely to be detectable later in the illness so that the lower sensitivity recorded in this study probably reflects the later collection of some of the samples. Indeed, review of current literature supports the idea that if plasma HIV RNA is not detected in a patient with suspected PHI then an alternative diagnosis should be considered (Kassutto and Rosenberg, 2004).

5.7. Treatment

Specific therapy in terms of antiretroviral drugs with or without immunomodulatory treatments for PHI is controversial. There are arguments and data that can be advanced to support or refute such treatment. Theoretically, there is the opportunity to lower what is known as the viral set-point and possibly to limit the dissemination of HIV to sanctuary sites such as the central nervous system (CNS). At the time of PHI there is an initial rise in plasma HIV viremia, followed within a few weeks by a marked reduction to a steady state level of replication, the so called viral set-point. The higher the viral set-point the more rapid the rate of progression of HIV-1 disease (Mellors et al., 1996). While lowering the viral set-point is an attractive concept there is as yet no proof that institution of antiretroviral therapy at PHI makes any impact. However, there is evidence from the animal model of HIV that at least short-term partial blockade of SIV replication in the setting of acute infection reduces the magnitude of virus specific CD^8 T cell responses and significantly alters the antigen specific B cell response (Deeks and Walker, 2004). Ideally, some balance between these opposing effects of PHI treatment is needed.

If there are conflicting theoretical and animal data are there any clinical trial data to settle the issue? Unfortunately, there has only been one properly conducted randomized placebo-controlled clinical trial. Some years ago a randomized study of zidovudine monotherapy showed modest benefit with reduction in subsequent opportunistic infections and slower CD^4 cell count decline, but these benefits were not sustained at two-year follow-up (Kinloch-de Loes and Perneger, 1997). Other studies have been observational. Nonetheless, the results thus far have been singularly unremarkable despite the use of highly active antiretroviral therapy (Desquilbet et al., 2004; Kaufman et al., 2004; Smith et al., 2004). Indeed, none of the three major published guidelines from the International AIDS Society USA, the United States Department of Health and Human Services or the British HIV Association unequivocally recommend antiretroviral therapy. However, there is still reason to think that such therapy may be beneficial, perhaps in a subset of PHI patients. The data thus far have included small numbers of patients, the trials have almost exclusively been observational and the trials have not targeted patients who are known to have risk factors for more rapid progression to AIDS.

This last issue is the very point that is most critical to patients with neurological complications of PHI. As mentioned in Section 5.3, neurological complications are associated with a worse prognosis for subsequent AIDS development, as well as more rapid development of cognitive impairment. Furthermore, such patients often have the other poor prognostic factors, namely longer duration and severity of PHI. It is therefore recommended that patients with neurological complications and indeed others with poor prognostic factors be treated with antiretroviral drugs (Apoola et al., 2002).

In those situations where it has been decided to use antiretroviral drugs there are several factors that must first be considered. Because transmission of drug-resistant strains of HIV is increasingly being reported, most clinicians would now recommend that resistance testing be performed especially if the source patient was taking antiretroviral drugs. Indeed this is the recommendation of the International AIDS Society USA. The results of such testing may take several weeks at which point the antiretroviral drug regimen can be modified appropriately. The optimal components of the antiretroviral drug regimen are still under evaluation but triple nucleosides should be avoided because of diminished efficacy (Kassutto and Rosenberg, 2004). Usually a protease inhibitor or non-nucleoside reverse transcriptase inhibitor on a nucleoside/nucleotide backbone is administered (Kassutto and Rosenberg, 2004). In those patients with neurological complications an argument can be made to use antiretroviral drugs with good CNS penetration (Cysique et al., 2004). Finally, the duration of therapy is contentious with some clinicians recommending only a few weeks to months followed by close clinical and laboratory monitoring. Any consideration of treatment must include an appraisal of the ability of the patient to comply with the sometimes complicated dosing schedules. Poor compliance has a high chance of leading to the development of resistance to some or all of the antiretroviral drugs in the combination.

The utility of immunomodulatory therapies alone or in combination with antiretroviral drugs is currently under assessment. Such treatments include mycophenolate mofetil, cyclosporine and interleukin-2 (Rizzardi et al., 2002; Kassutto and Rosenberg, 2004). The utility of corticosteroids in addition to highly active antiretroviral therapy (HAART) has not been explored but their use alone is not recommended in view of the case reports of more rapid progression of HIV disease (Apperley et al., 1987; Pedersen et al., 1989).

5.8. Neurologic clinical presentations

5.8.1. Principles

There are three principles relating to the neurologic sequelae of PHI: (i) all parts of the nervous system may be affected (see Table 5.4), (ii) there is parallel tracking — more than one part of the nervous system is frequently affected, a concept that is fundamental to the understanding of the clinical aspects of HIV neurology (Brew, 2001b), and (iii) the clinical features always occur in the context of an acute febrile illness — either with the illness or shortly afterwards.

Table 5.4

Nervous system involvement during primary HIV infection

Headaches
Meningoencephalitis
Seizures
Paroxysmal dyskinesias
Myelopathy
Cranial neuropathies (especially VII)
Guillain–Barré-like neuropathy
Sensory polyneuropathy
Brachial neuritis
Acute rhabdomyolysis

5.8.2. Epidemiology

The combined incidence of the neurological manifestations of PHI varies widely from 8 to 60% (Pedersen et al., 1989; Clark et al., 1991; Boufassa et al., 1995; Schacker et al., 1996) probably as a consequence of the differing numbers of patients in the studies, and the definition as to what exactly is neurological. For example, some series have included headache without other neurological complaints.

The specific incidence for each complication also varies according to sample size. Nonetheless, in several large reviews, encephalopathy and neuropathy were each reported in approximately 8% of cases (Pedersen et al., 1989; Clark et al., 1991). Headache is much more common, in 30–70% of patients (Gaines et al., 1988; Clark et al., 1991; Kaufman et al., 1998), largely because of its nonspecific association with a febrile illness. Excluding the common finding of myalgia (Pedersen et al., 1989; Quinn, 1997; Kaufman et al., 1998) significant PHI-related muscle disease is very unusual.

5.8.3. Natural history

In the short term, almost all patients make a complete recovery over several days to weeks. In the long term, there is evidence to suggest that patients who develop neurological complications progress to AIDS more rapidly than those who do not (Boufassa et al., 1995). Whether such patients have a greater probability of developing neurological complications later, with more advanced HIV disease, is unclear. As mentioned previously, Wallace et al. (2001) found that patients with symptoms of PHI more rapidly progressed to cognitive impairment; however, only one of the 29 patients in the study had definite neurological complications at PHI. The study by Boufassa et al. (1995) paradoxically showed that patients with a neurological PHI were less likely to develop AIDS dementia complex. As the authors themselves point out, this may be an artifact related to the small numbers of patients in the study. Nonetheless, it is theoretically possible as the clinical illness may be the result of an immune system effectively clearing HIV from the nervous system.

5.9. Clinical features

5.9.1. CNS syndromes

CNS manifestations of primary HIV infection cover a wide spectrum from mild headache to severe meningoencephalitis.

5.9.1.1. Headaches

Headaches frequently occur in PHI as previously noted. Retrobulbar pain, particularly on eye movement (Pedersen et al., 1989; Clark et al., 1991), and photophobia in some patients suggests that the headache is more than one would expect from fever alone. Indeed, it may represent a mild form of meningoencephalitis.

5.9.1.2. Meningoencephalitis

PHI-related meningoencephalitis (Carne et al., 1985; Ho et al., 1985) has a wide spectrum of presentations. At one extreme, patients have the rapid onset of confusion, progressing to obtundation and coma, usually within two weeks of onset of an acute febrile illness (Carne et al., 1985; Ho et al., 1985a; Biggar et al., 1986; Hardy et al., 1991). Other patients present with focal symptoms such as dysphasia (Brew et al., 1989). At the other extreme, patients have a mild course with lethargy and disorientation (Scully et al., 1989). Resolution typically occurs within two weeks of presentation (Biggar et al., 1986; Brew et al., 1989); significant persistent morbidity or death are very unusual.

5.9.1.3. Seizures

Seizures may also occur in PHI, but they do so in the context of the other clinical features of acute meningoencephalitis (Carne et al., 1985). This is in contradistinction to seizures later in the course of HIV infection, where they may be the only neurological abnormality (Wong et al., 1990; Dore et al., 1996).

5.9.1.4. Paroxysmal dyskinesias

Mirsattari et al. (1999) have described a patient who developed generalized paroxysmal nonkinesigenic dyskinesia in the context of PHI.

5.9.1.5. Spinal cord

The clinical features of primary HIV infection are known only by virtue of the reporting of a few cases (Denning et al., 1987; Gaines et al., 1988; Zeman and Donaghy, 1991; Silver et al., 1997) and variations clearly may exist. The most consistent feature is involvement of the thoracic cord and possibly conus. Patients complain of leg weakness with variable sensory disturbance and bladder dysfunction, usually within one to two weeks of the acute febrile illness that characterizes PHI (Kahn and Walker, 1998). A sensory level and spasticity are uncommon. Sometimes there are additional features consistent with more widespread pathology such as mild mental slowing and the presence of a snout reflex. More definite signs of cerebral involvement are unusual. One recently reported case presented with quadriplegia in association with confusion and cranial nerve abnormalities (Silver et al., 1997).

5.9.2. Peripheral nervous system

5.9.2.1. Neuropathies

Both cranial and peripheral neuropathies have been reported in association with PHI. While some patients may develop a cranial neuropathy in the context of a generalized neuropathy, others have developed isolated cranial neuropathies.

The most common reports have been of a facial nerve palsy (Anonymous, 1984; Piette et al., 1986; Wiselka et al., 1987; Paton et al., 1990). Although some of the cases have documented clear evidence of axonal damage, in others it is not apparent whether there is central involvement of the facial nerve nucleus, such as may occur in aseptic meningoencephalitis, or cranial nerve inflammatory neuropathy per se (Parry, 1988).

Several cases of severe and rapidly progressive neuropathy resembling Guillain–Barré syndrome have been described (Hagberg et al., 1986; Piette et al., 1986; Vendrell et al., 1987; Beytout et al., 1989; Paton et al., 1990). In general, the clinical features have been consistent with Guillain–Barré syndrome.

Acute bilateral brachial neuritis has also been reported (Calabrese et al., 1987; Brew et al., 1989). Neurologic symptoms begin approximately two weeks after onset of an acute febrile illness consistent with PHI and resolve within several weeks.

A subacute lumbosacral polyradiculoneuropathy may also occur. It is characterized by symmetrical or asymmetrical leg weakness, sacral paresthesias, sphincter disturbance and areflexia (Steiner et al., 1999).

Other reports of involvement of the peripheral nervous system during PHI include a sensory neuropathy of the limbs with areflexia, and ataxic neuropathy (Elder et al., 1986; Castellanos et al., 1994), as well as a symmetrical polyneuropathy (Hughes et al., 1992).

5.9.2.2. Myopathies

Several cases of rhabdomyolysis affecting all limbs with very high serum creatine kinase (CK) levels have been reported (McDonagh and Holman, 2003). The symptoms consisted of muscle pain and weakness developing over days. The CK levels resolved within a few days to several weeks. These cases may simply indicate a nonspecific, post-viral, infection-type myopathy.

5.10. Investigations

As expected, investigations are consistent with mild to moderate aseptic meningoencephalitis. This presumably is also true of patients who develop paroxysmal dyskinesias. Examination of cerebrospinal fluid (CSF) shows a mild to moderate mononuclear pleocytosis (Carne et al., 1985; Ho et al., 1985a, b; Ruutu et al., 1987) which improves and sometimes resolves within two weeks. Protein levels may be increased (Ho et al., 1985b), and may remain so for up to six weeks (Ruutu et al., 1987). HIV has been isolated from the CSF of some subjects with meningoencephalitis, sometimes within a day of presentation (Ho et al., 1985a, b; Ruutu et al., 1987). More recently, Tambussi et al. (2000) have shown that HIV RNA concentrations in the CSF are significantly elevated above those without neurologic complications. Electroencephalographic (EEG) studies show generally diffuse changes, consistent with the diagnosis of an aseptic meningoencephalitis (Carne et al., 1985; Ruutu et al., 1987). CT brain scan has not shown abnormalities, but the MRI appearance has not been reported.

In spinal cord syndromes, imaging investigations are typically unremarkable, but CSF analyses usually reveal a mild mononuclear pleocytosis with normal protein. These studies antedated the introduction of CSF HIV viral load. Patients spontaneously improved over several weeks usually with no residual deficits.

In patients with one of the neuropathies, the CSF usually shows a mild to moderate mononuclear pleocytosis with mild protein elevation as it does in patients with CNS complications. In those with Guillain–Barré syndrome, testing the cerebrospinal fluid does not show the classical finding of "albumino-cytologique" dissociation, but rather a raised protein and a mononuclear pleocytosis in at least one half of cases (Cornblath et al., 1987).

5.11. Pathogenesis

The neurologic complications of PHI are most likely caused by both direct and indirect mechanisms. The evidence for direct HIV involvement comes from the observation that patients with neurologic complications of PHI have higher concentrations of HIV RNA in the CSF (Tambussi et al., 2000). Secondly, there is the report of the patient who was accidentally infected by the intravenous route who experienced the rapid onset of confusion progressing to coma and death. Autopsy revealed evidence of productive brain infection (Davis et al., 1992). However, it is curious that the neurologic sequelae sometimes occur several weeks after, and not during, PHI suggesting that there may be a parallel with diseases such as acute disseminated encephalomyelitis that are immune mediated. The CSF abnormalities previously mentioned might reflect trafficking of infected cells through the nervous system, the Trojan horse mechanism (Peluso et al., 1985), or productive infection of the meningeal compartment, without necessarily implicating brain involvement.

As to the cellular mechanisms of tissue damage, it is likely that the processes are similar to those that operate later in HIV-1 disease. In the rhesus monkey, primary SIV infection has been shown to be associated with a significant increase in quinolinic acid levels, and animals who maintained high levels after the acute period appeared to have worse prognoses (Jordan and Heyes, 1993). The increased levels of quinolinic acid are presumably related to increased levels of virally mediated cytokines during PHI (Sonnerborg et al., 1989; von Sydow et al., 1991; Sinicco et al., 1993).

5.12. Treatment

While there is no evidence basis for antiretroviral therapy in patients with neurological complications of PHI, there is certainly reason to think that it may be beneficial acutely by lessening the symptoms and possibly in the longer term by ameliorating the severity of the illness. As mentioned in the previous section on treatment of PHI, there are several reasons for thinking that antiretroviral therapy would be helpful. PHI patients with neurological complications have a worse prognosis for subsequent AIDS development, and more rapid development of cognitive impairment.

The general nature of the antiretroviral combination has already been discussed in Section 5.7. The specific components probably should include at least three drugs with known ability to penetrate the CNS: zidovudine, stavudine, abacavir, lamivudine, nevirapine, efavirenz, indinavir and possibly ritonavir-boosted lopinavir (Brew, 2001b; Capparelli et al., 2005).

In those patients with Guillain–Barré syndrome, treatment with plasmapheresis and the indications for plasmapheresis are no different than those for non-HIV-related Guillain–Barré syndrome (Miller et al., 1988). As best as can be determined from the very few studies in this area, there does not appear to be a higher incidence of complications in patients with HIV disease (Berger et al., 1987), nor does the occurrence of Guillain–Barré syndrome render the patient more likely to develop other neurological complications at a later time.

References

Anonymous (1984). Needlestick transmission of HTLV-III from a patient infected in Africa. Lancet ii: 1376–1377.

Apoola A, Ahmad S, Radcliffe K (2002). Primary HIV infection. Int J STD AIDS 13(2): 71–78.

Apperley JF, Rice SJ, Hewitt P, et al (1987). HIV infection due to platelet transfusion after allogeneic bone marrow transplantation. Eur J Haematol 39: 185–189.

Berger JR, Difini JA, Swedloff MA, et al (1987). HIV seropositivity in Guillain-Barré syndrome. Ann Neurol 22: 393–394.

Beytout J, Llory JF, Clavelou P, et al (1989). Meningoradiculite a la phase de primoinfection a VIH. Interet de la plasmapherese. La Presse Medicale 18: 1031–1032.

Biggar RJ, Johnson BK, Musoke SS, et al (1986). Severe illness associated with the appearance of antibody to human immunodeficiency virus in an African. Br Med J 293: 1210–1211.

Boag FC, Dean R, Hawkins DA, et al (1992). Abnormalities of liver function during HIV-1 seroconversion illness. Int J STD AIDS 3: 46–48.

Boufassa F, Bachmeyer C, Carne C, et al (1995). Influence of neurologic manifestations of primary human immunodeficiency virus infection on disease progression. SEROCO study group. J Infect Dis 171: 1190–1195.

Brew BJ (2001a). Sequelae of primary HIV infection and non-focal complications of moderately advanced HIV disease. In: BJ Brew (Ed.), HIV Neurology. Oxford University Press, New York, pp. 74–77.

Brew BJ (2001b). Principles of HIV neurology. In: BJ Brew (Ed.), HIV Neurology. Oxford University Press, New York, pp. 32–35.

Brew BJ, Perdices M, Darveniza P, et al (1989). The neurologic features of early and 'latent' human immunodeficiency virus. Aust NZ J Med 19: 700–705.

Busch MP, Satten GA (1997). Time course of viraemia and antibody seroconversion following human immunodeficiency virus exposure. Am J Med 102(Suppl 5B): 117–124.

Busch MP, Lee LL, Satten GA, et al (1995). Time course of detection of viral and serologic markers preceding human immunodeficiency virus type 1 seroconversion: implications for screening of blood and tissue donors. Transfusion 35: 91–97.

Calabrese LH, Proffitt MR, Levin KH, et al (1987). Acute infection with the human immunodeficiency virus (HIV-1) associated with acute brachial neuritis and exanthematous rash. Ann Intern Med 107: 849–851.

Capparelli EV, Holland D, Okamoto C, et al (2005). Lopinavir concentrations in cerebrospinal fluid exceed the 50% inhibitory concentration for HIV. AIDS 19: 949–952.

Carne CA, Tedder RS, Smith A, et al (1985). Acute encephalopathy coincident with seroconversion for anti-HTLV-III. Lancet 2: 1206–1208.

Castellanos F, Mallada J, Ricart C, et al (1994). Ataxic neuropathy associated with human immunodeficiency virus infection. Arch Neurol 51: 236.

Clark SJ, Saag MS, Decker WD, et al (1991). High titers of cytopathic virus in plasma of patients with symptomatic primary HIV-1 infection. N Engl J Med 324: 954–960.

Cornblath DR, McArthur JC, Kennedy PG, et al (1987). Inflammatory demyelinating peripheral neuropathies associated with human T-cell lymphotropic virus type III infection. Ann Neurol 21: 32–40.

Cysique LA, Maruff P, Brew BJ (2004). Antiretroviral therapy in HIV infection: are neurologically active drugs important? Arch Neurol 61: 1699–1704.

Davis LE, Hjelle BL, Miller VE, et al (1992). Early viral brain invasion in iatrogenic human immunodeficiency virus infection. Neurology 42: 1736–1739.

Deeks SG, Walker BD (2004). The immune response to AIDS virus infection: good, bad or both? J Clin Invest 113(6): 808–810.

Denning DW, Anderson J, Rudge P, et al (1987). Acute myelopathy associated with primary infection with human immunodeficiency virus. Br Med J 294: 143–144.

Desquilbet L, Goujard C, Rouzioux C, et al (2004). Does transient HAART during primary HIV-1 infection lower the virological set-point? AIDS 18: 2361–2369.

Dore GJ, Law M, Brew BJ (1996). Prospective analysis of seizures occurring in human immunodeficiency virus type 1 infection. J NeuroAIDS 1(4): 59–70.

Elder G, Dalakas M, Pezeshkpour G, et al (1986). Ataxic neuropathy due to ganglioneuritis after probable acute human immunodeficiency virus infection. Lancet ii: 1275–1276.

Gaines H, von Sydow M, Pehrson PO, et al (1988). Clinical picture of primary HIV-1 infection presenting as a glandular-fever-like illness. Lancet 297: 1363–1368.

Gaines H, von Sydow MAE, von Stedingk LV, et al (1990). Immunological changes in primary HIV-1 infection. AIDS 4: 995–999.

Hagberg L, Malmvall B-E, Svennerholm L, et al (1986). Guillain-Barré syndrome as an early manifestation of HIV-1 central nervous system infection. Scan J Infect Dis 18: 591–592.

Hardy WD, Daar ES, Sokolov RT Jr, et al (1991). Acute neurologic deterioration in a young man. Rev Infect Dis 13: 745–750.

Hecht FM, Busch MP, Rawal B, et al (2002). Use of laboratory tests and clinical symptoms for identification of primary HIV infection. AIDS 16: 1119–1129.

Henrard DR, Phillips J, Windsor I, et al (1994). Detection of human immunodeficiency virus type 1 p24 antigen and plasma RNA: relevance to indeterminate serologic tests. Transfusion 34: 376–380.

Ho DD, Sarngadharan MG, Resnick L, et al (1985a). Primary human T-lymphotropic virus type III infection. Ann Intern Med 103: 880–883.

Ho DD, Rota TR, Schooley RT, et al (1985b). Isolation of HTLV-III from cerebrospinal fluid and neural tissues of patients with neurologic syndromes related to the acquired immunodeficiency syndrome. N Engl J Med 313: 1493–1497.

Hughes PJ, McLean KA, Lane RJM (1992). Cranial polyneuropathy and brainstem disorder at the time of seroconversion in HIV-1 infection. Int J STD AIDS 3: 60–61.

Jordan EK, Heyes MP (1993). Virus isolation and quinolinic acid in primary and chronic simian immunodeficiency virus infection. AIDS 7: 1173–1179.

Kahn JO, Walker BD (1998). Acute human immunodeficiency virus type 1 infection. N Engl J Med 339: 33–39.

Kassutto S, Rosenberg ES (2004). Primary HIV type 1 infection. Clin Infect Dis 38(10): 1447–1453.

Kaufman DE, Lichterfeld M, Altfield M, et al (2004). Limited durability of viral control following treated acute HIV infection. PLoS Med 1: 1–12.

Kaufman G, Duncombe C, Zaunders J, et al (1998). Primary HIV-1 infection: a review of clinical manifestations, immunologic and virologic changes. AIDS Pat Care STDs 12: 759–767.

Keet IP, Krijnen P, Koot M, et al (1993). Predictors of rapid progression to AIDS in HIV-1 seroconverters. AIDS 7: 51–57.

Kinloch-de Loes S, Perneger TV (1997). Primary HIV infection: follow up of patients initially randomized to zidovudine or placebo. J Infect 333: 408–413.

Kinloch-de Loes S, de Saussure P, Saurat JH, et al (1993). Symptomatic primary infection due to human immunodeficiency virus type 1: review of 31 cases. Clin Infect Dis 17(1): 59–65.

Lindbacl S, Brostrom C, Karlsson A, et al (1994). Does symptomatic primary HIV-1 infection accelerate progression to CDC stage IV disease, CD4 count below 200 × 10^6/l. Br Med J 309: 1535–1537.

McDonagh CA, Holman RP (2003). Primary human immunodeficiency virus type 1 infection in a patient with acute rhabdomyolysis. South Med J 96(10): 1027–1030.

Mellors JW, Rinaldo CR, Gupta P, et al (1996). Prognosis in HIV-1 infection is predicted by the quantity of virus in plasma. Science 272: 1167–1170.

Miller RG, Parry G, Pfaeffl W, et al (1988). The spectrum of peripheral neuropathy in the acquired immune deficiency syndrome. Muscle Nerve 11: 857–863.

Mirsattari SM, Roke Berry ME, Holdan JK, et al (1999). Paroxysmal dyskinesias in patients with HIV infection. Neurology 52: 109–114.

Parry GJ (1988). Peripheral neuropathies associated with human immunodeficiency virus infection. Ann Neurol 23 (Suppl): S49–S53.

Paton P, Poly H, Gonnaud P-M, et al (1990). Acute meningoradiculitis concomitant with seroconversion to human immunodeficiency virus type 1. Res Virol 141: 427–433.

Pedersen C, Lindhardt B, Jensen BL, et al (1989). Clinical course of primary HIV-1 infection: consequences for subsequent course of infection. Br Med J 299: 154–157.

Peluso R, Haase A, Stowring L, et al (1985). A Trojan horse mechanism for the spread of visna virus in monocytes. Virology 147: 231–236.

Piette AM, Tusseau F, Vignon D, et al (1986). Acute neuropathy coincident with seroconversion for anti-LAV/HTLV-III. Lancet 1: 852.

Quinn TC (1997). Acute primary HIV infection. JAMA 278: 58–62.

Rizzardi GP, Harari A, Capiluppi B, et al (2002). Treatment of primary HIV-1 infection with cyclosporine A coupled with highly active antiretroviral therapy. J Clin Invest 109: 681–688.

Ruutu P, Suni J, Oksanen K, et al (1987). Primary infection with HIV-1 in a severely immunosuppressed patient with acute leukemia. Scand J Inf Dis 1987; 19: 369–372.

Schacker T, Collier AC, Hughes J, et al (1996). Clinical and epidemiologic features of primary HIV infection. Ann Intern Med 125: 257–264.

Scully RE, Mark EJ, McNeely WF, et al (1989). Case records of the Massachusetts General Hospital, Case 33 — 1989. N Engl J Med 321: 454–463.

Silver B, McAvoy K, Mikesell S, et al (1997). Fulminating encephalopathy with perivenular demyelination and vacuolar myelopathy as the initial presentation of human immunodeficiency virus infection. Arch Neurol 54: 647–650.

Sinicco A, Biglino A, Sciandra M, et al (1993). Cytokine network and acute primary HIV-1 infection. AIDS 7: 1167–1172.

Smith DE, Walker BD, Cooper DA, et al (2004). Is antiretroviral treatment of primary HIV infection clinically justified on the basis of current evidence? AIDS 18: 709–718.

Sonnerborg AB, von Stedingk L-V, Hansson L-O, et al (1989). Elevated neopterin and beta$_2$-microglobulin levels in blood and cerebrospinal fluid occur early in HIV-1 infection. AIDS 3: 277–283.

Steiner I, Cohen O, Leker RR, et al (1999). Subacute painful lumbosacral polyradiculoneuropathy in immunocompromised patients. J Neurol Sci 162(1): 91–93.

Taiwo BO, Hicks CB (2002). Primary human immunodeficiency virus. South Med J 95: 1312–1317.

Tambussi G, Gori A, Capiluppi B, et al (2000). Neurological symptoms during primary human immunodeficiency virus (HIV) infection correlate with high levels of HIV RNA in cerebrospinal fluid. Clin Infect Dis 30(6): 962–965.

Tindall B, Hing M, Edwards P, et al (1989). Severe clinical manifestations of primary HIV infection. AIDS 3: 747–749.

Vendrell J, Heredia C, Pujol M, et al (1987). Guillain-Barré syndrome associated with seroconversion for anti-LAV HTLV-III. Neurology 37: 544.

Von Sydow M, Sonnerborg A, Gaines H, et al (1991). Interferon-alpha and tumor necrosis factor-alpha in serum of patients in varying stages of HIV-1 infection. AIDS Res Hum Retrovir 7: 375–380.

Wallace MR, Nelson JA, McCutchan JA, et al (2001). Symptomatic HIV seroconverting illness is associated with more rapid neurological impairment. Sex Transm Inf 77: 199–201.

Wilson KM, Johnson EIM, Croom HA, et al (2004). Incidence immunoassay for distinguishing recent from established HIV-1 infection in therapy-naïve populations. AIDS 18: 2253–2259.

Wiselka MJ, Nicholson KG, Ward SC, et al (1987). Acute infection with human immunodeficiency virus associated with facial nerve palsy and neuralgia. J Infect 15: 189–194.

Wong MC, Suite NA, Labar DR (1990). Seizures in human immunodeficiency virus infection. Arch Neurol 47: 640–642.

Zeman A, Donaghy M (1991). Acute infection with human immunodeficiency virus presenting with neurogenic urinary retention. Genitourin Med 67: 345–347.

Zhu T, Wang N, Carr A, et al (1996). Genetic characterization of human immunodeficiency virus type 1 in blood and genital secretions: evidence for viral compartmentalisation and selection during sexual transmission. J Virol 70: 3098–3107.

Handbook of Clinical Neurology, Vol. 85 (3rd series)
HIV/AIDS and the Nervous System
P. Portegies, J. R. Berger, Editors

Chapter 6

AIDS dementia complex

BRUCE J. BREW*

Departments of Neurology, HIV Medicine, Centre for Immunology, and National Centre in HIV Epidemiology and Clinical Research, St Vincent's Hospital, University of New South Wales, Sydney, Australia

6.1. Introduction

HIV may directly involve the brain leading to AIDS dementia complex (ADC). It is perhaps the most important neurological complication of HIV disease, because it is so disabling but at the same time potentially treatable. While the introduction of highly active antiretroviral therapy (HAART) has lessened its incidence, the prevalence has increased. Moreover, the disorder appears to be changing and there are new potential confounds to the diagnosis.

6.2. Prevalence and incidence

Prior to the introduction of HAART, the prevalence of ADC, including mild severity, was approximately 20–30% of patients with advanced HIV disease (McArthur et al., 1993), possibly with an increase in females and in injecting drug users (Chiesi et al., 1996). The incidence in patients with advanced HIV disease was approximately 7% per year (McArthur et al., 1993; Dore et al., 1997). In the HAART era, the prevalence has approximately doubled because of an increased life span, while the incidence has halved (Dore et al., 1999; Maschke et al., 2000). These and other data (Bacellar et al., 1994; Brew et al., 1996) now strongly suggest that ADC is not an inevitable consequence of HIV disease: it only occurs in some patients because of both host and viral factors (see Section 6.7).

6.3. Clinical features

Until it is severe, ADC is a subcortical dementia — aphasia, alexia and agraphia are absent — and motor disturbance is prominent (Navia et al., 1986). In severe ADC these differentiating features are lost and there is global impairment. Before HAART, ADC symptomatology developed subacutely over several weeks to months. It is an impression yet to be defined more rigorously that ADC in HAART-treated patients unfolds over a much longer time period. The clinical features of ADC are evident in three areas: cognition, motor function and behavior. In a prospectively studied series of 549 patients, examined by both myself and R. W. Price, of whom 111 had ADC, we found the following frequencies of symptoms as delineated in Table 6.1.

It should be noted that in mild ADC there may be few if any abnormalities on examination. In severe ADC, patients are globally demented, mute, paraparetic and incontinent of urine and feces. The myelopathy that sometimes occurs is characterized by a spastic paraparesis usually without a definite sensory level (Petito et al., 1985; Dal Pan et al., 1994). Diminished proprioception and vibration sense are dominant over pinprick or light touch deficits. Changes are largely confined to the legs, and the onset is usually over weeks to months. Rarely, ADC patients may present with movement disorders such as myoclonus, chorea (Nath and Jankovic, 1989), hypomania (Everall, 1995), transient neurologic deficits (Brew and Miller, 1996) or paroxysmal nonkinesigenic dyskinesia (Mirsattari et al., 1999). Rarely ADC may mimic Huntington's disease (Sevigny et al., 2005).

ADC occurring in HAART-treated patients may be different. This is an evolving area. To date ADC in such patients seems to be milder. However, it is not clear at present whether there is a "phenotypic" change. There is reason to suspect that some clinical features may be different in view of the changes seen on neuropsychological assessment (see below). Indeed, there may be more

*Correspondence to: Professor B. J. Brew, Department of Neurology, Level 4 Xavier Building, St Vincent's Hospital, Victoria St Darlinghurst, Sydney 2010, Australia. E-mail: B.Brew@unsw.edu.au, Tel: +61-2-8382-4100, Fax: +61-2-8382-4101.

Table 6.1

Symptoms and signs of AIDS dementia complex

Symptom classification	Symptom	Frequency (%)
Cognition	Decreased concentration	80
	Forgetfulness	80
	Mental slowing	80
Motor	Gait unsteadiness	75
	Clumsiness	50
	Tremor	30
Behavior	Apathy	50
Other	Urinary urgency/hesitancy	30
	Headache	30
	Seizures	11

The main examination features are detailed below.

Sign	Frequency (%)
Slowing of fine finger movements	100
Hyperreflexia of the legs	100
Inaccuracy of pursuit and saccadic eye movements	75
Impaired tandem gait	75
Mask-like facies	50
Action tremor	50
Peripheral neuropathy	30
Myelopathy	30
Positive glabella response	25
Snout response	25

cortical features developing as a result of an interaction with the effects of aging and possibly an increased susceptibility to Alzheimer's disease (Brew, 2004). Further studies are needed.

The severity of ADC can be staged according to the criteria set forth in the Memorial Sloan Kettering scale with a range from 0 to 4 (Price and Brew, 1988) (Table 6.2). In essence it is a modified neurological Karnofsky score. Recently, it has been operationalized (Marder et al., 2003). The Neuropsychiatric AIDS Rating Scale (NARS) is another more complicated scale also specifically developed for ADC (Boccellari and Dilley, 1992).

The HIV Dementia Scale (Power et al., 1995) is a reasonable screen for ADC. In "mild" ADC the screen had a false negative rate of 17% and a false positive rate of

Table 6.2

Modified staging scheme for AIDS dementia complex

Stage 0 (normal)	Normal mental and motor function
Stage 0.5 (subclinical)	Minimal or equivocal symptoms without impairment of work or activities of daily living (ADL). "Background" neurological signs, such as slowed fine finger movements or primitive reflexes may be present
Stage 1 (mild)	Cognitive deficit that compromises the performance of the more demanding aspects of work or ADL
Stage 2 (moderate)	Cognitive deficit makes the patient unable to perform work or the more demanding aspects of ADL
Stage 3 (severe)	Cognitive deficit makes it possible for the patient to perform only the most rudimentary tasks. Assistance with walking is often required
Stage 4 (end-stage)	Cognitive deficit has reached the point where the patient has virtually no understanding of his or her surroundings and is virtually mute. There is often paraparesis with double incontinence

16%. Testing antisaccade eye movements and recording the antisaccade errors has limited utility (Currie et al., 1988): approximately 36% of normal HIV-infected patients and 86% of patients with AIDS but without ADC are abnormal.

6.4. Natural history

Pre-HAART the mean time to death for patients with varying degrees of severity of ADC was approximately six months (median 3.8 months) (Bouwman et al., 1998): 10 months for stage 1, 4.6 months for stage 2 and 1.4 months for stages 3 and 4 combined (Dore et al., 2003). Some patients do progress more rapidly: injecting drug users, patients with low CD^4 cell count especially below 200 cells/μl, those with a history of AIDS defining illnesses (Bouwman et al., 1998) and those with markedly elevated cerebrospinal fluid (CSF) concentrations of S100 (Pemberton and Brew, 2001). Pre-HAART, ADC occurred usually in advanced disease: mean CD^4 cell count of 109/μl (Portegies et al., 1993), while the median CD^4 cell count was 50/μl (Dore et al., 1999). With HAART, a normal or near-normal CD^4 cell count is found (McArthur et al., 2004; Cysique et al., in press).

6.5. Investigations

Currently, no investigation is diagnostic of ADC; it remains a disorder of exclusion in conjunction with supportive test results (Table 6.3). With the introduction of HAART, some aspects of investigations have changed; however, this is an evolving area that still needs clarification (Brew, 2004).

6.5.1. Metabolic investigations

The usual metabolic investigations that are performed as part of a dementia workup should also be performed in this situation: B12, red cell folate concentrations and thyroid function tests (Kieburtz et al., 1991).

6.5.2. Imaging investigations

Either computed tomographic (CT) scanning or magnetic resonance imaging (MRI) of the brain should be performed to exclude a mass lesion such as toxoplasmosis or lymphoma, as well as address the likelihood of disorders such as cytomegalovirus (CMV) encephalitis and progressive multifocal leukoencephalopathy (PML). MRI is, however, superior to CT generally and especially for the delineation of the latter two entities.

Table 6.3

Investigations for the diagnosis of ADC

Investigation	Result
CT brain scan	Cerebral atrophy Exclude mass lesion, PML
MRI brain scan	Cerebral atrophy Exclude mass lesion, PML (MRI more sensitive than CT) Possibly diffuse periventricular T2 abnormalities
MRS	Reduced levels of N-acetyl aspartate Increased levels of choline, myoinositol
CSF	Increased concentrations of immune activation markers Increased HIV RNA load Antiretroviral drug resistance Exclude cryptococcus, cytomegalovirus, syphilis
Neuropsychological assessment	Psychomotor slowing and decreased mental flexibility

The following findings that are supportive of ADC were defined in patients who were either naive to antiretroviral therapy or who were studied in the pre-HAART era. The abnormalities presumably are still relevant to those patients developing ADC on HAART.

The supportive imaging findings for an ADC diagnosis are as follows. CT scanning of the brain usually demonstrates general cerebral atrophy especially of the caudate nuclei (Dal Pan et al., 1992). MRI additionally often has T2-weighted patchy or diffuse periventricular abnormalities (Jarvik et al., 1988).

Similarly, single photon emission computed tomography (SPECT) often shows multifocal cortical and subcortical areas of hypoperfusion, although the changes are very nonspecific (Schielke et al., 1990). Positron emission tomography, however, has been found to show areas of hypometabolism in the basal ganglia that are relatively specific for ADC (Rottenberg et al., 1987).

Magnetic resonance spectroscopy (MRS) appears helpful in ADC. There are reduced levels of N-acetyl aspartate, a marker of neuronal function, in the deep frontal white matter and basal ganglia along with elevated levels of myoinositol (found only in glial cells) and choline, a marker that is present in higher concentrations in glial cells reflecting changes in cell membrane injury, turnover or glial cell activation (Chang et al., 1999).

6.5.3. CSF analyses

CSF analysis may reveal "background" abnormalities that can occur in asymptomatic HIV-infected individuals: a mild mononuclear pleocytosis, a mildly raised CSF protein, intrathecal IgG synthesis and oligoclonal bands (Marshall et al., 1988). Again, the influence of HAART on these abnormalities has not been fully addressed.

CSF analyses primarily serve the purpose of excluding other illnesses. Cryptococcal meningitis (cryptococcal antigen), cytomegalovirus encephalitis (cytomegalovirus DNA) and syphilis (VDRL) should especially be considered.

In the pre-HAART era, the number of HIV RNA copies in the CSF (CSF viral load) was helpful. It is the best correlate of severity (Brew et al., 1997; McArthur et al., 1997) in patients with advanced HIV disease — nadir CD^4 cell counts below 200/µl. Like other tests, however, it is often abnormal in patients who are systemically well and naive to antiretroviral agents. There are other caveats in the interpretation of CSF viral load. Central nervous system infections such as cryoptococcal meningitis can lead to significant elevations in CSF viral load in the absence of ADC which fall with treatment. Indeed, there are four sources of CSF viral load: (i) "leak" from the plasma through a disturbed blood–brain barrier, (ii) "spill over" from brain infection, (iii) "shedding" from productive infection of the meningeal compartment (meningeal macrophages) and (iv) "shedding" of HIV from the trafficking of activated T cells through the central nervous system as part of their normal regulatory immune function.

Other virological markers are only discussed here briefly for the sake of completeness as they have little clinical utility. The core protein of HIV-1, p24 antigen, is infrequently found in the CSF of ADC patients and is rarely present in non-demented patients (Brew et al., 1994). HIV-1 antibodies in CSF, and culture of CSF for HIV-1, are often found in both ADC and non-ADC patients (Chiodi et al., 1988).

Measures of immune activation within the CSF may be helpful as adjuncts to the diagnosis of ADC in patients naive to antiretroviral drugs or unexposed to HAART. Elevated CSF concentrations of $\beta_2$microglobulin (Brew et al., 1992), neopterin (Brew et al., 1990), quinolinic acid (Heyes et al., 1991) and monocyte chemotactic protein-1 (MCP-1) (Cinque et al., 1998) parallel the severity of ADC in patients without other confounding neurological illnesses. Other markers of immune activation such as tumor necrosis factor (TNF) concentrations and interleukin (IL)-1 and IL-6 are not as well correlated (Gallo et al., 1989; Grimaldi et al., 1991). Matrix metalloproteinase (MMP)-2, MMP-7 and MMP-9 are raised in approximately half of ADC patients (Conant et al., 1999) indicating blood–brain barrier disruption. At present it is unclear how these assays assist in clinical management as blood–brain barrier impairment occurs even in the absence of ADC (Petito and Cash, 1992).

CSF analysis is useful for antiretroviral therapy decisions. The patterns of resistance to antiretroviral drugs in the CSF and blood are discordant in up to one-third of patients: some are resistant to certain drugs in the CSF, but sensitive in the blood and vice versa (Cunningham et al., 2000). At present it is not clear whether measuring antiretroviral drug concentrations in the CSF is useful, although this is the case with the serum.

6.5.4. Neurophysiological investigations

Neurophysiological investigations have little value in the assessment of the ADC patient, because the changes frequently overlap with those in asymptomatic patients: disturbances in pursuit and saccadic eye movements (Sweeney et al., 1991), brainstem auditory evoked potentials (Koralnik et al., 1990), long latency event-related potentials (Goodin et al., 1990) and electroencephalography (EEG) (Nuwer et al., 1992).

6.5.5. Neuropsychological investigations

Neuropsychological testing, in contrast, is useful. It can validate the history of cognitive decline by comparing performance with estimates of premorbid intellect and it can assist ADC diagnosis by the pattern of deficits that is found. It is also useful in optimizing the day-to-day management of ADC patients by identifying which cognitive domains are particularly weak. Finally, neuropsychological tests can be used to monitor response to therapy.

Neuropsychological disturbances are characterized by subcortical deficits suggestive of ADC: impaired attention, slowing of intellectual processes, especially motor based, deficits of executive or frontal lobe type function (for example, mental flexibility), only relatively mild disturbances of memory with frequent sparing of recognition memory and less commonly visuospatial abnormalities (Maruff et al., 1994). In patients who develop ADC on HAART there is emerging evidence that there is a change to the pattern of abnormalities with fewer deficits in motor-based tests (Cysique et al., in press).

A full discussion of the neuropsychological tests used in ADC is beyond the scope of this review (but see Chapter 7). Suffice to say that most researchers employ a combination of the following: timed gait, Trail Making Tests A and B, finger tapping dominant and non-dominant, grooved pegboard dominant and non-dominant, digit symbol test, the Rey Auditory Verbal Learning Test, Rey-Osterreith Complex Figure, symbol digit modalities test, simple and choice reaction times, and California computerized assessment package (CAL-CAP). The issue of practice effect because of repeat assessments is important. The tests with the least practice effect are the grooved pegboard and symbol digit modalities test (Selnes et al., 1995). The grooved pegboard has been shown to be the earliest indicator of ADC at least in the pre-HAART era.

There are caveats to the interpretation of neuropsychological assessment. Firstly, confounding illnesses such as opportunistic infections or tumors, fatigue, anxiety, depression or substance abuse must be excluded. Secondly, impaired test performance does not always equal ADC: neuropsychological assessment should be linked to a clinical neurological examination.

The issue of neuropsychological impairment in otherwise healthy HIV-positive individuals is complex. In the pre-HAART era, large studies had not shown any significant impairment. Indeed, in this group of patients subjective cognitive complaints alone most often relate to the effects of depression (Wilkins et al., 1991). However, there was some evidence for neuropsychological dysfunction in a subgroup of patients with advanced HIV disease that has no clinical correlate and no impact on functioning (Cysique et al., 2004a). The issue is different in HAART-treated patients. While the prevalence of deficits has not significantly declined (in contradistinction to the decline in ADC in HAART versus pre-HAART eras) the pattern of neuropsychological impairment has changed with a reduction in attention, visuoconstruction deficits but a deterioration of learning efficiency and some aspects of complex attention. This change remained even in patients with an undetectable plasma viral load (Cysique et al., 2004a). Moreover, these abnormalities occurred in patients with only mildly depressed CD^4 cell counts.

6.6. Neuropathology

The neuropathological features of ADC chiefly relate to the brain; the oft-associated myelopathic findings will be only briefly mentioned.

6.6.1. Neuropathology: brain

Macroscopically, there is cerebral atrophy in virtually all patients (Brew et al., 1995), predominantly involving the frontal lobes (Gelman and Guinto, 1992). Leptomeningeal fibrosis occurs in approximately one-third of patients. Histopathological changes of perivascular mononuclear infiltrates are predominant in the deeper parts of the brain especially the basal ganglia (Brew et al., 1995). Microglial cells are frequently activated forming microglial nodules or multinucleated giant cells (Glass et al., 1993). HIV encephalitis is the term used for the combination of infiltrating mononuclear cells and multinucleated giant cells. Astrocytes are commonly increased in number and in patients who rapidly progress there is increased apoptosis (Thompson et al., 2001). Oligodendrocyte numbers are occasionally increased (Esiri and Morris, 1996). White matter pallor is common and is more probably related to blood brain–barrier-induced myelin damage than demyelination (Power et al., 1993). Rarely, there may be vacuolation throughout the white matter which has been termed vacuolar leukoencephalopathy (Budka et al., 1991). Finally, there may be neuronal loss: the large pyramidal neurons and interneurons in the cortex, spiny neurons in the putamen, medium size neurons in the globus pallidus and occasionally the interneurons in the CA3 region of the hippocampus (Giometto et al., 1997; Masliah et al., 1997). There are also frequent alterations in ultrastructure, again mainly in the frontal lobes and the basal ganglia. There is dendritic pruning and simplification of synaptic contacts and axonal damage with the accumulation of beta amyloid precursor protein (Giometto et al., 1997). The aforementioned neuropathological features relate to the pre-HAART era. The few studies that have been performed in the HAART era have usually shown a decrease in ADC frequency but with four additional findings. There is attenuation of the inflammatory response in most, but in some patients there can be a fulminant inflammatory leukoencephalopathy perhaps related to HAART-induced immune reconstitution (Gray et al., 2003), there is early evidence for more prominent involvement of the hippocampus (Anthony et al., 2004) and finally there is increased cerebral atherosclerosis (Morgello et al., 2002). It is likely that the neuropathological features will continue to evolve over the next few years as patients come to autopsy who have been on HAART for extended periods of time.

In the pre-HAART era, the best neuropathological correlate of ADC was the presence of activated microglia. White matter pallor and multinucleated giant cells are only found in half the patients and HIV encephalitis in only one-quarter to one-half (Glass et al., 1993; Brew et al., 1995). Importantly, neuronal loss is more a feature of severe ADC (Gray et al., 2001). Neither beta amyloid precursor protein accumulation nor the degree of dendritic change correlates with ADC severity.

Some of these neuropathological changes may also occur in the absence of ADC. Cerebral atrophy, leptomeningeal fibrosis, white matter pallor, microgliosis and micoglial nodules occur in approximately one-third to one-half of patients with advanced HIV disease (Bell et al., 1993; Gelman, 1993; Glass et al., 1993; Brew et al., 1995). Astrocytosis is very common in advanced HIV disease. Conversely, mononuclear infiltrates in the meninges, choroid plexus and perivascular spaces in the brain have been found in HIV-infected patients but not in ADC patients (Bell et al., 1993; Gelman, 1993; Glass et al., 1993; Brew et al., 1995).

6.6.2. Neuropathology: spinal cord

In approximately half of ADC patients there is evidence of vacuolar myelopathy and varying degrees of multinucleated giant cell myelitis (Petito et al., 1985; Dal Pan et al., 1994). Vacuolar myelopathy is characterized by multiple non-tract-associated vacuoles in the white matter of the posterior and lateral columns, especially in the thoracic area, infrequent lipid-laden macrophages and separation of the myelin lamellae on electron microscopy. Axonal damage occurs only in more severe forms (Rottnek et al., 2002). Multinucleated cell myelitis is characterized by multinucleated cell infiltrates without a predilection for any particular part of the cord (Rosenblum et al., 1989).

6.6.3. Neurovirological aspects of the neuropathology

The only intrinsic neural cell that can consistently support productive HIV infection is the microglial cell (Brew et al., 1995). Endothelial cells may also be able to support productive infection, but this seems to be uncommon and its significance is unknown (Moses et al., 1996). Astrocytes can only support restricted infection, where the production of viral particles stops at the stage of regulatory proteins (especially nef) and whole virions are not produced. This is especially important in children (Tornatore et al., 1994; Takahashi et al., 1996). Oligodendrocytes and neurons cannot support productive or restricted infection (Esiri and Morris, 1996). Latent infection, however, where the viral RNA has been reverse transcribed into the host DNA, has been described in neurons and not unexpectedly in microglial cells (Bagasra et al., 1996).

Whilst there is still controversy as to what constitutes neurotropism and neurovirulence as they pertain to HIV, the dominant viral strain that can be isolated from the brain is R5. This relates to the dichotomous classification of HIV into R5 (that is, it utilizes the CCR5 receptor and is almost always macrophage tropic) and X4 (that is, it utilizes the CXCR4 receptor and is almost always T cell tropic). It is not surprising that the dominant viral strain is R5 given that the only intrinsic neural cell that can be productively infected is the microglial cell. The X4 strain seems to be important in astrocyte infection.

6.6.4. Neuroimmunological aspects of the neuropathology

In general, the brains of ADC patients have diffuse upregulation of expression of MHC class I and II on both the infiltrating inflammatory cells and intrinsic neural elements. Also, there is increased expression of TNFα and decreased levels of IL-4 in the brain parenchyma, which correlate with the severity of ADC (Wesselingh et al., 1993). The latter findings strongly suggest that there is a degree of immune dysregulation within the brain leading to unchecked immune activation.

As mentioned in the previous section, the chemokine receptors CCR5 and CXCR4 are important. CCR5 is localized to macrophages, microglia, astrocytes and neurons. CXCR4 is expressed on astrocytes and hippocampal neurons. Chemokines have been found to be elevated in the brain parenchyma, predominantly in areas of inflammatory infiltrates but also in normal-appearing brain tissue: MIP-1α, MIP-1β, RANTES, IL-8 and IP-10 (Sanders et al., 1998). These probably amplify HIV and associated inflammatory changes by recruiting mononuclear cells into the brain.

6.7. Pathogenesis

6.7.1. Mechanism of HIV entry into the brain

The most widely held theory is that HIV crosses the blood–brain barrier in infected mononuclear cells either because the barrier is disturbed by HIV infection of the endothelial cells, or because activated and infected lymphocytes traffic through the brain as part of their normal immune surveillance function (Brew, 2001). A more recent aspect to this theory is the increasing realization that monocytes derived from the bone marrow may be important in the turnover of perivascular macrophages in the brain (Gartner, 2000). This would explain the perivascular distribution of the neuropathology of ADC. Additionally, some consider that HIV enters the brain through the CSF compartment, which in turn is involved by infection of the choroid plexus and the meningeal macrophages (Brew, 2001). This would partly explain the dominance of the pathological changes in the deeper parts of the brain. Most probably, there is a dynamic interplay between these entry points in vivo.

6.7.2. The timing of HIV entry into the brain

Most investigators now agree that HIV probably enters the brain soon after seroconversion but probably not in every case. This is based on patient anecdotes and the fact that latent infection of the brain has not been consistently found in those with moderately advanced HIV disease (Sinclair et al., 1994). The most convincing data come from the macaque infected with a hyperneurovirulent strain of SIV: there is universal brain involvement at seroconversion which is then quelled by the immune system leaving only evidence of latent infection. Later, in the course of immunodeficiency there is resurgence of productive intracerebral infection (Zink and Clements, 2002). This would suggest that immunodeficiency is important in the recrudescence of cerebral infection but another model advanced by Narayan has demonstrated that encephalitis can occur with R5 strains in the absence of immunodeficiency (Narayan, personal communication). At present it is not clear whether HAART at the time of seroconversion diminishes the chance of subsequent development of ADC even though intuitively this would be reasonable.

6.7.3. The mechanism of HIV brain damage

HIV damages the brain by a variety of toxins elaborated by both the host and the virus itself: quinolinic acid, TNFα, platelet activating factor, nitric oxide, neopterins, peroxynitrite, Ntox, chemokines, gp120, tat and nef (Brew, 2001). Interestingly, the viral toxins mediate their toxicity through secondarily activating the production of some of the latter mentioned host toxins. Most host and viral toxins effect neuronal death and dysfuncton by activation of the N-methyl-D-aspartate receptor (Lipton and Gendelman, 1995). Once neurons are killed, it seems likely that neurotransmitters such as glutamate are released in concentrations that initially are not neurotoxic but because of HIV and TNFα-induced astrocyte dysfunction (Fine et al., 1996) they cannot be cleared from the microenvironment, and so accumulate leading to further neuronal death. Chemokines may directly damage neurons or indirectly through activation of microglia (Cartier et al., 2005). Perhaps one of the most interesting chemokines is SDF1 which is overproduced by astrocytes in HIV infection and which can be cleaved to a highly neurotoxic fragment by MMP2, one of the matrix metalloproteinases (Zhang et al., 2003). Furthermore, there is some evidence that they may lead to neurodegeneration through inhibition of neural stem cell proliferation (Krathwohl and Kaiser, 2004). It should not be thought that all chemokines have neurotoxic potential. There is increasing evidence that fractalkine may be neuroprotective and capable of antagonizing the neurotoxicity of some chemokines (Mizuno et al., 2003). Moreover, other factors such as brain-derived neurotrophic factor may also have similar potential (Bachis et al., 2003).

A metabolic aspect to neuropathogenesis is suggested by the observation that neopterin may lead to downregulation of GTP cyclohydrolase I, which is essential in the synthesis of tetrahydrobiopterin, critical for synthesis of dopamine, serotonin and noradrenaline (Shen et al., 1988).

The non-neuronal damage is most likely related to the elaboration of a variety of cytokines and perhaps neopterin.

The most important aspect of pathogenesis is the discordancy between productive viral burden in the brain and clinical deficit (Brew et al., 1995). ADC is not simply correlated with the amount of virus in the brain parenchyma. There is, however, a relationship between cytokine excess in the brain and CSF and the severity of ADC (Tyor et al., 1992). These data suggest, therefore, that indirect mechanisms are important.

HIV-related vacuolar myelopathy is probably immune mediated though the details are not apparent. In vitro TNFα is associated with similar changes to those of vacuolar myelopathy but it is not clear if the association is causal (Weidenheim et al., 1996). A metabolic pathogenesis such as a methylation defect has been identified in HIV infection, leading to the hypothesis that there is reduced production of S adenosyl methionine, perhaps because of macrophage-derived nitric oxide. This in turn inhibits methionine synthase, and increases consumption of S adenosyl methionine in an attempt to repair cytokine and free radical-induced myelin damage. A recent trial, however, with S adenosyl methionine did not lead to any improvement (di Rocco et al., 2004) suggesting either that the drug does not work or that the trial may have been compromised by including patients with fixed damage. Multinucleated cell myelitis is essentially the spinal cord equivalent of HIV encephalitis.

6.8. Management

The management of the patient with ADC is evolving. The following is a personal approach that I found helpful. The first step is to determine whether the disorder is active or inactive as discussed previously. The chief tool by which this is assessed is the history from the patient, loved one and family. The second step in management then hinges on the nature of the activity: it should be possible to assess whether it is progressive, regressive or stable. However, at present better tools are needed to distinguish between inactive ADC and stable ADC — the latter seems to be characterized by evidence of CSF viral and immunological activity that presumably are just keeping in each other in check, while the former has no evidence of viral or immunological activity in the CSF. Active progressive and stable ADC should have their antiretroviral regimen altered if possible. Inactive ADC and perhaps regressive ADC should have aspects of palliative management incorporated into their clinical management.

6.8.1. Antiretroviral drugs

The treatment of ADC at present is centered on antiretroviral drugs. Adjunctive therapy in the form of NMDA receptor and cytokine antagonists as well as inhibitors of apoptosis are yet to have a well-defined and proven role. Evidence for efficacy of individual antiretroviral agents is checkered and despite the current proven practice of combination therapy in systemic disease, the relative neurological potencies (either additive or synergistic) of different combinations of antiretroviral drugs are unknown. Antiretroviral drugs can be divided into the following classes: nucleoside reverse transcriptase inhibitors (NRTIs), ribonucleotide reductase inhibitors, nucleotide reverse transcriptase inhibitors, non-nucleoside reverse transcriptase inhibitors (NNRTIs), protease inhibitors and entry inhibitors.

There are three characteristics of an antiretroviral drug that theoretically at least should make it likely to be effective in ADC: (i) it should be able to enter the brain in efficacious concentrations, (ii) it should be effective against HIV infection in the infiltrating cells — the lymphocytes and monocytes, and (iii) it should be effective against the intrinsic brain cells — the perivascular macrophages and microglia. The significance of astrocytic infection and therefore the need of a drug to be able to work in astrocytes are under evaluation. Currently, the CSF concentration of a particular drug is the only practical method of addressing the first criterion. There are standard features to a drug that make it more likely to be able to enter the brain: small size, increased lipophilicity and more acidic composition. But in relation to antiretroviral drugs, there are active transporter systems that pump drugs out of the brain that must be considered. Of especial importance are the multiple resistance associated transporters (MRPs) and the P-glycoprotein system: the first is important for zidovudine and possibly other nucleosides while the second is important for the protease inhibitors (Kim et al., 1998; Konig et al., 1999). Ritonavir, a protease inhibitor, is not only a substrate but also a potent inhibitor of the P-glycoprotein system. Consequently, combining ritonavir with another protease inhibitor may facilitate brain entry but definitive data are needed. The fact that there is neuropathological evidence of increased P-glycoprotein expression in HIV encephalitis further supports the importance of adding ritonavir to a protease inhibitor (Langford et al., 2004). The ability of a particular drug to be effective in lymphocytes, macrophages and microglia is critical to ADC efficacy and involves two important types of infection: acute (in lymphocytes) and chronic (in monocytes, macrophages and microglia). All antiretroviral drugs are effective against acute infection but only some (abacavir, lamivudine, didanosine, dideoxycytidine and possibly the NNRTIs and protease inhibitors) are effective against chronic infection. Comparative efficacy data are limited (Aquaro et al., 1997; Saavedra-Lozano et al., 2004).

Most of the NRTIs are useful in ADC. Zidovudine (ZDV) is the only antiretroviral medication with proven efficacy in ADC (Sidtis et al., 1993), but higher than usual doses are required with the resultant risk of hematological intolerance manifesting as anemia. Thus the highest tolerable dose of ZDV should be given. Didanosine (ddI), another NRTI, has some efficacy in HIV-1-infected children, but it may not be effective in adults (Gisslen et al., 1997). Zalcitabine (ddC) probably has no role in ADC (Brew, 2001). Stavudine (d4T) appears effective in ADC based on its ability to penetrate into the CSF in efficacious concentrations and to clear CSF viral load in a small number of patients (Foudraine et al., 1998). Lamivudine can also reach efficacious steady state concentrations in the CSF and may be useful (Foudraine et al., 1998). Abacavir, a relatively new NRTI with CSF penetration comparable to zidovudine, should be beneficial although a large randomized double blind placebo-controlled trial did not show this, despite superior CSF viral load clearance. The failure may have been related to background combination therapy efficacy as well as resistance to abacavir and some patients having inactive disease (Brew, 2001).

Hydroxyurea is a ribonucleotide reductase inhibitor that effectively inhibits HIV replication (Lori et al., 1994), especially in combination with d4T and ddI. The drug has good brain penetration but firm evidence for ADC efficacy is lacking. Recently, it has fallen out of fashion because of the risk of myelosuppression, and neuropathy when used with d4T or ddI.

The only nucleotide reverse transcriptase inhibitor currently available is tenofovir. Data on its potential neurological efficacy are lacking.

The NNRTIs nevirapine and efavirenz are likely to be effective in ADC because of potentially efficacious steady state CSF concentrations. Neurologists, however, should be aware of the side effects of efavirenz that may occur in the first few weeks of treatment, especially mild confusion and somnolence (Brew, 2001). Usually these settle spontaneously.

The only protease inhibitor with the potential for ADC efficacy is indinavir because of favorable CSF steady state concentrations (Martin et al., 1999). This is probably improved by the addition of ritonavir through its inhibition of the P-glycoprotein efflux transporter system.

The entry inhibitors are a new class of antiretroviral drug. Their mode of action is the inhibition of entry of HIV into the cell either by stopping fusion of HIV with the cell membrane (T20; also known as fuzeon) or by blocking one of the chemokine receptors (either CCR5 or CXCR4). At present, neurological efficacy data for this class are lacking but given the large molecular weight of T20 it is unlikely to be able to penetrate into the brain.

In clinical practice, ADC patients should have their blood and CSF analyzed for any evidence of resistance mutations to the latter mentioned antiretroviral drugs. A viral load of approximately 1000 copies/ml is usually required to enable the testing to be performed. The results of the resistance testing can then be used to guide which antiretrovirals would be appropriate to use. At present, however, it is still not apparent how many and which should be used. Nonetheless, the guiding principles are that at least three neurologically active NRTIs (zidovudine, stavudine, abacavir and lamivudine) should be used together (except for zidovudine and stavudine) along with one of the NNRTIs (nevirapine or efavirenz) and a protease inhibitor (indinavir) preferably boosted by ritonavir to aid in brain penetration and compliance.

The latter recommendations are based upon evolving evidence of the need for three neuroactive drugs in the regimen (Cysique et al., 2004b). However, there are some clinicians who hold that it does not matter whether the antiretroviral regimen has neuroactive drugs in it or not. For a critical review of this complex issue the interested reader is referred to Cysique et al. (2004b).

The potential toxicity of these regimens should not be forgotten. NRTIs can be associated with tissue-specific mitochondrial toxicity, for example zidovudine with myopathy and didanosine with peripheral neuropathy. There is also the potential for additive and perhaps synergistic mitochondrial toxicity when several NRTIs are used together.

Response to HAART may take weeks to months with the vast majority showing some improvement by six months (Brew, 2001). Clinical response is usually mirrored in the CSF by a decline in markers of immune activation, β_2 microglobulin, neopterin, quinolinic acid and possibly viral load, but it is not yet clear whether these changes can predict response. Chang et al. (2003) have shown that HAART can reverse abnormalities seen on MRS in both the frontal lobe and basal ganglia after approximately nine months.

6.8.2. Adjunctive treatments

Memantine, an open channel NMDA antagonist, showed modest efficacy in ADC patients, while pentoxfylline, nimodipine and lexipafant were less helpful. Recently, the monoamine oxidase inhibitor deprenyl showed promise, probably through its anti-apoptotic effect (Brew, 2001). Minocycline is effective in inhibiting HIV infection of microglia in vitro and has anti-apoptotic and immune regulating properties thereby making it an attractive candidate for adjunctive therapy in ADC (Si et al., 2004).

6.8.3. Palliative management

Palliative management of patients with ADC should be seen as an integral component to the whole management process: its role should vary according to the patient's needs. It is useful in assisting with symptom-based aspects of management even in patients who are doing well. It becomes more important in those patients who do not respond to antiretroviral therapy.

Some patients have prominent extrapyramidal symptomatology that may be assisted by the use of low doses of dopaminergic agents. Disinhibited or even psychotic behavior may benefit from low doses of one of the neuroleptic drugs but caution should be used given that ADC patients are more susceptible to the extrapyramidal effects. Where possible, pimozide or olanzapine should be used to minimize such effects (Brew, 2001).

Psychosocial aspects of palliative management are also important. Appropriate support facilitates early diagnosis which in turn allows participation in identifying power of attorney and decisions on treatment limitations. Modifications to the home can be discussed, such as adding stair rails and extra bathroom fittings. Because response to treatment may take weeks to months, the patient and important others need reassurance to avoid frustration or despair at the lack of response in the early stage of treatment. Additionally, loved ones and staff need support in coping with the apathy that ADC patients display; at times it can lead to frustration and unrealistic expectations in terms of rehabilitation at least in the early stages of treatment. A more complete discussion of the role of palliative management can be found in the monograph by Maddocks et al. (2005).

References

Anthony IC, Ramage S, Brannan FW, et al (2004). Premature neurodegeneration, inflammation and opiate misuse in HIV/AIDS before and after HAART. J Neurovirol 10(Suppl 3): 52.

Aquaro S, Perno CF, Balestra E, et al (1997). Inhibition of replication of HIV in primary monocyte/macrophages by different antiviral drugs and comparative efficacy in lymphocytes. J Leukoc Biol 62(1): 138–143.

Bacellar H, Munoz A, Miller EN, et al (1994). Temporal trends in the incidence of HIV-1 related neurologic diseases: Multicentre AIDS Cohort Study 1985–1992. Neurology 44: 1892–1900.

Bachis A, Major EO, Mocchetti I (2003). Brain-derived neurotrophic factor inhibits human immunodeficiency virus-1/gp120-mediated cerebellar granule cell death by preventing gp120 internalization. J Neurosci 23(13): 5715–5722.

Bagasra O, Lavi E, Bobroski L, et al (1996). Cellular reservoirs of HIV-1 in the central nervous system of infected individuals: identification by the combination of in situ polymerase chain reaction and immunochemistry. AIDS 10(6): 573–586.

Bell JE, Busuttil A, Ironside JW, et al (1993). Human immunodeficiency virus and the brain: investigation of virus load and neuropathologic changes in pre-AIDS subjects. J Infect Dis 168: 818–824.

Boccellari AA, Dilley JW (1992). Management and residential placement problems of patients with HIV related cognitive impairment. Hosp Community Psych 43: 32–37.

Bouwman FH, Skolasky R, Hes D, et al (1998). Variable progression of HIV-associated dementia. Neurology 50: 1814–1820.

Brew BJ (2001). HIV Neurology. Contemporary Neurology Series. Oxford University Press, New York, pp. 53–90.

Brew BJ (2004). Evidence for a change in AIDS dementia complex (ADC) in the era of highly active antiretroviral therapy and the possibility of new forms of ADC. AIDS 18(Suppl 1): S75–S78.

Brew BJ, Miller J (1996). Human immunodeficiency virus type 1 related transient neurologic deficits. Am J Med 101(3): 257–261.

Brew BJ, Bhalla RB, Paul M, et al (1990). Cerebrospinal fluid concentrations of neopterin in human immunodeficiency virus infection. Ann Neurol 28: 556–560.

Brew BJ, Bhalla RB, Paul M, et al (1992). Cerebrospinal fluid β_2 microglobulin in patients with AIDS dementia complex: an expanded series including response to zidovudine treatment. AIDS 6: 461–465.

Brew BJ, Paul M, Khan A, et al (1994). Cerebrospinal fluid HIV-1 p24 antigen and culture: sensitivity and specificity for AIDS dementia complex. J Neurol Neurosurg Psych 57: 784–789.

Brew BJ, Rosenblum M, Cronin K, et al (1995). The AIDS dementia complex and human immunodeficiency virus type 1 brain infection: clinical–virological correlations. Ann Neurol 38: 563–570.

Brew BJ, Evans L, Byrne C, et al (1996). The relationship between AIDS dementia complex and the presence of macrophage tropic and non syncytium inducing isolates of human immunodeficiency virus type 1 in the cerebrospinal fluid. J Neurovirol 2: 152–157.

Brew BJ, Pemberton L, Cunningham P, et al (1997). Levels of HIV-1 RNA correlate with AIDS dementia. J Infect Dis 175: 963–966.

Budka H, Wiley CA, Kleihues P, et al (1991). HIV-associated disease of the nervous system: review of nomenclature and proposal for neuropathology based terminology. Brain Pathol 1: 143–152.

Cartier L, Hartley O, Dubois-Dauphin M, et al (2005). Chemokine receptors in the central nervous system: role in brain inflammation and neurodegenerative diseases. Brain Res Brain Res Rev 48(1): 16–42.

Chang L, Ernst T, Leonido-Yee M, et al (1999). Cerebral metabolites correlate with clinical severity of HIV-cognitive motor complex. Neurology 52: 100–108.

Chang L, Ernst T, Witt MD, et al (2003). Persistent brain abnormalities in antiretroviral-naive HIV patients 3 months after HAART. Antivir Ther 8(1): 17–26.

Chiesi A, Vella S, Dally LG, et al (1996). Epidemiology of AIDS dementia complex in Europe. HNRC Group. HIV Neurobehavioral Research Center. J Acq Imm Def Syn Hum Retrov 11: 39–44.

Chiodi F, Albert J, Olausson E, et al (1988). Isolation frequency of human immunodeficiency virus from cerebrospinal fluid and blood of patients with varying severity of HIV infection. AIDS Res Hum Retrov 4: 351–358.

Cinque P, Vago L, Mengozzi M, et al (1998). Elevated cerebrospinal fluid levels of monocyte chemotactic protein-1 correlate with HIV-1 encephalitis and local viral replication. AIDS 12(11): 1327–1332.

Conant K, McArthur JC, Griffin DE, et al (1999). Cerebrospinal fluid levels of MMP-2, 7 and 9 are elevated in association with Human Immunodeficiency Virus Dementia. Ann Neurol 46: 391–398.

Cunningham P, Smith D, Satchell C, et al (2000). Evidence for independent development of reverse transcriptase inhibitor resistance patterns in the cerebrospinal fluid. AIDS 14(13): 1949–1954.

Currie J, Benson E, Ramsden B, et al (1988). Eye movement abnormalities as a predictor of the acquired immunodeficiency syndrome dementia complex. Arch Neurol 45: 949–953.

Cysique LA, Perdices M, Maruff P, et al (2004a). Prevalence and pattern of neuropsychological impairment in HIV/AIDS infection across pre-HAART and HAART eras: a combined study of 2 cohorts. J Neurovirol 10: 350–357.

Cysique LA, Maruff P, Brew BJ (2004b). Antiretroviral therapy in HIV infection: are neurologically active drugs important? Arch Neurol 61: 1699–1704.

Cysique LA, Bain MP, Wright E, et al (2007). Changes to the neuropsychological profile of AIDS dementia complex across pre-HAART and HAART eras and its relation to plasma and cerebropsinal fluid markers of

virological and immunological activity. J Neurol, (in press).
Dal Pan GJ, McArthur JH, Aylward E, et al (1992). Patterns of cerebral atrophy in HIV-1-infected individuals: results of a quantitative MRI analysis. Neurology 42(11): 2125–2130.
Dal Pan GJ, Glass JD, McArthur JC (1994). Clinicopathological correlations of HIV-1 associated vacuolar myelopathy: an autopsy based case control study. Neurology 44: 2159–2164.
Di Rocco A, Werner P, Bottiglieri T, et al (2004). Treatment of AIDS-associated myelopathy with L-methionine: a placebo-controlled study. Neurology 63(7): 1270–1275.
Dore GJ, Hoy JF, Mallal SA, et al (1997). Trends in incidence of AIDS illnesses in Australia from 1983 to 1994: the Australian AIDS Cohort. J Acquir Immun Defic Syndr Hum Retrovirol 6(1): 39–43.
Dore G, Correll P, Kaldor J, et al (1999). Changes to the natural history of AIDS dementia complex in the era of HAART. AIDS 13: 1249–1253.
Dore GJ, McDonald A, Li Y, et al (2003). Marked improvement in survival following AIDS dementia complex in the era of highly active antiretroviral therapy. AIDS 17(10): 1539–1545.
Esiri MM, Morris CS (1996). The cellular basis of HIV infection of the CNS and the AIDS dementia complex: oligodendrocyte. In: RW Price, JJ Sidtis (Eds.), The Cellular Basis of Central Nervous System HIV-1 Infection and the AIDS Dementia Complex. Haworth Press, New York, pp. 133–160.
Everall IP (1995). Neuropsychiatric aspects of HIV infection. J Neurol Neurosurg Psych 58: 399–402.
Fine SM, Angel RA, Perry SW, et al (1996). Tumor necrosis factor α inhibits glutamate uptake by primary human astrocytes: implications for pathogenesis of HIV-1 dementia. J Biol Chem 271: 15303–15306.
Foudraine NA, Hoetelmans RM, Lange JM, et al (1998). Cerebrospinal-fluid HIV-1 RNA and drug concentrations after treatment with lamivudine plus zidovudine or stavudine. Lancet 351: 1547–1551.
Gallo P, Frei K, Rordorf C, et al (1989). Human immunodeficiency virus type 1 (HIV-1) infection of the central nervous system: an evaluation of cytokines in cerebrospinal fluid. J Neuroimmunol 23: 109–116.
Gartner S (2000). HIV infection and dementia. Science 287 (5453): 602–604.
Gelman BB (1993). Diffuse microgliosis associated with cerebral atrophy in the acquired immunodeficiency syndrome. Ann Neurol 34: 65–70.
Gelman BB, Guinto FC (1992). Morphometry, histopathology and tomography of cerebral atrophy in the acquired immunodeficiency syndrome. Ann Neurol 32: 31–40.
Giometto B, An SF, Groves M, et al (1997). Accumulation of β amyloid precursor protein in HIV encephalitis: relationship with neuropsychological abnormalities. Ann Neurol 42: 34–40.
Gisslen M, Norkrans G, Svennerholm B, et al (1997). The effect on human immunodeficiency virus type 1 RNA levels in the cerebrospinal fluid after initiation of zidovudine or didanosine. J Infect Dis 175: 434–437.
Glass JD, Wesselingh SL, Selnes OA, et al (1993). Clinical–neuropathologic correlation in HIV-associated dementia. Neurology 43: 2230–2237.
Goodin DS, Aminoff MJ, Chernoff DN, et al (1990). Long latency event-related potentials in patients infected with human immunodeficiency virus. Ann Neurol 27: 414–419.
Gray F, Adle-Biassette H, Chretien F, et al (2001). Neuropathology and neurodegeneration in human immunodeficiency virus infection. Pathogenesis of HIV-induced lesions of the brain, correlations with HIV-associated disorders and modifications according to treatments. Clin Neuropathol 20(4): 146–155.
Gray F, Chretien F, Vallat-Decouvelaere AV, et al (2003). The changing pattern of HIV neuropathology in the HAART era. J Neuropathol Exp Neurol 62(5): 429–440.
Grimaldi LM, Martino GV, Franciotta DM, et al (1991). Elevated alpha-tumor necrosis factor levels in spinal fluid from HIV-1-infected patients with central nervous system involvement. Ann Neurol 29(1): 21–25.
Heyes MP, Brew BJ, Martin A, et al (1991). Cerebrospinal fluid quinolinic acid concentrations are increased in acquired immune deficiency syndrome. Ann Neurol 29: 202–209.
Jarvik JG, Hesselink JR, Kennedy C, et al (1988). Acquired immunodeficiency syndrome. Magnetic resonance patterns of brain involvement with pathologic correlation. Arch Neurol 45(7): 731–736.
Kieburtz KD, Giang DW, Schiffer RB, et al (1991). Abnormal vitamin B12 metabolism in human immunodeficiency virus infection. Association with neurological dysfunction. Arch Neurol 48: 312–314.
Kim RB, Fromm MF, Wandel C, et al (1998). The drug transporter P glycoprotein limits oral absorption and brain entry of the HIV-1 protease inhibitors. J Clin Invest 101: 289–294.
Konig J, Nies AT, Cui Y, et al (1999). Conjugate export pumps of the multidrug resistance protein (MRP) family: localization, substrate specificity, and MRP2-mediated drug resistance. Biochim Biophys Acta 1461: 377–394.
Koralnik IJ, Beaumanoir A, Hausler R, et al (1990). A controlled study of early neurologic abnormalities in men with asymptomatic human immunodeficiency virus infection. N Engl J Med 323(13): 864–870.
Krathwohl MD, Kaiser JL (2004). HIV-1 promotes quiescence in human neural progenitor cells. J Infect Dis 190 (2): 216–226.
Langford D, Grigorian A, Hurford R, et al (2004). Altered P-glycoprotein expression in AIDS patients with HIV encephalitis. J Neuropathol Exp Neurol 63(10): 1038–1047.
Lipton SA, Gendelman HE (1995). Dementia associated with the acquired immunodeficiency syndrome. N Engl J Med 332(14): 934–940.
Lori F, Malykh A, Cara A, et al (1994). Hydroxyurea as an inhibitor of human immunodeficiency virus type 1 replication. Science 266: 801–805.
Maddocks I, Waddy H, Brew BJ (2005). Palliative Management in Neurological Disorders: A Practical Guide. Cambridge University Press, London.

Marder K, Albert SM, McDermott MP, et al (2003). Interrater reliability of a clinical staging of HIV-associated cognitive impairment. Neurology 60: 1467–1473.

Marshall DW, Brey RL, Cahill WT, et al (1988). Spectrum of cerebrospinal fluid findings in various stages of human immunodeficiency virus infection. Arch Neurol 45: 954–958.

Martin C, Sonnerborg A, Svensson JO, et al (1999). Indinavir-based treatment of HIV-1 infected patients: efficacy in the central nervous system. AIDS 13: 1227–1232.

Maruff P, Currie J, Malone V, et al (1994). Neuropsychological characterisation of the AIDS dementia complex and rationalisation of a test battery. Arch Neurol 51: 689–695.

Maschke M, Kastrup O, Esser S, et al (2000). Incidence and prevalence of HIV-associated neurological disorders since the introduction of highly active antiretroviral therapy (HAART). J Neurol Neurosurg Psych 69(3): 376–380.

Masliah E, Heaton RK, Marcotte TD, et al (1997). Dendritic injury is a pathological substrate for human immunodeficiency virus related cognitive disorders. Ann Neurol 42 (6): 963–972.

McArthur JC, Hoover DR, Bacellar H, et al (1993). Dementia in AIDS patients: incidence and risk factors. Multicenter AIDS Cohort Study. Neurology 43: 2245–2252.

McArthur JC, McClernon DR, Cronin MF, et al (1997). Relationship between human immunodeficiency virus-associated dementia and viral load in cerebrospinal fluid and brain. Ann Neurol 42: 689–698.

McArthur JC, McDermott MP, McClernon D, et al (2004). Attenuated central nervous system infection in advanced HIV/AIDS with combination antiretroviral therapy. Arch Neurol 61(11): 1687–1696.

Mirsattari SM, Roke Berry ME, Holden JK, et al (1999). Paroxysmal dyskinesias in patients with HIV infection. Neurology 52: 109–114.

Mizuno T, Kawanokuchi J, Numata K, et al (2003). Production and neuroprotective functions of fractalkine in the central nervous system. Brain Res 979(1–2): 65–70.

Morgello S, Mahboob R, Yakoushina T, et al (2002). Autopsy findings in a human immunodeficiency virus-infected population over 2 decades: influences of gender, ethnicity, risk factors, and time. Arch Pathol Lab Med 126(2): 182–190.

Moses AV, Stenglein SG, Nelson JA (1996). HIV infection of the brain microvasculature and its contribution to the AIDS dementia complex. J NeuroAIDS 1: 85–100.

Nath A, Jankovic J (1989). Motor disorders in patients with human immunodeficiency virus infection. In: H Rotterdam, SC Sommers, P Racz, PR Meyer (Eds.), Progress in AIDS Pathology Vol. 1. Field and Wood Medical, New York, pp. 159–166.

Navia B, Jordan BD, Price RW (1986). The AIDS dementia complex: I. Clinical features. Ann Neurol 19: 517–524.

Nuwer MR, Miller EN, Visscher BR, et al (1992). Asymptomatic HIV infection does not cause EEG abnormalities: results from the Multicentre AIDS Cohort Study (MACS). Neurology 42: 1214–1219.

Pemberton L, Brew BJ (2001). CSF S-100b and its relationship with AIDS dementia complex. J Clin Virol 22: 249–253.

Petito CK, Cash KS (1992). Blood brain barrier abnormalities in the acquired immunodeficiency syndrome: immunohistochemical localization of serum proteins in postmortem brain. Ann Neurol 32: 658–666.

Petito CK, Navia BA, Cho ES, et al (1985). Vacuolar myelopathy pathologically resembling subacute combined degeneration in patients with acquired immunodeficiency syndrome. N Engl J Med 312: 874–879.

Portegies P, Enting RH, de Gans J, et al (1993). Presentation and course of AIDS dementia complex: 10 years of follow-up in Amsterdam, The Netherlands. AIDS 7: 669–675.

Power C, Kong PA, Crawford TO, et al (1993). Cerebral white matter changes in acquired immunodeficiency syndrome dementia: alterations of the blood brain barrier. Ann Neurol 34: 339–359.

Power C, Selnes OA, Grim JA, et al (1995). HIV dementia scale: a rapid screening test. J Acquir Immun Defic Syndr Hum Retrovirol 8: 273–278.

Price RW, Brew BJ (1988). The AIDS dementia complex. J Infect Dis 158: 1079–1083.

Rosenblum M, Scheck AC, Cronin K, et al (1989). Dissociation of AIDS related vacuolar myelopathy and productive HIV-1 infection of the spinal cord. Neurology 39: 892–896.

Rottenberg DA, Moeller JR, Strother SC, et al (1987). The metabolic pathology of the AIDS dementia complex. Ann Neurol 22: 700–706.

Rottnek M, Di Rocco A, Laudier D, et al (2002). Axonal damage is a late component of vacuolar myelopathy. Neurology 58: 479–481.

Saavedra-Lozano J, McCoig CC, Cao Y, et al (2004). Zidovudine, lamivudine, and abacavir have different effects on resting cells infected with human immunodeficiency virus in vitro. Antimicrob Agents Chemother 48(8): 2825–2830.

Sanders VJ, Pittman CA, White MG, et al (1998). Chemokines and receptors in HIV encephalitis. AIDS 12(9): 1021–1026.

Schielke E, Tatsch K, Pfister HW, et al (1990). Reduced cerebral blood flow in early stages of human immunodeficiency virus infection. Arch Neurol 47(12): 1342–1345.

Selnes OA, Galai N, Bacellar H, et al (1995). Cognitive performance after progression to AIDS: a longitudinal study from the multicenter AIDS cohort study. Neurology 45: 267–275.

Sevigny JJ, Chin SS, Milewski Y, et al (2005). HIV encephalitis simulating Huntington's disease. Mov Disord 20(5): 610–613.

Shen R, Alam A, Zhang Y (1988). Inhibition of GTP cyclohydrolase I by pterins. Biochem Biophys Acta 965: 9–15.

Si Q, Cosenza M, Kim MO, et al (2004). A novel action of minocycline: inhibition of human immunodeficiency virus type 1 infection in microglia. J Neurovirol 10(5): 284–292.

Sidtis JJ, Gatsonis C, Price RW, et al (1993). Zidovudine treatment of the AIDS dementia complex: results of a placebo controlled trial. AIDS Clinical Trials Group. Ann Neurol 33: 343–349.

Sinclair E, Gray F, Giardi A, et al (1994). Immunohistochemical changes and PCR detection of HIV provirus DNA in brains of asymptomatic HIV positive patients. J Neuropathol Exp Neurol 53: 43–50.

Sweeney JA, Brew BJ, Keilp JG, et al (1991). Pursuit eye movement dysfunction in HIV-1 seropositive individuals. J Psych Neurosci 16(5): 247–252.

Takahashi K, Wesselingh SL, Griffin DE, et al (1996). Localization of HIV-1 in human brain using polymerase chain reaction/in situ hybridization and immunohistochemistry. Ann Neurol 39: 705–711.

Thompson KA, McArthur JC, Wesselingh SL (2001). Correlation between neurological progression and astrocyte apoptosis in HIV-associated dementia. Ann Neurol 49(6): 745–752.

Tornatore C, Chandra R, Berger JR, et al (1994). HIV-1 infection of subcortical astrocytes in the pediatric central nervous system. Neurology 44: 481–487.

Tyor WR, Glass JD, Griffin JW, et al (1992). Cytokine expression in the brain during the acquired immune deficiency syndrome. Ann Neurol 31: 349–360.

Weidenheim KM, Ausubel M, Lyman WD (1996). Quantification of microglia in vacuolar myelopathy is a more sensitive measure of pathology than changes in myelin protein immunoreactivity. J Neuropathol Exp Neurol 55: 660.

Wesselingh SL, Power C, Glass JD, et al (1993). Intracerebral cytokine messenger RNA expression in acquired immunodeficiency syndrome dementia. Ann Neurol 33(6): 576–582.

Wilkins JW, Robertson KR, Snyder CR, et al (1991). Implications of self reported cognitive and motor dysfunction in HIV positive patients. Am J Psych 148: 641–643.

Zhang K, McQuibban GA, Silva C, et al (2003). HIV-induced metalloproteinase processing of the chemokine stromal cell derived factor-1 causes neurodegeneration. Nat Neurosci 6(10): 1064–1071.

Zink MC, Clements JE (2002). A novel simian immunodeficiency virus model that provides insight into mechanisms of human immunodeficiency virus central nervous system disease. J Neurovirol 8(Suppl 2): 42–48.

Handbook of Clinical Neurology, Vol. 85 (3rd series)
HIV/AIDS and the Nervous System
P. Portegies, J. R. Berger, Editors

Chapter 7

Neurocognitive assessment of persons with HIV disease

SHARRON DAWES AND IGOR GRANT*

HIV Neurobehavioral Research Center, University of California, San Diego, CA, USA

7.1. Introduction

The role of neuropsychological assessment is to examine change in cognition and emotion in individuals after brain dysfunction related to trauma, psychiatric/neurological disorders or general medical conditions (Franzen, 2000). Within the area of HIV/AIDS research, neuropsychological assessment seeks to identify cognitive strengths and weaknesses, with relevance for treatment planning or monitoring as well as identification of the early symptomatology of HIV-related cognitive deficits (Orr and Pinto, 1993).

Brain pathology can produce variable behavioral impairments whose accurate delineation can be helpful in diagnosis, assessment of treatment outcomes and research. The behaviors whose alteration most often suggests central nervous system disturbance are usually termed neurocognitive functions and abilities. Examples of neurocognitive processes include: psychomotor speed and attentional capabilities; neurocognitive abilities are often grouped in terms of verbal-language skills; executive functioning and abstraction ability; complex perceptual motor skills; memory; motor skills; and sensory functions.

Systematic assessment of these various functions and abilities can be accomplished through neuropsychological testing. Such tests involve use of stimuli and procedures that engage the neurocognitive function or ability in question in a manner that is valid and reliable. For some tests, demographically adjusted age-education (and in some cases, racial-, gender- and language-based) norms have been developed, which permit classification of a person's performance into normal versus abnormal ranges.

In this chapter, we review some of the common neuropsychological testing approaches to the various functions and abilities as they apply in the context of HIV infection. In each section, we provide a brief definition of the neuropsychological construct, review some of the tests commonly used and discuss how such tests have been applied in studies with HIV disease. Most studies of HIV/AIDS-infected individuals give a comprehensive neuropsychological battery that covers a number of domains. Therefore, some studies are discussed in a number of areas.

7.2. General intellectual abilities

7.2.1. Definition of the construct

The concept of intelligence evolved from attempts to capture diverse aspects of cognitive adaptation into a summary measure. Commonly intelligence has been subdivided into two clusters: crystallized intelligence, which embraces well-learned skills and abilities, e.g. knowledge of facts or concepts; and fluid intelligence, thought to represent the ability to address novel problems. Over the past century, various standardized tests have evolved. Some of these had enough data to propose norms (e.g. intelligence quotient (IQ) scores) based on the distribution of results from large population studies allowing more appropriate scoring.

Generalized intelligence is not a unitary concept but is a multiplicity of abilities and the traditional measures are influenced by quality of education, age, cultural background and gender. IQ scores tend to reflect the level of achievement attained at school unless the scores are demographically adjusted (Spreen and Strauss, 1998). Testing conditions can also influence results, e.g. poor rapport, the room being too cold or if a fire alarm sounds during testing may all lead to artificially poorer results. Despite a number of drawbacks, IQ

*Correspondence to: Igor Grant, MD, FRCP (C), Department of Psychiatry, University of California, San Diego, 9500 Gilman Drive, La Jolla, CA, 92093-0680, USA. E-mail: igrant@ucsd.edu, Tel: +1-858-534-3652, Fax: +1-858-534-7723.

measures can be helpful in interpreting the results of other neuropsychological tests. For example, people with low IQ may have difficulty with tests of verbal learning, and such difficulty should not be interpreted as meaning that there is a new, acquired brain disease.

7.2.2. Review of the tests

Tests that measure the construct of intelligence may be based on one measure, usually verbal or visual, or they may be based on a variety of measures that are then joined in a composite "intelligence" score or quotient. Some of the verbally based tests are the Shipley Institute of Living Scale (Zachary, 1991), which has two subtests, a multiple choice vocabulary test and a sequencing task; and the Peabody Picture Vocabulary Test (Dunn and Dunn, 1997), where the person matches pictures to words read by the examiner. A popular visually based intelligence test is Raven's Standard Progressive Matrices where in completing 60 items, the person matches one of six or eight choices to a pattern that is missing a piece. This test is less language, educationally and racially biased and might therefore be useful when testing persons with lower education attainment, who speak English as a second language or who are from a different culture.

The Wechsler Scales, including the original, revised and third editions, as well as the children's versions, are some of the most common tests in use today, in many different settings (Sharpley and Pain, 1988; Piotrowski and Keller, 1989; Butler et al., 1991; Lees-Haley et al., 1996; Sullivan and Bowden, 1997; Camara et al., 2000). The Wechsler Adult Intelligence Scale, third edition (The Psychological Corporation, 1997a), the latest incarnation of the adult intelligence test, consists of 13 subtests, which measure a variety of cognitive abilities. The test is able to yield a number of scores. The total score overall is the Full Scale IQ, which gives a performance level of the person across the whole test. The next two IQ scores that may be calculated are made up of either the verbally based subtests (Verbal IQ) or the visually based subtests (Performance IQ). These subtests and their description, according to the WAIS-III Administration and Scoring manual (WAIS-III; Psychological Corporation, 1997a), include: Vocabulary (respondent gives oral dictionary-like definitions of words), Similarities (explanation of how two words are alike), Information (general knowledge questions, orally administered and answered about common events, objects, places and people), Comprehension (situations and scenarios given orally to assess respondents' understanding of everyday mores and solutions to problems), Arithmetic (orally presented arithmetic questions that the person mentally calculates), Digit Span (series of number sequences that the respondent repeats verbatim for the forwards trials and in reverse for the backwards trials), Letter-Number Sequencing (a series of number and letter sequences that the person must rearrange and orally give back ordered numbers first in ascending order and then letters in alphabetical order and is only on the third edition), Picture Arrangement (respondent organizes a mixed-up series of cartoon-like picture cards of scenarios into logical order), Picture Completion (set of colored pictures that have an important piece missing which the respondent orally tells the examiner what it is), Block Design (respondent rearranges two-tone blocks to match patterns on stimulus card), Matrix Reasoning (new to the third edition; consists of incomplete grid patterns where the person must choose which of five possible choices fits the pattern to complete the grid), Symbol Search (also new to the third edition; requires a respondent to verify sets of symbols as to whether they contain one of two target symbols) and Digit Symbol-Coding (where respondent matches symbols to numbers based on a key of paired numbers and symbols). Supplementing the traditional Full Scale IQ, Verbal and Performance IQ are several factor or Index scores. These factor indices include Verbal Comprehension (VC, composed of Vocabulary, Similarities and Information), Working Memory (WM, which includes Arithmetic, Digit Span and Letter-Number Sequencing), Perceptual Organization (PO, made up of Picture Completion, Block Design and Matrix Reasoning) and Processing Speed (PS, including Digit Symbol and Symbol Search). Administration of the test in its entirety is not necessary to generate the IQ Composites and/or the factor scores. Nonetheless, it is necessary to administer all 13 subtests if both the Verbal and Performance IQs and the Index scores are desired. The children's scales are similar but with tests that are modified to be more appropriate for children according to their age or developmental capacities. These tests have been used for many years in many different settings.

Raven's Colored Progressive Matrices is a test of reasoning in visual modality that can be utilized to derive an intelligence score based on a regression equation. This test may be more "culture-free" as it is presented in a non-verbal format; however, extensive cross-cultural validations have not been completed. This test has 60 items in 5 sets (12 items per set) and the respondent must work out the rule related to the series and then choose the next progression for the series. The problems get harder as the respondent progresses through the test. Each item has a pattern problem with one part removed and 6–8 inserts of which the person has to choose the correct insert.

Another of the intelligence batteries is the Kaufman assessment batteries (Kaufman and Kaufman, 1983), a battery of tests designed for children aged 2½–12½ years old. There are 16 subtests that cover the domains of intelligence, both verbal and visual and educational abilities. The measures of intelligence are determined by assessing simultaneous and sequential processing abilities, and are reportedly more culture-free.

7.2.3. Application to HIV research

Most of the research using intelligence measures related to HIV has been with the Wechsler scales. These scales for both children and adults were used in a number of different studies involving those infected with HIV.

The impact of HIV on overall intelligence is difficult to evaluate because many studies purposely attempt to match HIV-positive and HIV-negative subjects at least on verbal intelligence in order to determine "premorbid" (i.e. pre-infection) cognitive status from post-infection. Thus, most studies of intelligence in adults have focused more on associations of IQ with commonly co-occurring conditions, such as psychiatric and substance-use disorders. In this vein, Honn and Bornstein (2002) found that in 217 HIV-positive men, higher intelligence correlated negatively with stressful life events and depressive symptomatology, but positively correlated with social contacts. In another study assessing the impact of past alcohol abuse and HIV on neuropsychological function, it was found that those with past alcohol abuse and HIV infection were more likely to have lower Verbal IQ scores (Green et al., 2004). In a recent finding this was not supported, however (Young et al., 2005). Further, it was found that in those who were HIV positive and drug abusers, such substances, and not HIV, were associated with reduced intellectual abilities (Margolin et al., 1998).

In children, intelligence is often measured by the Wechsler Intelligence Scales for Children, either revised or third editions. In one study, findings showed that there was no difference between HIV positive and negative serostatus children on these measures of intelligence (Blanchette et al., 2002). However, it has been assessed that there was a relationship between intelligence and CD^4 counts, with significant differences noted in intelligence based on CD^4 counts. In this study those children with higher counts had corresponding lower intelligence (Brouwers et al., 1995). A larger study of 40 children (aged 3 to 5 years) was conducted to assess the cognitive effects of HIV in children (Fishkin et al., 2000). At the time of testing, with the revised preschool version of the Wechsler intelligence scales, 21 children had AIDS symptomatology including eight with sub-syndromic AIDS symptomatology, and eight asymptomatically HIV positive. Compared to a group of 40 control participants, no significant differences were found on any demographic variables or on overall IQ scores, even though both groups were approximately one standard deviation below the published norms for this test. The authors concluded that although there were no gross cognitive deficits found, this may be related to the age of the children (too young to discriminate deficits) or that the tests were not sensitive enough.

Another pediatric study examining the effects of HIV on brain development utilized the Kaufman ABC test (Diamond et al., 1987). In this study 12 children (median age of 5 years, 3 months; seven with AIDS-related complex and five with AIDS) were followed for 3 years with two being diagnosed as mildly retarded, six as borderline and four with average intelligence. Irrespective of the level of intellectual functioning, the children were found to have a pattern of selective impairment related to their visual abilities measured as part of their overall intellectual functioning.

However, there were some studies that used other methods of measuring overall intellectual abilities. These studies utilized Raven's Colored Progressive Matrices and found in both children infected with vertically transmitted HIV compared to their non-infected siblings (Bisiacchi et al., 2000), and in adult intravenous, drug use populations, with or without concurrent HIV, that there were no differences between these subjects on Raven's Progressive Matrices as a measure of intelligence (Pakesch et al., 1992a, b; Bono et al., 1996).

The Shipley Institute of Living Scale was used as part of a larger neuropsychological battery to determine impairment compared to brain atrophy as measured by an MRI scan to neuropsychological impairment (Di Sclafani et al., 1997). The findings from this study indicated that MRI-detectable brain atrophy that was related to HIV infection did not appear to be the main cause of progressive neuropsychological impairment in this group of 31 HIV-infected men.

Treatment studies are common in the era of highly active antiretroviral therapy (HAART) and with new antiretroviral medications. In an assessment of intelligence on intravenous drug users with the WAIS-R, it was found that Lamotrigine, a medication used to treat HIV neuropathic pain, did not affect cognition (Grassi et al., 1993). Four children (aged 2.5 to 12.5), were assessed longitudinally over the 12-month period after initialization of HAART to determine the effect of this treatment on their cognition (Toledo Tamula et al., 2003). It was found that although all four of these children had apparent stable disease and treatment states, there was a significant decline in neurocognitive functioning with a significant drop in IQ scores, as measured by the WISC-III or the MacCarthy Scales, within 6–12 months of starting HAART. Therefore, the authors concluded that the reduction in IQ scores, which was considered to be unusual after the commencement of HAART, was more likely to be related to the effects of HIV on brain development.

Most studies appear to show that there is possibly a detrimental effect on neurocognitive functioning that may be attributable to HIV, as shown particularly by the relationship between CD^4 counts and intelligence. However, most also indicate that other factors, including drug and alcohol abuse, seem to affect cognitive abilities more than HIV.

7.3. Premorbid functioning

7.3.1. Definition of the construct

Premorbid functioning is an estimate of one's cognitive functioning before injury, illness, trauma or any event that may have an effect on the person's brain (Lezak, 1995). Premorbid intellectual functioning influences performance on neuropsychological tests that are designed to detect effects of disease on the brain (such as HIV). Therefore, estimates of pre-illness abilities become important for interpreting neuropsychological data correctly.

7.3.2. Review of the tests

Most of the tests that are utilized to estimate premorbid functioning are very educationally biased and depend on the level of educational attainment of the person being tested. However, while these tests and methods are usually the most resistant to brain dysfunction, they may disadvantage those who are not in the mainstream socioeconomic groups, e.g. in the USA: middle-class Caucasians, with at least a high-school level of education (Groth-Marnat, 1997). There is little empirical evidence to suggest that premorbid measures are able to accurately predict functioning before illness onset in those people with very high or very low premorbid attainment. Because such tests are usually verbally based, they may disadvantage those people who have low educational levels for reasons other than inability to perform at school; furthermore there are often no ethnicity-based adjustments. However, these tests and methods have often shown that those with lowered cognitive reserves based on premorbid estimates usually do worse on other cognitive tests than those with higher cognitive reserves (Stern et al., 1996).

There are four common methods of assessing premorbid functioning (Spreen and Strauss, 1998). These are the best-performance method, the use of "hold" tests, tests of over-learned skills and demographic equations.

The best-performance method utilizes the highest test score in a person's result as the premorbid estimate, and therefore as the benchmark to which to compare all other scores. For example, if an individual scored at one standard deviation above the mean on the Object Assembly test on the WAIS-R and this was the highest score, then all other scores would be compared to the Object Assembly score as a benchmark for premorbid abilities. If this same person's scores on all other subtests of the WAIS-R fell at the 50th percentile (i.e. he/she did as well as or better than 50% of his/her age, education-matched peers), in comparison to the Object Assembly subtest score, this may indicate a decline in the individual's cognitive abilities. The best-performance method is considered a poor method of estimating premorbid skills as the score may reflect an area of cognitive strength for the person, therefore overestimating the person's abilities with no individual test performance variability taken into account. This method has little empirical validation supporting it.

The second method is the use of so called "hold" tests which are usually thought to be relatively insensitive to brain injury. Most of these tests are verbally based with some common examples being the Information subtest from the WAIS-III where the person is required to verbally answer general knowledge questions about famous events and people, or the Vocabulary subtest from the WAIS-III where the person is required to give oral dictionary-like definitions of words. This was suggested as part of the National Institute of Mental Health (NIMH) battery for the assessment of premorbid abilities in HIV/AIDS, as well as one of the tests in the following section of over-learned skills, the National Adult Reading Test (Butters et al., 1990). While such tests are less likely to be impacted by brain dysfunction, unless there is a lesion in the areas of the brain mediating language, they may underestimate premorbid abilities of poorly educated persons, or persons from different cultural backgrounds.

The third method of assessing premorbid abilities is that of utilizing over-learned skills, usually reading or other academic skills. These tests may be tests of reading irregularly spelt English words like drachm (pronounced "dram") or regular reading tests where the words become harder as the test progresses (e.g. "in" through to "contagious"). The person is given a card with the words on it and they simply read it back to the examiner. Their score is the number of words pronounced correctly, which is then placed into a regression equation usually including demographic variables which gives an IQ score. The tests of irregularly pronounced English words have many versions including the National Adult Reading Test (NART), National Adult Reading Test-2 (NART-2), American National Adult Reading Test (AMNART) and the North American Adult Reading Test (NAART). The version used depends on the country in which it is being used; for example, Australia uses the NART-2 where as the NAART and AMNART are more popular in the USA. Each version has slightly different words included. Achievement tests are usually measures of how well a person performed in school. They can be used as general screening measures to assess for difficulties that the person may be having due to a learning disability that is not related to any distinct cognitive impairment, or they may often be used in the assessment of premorbid abilities (Spreen and Strauss, 1998).

Some of the most popular regular reading tests to be used for this purpose are the Wide Range Achievement Test Third Edition — Reading (WRAT-3 Reading) and the new Wechsler scale reading test, the Wechsler Test of Adult Reading (The Psychological Corporation, 2001). Other achievement tests may be large batteries like the Woodcock-Johnson Psychoeducational Battery containing 12 standard tests and 10 extra for the extended version, or the Wechsler Individual Achievement Test (WIAT) which has nine subtests. Simpler tests like the Wide Range Achievement Tests, which have only three subtests, namely reading, spelling and arithmetic, are also utilized for this purpose. If both the WIAT and the WAIS-III are given, a discrepancy score can be generated between overall intelligence and educational achievement that can assist in the assessment of individuals markedly. These tests usually target specific areas of learning and when scored are usually able to give a score based on the level of actual attainment at school and based on the age of the person. These tests are usually easy to administer but can be time consuming depending on the amount of testing required. These tests are very useful as measures of premorbid functioning but still disadvantage those with learning disabilities like reading and spelling deficits, unrelated to acquired brain damage, as they may underestimate their levels of premorbid functioning.

The fourth method of estimating premorbid functioning utilizes regression equations that attempt to account for the effects of race, education and occupation, as these variables have been found to have the most influence on premorbid abilities. The main demographic equations are the Oklahoma Premorbid Intelligence Estimates (Krull et al., 1995), which utilizes both demographics and Wechsler intelligence scores, and the Barona adjustments (Barona et al., 1984), which only accounts for demographic variables. While this approach is arguably the most accurate, it shares with the other approaches limitations when dealing with educational extremes, or lack of acculturation to the dominant culture.

7.3.3. Application to HIV research

1. Best-performance method is rarely used in the literature and no research was found that utilizes this method with relation to individuals with HIV or AIDS.

2. Hold tests (Margolin et al., 2002) utilized the WAIS-R Information subtest score as an estimate of premorbid functioning. This test was used to estimate the premorbid level of educational attainment, which was then entered into a hierarchical multiple regression analysis in order to predict neuropsychological performance based on serostatus. They reported that although the majority of their 90 HIV-positive intravenous drug users reported 12 or more years of education, scores on the WAIS-R Information subtest suggested that premorbid educational attainment was generally below average. However, it was also noted that the Information subtest scores contributed significantly to the prediction of neuropsychological performance as a measure of actual educational achievement.

3. Over-learned skills: The NART has been used in four studies as a measure of premorbid abilities (Dana Consortium on the Therapy of HIV Dementia and Related Cognitive Disorders, 1996a, b; Lopez et al., 1998; Marder et al., 2003; Cysique et al., 2004). Benedict et al. (2000) have used NAART, which is the North American version of the NART and has some different words from the NART. The most popular test that has been used in this area appears to be the WRAT-3 Reading subtest. Moser et al. (2002) administered this subtest of the WRAT-3 to obtain a gross estimate of premorbid cognitive functioning in 25 schizophrenic and 25 HIV-positive men and women. In a comparison between HIV-infected patients and schizophrenic patients, there were no differences found between the two groups (Moser et al., 2002). Other studies have utilized the WRAT as part of a battery of tests that make up a global impairment index and have found that in dependent users of methamphetamine who have HIV, their impairment is greater than in those who have only methamphetamine dependence or HIV (Cherner et al., 2002; Rippeth et al., 2004). In a small sample of children with vertically transmitted HIV, who had been assessed using the WRAT (revised and third editions), it was found that they performed within the Low Average to Borderline ranges of functioning but that this was not different from their siblings who did not have HIV (Blanchette et al., 2002). Most of these studies utilize the premorbid estimate as a measure of comparing

groups' demographics on educational achievement or premorbid IQ as a baseline to compare to other measures of neuropsychological functioning.

4. Demographic equations. Researchers have found that in HIV, estimated premorbid intelligence calculated by demographic equations influences the rate of decline in cognitive functioning relative to IQ. Therefore, those with higher estimated premorbids have purportedly greater cognitive reserve capacities, which affects the rate of decline in their cognitive abilities related to disease progression (Basso and Bornstein, 2000).

In summary, the use of estimated premorbid abilities is consistently utilized as a method of determining overall cognitive functioning. Findings of estimated premorbid abilities within the HIV population are mixed. Many of the studies reviewed indicate that the level of premorbid attainment of people infected with HIV is lower than in those without HIV. However, there have been some studies that indicate that there is high variability in the premorbid functioning of those with HIV. Results indicate that some people, who are infected with the virus, premorbidly did function at the above average level.

7.4. Tests of specific ability areas

7.4.1. Attention and speed of information processing

7.4.1.1. Definition of the construct

There are many differing ways that attention may be described or broken into components. Attention is the ability to filter excess material and concentrate only on relevant sources balanced with the ability to move the focus of attention when required. Therefore, another presentation of respondents with attentional problems may be perseveration, or the inability to shift their attentional focus to other stimuli (Groth-Marnat, 1997). Others indicate that attention has three main processes, namely information selection, vigilance or maintenance of information and information processing or control (Parasuraman et al., 2000). However, attentional deficits are often one of the most frequent complaints in neuropsychology, and therefore there are many different tests used in the assessment of this construct.

7.4.1.2. Review of the tests

Types of tests used range from simple reaction time tests to complex measures using arithmetic and set shifting. The more complexity that is built into the tests, the more sensitive they are to impairment. Most of the tests are timed, and therefore they are tests of reaction time or speed as well as attention. Approaches include recall of increasing numbers of digits, letters or divided attentional tasks (Spreen and Strauss, 1998).

The Wechsler tests of attention include Letter-Number Sequencing, which is a series of number and letter sequences that are given orally that the person must rearrange and orally report to the examiner, with the numbers first in ascending order and then letters in alphabetical order. Arithmetic is a subtest that ranges from counting of the number of blocks presented, to complex cognitive processing of verbally presented mathematical problems. Digit Span is the third of these tests and consists of two parts. The first is where the person repeats sequences of three to nine digits that are given orally verbatim, and the second part is where the person must repeat back digit strings of two to eight digits in length, in reverse order.

The Symbol Digit Modalities Test (Smith, 1982) is similar to the Digit Symbol subtest from the Wechsler intelligence scales, but the key has the symbols at the top and numbers matched to the symbols. Therefore, the person has to fill in the numbers as quickly as possible in the first trial, and then read them out loud in the second trial. This test is therefore capable of removing the graphomotor component of the assessment.

In the Paced Auditory Serial Addition Test (Gronwall, 1977), the person must add each subsequent single-digit number to the previous single-digit number that has been given orally, stating the answer out loud, while attending to the next number to add. This test is used to assess the capacity and rate of information processing, and also the person's ability to both sustain and divide their attentional abilities.

The Visual Search and Attention Test (Trenerry et al., 1990) is utilized to assess visual scanning, and sustained attention in adults. It is essentially a cancellation task, and has four different trials where the person must scan the page and cancel out the target symbols as quickly as possible.

The Trail Making Tests (Army Individual Test Battery, 1944) are tests of speed and attention, sequencing and mental flexibility, as well as visual search and motor function. Part A — connect circles labeled 1–25 sequentially, as quickly as possible. Part B — connect circles alternating between numerals and letters in sequential order from 1-A-2-B-...13. Because of the "parallel processing" that is involved in the Trail B task, it is usually grouped with the tests of executive functioning.

7.4.1.3. Application to HIV research

HIV research has indicated that attention is often one of the cognitive areas that is affected as the disease process progresses. Reger et al. (2002) completed a meta-analysis of 39 studies related to the neuropsychological sequelae of HIV/AIDS published from 1984 through the first quarter of 2000. In this meta-analysis when attention/concentration was assessed, the main tests given in this area were Digit Span followed by Arithmetic, and then a Cancellation test (given by 54, 12 and 10% of studies reviewed, respectively). In another review of HIV neuropsychological features, simple attention is indicated to only become impaired in the later stages of dementia, delirium or acutely physically ill (Hinkin et al., 1998). These authors also indicate that HIV-infected patients often have difficulties on the PASAT and other tasks of divided attention, but that tasks of sustained concentration like the Continuous Performance Task and the first two trials of the Stroop are normally not impaired. A large study indicated that there were no significant differences between HIV-negative and HIV-positive participants on the WAIS-R attentional subtests of Digit Span and Arithmetic, nor the Rhythm, Speech Sounds Perception or Digit Vigilance–Errors tests from the expanded Halstead-Reitan Neuropsychological Test Battery (Heaton et al., 1995). However, they did find that participants with a history of AIDS-defining illness did perform much worse on the PASAT. Another of the Wechsler Intelligence scales subtests, Digit Symbol, has been shown as a sensitive measure of speed of processing and attention. Becker and colleagues, in their 1997 study, found that there were differences between those who were HIV positive, and those who were negative for the virus. Findings indicated that those who were HIV negative performed better than those who were HIV positive.

Many of the studies that involve tests of attention and speed of processing use these measures to predict performance on everyday activities like driving. The PASAT along with Digit Span and Arithmetic from the Wechsler Intelligence scales were used as measures of attention and speed of information processing in driving (Marcotte et al., 1999). In this study, the authors found that those with neuropsychological impairment performed significantly worse on the PASAT than the non-neuropsychologically impaired persons and that these people were also impaired in their performance on the driving simulator. Studies utilizing the Wechsler subtests have found differing results. While some authors have found no differences between those people infected with HIV and those who were not (Becker et al., 1997) others have found that some people do have deficits on these tests (Manly et al., 1997). The Digit Symbol subtest from the Wechsler scales appears to be good at detecting cognitive improvement correlated to increased CD^4 levels (Stern et al., 1995), and HAART treatment (Ferrando et al., 1998).

Two studies that used the PASAT as a measure of attention related the attentional performances of people to everyday stressors and their perceived level of illness-related disability. In the first study, poorer performances on the PASAT were associated with more negative stressors reported by patients (Pukay-Martin et al., 2003), and better performances in a path analysis contributed to less illness-related disability (Honn and Bornstein, 2002). Therefore, poorer attentional performances may be related to the negative impact of stress in a person's life, and people who perceive that they have less illness-related disability have better attentional abilities.

The Trail Making Test — Part A (TMT-A) has been utilized heavily in the HIV literature and sometimes in conjunction with the Digit Symbol subtest (Becker et al., 1997). Some of the other studies have utilized this test as a measure of processing speed and attentional capacity, with the Symbol Digit Modalities Test, when assessing the relationship to medication regime adherence. In this study it was found that people with better processing speed and attentional capacity are more likely to be adherent than those with poorer processing speed and attentional capacities (Hinkin et al., 2002). The TMT-A has also been used when assessing coping, finding that those who are able to cope better usually have better attentional and speed of processing abilities, and are less likely to revert to impulsive forms of coping (Manly et al., 1997). Other authors have utilized this test as an early indicator for global impairment, finding that attentional performance is one of the more sensitive measures in the indication of impending cognitive impairment in HIV (Sarter and Podell, 2000).

In an assessment of risk of dementia (Stern et al., 2001) attentional tasks like Digit Symbol Coding have been shown to be significantly worse in people not only currently diagnosed with dementia related to HIV but also in those who developed dementia later. Therefore, poorer performances on attentional tasks, especially Digit Symbol Coding, may be an early indicator again, but this time of dementia. The Visual Search and Attention Test indicated significant disturbances in attention in individuals who were HIV-positive drug-users with co-morbid depression (Váquez Justo et al., 2003).

Although many deficits are found in the area of attention and speed of information processing in the HIV/AIDS population, it is interesting to note that individuals' performances on specific tests of attention may not be impaired all of the time. It is also apparent that some tests are more indicative of a possible dementing process rather than others (e.g. Digit Symbol Coding). Deficits in attentional abilities may also be linked to poorer medication adherence and coping and increased perceived stress in the patient's life. It also appears that tests of simple attention are more preserved in HIV, and that impairment is usually noted first in the more demanding tasks of attention and speed, especially the dual tasks (e.g. PASAT).

7.4.2. Verbal language

7.4.2.1. Definition of the construct

This construct itself encompasses many differing aspects including verbal fluency (ability to express words orally), word knowledge, naming, reading, writing and comprehension of verbal material. The construct of verbal language abilities are most closely associated with the left hemisphere of the brain, in most humans, and therefore damage to this hemisphere may lead to deficits in this area.

7.4.2.2. Review of the tests

There are many different tests of verbal abilities; however, focus shall only be on those that have been utilized in HIV research. The first is the Boston Diagnostic Aphasia Exam (Goodglass and Kaplan, 1983), which has two parts. The first is an assessment of the person's conversational speed by the use of seven brief open-ended questions, which are scored on a 5-point aphasia rating scale. The second part is a formal assessment of a number of areas including auditory comprehension. The Boston Naming Test is one of the subtests of the Boston Diagnostic Aphasia Examination (Kaplan et al., 2001), and is often administered alone. This test assesses the person's ability to name 60 line drawings of objects that become increasingly harder. Performance is scored based on the number of items that are correct, with and without semantic cues.

The Controlled Oral Word Association Test (COWAT; Benton and Hamsher, 1976) is one of the most popular tests of verbal fluency. The person must generate as many words as possible (excluding proper nouns) starting with the letters F, A or S in turn, or alternatively C, F and L for English speakers and the letters P, M and R for Spanish speakers. For each letter, the person has 60 seconds to generate as many words as possible. This is completed over three one-minute trials with the dependent variable being the total number of words generated (Gladsjo et al., 1999). Other tests of verbal fluency include semantic naming, for example animals, furniture or fruit and vegetables.

The Peabody Picture Vocabulary Test (Revised or Third edition) is a test of vocabulary where the person is required to choose which one of four pictures matches the word that the examiner just said. This is a very good test of vocabulary for people who may not be able to give dictionary-like definitions of words (e.g. the WAIS-III Vocabulary subtest), but who have contextual knowledge of the word. This test is also available in Spanish for those who work with Spanish-speaking populations.

Another vocabulary test that does not rely on giving dictionary-like definitions is the Shipley Institute of Living Scale — Vocabulary subtest (Zachary, 1991), which is a multiple choice vocabulary test. The person is asked to choose which word from four options matches a target word. This test is also able to be used as a quick measure of intelligence, if the second subtest of verbal abstraction is given as well (administration takes no more than 20 minutes). This test has 20 items with each one paired in difficulty with two vocabulary items, where the person must fill in sequences (e.g., A, B, C, D, _).

Wechsler Intelligence scales have four tests of verbal abilities. They are Vocabulary, Similarities, Information and Comprehension (refer to Section 7.2.2 for descriptions).

The Token Test (De Renzi and Vignolo, 1962) is used to assess how well the person is able to follow commands of increasing difficulty. Respondents are given 20 plastic tokens that are of differing sizes and colors. They are given increasingly longer commands that they have to follow by arranging the tokens in particular ways.

7.4.2.3. Application to HIV research

In the meta-analytic study by Reger et al. (2002), the main verbal tests that were found to be prevalent in HIV literature were the Wechsler Intelligence Scale subtests of Comprehension, Information, Similarities and Vocabulary. Other common tests were the Boston Naming Test, Aphasia Screening Test, Speech Sounds Test, Shipley Vocabulary, Semantic Processing and the Receptive and Expressive Speech subtests from the Luria-Nebraska. In a study that utilized a number of these measures including the Boston Naming Test, the Controlled Oral Word Association Test (COWAT) and the Vocabulary subtest from the WAIS-R, it was found that those who were seronegative performed better on all measures of verbal language than those who were seropositive (Levin et al., 1992). Similarly, Stern et al. (1996) used the COWAT as a measure of verbal fluency, and established also that those who were seropositive performed worse than those who were seronegative. In a sample of HIV-positive drug users with and without depression, results indicated that those with depression performed worse on verbal fluency as measured by the COWAT. Green et al. (2004) found that overall, although people with HIV did not differ in their performance on verbal measures (WAIS-R verbal subtests) compared to those without HIV, a sub-sample of the those with HIV and past alcohol abuse performed worse on these measures compared to others with HIV but no alcohol abuse history. In an investigation of verbal abilities of children infected with HIV compared to their seronegative siblings, as measured by the Wechsler intelligence scale subtests of Vocabulary, Comprehension and Similarities and on the verbal fluency measure of the COWAT, the HIV-positive children were not impaired compared with the control group (Blanchette et al., 2002).

Verbal language scores may also be used as predictive measures of generalized neuropsychological abilities. For example, Meyerhoff et al. (1999) utilized the COWAT as well as the Shipley Institute of Living Scale: Vocabulary subtest, as measures of verbal language, finding that verbal abilities accounted for 13% of the variance for the differences in global scores in HIV-positive compared to HIV-negative participants, with the seropositive participants performing worse. Using the Boston Naming Test, WAIS-R Vocabulary and phonemic and category fluency as measures of verbal abilities, Carter et al. (2003) found that cognitive complaints were able to predict neuropsychological skills including the scores from the tests of verbal ability. This prediction was independent of mood and medical symptomatology, based on the HIV-positive participants' cognitive complaints. In an assessment of the predictive ability of neuropsychological tests on overall neuropsychological outcomes, the WAIS-R verbal subtests, which were included as a verbal abilities factor, contributed most to the prediction of overall cognitive outcomes in both HIV-positive and AIDS participants (Becker et al., 1997).

In summary, the verbal area does appear to be variably affected by HIV. Although some studies indicate that there are no differences between seronegative and seropositive groups on tests within this domain, others indicate that there are significant differences and that those who are seropositive are less likely to achieve the same results as their seronegative counterparts on these tests. Some of the literature indicates that the area of verbal fluency is often more affected than other verbal functioning areas, like naming, comprehension and reading, and verbal fluency appears to be the area that is most affected, suggesting a possible pattern of frontal-striatal deficits.

7.4.3. Executive-abstraction

7.4.3.1. Definition of the construct

Executive functioning and measures of abstract reasoning measure what used to be known as frontal lobe functioning. Therefore, this construct applies to the functions of switching set rapidly, learning new rules and the ability to control responses without perseveration (Spreen and Strauss, 1998). As this construct is a measure of a person's ability to essentially self-regulate and direct their own behavior, it has many subparts. These include: the ability to make informed choices, planning and being able to act purposively and effectively (Groth-Marnat, 1997). This may be difficult for some people with deficits in this area as some have very little insight into their difficulties, which may lead to social problems due to their disinheriting and inappropriate behavior. Some of the classic behaviors in people with this type of cognitive dysfunction according to Lezak (1995) are irritability, amotivation and lability, carelessness and rigidity in their thinking.

7.4.3.2. Review of the tests

There are many tests that purportedly measure these functions and some of the most widely used are tests of set shifting (Trail Making Test: Part B), rule learning (e.g. Wisconsin Card Sorting Test) and categorization of objects (e.g. Booklet Category Test and Delis-Kaplan Executive Function Scale: Sorting Task). Other tests that assess this type of functioning are reasoning tasks that may be verbal (Shipley Institute of Living Scale — Abstraction; WAIS-III — Similarities) or visual (WAIS-III: Matrix Reasoning; Raven's Progressive Matrices). All of these tasks are said to tap into the higher level reasoning abilities, but most are only able to be given once, otherwise the nature of the task is changed, especially for sorting tasks and reasoning tasks where the person is able to benefit from previous exposure.

The Wisconsin Card Sorting Test (WCST) is one of the most common tests that is used in this area and purportedly measures the ability to form abstract concepts, and learn and follow new rules by utilizing feedback when necessary. The person must sort cards into different categories (color, number, shape), where the rules or categories change without any warning or instruction. The original test has a total of 128 trials for the person to complete, while the shortened version has only 64 trials.

The Design Fluency Test requires one to draw as many abstract designs as possible within a time limit. This test was developed as an equivalent test to word fluency tasks but in the non-verbal domain (Jones-Gotman and Milner, 1977).

The Category Test is part of the Halstead-Reitan Neuropsychological Battery, and is purported to measure a person's abstraction abilities, how easily they form concepts and problem solve, and learn from previous experience (Spreen and Strauss, 1998). The person works through seven sets of items (208 individual images), which are organized, related to the number of objects, and their position, etc., on the page. The person points to the relevant number (1 through 4) on a card that is laid in front of them and is given feedback as to whether their choice was correct or incorrect.

The Stroop Color-Word Test (Golden, 1978) is made up of three trials, and is a measure of how well a person can suppress a habitual response for a novel one. In the first two trials, the person reads the names of colors and then the actual colors on the card. In the third trial, the person must ignore the word (name of a color) that is written, and report the color of the ink that the word is written in, therefore suppressing a habitual response for a novel one. For example, the word "blue" may be written in green ink and the correct response for this item would be "green". Each of the three trials is timed and the person must complete the task as quickly as possible.

The Trail Making Test — Part B (TMT-B) is one of the most commonly utilized tests in this area within HIV research (Reger et al., 2002). It is a test of mental flexibility, attention and sequencing (Spreen and Strauss, 1998), as described previously in the section on attentional abilities.

The Abstraction subtest of the Shipley Institute of Living Scale is a verbally based sequencing task where the person must complete as many of the 20 problems within a 10-minute time limit. This task is thought to measure verbal abstract reasoning abilities.

7.4.3.3. Application to HIV research

Reger et al. (2002) in their meta-analysis of the HIV research found that the most commonly given tests in this area of the 41 journal articles that they reviewed were: the TMT-B, Word Fluency, Category Fluency, Stroop Color-Word, Halstead Category Test and then the WCST.

The TMT-B is one of the most frequently used tests and has been applied in a number of papers within the HIV research. In some of its applications within this research it has been applied to studies assessing generalized cognitive impairment within HIV (Silberstein et al., 1993; Becker et al., 1997; Lopez et al., 1998), the progression to HIV-related dementia (Luo et al., 2003), and the effectiveness of HAART (Ferrando et al., 1998; Stankoff et al., 2001). It has also been used to evaluate cognitive functioning in longitudinal studies (Bono et al., 1996; Stankoff et al., 1999), and in the assessment of medication compliance (Hinkin et al., 2002), the effects of stress (Pukay-Martin et al., 2003), self-impairment ratings (Knippels et al., 2002) and drug use (Margolin et al., 2002). In many of these studies, performance on this frequently used test has been found to be impaired in the seropositive population.

Between a seropositive and control group, there were no differences on the performance on the WCST (Levin et al., 1992) but in a study of well-educated seropositive males, the cohort that exercised regularly scored less on the categories completed index of this test than those who did not (Honn et al., 1999). In a different HIV-positive cohort with or without alcohol abuse, those who abused alcohol performed worse than those who did not (Green et al., 2004). In another cohort of HIV-positive substance users, those with depression performed worse than those without depression on both the WCST and on the Stroop Color-Word Test (Váquez Justo et al., 2003). Carter et al. (2003) utilized the WCST as one measure of executive functioning in their calculation of a global clinical rating scale finding that it was sensitive to impairment related to HIV.

Another test that has been utilized as the measure of executive functioning is the Category Test from the Halstead-Reitan. It has been used in calculations of global impairment scores in a number of articles from the HIV Neurobehavioral Research Center, University of California, San Diego (e.g. Grant et al., 1987, 1988, 1990; Heaton et al., 1995; Dikmen et al., 1999; Cherner et al., 2002; Carey et al., 2004). These studies indicate that the Category Test performance in those with HIV is usually impaired compared to those who are seronegative.

The Stroop Color-Word Test has been used in a number of other studies, by itself and in conjunction with other tests. The Stroop, Shipley Institute of Living Scale: Abstraction and the Design Fluency tests were found not to discriminate between asymptomatic persons who were seropositive and those who were seronegative (Levin et al., 1992). In a study assessing the impact of methamphetamine abuse on those with HIV, it was found that those who abused methamphetamines performed worse on the Stroop Color-Word than those who did not (Rippeth et al., 2004), and in a study that assessed those who were employed versus those who were not, it was found that the unemployed participants performed worse on the response inhibition on Stroop Color-Word and the cognitive flexibility task TMT-B (van Gorp et al., 1999).

The prediction of the effects of HAART treatment on executive functioning were assessed by Tozzi et al. (1999) in their study with the Stroop Color-Word Test, and the TMT-B. They found significant improvements at both six and fifteen month follow-ups on both tests. However, previously in another study that compared subjects who were on HAART to those who were not, it had conflicting results with scores on the Stroop

Color-Word Test found to be no different between these groups (Ferrando et al., 1998).

Although, again, there are mixed findings in the literature, it seems that more often than not, research involving executive and abstraction tests usually finds deficits in this construct when the research is conducted with HIV seropositive participants. Therefore, this generalized finding leads again to the possibility of frontal-striatal impairment.

7.4.4. Complex perceptual motor

7.4.4.1. Definition of the construct

This construct is usually divided into three distinct areas, according to Groth-Marnat (1997). The first area is that of visuoperceptual abilities, and assesses the discernment of visual stimuli, color and recognition. The second area is visuospatial abilities relating to the orientation and location of points in space, as well as distance measurement and direction. The last area to be covered by this construct is that of visuomotor abilities, which covers not only manual dexterity and construction but also eye movement. Deficits within this cognitive construct may be within only one area or may fall into overlapping areas. It is generally thought that when constructional tasks show fragmentation and lack of organization with an apparent loss of the gestalt, then the right hemisphere is damaged, whereas left hemisphere injury is associated with intact gestalt but with loss of detail.

7.4.4.2. Review of the tests

Picture Arrangement, Object Assembly, Block Design and Picture Completion, subtests from the Wechsler scales, are commonly used tests of complex perceptual motor functioning. All of these are described in Section 7.2.

The Tactual Performance Test from the Halstead-Reitan Battery is a test of tactile form recognition, as well as shape and spatial location memory, and psychomotor problem-solving (Spreen and Strauss, 1998). It was originally developed as a test of visuospatial performance. However, the memory component was added later by the addition of a blindfolded tactile subtest where the person is blindfolded, and required to place different shaped blocks into the correctly shaped holes with firstly their dominant, non-dominant, and then both hands. The final part of the test is a drawing recall task where the person has to draw from memory the position of the board, and where each of the blocks were placed from memory (Lezak, 1995).

The Developmental Test of Visual Motor Integration (Beery and Buktenica, 1989) was originally a test for children where subjects were instructed to copy 24 line drawings which start very simply and end with some quite complex, three-dimensional drawings. The 24 designs follow a developmental gradient from 2 years to 15 years old, and have no time limit for completion.

The Rey-Osterreith Complex Figure Test — copy subtest (Rey, 1941) was designed to assess visuospatial constructional ability as it allows the assessment of planning, organization and problem solving. The person must copy a line drawing of a complex figure from a sheet. The scoring is both quantitative and qualitative, with a total of 36 points being awarded for a fully correct copy. There is also an informal assessment of style, approach to the problem, organizational skills and the overall gestalt of the figure.

Another test that is used in this area is that of Judgment of Line Orientation. The person has to match two short line segments to the key, which has eleven numbered radii forming a semicircular cluster. Therefore, this test assesses the person's ability to estimate angular relationships, and visual matching (Lezak, 1995). Another test that assesses visual judgment of stimuli is Visual Form Discrimination. This is a multiple choice test of visual recognition, where the person has to nominate which of the four stimuli (one match and three distracters) match the target. The three distracters are variations of the target stimuli. The three variations are a displacement of the small figure or a rotation or distortion of the set of figures.

7.4.4.3. Application to HIV research

Reger et al. (2002) in their meta-analysis of the HIV research found that the most commonly given tests in this area of the 41 journal articles that they studied were Block Design, Picture Completion, Tactual Performance Test, Object Assembly, and Rey-Osterrieth Complex Figure.

The Tactual Performance Test from the Halstead-Reitan Neuropsychological Battery has been used repeatedly in the investigation of HIV related neuropsychological performance (Butters et al., 1990; Grant et al., 1992; Heaton et al., 1995; Goggin et al., 1997; Cherner et al., 2002). The findings from these studies suggest this test often reveals impaired performance in those who were positive for HIV.

Other tests that have been used frequently in the literature are those from the Perceptual Organization Factor from the Wechsler scales. In one study that assessed the overall cognitive effects of HIV, the Trail Making Test was paired with Block Design (Becker and Salthouse, 1999). In this study they found that those with AIDS performed worse on Block Design and Trail Making Test – Part B but not on Trail Making

Test – Part A compared to HIV negative controls. In a comparison between children who had been vertically infected with HIV and their siblings, Blanchette et al. (2002) utilized the four perceptual tests from the Wechsler children's scales (Block Design, Object Assembly, Picture Completion and Picture Arrangement) as well as the Beery Visual Motor Integration test finding no differences between the groups.

In two treatment studies, Heseltine et al. (1998) found that there was no difference with treatment by Peptide T in HIV-positive participants with cognitive impairment on Block Design, and using the Trail Making Test paired with Symbol Digit Modalities in the assessment of an antioxidant treatment on HIV cognitive decline found no difference between the group treated and those on placebo on these measures (Clifford et al., 2002). Gruzelier et al. (1996) utilized the Block Design test as part of a larger neuropsychological battery finding that neuropsychological performance predicts immune competence and compromise in those infected with HIV but who were otherwise healthy.

The Trail Making Test, described earlier in Section 7.4.3, has been viewed by some authors as a complex-perceptual motor measure. The Trail Making Test and the Symbol Digit Modalities Test have been paired in the assessment of depression (Harker et al., 1995), AIDS-associated dementia (Di Rocco et al., 2000) and the effects of cognition on unemployment (van Gorp et al., 1999). Harker et al. (1995) found that there was no link between depression in those with HIV, and their neuropsychological performance on these tests. Depressed dopaminergic function in those with AIDS-associated dementia did not appear to relate to these tests' performances compared to HIV-negative controls (Di Rocco et al., 2000). These tests, however, did discriminate between participants with HIV who were unemployed, and those who were employed, with the employed group performing better (van Gorp et al., 1999). The performance on the Trail Making Test paired with Symbol Digit Modalities has been correlated with brain atrophy in HIV on MRI scans (Di Sclafani et al., 1997). Findings from this study indicated that MRI detectable brain atrophy did not appear to be the primary reason for HIV-associated neuropsychological decline.

Only two of the studies reviewed utilized the Rey Complex Figure – Copy trial as the measure of complex perceptual motor abilities in an assessment of overall cognitive abilities (Levin et al., 1992; Pereda et al., 2000). Levin et al. (1992) utilized this test along with a number of other visually based tests including the Judgment of Line Orientation test, Block Design and the Design Fluency test. This study indicated no differences in the performance between HIV-positive asymptomatic individuals and matched HIV-negative individuals on any of these measures. The Rey Complex Figure – Copy trial was also shown not to discriminate between HIV-positive and HIV-negative participants in another study of the assessment of overall cognitive abilities (Pereda et al., 2000).

Therefore, within this area of neuropsychology, most studies indicate no differences are found between those who are seropositive and those who are seronegative. However, it seems that with advanced AIDS there are sometimes deficits in the area of complex perceptual motor skills.

7.4.5. Memory

7.4.5.1. Definition of the construct

There are three main delimiters in memory functioning. These are working memory, learning and encoding, and recall. Working memory is utilized when information is required to be manipulated and processed. It is sometimes linked to attentional resources and is an active memory function, as the person actively concentrates on remembering and manipulating information. This type of memory is usually measured by tasks of arithmetic, or memory for number strings, etc.

The second part of memory, per se, is the learning and encoding of information into mental representations that are able to be retrieved into conscious awareness at a later time that is remembering. Encoding refers to the transfer of information into mental representations. Tests of encoding usually involve several presentations (trials) in order to gauge the efficiency of this process. Several categories of learning and encoding have been delineated. The first is implicit learning, of which there is no conscious recollection, and is akin to an information snapshot that is taken without overt effort of awareness. For example, a person sees a red car drive past, has no conscious recollection of it, but later, when he/she is asked to choose between the colors green and red, he/she chooses red.

The next type of learning is explicit learning, an active learning process with conscious recollection of the event where the learning took place. Explicit learning can be tested by presenting novel information followed by recognition or recall tasks. Recognition tasks involve selecting previously presented information from among a number of distracter items. These are usually considered to be easier to perform than free recall tasks.

The last type of learning and encoding of information is incidental learning, where again the process of learning is unconscious. While incidental learning is sometimes thought of as a type of explicit learning, the encoding of information is done unintentionally. One of the best examples of incidental learning and encoding is the Rey Complex Figure Test. The person is asked initially to copy a figure, and then after a 3- and then 30-minute delay is asked to redraw the figure from memory. There is no warning given to the person that this is a memory task, and therefore it relies on the incidental learning of the person. Incidental learning, although an unintentional learning process like implicit learning, is different because the person has usually done something active to encode the information, even if unintentional.

Recall is the act of actually trying to retrieve information from memory. There are two types of memory from which information can be retrieved, short-term memory and long-term memory. Short-term memory is considered to be a temporary storage unit that only holds information for one to two minutes to be given back in the form that it was presented in. It is passive memory, and the information held in short-term memory is subjected to one of two outcomes. The information may be encoded into long-term memory or it may be forgotten. This is not to be confused with working memory where, although the information is only held for a short amount of time, it is manipulated and actively "worked on" to derive an answer and then forgotten, whereas short-term memory is more of a passive process of just giving back the initial information that was given, and is more likely to be remembered for longer. Information is usually held in short-term memory by rehearsal, and is subjected to primacy and recency effects. Primacy reflects the tendency to rehearse the items presented first, so they are remembered better. Recency refers to recalling the last presented items better.

Long-term memory is usually permanent and more stable, boasting an unlimited capacity and duration for recall. There are a number of different categories of long-term memory. The main theories about long-term memory often delineate long-term memory as either procedural or declarative memory. Procedural memory is the memory for automatic processes, for example driving a car, or riding a bike. Declarative memory, which is related to specific knowledge, is usually broken down into two categories, i.e. semantic memory (storage of general knowledge and concepts, e.g. who is Tchaikovsky?) and episodic memory (situation- or context-specific memory, e.g. what did you have for dinner last night?). These tasks are usually tapped in delayed recall tasks.

The last type of memory to be discussed is that of prospective memory which is a form of episodic memory and is the measurement of intention, i.e. the act of remembering to remember. Therefore, this encompasses tasks like remembering future appointments and remembering to take medication.

7.4.5.2. Review of the tests

The first tests to be reviewed are the Wechsler Memory Scales (first, revised and third editions). These measures may assess various aspects of memory functioning including overall immediate and delayed memory, working memory and verbal and visual memory. There are many subtests within these three batteries; however, the main subtests are: Logical Memory (story recall), Verbal-Paired Associates (word pair recall), Visual Reproduction (simple line drawing recall), Faces (facial recognition), Digit Span (working memory for lists of digits given back by the examinee verbatim and in reverse), Mental Control (repetition of counting, days of week, year, etc.) and Visual Memory Span (memory for block tapping sequences both verbatim and in reverse). A number of other tests are also available depending on the version including a verbal list-learning task with 12 items and four learning trials (Word Lists), complex visual memory (Family Pictures), memory for simple line drawings paired with colors (Visual Paired Associates), complex working memory task (Letter-Number Sequencing) and an information and orientation task. The normative data for each of these tests are robust but the norms for the latest version (Psychological Corporation, 1997b) are very forgiving for the elderly.

The California Verbal Learning Test (CVLT) was designed to provide an assessment of verbal learning and memory including the way that people learn and retrieve verbal information (Delis et al., 1987). The person being tested must learn and remember two shopping lists containing 16 words from four semantic categories. There are five learning trials in total for the "Monday" shopping list, and then they are given the "Tuesday" shopping list which has 16 different words. After these trials the person is asked to recall "Monday's" list immediately, with no prompts. There is then a delayed recall trial after a 20-minute delay where the person is again required to give "Monday's" list without prompts. Once the delayed free recall is administered, a recognition trial is given with words from both lists, plus phonemic and semantic distracters.

Another verbal list learning task that is often used is the Rey Auditory Verbal Learning Test, which again is designed to test learning and recall of verbal materials (Rey, 1964). There are five learning trials administered, each with 15 unrelated words, then an interference trial of 15 different words, followed by the immediate recall trial of the initial 15 words. After a 25-minute delay the person is then asked to recall the words from the first list again, and then is given a recognition trial, varying from 30 to 50 words depending on the version of the test. The recognition trial may ask the person to recall the words from just the original list, not words from the second list or the words that are either semantically or phonemically similar.

Another verbal list learning and memory task is the Hopkins Verbal Learning Task (Brandt, 1991). It utilizes 12 words from three semantic categories. The list is given over three learning trials, in no particular order, so as to decrease the primacy and recency effects, and then a 24-word recognition test is given. The recognition trial contains all of the 12 target words with six semantic and six unrelated distracter items.

The Selective Reminding Test is again a measure of verbal and memory by list learning task over twelve trials. There are three different cards. The first has the list of 12 words that are shown for immediate recall. From then on, only the words that the person neglected to recall are told to the subject until they are able to recall all of the words on three consecutive trials or until 12 trials have been given. There is then a cued recall trial with the second card that has the first two to three letters of each word on it. The last card is the stimulus for a multiple choice recognition trial.

The Story Memory Test has five learning trials and a delayed recall of four hours after the person has completed the last learning trial. The story contains 29 items and is told to the person via a tape recording (Heaton et al., 2004).

The Benton Visual Retention Test (Benton and Spreen, 1961) is a visual memory test that also assesses perception and constructive abilities. There are three alternative forms of the test each with ten designs presented. In each of the alternate forms, the first two cards have only two major figures on them with the other eight cards having two major figures with one smaller peripheral figure. Each of the designs is displayed for ten seconds and is then removed for immediate recall by the subject by drawing. Then, each of the cards is displayed for five seconds with the subject then drawing them from memory again. In the final trial the cards are displayed while the subject copies the designs directly from them.

The Brief Visuospatial Memory Test is used to assess visuospatial memory and is similar to the Visual Reproduction subtest from the Wechsler Memory Scales. This test has six alternative forms each with an immediate recall trial, a rate of acquisition measure, a delayed recall trial and finally a recognition trial. The person is shown six simple geometric designs in a two by three matrix for ten seconds, and is then asked to reproduce on a blank sheet as many of the designs as possible. When reproducing the designs, they are asked to try to place them in the same location as they were displayed, and there is no time limit for recall. The person then completes two more learning trials. There is a delayed recall trial, which is presented after 25 minutes' delay, where the person is asked to recall all of the designs without prompts.

The Warrington Recognition Memory Test is, in essence, two tests. The first is a test of verbal memory while the second is a test of non-verbal memory abilities. Each test has a total of 100 items, 50 stimuli and 50 distracters. The verbal test contains one-syllable words that are of high frequency within the English language, which in the recognition trial are placed to one side of a distracter item. The visual test is similar but the stimuli are all male faces instead of words.

The Rey-Osterrieth Complex Figure test (Rey, 1941) has a number of alternative forms and by using a copy, immediate and delayed recall trial, with a recognition trial at the end, is a test of complex visual memory. The person initially is asked simply to copy the complex figure that is placed in front of them. Then, usually three minutes later they are asked to recall the figure with no cues. After another delay of between 30 minutes to one hour the person is then asked to recall the figure, and then complete the recognition trial, which contains 18 pieces of figure, nine from the stimulus and nine distracter items.

Another visual memory test is the Figural Memory test which was developed to assess visual learning and delayed recall with the figures from the WMS Visual Reproduction subtest. The person is shown the three cards with the figures on them, for five learning trials, with a delayed recall trial four hours after the last learning trial (Heaton et al., 2004).

The prospective memory task that has been utilized in HIV research is that of the Memory for Intentions Screening Test (Raskin, 2004). This task consists of eight prospective memory tasks given over a 30-minute time period where the person is given the tasks either 2 or 15 minutes after being told the instructions. The delay between the instructions, like "In two minutes, ask me what time the session ends", is filled with a word search puzzle distracter task.

7.4.5.3. Application to HIV research

Reger et al. (2002) in their meta-analysis of the HIV research found that the most commonly given tests in this area for verbal measures of memory included Logical Memory, Associate Learning, California Verbal Learning Test and Rey Auditory Verbal Learning Test. The main visual memory tests were Visual Reproduction, Rey-Osterreith Complex Figure Test, Visual Paired Associates and the Benton Visual Retention Test.

The California Verbal Learning Test has been utilized in characterizing the "unique" learning and memory profiles associated with HIV/AIDS, which were similar to the "subcortical" pattern of Huntington's chorea reflected by relatively poor acquisition and retention of words lists, with relatively spared recognition memory (Becker et al., 1995). Other studies that have utilized this test have assessed the relationship between HIV and unemployment (van Gorp et al., 1999), drug use (Margolin et al., 2002) and alcohol abuse (Green et al., 2004), with the main findings being differences between those HIV positive and those HIV negative who had a history of drug use, but not with those who abuse alcohol. However, HIV-positive individuals who were unemployed performed worse than those who were employed on this test.

The Rey Auditory Verbal Learning Test was used by Selnes et al. (1995). In this study they found no decline in functioning on this test except in advanced HIV-associated dementia complex. Two other studies used the Rey Auditory Verbal Learning Test in the assessment of HIV-related dementia, finding that dopamine and changes in monocyte functioning preceded the development of dementia and that this affected memory (Di Rocco et al., 2000; Luo et al., 2003). Other authors have used the Rey Auditory Verbal Learning Test in an assessment of abnormal brain activity in HIV (Ernst et al., 2002), the implications of the use of HAART (Ferrando et al., 1998) and neuropsychological performance and its relationship to apathy for medication adherence (Rabkin et al., 2000). The main findings of these studies are that people with HIV tend to show slower learning, but retention and recall of the information learned is usually intact. The Warrington Recognition Memory Test has also been utilized to assess memory functioning in HIV and medication adherence, finding no differences between zidovudine versus placebo treatment groups (Gruzelier et al., 1996).

The Benton Visual Retention Test was utilized to assess the additive impact on HIV neurocognitive dysfunction of drug use, finding that there was an effect (Pakesch et al., 1992a, b). Therefore, those who were both seropositive and positive for drug use performed worse than those who were just seropositive.

Hinkin et al. (1998) used a number of memory tasks including the California Verbal Learning Test, Hopkins Verbal Learning Test and Story and Figural Memory to assess for the neuropsychological impact of HIV, finding that those with poor medication adherence had poorer memory scores. Marcotte et al. (1999) used these tests, except for the Hopkins Verbal Learning test, to assess neuropsychological impacts on driving, finding significant differences between those with neuropsychological impairment on memory scores, and that these impaired participants performed worse on the driving simulator.

Using two of the major recall tasks, the California Verbal Learning Test and Rey Complex Figure Test, Di Sclafani et al. (1997) found no decline in performance related to brain atrophy, as assessed by MRI and HIV infection. Another study using the Rey Auditory Verbal Learning Test paired with the Rey Complex Figure Test (Pereda et al., 2000) found that cognitive abilities in HIV at diagnosis were usually intact. Many other articles have assessed both verbal and visual memory in HIV at once. One study (Manly et al., 1997) utilized the Story and Figural Memory tests by Heaton et al. (2004) to assess the neuropsychological relationship to coping. Others have used Visual Reproduction and Logical Memory from the Wechsler Memory Scales to assess the long-term impact of HIV on cognition (Bono et al., 1996), cognitive abilities and medication management (Albert et al., 1995) and general neuropsychological test performance (Becker and Salthouse, 1999). These last two studies also employed other memory tests including the Rey-Osterreith Complex Figure Test (Albert et al., 1995) and Digit Span and Verbal Free Recall (Becker and Salthouse, 1999).

In a study that assessed people's complaints of neuropsychological impairment, it was found that the number of cognitive complaints that the person had was related to their level of actual neuropsychological impairment (Lopez et al., 1998). This battery of tests included Visual Memory Span and the Selective Reminding Task. Another study that assessed the person's perceived complaints to actual neuropsychological functioning utilized Digit Span, Visual Reproduction and the Rey Auditory Verbal Learning Task (Knippels et al., 2002).

Two studies used the Brief Visuospatial Memory Test in conjunction with the Hopkins Verbal Learning Test, finding that there was little effect on cognition by treatment with an antioxidant (Clifford et al., 2002), and that both HIV infection and methamphetamine abuse were related to poor neurocognitive performance (Rippeth et al., 2004). The last article to be cited involved a comprehensive memory assessment as measured by Digit Span, Visual Memory Span, Benton Visual Retention Test, California Verbal Learning Test, Visual Reproduction and the Rey-Osterreith Complex Figure Test to see the cognitive impact of Peptide T on overall functioning in HIV, with results indicating no difference between treatment with the peptide over a placebo (Heseltine et al., 1998).

The one prospective memory study that has been completed with the HIV population (Carey et al., 2006) found that the seropositive individuals performed worse than the seronegative people on the tasks of prospective memory with relation to both time- and event-based tasks. The HIV-positive group also performed worse on the 24-hour delay, and made more task substitution errors. However, it was found that the HIV group was not impaired in their recognition performance.

Therefore, research into HIV memory deficits indicates variable patterns, with deficits associated with free recall tasks, high rates of interference effects, recall inconsistencies across trials and repetition errors. The memory functions that are usually spared in HIV are those of retention and discrimination. Therefore, although people who are seropositive have difficulty in getting the information into their memory, they are able to recall what they do learn and this is enhanced by cues.

7.4.6. Motor

7.4.6.1. Definition of the construct

In the neuropsychological framework, testing within the area of the motor construct is usually limited to the testing of hand strength and dexterity, coordination and gait. Other tests of motor function may include everyday activities. Testing of other aspects of motor functioning generally occurs during neurological examination. Tests of motor function are often used to assess the lateralization of a lesion, and tend to have a volitional component to them. This is to ensure that they are not mistaken for reflex or random movement.

7.4.6.2. Review of the tests

The Purdue Pegboard was developed in the 1940s, and is used to assess both finger and hand dexterity. The actual pegboard has two parallel rows of 25 holes, and the pins that are to be placed in the holes are in small cups at the top left and right hand corners of the board. The set also contains collars and washers, which sit in two middle cups and are utilized in the fourth subtest. The first three subtests require the person to place the pins into the holes as quickly as possible with first the dominant, then non-dominant and then both hands. The fourth subtest requires making assemblies utilizing alternate hands to construct them. Each assembly consists of a pin, washer, collar and another washer, with the score for the individual being how many assemblies were constructed within one minute.

Another pegboard test is the Grooved Pegboard. In this test the person must rotate the pegs so that they may be correctly inserted into slotted holes. This test is more sensitive than the Purdue Pegboard test because of the increased complexity associated with the rotation of the peg.

The second type of test is Finger Tapping, and it was originally part of the Halstead-Reitan Neuropsychological Battery under another name, the Finger Oscillation Test. In this test, the person must tap as rapidly as possible with the index finger of their preferred and then non-preferred hand. The mean score of five consecutive 10-second trials within 5 taps is the recorded score for the person. There is also a similar test that is utilized for measuring foot tapping.

Another test is the Hand Dynamometer, which is also called the Grip-Strength test. The main purpose of this test is to measure the voluntary grip strength and intensity of each hand of the person. The person holds the top part of the dynamometer in the palm of their hand, and is then required to squeeze the stirrup as hard as possible with fingers. The force, measured in kilograms, which the person squeezed, is then registered.

7.4.6.3. Application to HIV research

Motor tests are abundantly utilized within the HIV research literature. These tests have been popular within this area as HIV neurological complications have long been thought to contain a "motor" component (Brew et al., 1988; Navia and Price, 1987; Butters et al., 1990; American Academy of Neurology, 1991, 1996). Motor tests are also thought to be more culture-free and easy to administer, in what are often diverse and international settings, as HIV is a global problem. Thus, they are often utilized in brief batteries or even on their own in clinical trials, as well as being utilized as part of a larger battery of neuropsychological tests.

Reger et al. (2002) in their meta-analysis of the HIV research found that the most commonly given test in this area of the 43 journal articles that they studied was the Grooved Pegboard (49% of the studies). Other tests that were frequently used were the Finger Tapping, Grip Strength, Finger Sequencing, Matthews-Klove Motor Steadiness, Kimura Box Test, Graphesthia Test, Foot Tapping, and Luria-Nebraska Motor examination.

The Grooved Pegboard test has been used in many prevalence studies. One assessed the cognitive functioning and activities of daily living in those with HIV, discriminating well between those people falling within the normal, mild and moderate–severely neuropsychologically impaired groups (Benedict et al., 2000). This test was also utilized in the assessment of cognitive reserve in HIV-positive individuals (Pereda et al., 2000). In this study, it was found that the non-dominant hand trial discriminated between those with high and low cognitive reserve in those who were seropositive, while the dominant hand trial did not. Heaton et al. (1995) found that HIV-negative controls performed better compared to HIV-positive participants on tests of motor functioning including Finger Tapping, Grip Strength and Grooved Peg Board. Within the HIV-positive groups, it was found that those who were more advanced in their illness (CDC-B and CDC-C) performed significantly worse than those who were not as advanced (CDC-A) in both Finger Tapping and Grip Strength. Studies using Grooved Pegboard and Hand Dynamometer (Blanchette et al., 2002) and other motor tests (Pulsifer and Aylward, 2000) have also noted deficits in school-age children with vertically transmitted HIV.

The Grooved Pegboard test has been used in a number of predictive studies assessing many constructs including the relative risk for AIDS-related dementia in seropositive drug users (Selnes et al., 1997; Margolin et al., 2002; Rippeth et al., 2004), seropositive persons with depression (Rabkin et al., 2000) and people who were just seropositive (Stern et al., 2001). These studies indicated that Grooved Pegboard performance was able to assist in the prediction of dementia progression. This is corroborated by a study that found that, as screening measures, the Grooved Pegboard, when paired with the Hopkins Verbal Learning Test (total recall score), was the most sensitive test combination when predicting neuropsychological impairment versus non-impairment, from more extensive neuropsychological testing (Carey et al., 2004). It has also been found that test performance on the Grooved Pegboard has also been related to MRI findings of cortical atrophy in those with HIV (Hestad et al., 1993; Heaton et al., 1995).

The other pegboard test, the Purdue Pegboard, although not utilized in as many manuscripts, still appears to be useful in the assessment of cognitive complaints in HIV. The Purdue Pegboard was used in a study to validate a brief screening measure for the assessment of HIV-related cognitive impairment in France (Desi et al., 1995), and has also been used in treatment studies and was noted to be positively affected by the administration of HAART (Suarez et al., 2001) with better performances noted after HAART treatment had been implemented. Nevertheless, despite treatment with HAART, HIV-positive persons tend to perform worse than HIV-negative controls (Deutsch et al., 2001; Cysique et al., 2004). Grooved Pegboard and Timed gait were used to assess the motor functioning in HIV-positive participants with HIV-associated cognitive-motor impairment, while taking an antioxidant. However, it was noted that only the Grooved Pegboard performance improved with the ingestion of the antioxidant (Clifford et al., 2002).

Combinations of motor tests are often considered when exploring the relationship of neuromotor changes and other effects of HIV. For example, Finger Tapping and Grooved Pegboard were placed into a Structural Equation Model (SEM) as the motor part of the neuropsychological performance (Carter et al., 2003). The findings from the SEM indicated that the number of cognitive complaints a person made was associated with poorer neuropsychological performance. In a study of cerebrospinal fluid dopamine reduction in HIV, it was found that motor performance was better in those with higher dopamine metabolite levels in the cerebrospinal fluid (Di Rocco et al., 2000).

The motor abilities, along with other neuropsychological constructs, were assessed in HIV-positive participants with the Grooved Pegboard Test, Hand Dynamometer, and Finger Tapping test in an attempt to ascertain premortem the postmortem incidence of HIV-related encephalitis (Cherner et al., 2002). This study found that large neuropsychological batteries are able to predict the postmortem incidence of HIV-related encephalitis. Finger Tapping and Grip Strength were used in a study to represent motor abilities that was examining the relationship between MRI, CD^4 count and neuropsychological impairment outcomes, finding that neuropsychological impairment outcomes were not as important in predicting brain atrophy on MRI as was CD^4 counts (Di Sclafani et al., 1997).

Therefore, most of the findings within the HIV positive groups seem to indicate that there is motor impairment within the gross and fine motor domains. Some of these deficits can be remediated with particular drug therapies, including HAART or antioxidant interventions.

7.4.7. Sensory

7.4.7.1. Definition of the construct

Detailed sensory testing is normally part of a neurological evaluation. However, some neuropsychological batteries include assessment of selected sensory-perceptual functions such as finger agnosia, tactile form recognition, graphesthesia and tests for neglect.

7.4.7.2. Review of the tests

The Sensory Perceptual Examination is part of the Halstead-Reitan Battery. The first subtest involves the assessment of tactile inattention, and requires the examiner to touch points on either side of the body (usually on the face or the hands), singly and then simultaneously (Lezak, 1995). The Auditory subtest requires the examiner to stand behind the subject, and make a barely audible finger snap next to one or both of the respondent's ears (Reitan and Wolfson, 1985). The subject is then required to identify where the noise was made. The visual part of this test is along the same principles, but the examiner sits in front of the person, with their arms extended to the side. The subject is to look at the examiner's nose while identifying which fingers the examiner is moving (Reitan and Wolfson, 1985).

Another test of sensory abilities is the Fingertip Number-Writing Perception test. This test is designed to assess for sensory abilities in identification of numbers and letters (Lezak, 1995). The examiner writes with a pencil on each of the subject's fingertips to assess for sensory deficits. Another test involving finger recognition is the Finger Agnosia test where the subject must identify, by touch, with their eyes closed, which fingertip the examiner just touched (Reitan and Wolfson, 1985).

7.4.7.3. Application to HIV research

Because various sensory disturbances do complicate more advanced HIV disease and can also be a byproduct of HIV treatment (especially dideoxynucleoside antiretroviral drugs), comprehensive neurobehavioral studies of HIV usually include neurological evaluations of sensory functions. Typically about 20% of HIV cases have sensory abnormalities, due usually to distal sensory polyneuropathy. Neuropsychological studies themselves rarely concentrate on sensory examinations because of availability of concurrent neurological exams. However, a few studies have reported data. For example, a large study assessing neuropsychological performance across differing levels of disease stage found that although Tactile Form Recognition discriminated well between those who were most ill (CDC-C) and other seropositive groups (CDC-A and CDC-B), as well as HIV-negative subjects, other sensory measures did not. Among the subjects who were globally impaired in this study, 28% were found to have sensory deficits. This finding was not unlike the finding in a neurological paper (Marra et al., 1998) which found that 21% of people with global deficits had sensory impairment.

Three other studies with sensory measures included (Hall et al., 1996; Hinkin et al., 1998; Marcotte et al., 1999) used the Sensory Perceptual Exam, Spatial Relations, Fingertip Writing, Finger Agnosia, Tactile Form Recognition and Tactile Perception. Hinkin et al. (1998) indicate that although sensory functions are rarely affected in early to middle stages of HIV infection, they are more likely to be impaired in late and end stages of infection. Marcotte et al. (1999) found that in 68 seropositive participants, sensory measures discriminated between those who were neuropsychologically normal ($n = 36$) and those who were impaired ($n = 32$). In an MRI study (Hall et al., 1996), it was found that there was a correlation between the level of performance in tactile perception, and quantitative changes in brain matter over time. Those authors went on to conclude that this indication of structural changes in the brain of those who were seropositive leads to poorer sensory performances.

7.4.8. Summary measures

7.4.8.1. Definition of the construct

Although there is some predilection for HIV to affect basal ganglia and white matter preferentially to other brain regions, there are also many cases of cortical involvement, and patchy, irregularly distributed pathology may be seen. Therefore, the pattern of neurocognitive deficits is not always predictable, and while many cases present with difficulties in attention, speeded information processing, learning and executive functioning, others may have perceptual-motor, language or other problems. In order to ascertain if cognitive impairment exists, it is thus necessary to evaluate multiple ability areas.

Once this is accomplished, it is often useful to arrive at a composite or global judgment of presence and severity of impairment. The approaches have included creation of impairment indices based on proportion of tests or ability areas falling into the impaired range; calculation of standardized scores (z- or t-scores); calculation of deficit scores; and ratings based on clinical judgments.

7.4.8.2. Approaches to creating summary measures

7.4.8.2.1. Average score

One common approach to summarizing the results of evaluation when several tests are involved is to create an average score. Since tests usually employ different approaches in scoring, it is first necessary to convert such diverse results to a common metric. Typically, standard scores are computed, and expressed in standard deviation units from the mean of a reference group's performance on that test. Two of the most common methods of standardizing tests are z-scores

$$z = \frac{x - \mu}{\sigma} \tag{1}$$

where x = the sample mean, μ = the population mean and σ = the population standard deviation, and t-scores

$$t = 10z + 50 \tag{2}$$

which permit easier manipulation of data by converting z to a positive number. Once the results of each test are expressed in standard units, performance on a group of tests can be expressed as an average.

7.4.8.2.2. Proportion of tests in the impaired range

When neurocognitive impairment in a patient group is expected to be mild, as may be the case in the earlier phases of HIV disease, performance of the patient group may be somewhat variable, with good performances on some tests in some patients offsetting worse performances by other patients on the same tests. This may obscure differences between patients and controls when using group mean comparisons. This is especially likely to happen when patient performance falls into the very mildly impaired range on any particular test. If suitable norms exist, or if there is an adequate comparison group, it is possible to classify performance on a particular test as impaired or not based on some cut point, e.g. a score of one standard deviation or less below the mean of the reference group. One can then count how many tests a patient has "failed" and create a percent tests impaired score. This is the basis for some well-known indices of impairment for neuropsychological test batteries such as the Halstead Impairment Index and the Average Impairment Rating (AIR) of Russell et al. (1970).

7.4.8.2.3. Deficit scores

A potential difficulty in using an average score approach is that good and bad performances are given equal weight, when in fact the interest may be more in analyzing who, and under what circumstances, performs worse than expected. To address this issue, the deficit score approach uses a scaling procedure wherein progressively higher scores are given to worse performance, but normal or exceptionally good scores are collapsed into one scale. For example, when a person is given a comprehensive neuropsychological battery, the raw scores are converted to standardized scores which are weighted, and converted to a single deficit score for each measure. More weight is given to those scores that are impaired. Once this is done, the standardized scores are averaged to create an overall deficit score for that assessment (Heaton et al., 1994, 1995; Carey et al., 2004). These deficit scores range from 0 (no impairment) to 5 (severe impairment). The deficit score usually classifies neuropsychological impairment with a cut-point score of 0.5, which yields the best balance between sensitivity and specificity of detection of impairment (Carey et al., 2004).

7.4.8.3. Clinical ratings of impairment

Historically, the approach to summary data that has shown greatest sensitivity and specificity has involved blinded ratings of test results by suitably trained clinicians (Heaton et al., 1981). In this approach the rater utilizes norms adjusted for age, education and other factors, as available, to determine what tests were in the impaired range. Furthermore, tests are grouped into ability areas (for example, speed of information processing, verbal abilities, learning, memory, abstract reasoning, attention, working memory and motor abilities) and then a judgment is made whether a particular ability area falls into the impaired range. Generally speaking, a person's performance is judged to be "globally" impaired if at least two ability areas fall into the impaired range. The deficit score approach described above can also be used as a surrogate for clinical ratings by similar grouping of tests, and determining whether deficit scores for ability areas exceed some criterion.

7.4.8.4. Review of the tests and application to HIV

One of the major batteries used in the assessment of impairment is the Halstead–Reitan battery (Reitan and Wolfson, 1993). The original Halstead–Reitan battery yielded 10 scores that could be classified as impaired or not. The Halstead Impairment Index (HII) is the fraction of the tests impaired (range 0–1). The current HRB has seven scores, so the HII is the fraction impaired of seven. The HRB contains a core battery of five tests which yield an impairment index. The five tests contributing to the seven scores are the Category Test, Tactual Performance Test (3 scores: Total Time, Memory, and Localization), Rhythm Test, Speech Sounds Perception Test and the Finger Oscillation or Finger Tapping Test. An HII of 0.6 or greater was considered to be indicative of definite impairment. Other tests that may be added into this battery are the TMT and the Aphasia Screening Test, as well as the Hand Dynamometer. All of these tests have been described previously.

Other measures are constructed from a number of different tests from different sources, put together in a way that gives an overall or global performance rating. This may be constructed by just a few tests or by many measures across many different areas. For example, Cherner et al. (2002) utilized a global clinical rating scale that was developed by Heaton et al. (1995), in their assessment of antemortem cognitive functioning and its relationship to postmortem evidence of HIV-related encephalitis. This global clinical rating is scaled on a 9-point Likert scale, and is based on performance on a number of cognitive domains including verbal, abstraction, psychomotor, learning, memory, motor and attention. These scores are grouped together into the global score based on T-scores with those who are impaired on neuropsychological performance being rated according to how severe the impairment is and how many constructs they are impaired on. This global performance rating was also utilized in assessment of the impact of neuropsychological dysfunction on driving behaviors (Marcotte et al., 1999), quality of life (Kaplan et al., 1997) and coping (Manly et al., 1997). This method was also used to assess the added impact of methamphetamine abuse on those who are HIV seropositive (Rippeth et al., 2004).

The Dana Consortium (Dana Consortium on the Therapy of HIV Dementia and Related Cognitive Disorders, 1996a, b) developed criteria for the measurement of global impairment in HIV with people who scored two standard deviations below demographically adjusted means on one test or who scored one standard deviation below demographically adjusted means on two or more tests. They covered six cognitive constructs, including verbal and visual memory, constructional, psychomotor and motor skills, and executive functioning.

Another way of generating Global Impairment scores was to convert *z*-scores derived from nine domains (attention, verbal, abstraction, spatial processing, psychomotor, memory, learning, motor and reaction time) into percentile ranks and then convert these to clinical ratings based on a three-point scale (0: unimpaired; to 2: severely impaired), which was then summed. The participants were then graded according to these impairments with the severely impaired group having ratings of more than six points, the mild-to-moderately impaired group with ratings of two to five points and the unimpaired group of less than two points. Two studies have utilized this method of derivation for global impairment scores in HIV (Fletcher et al., 1997; Nielsen-Bohlman et al., 1997). The first assessed inter- and intra-hemispheric EEG coherence, and the second, semantic priming effects. Another study that has utilized this method assessed the relationship between HIV neurocognitive status and semantic and repetition priming (Jasiukaitis and Fein, 1999).

Similar to this method was the one that was employed by Hinkin et al. (2002) in their assessment of medication adherence in HIV-positive subjects. However, the method utilized appears to be simpler, as an impaired global deficit score is classified if the mean *t*-scores across domains were less than one standard deviation below the mean.

Another study rated individuals as having neuropsychological impairment if they performed at least two standard deviations below the mean on at least two neuropsycholological measures based on appropriate normative data (Ferrando et al., 1998). In this study however, the participants were given a smaller number of tests (Trail-Making Tests, the California Verbal Learning Test, the Grooved Pegboard Test, the Digit Symbol subtest from the WAIS-R and the Stroop Color-Word Test). Another study that utilized these criteria also had a small number of measures (De Ronchi et al., 2002). The tests included were the verbal fluency, Digit Span, Vocabulary, Digit Symbol and Block Design from the WAIS, and the short and long delay measures from a 15-word list-learning task. This impairment criterion of at least two standard deviations below the appropriate mean on at least two tests was also used in a study assessing the effects of HAART on neurocognitive abilities (Cysique et al., 2004). The study utilized 13 different tests, covering five different constructs including attention, memory, psychomotor, motor and visuoconstruction, whose composite scores were made from the average of the tests that fell within that construct. This method of assessing overall neurocognitive impairment was also used to evaluate the relationship between HIV neuropsychological impairment and employment status (van Gorp et al., 1999).

Margolin et al. (2002) utilized a neuropsychological impairment score for each respondent as the dependent variable, in a multiple regressional analysis to predict the impact on performance by drug use, CD^4 counts, educational achievement and other variables. The neuropsychological impairment score was based on the demographically adjusted means on each test, which were summed. Pukay-Martin et al. (2003) also appeared to utilize a fairly simple method of deriving summary scores. This method involved representing the number of neuropsychological measures in which the participants' performance fell at least one standard deviation below that of the controls. This method was utilized in their assessment of the impact of stressful life events on neurocognitive functioning in HIV-positive participants.

In a number of reviews of the main neurocognitive effects of HIV, it has been noted that there is little consistency in the findings of deficits. This is mainly related to the different tests that are being utilized, the differing co-morbidities that are affecting those who are infected, the different levels of disease state and differing methods of analysis. This was reported by Grant and Heaton (1990), and 5 years later it was reported in a meta-analytic review of the HIV literature that these areas as well as sample size, test battery size and type and mode of infection also contributed to differing deficit findings.

In 1990, it was noted that persons with AIDS appeared to do worse on many differing cognitive constructs including motor, sensation, memory and verbal tasks. It was then noted in a large study involving 500 individuals that the main areas of deficit appeared to be in attention, speed of processing and in learning

(Heaton et al., 1995). In two recent meta-analyses, one in adults (Reger et al., 2002) and the other in children (Wachsler Felder and Golden, 2002), the main deficits that were concurrent with HIV infection were those of motor abilities, executive functioning, attentional processing and speed.

The summary measures indicate that there is large variability within the HIV population. It has been found that some people with HIV are impaired according to these measures while others are not. There are many variables that affect these results including: the size of the test battery, the tests utilized in the measure, the population, sample size, mode of infection and any other co-morbid events (e.g. substance use). There is general agreement from a variety of studies that likelihood and severity of impairment increases with advancement of disease (see Heaton et al., 1995). While no single biomarker has proven to be an extremely reliable predictor of impairment, lower CD^4, higher plasma and CSF viral load and plasma and CSF indicators of immune activation such as a rise in beta-2-microglobulin or reduced hemoglobin are associated with neuropsychological impairment (Woods and Grant, 2005).

7.4.9. Bedside screening examinations

7.4.9.1. Definition of the construct

The measures that are incorporated in this section are those that are utilized to assess quickly for cognitive deficits. Screening measures are usually considered to be preliminary and are, therefore, brief and easy to administer. However, due to the briefness of these tests, they often do not cover many constructs and have a very narrow focus (Groth-Marnat, 1997). Therefore, deficits may be missed if they do not fall into the areas that are being assessed by the screening measure. Another problem that may occur with these measures is that they are usually not very sensitive, and ceiling effects are common. This is usually due to the relatively low level of difficulty that are assessed by these tests, and the single summary score that is normally given (Fogel, 1991). Therefore, one of the major limitations of these screening techniques is usually when an individual performs "normally" due to the brevity and structure of the test. However, as a quick measure of cognitive abilities these tests still are very useful.

7.4.9.2. Review of the tests and application to HIV research

The Mattis Dementia Rating Scale is a brief cognitive screening test that examines five areas that are purported to be sensitive to cognitive impairment in Alzheimer's dementia (Lezak, 1995). This test was designed to be used as an index of functioning when a person is suspected to have dementia (Spreen and Strauss, 1998). The test covers the domains of attention, initiation and perseveration, construction, conceptual abilities and memory. The most difficult items are given first, which is contrary to the way that most tests are administered (i.e. easiest items first). The Mattis Dementia Rating Scale has been used as a brief screening measure to detect dysfunction in HIV (Kovner et al., 1992). This test was also paired with the Mini-Mental Status Exam as a brief neuropsychological battery in two studies (Suarez et al., 2000; Stankoff et al., 2001). In these studies, the short screening measures showed utility as they were able to discern HIV-related deficits in cognitive functioning.

The Mini-Mental State Examination is one of the most widely used brief screening measures (Lezak, 1995). According to Spreen and Strauss (1998) it is mainly designed to be used with the elderly as a measure of cognitive impairment. This test has 11 items (e.g. what is the year, season, month, date, and day?), and is most effective at discriminating between those who are normal and those who are moderately to severely impaired (Lezak, 1995).

The HIV Dementia Scale was developed because of the lack of testing of the subcortical functions by other brief screening measures. Most cognitive screening measures are designed to assess the cortical functionality of the brain. However, most of the difficulties associated with HIV dementia are related to subcortical difficulties (van Harten et al., 2004). It is designed to be a brief, clinician-rated test that is considered to be more sensitive than other brief screening measures of cognitive functioning (e.g. Dementia Rating Scale and Mini-Mental Status). This is because it is purportedly able to identify mild levels of dementia in those with HIV (Pessin et al., 2003). The HIV Dementia Scale has been used to select neurocognitively intact participants for inclusion in research (Fernandez et al., 2004), and the scores on this measure have been shown to be better with HAART treatment (Chang et al., 1999).

The HIV Dementia Scale was developed solely to assess HIV-related dementia. Some research has been found to be effective in classifying cases as being neuropsychologically impaired (van Harten et al., 2004). In that study, the cases all had normal Mini-Mental State Examination scores. However, other attempts at validation were not as successful. The HIV Dementia Scale's clinical utility was assessed in a study with over 200 HIV-positive participants, and was found to be reasonably insensitive unless score cut-off was less than eleven (Childers et al., 2002).

The HNRC–UCSD developed a screening battery for the assessment of HIV-related neuropsychological impairment that would detect which participants required further testing (Carey et al., 2004). The selected tests, from a larger battery, were paired into combinations and blind clinical ratings of the neuropsychological results were assessed. The most sensitive combinations to impairment were found to be (i) the Hopkins Verbal Learning Test — Revised (Total Recall, HVLT-R: TR) and the Grooved Pegboard Test non-dominant hand pair, and (ii) the HVLT-R: TR and the WAIS-III Digit Symbol subtest pair. These combinations were found to be more sensitive than the HDS in the classification of neuropsychological impairment.

7.4.9.3. Some special considerations for repeated measures; e.g. monitoring the effects of disease progress or treatments

An important application of neuropsychology in HIV research and treatment has to do with repeated evaluations to detect change in performances over time. A challenge is how to actually determine meaningful change. For example, improvement in test scores may be due to recovery of cognitive functions, statistical regression to the mean or practice effects. Although statistical regression to the mean may be a problem, practice effects or recovery in functioning are more likely to occur within clinical populations. For the determined change to be related to practice effects, assumptions need to be made, for example, that such effects are equal across both unimpaired (i.e. normative sample) and impaired individuals. However, it is undetermined as to whether impaired individuals actually gain as much benefit from repeated test exposure as unimpaired individuals. It has also been noted that those people with higher cognitive reserves or "intelligence" tend to benefit more from practice effects than those who do not have such good cognitive reserves or "intelligence" (Rapport et al., 1997).

Many differing methods of assessing neuropsychological change have been proposed. Chelune (2000) utilizes multiple regression to predict a retest score based upon a person's test score and their relevant demographic characteristics. Standardized *z*-scores of change are computed to assess this. The standardized *z*-scores are the person's observed score on retest minus their predicted retest score divided by the standard error of the regression. In another approach, Iverson (2001) utilizes the standard error of the difference between group test and retest scores. The standard error of difference was chosen as it provides an estimate based upon test and retest scores. A critical difference score between test and retest can be computed at the 90% confidence level by multiplying the standard error of difference by 1.64. Scores that fall below this cut-score are likely to fall within the expected range of practice effects for the group from which the standard error was derived. In two comprehensive reviews of methods of determining change, Heaton et al. (2001) and Temkin et al. (1999) assessed the Reliable Change Index (Jacobson and Truax, 1991), and other methods of change score prediction. The Reliable Change Index indicates significant change based on the discrepancy between test and retest scores in a normative sample. When compared to other measures of assessing reliable change, this index performed worst demonstrating the highest error rates and largest confidence intervals. However, by adding to this method correction for practice effects, and the addition of simple or multiple regression analyses, the ability to predict change scores accurately was increased.

Another problem when assessing practice effects is the interval at which the practice effect is determined. For example, most reported retest intervals in the manuals are approximately 12 weeks but some of the actual retest times in research and in clinical practice can be more or less than this depending on the retest purpose. For example, three studies utilized interval testing (Villa et al., 1996; Basso and Bornstein, 2000; McCaffrey et al., 2000). For each of the studies the intervals differed with the first being 2 weeks; 6 months, 12 months and 18 months for the second; and the third was 12–18 months. The studies by Basso and Bornstein and Villa et al. were both studies that were assessing the cognitive effects of HIV over time and therefore the subjects were only tested twice. But the study by McCaffrey et al. was assessing the efficacy of a test for the utilization in HIV, hence the utilization of more frequent and a larger number of assessments. It is also noted that with this population the sicker the participants are the frequency of assessments is often more rapid.

When assessing the cognitive impact over time of a degenerative illness, often the time periods involved are long. This was demonstrated in a study that followed seronegative participants for a mean of approximately 38 months and seropositive for approximately 27 months to assess the cognitive impact of HIV (Selnes et al., 1997). In this study the participants were tested every 6 months and the authors noted a comparable general positive trend between the groups in the neuropsychological evaluation outcomes, which they attributed to practice effects. In one of their previous studies, Selnes et al. (1995) assessed the progression of neurocognitive impairment in the development of AIDS and found that there was no real decline in cognitive functioning before AIDS or in those with non-AIDS-defining illnesses, unless overt dementia was present. These participants were again tested every 6 months over at least a 2-year period. Again these authors suggested that the practice effects of the retests were accounted for by factoring in the amount of increase in scores, for example in verbal memory, before the diagnosis of AIDS, and then adding the decline in scores after the diagnosis. Although these authors appear to have accounted for the change in scores, the manner in which they accounted for them did not seem very empirically robust.

For recovery of functioning to be determined, there needs to be an increase in scores that is not attributable to practice effects. In studies that have assessed the cognitive impact of HAART, most have concluded that there has been improvement from the treatment. In one article (Schifitto et al., 1999), for neuropsychological assessment using the Rey Auditory Verbal Learning Test, Digit Symbol Test, Grooved Pegboard (dominant and non-dominant hands), Timed Gait and the California Computerized Assessment Package, a reaction time test was given at baseline, six and then ten weeks. The issue of practice effects was not noted in the article, but the authors concluded that there had been a significant increase in cognitive abilities on HAART. However, Lezak (1995) notes that the California Verbal Learning Test is susceptible to large practice effects. This is also the case for Digit Symbol (The Psychological Corporation, 1997). Suarez et al. (2000) utilized the Purdue Pegboard and the Trail Making Test in their assessment of the neurocognitive impact of HAART, in one to six serial neuropsychological assessments over a mean of 12 months. The median interval between assessments was approximately 161 days, with the shortest being 49 days. Both of these tests are affected by large practice effects (Lezak, 1995), and these authors did not mention how they adjusted for practice effects or even if this issue was taken into account (Suarez et al., 2000).

Other tasks that are susceptible to practice effects and that have been utilized in the HIV literature frequently are the Paced Auditory Serial Addition Test, the Stroop Color-Word Test, the Benton Visual Memory Test, the Hand Dynamometer (but only in males) and Object Assembly (Lezak, 1995). Therefore, caution should be taken when using these tests over serial assessments, and when interpreting findings of studies with short or multiple retest intervals. One other task that is often used for the assessment of HIV impairment is the Wisconsin Card Sorting Test. This test should never be used twice on the same person, as it changes the nature of the construct being measured (Lezak, 1995). Once a person is able to work out the nature and requirements of the test, they shall always perform at a much higher level than they did previously, unless they have severe memory problems and do not remember the task. Therefore, this test and others that are similar to it, like the Category test from the Halstead-Reitan, should never be used more than once on any person.

Another consideration when retesting persons is that of the age of cohort effect. This refers to a person being tested moving from one age cohort to another between testings. For example, if someone were 64 years old and 11 months and were tested with the WAIS-III subtest of Digit Symbol, and with a raw score of 63, their converted scaled score would therefore be a 10. If this person, however, had been tested one month later and was now 65 years and 0 months old, and had scored exactly the same raw score on the test, they would have been given a scaled score of 12. This is the edge of cohort effect, where scoring the person on one table at testing Time 1, and then on a different table at testing Time 2, leads to the person appearing to have a recovery of functioning. The easiest way to combat the edge of cohort problem is by scoring the person's second testing scores using the initial age category. Ultimately, the development of norms for change scores will provide a more satisfactory solution.

7.5. Summary

Neuropsychological testing that is correctly performed represents a valid and reliable indicator of brain injury in HIV. It has advantages of being reasonably sensitive, noninvasive, and measuring abilities that relate to the demands of everyday life. Depending on the length and comprehensiveness of batteries, results vary in their sensitivity and specificity. Screening measures reduce the cost of time and difficulties may be picked up by moderate to relatively short batteries. However, these usually are good for the assessment of severe impairment but are less sensitive to the mild to moderate impairment. The detection of impairment is important; especially if one thinks that a treatment may arrest the course of the central nervous system disease, or even reverse it.

The use of neuropsychological testing has to be performed carefully, however. These tests need to have good reliability, specificity, sensitivity and validity. The tests also should have appropriate and relevant normative data for the person being tested. The use of demographically adjusted norms is important, and assists in the correct diagnosis of neurocognitive pathology. Without the use of demographically corrected and appropriate norms, over- or under-pathologization of cognitive abilities may occur. Education, age, sex and ethnically adjusted norms, where appropriate, enable the correct diagnosis to be made and reduce the risk of incorrect detection or apparent lack of pathology.

Linking neuropsychological performance with the person's ability to perform their activities of daily living, for example their ability to work, socialize, drive, handle their financial affairs and plan, prepare and cook meals, is important as it assists in the diagnostic process. Those with moderate to severe neuropsychological impairment who also are unable to carry out their activities of daily living as adequately as they have been able to previously may be given the classification of having Minor Cognitive Motor Disorder according to the American Academy of Neurology Consortium guidelines. If the person has severe impairment in their neuropsychological functioning and also has severe impediment to their activities of daily living they may be diagnosed, according to these guidelines, as having HIV-associated dementia. Therefore, the additional information provides a more comprehensive picture of an individual's abilities and functions, thereby allowing a more formalized diagnosis to be made.

Both HIV and AIDS affect the cognition of an individual in many different ways. Neuropsychological impairment among those with the disorder is varied, and appears to be reasonably dependent on the age at which the individual was infected, co-morbid diagnoses, alcohol and other substance abuse and dependence, as well as the progression of the disease at the time of testing (e.g. severity of immunosuppression; viral load). The advent of HAART has reduced incidence of neurocognitive complications. However, prevalence has not changed, perhaps reflecting longer term survival of advanced cases. Beyond the benefits of HAART, the search continues for specific neuroprotective agents. No clearly effective neuroprotectants have, as yet, emerged.

Despite HAART, general cognitive impairment as well as specific deficits continue to be prevalent in HIV-infected persons, and these may have significant implications for everyday functioning, medication management and ultimately disease outcome.

References

Albert SM, Marder K, Dooneief G, et al (1995). Neuropsychologic impairment in early HIV infection: a risk factor for work disability. Arch Neurol 52(5): 525–530.

American Academy of Neurology (1991). Nomenclature and research case definitions for neurologic manifestations of human immunodeficiency virus-type 1 (HIV-1) infection. Neurology 41: 778–785.

American Academy of Neurology (1996). Clinical confirmation of the American Academy of Neurology algorithm for HIV-1-associated cognitive/motor disorder. Neurology 47 (5): 1247–1253.

Army Individual Test Battery (1944). Manual of Directions and Scoring, War Department, Adjutant General's Office, Washington DC.

Barona A, Reynolds C, Chastain R (1984). A demographically based index of premorbid intelligence for the WAIS-R. J Consult Clin Psychol 52: 885–887.

Basso MR, Bornstein RA (2000). Estimated premorbid intelligence mediates neurobehavioral change in individuals infected with HIV across 12 months. J Clin Exp Neuropsychol 22(2): 208–218.

Becker JT, Salthouse TA (1999). Neuropsychological test performance in the acquired immunodeficiency syndrome: independent effects of diagnostic group on functioning. J Int Neuropsychol Soc 5(1): 41–47.

Becker JT, Caldararo R, Lopez OL, et al (1995). Qualitative features of the memory deficit associated with HIV infection and AIDS: cross-validation of a discriminant function classification scheme. J Clin Exp Neuropsychol 17(1): 134–142.

Becker JT, Sanchez J, Dew MA, et al (1997). Neuropsychological abnormalities among HIV-infected individuals in a community-based sample. Neuropsychology 11(4): 592–601.

Beery KE, Buktenica NA (1989). Developmental Test of Visual-motor Integration. Psychological Assessment Resources, Odessa, FL.

Benedict RHB, Mezhir JJ, Walsh K, et al (2000). Impact of human immunodeficiency virus type-1-associated cognitive dysfunction on activities of daily living and quality of life. Arch Clin Neuropsychol 15(6): 535–544.

Benton AL, Hamsher K (1976). Multilingual Aphasia Examination, 2nd edn. AJA Associates, Iowa City, IA.

Benton AL, Spreen O (1961). Visual memory test: the simulation of mental incompetence. Arch Gen Psych 4: 74–83.

Bisiacchi PS, Suppiej A, Laverda A (2000). Neuropsychological evaluation of neurologically asymptomatic HIV-infected children. Brain Cogn 43(1–3): 49–52.

Blanchette N, Smith ML, King S, et al (2002). Cognitive development in school-age children with vertically transmitted HIV infection. Develop Neuropsychol 21(3): 223–241.

Bono G, Mauri M, Sinforiani E, et al (1996). Longitudinal neuropsychological evaluation of HIV-infected intravenous drug users. Addiction 91(2): 263–268.

Brandt J (1991). The Hopkins verbal learning test: development of a new verbal memory test with six equivalent forms. Clin Neuropsychol 5: 125–142.

Brew BJ, Rosenblum M, Price RW (1988). AIDS dementia complex and primary HIV brain infection. J Neuroimmunol 20(2–3): 133–140.

Brouwers P, Tudor Williams G, DeCarli C, et al (1995). Relation between stage of disease and neurobehavioral measures in children with symptomatic HIV disease. AIDS 9(7): 713–720.

Butler M, Retzlaff PD, Vanderploeg R (1991). Neuropsychological test usage. Prof Psychol Res Pr 22(6): 510–512.

Butters N, Grant I, Haxby J, et al (1990). Assessment of AIDS-related cognitive changes: recommendations of the NIMH workshop on neuropsychological assessment approaches. J Clin Exp Neuropsychol 12(6): 963–978.

Camara WJ, Nathan JS, Puente AE (2000). Psychological test usage: implications in professional psychology. Prof Psychol Res Pr 31(2): 141–154.

Carey CL, Woods SP, Rippeth JD, et al (2004). Initial validation of a screening battery for the detection of HIV-associated cognitive impairment. Clin Neuropsychol 18(2): 234–248.

Carey CL, Woods SP, Rippeth JD, et al (2006). Prospective memory in HIV-1 infection. J Clin Exp Neuropsychol 28: 536–548.

Carter SL, Rourke SB, Murji S, et al (2003). Cognitive complaints, depression, medical symptoms, and their association with neuropsychological functioning in HIV infection: a structural equation model analysis. Neuropsychology 17(3): 410–419.

Chang L, Ernst T, Leonido-Yee M, et al (1999). Highly active antiretroviral therapy reverses brain metabolite abnormalities in mild HIV dementia. Neurology 53(4): 782–789.

Chelune GJ (2000). Interpreting change scores on the WAIS-III and WMS-III in the context of serial assessments. Paper presented at the 20th Annual Conference of the National Academy of Neuropsychology, Orlando, FL.

Cherner M, Masliah E, Ellis RJ, et al (2002). Neurocognitive dysfunction predicts postmortem findings of HIV encephalitis. Neurology 59(10): 1563–1567.

Childers M, Ellis R, Deutsch R, et al (2002). The utility and limitations of the HIV dementia scale. J Int Neuropsychol Soc 8(2): 160.

Clifford DB, McArthur JC, Schifitto G, et al (2002). A randomized clinical trial of CPI-1189 for HIV-associated cognitive-motor impairment. Neurology 59(10): 1568–1573.

Cysique LAJ, Maruff P, Brew BJ (2004). Antiretroviral therapy in HIV infection: are neurologically active drugs important? Arch Neurol 61: 1699–1703.

Dana Consortium on the Therapy of HIV Dementia and Related Cognitive Disorders (1996a). Clinical confirmation of the American Academy of Neurology algorithm for HIV-1-associated cognitive/motor disorder. Neurology 47 (5): 1247–1253.

Dana Consortium on the Therapy of HIV Dementia and Related Cognitive Disorders (1996b). Safety and tolerability of the antioxidant OPC-14117 in HIV-associated cognitive impairment. Neurology 49(1): 142–146.

De Renzi E, Vignolo LA (1962). The token test: a sensitive test to detect receptive disturbances in aphasics. Brain 85: 665–678.

De Ronchi D, Faranca I, Berardi D, et al (2002). Risk factors for cognitive impairment in HIV-1-infected persons with different risk behaviors. Arch Neurol 59(5): 812–818.

Delis DC, Kramer JH, Kaplan E, et al (1987). California Verbal Learning Test. Psychological Corporation, San Antonio, TX.

Desi M, Seibel N, Korezlioglu J, et al (1995). [Cognitive impairment in HIV infection: validation of a brief battery for neuropsychological assessment/troubles]. Encephale 21 (4): 289–294.

Deutsch R, Ellis RJ, McCutchan A, et al (2001). AIDS-associated mild neurocognitive impairment is delayed in the era of highly active antiretroviral therapy. AIDS 15 (14): 1898–1899.

Di Rocco A, Bottiglieri T, Dorfman D, et al (2000). Decreased homovanilic acid in cerebrospinal fluid correlates with impaired neuropsychologic function in HIV-1-infected patients. Clin Neuropharmacol 23(4): 190–194.

Di Sclafani V, MacKay RDS, Meyerhoff DJ, et al (1997). Brain atrophy in HIV infection is more strongly associated with CDC clinical stage than with cognitive impairment. J Int Neuropsychol Soc 3(3): 276–287.

Diamond GW, Kaufman J, Belman A, et al (1987). Characterization of cognitive functioning in a subgroup of children with congenital HIV infection. Arch Clin Neuropsychol 2: 245–256.

Dikmen SS, Heaton RK, Grant I, et al (1999). Test-retest reliability and practice effects of expanded Halstead-Reitan neuropsychological test battery. J Int Neuropsychol Soc 5(4): 346–356.

Dunn LM, Dunn LM (1997). Peabody Picture Vocabulary Test, Third Edition. AGS Publishing, Circle Pines, MN.

Ernst T, Chang L, Jovicich J, et al (2002). Abnormal brain activation on functional MRI in cognitively asymptomatic HIV patients. Neurology 59(9): 1343–1349.

Fernandez MI, Collazo JB, Hernandez N, et al (2004). Predictors of Hiv risk among Hispanic farm workers in south Florida: women are at higher risk than men. AIDS Behav 8(2): 165–174.

Ferrando S, van Gorp W, McElhiney M, et al (1998). Highly active antiretroviral treatment in HIV infection: benefits for neuropsychological function. AIDS 12(8): F65–F70.

Fishkin PE, Armstrong FD, Routh DK, et al (2000). Brief report: relationship between HIV infection and WPPSI-R performance in preschool-age children. J Pediatr Psychol 25(5): 347–351.

Fletcher DJ, Raz J, Fein G (1997). Intra-hemispheric alpha coherence decreases with increasing cognitive impairment of HIV patients. Electroenceph Clin Neurophysiol 102(4): 286–294.

Fogel BS (1991). The high sensitivity cognitive screen. Int Psychogeriat Special Delir Adv Res Clin Pract 3(2): 273–288.

Franzen MD (2000). Neuropsychological assessment in traumatic brain injury. Crit Care Nurs Q 23(3): 58–64.

Gladsjo JA, Schuman CC, Evans JD, et al (1999). Norms for letter and category fluency: demographic corrections for age, education, and ethnicity. Assessment 2: 147–178.

Goggin KJ, Zisook S, Heaton RK, et al (1997). Neuropsychological performance of HIV-1 infected men with major depression. HNRC Group. HIV Neurobehavioral Research Center. J Int Neuropsychol Soc 3(5): 457–464.

Golden JC (1978). Stroop Color and Word Test. Stoelting, Chicago, IL.

Goodglass H, Kaplan E (1983). The Assessment of Aphasia and Related Disorders, 2nd edn. Lea and Febiger, Philadelphia, PA.

Grant I, Heaton RK (1990). Human immunodeficiency virus-type 1 (HIV-1) and the brain. J Consult Clin Psychol 58 (1): 22–30.

Grant I, Atkinson JH, Hesselink JR, et al (1987). Evidence for early central nervous system involvement in the acquired immunodeficiency syndrome (AIDS) and other human immunodeficiency virus (HIV) infections: studies with neuropsychologic testing and magnetic resonance imaging. Ann Intern Med 107(6): 828–836.

Grant I, Atkinson JH, Hesselink JR, et al (1988). Human immunodeficiency virus-associated neurobehavioral disorder. J R Coll Physicians Lond 22(3): 149–157.

Grant I, Hesselink JR, Kennedy CJ, et al (1990). HIV disease: brain–behavior relationships. In: DG Ostrow (Ed.), Behavioral Aspects of AIDS. Plenum Medical/Plenum Press, New York, pp. 247–266.

Grant I, Heaton RK, Atkinson JH, et al (1992). HIV-1 associated neurocognitive disorder. Clin Neuropharmacol 15 (Suppl 1): 364A–365A.

Grassi MP, Perin C, Clerici F, et al (1993). Neuropsychological performance in HIV-1-infected drug abusers. Acta Neurol Scand 88(2): 119–122.

Green JE, Saveanu RV, Bornstein RA (2004). The effect of previous alcohol abuse on cognitive function in HIV infection. Am J Psych 161(2): 249–254.

Gronwall DMA (1977). Paced auditory serial-addition task: a measure of recovery from concussion. Percept Motor Skills 44: 367–373.

Groth-Marnat G (1997). Handbook of Psychological Assessment, 3rd edn. John Wiley, New York.

Gruzelier J, Burgess A, Baldeweg T, et al (1996). Prospective associations between lateralised brain function and immune status in HIV infection: analysis of EEG, cognition and mood over 30 months. Int J Psychophysiol 23(3): 215–224.

Hall M, Whaley R, Robertson K, et al (1996). The correlation between neuropsychological and neuroanatomic changes over time in asymptomatic and symptomatic HIV-1-infected individuals. Neurology 46(6): 1697–1702.

Harker JO, Satz P, Del Jones F, et al (1995). Measurement of depression and neuropsychological impairment in HIV-1 infection. Neuropsychology 9(1): 110–117.

Heaton RK, Grant I, Anthony WZ, et al (1981). A comparison of clinical and automated interpretation of the Halstead-Reitan battery. J Clin Neuropsychol 3(2): 121–141.

Heaton RK, Kirson D, Velin R, et al (1994). The utility of clinical ratings for detecting cognitive change in HIV infection. In: I Grant, A Martin (Eds.), Neuropsychology of HIV Infection. Oxford University Press, New York, pp. 188–206.

Heaton RK, Grant I, Butters N, et al (1995). The HNRC 500: neuropsychology of HIV infection at different disease stages. J Int Neuropsychol Soc 1: 231–251.

Heaton RK, Temkin N, Dikmen S, et al (2001). Detecting change: a comparison of three neuropsychological methods, using normal and clinical samples. Arch Clin Neuropsychol 16: 75–91.

Heaton RK, Miller SW, Taylor MJ, et al (2004). Revised Comprehensive Norms for an Expanded Halstead-Reitan Battery: Demographically Adjusted Neuropsychological Norms for African American and Caucasian Adults, Psychological Assessment Resources, Lutz, FL.

Heseltine PNR, Goodkin K, Atkinson JH, et al (1998). Randomized double-blind placebo-controlled trial of peptide t for HIV-associated cognitive impairment. Arch Neurol 55 (1): 41–51.

Hestad K, McArthur JH, Dal Pan GJ, et al (1993). Regional brain atrophy in HIV-1 infection: association with specific neuropsychological test performance. Acta Neurol Scand 88(2): 112–118.

Hinkin CH, Castellon SA, van Gorp WG, et al (1998). Neuropsychological features of HIV disease. In: WG van Gorp, SL Buckingham (Eds.), Practitioner's Guide to the Neuropsychiatry of HIV/AIDS. Guilford Press, New York, pp. 1–41.

Hinkin CH, Castellon SA, Durvasula RS, et al (2002). Medication adherence among HIV+ adults: effects of cognitive dysfunction and regimen complexity. Neurology 59(12): 1944–1950.

Honn VJ, Bornstein RA (2002). Social support, neuropsychological performance and depression in HIV infection. J Int Neuropsychol Soc 8(3): 436–447.

Honn VJ, Para MF, Whitacre CC, et al (1999). Effect of exercise on neuropsychological performance in asymptomatic HIV infection. AIDS Behav 3(1): 67–74.

Iverson GL (2001). Interpreting change on the WAIS-III/WMS-III in clinical samples. Arch Clin Neuropsychol 16 (2): 183–191.

Jacobson NS, Truax P (1991). Clinical significance: a statistical approach to defining meaningful change in psychotherapy research. J Consult Clin Psychol 66(22): 400–410.

Jasiukaitis P, Fein G (1999). Differential association of HIV-related neuropsychological impairment with semantic versus repetition priming. J Int Neuropsychol Soc 5(5): 434–441.

Jones-Gotman M, Milner B (1977). Design fluency: the invention of nonsense drawings after focal cortical lesions. Neuropsychologia 15: 653–674.

Kaplan E, Goodglass H, Weintraub S (2001). Boston Naming Test: Record Booklet, 2nd edn. Lippincott, Williams and Wilkins, Philadelphia, PA.

Kaplan RM, Patterson TL, Kerner DN, et al (1997). The quality of well-being scale in asymptomatic HIV-infected patients. Qual Life Res 6(6): 507–514.

Kaufman AS, Kaufman NL (1983). K-abc, Kaufman Assessment Battery for Children Interpretive Manual. AGS Publishing, Circle Pines, MN.

Knippels HMA, Goodkin K, Weiss JJ, et al (2002). The importance of cognitive self-report in early HIV-1 infection: validation of a cognitive functional status subscale. AIDS 16(2): 249–267.

Kovner R, Lazar JW, Lesser M, et al (1992). Use of the dementia rating scale as a test for neuropsychological dysfunction in HIV-positive IV drug abusers. J Subst Abuse Treat 9(2): 133–137.

Krull K, Scott JG, Sherer M (1995). Estimation of premorbid intelligence from combined performance and demographic variables. Clin Neuropsychol 9: 83–87.

Lees-Haley PR, Smith HH, Williams CW, et al (1996). Forensic neuropsychological test usage: an empirical survey. Arch Clin Neuropsychol 11(1): 45–51.

Levin BE, Berger JR, Didona T, et al (1992). Cognitive function in asymptomatic HIV-1 infection: the effects of age, education, ethnicity, and depression. Neuropsychology 6 (4): 303–313.

Lezak M (1995). Neuropsychological Assessment, 3rd edn. Oxford University Press, New York.

Lopez OL, Wess J, Sanchez J, et al (1998). Neurobehavioral correlates of perceived mental and motor slowness in HIV infection and AIDS. J Neuropsych Clin Neurosci 10(3): 343–350.

Luo X, Carlson KA, Wojna V, et al (2003). Macrophage proteomic fingerprinting predicts HIV-1-associated cognitive impairment. Neurology 60(12): 1931–1937.

Manly JJ, Patterson TL, Heaton RK, et al (1997). The relationship between neuropsychological functioning and coping activity among HIV-positive men. AIDS Behav 1(2): 81–91.

Marcotte TD, Heaton RK, Wolfson T, et al (1999). The impact of HIV-related neuropsychological dysfunction on driving behavior. J Int Neuropsychol Soc 5(7): 579–592.

Marder K, Albert SM, McDermott MP, et al (2003). Interrater reliability of a clinical staging of HIV-associated cognitive impairment. Neurology 60(9): 1467–1473.

Margolin A, Avants SK, DePhilippis D, et al (1998). A preliminary investigation of lamotrigine for cocaine abuse in HIV-seropositive patients. Am J Drug Alcohol Abuse 24 (1): 85–101.

Margolin A, Avants K, Warburton LA, et al (2002). Factors affecting cognitive functioning in a sample of human immunodeficiency virus-positive injection drug users. AIDS Pat Care STDs 16(6): 255–267.

Marra CM, Boutin P, Collier AC (1998). Screening for distal sensory peripheral neuropathy in HIV-infected persons in research and clinical settings. Neurology 51(6): 1678–1681.

McCaffrey RJ, Duff K, Westervelt HJ, et al (2000). Preliminary serial assessment findings with the brief NIMH neuropsychological battery for HIV infection and AIDS. NYS Psychol 12(1): 51–53.

Meyerhoff DJ, Bloomer C, Cardenas V, et al (1999). Elevated subcortical choline metabolites in cognitively and clinically asymptomatic HIV+ patients. Neurology 52(5): 995–1003.

Moser DJ, Schultz SK, Arndt S, et al (2002). Capacity to provide informed consent for participation in schizophrenia and HIV research. Am J Psych 159(7): 1201–1207.

Navia BA, Price RW (1987). The acquired immunodeficiency syndrome dementia complex as the presenting or sole manifestation of human immunodeficiency virus infection. Arch Neurol 44(1): 65–69.

Nielsen-Bohlman L, Boyle D, Biggins C, et al (1997). Semantic priming impairment in HIV. J Int Neuropsychol Soc 3 (4): 348–358.

Orr DA, Pinto PF (1993). The clinical management of HIV-related dementia and other memory disorders in the residential drug treatment environment. J Subst Abuse Treat 10(6): 505–511.

Pakesch G, Loimer N, Grnberger J, et al (1992a). Neuropsychological findings and psychiatric symptoms in HIV-1 infected and noninfected drug users. Psych Res 41(2): 163–177.

Pakesch G, Pfersmann D, Loimer N, et al (1992b). Neuropsychological assessment and psychopathological status in HIV-1 patients of different risk groups/Noopsychische vernderungen und psychopathologische auffligkeiten bei HIV-1 patienten unterschiedlicher risikogruppen. Fortschritte der Neurologie, Psychiatrie 60(1): 17–27.

Parasuraman R, Greenwood PM, Alexander GE (2000). Alzheimer disease constricts the dynamic range of spatial attention in visual search. Neuropsychologia 38: 1126–1135.

Pereda M, Ayuso Mateos JL, del Barrio AG, et al (2000). Factors associated with neuropsychological performance in HIV-seropositive subjects without AIDS. Psychol Med 30(1): 205–217.

Pessin H, Rosenfeld B, Burton L, et al (2003). The role of cognitive impairment in desire for hastened death: a study of patients with advanced AIDS. Gen Hosp Psych 25(3): 194–199.

Piotrowski C, Keller JW (1989). Psychological testing in outpatient mental health facilities: a national study. Prof Psychol Res Pr 20(6): 423–425.

Psychological Corporation (1997a). WAIS-III: Administration and Scoring Manual. Harcourt Brace, San Antonio, TX.

Psychological Corporation (1997b). WMS-III: Wechsler Memory Scale Administration and Scoring Manual. Psychological Corporation, San Antonio, TX.

Psychological Corporation (2001). Wechsler Test of Adult Reading Manual. Psychological Corporation, San Antonio, TX.

Pukay-Martin ND, Cristiani SA, Saveanu R, et al (2003). The relationship between stressful life events and cognitive function in HIV-infected men. J Neuropsych Clin Neurosci 15(4): 435–441.

Pulsifer MB, Aylward EH (2000). Human immunodeficiency virus. In: KO Yeates, MEA Ris (Eds.), Pediatric Neuropsychology: Research, Theory, and Practice. Guilford Press, New York, pp. 381–402.

Rabkin JG, Ferrando SJ, van Gorp W, et al (2000). Relationships among apathy, depression, and cognitive impairment in HIV/AIDS. J Neuropsych Clin Neurosci 12(4): 451–457.

Rapport LJ, Axelrod BN, Theisen ME, et al (1997). Relationship of IQ to verbal learning and memory: test and retest. J Clin Exp Neuropsychol 19(5): 655–666.

Raskin S (2004). Memory for intentions screening test [abstract]. J Int Neuropsychol Soc 10(Suppl 1): 110.

Reger M, Welsh R, Razani J, et al (2002). A meta-analysis of the neuropsychological sequelae of HIV infection. J Int Neuropsychol Soc 8(3): 410–424.

Reitan RM, Wolfson D (1985). The Halstead-Reitan Neuropsychological Test Battery: Theory and Clinical Interpretation. Neuropsychology Press, Tucson, AZ.

Reitan RM, Wolfson D (1993). The Halstead-Reitan Neuropsychological Test Battery: Theory and Clinical Interpretation. Neuropsychology Press, Tucson, AZ.

Rey A (1941). L'examen psychologique dans les cas d'encephalopathie traumatique. Archives de Psychologie 28: 286–340.

Rey A (1964). L'examen clinique en psychologie. Presses Universitaires de France, Paris.

Rippeth JD, Heaton RK, Carey CL, et al (2004). Methamphetamine dependence increases risk of neuropsychological impairment in HIV infected persons. J Int Neuropsychol Soc 10(1): 1–14.

Russell EW, Neuringer C, Goldstein G (1970). Assessment of Brain Damage: A Neuropsychological Key Approach. Wiley-Interscience, Oxford.

Sarter M, Podell M (2000). Preclinical psychopharmacology of AIDS-associated dementia: lessons to be learned from the cognitive psychopharmacology of other dementias. J Psychopharmacol 14(3): 197–204.

Schifitto G, Sacktor N, Marder K, et al (1999). Randomized trial of the platelet-activating factor antagonist lexipafant in HIV-associated cognitive impairment. Neurology 53 (2): 391–396.

Selnes OA, Galai N, Bacellar H, et al (1995). Cognitive performance after progression to AIDS: a longitudinal study from the Multicenter AIDS Cohort Study. Neurology 45 (2): 267–275.

Selnes OA, Galai N, McArthur JC, et al (1997). HIV infection and cognition in intravenous drug users: long-term follow-up. Neurology 48(1): 223–230.

Sharpley CF, Pain MD (1988). Psychological test usage in Australia. Austral Psychol 23(3): 361–369.

Silberstein CH, O'Dowd MA, Chartock P, et al (1993). A prospective four-year follow-up of neuropsychological function in HIV seropositive and seronegative methadone-maintained patients. Gen Hosp Psych 15(6): 351–359.

Smith A (1982). Symbol Digit Modalities Test (SDMT), revised edn. Western Psychological Services, Los Angeles, CA.

Spreen O, Strauss E (1998). A Compendium of Neuropsychological Tests: Administration, Norms, and Commentary, 2nd edn. Oxford University Press, New York.

Stankoff B, Calvez V, Suarez S, et al (1999). Plasma and cerebrospinal fluid human immunodeficiency virus type-1 (HIV-1) RNA levels in HIV-related cognitive impairment. Eur J Neurol 6(6): 669–675.

Stankoff B, Tourbah A, Suarez S, et al (2001). Clinical and spectroscopic improvement in HIV-associated cognitive impairment. Neurology 56(1): 112–115.

Stern Y, Liu X, Marder K, et al (1995). Neuropsychological changes in a prospectively followed cohort of homosexual and bisexual men with and without HIV infection. Neurology 45(3, Pt 1): 467–472.

Stern RA, Silva SG, Chaisson N, et al (1996). Influence of cognitive reserve on neuropsychological functioning in asymptomatic human immunodeficiency virus-1 infection. Arch Neurol 53(2): 148–153.

Stern Y, McDermott MP, Albert S, et al (2001). Factors associated with incident human immunodeficiency virus-dementia. Arch Neurol 58(3): 476–479.

Suarez SV, Stankoff B, Conquy L, et al (2000). Similar subcortical pattern of cognitive impairment in AIDS patients with and without dementia. Eur J Neurol 7(2): 151–158.

Suarez S, Baril L, Stankoffa B, et al (2001). Outcome of patients with HIV-I-related cognitive impairment on highly active antiretroviral therapy. AIDS 15: 195–200.

Sullivan K, Bowden SC (1997). Which tests do neuropsychologists use? J Clin Psychol 53(7): 657–661.

Temkin NR, Heaton RK, Grant I, et al (1999). Detecting significant change in neuropsychological test performance: a comparison of four models. J Int Neuropsychol Soc 5(4): 357–369.

Toledo Tamula MA, Wolters PL, Walsek C, et al (2003). Cognitive decline with immunologic and virologic stability in four children with human immunodeficiency virus disease. Pediatrics 112: 679–684.

Tozzi V, Balestra P, Galgani S, et al (1999). Positive and sustained effects of highly active antiretroviral therapy on HIV-1-associated neurocognitive impairment. AIDS 13(14): 1889–1897.

Trenerry MR, Crosson B, DeBoe J, et al (1990). Visual Search and Attention Test. Psychological Assessment Resources, Odessa, FL.

van Gorp WG, Baerwald JP, Ferrando SJ, et al (1999). The relationship between employment and neuropsychological impairment in HIV infection. J Int Neuropsychol Soc 5(6): 534–539.

van Harten B, Courant MNJ, Scheltens P, et al (2004). Validation of the HIV dementia scale in an elderly cohort of patients with subcortical cognitive impairment caused by subcortical ischaemic vascular disease or a normal pressure hydrocephalus. Dement Geriat Cogn Disord 18(1): 109–114.

Váquez Justo E, Alvarez MR, Otero MJF (2003). Influence of depressed mood on neuropsychologic performance in HIV-seropositive drug users. Psychiatry Clin Neurosci 57 (3): 251–258.

Villa G, Solida A, Moro E, et al (1996). Cognitive impairment in asymptomatic stages of HIV infection. A longitudinal study. Eur Neurol 36(3): 125–133.

Wachsler Felder JL, Golden CJ (2002). Neuropsychological consequences of HIV in children: a review of current literature. Clin Psychol Rev 22(3): 441–462.

Woods SP, Grant I (2005). Neuropsychology of HIV. In: HE Gendelman, I Grant, I Everall, SA Lipton, S Swindells (Eds.), The Neurology of AIDS. 2nd edn. Oxford University Press, London, pp. 607–616.

Young C, Atkinson H, Lazzeretto D, et al (2005). Alcohol use disorders and risk of neurocognitive impairment associated with HIV: is there a relationship? Psychosom Med 67(1): A-43.

Zachary RA (1991). Shipley Institute of Living Scale Revised Manual. Western Psychological Services, Los Angeles, CA.

Further Reading

Basso MR, Bornstein RA (2000). Effects of immunosuppression and disease severity upon neuropsychological functioning in HIV infection. J Clin Exp Neuropsychol 22(1): 104–114.

Carey CL, Woods SP, Gonzalez R, et al (2004). Predictive validity of global deficit scores in detecting neuropsychological impairment in HIV infection. J Clin Neuropsychol 26(3): 307–319.

Grant I, Sacktor N, McArthur J (2005). HIV neurocognitive disorders. In: HE Gendelman, I Grant, I Everall, SA Lipton, S Swindells (Eds.), The Neurology of AIDS. 2nd edn. Oxford University Press, London, pp. 357–373.

Handbook of Clinical Neurology, Vol. 85 (3rd series)
HIV/AIDS and the Nervous System
P. Portegies, J. R. Berger, Editors

Chapter 8

HIV myelopathy

ALESSANDRO DI ROCCO*

New York University School of Medicine, New York, NY, USA

8.1. Introduction

The primary myelopathy associated with HIV is usually referred to as vacuolar myelopathy (VM) for its typical pathologic appearance. It is a common complication of AIDS, although it is often unrecognized. In many patients the clinical manifestations of VM are in fact limited to non-specific sphincter or sexual symptoms, and may remain completely asymptomatic. Even when motor and sensory symptoms become evident, the diagnosis is often complicated by a concomitant neuropathy. The subtle clinical presentation and overlapping signs and symptoms with other neurological complications may explain why it is often unrecognized in the clinical setting (Jabarri et al., 1993; Iragui et al., 1994). Although VM may manifest early in the course of the infection, it usually becomes clinically evident only in the late stages of the disease.

In the pre-highly active antiretroviral therapy (HAART) era autopsy series showed vacuolization of the thoracic white matter cord consistent with VM in 20–55% of all autopsied AIDS patients (Petito et al., 1985; Artigas et al., 1990; Dal Pan et al., 1994). There is remarkably little clinical or pathological data on the incidence and prevalence of VM after the widespread introduction of HAART. As with other neurological complications of AIDS, the disease has probably become less common, although it is unknown whether HAART has effectively reduced the incidence of spinal cord vacuolization or just mitigated the severity and progression of the demyelinating process.

As discussed in another chapter of this book, the spectrem of neurological and non-neurological complications of HIV infection in resource-poor countries is very different. Spinal cord disease is commonly observed, but it is usually due to concomitant infections with other pathogens, including tuberculosis, HTLV-1 co-infection, syphilis, or other opportunistic infections (Zenebe, 1995; Bhigjee et al., 2001) rather than VM. Whether this is due to different risk factors, nutritional status, possible genetic susceptibility, specific viral genomics and, above all, lack of widespread use of antiretroviral therapy remains unknown.

8.2. Pathology

The first large autopsy-based observation of spinal cord disease by Petito and colleagues, in 1985, reported vacuolization of the spinal cord in 20 of 89 consecutive AIDS patients (Petito et al., 1985). The high prevalence of spinal cord vacuolization in AIDS was later confirmed in other large pathologic series (Artigas et al., 1990; Dal Pan et al., 1994).

VM usually affects the mid portions of the thoracic cord. Less frequently the vacuolization may be present in other segments of the spinal cord, although cervical or lumbar vacuolization is rarely observed.

The histopathological hallmark of VM is a patchy vacuolization of white matter that is mostly confined to the lateral and posterior columns, with sparse macrophage infiltration surrounding the areas of vacuolization (Petito et al., 1985). Children with AIDS usually do not have vacuolization, but often have diffuse demyelination, axonal loss, multinucleated giant cells and prominent inflammatory infiltrates (Dickson et al., 1989; Sharer et al., 1990).

At a microscopic level, the disease is characterized by the presence of peri-axonal and intramyelinic vacuoles filled with foamy macrophages in the lateral and posterior columns of the spinal cord. Macrophages may be found either in the thin myelin sheath that surrounds the vacuole or, less frequently, within the

*Correspondence to: Alessandro Di Rocco, MD, Chief, Division of Movement Disorders, New York University School of Medicine, 650 First Avenue, New York, NY 10016, USA. E-mail: alessandro.dirocco@med.nyu.edu, Tel: +1-212-263-4838, Fax: +1-212-263-4837.

vacuole. Multinucleated giant cells and microglial nodules are occasional findings in the proximity of the vacuoles, without, however, lymphocytic infiltrates (Budka et al., 1988; Maier et al., 1989; Tan et al., 1995).

The spinal cord lesions may show variable evolution. Often early lesions progress to complete demyelination of axons which then become surrounded by fibrillary astrocytic processes. Some of these lesions can be partially remyelinated by oligodendrocytes while other lesions progress to necrosis with replacement by macrophages and astrocytic gliosis. The axons can preserve their integrity in areas of mild or moderate vacuolization, but in cases of more severe vacuolization the axons are disrupted (Petito et al., 1985; Budka et al., 1988; Maier et al., 1989; Artigas et al., 1990; Dal Pan et al., 1994, Tan et al., 1995; Rottnek et al., 2002).

A grading system has been developed to illustrate the pathological progression of VM that is based on the degree of vacuolization and correlates the severity of the vacuolization to the severity of symptoms of myelopathy (Petito et al., 1985; Artigas et al., 1990; Dal Pan et al., 1994; Tan et al., 1995) (Table 8.1).

8.3. Pathogenesis

The pathogenesis of VM remains elusive. It is debated whether immunological and biochemical mechanisms rather than direct viral infection may be responsible for the white matter disease.

A number of observations suggest that the direct infection of the spinal cord by HIV may not be the primary mechanism of disease. HIV is usually found only within macrophages but not in neuronal cells or microglia and there is no correlation between the presence of HIV in the spinal cord and the development of myelopathy (Eilbott et al., 1989; Rosenblum et al., 1989; Petito et al., 1994). Neither is there a correlation between the clinical severity of myelopathy and a number of markers of HIV infection including CD^4 count or presence of HIV in plasma and cerebrospinal fluid (CSF) (Rosenblum et al., 1989; Geraci et al., 2000).

A possible pathogenetic model implying a metabolic dysfunction of the trans-methylation pathway has been proposed. VM has striking pathological resemblance with the subacute combined degeneration associated with vitamin B12 deficiency (Petito et al., 1985), but the vacuolization in the course of HIV infection cannot be explained by cobalamin or folate deficiency, nor has treatment with vitamin B12 any effect on the myelopathy (Keibutz et al., 1991). It is possible however that HIV or the host response induces a metabolic interference with the cobalamin-dependent trans-methylation pathway, crucial for the synthesis of S-adenosyl-methionine (SAM), the major methyl group donor in the nervous system (Scott et al., 1981). In the white matter, methylation of myelin basic protein is essential in myelin formation, stabilization and repair, and any metabolic interference in the pathway leading to SAM production can cause spinal cord vacuolization (Scott et al., 1994; Castagna et al., 1995; Kim et al., 1997).

There is significant evidence that in VM, extracellular methyl group donors are severely depressed even in the absence of vitamin B12 and folate deficiency, and that the severity of the biochemical deficit is associated with the severity of myelopathy (Di Rocco et al., 2002). The cause of this metabolic abnormality is unknown, but it is possible that infected macrophages in areas of vacuolization may express cytokines that may inhibit trans-methylation, or that the virus itself may inhibit the trans-methyl cycle.

A number of functional polymorphisms affecting the methylation cycle have also been implied, but there is, at the moment, no obvious genetic vulnerability to the disease (Bottiglieri et al., 2002; Diaz-Arrastia et al., 2004).

Table 8.1

Clinico-pathological correlation in HIV-associated vacuolar myelopathy

	Pathological findings	Clinical findings
Grade I	<15 vacuoles per transverse section	Asymptomatic
Grade II	Numerous, non-confluent vacuoles	Rare myelopathy (20%)
Grade III	Areas of confluent vacuolization	Frequent myelopathy (80%)

Adapted from Dal Pan et al. (1994).

8.4. Clinical manifestations

Vacuolar myelopathy is usually a late complication of HIV infection (Petito et al., 1985; Di Rocco, 1999), although it has been occasionally described in early stages of the diseases or as the first manifestation of HIV infection (Eyer-Silva et al., 2001).

The clinical course of the disease is insidious, with symptoms developing slowly over weeks or months. Initial symptoms are often subtle and nonspecific, with fatigue and sphincter abnormalities such as constipation and urinary urgency and frequency, or erectile dysfunction in men (Di Rocco et al., 1998; Berger and Sabet, 2002). A spastic paraparesis may eventually develop, with slowly progressing weakness in the lower extremities that is typically asymmetric, and gait abnormalities that can progress to severe paraparesis often accompanied by painful spasms. Walking becomes progressively more difficult with a spastic or ataxo-spastic gait that can lead to complete ambulatory disability. Sensory abnormalities, in the absence of a concomitant neuropathy, are usually mild, with mild dysesthesias or numbness in the feet. Sensory symptoms such as paresthesias, numbness, burning and tingling in the lower extremities are rarely severe in VM.

Proprioceptive deficits may however result in ataxia of the lower limbs, and further impair the gait (Petito et al., 1985; Di Rocco et al., 1998; Berger and Sabet, 2002). On examination the spasticity of the legs is accompanied by hyper-reflexia at the knees and ankles and extensor plantar responses with clonus. The abnormalities on sensory examination typically are limited to diminished or loss of vibration and proprioception in the lower extremities, more profound distally, with relative preservation of pain and temperature sensation. A discrete sensory level is usually absent in VM, and its presence strongly suggests an alternative cause of spinal cord disease, especially focal and structural lesions.

As the spinal cord vacuolization is in general limited to the thoracic segments, the arms are usually spared, except in rare or very advanced cases, when the vacuolization spreads to the cervical segments.

It is important to stress that in many patients the myelopathy develops in concomitance with distal sensory polyneuropathy (DSP), which may mask some of the symptoms of VM and complicate the interpretation of the neurological examination. The overlapping peripheral neuropathy defines a frequent clinical picture of myeloneuropathy, with brisk reflexes at the knees in the presence of reduced or absent ankle reflexes and complex patterns of sensory deficit. Typically however, abnormal plantar reflexes can be detected even in cases of severe neuropathy. VM is not usually associated with pain, and thermal and pain sensations are therefore suggestive of neuropathy.

Cognitive impairment is also usually absent in VM, although the relationship between VM and AIDS dementia has not been defined. It has been postulated that VM may be associated with white matter changes in the central nervous system that may contribute to cognitive deficits (Budka et al., 1991; Schmidbauer et al., 1992; Power et al., 1993; Price, 1994).

8.5. Diagnosis

The diagnosis of VM is mostly by exclusion, based primarily on the slowly evolving typical symptoms and signs of myelopathy (or myeloneuropathy) and a diagnostic work aimed at excluding other pathologies affecting the spinal cord.

Specific clinical findings that suggest other causes of myelopathy include the presence of a discrete sensory level, a rapid progression of myelopathy over days or weeks, CSF pleocytosis greater than 30 cells/ml and back pain, all findings that are not associated with VM.

The differential diagnoses (summarized in Table 8.2) include myelopathy due to HTLV-1 or HTLV-2, acute HIV myelitis, toxoplasmosis, lymphoma, tuberculosis, cytomegalovirus infection or other opportunistic infections affecting the spinal cord or causing cord compression by affecting vertebral and paraspinal structures. In nutritionally deficient patients, a true subacute combined degeneration of the spinal cord due to cobalamin or folate deficiency can be present (Harriman et al., 1989).

Table 8.2

Differential diagnosis of most common forms of myelopathy associated with HIV infection

	Type of progression	Clinical manifestations	Laboratory findings
Vacuolar myelopathy	Slow	Transverse myelitis (TM) (no sensory level)	Normal or nonspecific elevated CSF proteins
HIV myelitis	Acute	TM (frequent sensory level; back pain)	CSF pleocytosis (>30 cell/ml)
HTLV-1 or -2 myelitis	Slow	TM (frequent sensory level)	CSF anti-HTLV Ab +
Toxoplasmosis	Slow	TM	Toxo +; MRI lesions
Tuberculosis	Slow	TM	PCR; mononuclear pleocytosis; hypoglycorrhachia
Other opportunistic infections	Acute or subacute	TM (radiculomyelopathy or cone syndrome in CMV)	CSF anti-virus Ab + or PCR +
Lymphoma	Slow or subacute	TM or myeloradiculitis	CSF neoplastic cells; MRI +

Serologic and imaging studies are therefore indicated in all patients with HIV myelopathy to rule out secondary causes of spinal cord disease.

Magnetic resonance imaging (MRI) is usually normal. Mild atrophy of the thoracic cord or, less frequently, nonspecific areas of increased signal on T-2 images (Barakos et al., 1990; Santosh et al., 1995; Sartoretti-Schefer et al., 1997; Chong et al., 1999) have been reported, but these findings are nonspecific and do not correlate with the severity of the disease. CSF studies are also usually normal or reveal mild pleocytosis (<20 cells/mm^3) and nonspecific increase in protein content.

Electrophysiological tests are essential to confirm the diagnosis of VM. Somatosensory evoked potentials (SEP) can reliably measure abnormal conduction of bioelectrical signals within the spinal cord and may also be useful to detect subclinical myelopathy (Jabarri et al., 1993; Iragui et al., 1994) or to monitor disease progression and therapeutic outcome. They typically demonstrate abnormal central conduction time (CCT), often associated with electrophysiological evidence of peripheral nerve disease (Helweg-Larsen et al., 1988; Tagliati et al., 2000). When the peripheral neuropathy limits the ability to effectively stimulate distally, posterior tibial stimulation increases the diagnostic yield of SEP, bypassing the distal peripheral nerve abnormalities (Tagliati et al., 2000). A more precise electrophysiological measure of spinal cord dysfunction may be derived by focusing the evaluation on the thoracolumbar segments of the spinal cord, which are primarily affected in VM, calculating a derived spinal conduction time by subtracting the median nerve CCT from the posterior tibial or peroneal nerve CCT (Tagliati et al., 2000).

8.6. Treatment

There is no specific treatment for VM. The role of antiretroviral agents in the treatment of VM is also not well defined. It is estimated that, as in other neurological complications of HIV, the incidence of the disease has decreased with the introduction of HAART. However, except for a small uncontrolled study reporting remission of HIV myelopathy with zidovudine (Osenhendler et al., 1990) and a few case reports of remission of myelopathy after treatment with HAART (Staudinger and Henry, 2000; Fernandez-Fernandez et al., 2004), the therapeutic efficacy of specific antiretroviral agents or HAART in VM has not been studied, and it is still unknown whether these agents can prevent or modify the progression of myelopathy. It is established, however, that VM continues to present even in patients who have been on chronic treatment with HAART and have no other HIV-related neurological problems.

There have been very few clinical trials for VM. Small open-label trials with intravenous gammaglobulins or corticosteroids have proven ineffective (Di Rocco, 1999). The pathological similarity of VM with the subacute combined degeneration of the spinal cord prompted a study with vitamin B12 supplementation, which, however, did not produce any significant improvement (Keibutz et al., 1991).

The only double-blind placebo-controlled study of treatment of VM used high doses of L-methionine (6 g/day), the direct metabolic precursor of SAM. Although patients treated with L-methionine exhibited a non-significant improvement in electrophysiological parameters, overall the study did not demonstrate any significant clinical benefit. Further, in some of the patients treated with L-methionine there was an increase in serum homocysteine (Di Rocco et al., 2004). Despite the negative results of the study, it is possible that more potent methyl donors, including direct supplementation with SAM, may be effective in the treatment of VM.

In the absence of a reliable treatment, symptomatic management of spasticity and sphincter dysfunction remains an important therapeutic goal. Baclofen, dantrolen, tizanidine and botulin toxin are useful in the management of spasticity, while bladder training and anticholinergic drugs or alpha-adrenergic blockers may be used for sphinteric dysfunction. Physical and occupational therapy also play an important role in limiting the effect of progressive disability.

References

Artigas J, Grosse G, Niedobitek F (1990). Vacuolar myelopathy in AIDS. A morphological analysis. Pathol Res Pract 186: 228–237.

Barakos JA, Mark AS, Dillon WP, et al (1990). MR imaging in acute transverse myelitis and AIDS myelopathy. J Comput Assist Tomogr 14: 45–50.

Berger JR, Sabet A (2002). Infectious myelopathies. Semin Neurol 22(2): 133–142.

Bhigjee AI, Madurai S, Bill PL, et al (2001). Spectrum of myelopathies in HIV seropositive South African patients. Neurology 57: 348–351.

Bottiglieri T, Ozelius L, Godbold J, et al (2002). Frequency of the MTHFR (C677T) and MetSyn (A2756G) functional polymorphisms in patients with AIDS myelopathy. Neurology 56(Suppl 3): A475.

Budka H, Maier H, Pohl P (1988). Human immunodeficiency virus in vacuolar myelopathy of the acquired immunodeficiency syndrome. N Engl J Med 319: 1667–1668.

Budka H, Wiley CA, Kleihues P, et al (1991). HIV associated disease of the nervous system: review of nomenclature and proposal for neuropathology based terminology. Brain Pathol 1: 143–152.

Castagna A, Le Grazie C, Accordini A, et al (1995). Cerebrospinal fluid S-adenosylmethionine (SAMe) and glutathione concentrations in HIV infections: effect of parenteral treatment with SAMe. Neurology 45: 1678–1683.

Chong J, Di Rocco A, Danisi F, et al (1999). MR abnormalities in AIDS-associated vacuolar myelopathy. AMJNR Am J Neuroradiol 20: 1412–1416.

Dal Pan GJ, Glass JD, McArthur JC (1994). Clinicopathologic correlation of HIV-1 associated vacuolar myelopathy: an autopsy based case-control study. Neurology 44: 2159–2164.

Di Rocco A (1999). Diseases of the spinal cord in human immunodeficiency virus infection. Semin Neurol 19: 151–155.

Di Rocco A, Tagliati M, Danisi F, et al (1998). L-methionine for AIDS-associated vacuolar myelopathy. Neurology 51: 266–268.

Di Rocco A, Bottiglieri T, Werner P, et al (2002). Abnormal cobalamin-dependent transmethylation in AIDS-associated myelopathy. Neurology 58(5): 730–735.

Di Rocco A, Werner P, Bottiglieri T, et al (2004). Treatment of AIDS-associated myelopathy with L-methionine: a placebo controlled study. Neurology 63: 1270–1275.

Diaz-Arrastia R, Gong Y, Kelly CJ, et al (2004). Host genetic polymorphisms in human immunodeficiency virus-related neurologic disease. J Neurovirol 10(Suppl 1): 67–73.

Dickson DW, Belman AL, Kim TS, et al (1989). Spinal cord pathology in pediatric acquired immunodeficiency syndrome. Neurology 39: 227–235.

Eilbott DJ, Peress N, Burger H, et al (1989). Human immunodeficiency virus type 1 in spinal cords of acquired immunodeficiency syndrome patients with myelopathy: expression and replication in macrophages. Proc Natl Acad Sci USA 86: 3337–3341.

Eyer-Silva WA, Auto I, Pinto JF, et al (2001). Myelopathy in a previously asymptomatic HIV-1-infected patient. Infection 29(2): 99–102.

Fernandez-Fernandez FJ, de la Fuente Aguardo J, Ocampo-Hermida A, et al (2004). Remission of HIV-associated myelopathy after highly active antiretroviral therapy. J Postgrad Med 50: 195–196.

Geraci A, Di Rocco A, Liu M, et al (2000). AIDS myelopathy is not associated with elevated HIV viral load in cerebrospinal fluid. Neurology 55: 440–442.

Harriman GR, Smith PD, Horne MK, et al (1989). Vitamin B12 malabsorption in patients with acquired immunodeficiency syndrome. Arch Intern Med 149: 2039–2041.

Helweg-Larsen S, Jakobsen J, Boesen F, et al (1988). Myelopathy in AIDS. A clinical and electrophysiological study of 23 Danish patients. Acta Neurol Scand 77: 64–73.

Iragui VJ, Kalmijin J, Thal L, et al (1994). Neurological dysfunction in asymptomatic HIV-1 infected men: evidence from evoked potentials. Electroenceph Clin Neurophysiol 2: 1–10.

Jabarri B, Coats M, Salazar A, et al (1993). Longitudinal study of EEG and evoked potentials in neurologically asymptomatic HIV infected subjects. Electroenceph Clin Neurophysiol 86: 145–151.

Keibutz KD, Giang DW, Schiffer RB, et al (1991). Abnormal vitamin B12 metabolism in human immunodeficiency virus infection. Arch Neurol 48: 312–314.

Kim S, Lim IK, Park GH, et al (1997). Biological methylation of myelin basic protein: enzymology and biological significance. Int J Biochem Cell Biol 29: 743–751.

Maier H, Budka H, Lassmann H, et al (1989). Vacuolar myelopathy with multinucleated giant cells in the acquired immune deficiency syndrome (AIDS). Light and electron microscopic distribution of human immunodeficiency virus (HIV) antigens. Acta Neuropathol 78: 497–503.

Osenhendler E, Ferchal F, Cadranel J, et al (1990). Zidovudine for HIV-related myelopathy. Am J Med 88: 565N–566N.

Petito CK, Navia BA, Cho ES, et al (1985). Vacuolar myelopathy pathologically resembling subacute combined degeneration in patients with the acquired immunodeficiency syndrome. N Engl J Med 312: 874–879.

Petito CK, Vecchio D, Chen YT (1994). HIV antigen and DNA in AIDS spinal cords correlate with macrophages infiltration but not with vacuolar myelopathy. J Neuropathol Exp Neurol 53: 86–94.

Power C, Kong PA, Crawford TO, et al (1993). Cerebral white matter changes in acquired immune deficiency syndrome dementia: alterations of the blood brain barrier. Ann Neurol 34: 339–350.

Price RW (1994). Understanding the AIDS dementia complex (ADC). The challenge of HIV and its effects on the central nervous system. Res Publ Assoc Res Nerv Ment Dis 72: 1–45.

Rosenblum M, Scheck AC, Cronin K, et al (1989). Dissociation of AIDS-related vacuolar myelopathy and productive HIV-1 infection of the spinal cord. Neurology 39: 892–896.

Rottnek M, Di Rocco A, Laudier D, et al (2002). Axonal damage is a late component of vacuolar myelopathy. Neurology 58: 479–481.

Santosh CG, Bell JE, Best JJ (1995). Spinal tract pathology in AIDS: postmortem MRI correlation with neuropathology. Neuroradiology 37: 134–138.

Sartoretti-Schefer S, Blattler T, Wichmann W (1997). Spinal MRI in vacuolar myelopathy, and correlation with histopathological findings. Neuroradiology 39: 865–869.

Schmidbauer M, Huemer M, Cristina S, et al (1992). Morphological spectrum, distribution and clinical correlation of white matter lesions in AIDS brains. Neuropathol Appl Neurobiol 18: 489–501.

Scott JM, Dinn JJ, Wilson P, et al (1981). Pathogenesis of subacute combined degeneration: a result of methyl group deficiency. Lancet 2(8242): 334–337.

Scott JM, Molloy AM, Kennedy DG, et al (1994). Effects of destruction of transmethylation in the central nervous system: an animal model. Acta Neurol Scand 154(Suppl): 27–31.

Sharer LR, Dowling PC, Michaels J, et al (1990). Spinal cord disease in children with HIV-1 infection: a combined molecular biology and neuropathological study. Neuropathol Appl Neurobiol 16: 317–331.

Staudinger R, Henry K (2000). Remission of HIV myelopathy after highly active antiretroviral therapy. Neurology 54: 267–268.

Tagliati M, Di Rocco A, Danisi F, et al (2000). The role of somatosensory evoked potentials in the diagnosis of AIDS-associated myelopathy. Neurology 54: 1477–1482.

Tan SV, Guiloff RJ, Scaravilli F (1995). AIDS-associated vacuolar myelopathy. A morphometric study. Brain 118: 1247–1261.

Zenebe G (1995). Myelopathies in Ethiopia. East Afr Med J 72: 42–45.

Handbook of Clinical Neurology, Vol. 85 (3rd series)
HIV/AIDS and the Nervous System
P. Portegies, J. R. Berger, Editors

Chapter 9

Peripheral neuropathy in HIV infection

SUSAMA VERMA AND DAVID M. SIMPSON*

Mount Sinai Medical Center, New York, NY, USA

9.1. Introduction

In the context of a global pandemic, AIDS has affected millions of people worldwide. Following the introduction of highly active antiretroviral therapy (HAART) in 1996, the epidemiology of HIV and AIDS has changed, with a dramatic impact on morbidity and mortality. Improved use of chemoprophylaxis has also contributed to epidemiologic changes in the neurological complications. There has been a decline in the incidence of many of the neurological manifestations of HIV infection, but the prevalence of some of these complications have increased as a result of a larger pool of long-term survivors, providing the opportunity for patients to develop late-stage neurological disorders. Thus, in the HAART era, HIV is now considered to be a chronic illness. The incidence of HIV infection is also increasing among the late middle-aged and older adults (Savasta, 2004; Verma et al., 2004).

HIV-1 infection is associated with a variety of central and peripheral nerve diseases. The most frequent neurologic complication in the context of HIV infection is peripheral neuropathy. Distal symmetrical sensory polyneuropathy (DSP) is the most common form of neuropathy in HIV infection. Other patterns of neuropathy include mononeuropathy multiplex, inflammatory demyelinating polyneuropathy and progressive polyneuropathy. Pain is the major symptom of HIV peripheral neuropathy, with major impact on activities of daily living and quality of life. HIV-associated neuropathies pose a challenge in diagnosis and management. This chapter reviews the various forms of neuropathy associated with HIV infection, and their pathogenetic mechanisms and treatment strategies.

9.2. Types of HIV neuropathy

Peripheral neuropathy is subdivided into different types. These include distal symmetrical peripheral neuropathy (DSP), toxic neuropathies (TN), mononeuropathy multiplex (MM; including brachial and lumbar plexopathy), inflammatory demyelinating polyneuropathy (IDP; acute or chronic type) and progressive polyradiculopathy (PP).

The types of peripheral neuropathy and their symptoms, signs, mechanisms, diagnoses and therapy are summarized in Table 9.1.

9.2.1. Distal symmetrical polyneuropathy

9.2.1.1. Epidemiology

DSP is the most commonly reported type of neuropathy in HIV infection, affecting approximately one-third of individuals. Concomitant conditions such as diabetes mellitus, alcohol abuse (Lopez et al., 2004), uremia, vitamin B12 or thiamine deficiencies, weight loss and low albumin or hemoglobin levels may increase the risk of development of HIV-associated neuropathy.

Pre-HAART data indicate that advanced immunosuppression, as reflected by depressed CD^4 lymphocyte count, and increased plasma HIV RNA, is a risk factor for development of DSP. Childs et al. (1999) have reported a 2.3-fold greater risk for acquiring DSP if plasma HIV RNA is greater than 10,000 copies/ml versus less than 500 copies/ml prior to the initiation of antiretroviral therapy. Data from the Johns Hopkins HIV clinical cohort study indicate that the prevalence of both HIV-associated and toxic neuropathy has increased through 2000, which may be a result

*Correspondence to: David M. Simpson, MD, Professor of Neurology, Director, Clinical Neurophysiology Laboratories, Director, Neuro-AIDS Program, Mount Sinai Medical Center, One Gustave Levy Place, Box 1052, New York, NY 10029, USA. Email: david.simpson@mssm.edu, Tel: +1-212-241-8748, Fax: +1-212-987-3301.

Table 9.1

Peripheral neuropathies in HIV disease

Type of neuropathy	Presenting symptoms	Neurological signs	Mechanism	Diagnosis studies	Therapy
Distal symmetrical polyneuropathy (DSP)	Gradual in onset	Decreased/absent ankle reflexes with intact knee reflexes	Immune dysfunction	Rule out metabolic, infectious and nutritional causes	Symptomatic relief of pain management as demonstrated by the WHO ladder
	Distal symmetrical (glove and stocking distribution) sensory loss, paresthesia and burning pain	Pinprick, temperature and vibration impaired	Macrophage mediated	Nerve conduction studies: distal, symmetrical axonopathy with diminished SNAP	Trycyclic antidepressants (nortriptyline, desipramine, amitriptyline)
			Axonal injury	QST helpful	Anticonvulsants (phenytoin, carbamazepine, lamotrigine, gabapentin)
				Skin biopsy	Topical lidocaine and capsaicin
Toxic neuropathy (TN)	Neuropathic pain	Sensory and motor impairment	Mitochondrial damage from d-drugs	Same as DSP	Discontinue the offending agent Same as DSP
Mononeuropathy multiplex (MM)	Facial or extraocular muscle weakness, wrist drop, foot drop	Cranial neuropathies, wrist drop, foot drop	Immune mediated	Electrophysiologic studies	Immunomodulating agents: corticosteroids, plasmapheresis, IVIG
			Immune-complex deposition	CSF analysis	Empiric therapy for CMV
			Vasculitic neuropathy	Sural nerve biopsy	
			CMV: demyelinating neuropathy		
Inflammatory demyelinating polyneuropathy (IDP)	Progressive weakness in lower extremities and then upper extremities, paresthesia	Motor weakness, arreflexia and mild sensory loss	Immune dysfunction	CSF with mild increase WBC and moderate increase protein	Plasmapheresis, steroids, IVIG
			Demyelinating neuropathy	NCS: demyelination	
Progressive polyradiculopathy (PP)	Lower extremities weakness, paresthesia and urinary dysfunction	Flaccid paraparesis, decreased reflexes and urinary retention	CMV infection	CSF with pleocytosis and positive CMV culture	Anti-CMV: ganciclovir, foscarnet
			Necrotizing neuropathy	MRI scan of lumbosacral spine may show enhancement of nerve roots Electrophysiologic studies: polyradiculopathy	

of prolonged survival (McArthur, 2003). The DANA (pre-HAART) prospective cohort study, which enrolled subjects with CD^4 lymphocyte <200 cells/mm^3 (Schifitto et al., 2002), revealed that the probability of developing symptomatic HIV sensory neuropathy is 36% after 12 months and 52% after 24 months. Recent data from HAART era cohorts indicate that the one-year incidence of neuropathy is 21%, which is a 15% decrease from that observed in the pre-HAART DANA cohort, with similar inclusion and exclusion criteria (Schifitto et al., 2005).

Simpson et al. (2002) reported a correlation between plasma HIV-1 RNA levels and severity of pain and quantitative sensory test (QST) results in HIV-associated DSP. However, recent data indicate that previously established risk factors for HIV PN, including plasma HIV RNA and CD^4 count, no longer correlate with either the presence or development of peripheral neuropathy (Simpson et al., 2005).

9.2.1.2. Clinical features

The symptoms of DSP are usually gradual in onset. They consist mainly of sensory disturbances, beginning symmetrically in the distal portions of the extremities (i.e. toes and feet) and ascending proximally to involve the legs, and in advanced or severe stages, the fingers and hands; thus, the terminology "glove and stocking" distribution. The most common sensory symptoms are numbness, paresthesias and burning or lancinating pain, often worse at night with resultant sleep interference. Other complaints include gait instability, heightened perception of pain (hyperesthesia), burning sensation (dysesthesias), pain from non-noxious stimuli (allodynia) or contact sensitivity with bedclothes, shoes and socks (Keswani et al., 2002).

DSP is frequently asymptomatic (Marra et al., 1998), and thus the condition is detected only on neurological examination. Marra et al. (1998) have reported a single site study, comparing the sensitivity and specificity of a screening examination for HIV DSP by non-neurologist clinicians, as compared with neurologists. They found a good correlation between diagnoses established by the clinicians and neurologists, and a high rate of asymptomatic DSP, with 71% of DSP patients lacking symptoms.

The most common neurological signs in DSP are reduced or absent ankle reflexes with intact patellar (knee) reflexes. It is not uncommon in HIV infection to note the combination of depressed ankle reflexes and hyperactive knee reflexes, indicating the presence of peripheral and central nervous system diseases. HIV may preferentially affect small nerve fibers, in which ankle reflexes may be relatively preserved. Pinprick, temperature and vibration modalities are impaired or reduced in a distal distribution. Proprioception may be relatively spared, except in long-standing and severe cases.

9.2.1.3. Laboratory features

DSP is distinguished from other forms of neuropathy based on the clinical history, neurological examination and appropriate laboratory studies. Results of electrophysiologic studies suggest a dying-back axonopathy. Nerve conduction studies show characteristics of distal and symmetrical sensorimotor axonopathy with a reduction in sensory nerve action potential (SNAP) and reduced sensory and motor conduction velocities. Electromyography may show active and chronic denervation with reinnervation, particularly in distal muscles.

When symptoms are predominantly due to small nerve fiber involvement, quantitative sensory test (QS) is a helpful diagnostic test to assess the thresholds for cooling, and heat pain. Cerebrospinal fluid (CSF) analysis is not routinely necessary for the diagnosis of DSP, unless there is a suspicion of rare infectious, inflammatory or neoplastic etiologies. Skin biopsy with epidermal nerve fiber density analysis is a diagnostically useful and simple test (Figs 9.1 and 9.2). In subjects without

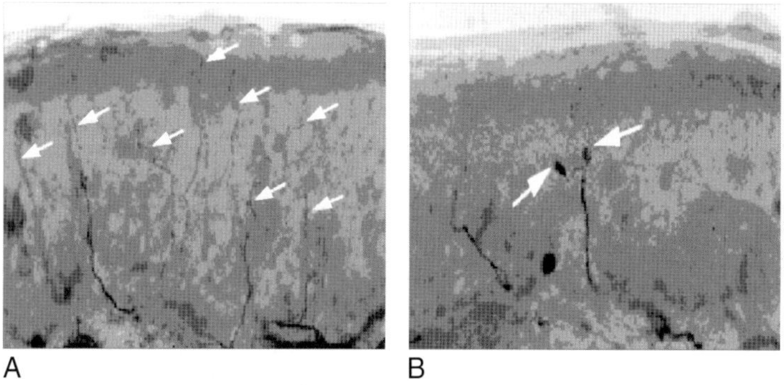

Fig. 9.1. HIV sensory neuropathy. Skin biopsy to assess unmyelinated nerve fibers. A. Thigh: normal density; B. distal leg: reduced density and nerve fiber swellings. Courtesy of Justin C. McArthur MBBS MPH.

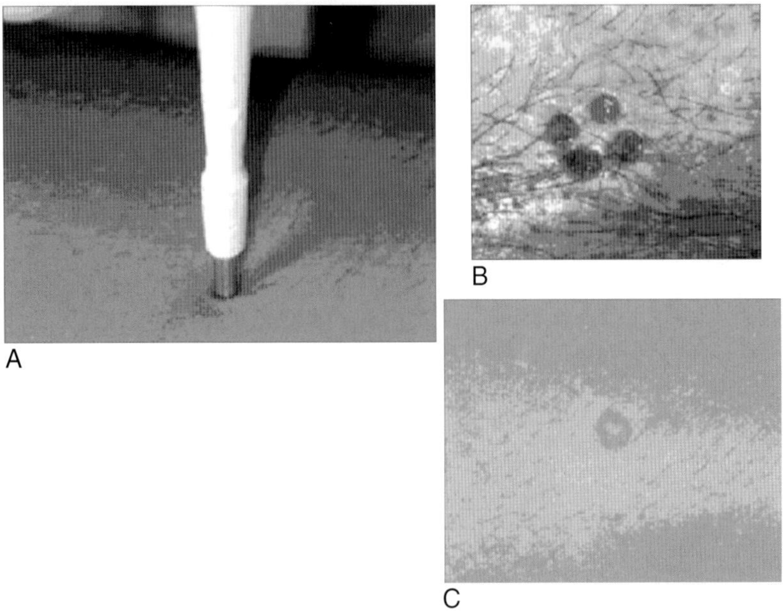

Fig. 9.2. Skin biopsy technique. A. 3 mm punch; B. fresh punch site; C. healed scar after 2 weeks. Courtesy of Justin C. McArthur MBBS MPH.

HIV neuropathy or with asymptomatic DSP, epidermal nerve fiber densities are similar to those of HIV seronegative controls. However, subjects with symptomatic DSP have significant reduction in epidermal nerve fiber density (Schifitto et al., 2002).

9.2.1.4. Pathogenesis

The pathogenesis of primary HIV-associated DSP remains undetermined but may involve both viral-associated neurotoxicity and immune activation (Yoshioka et al., 1994). There is no evidence to support HIV infection of nerve roots, dorsal root ganglion (DRG) or peripheral nerve. The predominant pathology in HIV neuropathy is axonal degeneration and loss of the distal region of the long axons of the sensory neurons, termed the "dying back" phenomenon (Pardo et al., 2001). Loss of unmyelinated axons in the distal regions of the sensory nerves is followed by the Wallerian degeneration of the distal myelinated fibers.

Nerve damage in HIV disease is associated with immune dysregulation and macrophage activation within the peripheral nervous system (Tyor et al., 1995). Immunohistochemistry reveals activation of macrophages in the epineurium (Griffin et al., 1998) of peripheral nerves and dorsal nerve ganglia. These reactive macrophages in the areas of axonal degeneration cause local release of proinflammatory neurotoxic cytokines with expression of TNF-alpha, interleukin (IL)-1 and IL-6, as detected by in situ PCR. Appropriate cytokine response is important for the clearance of the pathogen but a prolonged or excessive response may result in pathology (Wesselingh et al., 1994).

Studies in neuritic cell culture have shown evidence that HIV virus envelope glycoprotein, gp120, induces mitochondrial pathology. Mitochondrial membrane depolarization in the DRG sensory neurons has been noted six hours after exposure to the neurotoxicity secreted by gp120 (Keswani et al., 2003). Data are lacking for the in vivo mitochondrial toxicity directly by HIV virus.

9.2.2. Toxic neuropathy

Toxic neuropathy (TN) commonly occurs in HIV infection. TN is associated with neurotoxic adverse effects of the dideoxynucleoside analogues of reverse transcriptase inhibitors (ddN). Early clinical trials showed that the effects of specific ddN, including didanosine (ddI), stavudine (d4T) and zalcitabine (ddC), are dose-dependent (Simpson and Tagliati, 1995), and may trigger, aggravate or unmask painful DSP (Yarchoan et al., 1990). ddC is more potent than ddI, which is more potent than d4T in causing sensory neurotoxicity. This hierarchy correlates clinically with the incidence of peripheral neuropathy associated with the use of these drugs (Keswani et al., 2004). Other studies indicate that d-drug regimens may be used with little potential for neurotoxicity (Gulick, 1998). These results were based on trials where the enrolled patients had relatively early

HIV disease. Recent studies performed in HAART-exposed cohorts reveal that d-drug ARV therapy does not appear to increase the risk of development of DSP (Schifitto et al., 2002; Lichtenstein et al., 2005). However, one must be cautious in interpreting these results, since these studies were not designed specifically to determine the neurotoxicity of d-drug ARV, and ascertainment bias in these cohort studies may confound the results.

TN is clinically and electrophysiologically indistinguishable from HIV DSP. The onset and worsening of painful symptoms of TN tend to be more abrupt and progressive as compared with HIV-associated DSP (Moyle and Sadler, 1998). Incidence rates of neuropathy increase by 28-fold with the combination use of ddI, d4T and hydroxyurea. Moore et al. (2000) provided an explanation that hydroxyurea supports the integration of ddI and d4T into the mitochondrial matrix, thus increasing the neurotoxicity and risk of neuropathy.

The symptoms of TN may have a temporal (Keswani et al., 2002) relationship with the initiation of the therapy. A decrease or resolution of that component of neuropathic symptoms should occur with discontinuation or decrease of the dosage of drug. There may be a "coasting phenomenon" during which neuropathic symptoms may continue to intensify 2–4 weeks following the withdrawal of neurotoxic drug. Improvement or resolution may follow within 4 to 8 weeks (Blum et al., 1996) after drug withdrawal.

It is not clear to what extent mitochondrial toxicity contributes to the pathogenesis of ddN neurotoxicity (Dalakas et al., 1990; Schifitto et al., 2002). The mechanism of neurotoxicity of peripheral nerve due to d-drug ARV is believed to be secondary to mitochondrial damage. Toxic metabolites of d-drugs inhibit human gamma DNA polymerase required for mitochondrial DNA (mtDNA) replication, leading to disruption of cristae (Lewis and Dalakas, 1995). The metabolites of d-drugs compete with the substrates in the synthesis of neuronal mitochondrial DNA, resulting in depletion of mtDNA (Estanislao et al., 2004). Brew et al. (2003) reported a significant elevation of serum lactate levels in patients with d-drug neuropathy. ddC and ddI decrease mtDNA levels, alter mitochondrial morphology and increase intracellular lactate levels (Keswani et al., 2002). Mitochondrial alteration and DNA depletion have been observed in peripheral nerves of AIDS patients treated with ddI (Dalakas et al., 2001).

While ddN antiretroviral drugs are the major cause of TN in HIV patients, there are many other neurotoxic medications used in this population; i.e. dapsone, metronidazole and antineoplastic agents. The primary treatment of ddN toxic neuropathy is withdrawal of the offending drug. However, this should be done only after a cost–benefit analysis, weighing the need for the ddN drug for HIV virological control, versus the ability to stop the drug, and switch to a non-toxic ARV medication. There are still some patients with multi-drug resistance that must remain on ddN, even in the setting of peripheral neuropathy. Therapeutic agents under study in ddN-associated TN include acetyl-L-carnitine and coenzyme Q (Christensen et al., 2004).

9.3. Management of DSP

Pain is now considered to be the fifth vital sign, and remains one of the most common reasons for consultation with a physician. Because of the variability and subjectivity inherent in patients' reports of pain, as well as evidence that a majority of physicians have inadequate training in pain management, pain is often suboptimally treated. There are significant barriers to optimal pain management, which exist both in patients and health care providers. Some of the major barriers include patients' reluctance to report pain and their fear of addiction. Health care providers have fear of substance abuse, prescription scam and concerns of regulatory oversight, especially with the use of narcotic analgesics. A heightened awareness of a systematic approach for successful pain control consists of disease modification and local or regional measures to achieve an improved level of comfort, function and quality of life.

Pain contributes to considerable psychological and functional morbidity and leads to a diminished quality of life in HIV-infected individuals. Similarly, psychological and emotional features also contribute to pain. Appropriate pain management should be approached from both medical and psychosocial levels. Evans and co-workers noted a role of cognitive-behavioral treatments for HIV-related peripheral neuropathic pain (Evans et al., 2003a), whereas supportive psychotherapy has shown less benefit (Evans et al., 2003a, b). The possible etiologies and risk factors of neuropathy should be initially identified. Metabolic, infectious, malignant and nutritional causes of neuropathic pain may be amenable to treatment. However, the primary approach to DSP remains pain management (Wulff and Simpson, 1998).

As elevated plasma HIV-1 viral load is associated with the intensity of neuropathic pain due to HIV-associated DSP (Simpson et al., 2002), it is important to aggressively optimize HIV virologic control. Martin et al. (2000) have reported improvement of thermal quantitative sensory test (QST) results in HIV neuropathy, in patients that achieved good HIV control. However, it remains unclear whether reduction of HIV viral load results in clinical improvement in neuropathy.

There is no current pharmacological management available to reverse the pathology of HIV-associated

DSP. Thus the primary focus of management is on the relief of symptoms. Multidisciplinary, comprehensive approaches to pain management modalities include non-narcotic and narcotic analgesics, anticonvulsants, tricyclic antidepressants, antidysrhythmics, physical therapy and psychological techniques.

The World Health Organization developed an "analgesic ladder", with recommended approaches to pharmacological management of escalating pain (World Health Organization, 1990, 1996; Fig. 9.3). While this schema was proposed for cancer-related pain, it may be adapted to management of painful HIV neuropathy. Non-narcotic analgesics are recommended for mild to moderate pain. Examples include acetaminophen, nonsteroidal anti-inflammatory agents and topical analgesics, e.g. lidocaine ointment or patch. Capsaicin, which is the pungent ingredient of the hot chili pepper, is currently being investigated as an adjuvant topical analgesic for DSP. Capsaicin's mechanism of action includes binding to the excitatory vanilloid receptor (TRPVR1) in the epidermal and dermal nerve fibers with selective desensitization and reduction in the hyperactivity of the primary afferent nociceptors of the peripheral nerve (Verma et al., 2004). In its low concentration form, topical capsaicin is ineffective in the treatment of painful HIV neuropathy (Paice et al., 2000). An experimental high concentration topical capsaicin patch has been developed and is applied directly to the affected region of the feet. The high concentration capsaicin patch has shown benefit in open label studies of painful HIV neuropathy (Simpson et al., 2004), and the results of a large, double-blind, controlled study are under analysis.

Anticonvulsants may be used as adjuvant analgesics. These include gabapentin, carbamazepine and lamotrigine (Beckonja, 2002). Gabapentin has the advantage of a low side effect profile and minimal drug–drug interactions. Gabapentin is effective and is the most commonly used anticonvulsant in painful diabetic neuropathy and postherpetic neuralgia. Recent data based on the results of a multicenter, prospective, randomized, double-blind, placebo-controlled trial showed gabapentin to be more effective than placebo in reducing pain in HIV-associated sensory neuropathy (Hahn et al., 2004). Pregabalin is a gamma-aminobutyric acid (GABA) analogue with structural similarity and a similar mechanism of action as gabapentin on the alpha-2-delta ligand of voltage-activated calcium channels. In comparison to gabapentin, pregabalin has better bioavailability and more convenient titration. Pregabalin is US Food and Drug Administration (FDA)-approved for the treatment of painful diabetic neuropathic pain, postherpetic neuralgia and partial seizures. A placebo-controlled trial is underway to evaluate pregabalin for the treatment of painful HIV neuropathy.

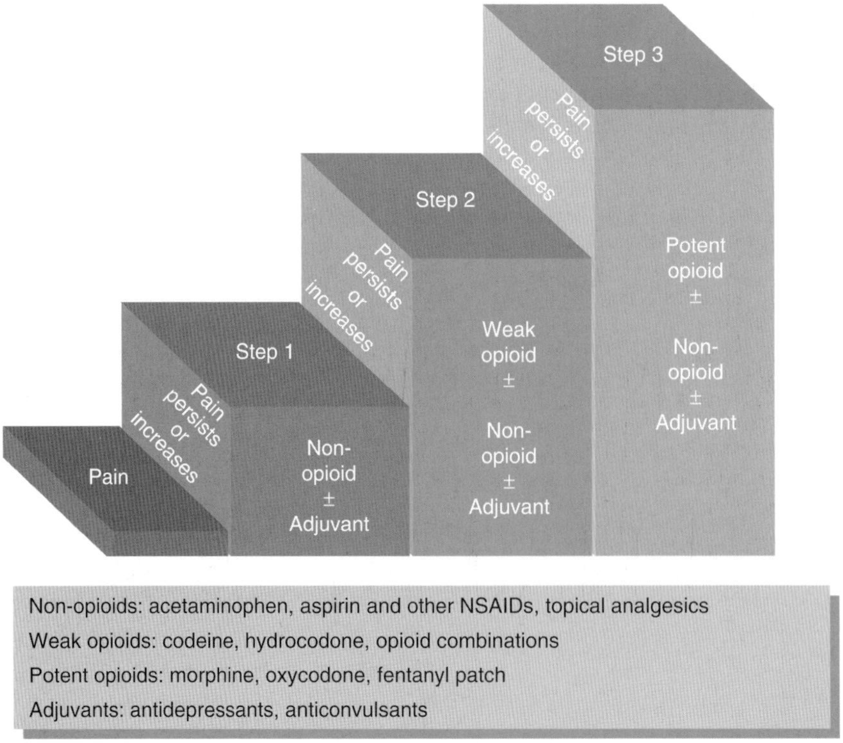

Fig. 9.3. WHO three-step analgesic ladder. (World Health Organization. Copyright free.)

Two randomized, double-blind, placebo-controlled trials of lamotrigine for HIV-associated painful sensory neuropathy showed significant difference in the primary measurement of pain using standardized scales (Simpson et al., 2000, 2003). Notably, in the larger lamotrigine study (Simpson et al., 2003), the beneficial effect was observed only in that group receiving ddN ARV. Most other antidepressants and anticonvulsants have not been adequately studied in HIV PN. Opiate analgesics may also be helpful for optimal control of neuropathic pain. When such treatment modalities fail to adequately control intractable neuropathic pain, specialists in pain management may be consulted.

Neuro-regenerative agents are of great interest in the management of peripheral neuropathy. While recombinant human nerve growth factor reduced pain in HIV neuropathy, it did not show evidence of nerve regeneration in diabetic or HIV neuropathies (McArthur et al., 2000). Acetyl-L-carnitine treatment resulted in increase in epidermal nerve fiber density in a study of ddN neuropathy (Hart et al., 2004). Studies are underway in an effort to replicate these results. Administration of the hormone erythropoietin may also be useful in the treatment of HIV DSP and TN (Keswani et al., 2004), with clinical trials under development.

9.4. Mononeuropathy multiplex

Mononeuropathy multiplex (MM) manifests with multifocal motor and sensory abnormalities in an asymmetric distribution. It may develop within days or weeks of acute HIV infection (Leger et al., 1992). MM has a bimodal pattern of occurrence. It may occur in relatively mild immunosuppression with a CD^4 lymphocyte count >200 cells/mm^3 or in advanced stages of immunosuppression with CD^4 lymphocyte count <50 cells/mm^3. Mechanisms implicated in MM include vasculitis, occasionally associated with hepatitis C, cryoglobinemia and dysimmune mechanisms (Simpson and Olney, 1992; Caniatti et al., 1996; Schifitto et al., 1997). In advanced immunosuppression, cytomegalovirus (CMV) may cause MM (Manji, 2000).

Multifocal cranial or peripheral nerves may be involved in MM. Cranial neuropathies may present as facial or extraocular muscle weakness. The most commonly involved peripheral nerves in MM are wrist drop (radial nerve palsy) and foot drop (peroneal neuropathy). MM may also affect cutaneous nerves, with paresthesias and numbness in a patchy distribution.

Electrophysiologic assessments in MM demonstrate asymmetric axonal degeneration with reduced amplitude of motor nerve compound muscle action potentials (CMAP) and sensory nerve action potentials (SNAP). EMG shows evidence for denervation and neuropathic motor unit recruitment patterns. CSF analysis may show elevated protein and mild mononuclear pleocytosis. CSF analysis for CMV by PCR (polymerase chain reaction) assay should be performed in patients with advanced immunosuppression. Definitive etiologic diagnosis may be established by biopsy of an affected sensory nerve, most commonly the sural nerve. Findings range from axonal degeneration, with perivascular infiltrates in the endoneurium and epineurium, to numerous polymorphonuclear infiltrates with mixed axonal and demyelinative lesions.

Patients with MM in early HIV disease often experience spontaneous improvement. For those with persistent or progressive deficits, while there are no controlled trials, treatment may include immunomodulating agents, such as corticosteroids, plasmapheresis or intravenous immunoglobulin. Empiric therapy for CMV may be considered in MM in advanced HIV infection, especially if CSF, CMV, PCR, or nerve biopsy results are supportive.

9.5. Inflammatory demyelinating polyneuropathy

Similar to MM, inflammatory demyelinating polyneuropathy (IDP) has a bimodal pattern, and can occur during the early stages with a CD^4 lymphocyte count >500 cells/mm^3 or in the advanced stages of the infection with CD^4 lymphocyte count <50 cells/mm^3. Pathogenetic mechanisms of IDP include autoimmune mechanisms (Cornblath et al., 1987) in early-stage HIV infection, and opportunistic infections such as CMV in late stages of HIV.

Clinical features of IDP include progressive weakness of the distal extremities, followed by involvement of the upper extremities, paresthesias, mild sensory loss and generalized areflexia. CSF analysis reveals markedly elevated protein and a mild lymphocytic pleocytosis, in comparison to the acellular CSF in HIV-negative IDP. Electrodiagnostic findings are consistent with primary demyelinating neuropathy, with markedly decreased nerve conduction velocities, conduction block and prolonged latencies. There may be superimposed axonal loss particularly in chronic inflammatory demyelinating polyneuropathy (CIDP).

Treatment of IDP in early-stage HIV infection includes supportive care, particularly in the intensive care unit setting. Similar to HIV-negative IDP, treatment in HIV-infected patients in early-stage HIV illness may include intravenous immunoglobulin, plasmapheresis or corticosteroids, although no controlled studies have been done in HIV-infected subjects. For IDP in late-stage HIV, ganciclovir, foscarnet or cidofovir may be considered, particularly if there is evidence of CMV infection.

9.6. Progressive polyradiculopathy

Progressive polyradiculopathy (PP) usually occurs in advanced stages of immunosuppression, with CD^4 lymphocyte count <50 cells/mm^3. The most common etiology is CMV infection of the lumbosacral nerve roots. PP is usually rapidly progressive and is fatal if not rapidly treated. With the advent of HAART, the incidence of CMV-related PP has dramatically declined. Less common causes of PP include herpes viruses (HSV and VZV), tuberculosis, neurosyphilis and lymphoma (Lanska et al., 1988; Leger et al., 1992).

PP characteristically presents with lumbosacral pain and paresthesias, usually in the cauda equina distribution, with "saddle anesthesia" and radicular leg involvement. Lower motor neuron signs are present, with rapidly progressive flaccid paraparesis, areflexia of the distal extremities and bowel and bladder dysfunction. Myelopathic features may be noted. PP rarely involves cervical or the thoracic roots.

Magnetic resonance scan of the lumbosacral spine with gadolinium contrast is important to exclude other etiologies of PP, including structural lesions. Most commonly, MRI in CMV PP may show enhancement of nerve roots. Electrophysiological studies reveal evidence of polyradiculopathy in the lower extremities, including reduced motor compound muscle action potentials, relatively normal sensory potentials, delayed late responses and evidence of denervation. Lumbar puncture is a critical diagnostic test in PP, with CSF usually containing a marked polymorphonuclear pleocytosis, moderately elevated protein and marked hypoglycorrhachia. The detection of CSF CMV by PCR has 92% sensitivity and 94% specificity in the diagnosis of CMV PP. Treatment of CMV polyradiculopathy includes anti-CMV therapy with ganciclovir, foscarnet or cidofovir, alone or in combination.

9.7. Conclusion

Peripheral neuropathy is one of the most common neurological complications seen in the HIV infection. Peripheral nerve damage in HIV-infected patients can occur from the time of seroconversion to the time of significant immunodeficiency. DSP is the most common form of neuropathy in HIV infection. The pathogenesis of DSP is still unclear but the favored hypothesis is that of immune dysregulation. The most common symptoms of DSP are painful dysesthesias in a stocking and glove distribution. Currently, there is no FDA-approved treatment for HIV-associated peripheral neuropathy. Clinical trials are ongoing in search of better options for management of this disabling disorder.

References

Beckonja MM (2002). Use of anticonvulsants for treatment of neuropathic pain. Neurology 59(5 Suppl 2): S14–S17.

Blum AS, Dal Pan CI, Feinberg I, et al (1996). Low dose zalcitabine-related toxic neuropathy. Neurology 46(4): 999–1003.

Brew BJ, Tisch S, Law M (2003). Lactate concentrations distinguish between nucleoside neuropathy and HIV neuropathy. AIDS 17(7): 1094–1096.

Caniatti LM, Tugnoli V, Eleopra R, et al (1996). Cryoglobinemic neuropathy related to hepatitis C virus infection. Clinical, laboratory and neurophysiological study. J Peripher Nerv Syst 1: 131–138.

Childs EA, Lyles RH, Selnes OA, et al (1999). Plasma viral load and CD4 lymphocytes predict HIV-associated dementia and sensory neuropathy. Neurology 52(3): 607–613.

Christensen ER, Stegger M, Jensen-Fangel S, et al (2004). Mitochondrial DNA levels in fat and blood cells from patients with lipodystrophy or peripheral neuropathy and the effect of 90 days of high dose coenzyme Q treatment: a randomized, double-blind, placebo-controlled pilot study. Clin Infect Dis 39: 1371–1379.

Cornblath DR, McArthur JC, Kennedy PG, et al (1987). Inflammatory demyelinating peripheral neuropathies associated with human T-cell lymphotropic virus type III infection. Ann Neurol 21: 32–40.

Dalakas MC, Illa I, Pezeshkpour GH, et al (1990). Mitochondrial myopathy caused by long-term zidovudine therapy. N Engl J Med 322: 1098–1105.

Dalakas MC, Semino-Mora C, Leon-Monzon M (2001). Mitochondrial alterations with mitochondrial DNA depletion in the nerves of AIDS patients with peripheral neuropathy induced by 2′3′ dideoxycytidine (ddC). Lab Invest 81: 1537–1544.

Estanislao L, Thomas D, Simpson DM (2004). HIV neuromuscular disease and mitochondrial function. Mitochondrion 4: 131–139.

Evans S, Fishman B, Spielman L, et al (2003a). Randomized trial of cognitive behavior therapy versus supportive psychotherapy for HIV-related peripheral neuropathic pain. Psychosomatics 44(1): 44–50.

Evans S, Weinberg BA, Spielman L, et al (2003b). Assessing negative thoughts in response to pain among people with HIV. Pain 105(1–2): 239–245.

Griffin J, Crawford T, McArthur J (1998). Peripheral Neuropathies Associated with HIV Infection. The Neurology of AIDS. Chapman and Hall, New York.

Gulick R (1998). Combination therapy for patients with HIV-1 infection: the use of dual nucleoside analogues with protease inhibitors and other agents. AIDS 12(Suppl 3): S17–S22.

Hahn K, Arendt G, Braun JS, et al (2004). A placebo-controlled trial of gabapentin for painful HIV-associated sensory neuropathies. J Neurol 251: 1260–1266.

Hart AM, Wilson AD, Montovani C, et al (2004). Acetyl-l-carnitine: a pathogenesis based treatment for HIV-associated antiretroviral toxic neuropathy. AIDS 18(11): 1549–1560.

Keswani S, Pardo C, Cherry C, et al (2002). HIV-associated sensory neuropathies. AIDS 16(16): 2105–2117.

Keswani S, Polley M, Pardo C, et al (2003). Schwann cell chemokine receptors mediate HIV-1 gp120 toxicity to sensory neurons. Ann Neurol 54: 287–296.

Keswani S, Leitz G, Hoke A (2004). Erythropoietin is neuroprotective in models of HIV sensory neuropathy. Neurosci Lett 37: 102–105.

Lanska MJ, Lanska DJ, Schmidley JW (1988). Syphilitic polyradiculopathy in an HIV-positive man. Neurology 38: 1297–1301.

Leger JM, Henin D, Belec L, et al (1992). Lymphoma-induced polyradiculopathy in AIDS: two cases. J Neurol 239: 132–134.

Lewis W, Dalakas M (1995). Mitochondrial toxicity of antiviral drugs. Nat Med 1: 417–422.

Lichtenstein KA, Armon C, Baron A, et al (2005). Modification of the incidence of drug-associated symmetrical peripheral neuropathy by host and disease factors in the HIV outpatient study cohort. Clin Infect Dis 40(1): 148–157.

Lopez OL, Becker JT, Dew MA, et al (2004). Risk modifiers for peripheral sensory neuropathy in HIV infection. Eur J Neurol 11(2): 97–102.

Manji H (2000). Neuropathy in HIV infection. Curr Opin Neurol 13: 589–592.

Marra CM, Boutin P, Collier AC (1998). Screening for distal sensory peripheral neuropathy in HIV-infected persons in research and clinical settings. Neurology 51(6): 1678–1681.

Martin C, Solders G, Sonnerborg A, et al (2000). Antiretroviral therapy may improve sensory function in HIV-infected patients: a pilot study. Neurology 54: 2120–2127.

McArthur JC (2003). Emerging neurological complications of HIV. Paper presented at 6th Annual Controversies in the Management of HIV. New York, November 7, 2003.

McArthur JC, Yiannoutsos C, Simpson DM, et al (2000). A phase II trial of nerve growth factor for sensory neuropathy associated with HIV infection. AIDS clinical trial group team 291. Neurology 54(5): 1080–1088.

Moore RD, Wong WE, Keruly JC, et al (2000). Incidence of neuropathy in HIV-infected patients on monotherapy versus those on combination therapy with didanosine, stavudine and hydroxyurea. AIDS 14(3): 273–278.

Moyle GJ, Sadler M (1998). Peripheral neuropathy with nucleoside antiretrovirals: risk factors, incidence, and management. Drug Safety 19(6): 481–494.

Paice JA, Ferrans CE, Lashley FR, et al (2000). Topical capsaicin in the management of HIV-associated peripheral neuropathy. J Pain Symptom Manage 19(1): 45–52.

Pardo CA, McArthur JC, Griffin JW (2001). HIV neuropathy: insights in the pathology of HIV peripheral nerve disease. J Periph Nervous Syst 1: 21–27.

Savasta AM (2004). HIV associated transmission risks in older adults: an integrative review of the literature. J Assoc Nurses AIDS Care 15(1): 50–59.

Schifitto G, Barbano RL, Kieburtz KD, et al (1997). HIV related vasculitic mononeuropathy multiplex: a role for IVIg? J Neuro Neurosurg Psych 63: 255–256.

Schifitto G, McDermott MP, McArthur JC, et al (2002). Incidence of and risk factors for HIV-associated distal symmetrical sensory polyneuropathy. Neurology 58: 1764–1768.

Schifitto G, McDermott MP, McArthur JC, et al (2005). Markers of immune activation and viral load in HIV-associated sensory neuropathy. Neurology 64(5): 842–848.

Simpson D, Olney R (1992). Peripheral neuropathies associated with human immunodeficiency virus infection. Periph Neuropathies: New Concepts Treat 685–711.

Simpson D, Tagliati M (1995). Nucleoside analogue-associated peripheral neuropathy in human immunodeficiency virus infection. JAIDS 9: 153–161.

Simpson DM, Haidich A, Schifitto G, et al (2002). Severity of HIV-associated neuropathy is associated with plasma HIV-1 RNA levels. AIDS 16(3): 407–412.

Simpson DM, McArthur JC, Olney R, et al (2003). Lamotrigine for HIV-associated painful sensory neuropathies: a placebo controlled trial. Neurology 60(9): 1508–1514.

Simpson D, Brown S, Sampson J, et al (2004). A single application of high concentration trans-capsaicin leads to 12 weeks of pain relief in HIV DSP: results of an open label trial. Platform presentation, 56th American Academy of Neurology Annual Meeting. San Diego, CA.

Simpson D, Evans S, Kitck D, et al (2005). HIV neuropathy natural history cohort study: clinical and laboratory features and risk factors for progression. 12th Conference on Retroviruses & Opportunistic Infection. Section 75 #397 (Poster): 1.

Tyor WR, Wesselingh SL, Griffin JW, et al (1995). Unifying hypothesis for the pathogenesis of HIV-associated dementia complex, vacuolar myelopathy, and sensory neuropathy. J Acquir Immune Defic Syndr Hum Retrovir 9: 379–388.

Verma S, Estanislao L, Mintz L, et al (2004). Controlling neuropathic pain in HIV. Curr Infec Dis Rep 6: 237–242.

Wesselingh SL, Glass J, McArthur J, et al (1994). Cytokine dysregulation in HIV-associated neurological disease. Adv Neuroimmunol 4: 199–206.

World Health Organization (1990). Cancer Pain Relief and Palliative Care. WHO, Geneva.

World Health Organization (1996). Cancer Pain Relief, 2nd edn, with a guide to opioid availability. WHO, Geneva.

Wulff EA, Simpson DM (1998). HIV-associated neuropathy recent advances in management. HIV Adv Res Ther 8: 23–29.

Yarchoan R, Pluda JM, Thomas RV, et al (1990). Long-term toxicity/activity profile of 2′,3′-dideoxyinosine in AIDS or AIDS-related complex. Lancet 336(8714): 526–529.

Yoshioka M, Shapshak P, Srivastava AK, et al (1994). Expression of HIV-1 and interleukin-6 in lumbosacral dorsal root ganglia of patients with AIDS. Neurology 44: 1120–1130.

Further Reading

Leger JM, Bouche P, Bolgert F, et al (1989). The spectrum of polyneuropathies in patients infected with HIV. J Neurol Neurosurg Psych 52: 1369–1374.

Handbook of Clinical Neurology, Vol. 85 (3rd series)
HIV/AIDS and the Nervous System
P. Portegies, J. R. Berger, Editors

Chapter 10

Myopathy in HIV infection

ADRIAN TIEN-AUH CHAN, CARL KIRTON, LYDIA ESTANISLAO, AND DAVID M. SIMPSON*

Mount Sinai Medical Center, New York, NY, USA

10.1. Introduction

Neurological complications are common in patients with HIV infection. Myopathy may be a consequence of HIV infection itself or associated with antiretroviral (ARV) therapy use or both (Table 10.1). Secondary myopathy may also occur, due to metabolic, infectious, neoplastic and vasculitic processes. Secondary myopathies are relatively rare in HIV-infected individuals.

In one of the earliest reports of neurological complications in HIV infection in 1983, HIV-associated myopathy was described as a polymyositis-type syndrome (Snider et al., 1983b). Later, other patients with polymyositis associated with AIDS were published (Dalakas et al., 1986). These cases were histologically indistinguishable from seronegative polymyositis. There were unusual pathological features in these cases, and relatively few patients satisfied the full diagnostic criteria for polymyositis. This led to the descriptive term "HIV-associated myopathy" as a distinctive entity unique to HIV infection.

With the advent of ARV therapy in the late 1980s, cases of zidovudine (ZDV or AZT)-related myopathy were reported (Bessen et al., 1988). More recently, cases of myopathy have been noted in association with other nucleoside reverse transcriptase inhibitors (NRTIs), occasionally in the setting of lactic acidosis. We have termed this disorder the HIV-associated neuromuscular weakness syndrome (HANWS). The distinguishing feature of some of these cases is believed to be mitochondrial pathology (Miller et al., 2000; Mokrzycki et al. 2000; Galera et al., 2001; Marcus et al., 2002; Cossarizza and Moyle, 2004; Estanislao et al., 2004; Simpson et al., 2004).

The focus of this chapter is primary muscle disease, i.e. HIV-associated and ARV-related myopathy. It reviews the clinical, electrophysiological and histopathological features of these two major forms of myopathy, as well as theories of pathogenesis and management options.

10.2. Epidemiology

Prior to the availability of highly active antiretroviral therapy (HAART) in HIV disease, numerous reports indicated a wide variety of neurological disorders in association with HIV infection. In the first decade of the HIV epidemic, about 40–70% of people with AIDS experienced some type of neurologic complication. Abnormal neuropathologic findings were found in more than 90% of AIDS patients at autopsy. Muscle disease in AIDS was first described in 1983 (Snider et al., 1983b). In this paper, one case of HIV-associated polymyositis was described in a cohort of 50 patients with neurological complications of AIDS. Subsequently, isolated cases of myopathy were reported.

Myopathy is often the initial clinical manifestation of HIV infection in its asymptomatic stage (Dalakas et al., 1986; Simpson and Bender, 1988). The incidence of myopathy in HIV infection has not been determined in large clinical studies. In one series of 101 HIV patients, two were found to have polymyositis (Berman et al., 1988). Following the use of the first available ARV, AZT, cases of "AZT myopathy" were reported. In 1988, Bessen et al. described polymyositis in four HIV seropositive patients treated with AZT, three of whom reportedly experienced improvement after AZT withdrawal (Bessen et al., 1988). Several case reports followed, describing myopathies in AZT-treated patients, whose symptoms improved after drug withdrawal (Helbert et al., 1988; Gertner et al., 1989). A retrospective study of a large primary antiretroviral protocol (ACTG 016) comparing the efficacy safety of zidovu-

*Correspondence to: David M. Simpson, MD, Professor of Neurology, Director, Clinical Neurophysiology Laboratories, Director, Neuro-AIDS Program, Mount Sinai Medical Center, One Gustave Levy Place, Box 1052, New York, NY 10029, USA. Email: david.simpson@mssm.edu, Tel: +1-212-241-8748, Fax: +1-212-987-3301.

Table 10.1

Myopathies encountered in HIV-infected individuals

HIV-associated myopathy	Antiretroviral (ARV)-related myopathy	Secondary myopathies
Polymyositis	AZT-associated myopathy	Metabolic, infectious, neoplastic and vasculitic myopathies
Structural myopathy (e.g. nemaline rod myopathy)	Other nucleoside (d4T, ddI)-associated myopathy	
	HIV neuromuscular weakness syndrome	Toxic myopathy (e.g. cholesterol lowering agent myopathy)

dine (AZT) found a 0.4% incidence of myopathy in the AZT-exposed group (Simpson et al., 1997).

With the introduction of HAART, the incidence of some HIV-associated neurological disease has decreased. Little is known about the incidence of primary myopathy in the HAART era, as there have been no large clinical and epidemiological studies done. Manfredi et al. (2002) reported that 15% of HIV-infected subjects have elevations of muscle enzyme creatine phosphokinase. In one longitudinal study of outcome of HIV myopathy, 13 of 64 had biopsy-proven polymyositis, 9 of whom were on HAART (Johnson et al., 2003). Recently, a clinical syndrome has been reported, characterized by progressive limb weakness in the setting of NRTI use and associated with lactic acidosis (Marcus et al., 2002; Simpson et al., 2004). We termed this syndrome HIV-associated neuromuscular weakness syndrome (HANWS) (see below). Of the 69 patients identified, 3 (4%) had clinical presentation and electrophysiologic features consistent with myopathy.

10.3. Clinical features

HIV-infected patients with myopathy (HIV-associated or ARV-related) classically present with subacute progressive weakness of the proximal muscles. Patients report difficulty in climbing stairs or difficulty rising from the seating position. They may also describe difficulty in lifting objects or performing tasks requiring proximal arm strength, such as combing hair, brushing teeth and reaching for objects above their heads. Myalgias, particularly of the thighs, is present in 25 to 50% of patients, but is not specific to myopathy (Simpson and Bender, 1988). Fatigue, muscle cramps and dysphagia have also been reported. Physical findings include weakness that is symmetrical, and predominately involving neck flexors and proximal upper and lower limb muscles. Weakness is usually greater in the lower than the upper extremities. Deep tendon reflexes are often preserved, unless there is superimposed peripheral neuropathy. Functional testing may demonstrate inability to sustain the arms above the head and difficulty rising from a seated or squatted position (Snider et al., 1983a; Dalakas et al., 1986; Simpson and Bender, 1988). Clinical findings, together with laboratory and electrodiagnostic testing, are necessary for the diagnosis.

10.4. Laboratory studies

An important laboratory test for myopathy is serum creatine kinase (CK) level, which can be elevated roughly 2–4 times the normal values. In a retrospective series by Simpson et al. (1993), 92% of patients with myopathy had CK elevations. The degree of CK elevation is highly variable, but they are sometimes elevated to >1000 IU/L. The CK level correlates with the degree of myonecrosis seen in muscle biopsy, but does not correlate well with the degree of muscle weakness (Simpson and Wolfe, 1991).

10.5. Electrodiagnostic features

Electromyography (EMG) is a sensitive diagnostic test for HIV-associated myopathies, with up to 94% diagnostic yield in one series of 50 patients with HIV-associated myopathy (Simpson et al., 1993). EMG reveals a typical myopathic pattern, similar to that in seronegative polymyositis, with short, brief motor unit action potentials, early recruitment, with full interference patterns. Frequently, abnormal irritative activity, such as fibrillation potentials, positive sharp waves and complex repetitive discharges, is found (Simpson et al., 1993). The iliopsoas is the most sensitive muscle for EMG diagnosis.

10.6. Histopathology and pathogenesis

10.6.1. HIV-associated myopathy

In the earliest reported cases of HIV-associated myopathy, muscle biopsy revealed histopathological changes indistinguishable from seronegative polymyositis, although the inflammatory cell infiltrate was relatively mild (Stern et al., 1987). There was a later report (Dalakas et al., 1986) of two ARV-naive AIDS patients with clinical myopathy who displayed myofiber necrosis and phagocytosis, as well as interstitial and interfascicular mononuclear cell inflammation in the muscle biopsies (Table 10.2). HIV was detected immunohistochemically within CD^{4} cells in the inflammatory cell infiltrates (Dalakas et al., 1986). Subsequent studies showed that approximately 49% of the endomyseal infiltrates were composed of CD^{8} cells, 38% macrophages and 13% CD^{4} cells (Illa et al., 1991). There were reports of structural myofiber abnormalities, including rod and cytoplasmic bodies, identified in muscle biopsies of AZT-naive patients with myopathy (Gonzales et al., 1988). However, HIV could not be detected in these biopsies by either immunohistochemistry or in situ hybridization.

The pathogenesis of these two histologic types of myopathies is still unknown. HIV DNA has been identified with the use of *in situ reverse-transcriptase polymerase chain reaction (RT-PCR)* within endomyseal macrophages and myocyte nuclei, raising the possibility that myopathies in ARV-naive patients may be due in part to *direct* HIV infection of myocytes (Seidman et al., 1994). An autoimmune mechanism has also been implicated because of the inflammatory nature of some biopsies. The presence of CD^{8} lymphocytes, together with diffuse expression of major histocompatibility complex (MHC) class I antigens on myofibers, has led to speculation that HIV-associated myopathy is a T-cell mediated and MHC class I restricted cytotoxic process. This process is also found in seronegative polymyositis (Dalakas et al., 1986; Illa et al., 1991). Tubuloreticular inclusions, which are thought to be typical of autoimmune phenomena, have been identified in capillary endothelial cells of HIV-positive myopathic patients (Lane et al., 1993). More recently, HIV-associated myopathy has been reported during immune reconstitution with HAART (Sellier et al., 2000), suggesting that myopathy can occur as an immunological phenomenon in the context of a rapid restoration of T-cell subsets. Furthermore, clinical response to corticosteroids and intravenous immunoglobulins (IVIG) is evident in many cases, supporting an autoimmune etiology.

10.6.2. Antiretroviral-related myopathy

AZT is a dideoxynucleoside analogue of thymidine, which has affinity for mitochondrial DNA polymerase-

Table 10.2

Histopathological patterns of HIV-associated and ARV-related myopathy

HIV-associated myopathy		
Inflammatory myopathy	Structural myopathy	ARV-related myopathy
Presence of myofiber necrosis, phagocytosis and mononuclear cell inflammation in the interstitial and interfascicular areas	Abnormal myofiber structure, with rod and cytoplasmic bodies, and basophilic granular material (Simpson and Bender, 1988)	Non-necrotic moderately atrophic myofiber degeneration, with nemaline rod bodies or cytoplasmic inclusion bodies; red ragged fibers may be present (Simpson et al., 1993)
CD^{8} cells comprised the majority of cells in the infiltrate (Illa et al., 1991)	Inflammatory cell infiltrates were often mild or absent	Variable degree of inflammatory infiltrates (from mild to marked) indistinguishable from biopsies obtained from NRTI-naive patients
Immunohistochemistry revealed HIV positive cells and macrophages within the infiltrate	HIV genome cannot be demonstrated within myofibers by immunohistochemistry and in situ hybridization	HIV genome has not been demonstrated within myofibers
Proliferation and enlargement of abnormal mitochondria in subsarcolemmal space.		
Mitochondrial abnormalities seen under electron microscopy: double membrane bounded profiles containing tubular, circinate and otherwise convoluted cristae and amorphous dense inclusions		
Amount of mitochondrial DNA is diminished in muscles		

gamma, the enzyme required for mitochondrial DNA replication. In vitro studies demonstrate that AZT and other NRTIs, particularly the so-called "d-drugs" (ddC or zalcitabine, ddI or didanosine, d4T or stavudine), may induce mitochondrial dysfunction or toxicity via its inhibition of DNA polymerase-gamma. Their effects on the mitochondrial enzyme system may be more important (Cossarizza and Moyle, 2004), as defects of mitochondrial enzymes such as cytochrome C oxidase and reductase have been described in AZT-treated patients (Mhiri et al., 1991; Chariot et al., 1993).

There are numerous reports indicating that mitochondrial toxicity is at least partially involved in the pathogenesis of NRTI-associated myopathy. Electron microscopy (EM) from biopsies of patients with AZT myopathy revealed a proliferation and enlargement of abnormal mitochondria in subsarcolemmal spaces. Such mitochondrial abnormalities include paracrystalline inclusions, giant double membrane-bounded forms with convoluted tubular cristae and amorphous dense inclusions (Dalakas et al., 1990; Simpson et al., 1993). Muscle mitochondrial DNA also appears to be depleted in some patients exposed to AZT (Dalakas, 1991).

Like the familial forms of mitochondrial myopathy, ragged red fibers (RRF) were demonstrated on modified Trichrome staining of muscle, considered to be a pathological hallmark of mitochondrial dysfunction. Dalakas et al. (1990) reported that muscle biopsies taken from all patients with high dose AZT (1000–1200 mg) showed RRF, whereas in subsequent studies when lower doses were taken, less RRF were seen in biopsies (Mhiri et al., 1991; Simpson et al., 1993). The percentage of RRF was also reported to correlate with the severity of clinical myopathy (Dalakas et al., 1990).

AZT-induced myopathy also appears to be dose-dependent. Mhiri et al. showed that AZT-associated myopathy as well as mitochondrial and structural abnormalities in biopsies were commonly seen among patients taking doses above 250 mg (Mhiri et al., 1991). A T-cell-mediated inflammatory change that is indistinguishable from those seen in HIV-associated myopathy was present in the biopsies of these patients, suggesting that mitochondrial toxicity alone may not account for myopathy seen in AZT-associated myopathy (Dalakas et al., 1990). Myofiber degeneration and structural abnormalities similar to those seen in primary HIV-associated myopathy can also be seen in myopathy occurring in AZT-treated patients (see Table 10.2).

10.6.2.1. Effects of HIV on muscle mitochondrial function

There is controversy on the specificity of mitochondrial abnormalities in AZT myopathy. In earlier studies, mitochondrial depletion and abnormalities were thought to be unique among HIV-infected patients treated with AZT (Dalakas et al., 1990). However, in 1993, our group reported a study of 50 HIV-infected patients with myopathy, diagnosed by clinical and laboratory criteria, in which pathological findings failed to distinguish myopathy in patients exposed to AZT, from those naive to AZT (Simpson et al., 1993). Clinical features did not differ between AZT-treated and AZT-naive patients with myopathy. Muscle biopsy findings, including myofiber degeneration, inflammatory infiltrates, inclusion bodies and mitochondrial abnormalities, were present in both groups.

A more detailed pathologic characterization of 18 AZT-treated and 9 AZT-naive patients reached similar conclusions (Morgello et al., 1995). In this study, the mitochondrial abnormalities correlated with the degree of myofiber degeneration present in muscle biopsies, regardless of AZT exposure history. The conclusion was that mitochondrial abnormalities are not specific for AZT myopathy. Furthermore, HIV infection alone appears to have direct effects on mitochondrial function. Cote et al. (2002) reported depletion of mitochondrial DNA in HIV-infected, ARV-naive patients. These changes may reflect a direct effect of HIV or cytokines released in response to HIV or immune reconstitution, which may cause mitochondrial damage, potentially making them more vulnerable to the toxicity of NRTI. Thus, both the direct effect of HIV infection via immune or mitochondrial mechanism and the concomitant use of NRTI via mitochondrial toxicity predisposes patients to the development of myopathy. This proposed mechanism is similar to that in NRTI-related peripheral neuropathy, in which synergistic effects of HIV and toxic nucleoside analogue ARV likely cause symptomatic neuropathy.

10.6.2.2. HIV-associated neuromuscular weakness syndrome

Other NRTIs, in addition to AZT, have been implicated in muscle disease. In 2000, there were reports of hepatic steatosis and lactic acidosis with the use of stavudine (d4T) that were accompanied by mitochondrial abnormalities in muscle, including ragged-red fibers, and cytochrome C oxidase-negative fibers (Miller et al., 2000; Mokrzycki et al., 2000). These were followed by reports of progressive weakness occurring in the setting of NRTI use and lactic acidosis (Galera et al., 2001; Marcus et al., 2002; Simpson et al., 2004). This disorder, termed *HIV-associated neuromuscular weakness syndrome* (HANWS), encompasses cases of progressive neuromuscular manifestations, including subacute progressive myopathy and rapidly progressive sensorimotor polyneuropathy. In our report of 69 patients with this syndrome, 3 of 15 muscle biopsies revealed inflammation, and 4 showed mitochondrial abnormalities, including ragged red fibers, abnormalities in respiratory chain enzymes and mitochondrial DNA depletion (Simpson et al., 2004). The pathophysiological mechanism underlying HANWS, and its relationship to lactic acidosis syndrome and NRTI therapy, is not clear. Mitochondrial toxicity by NRTIs, together with HIV-related immune-mediated neuromuscular disease, may explain at least some of the components of this syndrome.

10.7. Diagnosis

There are no clinical criteria specifically developed for the diagnosis of HIV myopathy. Because the characteristics of HIV myopathy are similar to, if not indistinguishable from, those of classic seronegative polymyositis, the diagnostic criteria used for the latter (Bohan and Peter, 1975a, b; Mastaglia and Ojeda, 1985a, b; Dalakas, 1991) may be employed for HIV-associated myopathy, as follows:

- Progressive, symmetrical weakness of limb-girdle muscles and neck flexors.
- Elevation of serum skeletal muscle enzymes, particularly CK.
- Electromyographic abnormalities with short, brief, polyphasic motor unit action potentials, that recruit with early and full interference patterns, with or without associated irritative activity.
- Muscle biopsy evidence of myofibrillar necrosis, phagocytosis, variation in fiber size, regeneration and degeneration, with or without endomyseal inflammatory infiltrates.

The diagnosis of HIV myopathy is considered *definite* if all four criteria are fulfilled and *probable* if three criteria are present. In clinical practice it may not always be possible to obtain muscle biopsies in all patients and so a *possible* diagnosis can be made when only two criteria are present without a muscle biopsy. With the advent of NRTIs, particularly AZT, it is often difficult to distinguish between myopathy due to HIV infection from that due to NRTIs. A history of a temporal relationship between the initiation of NRTIs and development of myopathy, as well as improvement of myopathy after withdrawal of NRTIs, helps to support the supposition that ARVs are myotoxic in such patients. However, there are many cases where no improvement in myopathic symptoms occurs following withdrawal of NRTIs (Espinoza et al., 1991; Manji et al., 1993; Simpson et al., 1993; Prime et al., 2003).

10.8. Therapeutic interventions

The initial management of patients with myopathy in the setting of use of AZT, and other NRTIs, is withdrawal of the drug if it is not essential for virological control. Early reports noted a wide variety of objective responses to this strategy. In different studies, improvement of myopathy following AZT withdrawal was noted in highly variable fashion, ranging from 18 to 100% of subjects (Bessen et al., 1988; Dalakas et al., 1990; Chalmers et al., 1991; Grau et al., 1993; Manji et al., 1993).

In patients with HANWS, prompt discontinuation of NRTIs and supportive treatment are essential. Treatment for the neuromuscular disorder is based on the specific pathology, and may include IVIG, plasmapheresis or corticosteroids. While no controlled studies are available, some investigators suggest mitochondrial protective agents, including acetyl-carnitine, riboflavin and coenzyme Q.

While there are very few clinical trials available upon which to base recommendations, the usual treatment strategy for HIV-associated myopathy is immune modulation. For those with moderate to advanced immunosuppression, a trial of intravenous immunoglobulins (1 mg/kg/day for 2 days monthly) may be tried. If no response is noted after 2 or 3 infusion cycles, and for those who are not profoundly immunosuppressed, corticosteroids may be used. To date, only one placebo-controlled trial of prednisone in HIV-associated myopathy has been conducted (Simpson et al., 1993). A high dose of at least 1 mg/kg or 60–80 mg/day is administered and tapered according to the clinical response (improved strength with or without CK normalization). Anabolic steroids, such as oxandralone, may lead to improvement in muscle bulk and body weight in patients with AIDS.

10.9. Conclusion

Myopathy in HIV-infected individuals may be related to HIV infection itself, nucleoside antiretroviral therapy, or both. Clinical manifestations are similar and, in most cases, indistinguishable in these two forms. Contributory mechanisms include immune-mediated in HIV myopathy and mitochondrial dysfunction in ARV myopathy. It is likely that in most patients, simultaneous or sequential occurrence of these phenomena leads to synergistic effects on muscle. The diagnosis of myopathy is based on clinical features, electrophysiologic and histopathologic findings. Management includes discontinuation of the NRTI, if the virologic and immune status permits this approach, and immune modulation with IVIG or corticosteroids.

References

Berman A, Espinoza LR, Diaz JD, et al (1988). Rheumatic manifestations of human immunodeficiency virus infection. Am J Med 85(1): 59–64.

Bessen LJ, Greene JB, Louie E, et al (1988). Severe polymyositis-like syndrome associated with zidovudine therapy of AIDS and ARC. N Engl J Med 318(11): 708.

Bohan A, Peter JB (1975a). Polymyositis and dermatomyositis (part 1). N Engl J Med 292(7): 344–347.

Bohan A, Peter JB (1975b). Polymyositis and dermatomyositis (part 2). N Engl J Med 292(8): 403–407.

Chalmers AC, Greco CM, Miller RG (1991). Prognosis in AZT myopathy. Neurology 41(8): 1181–1184.

Chariot P, Monnet I, Gherardi R (1993). Cytochrome c oxidase reaction improves histopathological assessment of zidovudine myopathy. Ann Neurol 34(4): 561–565.

Cossarizza A, Moyle G (2004). Antiretroviral nucleoside and nucleotide analogues and mitochondria. AIDS 18: 137–151.

Cote HC, Brumme ZL, Craib KJ, et al (2002). Changes in mitochondrial DNA as a marker of nucleotide toxicity in HIV-infected patients. N Engl J Med 346(11): 811–820.

Dalakas MC (1991). Polymyositis, dermatomyositis and inclusion-body myositis. N Engl J Med 325(21): 1487–1498.

Dalakas MC, Pezeshkpour GH, Gravell M, et al (1986). Polymyositis associated with AIDS retrovirus. JAMA 256(17): 2381–2383.

Dalakas MC, Illa I, Pezeshkpour GH, et al (1990). Mitochondrial myopathy caused by long-term zidovudine therapy. N Engl J Med 322(16): 1098–1105.

Espinoza LR, Aguilar JL, Espinoza CG, et al (1991). Characteristics and pathogenesis of myositis in human immunodeficiency virus infection: distinction from azidothymidine-induced myopathy. Rheum Dis Clin North Am 17(1): 117–129.

Estanislao L, Thomas D, Simpson D (2004). HIV neuromuscular disease and mitochondrial function. Mitochondrion 4: 131–139.

Galera C, Redondo C, Pozo G (2001). Symptomatic hyperlactatemia and lactic acidosis syndrome in HIV patients treated with nucleoside analogue reverse transcriptase inhibitors. First Conference on HIV Pathogenesis and Treatment, Buenos Aires, Argentina 2001.

Gertner E, Thurn JR, Williams DN, et al (1989). Zidovudine-associated myopathy. Am J Med 86(6 Pt 2): 814–818.

Gonzales MF, Olney RK, So YT, et al (1988). Subacute structural myopathy associated with human immunodeficiency virus infection. Arch Neurol 45(5): 585–587.

Grau JM, Masanes F, Pedrol E, et al (1993). Human immunodeficiency virus type 1 infection and myopathy: clinical relevance of zidovudine therapy. Ann Neurol 34(2): 206–211.

Helbert M, Fletcher T, Peddle B, et al (1988). Zidovudine-associated myopathy. Lancet 2(8612): 689–690.

Illa I, Nath A, Dalakas M (1991). Immunocytochemical and virological characteristics of HIV-associated inflammatory myopathies: similarities with seronegative polymyositis. Ann Neurol 29(5): 474–481.

Johnson R, Williams F, Kazi S, et al (2003). Human immunodeficiency virus-associated polymyositis: a longitudinal study of outcome. Arthritis Rheum 49(2): 172–178.

Lane RJ, McLean KA, Moss J, et al (1993). Myopathy in HIV infection: the role of zidovudine and the significance of tubuloreticular inclusions. Neuropathol Appl Neurobiol 19(5): 406–413.

Manfredi R, Motta R, Patrono D, et al (2002). A prospective case-control survey of laboratory markers of skeletal muscle damage during HIV disease and antiretroviral therapy. AIDS 16(14): 1969–1971.

Manji H, Harrison MJ, Round JM, et al (1993). Muscle disease, HIV and zidovudine: the spectrum of muscle disease in HIV-infected individuals treated with zidovudine. J Neurol 240(8): 479–488.

Marcus K, Truffa M, Boxwell D, et al (2002). Recently identified adverse events secondary to NRTI therapy in HIV-infected individuals: cased from the FDA's adverse event reporting system (AERS) [abstract LB14]. Paper presented at the Ninth Conference on Retroviruses and Opportunistic Infections, Seattle, WA.

Mastaglia FL, Ojeda VJ (1985a). Inflammatory myopathies: Part 1. Ann Neurol 17(3): 215–227.

Mastaglia FL, Ojeda VJ (1985b). Inflammatory myopathies: Part 2. Ann Neurol 17(4): 317–323.

Mhiri C, Baudrimont M, Bonne G, et al (1991). Zidovudine myopathy: a distinctive disorder associated with mitochondrial dysfunction. Ann Neurol 29(6): 606–614.

Miller KD, Cameron M, Wood LV, et al (2000). Lactic acidosis and hepatic steatosis associated with use of stavudine: report of four cases. Ann Intern Med 133(3): 192–196.

Mokrzycki MH, Harris C, May H, et al (2000). Lactic acidosis associated with stavudine administration: a report of five cases. Clin Infect Dis 30(1): 198–200.

Morgello S, Wolfe D, Godfrey E, et al (1995). Mitochondrial abnormalities in human immunodeficiency virus-associated myopathy. Acta Neuropathol 90(4): 366–374.

Prime KP, Edwards SG, Pakianathan MR, et al (2003). Polymyositis masquerading as mitochondrial toxicity. Sex Transm Infect 79(5): 417–418.

Seidman R, Peress NS, Nuovo GJ (1994). In situ detection of polymerase chain reaction-amplified HIV-1 nucleic acids in skeletal muscle in patients with myopathy. Mod Pathol 7(3): 369–375.

Sellier P, Monsuez JJ, Evans J, et al (2000). Human immunodeficiency virus-associated polymyositis during immune restoration with combination antiretroviral therapy. Am J Med 109(6): 510–512.

Simpson DM, Bender AN (1988). Human immunodeficiency virus-associated myopathy: analysis of 11 patients. Ann Neurol 24(1): 79–84.

Simpson DM, Wolfe DE (1991). Neuromuscular complications of HIV infection and its treatment. AIDS 5(8): 917–926.

Simpson DM, Citak KA, Godfrey E, et al (1993). Myopathies associated with human immunodeficiency virus and zidovudine: can their effects be distinguished? Neurology 43 (5): 971–976.

Simpson DM, Slasor P, Dafni U, et al (1997). Analysis of myopathy in a placebo-controlled zidovudine trial. Muscle Nerve 20(3): 382–385.

Simpson DM, Estanislao L, Evans S, et al (2004). HIV-associated neuromuscular weakness syndrome. AIDS 18(10): 1403–1412.

Snider WD, Simpson DM, Aronyk KE, et al (1983a). Primary lymphoma of the nervous system associated with acquired immune-deficiency syndrome. N Engl J Med 308(1): 45.

Snider WD, Simpson DM, Nielsen S, et al (1983b). Neurological complications of acquired immune deficiency syndrome: analysis of 50 patients. Ann Neurol 14(4): 403–418.

Stern R, Gold J, DiCarlo EF (1987). Myopathy complicating the acquired immune deficiency syndrome. Muscle Nerve 10(4): 318–322.

Further Reading

Gabbai AA, Schmidt B, Castelo A, et al (1990). Muscle biopsy in AIDS and ARC: analysis of 50 patients. Muscle Nerve 13(6): 541–544.

Wrzolek MA, Sher JH, Kozlowski PB, et al (1990). Skeletal muscle pathology in AIDS: an autopsy study. Muscle Nerve 13(6): 508–515.

Handbook of Clinical Neurology, Vol. 85 (3rd series)
HIV/AIDS and the Nervous System
P. Portegies, J. R. Berger, Editors

Chapter 11

Cerebral toxoplasmosis in AIDS

LISA M. CHIRCH AND BENJAMIN J. LUFT*

Stony Brook University School of Medicine, Stony Brook, NY, USA

11.1. Introduction

Toxoplasma gondii, an obligate intracellular protozoan parasite in human beings, has the potential to cause disease both in immunocompetent and immunocompromised individuals. Cerebral toxoplasmosis is a frequent cause of neurological pathology in HIV-infected patients. Toxoplasmic encephalitis (TE) is a life-threatening central nervous system (CNS) infection observed in advanced HIV infection, in particular in those patients with CD^4 cell counts of 100 cells/mm^3 or less. The introduction of protease inhibitors to the antiretroviral armamentarium and establishment of highly active antiretroviral therapy (HAART) as standard of care in developed countries have had profound effects on the incidence of HIV-related opportunistic infections, including toxoplasmosis, and have spectacularly diminished morbidity and mortality in these areas of the world. In addition, immune reconstitution related to HAART has led to the identification of new and unusual presentations of latent infections such as toxoplasmosis, cytomegalovirus (CMV) and atypical mycobacteria ("immune reconstitution inflammatory syndromes (IRIS)"). In developing countries where HAART is not readily available, TE remains a major cause of CNS pathology in HIV-infected patients, despite the potential utility of effective prophylactic medications, such as trimethoprim/sulfamethoxazole.

This chapter focuses on the epidemiology, both in developed and developing countries, pathogenesis, clinical manifestations, diagnosis and treatment of CNS toxoplasmosis in HIV-infected individuals.

11.2. The parasite

T. gondii is a coccidian in the family Sarcocystidae that was first identified in 1908 (Lainson, 1958; Katz et al., 1989). The organism exists in nature in different forms: the tachyzoite, which is the invasive, rapidly proliferating form; the tissue cyst, containing bradyzoites; and the oocyst, which produces infectious sporozoites. Sabin and Olitsky (1937) described the obligate intracellular location of *T. gondii*, and Dubey et al. (1970) reported experiments that described its life cycle. Domestic cats are the only definitive hosts for *T. gondii*. Cats become infected by eating animal tissues that contain cysts (usually rodents), or by incidentally ingesting oocysts passed in the feces of other cats. The sexual phase occurs only in the feline gut and begins when the cat ingests oocysts or tissue cysts. Sexual and asexual forms of the parasite begin to develop (gametogeny) when *T. gondii* penetrates the intestinal epithelial cells. The protozoa infect the epithelial cells of the small intestine and develop into merozoites, which in turn may infect more epithelial cells, resulting in an enteroepithelial cycle. Some of these merozoites then develop into gametocytes which fuse and form diploid oocysts. In humid, warm climates oocysts may remain infectious in the environment for a year or more. Oocysts become infectious after sporulating, which occurs 2 to 21 days after release into the environment (Sheffield and Melton, 1970).

The asexual phase of the life cycle is in most part extraintestinal, and occurs both in felines and in incidental hosts, such as humans. Oocysts are ingested with release of sporozoites into the small intestine, where they are able to penetrate the intestinal mucosa, replicate and

*Correspondence to: Benjamin J. Luft, MD, Edmund D. Pellegrino Professor, Chief, Divisioin of Infectious Diseases, University Hospital, Health Sciences Center T15, 080, Stony Brook, NY 11794-8153, USA. E-mail: bluft@notes.cc.sunysb.edu, Tel: +1-631-444-3490, Fax: +1-631-444-7518.

enter the bloodstream. *T. gondii* is unusual in that the organism is able to invade and multiply inside various types of mammalian cells; one notable exception includes erythrocytes. The organism then transforms into a tachyzoite intracellularly, proliferating by binary fission until the host cell ruptures. Tachyzoites are released and may subsequently invade adjacent host cells. This cycle continues until adequate host immunity ultimately develops to limit further multiplication. In addition, formation of tissue cysts may occur upon cell entry and replication; cysts grow and remain within the host cell cytoplasm, and bradyzoites continue to divide within the cysts. Tachyzoites may aggregate and form pseudocysts that accumulate in host cells over time; pseudocysts eventually form true tissue cysts. The complex outer membrane of the tissue cyst is immunologically inert, and it is unusual to find inflammatory reaction surrounding a cyst in chronically infected individuals (Frenkel, 1956).

T. gondii is able to survive within a variety of cells by evading the host immune response by several mechanisms (Murray, 1983; Luft, 1988): the organism is able to avoid triggering the oxidative burst of some phagocytes (Wilson et al., 1980), can prevent fusion with the lysosome (Jones et al., 1972) and acidification (Sibley et al., 1985) and produces catalase and superoxide dismutase (Murray et al., 1980). Organisms ingested by macrophages are therefore able to survive and multiply, as they do not trigger generation of reactive oxygen species (Wilson et al., 1980).

11.3. Epidemiology

Since the introduction of HAART, including protease inhibitors and non-nucleoside reverse transcriptase inhibitors (NNRTIs), epidemiological studies in industrialized nations such as the USA, Great Britain, France, Spain and Australia have shown dramatic reductions in HIV-related complications and opportunistic infections, including cerebral infections such as toxoplasmosis. Neurological complications, however, remain an important cause of morbidity and mortality in HIV-infected individuals. Gray et al. (2003) demonstrated a dramatic decrease in the number of autopsies performed on AIDS patients at two hospitals in France between 1997 and 2002. There was an overall decrease in cerebral toxoplasmosis, for which treatment is available. However, when these investigators looked specifically at pathology occurring after 2000, they found a decrease in infections associated with severe immunosuppression, with a relative increase in those infections that may occur in patients with more mild immune compromise, such as TE, and encephalitis associated with varicella zoster virus (VZV) and herpes simplex virus (HSV). Unusual presentations of previously identified diseases were also identified, including IRISs (Gray et al., 2003). In the Multicenter AIDS Cohort Study, a longitudinal cohort of homosexual or bisexual men in several US cities, there was a trend for a decreased incidence of CNS toxoplasmosis since the introduction of HAART which could not be accounted for by changes in the use of prophylactic medications, such as trimethoprim/sulfamethoxazole (TMP/SMX) (Sacktor, 2002). Similar trends were reported over this time period from the Australian National AIDS Registry (Sacktor, 2002). In Madrid, Spain, San-Andres et al. reported a significant decreasing trend in the incidence of cerebral toxoplasmosis from 1989 through 1997 in patients who adhered to antiretroviral therapy and *Pneumocystis carinii* (PCP) prophylaxis (San-Andres et al., 2003). The authors demonstrated an annual risk reduction of 18% (95% confidence interval) from 1993, with a 73% risk reduction for those patients adhering to PCP prophylaxis, $p < 0.001$ (San-Andres et al., 2003). Abgrall et al. and the Clinical Epidemiology Group of the French Hospital Database on HIV (2001) analyzed the incidence and risk factors for TE in two periods, before and after the introduction of HAART. The authors report a decline in the incidence of TE from 3.9 to 1 case per 100 person-years during the two time periods, respectively. In addition, discontinuation of TMP/SMX increased the risk of TE in both periods; $p < 0.001$. This increase was not observed in patients receiving HAART whose CD^4 cell counts increased to >200 cells/mm^3 during follow-up; $p = 0.45$ (Abgrall et al., 2001). In the USA, between 10 and 40% of adults with AIDS are latently infected, and it is estimated that 30–50% of these patients will develop TE (Clumeck et al., 1984; Levy et al., 1988).

The prevalence of cerebral toxoplasmosis in AIDS patients varies markedly from place to place, reflecting underlying seroprevalence rates in given populations. For example, in Africa, Europe, and Latin America, where the incidence of latent infection is much higher than in the United States, the proportion of individuals who will develop TE is three to four times higher (Clumeck et al., 1984). For example, in a study during the early 1990s in Austria, Glatt et al. reported that 47% of patients with AIDS seropositive for *T. gondii* developed TE (Glatt, 1992). These trends and paradigms, however, do not necessarily hold true in today's era of HAART, given its generally positive effects on levels of immunosuppression and incidence of opportunistic infections.

In contrast to recent trends in developed countries, HAART is almost universally unavailable in the rest of the world, where the overwhelming majority of HIV-infected individuals live. Areas with rapid recent increases in infection rates include sub-Saharan Africa, China and Eastern Europe (Sacktor, 2002). South Africa has more HIV cases than any other country, which has had devastating social, economic and political consequences (Sacktor, 2002).

The prevalence of neurological complications of HIV infection in developing countries is similar to that observed in the developed world prior to the advent of HAART. The epidemic is still expanding in most regions of Sub-Saharan Africa. According to the UNAIDS 2002 report on the global epidemic, 2.1 million Ethiopians were living with HIV at the end of 2001, third after Nigeria and South Africa (Bane et al., 2003). A one-year prospective study conducted at a tertiary teaching hospital in a major urban center in Ethiopia revealed CNS mass lesions "suggestive of toxoplasmosis" as the admitting diagnosis in 31% of HIV-infected patients admitted. Some 56% of those who died during hospitalization had a CNS mass lesion cited as the cause of death. CNS toxoplasmosis was usually diagnosed in this series using epidemiological information (80% seroprevalence in this community), suggestive CT scan findings and response to therapy (Bane et al., 2003).

Mathew and Chandy (1999) reported two cases of CNS toxoplasmosis in India, in which toxoplasmosis was not initially suspected, and the diagnosis was only made after surgical intervention. Despite the fact that only a few cases have been reported in India, the AIDS epidemic continues to surge in that region of the world. One would therefore expect the incidence of CNS toxoplasmosis to become increasingly common given the ubiquitous nature of this organism. At present all patients in India scheduled to undergo elective neurosurgery are tested for HIV antigen, and it is now recommended that any patient with multiple lesions on CT and positive serology be started on empiric therapy (Mathew and Chandy, 1999).

In rural Thailand, another region of rapid expansion of the AIDS epidemic, Inverarity et al. (2002) conducted a cross sectional study to determine the spectrum of HIV-related disease presenting to rural primary and secondary healthcare facilities between 1997 and 2000. Serology was not available for this study, but patients were presumed to have cerebral toxoplasmosis if there was a recent history of focal neurological abnormality or reduced level of consciousness, and a contrast-enhanced CT scan showing mass lesion(s) and successful response to therapy (Inverarity et al., 2002). Toxoplamosis was among the most common opportunistic infections, along with PCP, cryptococcal meningitis and tuberculosis, and accounted for 4% of admissions. The authors suggest that, in this region, where HAART is not readily available, the use of prophylactic medications such as cotrimoxazole for PCP as well as toxoplasmosis may result in substantial reductions in morbidity and mortality.

Although the AIDS epidemic began more than 2 decades ago, the large majority of individuals infected with HIV in the developing world do not have access to HAART, and thus continue to suffer from a wide range of opportunistic infections, including cerebral toxoplasmosis. Perhaps widespread institution of primary prophylactic medications that are more readily available and less expensive will help ameliorate this to a certain extent, but it is clear that more rigorous international efforts to provide effective antiretroviral medication to these parts of the world are sorely needed.

11.4. Pathogenesis and pathology

Cerebral toxoplasmosis in immunocompromised hosts most likely occurs as a result of reactivation of latent infection. There is evidence to support this presumption, although much of it is circumstantial. First, it is rare to identify CNS involvement with toxoplasmosis in acutely infected immunocompetent patients (Remington, 1974; Krick and Remington, 1978). Also, autopsy findings have demonstrated incidental findings of tissue cysts in various organs of immunocompetent patients (Frenkel, 1956). In addition, in patients with AIDS, cerebral toxoplasmosis rarely develops in the absence of IgG antibodies (Wong et al., 1984). All of this suggests that HIV-infected individuals who are seropositive for antibodies to *T. gondii* are at risk for recrudescence of latent disease and cerebral toxoplasmosis by virtue of significant compromise of cell-mediated immunity. The manifestations of acute infection in this population, however, are not well defined, and it could therefore be theoretically impossible to distinguish between reactivation and atypical manifestations of acute infection in these patients.

Cerebral toxoplasmosis is characterized pathologically by necrosis and thrombosis of involved vessels with perivascular inflammation, with frank vasculitis in extreme cases. There may be microglial response with formation of nodules, and tachyzoites or toxoplasmic antigens have been associated with these nodules. Conley et al. (1981) demonstrated this using the peroxidase-antiperoxidase staining method. Focal areas of encephalitis or abscess are common, but a diffuse necrotizing encephalitis in severely immunocompromised patients has been reported (Gray et al., 1989). The white matter, gray matter and other areas of the CNS may be involved, though there is a propensity to involve the basal ganglia, the corticomedullary junction, the thalamus and the white matter (Luft and Sivadas, 2002). The meninges are usually spared (Matheron et al., 1990).

In AIDS, the presence of cysts and extracellular tachyzoites is a hallmark of the disease process, and the most intense inflammation is histopathologically associated with these entities (Post et al., 1983; Luft et al., 1984; Millard, 1984; Farkash et al., 1986; Israelski and Remington, 1988). Post et al. (1983) described three distinct zones of a *Toxoplasma* abscess in patients with AIDS: in general, a middle zone consisting of severe perivascular inflammation is surrounded by lateral zones with coagulative necrosis and thrombosis of vessels. Huang and Chou (1988) also looked at ring-enhancing lesions and demonstrated tachyzoites within arterial walls associated with necrotizing arteritis in early lesions. In more advanced lesions, concentric fibrosis with necrosis and thrombosis was found associated with tachyzoites. An intermediate zone contained engorged vessels with patchy necrosis. Tachyzoites, along with variable numbers of cysts, were also identified. In outer zones necrosis and vascular lesions were rare and organisms more often appeared in the form of tissue cysts.

11.5. Clinical manifestations

Cerebral toxoplasmosis in HIV-infected individuals may have variable presentations, but typically progresses subacutely over a period of weeks. Initial neurological symptoms may be focal or nonfocal, depending on the location of the lesion(s). Presenting symptoms of hemianopsia, cranial nerve abnormalities, diplopia, hemiparesis, hemiplegia, aphasia, focal seizures or impaired equilibrium may provide insight into lesion location (Luft and Sivadas, 2002). Because TE tends to localize in the basal ganglia, movement disorders such as tremor, hemiballism, dyskinesias and parkinsonism have also been described (Cardoso, 2002). Hyperkinesias in patients with AIDS are most commonly caused by *T. gondii*, though *Treponema pallidum*, *Cryptococcus neoformans* and JC virus (the agent of progressive multifocal leukoencephalopathy) may also play a role. By and large these opportunistic pathogens create hyperkinesias by damaging the basal ganglia and its connections. Occasionally hemichorea or hemiballism may be the first manifestation of AIDS (Cardoso, 2002).

Initially, generalized signs and symptoms, including weakness, confusion, lethargy, altered mental status and headache, may predominate. More focal symptoms may later evolve. In addition, panhypopituitarism and hyponatremia (Conley et al., 1981), a result of inappropriate antidiuretic hormone secretion, can occur. Cerebral toxoplasmosis can be extremely difficult to distinguish from other CNS opportunistic infection, such as progressive multifocal leukoencephalopathy (PML) and the AIDS dementia syndrome (ADS) itself, in particular when patients present primarily with cognitive decline. Systemic symptoms such as fever and weight loss are common but are nonspecific.

Cerebral toxoplasmosis typically progresses in a subacute manner, over a period of weeks to months. Subtle, generalized symptoms eventually become more focal with evidence of mass effect. Generalized encephalitis may ensue with increasing confusion, dementia and stupor (Navia et al., 1986); some patients present with seizures, and this may be the initial manifestation of the disease (Levy et al., 1985). Cerebrospinal fluid (CSF) findings from a diagnostic lumbar puncture are generally nonspecific as well, and include a monocytic pleocytosis, elevated protein and a normal glucose (Mills, 1986).

By and large, toxoplasmosis in HIV-infected individuals manifests as CNS disease only; however, case reports have described disseminated infection of multiple organs. Rarely, ocular toxoplasmosis or retinochoroiditis may be a harbinger for cerebral toxoplasmosis (Alonso et al., 1984). Also, since the widespread availability of HAART in the developed world, one must consider atypical manifestations of toxoplasmosis as a consequence of immune reconstitution. For example, Ghosn et al. (2003) describe a case of toxoplasmic adenitis, an unusual manifestation of this infection, after discontinuation of secondary prophylaxis in the presence of significant immune recovery. The authors address the possibility of reinfection with an atypical strain of *T. gondii*, as this patient had lived in the French West Indies for many years. The other potential explanation is that this case of toxoplasmic inguinal adenitis occurring three years after an episode of toxoplasmic encephalitis represented reactivation disease associated with significant immune reconstitution on antiretroviral therapy.

Cerebral toxoplasmosis may occur in HIV-infected individuals as one of any number of opportunistic central nervous system infections. The differential diagnosis of central nervous system disease in AIDS patients includes cryptococcal meningitis, tuberculosis, nocardiosis, bacterial abscess, PML and lymphoma (Snider et al., 1983). Although empiric treatment for presumed toxoplasmosis, as is now common practice, could potentially delay treatment of other rare but treatable CNS infections such as tuberculosis or nocardiosis, the two disease entities that most closely resemble toxoplasmosis are the other most common causes of focal lesions in AIDS patients: lymphoma and PML. Both of these diagnoses carry poor prognoses and generally show minimal response to therapy. Luft et al. (1993) reported that more than 50% of patients treated empirically in a prospective trial demonstrated a quantifiable clinical improvement by the fifth day of therapy and more than 70% showed clinical improvement after one week of therapy. For all patients who ultimately failed to respond to therapy or who had lymphoma, baseline signs and symptoms worsened and new abnormalities appeared by day 10 of therapy. A total of 91% of patients who experienced clinical response also had a radiological response within 6 weeks.

11.6. Diagnosis

11.6.1. Serology

Because reactivation of chronic infection is the most common cause of toxoplasmosis in HIV-infected individuals, standard initial assessment should include assays for *T. gondii* IgG antibodies (Montoya, 2002). Those patients with positive toxoplasma serology are at risk for reactivation disease in the setting of advanced immunodeficiency (CD^4 count less than 200 cells/mm^3). More than 97% of patients with AIDS and TE have antibody titers against *T. gondii* (Luft and Sivadas, 2002). The predictive value of a positive serology in a patient with characteristic abnormalities on CT scan may be as high as 80% in the USA (Potasman et al., 1988). This leads many practitioners to initiate a trial of empiric anti-toxoplasma therapy in patients with AIDS and characteristic CNS lesions who are seropositive for *T. gondii*. The validity of this approach is dependent on multiple factors, including seroprevalence within a given population, and the relative frequencies of other CNS mass lesions in the same population, both of which affect the predictive value of serology.

Polymerase chain reaction (PCR) has been widely employed in the detection of pathogenic microorganisms. In the case of toxoplasmosis, however, there remains a general lack of standardized kits available, leading to the development of various in-house assays (Bretagne, 2003). Performances reported for these assays, both sensitivity and specificity, vary widely; in HIV-infected individuals with toxoplasmosis, the PCR on blood using the same DNA target is either of limited value with a sensitivity of 13%, or of excellent yield, with a sensitivity of 87.5% (Khalifa et al., 1994; Foudrinier et al., 1996). It is therefore difficult to know how reliable these assays are. However, the technique of detecting *T. gondii* DNA in blood by PCR may prove to be an essential tool in making rapid and definitive diagnoses in the near future, should standardized assays become available (Bretagne, 2003).

Contini et al. (2002) recently studied the role of stage-specific oligonucleotide primers in providing supportive information for the diagnosis of CNS toxoplasmosis in HIV-infected patients. Cerebral toxoplasmosis results from the stage conversion from the encysted bradyzoites located in tissue cysts to tachyzoites; therefore the recognition of individual stages could potentially aid in the diagnosis of TE. The authors used a nested PCR assay on CSF to assess the efficacy of primer sets which amplify target sequences expressed on tachyzoites (SAG1), bradyzoites (SAG4 and MAG1) or both stages (B1). Their results imply that specific primers to amplify genes expressed specifically by tachyzoites and bradyzoites may play a role in the diagnosis of relapse of toxoplasmosis in AIDS patients. Results were more accurate if lumbar puncture was performed before or shortly after therapy was initiated. CSF-PCR with the B1 (nonspecific stage) gene proved helpful in the diagnosis of acute disease when patients had not received more than one week of therapy.

11.6.2. Imaging

Both CT and MRI scans are utilized for the diagnosis of cerebral toxoplasmosis. In general MRI is more sensitive, and can identify lesions not detected by CT with contrast enhancement (Ramsey and Gean, 1997). As a rule, CT scans underestimate the number of lesions found at autopsy (Navia et al., 1986). As noted previously, toxoplasmosis typically involves the basal ganglia, corticomedullary junction and thalamus, although any area of the brain may be involved (Dietrich et al., 2000; Provenzale and Jinkins, 1997). CT scans with contrast enhancement characteristically show rounded isodense or hypodense ring enhancing lesions; ring enhancement seems to correlate with vascular proliferation and inflammation (Luft and Sivadas, 2002). A majority of patients have multiple lesions identified; 75% according to a study involving Haitian adults with AIDS and toxoplasmic encephalitis (Post et al., 1983). According to Ramsey and Gean (1997), focal lesions in TE can be divided into three types: necrotizing abscess, organizing abscess, and chronic abscess. Cystic lesions in patients with disseminated disease may not demonstrate enhancement with contrast; other stages, however, typically enhance.

Imaging features of TE are by no means pathognomonic and may resemble several other central nervous system disease processes, including lymphoma, PML, nocardial or bacterial abscess, primary or metastatic malignancy and tuberculoma. It is generally accepted that toxoplasmosis tends to manifest with multiple lesions more often, whereas patients with lymphoma more often have a single lesion. This criterion is suggestive at best, and more recently newer techniques have been investigated in order to assist in distinguishing the two disease processes. There are also specific findings that may be helpful if present. The eccentric target sign is highly suggestive of toxoplasmosis (specificity >90%), and consists of a small nodule associated with the wall of an abscess, or enhancing ring.

The location of the lesion(s) may also be suggestive, as lymphoma is much less often located subcortically, though both tend to affect the basal ganglia. The corpus callosum and leptomeninges are rarely involved in toxoplasmosis, and their involvement should suggest alternative diagnoses. Toxoplasmic abscesses typically demonstrate a uniform wall thickness of less than 3 mm, and tend to induce an intense surrounding inflammatory response.

Batra et al. (2004) describe the use of diffusion-weighted magnetic resonance (MR), MR spectroscopy and perfusion MRI features of toxoplasmosis in AIDS patients, in order to delineate specific characteristics that may be distinct from other disease processes. A total of 23 lesions in 8 adult male patients were found on MRI over a 14-month period; half of the patients had multiple lesions. All of the lesions examined in this study revealed predominant lipid peak on MR spectroscopy and were quite hypovascular on perfusion MR studies. The lesions examined generally demonstrated a liquefied central necrosis with increased diffusion and a lipid peak detectable on MR spectroscopy. In addition, a majority (13 of 17 lesions examined) revealed hypointensity on diffusion-weighted imaging, suggesting increased diffusion, which is consistent with the destructive nature of these lesions. The authors propose that these techniques may facilitate a more rapid and reliable diagnosis of cerebral toxoplasmosis.

Camacho et al. (2003) also studied diffusion coefficients in AIDS patients in an effort to distinguish between TE and CNS lymphoma. These authors also reported that TE lesions demonstrated significantly greater diffusion than that of lymphoma lesions ($p = 0.004$). Apparent diffusion coefficient ratios were higher in TE lesions, although there was considerable overlap for ratios between 1.0 and 1.6. Ratios above 1.6, however, accounted for more than half of the TE lesions, and were associated solely with toxoplasmosis.

Functional nuclear imaging may also prove useful in the near future in differentiating between TE and lymphoma (Bakshi, 2004). Ruiz et al. (1994) reported a 94% positive predictive value using thallium single photon emission CT (SPECT) in identifying lymphoma lesions. Early studies have also described the accuracy of FDG-PET scanning to differentiate between the two, 90% in one study (Villringer et al., 1995). Pomper et al. (2002) showed that patients with hypermetabolic lesions on PET and SPECT universally had lymphoma (100% specificity), and suggested these techniques may be useful as screening tools in AIDS patients with enhancing CNS mass lesions. It has been suggested that hypometabolic lesions are the best candidates for empiric anti-toxoplasma therapy with close follow-up (Bakshi, 2004).

Both MRI and CT scans are useful for monitoring and assessment of response to therapy in patients with TE. Those who respond to therapy will generally demonstrate radiographic response within 3 weeks, although complete resolution may take as long as 6 months (Luft and Sivadas, 2002). As with other disease entities, radiographic response lags behind clinical response (Luft et al., 1993). Because other processes, listed above, may occur concomitant to TE, the presence of a persistent or enlarging lesion despite adequate therapy should prompt further investigation, including a biopsy.

11.6.3. Organism identification

TE is generally diagnosed presumptively using clinical scenarios, serologies and imaging techniques. A definitive diagnosis rests on the identification of the organism in a clinical specimen, primarily by brain biopsy. The organism has been identified on occasion in bronchoalveolar lavage fluid, CSF and blood in patients with TE (Luft and Sivadas, 2002). The technique of inoculating laboratory animals in order to identify the organism from clinical specimens is labor-intensive and cannot be completed in a timely fashion, and thus is rarely utilized (Derouin et al., 1989). Selective amplification by PCR of specific *T. gondii* DNA products may prove a sensitive and specific manner of identifying the organism in clinical specimens, including CSF, though this remains to be demonstrated (Burg et al., 1989; Luft and Sivadas, 2002).

Stereotactic needle biopsy utilizing intraoperative sonography is currently the most common technique used to definitively diagnose TE. Immunohistochemical stains and other techniques including the use of monoclonal antitoxoplasmic antibodies (Sun et al., 1986) are used to identify characteristic changes associated with TE. These can sometimes, however, closely resemble those of viral infections, and therefore obtaining more tissue for additional studies is sometimes necessary if needle biopsy provides insufficient information. Tachyzoites are generally best identified using histopathologic stains at the periphery of necrotic lesions, or within normal brain (Luft and Sivadas, 2002).

11.7. Treatment

The therapy of choice for acute cerebral toxoplasmosis is pyrimethamine in combination with sulfadiazine (Klinker et al., 1996), although several alternative regimens have been examined in the recent past, or are currently under investigation. Pyrimethamine, a dihydrofolate reductase inhibitor, and sulfadiazine, a dihydrofolate synthetase inhibitor, act synergistically against

T. gondii by sequentially blocking folic acid metabolism (Kayhoe et al., 1957; Frenkel et al., 1960). They should therefore be used in combination when diffusion of the drug to the site of infection may be compromised (Luft and Sivadas, 2002).

Pyrimethamine has a half-life of approximately 96 hours and is well absorbed after oral administration. A recent study conducted by Klinker et al. (1996) demonstrated the variability of plasma levels of pyrimethamine in patients with AIDS and cerebral toxoplasmosis. In this study steady state plasma pyrimethamine levels were measured by gas chromatography in 74 adult patients with advanced AIDS receiving therapy for cerebral toxoplasmosis. Steady state levels were generally reached within 12 to 20 days, and concentrations increased with the weekly dosage given; however, there was wide variability among patients for each dosage given. Levels were particularly unpredictable in patients with diarrhea, who were on other medications that induced P-450 enzymes, or who were poorly compliant. A weekly dose of 175 mg (or 25 mg per day) was necessary to achieve mean serum levels of >75 ng/ml; when dosed at 100 mg to 15 mg per day plasma levels reached >3000 ng/ml. Based on these findings the authors suggest that during curative treatment of cerebral toxoplasmosis, a daily pyrimethamine dose of 100 to 150 mg as monotherapy or 25 mg in combination with sulfadiazine should be successful. The AIDS Clinical Trial Group (ACTG) evaluated the pharmacokinetics of pyrimethamine when administered with the nucleoside reverse transcriptase inhibitor zidovudine, and found steady state trough levels at a dose of 50 mg daily to be similar to those reported in patients not infected with HIV (Jacobson et al., 1996). Wide variability in serum concentrations was again demonstrated. It should be noted that the above combination be used with caution due to the potential of additive bone marrow toxicity; the same is true for several medications used relatively frequently in the HIV-infected population, such as trimethoprim/sulfamethoxazole and interferon.

Pyrimethamine is administered as an initial loading dose of 100 to 200 mg in order to reach steady state concentrations more rapidly, followed by a daily dose of 50 to 75 mg a day for the first 3 to 6 weeks of therapy (Luft and Sivadas, 2002). These doses have been shown to maintain serum levels within the therapeutic range. Sulfadiazine is rapidly absorbed from the gastrointestinal tract and peak levels are generally reached within 6 hours. Sulfadiazine is dosed at 4 to 6 g daily in four divided doses orally. Folinic acid (leucovorin) is given as a supplement at 10 to 20 mg daily in order to minimize the additive effects of these two agents on the bone marrow. This combination is highly effective against tachyzoites but has no effect on the cyst form of the organism; as such, prolonged maintenance therapy consisting of lower doses of the pyrimethamine and sulfadiazine is required in order to prevent relapse in severely immunocompromised patients. Haverkos (1987) from the Toxoplasma Encephalitis Study Group reported that 50% of patients with AIDS and TE relapse after hospital discharge. Maintenance therapy generally consists of pyrimethamine at 25 to 50 mg and sulfadiazine at 4 to 6 g, given daily or two to three times weekly (Leport et al., 1988). The most appropriate dosage and dosing interval long-term therapy for cerebral toxoplasmosis in AIDS has not yet been determined.

It has been reported that up to 60% of patients treated with the above regimen for cerebral toxoplasmosis experience significant toxicity, frequently prompting discontinuation of therapy (Haverkos, 1987). Skin rash is the most common dose-limiting adverse reaction, particularly during induction therapy, and may occur in up to 20% of patients (Luft and Sivadas, 2002). Bone marrow suppression usually occurs later in the course of therapy and may be exacerbated by concomitant use of other medications with similar effects. Because of the relative frequency of dose-limiting side effects with the use of pyrimethamine and sulfadiazine, other options for the treatment of cerebral toxoplasmosis are required.

There have been reports of several cases in which clindamycin was used successfully in combination with pyrimethamine for the treatment of cerebral toxoplasmosis (Rolston and Hoy, 1987; Podzamczer and Gudiol, 1988; Westblom and Belshe, 1988). Prior to the AIDS epidemic, Araujo and Remington (1974) demonstrated the efficacy of clindamycin in treating murine toxoplasmosis. A prospective study by the California University-wide Task Force on AIDS reported that intravenous clindamycin in combination with oral pyrimethamine was equal in efficacy to the latter in combination with sulfadizine for the treatment of AIDS-associated TE (1988). Luft et al. (1993) in an ACTG-sponsored trial demonstrated a 71% response rate in TE patients treated with pyrimethamine (200 mg loading dose followed by 75 mg daily) plus oral clindamycin 600 mg every 6 hours. Although it appears that the combination of clindamycin and pyrimethamine has equivalent efficacy to pyrimethamine/sulfadiazine for acute toxoplasmic encephalitis, it is less effective in the long-term treatment in patients who remain severely immunocompromised (Katlama et al., 1996).

To address the issue of CSF penetration of clindamycin and its metabolite *N*-demethylclindamycin, Gatti et al. (1998) measured concentrations after a single 1200 mg intravenous infusion beginning at 1 to 3 hours prior to lumbar puncture. The concentrations of clindamycin in CSF were well above the 50% inhibitory concentration of 0.001 mg/l and the parasiticidal concentration of 0.006 mg/l, thus supporting its use, at least intravenously, for the treatment of TE.

A few studies have examined the possibility of utilizing newer macrolide antibiotics for the treatment of TE. Trotta et al. (1997) reported a case of TE in an AIDS patient who was unable to tolerate sulfadiazine or clindamycin, but who responded to azithromycin 1000 mg daily for 21 days followed by 1500 mg weekly, in combination with pyrimethamine. After 50 days of follow-up a repeat CT scan showed complete resolution of multiple ring-enhancing lesions identified previously. Another scan performed one month late showed no evidence of recurrence. In a dose escalation, phase I/II trial of azithromycin and pyrimethamine for the treatment of TE, Jacobson et al. (2001) described the regimen as a possible alternative, though noted that use of the regimen as maintenance therapy may be associated with higher relapse rates. The role of clarithromycin in combination with minocycline in maintenance treatment of cerebral toxoplasmosis has also been addressed, though concrete data are still lacking (Sellal et al., 1996).

Trimethoprim-sulfamethoxazole (TMP-SMX) has also been evaluated as an alternative treatment of TE in AIDS, a particularly relevant question in developing countries where intravenous forms of other medications are not available or exceedingly expensive. A retrospective study of intravenous TMP-SMX showed a favorable outcome in 87% of treated patients (Francis et al., 2004). The same authors then conducted a prospective, multicenter, randomized pilot study comparing TMP-SMX to standard therapy (Torre et al., 1998). No statistically significant differences were identified between the two regimens in terms of efficacy, but patients who received TMP-SMX had fewer adverse reactions. Francis et al. (2004) conducted another prospective study examining the efficacy of oral TMP-SMX in the province of Kwa Zulu-Natal in South Africa. All 20 patients who completed the study showed complete or partial clinical and radiological improvement, although long-term follow-up was not available for all patients. The authors therefore conclude that TMP-SMX is likely to be a viable alternative for the treatment of acute TE in AIDS patients, in particular in developing countries and in the public sector; implications for maintenance therapy are less clear.

Lastly, atovaquone in combination with either pyrimethamine or sulfadiazine may prove to be another promising alternative for the treatment of TE in AIDS. A phase II ACTG trial (Chirgwin et al., 2002) examining the above regimens showed both to have sufficient antitoxoplasma activity to be considered for further study. Both regimens were well tolerated with similar side effect profiles to standard regimens. Further study including prospective randomized trials is necessary to more clearly define the roles of atovaquone, as well as other potential alternative agents for the treatment of AIDS-related cerebral toxoplasmosis.

11.8. Prophylaxis

In industrialized countries, primary prophylaxis with TMP-SMX is common practice for HIV-infected individuals with CD^4 counts of less than 200×10^6 cells/l, significantly reducing the incidence of PCP and TE (Maynart et al., 2001). Secondary prophylaxis is generally indicated in those patients who are treated for active TE and recover, but remain severely immunocompromised. The advent of potent antiretroviral therapy (ART) has prompted considerable investigation into the possibility of discontinuing prophylaxis in patients who are sufficiently immune reconstituted. Several studies (Soriano et al., 2000; Kirk et al., 2002) demonstrate the safety of interrupting maintenance therapy for TE in patients with CD^4 counts of greater than 200 cells/mm^3 for at least 6 months after beginning ART.

Intolerance to sulfonamides poses a difficult problem in these patients, in particular in developing countries where other agents are less readily available. Although neither dapsone nor pyrimethamine alone seems to be effective prophylaxis against TE, the two in combination may be (Clotet et al., 1991; Derouin et al., 1991; Girard et al., 1993). Girard et al. (1993) in an open label randomized study demonstrated that dapsone (50 mg daily) in combination with pyrimethamine (50 mg weekly) and leucovorin was more effective in preventing TE than aerosolized pentamidine (300 mg monthly). Another study demonstrated no significant difference in TE rates between patients receiving dapsone plus pyrimethamine or TMP-SMX (Podzamczer et al., 1993).

Several studies have examined the use of pyrimethamine alone for TE prophylaxis, although results have been less compelling. In a randomized, double-blinded placebo-controlled trial, Leport et al. (1996) studied the use of pyrimethamine (100 mg loading dose followed by 50 mg three times weekly) for TE prophylaxis. This trial failed to show a difference in the rates of TE between the treatment and placebo arms using intent-to-treat analysis (Chene et al., 1998). However, pyrimethamine demonstrated significant efficacy in the on-treatment analysis (HR 0.44).

According to the IDSA guidelines, clinicians may safely discontinue primary prophylaxis in patients on ART who experience a sustained rise in CD^4 count as previously described, for at least 3 months (MMWR,

2002). Lifelong suppressive therapy or chronic maintenance should be administered to patients who have been treated for acute TE, primarily with pyrimethamine plus either sulfadiazine (and leucovorin) or clindamycin. These regimens may also be discontinued, as per the IDSA guidelines, in patients who experience significant immune reconstitution for 6 months or more. Patients who have been adequately treated for acute TE and who have favorable immunologic responses to potent antiretroviral therapy should be at low risk for recurrence.

11.9. Future study

Several other agents from various classes have proven effective in vitro and in vivo models of TE. Macrolide antimicrobials, such as roxithromycin (Chan and Luft, 1986; Chang and Pechere, 1987), azithromycin (Araujo et al., 1992; Blais et al., 1993) and clarithromycin (Pfefferkorn et al., 1992; Beckers et al., 1995), have been studied. Luft (1986) reported the in vivo activity of arprinocid, a purine analogue and competitive inhibitor of hypoxanthine transmembrane transport. Arprinocid may prove particularly useful in that this mechanism is essential for *T. gondii*, which cannot synthesize purines de novo. Other experimental agents include piritrexim and trimetrexate, highly lipid-soluble dihydrofolate reductase inhibitors that have a higher affinity for toxoplasmic dihydrofolate reductase than pyrimethamine (Araujo et al., 1987; Kovacs et al., 1987). All of these agents require further study in humans in order to determine their potential role in TE therapy.

References

Abgrall S, Rabaud C, Costagliola D, et al (2001). Incidence and risk factors for toxoplasmic encephalitis in human immunodeficiency virus-infected patients before and during the highly active antiretroviral therapy era. Clin Infect Dis 33: 1747–1755.

Alonso R, Heiman-Patterson T, Mancall EL (1984). Cerebral toxoplasmosis in acquired immune deficiency syndrome. Arch Neurol 41: 321–323.

Araujo FG, Remington JS (1974). Effect of clindamycin on acute and chronic toxoplasmosis in mice. Antimicrob Agents Chemother 5: 647–651.

Araujo FG, Guptill DR, Remington JS (1987). Concise communications: in vivo activity of piritrexin against *Toxoplasma gondii*. J Infect Dis 156: 828–830.

Araujo FG, Guptill DR, Remington JS (1992). Azithromycin, a macrolide antibiotic with potent activity against *Toxoplasma gondii*. Antimicrob Agents Chemother 36: 1091–1096.

Bakshi R (2004). Neuroimaging of HIV and AIDS related illnesses: a review. Frontiers Biosci 9: 632–646.

Bane A, Yohannes AG, Fekade D (2003). Morbidity and mortality of adult patients with HIV/AIDS at Tikur Anbessa teaching hospital, Addis Ababa, Ethiopia. Ethiop Med J 41: 131–140.

Batra A, Tripathi RP, Gorthi SP (2004). Magnetic resonance evaluation of cerebral toxoplasmosis in patients with the acquired immune deficiency syndrome. Acta Radiologica 2: 212–221.

Beckers CJM, Roos DS, Donale RGK, et al (1995). Inhibition of cytoplasmic organellar protein synthesis in *Toxoplasma gondii*: implication for the target of macrolide antibiotics. J Clin Invest 95: 367–376.

Blais J, Garneau V, Chamberland S (1993). Inhibition of *Toxoplasma gondii* protein synthesis by azithromycin. Antimicrob Agents Chemother 37: 1701–1703.

Bretagne S (2003). Molecular diagnostics in clinical parasitology and mycology: limits of the current polymerase chain reaction (PCR) assays and interest of the real-time PCR assays. Clin Microbiol Infect 9: 505–511.

Burg JL, Grover CM, Pouletty P, et al (1989). Direct and sensitive detection of a pathogenic protozoan, *Toxoplasma gondii*, by polymerase chain reaction. J Clin Microbiol 27: 1787–1792.

Camacho D, Smith JK, Castillo M (2003). Differentiation of toxoplasmosis and lymphoma in AIDS patients by using apparent diffusion coefficients. Am J Neuroradiol 24: 633–637.

Cardoso F (2002). HIV-related movement disorders; epidemiology, pathogenesis and management. CNS Drugs 16 (10): 663–668.

Chan J, Luft BJ (1986). Activity of roxithromycin (RU28965), a macrolide, against *Toxoplasma gondii* infection in mice. Antimicrob Agents Chemother 30: 323–324.

Chang HR, Pechere JF (1987). Effect of roxithromycin on acute toxoplasmosis in mice. Antimicrob Agents Chemother 31: 1147–1149.

Chene G, Morlat P, Leport C, et al (1998). Intention-to-treat analysis vs. on-treatment analyses of clinical trial data: experience from a study of pyrimethamine in the primary prophylaxis of toxoplasmosis in HIV-infected patients. Control Clin Trials 19(3): 233–248.

Chirgwin K, Hafner R, Leport C, et al (2002). Randomized phase II trial of atovaquone with pyrimethamine or sulfadiazine for treatment of toxoplasmic encephalitis in patients with acquired immunodeficiency syndrome: ACTG 237/ANRS 039 Study. AIDS Clinical Trials Group 237/Agence Nationale de Recherche sur le SIDA, Essai 039. Clin Infect Dis 34(9): 1243–1250.

Clotet B, Sirera G, Romeu J, et al (1991). Twice weekly dapsone-pyrimethamine for preventing PCP and cerebral toxoplasmosis. AIDS 5: 601–602.

Clumeck N, Sonnet J, Taelman H, et al (1984). Acquired immunodeficiency syndrome in African patients. N Engl J Med 310: 492–497.

Conley FK, Jenkins KA, Remington JS (1981). *Toxoplasma gondii* infection of the central nervous system. Hum Pathol 12: 690–698.

Contini C, Cultrera R, Seraceni S, et al (2002). The role of stage-specific oligonucleotide primers in providing effective laboratory support for the molecular diagnosis of reactivated *Toxoplasma gondii* encephalitis in patients with AIDS. J Med Microbiol 51: 879–890.

Derouin F, Sarfati C, Beauvais B, et al (1989). Laboratory diagnosis of pulmonary toxoplasmosis in patients with acquired immune deficiency syndrome. J Clin Microbiol 27: 1661–1663.

Derouin F, Piketty C, Chastang C, et al (1991). Anti-toxoplasma effects of dapsosne alone and combined with pyrimethamine. Antimicrob Agents Chemother 35: 252–255.

Dietrich U, Maschke M, Dorfler A, et al (2000). MRI of intracranial toxoplasmosis after bone marrow transplantation. Neuroradiology 42: 14–18.

Dubey JP, Miller NL, Frenkel JK (1970). The *Toxoplasma gondii* oocyst from cat feces. J Exp Med 133: 636–662.

Farkash AE, Maccabee PJ, Sher JH, et al (1986). CNS toxoplasmosis in acquired immune deficiency syndrome: a clinical-pathological-radiological review of 12 cases. J Neurol 49: 744–748.

Foudrinier F, Aubert D, Puygauthier-Toubas D, et al (1996). Detection of *Toxoplasma gondii* in immunodeficient subjects by gene amplification: influence of therapeutics. Scand J Infect Dis 28: 383–386.

Francis P, Patel VB, Bill P, et al (2004). Oral trimethoprim-sulfamethoxazole in the treatment of cerebral toxoplasmosis in AIDS patients: a prospective study. S Afr Med J 94 (1): 51–53.

Frenkel JK (1956). Pathogenesis of toxoplasmosis and of infections with organisms resembling *Toxoplasma*. Ann NY Acad Sci 64: 215.

Frenkel JK, Weber RW, Lunde MN (1960). Acute toxoplasmosis: effective treatment with pyrimethamine, sulfadiazine, leucovorin calcium, and yeast. JAMA 173: 1471–1476.

Gatti G, Malena M, Casazza R, et al (1998). Penetration of clindamycin and its metabolite N-demethylclindamycin into cerebrospinal fluid following intravenous infusion of clindamycin phosphate in patients with AIDS. Antimicrob Agents Chemother 42(11): 3014–3017.

Ghosn J, Paris L, Ajzenberg D, et al (2003). Atypical toxoplasmic manifestation after discontinuation of maintenance therapy in a human immunodeficiency virus type 1-infected patient with immune recovery. Clin Infect Dis 37: e112–e114.

Girard PM, Landeman R, Gaudabout C, et al (1993). Dapsone-pyrimethamine compared with aerosolized pentamidine as primary prophylaxis against *Pneumocystis carinii* pneumonia and toxoplasmosis in HIV infection. N Engl J Med 328: 1514–1520.

Glatt A (1992). *Toxoplasma gondii* serologies in patients with human immunodeficiency virus infection. Infect Dis Clin Pract 1: 237.

Gray F, Gherardi R, Wingate E, et al (1989). Diffuse "encephalitic" cerebral toxoplasmosis in AIDS. J Neurol 236: 273–277.

Gray F, Chretien F, Vallat-Decouvelaere AV, et al (2003). The changing pattern of HIV neuropathology in the HAART era. J Neuropathol Exp Neurol 62(5): 429–440.

Haverkos HW (1987). Assessment of therapy for *Toxoplasma encephalitis*. The Toxoplasma Encephalitis Study Group. Am J Med 82: 907–914.

Huang TE, Chou SM (1988). Occlusive hypertrophic arteritis as the cause of discrete necrosis in CNS toxoplasmosis in the acquired immune deficiency syndrome. Hum Pathol 19: 1210–1214.

Inverarity D, Bradshaw Q, Wright P, et al (2002). The spectrum of HIV-related disease in rural central Thailand. Southeast Asian J Trop Med Public Health 33(4): 822–831.

Israelski DM, Remington JS (1988). Toxoplasma encephalitis patients with AIDS. Infect Dis Clin North Am 2: 429–445.

Jacobson JM, Davidian M, Rainey PM, et al (1996). Pyrimethamine pharmacokinetics in HIV-positive patients seropositive for *Toxoplasma gondii*. Antimicrob Agents Chemother 40: 1360–1365.

Jacobson JM, Hafner R, Remington J, et al (2001). Dose-escalation, phase I/II study of azithromycin and pyrimethamine for the treatment of toxoplasmic encephalitis in AIDS. AIDS 15(5): 583–589.

Jones TC, Yeh S, Hirsh JG (1972). The interaction between *Toxoplasma gondii* and mammalian cells: mechanism of entry and intracellular fate of the parasite. J Exp Med 136: 1157–1172.

Katlama C, De Wit S, O'Doherty S, et al (1996). Pyrimethamine-clindamycin vs. pyrimethamine-sulfadiazine as acute and long-term therapy for toxoplasmic encephalitis in patients with AIDS. Clin Infect Dis 22: 268–275.

Katz M, Despommier DD, Gwadz RW (1989). Parasitic Diseases. Springer-Verlag, New York.

Kayhoe DE, Jacobs L, Beye HK, et al (1957). Acquired toxoplasmosis: observations on two parasitologically proved cases treated with pyrimethamine and triple sulfonamides. N Eng J Med 257: 1247–1254.

Khalifa KS, Roth A, Roth B, et al (1994). Value of PCR for evaluating occurrence of parasitemia in immunocompromised patients with cerebral and extracerebral toxoplasmosis. J Clin Microbiol 32: 2813–9978; 298: 550–553.

Kirk O, Reiss P, Uberti-Foppa C, et al (2002). Safe interruption of maintenance therapy against previous infection with four common HIV-associated opportunistic pathogens during potent antiretroviral therapy. Ann Intern Med 137: 239–250.

Klinker H, Langmann P, Richter E (1996). Plasma pyrimethamine concentrations during long-term treatment for cerebral toxoplasmosis in patients with AIDS. Antimicrob Agents Chemother 40(7): 1623–1627.

Kovacs JA, Chabner BA, Lunde M, et al (1987). Potent effect of trimetraxate, a lipid-soluble antifolate, on *Toxoplasma gondii*. J Infect Dis 155: 1027–1032.

Krick JA, Remington JS (1978). Toxoplasmosis in the adult: an overview. N Engl J Med 298: 550–553.

Lainson R (1958). Observations on the development and nature of pseudocysts and cysts of *Toxoplasma gondii*. Trans R Soc Med Hyg 52: 396–407.

Leport C, Raffi F, Matheron S, et al (1988). Treatment of central nervous system toxoplasmosis with pyrimethamine/sulfadiazine combination in 35 patients with the acquired immune deficiency syndrome: efficacy of long-term continuous therapy. Am J Med 84: 94–100.

Leport C, Chene G, Morlat P, et al (1996). Pyrimethamine for primary prophylaxis of toxoplasmic encephalitis in patients with human immunodeficiency virus infection: a double-blind, randomized trial. J Infect Dis 173: 91–97.

Levy RM, Bredesen DE, Rosenblum ML (1985). Neurological manifestations of the acquired immune deficiency syndrome (AIDS): experience of UCSF and review of the literature. J Neurosurg 62: 475–495.

Levy RM, Janssen RS, Bush TJ, et al (1988). Neuroepidemiology of acquired immunodeficiency syndrome. J Acquir Immune Defic Syndr 1: 31.

Luft BJ (1986). Potent in vivo activity of aprinocid, a purine analogue, against murine toxoplasmosis. J Infect Dis 154: 692–694.

Luft BJ (1988). *Toxoplasma gondii*. In: PD Walzer (Ed.), Parasitic Infections in the Immunocompromised Host: Immunologic Mechanisms and Clinical Applications. Marcel Dekker, New York, pp. 179–279.

Luft BJ, Sivadas R (2002). Toxoplasmosis. In: WM Scheld, RJ Whitley, CM Marra (Eds.), Infections of the Central Nervous System. Lippincott, Williams and Wilkins, Philadelphia, PA, pp. 755–776.

Luft BJ, Brooks RG, Conley FK, et al (1984). Toxoplasmic encephalitis in patients with acquired immune deficiency syndrome. JAMA 252: 913–917.

Luft BJ, Hafner R, Korzun AH, et al (1993). Toxoplasmic encephalitis in patients with the acquired immune deficiency syndrome. N Engl J Med 329: 995–1000.

Matheron S, Dournon E, Garakhanian S, et al (1990). Prevalence of toxoplasmosis in 365 AIDS and ARC patients before and during zidovudine treatment. In: Programs and Abstracts of the 5th International Conference on AIDS, San Francisco, CA.

Mathew MJ, Chandy MJ (1999). Central nervous system toxoplasmosis in acquired immune deficiency syndrome: an emerging disease in India. Neurol India 47: 182–187.

Maynart M, Lievre L, Salif Sow P, et al (2001). Primary prevention with cotrimoxazole for HIV-1-infected adults: results of the pilot study in Dakar, Senegal. J AIDS 26: 130–136.

Millard PR (1984). AIDS: histopathological aspects. J Pathol 143: 223–239.

Mills J (1986). *Pneumocystic carinii* and *Toxoplasma gondii* infections in patients with AIDS. Rev Infect Dis 8: 1001–1011.

MMWR (2002). Guidelines for the prevention of opportunistic infections among HIV-infected persons. MMWR 51(RR08): 1–46.

Montoya JG (2002). Laboratory diagnosis of *Toxoplasma gondii* infection and toxoplasmosis. J Infect Dis 185(Supp 1): S73–S82.

Murray HW (1983). How protozoa evade intracellular killing. Ann Intern Med 98: 1016–1018.

Murray HW, Nathan CF, Cohn ZA (1980). Macrophage oxygen-dependent antimicrobial activity: IV. Role of endogenous scavengers of oxygen intermediates. J Exp Med 152: 1610–1624.

Navia BA, Petito CK, Gold JWM, et al (1986). Cerebral toxoplasmosis complicating the acquired immune deficiency syndrome: clinical and neuropathological findings in 27 patients. Ann Neurol 19: 224–238.

Pfefferkorn ER, Nothnagel RF, Borotz SE (1992). Parasiticidal effect of clindamycin on *Toxoplasma gondii* grown in cultured cells and selection of drug-resistant mutant. Antimicrob Agents Chemother 36: 1091–1096.

Podzamczer D, Gudiol F (1988). Clindamycin in cerebral toxoplasmosis. Am J Med 84: 800.

Podzamczer D, Satin M, Jimenez J, et al (1993). Thrice-weekly cotrimoxazole is better than weekly dapsone-pyrimethamine for primary prevention of Pneumocystis carinii pneumonia in HIV-infected patients. AIDS 7: 501–506.

Pomper M, Constantinides CD, Barker PB, et al (2002). Quantitative MR spectroscopic imaging of brain lesions in patients with AIDS: correlation with [11C-methyl]thymidine PET and thallium-201 SPECT. Acad Radiol 9: 398–409.

Post MJD, Chan JC, Hensley GT, et al (1983). Toxoplasma encephalitis in Haitian adults with acquired immune deficiency syndrome: a clinical-pathologic-CT correlation. AJR Am J Roentgenol 140: 861–868.

Potasman I, Resnick L, Luft BJ, et al (1988). Intrathecal production of antibodies against *Toxoplasma gondii* in patients with toxoplasmic encephalitis and the acquired immune deficiency syndrome. Ann Intern Med 108: 49–51.

Provenzale JM, Jinkins JR (1997). Brain and spine imaging findings in AIDS patients. Radiol Clin North Am 35: 1127–1166.

Ramsey RG, Gean AD (1997). Central nervous system toxoplasmosis. Neuroimag Clin North Am 7(2): 171–186.

Remington JS (1974). Toxoplasmosis in the adult. Bull N Y Acad Med 50: 211–227.

Rolston KV, Hoy J (1987). Role of clindamycin in the treatment of central nervous system toxoplasmosis. Am J Med 83: 551–554.

Ruiz A, Ganz W, Post MJ (1994). Use of thallium-201 brain SPECT to differentiate cerebral lymphoma from toxoplasma encephalitis in AIDS patients. Am J Neuroradiol 15: 1885–1894.

Sabin AB, Olitsky PK (1937). Toxoplasma and obligate intracellular parasitism. Science 85: 336–338.

Sacktor N (2002). The epidemiology of human immunodeficiency virus-associated neurological disease in the era of highly active antiretroviral therapy. J Neurovirol 8: 115–121.

San-Andres FJ, Rubio R, Castilla J, et al (2003). Incidence of acquired immunodeficiency syndrome-associated opportunistic diseases and the effect of treatment on a cohort of 1115 patients infected with human immundeficiency virus, 1989–1997. Clin Infect Dis 36: 1177–1185.

Sellal A, Rabaud C, Amiel C, et al (1996). Traitement d'entretien de la toxoplasmose cerebrale au cours du SIDA: place de l'association clarithromycine-minocycline. Presse Med 25(10): 509.

Sheffield HG, Melton ML (1970). *Toxoplasma gondii*: the oocyst, sporozoite, and infection of cultured cells. Science 167: 892–893.

Sibley LD, Weidner E, Krajenbuhl JL (1985). Phagosome acidification blocked by intracellular *Toxoplasma gondii*. Nature 315: 416–419.

Snider WD, Simpson DM, Nielson S, et al (1983). Neurologic complications of the acquired immune deficiency syndrome: analysis of 50 patients. Ann Neurol 14: 403–418.

Soriano V, Dona C, Rodriguez-Rosado R, et al (2000). Discontinuation of secondary prophylaxis for opportunistic infections in HIV-infected patients receiving highly active antiretroviral therapy. AIDS 14: 383–386.

Sun T, Greenspan J, Tenenbaum M, et al (1986). Diagnosis of cerebral toxoplasmosis using fluoroscein-labeled anti-toxoplasma monoclonal antibodies. Am J Surg Pathol 10: 312–316.

Torre D, Speranza F, Martegnani R, et al (1998). A retrospective study of treatment of cerebral toxoplasmosis in AIDS patients with trimethoprim-sulphamethoxazole. J Infect 37: 15–18.

Trotta M, Sterrantino G, Milo D, et al (1997). Azitomicina in associazione a pirimetamina nel trattamento della neurotoxoplasmosi, in un paziente con AIDS. Minerva Med 88: 117–119.

Villringer K, Jager H, Dichgans M, et al (1995). Differential diagnosis of CNS lesions in AIDS patients by FDG-PET. J Comput Assist Tomogr 19: 532–536.

Westblom TU, Belshe RB (1998). Clindamycin therapy of cerebral toxoplasmosis in an AIDS patient. Scand J Infect Dis 20: 561–563.

Wilson CB, Tsai V, Remington JS (1980). Failure to trigger the oxidative metabolic burst by normal macrophages: possible mechanism for survival of intracellular pathogens. J Exp Med 151: 328–346.

Wong B, Gold JWM, Brown AE, et al (1984). Central nervous system toxoplasmosis in homosexual men and parenteral drug abusers. Ann Intern Med 100: 36–42.

Further Reading

Anekthananon T, Techasathit W, Suwanagool S (2004). HIV infection/acquired immunodeficiency syndrome at Siriraj Hospital, 2002: time for secondary prevention. J Med Assoc Thai 87: 173–179.

Dannemann BR, Israelski DM, Remington JS (1988). Treatment of toxoplasmosis with intravenous clindamycin. Arch Intern Med 148: 2477–2482.

Del Rio-Chiriboga C, Orzechowski-Rallo A, Sanchez-Mejorada G (1997). Toxoplasmosis of the central nervous system in patients with AIDS in Mexico. Arch Med Res 28(4): 527–530.

Frenkel JK, Dubey JP, Miller NL (1970). *Toxoplasma gondii* in cats: fecal stages identified as coccidian oocysts. Science 167: 893–896.

Hofflin JM, Remington JS (1987). In vivo synergism of roxithromycin (RU28965) and interferon against *Toxoplasma gondii*. Antimicrob Agents Chemother 31: 346–348.

Julander I, Martin C, Lappalainen M, et al (2001). Polymerase chain reaction for diagnosis of cerebral toxoplasmosis in cerebrospinal fluid in HIV-positive patients. Scan J Infect Dis 33: 538–541.

Kirk O, Lundgren J, Pederson C, et al (1999). Can chemoprophylaxis against opportunistic infections be discontinued after an increase in CD4 cells induced by highly active antiretroviral therapy? AIDS 13: 1647–1651.

Langford TD, Letendre SL, Larrea GJ, et al (2003). Changing patterns in the neuropathogenesis of HIV during the HAART era. Brain Pathol 13: 195–210.

Naddaf SY, Akisik MF, Aziz M, et al (1998). Comparison between 201Tl-chloride and 99Tc(m)-sestamibi SPET brain imaging for differentiating intracranial lymphoma from non-malignant lesions in AIDS patients. Nucl Med Commun 19: 47–53.

Smadja D, Fournerie P, Cabre P, et al (1998). Efficacite et bonne tolerance du cotrimoxazole comme traitement de la toxoplasmose cerebrale au cours du SIDA. Presse Med 27: 1315–1320.

Vastava PB, Pradhan S, Jha S, et al (2002). MRI features of *Toxoplasma* encephalitis in the immunocompromised host: a report of 2 cases. Neuroradiology 44: 834–838.

Handbook of Clinical Neurology, Vol. 85 (3rd series)
HIV/AIDS and the Nervous System
P. Portegies, J. R. Berger, Editors

Chapter 12

Cryptococcal infection

ARTHUR JACKSON AND WILLIAM G. POWDERLY*

University College Dublin, Dublin, Ireland

12.1. Introduction

Cryptococcus neoformans (Fig. 12.1) was identified as a human pathogen in 1894. It has since been recognized as a major cause of morbidity and mortality, specifically in the host with compromised cell-mediated immunity. The advent of infection with the human immunodeficiency virus (HIV) and its associated acquired immunodeficiency syndrome (AIDS) has led to its emergence as a more important cause of disease. It is the most common life-threatening mycosis in HIV patients. In some studies of HIV-infected patients in sub-Saharan Africa it has been shown to account for 13–44% of deaths (Bicanic and Harrison, 2005). Over the past 50 years our understanding of the fungus and our knowledge of host factors have developed considerably. Paralleled with this has been a major improvement in treatment options.

12.2. Microbiology

Clinically relevant *Cryptococcus neoformans* is almost always in the form of an encapsulated yeast and reproduces by asexual budding. The sexual state (known as *Filobasidiella neoformans*) has not been described in clinical practice, and a hyphal form is extremely rarely found in tissue samples.

Cryptococcal cells are visualized on microscopy as spherical or ovoid and measure 4–6 μm in diameter without their capsule. Their main identifying feature is the presence of this polysaccharide capsule, the presence of which is implied by the characteristic halo effect of the India ink stain. There are also some non-polysaccharide, proteinaceous components to the capsule, including certain enzymes. The capsule size can vary from 1 to 50 μm, depending on environmental and genetic factors. In general the capsule size is greater during infection.

This polysaccharide capsule has four major antigenic serotypes, which is the basis of a sub-classification of the organism. These are known as serotypes A, B, C and D. Serotype A is classified as *C. neoformans grubbi*. Serotype D is classified as *C. neoformans* var. *neoformans*. Serotypes B and C are classified as *C. neoformans* var. *gattii*. A serotype AD has been described which has antigenic properties found in both A and D strains.

Inside the capsule, and external to the cell membrane, is a classic fungal cell wall, a meshwork predominantly consisting of polysaccharide. This structure gives shape and strength to the cells.

C. neoformans grows on agar plates, such as Sabouraud agar. Characteristic colonies appear in 36–72 hours and are creamy and mucoid. *C. neoformans* has the ability to replicate at 37°C, termed thermotolerance, unlike non-pathogenic cryptococci. Cryptococcal colonies grown on agar can appear brown given specific nutrient substrates, due to production of dark pigment, including melanin. This feature can be used as a diagnostic method, as it is more prominent in colonies of *C. neoformans* than other species.

12.3. Epidemiology

Serological classification has been used to study the epidemiology of cryptococcosis on a global level. *C. neoformans* var. *neoformans* and *grubii* are ubiquitous in the environment throughout the world and have been associated with pigeon and other bird droppings and soils contaminated with these droppings. *C. neoformans* var. *grubii* is more common than *C. neoformans* var. *neoformans* (Steenbergen and Casadevall, 2004).

*Correspondence to: William G. Powderly, MD, FRCPI, Professor of Medicine and Therapeutics, University College Dublin, Mater University Hospital, 44 Eccles Street, Dublin 7, Ireland. E-mail: bpowderly@mater.ie, Tel: +353-1-716-6366, Fax: +314-1-716-6335.

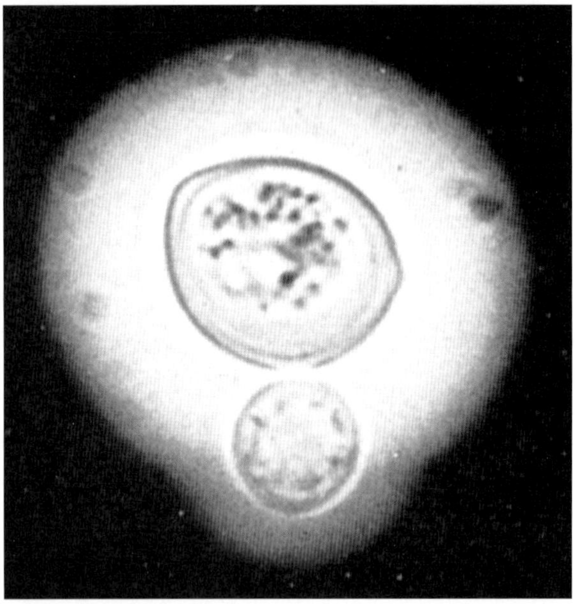

Fig. 12.1. *Cryptococcus neoformans.*

High seroprevalence has been encountered in many populations. A study in a large US urban area showed a majority of children above the age of 2 years have been infected. For children above the age of 5 years, the incidence was 70%. This reinforces the understanding that primary infection can be asymptomatic, or mistaken for a nonspecific viral illness, and that cryptococcus is a ubiquitous organism. There is a specific association between cryptococcal infection and pigeons, and therefore in urban areas, where pigeon populations are greater, these exposures are likely to be higher than average (Goldman et al., 2001).

C. neoformans var. *gattii* is usually found in tropical and subtropical areas. It is considered to be genetically and biochemically distinct to *C. neoformans* var. *neoformans*. It is most commonly found in southern Europe, southern California, Australia, areas of Africa, Southeast Asia and South America. There is an association in these areas with the river red and the forest red gum trees (*Eucalyptus camaldulensis* and *Eucalyptus tereticornis*), with a peak incidence of disease corresponding to the flowering period of these trees: November to February. The association with decaying wood suggests a potential endophytic existence. It affects mainly healthy individuals and is rare in HIV-infected patients (Sorrel, 2001).

Human-to-human transmission of cryptococcosis is extremely rare. It has been described following corneal transplant (Beyt and Waltman, 1978), and has been seen to be transmitted between neighboring patients in an intensive care unit setting on at least one occasion (Wang et al., 2005). Cases of mother-to-child transmission have been described rarely (Sirinavin et al., 2004).

Disseminated cryptococcosis has been the most common life-threatening fungal infection in patients with AIDS, affecting up to 8% of patients with advanced HIV infection (Dismukes, 1988; Powderly, 1993). With the use of highly active antiretroviral therapy (HAART) for the treatment of HIV infection and the widespread use of the azole antifungals, the incidence of invasive cryptococcosis in HIV-infected population has declined in the Western world (Hajjeh et al., 1999). However, cryptococcosis remains extremely important in other parts of the world. Cryptococcal meningitis is the second most common opportunistic infection (after tuberculosis) associated with AIDS in sub-Saharan Africa and southern Asia.

12.4. Pathogenesis

It is interesting and enigmatic that virulence factors enabling animal infection have developed despite the capacity to complete a saprophytic lifecycle in the environment. This may be a response to environmental selective pressures from predators on cryptococcus such as nematodes and amoebae (Casadevall et al., 2003). Cryptococcal species have the potential for a sexual cycle, allowing exchange of virulence factors, and development of more pathogenic strains (Nielsen et al., 2003).

Cryptococcal infection is usually acquired via the respiratory route, when either the yeast form or the basidospores are inhaled. Initial colonization and infection of the respiratory tract occurs, and there may or may not be subsequent dissemination. Skin involvement in cryptococcosis is usually a feature of dissemination. However, direct inoculum penetrating the skin may rarely lead to primary cutaneous cryptococcosis. *C. neoformans* is one of the few yeasts which is capable of reproducing at 37°C and above, which is considered as a virulence factor (Steen et al., 2002). At higher temperatures the reproductive capacity is reduced, and at 40°C the yeast is unable to multiply.

Cell-mediated immunity is critical in protection against cryptococcal infection. The absence of an intact cell-mediated immune response results in ineffective ingestion and killing of the organism leading to dissemination and increased cryptococcal burden. Macrophages have a critical initial role host defense against cryptococcal infection. In animal models macrophage dysfunction is associated with a decreased resistance to infection (Perfect et al., 1987). Granuloma are formed, localizing the infection, and these contain giant cells with ingested cryptococci. The role of giant cells is to engulf the cryptococcal cells which have increased in size beyond the capacity for phagocytosis by a single macrophage. In HIV-infected hosts, granulomatous formation is impaired with a consequent greater yeast burden (Neuville et al., 2002).

Complement-mediated phagocytosis appears to be the primary initial defense against cryptococcal invasion (Kwon-Chung et al., 1992). C5 deficiency in animals has been associated with increased susceptibility to infection and complement deficiency leads to increased mortality in animal models. Cytokines, including interleukin (IL)-1 and tumor necrosis factor (TNF)-alpha, are important in the enlisting of the inflammatory response. In both animal models and human models higher levels of interferon-gamma have been associated with better clearance of cryptococcal infection. It has been postulated that this may lead to a potential therapeutic use of interferon-gamma in cases of cryptococcal meningitis (Siddiqui et al., 2005).

Factors associated with virulence in *C. neoformans* include its polysaccharide capsule, melanin production, the mating type and growth at 37°C (thermotolerance). The cryptococcal capsule is of central importance to its pathogenic effects (Kwon-Chung and Rhodes, 1986). The potential mechanisms by which this acts include antiphagocytosis, complement depletion, antibody unresponsiveness, inhibition of leucocyte migration and dysregulation of cytokine secretion. Acapsular strains are shown to have markedly reduced virulence.

The release of cryptococcal capsular polysaccharide glucuronoxylomannan has been shown to be significantly decreased in the presence of capsule-binding monoclonal antibody. This supports a humoral immunity role for protection against cryptococcal disease. It also suggests future treatment options for cryptococcal infection (Martinez et al., 2004).

Cryptococcus has been noted to produce melanin from cathecolamine sources, e.g. dopa, dopamine, norepinephrine and epinephrine. The high concentration of these cathecolamines in the central nervous system (CNS) may explain the predilection of cryptococcus for the CNS. It is considered that melanin offers protection from oxidative stresses and reduces the killing effect of macrophages (Casadevall et al., 2000). Melanin also reduces the T-cell response and cytokine production, reduces phagocytosis and makes the organism less sensitive to amphotericin B and caspofungin (van Duin et al., 2002).

12.5. Clinical presentation

Involvement of the brain causing meningitis or meningoencephalitis is the most common presentation of invasive cryptococcal infection. Cryptococcal meningitis typically presents as an indolent syndrome of lethargy, personality change, memory difficulties and cognitive decline. Fever is a common feature. Specific meningeal symptoms of headache, neck stiffness and photophobia occur in 25–33% of patients (Haddad and Powderly, 2001). However, these symptoms are very nonspecific, and not reliably present especially in advanced HIV, where the inflammatory response is limited. There may be papilloedema or peripapillary retinal hemorrhages due to raised intracranial pressure. Visual disturbance or blindness can occur due to raised intracranial pressure. Rarely cases have presented with complications of deafness (Low, 2002) or dementia (Aharon-Peretz et al., 2004; Ala et al., 2004). Although in patients with AIDS cryptococcomas (which represent large granulomas) are uncommon, they can present with features of intracranial space occupying lesions. Central diabetes insipidus has been described as a complication of cryptococcal meningitis (Juffermans et al., 2002).

In patients with AIDS, disseminated disease is common. About half of HIV-positive patients with cryptococcal meningitis have evidence of pulmonary involvement at presentation, with clinical symptoms such as cough or dyspnea and abnormal chest radiographs (Cameron et al., 1991). The chest radiographic finding is usually diffuse interstitial infiltrates in immunocompromised patients or focal lesions in immunocompetent patients. Cutaneous involvement is common and suggests disseminated disease (Durden and Elewski, 1994). The most common skin involvement resembles that of molluscum contagiosum. As many as three-quarters of patients with cryptococcal meningitis have positive blood cultures. Infection of bone, eye, heart, adrenal glands, prostate and urinary tract has also been described. The prostate gland can represent as a reservoir of infection and potential source of reinfection after completion of therapy (Larsen et al., 1989).

12.6. Diagnosis

HIV-associated cryptococcal infection occurs when patients have advanced immunodeficiency — the median CD^4 lymphocyte count in most series is <50 cells/mm^3. A subacute history of meningeal signs in any HIV-positive patient with moderately advanced HIV immunodeficiency should warrant investigation for potential cryptococcal meningitis. Lumbar puncture should be performed and opening pressure measured unless there are specific contraindications. The cerebrospinal fluid (CSF) should be analyzed for protein, glucose and cell count. Serum and CSF cryptococcal antigen titers and routine bloods should be done.

The typical pattern in the CSF is a chronic meningitis with a lymphocytic pleocytosis. However, the CSF may appear normal in HIV-positive patients with cryptococcal meningitis, since the usual response to infection is usually markedly blunted. In fact, fewer than half of HIV-positive patients with cryptococcal meningitis have an elevated protein level, only about one-third have hypoglycorrhea, and only about 20% have more than 20 white blood cells per cubic millimeter of CSF (Darras-Jolly et al., 1996). The opening pressure is usually elevated in patients with cryptococcal meningitis (up to 70% of patients present with pressures greater than 20 cmH_2O) and is an important issue associated with therapy (Graybill et al., 2000). India ink stain of the CSF is positive, showing encapsulated yeast, in about 75% of cases. The capsule repels the ink, leaving a visible "halo" around the individual yeast cells.

Cryptococcal antigen (CrAg) can be measured in the serum and CSF with a latex agglutination test. When detected in the CSF it is 93–100% sensitive, with a specificity of 93–98%. Serum cryptococcal antigen (sCRAG) is usually elevated in 95% of patients with meningitis. A serum CrAg with a titer of >1:8 suggests disseminated disease (Feldmesser et al., 1996). Such patients should be evaluated for possible meningeal involvement. A large Ugandan study showed that antigenemia was detectable a median of 22 days before symptoms, and in 11% it was detectable over 100 days before symptoms (French et al., 2002). Measurement of serum cryptococcal antigen is probably the most sensitive and robust test used in practice for diagnosis of cryptococcosis. There is little value in the serial measurement of CSF and sCRAG in the routine management of cryptococcal meningitis. CSF cryptococcal antigen titers should decrease after effective therapy. There is no evidence to support following serial serum antigen, which shows no correlation with outcome of antifungal therapy (Powderly et al., 1994). Management should rely on clinical assessment not on the cryptococcal antigen alone (Aberg et al., 2000).

12.7. Prognostic indicators

In many clinical studies factors associated with outcome have included variables associated with the clinical illness and with the host response. In several studies, the presence of an increased opening pressure, presumably reflecting raised intracranial pressure, has been an indicator for the increased likelihood of headache, meningismus, papilledema, hearing loss and pathological reflexes. It is also associated with increased mortality (Graybill et al., 2000).

A more severe clinical presentation has also been associated with a worse outcome. A Thai study showed that early death with cryptococcal meningitis in patients treated with high-dose amphotericin B was associated with a previous history of weight loss, a Glasgow Coma Scale <13 and hypoalbuminaemia. Late mortality was associated with delayed CSF yeast clearance, and relapse of infection (Pitisuttithum et al., 2001). A CSF leucocyte count <20 cells/mm^3, high CSF cryptococcal antigen at baseline and fungemia are also poor prognostic indicators (Powderly, 2000).

12.8. Treatment

Cryptococcal meningitis is fatal if not treated. Standard therapy includes a combination of antimicrobial therapy and measures aimed towards the complications of raised intracranial pressure. Multiple studies have been conducted in HIV-infected patients with cryptococcal meningitis leading to published guidelines that recommend that antifungal therapy be divided into an induction phase of at least 2 weeks, a consolidation phase for at least 10 weeks, and a maintenance phase, which is often protracted and may be lifelong if the immunodeficiency is not corrected using antifungal therapy (Saag et al., 2000; Benson et al., 2004).

The standard induction phase for cryptococcal meningitis in HIV disease consists of co-administration of high-dose amphotericin B (0.7 mg/kg/day) and 5-flucytosine (100 mg/kg/day) for 2 weeks. In a large randomized clinical trial this was compared to amphotericin alone (Van der Horst et al., 1997). The acute mortality with this regimen was 6%. The addition of 5-FC to amphotericin B did not significantly improve the mortality and clinical course. However, 5-FC was well tolerated and there was a trend to a better CSF sterilization rate with its use (60% compared with 51% for negative culture in 2 weeks). Furthermore, the use of 5-FC as initial therapy has been associated with a decreased risk of later relapse of cryptococcal meningitis. More recent studies, in which CSF was sampled frequently in the initial phase of treatment, showed that this combination regimen was associated with a faster clearance of cryptococcus from CSF samples, than when compared to amphotericin B alone, or when fluconazole was added to these two agents (Brouwer et al., 2004).

Most studies have been associated with standard amphotericin B deoxycholate. This has the disadvantage of considerable infusion-related toxicity and a risk of nephrotoxicity. Small, randomized studies of liposomal amphotericin B (4 mg/kg) compared with conventional amphotericin B (0.7 mg/kg) noted an earlier CSF sterilization rate in patients in the liposomal preparation arm with also less nephrotoxicity (Leenders et al., 1997). Uncontrolled studies of amphoterin B lipid complex also suggest effectiveness for this compound (Baddour et al., 2005). There is less toxicity with the lipid-based therapy, but whether this justifies its considerable extra expense is unclear. Thus, the role of lipid preparations of amphotericin B remains uncertain, although they may be useful in patients with impaired renal function.

Once the induction phase is completed, it is followed by a consolidation phase of the triazole antifungal fluconazole, at a dosage of 400 mg/day for 8–10 weeks. After this the fluconazole can be reduced to 200 mg/day as maintenance therapy (Powderly, 1996). Lifelong maintenance therapy is required in AIDS patients with cryptococcal infection to prevent relapse of infection, if antiretroviral therapy is not provided or is ineffective. Relapse rates of 50–60% and a shorter life expectancy have been reported in patients who did not receive chronic suppressive therapy. Fluconazole 200 mg daily is the drug of choice. In a study comparing fluconazole weekly and amphotericin B infusions, the relapse rate in the fluconazole arm was 2% compared to 18% in the amphotericin B arm (Powderly et al., 1992). Fluconazole was also superior to itraconazole in a trial of maintenance therapy (Saag et al., 1999). Relapse is usually a clinically evident disease (e.g. fever and recurrence of headache). Routine monitoring by measurement of sCRAG has not been shown to predict relapse, although elevations of antigen in the CSF may predict recurrence (Powderly et al., 1994). Itraconazole is an alternative to fluconazole where the latter is not tolerated (Saag et al., 2000).

If antiretroviral therapy is administered with consequent reconstitution of the immune system, it is possible to discontinue maintenance therapy at some point after acute therapy is administered. A retrospective international study has suggested that once the CD^4 count has risen above 100 cells/mm^3 it is safe to discontinue maintenance therapy (Mussini et al., 2004). This assumes that adequate antifungal therapy (minimum of 6 months and probably one year) has also been given with successful sterilization of the CSF.

Although current guidelines for treatment of cryptococcal meningitis with AIDS recommend initial amphotericin B (with 5-FC if possible) followed by consolidation azole therapy, outcomes in clinical practice suggest that improvements in treatment are still needed. A recent study from Thailand using such an approach reported a mortality rate of 16 and 24% at 2 and 4 weeks (Pitisuttithum et al., 2001). Furthermore, in many parts of the world, neither amphotericin B nor flucytosine are readily available. In these settings, the only option is oral fluconazole therapy. The effectiveness of fluconazole is clearly more limited and only about 50% of patients respond (Powderly, 1996). Whether better response rates would be seen with higher doses of fluconazole is worthy of further study. Unfortunately the standard of care is rarely achievable in resource-poor countries. In a Zambian study there was a case fatality of 100% described for patients with cryptococcal meningitis treated with only azole therapy (Mwaba et al., 2001), presumably reflecting the progressive immunodeficiency in the absence of effective antiviral therapy.

A very important aspect of management of acute cryptococcal meningitis in AIDS is the recognition that clinical deterioration may be due to increased intracranial pressure and that this aspect may not respond rapidly to antifungal therapy. In a large treatment trial of cryptococcal meningitis, the median opening pressure at baseline for patients who cleared their CSF at 2 weeks was 22 cmH_2O and for those whose cultures remained positive the median was 28 cmH_2O (Van der Horst et al., 1997). Additionally, 13 of 14 patients who died had opening pressure more than 25 cmH_2O at their most recent prior lumbar puncture. Further analyses also showed a relationship between baseline opening pressure and long-term outcome, with the median survival in patients with the highest pressures being significantly less than that in patients whose pressure were normal (Graybill et al., 2000). These data do not prove a causal relationship between the pressure and outcome, nor do they prove that lowering pressure will improve the prognosis. Nevertheless, all patients with cryptococcal meningitis should have opening pressure measured when a lumbar puncture is performed and strong consideration should be given to reducing such pressure if the opening pressure is high (more than 20 cmH_2O). A CSF opening pressure of over 25 cmH_2O should be treated aggressively with repeated high volume CSF drainage. Uncontrollable elevation of intracranial pressure can be relieved with repeat lumbar punctures, or a ventriculoperitoneal shunt (Liliang et al., 2002). Therefore it is recommended to perform regular lumbar punctures initially to relieve this elevated pressure and assess the need for a definitive procedure. Lumbar puncture with removal of spinal fluid daily to reduce opening pressure to less than 20 cmH_2O or 50% of the initial opening pressure is recommended (Saag et al., 2000). Medical therapy using acetazolamide to reduce intracranial pressure is not recommended, following suggestions that metabolic acidosis occurs and is associated with an increased frequency of serious adverse events (Newton et al., 2002).

Cutaneous cryptococcosis, cryptococcal urinary tract disease and non-CNS-associated cryptococcaemia in HIV-infected patients should be treated as if they have disseminated infection. In all cases careful exclusion of CNS infection is required. HIV-positive patients who are asymptomatic with cryptococcal infection require treatment in all cases. It is not appropriate to monitor their course without antifungal therapy. Urinary or pulmonary disease should be treated with fluconazole 200–400 mg/day.

12.9. Prevention

Fluconazole 100–200 mg daily also prevents cryptococcal meningitis in patients with AIDS (Powderly et al., 1995). Weekly fluconazole (400 mg) is also effective (Havlir et al., 1998) as is itraconazole 200 mg daily (McKinsey et al., 1999). However, given alone fluconazole was not associated with increased survival in studies conducted in the developed world. Furthermore, there has been concern that prolonged usage of fluconazole may result in acquired resistance to fluconazole, especially in *Candida* species (Fichtenbaum and Powderly, 1998). However, the situation could be quite different in the developing world, especially where cryptococcal infection is highly prevalent. Preliminary data from a small, randomized trial of primary prophylaxis with fluconazole (400 mg once weekly) in Thailand suggested that there may be some survival benefit with daily fluconazole when CD^4 counts fall below 100, but a larger trial will be needed to assess the relevance of this (Chetchotisakd et al., 2004). A previous study of itraconazole use in advanced HIV as prophylactic measure showed no survival benefit (Chariyalertsak et al., 2002). These observations support large, randomized trials of primary prophylaxis in areas such as South and Southeast Asia and sub-Saharan Africa, where the incidence of cryptococcal infection is very high. It is important to consider that mass chemoprophylaxis could lead to increased prevalence of resistant fungi in the community and indeed preliminary reports of primary fluconazole-resistant cryptococcosis have emerged from these areas. A report from Cambodia indicated that the MIC of fluconazole changed significantly between 2000 and 2002. Some 2.5% of strains were resistant in the first year and this progressed to 14% of strains resistant to fluconazole in the second year (Sar et al., 2004).

In reality, although fluconazole has some effectiveness as a preventive agent, the best form of prevention is appropriate treatment of HIV, with commencement of HAART before severe immunosuppression occurs.

12.9.1. Use of antiretroviral therapy in patients with cryptococcal meningitis

Potent antiretroviral therapy reduces the incidence of opportunistic infections and is really the best prophylxis. All HIV-positive patients with cryptococcal meningitis should receive as potent antiretroviral therapy as possible. Efavirenz-based therapy has been shown to be well tolerated in patients receiving fluconazole for cryptococcal meningitis (Sungkanuparph et al., 2003).

A potential complication of the restoration of the immune system is the recognition of infection that may have been latent up to that point, or a more vigorous inflammatory response to an infection which was already evident. The recognition of antigen and the corresponding inflammatory response can lead to distinct clinical syndromes, the presentation of which depend on the site of the newly recognized infection. This effect is referred to as immune reconstitution inflammatory syndrome (IRIS) (DeSimone et al., 2000).

In the case of cryptococcal meningitis, IRIS may manifest as an apparent recurrence of meningitis, with all the features of the initial meningitis presentation. Lumbar puncture will typically show inflammation, but, by definition, will remain culture negative. Rarely, the syndrome may present outside the CNS as pulmonary infiltrates or hilar/mediastinal lymphadenitis due to extra-CNS cryptococcus. The definition of immune reconstitution disease relies on the culture of cryptococcus remaining negative, and the CSF antigen titer remaining at least fourfold less than the admission value. Typically IRIS presents following an initial clinical improvement. The majority of IRIS occurs within 30 days of starting antiretroviral therapy (Shelbourne et al., 2005). The frequency of IRIS following initiation of antiretroviral therapy in cryptococcal meningitis has varied from 10 to 50% in different series (Jenny-Avital and Abadi, 2002; Shelbourne et al., 2005; Singh et al., 2005).

IRIS may also lead to the initial diagnosis of previously latent cryptococcosis. This should be considered when meningeal symptoms occur within 180 days of commencing HAART, CSF cultures are negative, blood and CSF cryptococcal antigen titers are positive (CSF titers $<1{:}8$) and CSF white blood cell count is >50 cells/μl.

IRIS has been found more frequently in patients who are antiretroviral naive before HAART therapy. It is also more commonly seen in those with a higher CSF cryptococcal antigen, probably due to increased antigen producing a greater inflammatory response. Starting HAART within 30 days of diagnosis of cryptococcal meningitis was also associated with a higher likelihood of IRIS, presumably due to a similar antigenic burden (Shelbourne et al., 2005). The pathogenesis of IRS is unclear but thought to represent an over-exuberant immunological response to a high antigen load in the context of an immune response that is recovering but not yet fully normal. Clinically there can be considerable morbidity and even mortality. In a recent study of cryptococcosis from France, of 10 cases of IRIS, three were ultimately fatal (Singh et al., 2005).

If an immune reconstitution syndrome develops due to *C. neoformans*, patients should remain on antiviral therapy, and continue antifungal treatment. Anti-inflammatory treatment may be needed for symptom management, and in some cases, immunosuppressive therapy such as corticosteroids has been used. However, the indications for treatment of IRIS (and the duration) of such treatment have not been well delineated, and clearly further research is needed.

12.10. Future directions

Although there are clearly effective treatments available for infection with *C. neoformans*, mortality and morbidity of established infection in patients with AIDS remains high. Of the newer antifungals that have been or are being developed, both voriconazole (van Duin et al., 2004) and posaconazole (Pfaller et al., 2001) have activity against *C. neoformans* but are unlikely to have substantial advantage over fluconazole, except in the rare cases of resistant infection.

Immunity to cryptococcal infection is associated with a granulomatous inflammatory response, intact cell-mediated immunity and a T-helper 1 pattern of cytokine release. Interferon-gamma has been shown to be a prognostic indicator, with higher levels in the CSF associated with faster rate of cryptococcal clearance in both animal models and human models. It has been postulated that this may lead to a potential therapeutic use of interferon-gamma in future cases of cryptococcal meningitis (Siddiqui et al., 2005). Phase I studies have investigated the use of recombinant interferon-gamma 1b as adjunctive therapy for cryptococcal meningitis in patients with AIDS. Co-administration of interferon-gamma three times per week at doses of 100 and 200 mg was associated with faster conversion of CSF from culture positive to culture negative (Pappas et al., 2004). There was no appreciable deleterious effect on CD^4 counts and HIV virus load levels. These promising data suggest further studies are warranted.

There has also been ongoing interest in the adjunctive use of murine antibodies for passive immunotherapy in the treatment of cryptococcal meningitis. Monoclonal murine antibodies directed against the cryptococcal capsular polysaccharide coat have been tested in a phase I trial (Larsen et al., 2005) and high doses were associated with a decline in cryptococcal antigen titer. Again further studies are needed before the usefulness of this approach can be assessed.

Finally it may be possible to exploit antiretroviral therapy. It has been shown that certain antiviral agents, including indinavir, have direct antifungal effects. Indinavir can inhibit production of urease and protease, and capsule formation. Following indinavir exposure cryptococcal cells are more susceptible to the activity of natural effector cells. It is unclear as to the degree of clinical significance of this finding (Monari et al., 2005) but it would suggest that investigation of other more widely used protease inhibitors is appropriate.

References

Aberg JA, Watson J, Segal M, et al (2000). Clinical utility of monitoring serum cryptococcal antigen (sCRAG) titers in patients with AIDS-related cryptococcal disease. HIV Clin Trials 1(1): 1–6.

Aharon-Peretz J, Kliot D, Finkelstein R, et al (2004). Cryptococcal meningitis mimicking vascular dementia. Neurology 62(11): 2135.

Ala TA, Doss RC, Sullivan CJ (2004). Reversible dementia: a case of cryptococcal meningitis masquerading as Alzheimer's disease. J Alzheimer's Dis 6(5): 503–508.

Baddour L, Perfect J, Ostrosky-Zeichner L (2005). Successful use of amphotericin B lipid complex in the treatment of cryptococcosis. Clin Infect Dis 40(Suppl 6): S409–S413.

Benson CA, Kaplan JE, Masur H, et al (2004). Treating opportunistic infections among HIV-exposed and infected children: recommendations from CDC, the National Institutes of Health, and the Infectious Diseases Society of America. MMWR Recomm Rep 53(RR-15): 1–112.

Beyt BE Jr, Waltman WR (1978). Cryptococcal endophthalmitis after corneal transplantation. N Engl J Med 298: 825–826.

Bicanic T, Harrison TS (2005). Cryptococcal meningitis. Br Med Bull 72: 99–118.

Brouwer AE, Rajanuwong A, Chierakul W, et al (2004). Combination antifungal therapies for HIV-associated cryptococcal meningitis: a randomised trial. Lancet 363(9423): 1764–1767.

Cameron ML, Bartlett JA, Gallis HA, et al (1991). Manifestations of pulmonary cryptococcosis in patients with acquired immunodeficiency syndrome. Rev Infect Dis 13: 64–69.

Casadevall A, Rosas AL, Nosanchuk JD (2000). Melanin and virulence in *Cryptococcus neoformans*. Curr Opin Microbiol 3(4): 354–358.

Casadevall A, Steenbergen JN, Nosanchuk JD (2003). "Ready made" virulence and "dual use" virulence factors in pathogenic environmental fungi: the *Cryptococcus neoformans* paradigm. Curr Opin Microbiol 6(4): 332–337.

Chariyalertsak S, Supparatpinyo K, Sirisanthana T, et al (2002). A controlled trial of itraconazole as primary prophylaxis for systemic fungal infections in patients with advanced human immunodeficiency virus infection in Thailand. Clin Infect Dis 34(2): 277–284.

Chetchotisakd P, Sungkanuparph S, Thinkhamrop B, et al (2004). A multicentred, randomised, double-blind, placebo-controlled trial of primary cryptococcal meningitis prophylaxis in HIV-infected patients with severe immune deficiency. HIV Med 5(3): 140–143.

Darras-Jolly C, Chevret S, Wolff M, et al (1996). *Cryptococcus neoformans* infection in France: epidemiologic features of and early prognostic parameters for 76 patients who were infected with human immunodeficiency virus. Clin Infect Dis 15: 369–376.

DeSimone JA, Pomerantz RJ, Babinchak TJ (2000). Inflammatory reactions in HIV-infected persons after initiation of highly active antiretroviral therapy. Ann Intern Med 133: 447–454.

Dismukes WE (1988). Cryptococcal meningitis in patients with AIDS. J Infect Dis 157: 624.

Durden FM, Elewski B (1994). Cutaneous involvement with *Cryptococcus neoformans* in AIDS. J Am Acad Dermatol 30: 844–848.

Feldmeser M, Harris C, Reichberg S, et al (1996). Serum cryptococcal antigen in patients with AIDS. Clin Infect Dis 23: 827.

Fichtenbaum CJ, Powderly WG (1998). Refractory mucosal candidiasis in patients with human immunodeficiency virus infection. Clin Infect Dis 26(3): 556–565.

French N, Gray K, Watera C, et al (2002). Cryptococcal infection in a cohort of HIV-1 infected Ugandan adults. AIDS 16 (7): 1031–1038.

Goldman DL, Khine H, Abadi J, et al (2001). Serological evidence for *Cryptococcus neoformans* infection in early childhood. Pediatrics 107(5): E66.

Graybill JR, Sobel J, Saag M, et al (2000). Diagnosis and management of increased intracranial pressure in patients with AIDS and cryptococcal meningitis. The NIAID Mycoses Study Group and AIDS Cooperative Treatment. Clin Infect Dis 30(1): 47–54.

Haddad NE, Powderly WG (2001). The changing face of mycoses in patients with HIV/AIDS. AIDS Read 11(7): 365–378.

Hajjeh RA, Conn LA, Stephen DS, et al (1999). Cryptococcosis: population-based multistate active surveillance and risk factors in human immunodeficiency virus-infected persons. Cryptococcal Active Surveillance Group. J Infect Dis 179: 449–454.

Havlir DV, Dube MP, McCutchan JA, et al (1998). Prophylaxis with weekly versus daily fluconazole for fungal infections in patients with AIDS. Clin Infect Dis 27: 1369–1375.

Jenny-Avital E, Abadi M (2002). Immune reconstitution cryptococcosis after initiation of successful highly active antiretroviral therapy. Clin Infect Dis 35: 128–133.

Juffermans NP, Verbon A, Van der Poll T (2002). Diabetes insipidus as a complication of cryptococcal meningitis in an HIV-infected patient. Scand J Infect Dis 34(5): 397–398.

Kwon-Chung KJ, Rhodes JC (1986). Encapsulation and melanin formation as indicators of virulence in *Crytococcus neoformans*. Infect Immun 51: 218–224.

Kwon-Chung KJ, Kozel TR, Edman JC, et al (1992). Recent advances in biology and immunology of *Crytococcus neoformans*. J Med Vet Mycol 30: 133–138.

Larsen RA, Bozzette S, McCutchan JA, et al (1989). Persistent *Cryptococcus neoformans* infection of the prostate after successful treatment of meningitis. California Collaborative Treatment Group. Ann Intern Med 111: 125–128.

Larsen RA, Pappas PG, Perfect J, et al (2005). Phase I evaluation of the safety and pharmacokinetics of murine-derived anticryptococcal antibody 18B7 in subjects with treated cryptococcal meningitis. Antimicrob Agents Chemother 49(3): 952–958.

Leenders AC, Reiss P, Portegies P, et al (1997). Liposomal amphoterecin B (AmBisome) compared with amphoterecin B both followed by oral fluconazole in the treatment of AIDS-associated cryptococcal meningitis. AIDS 11: 1463–1471.

Liliang PC, Liang CL, Chang WN, et al (2002). Use of ventriculoperitoneal shunts to treat uncontrollable intracranial hypertension in patients who have cryptococcal meningitis without hydrocephalus. Clin Infect Dis 12: 64–68.

Low WK (2002). Cryptococcal meningitis: implications for the otologist. ORL J Otorhinolaryngol Relat Spec 64(1): 35–37.

Martinez LR, Moussai D, Casadevall A (2004). Antibody to *Cryptococcus neoformans* glucuronoxylomannan inhibits the release of capsular antigen. Infect Immun 72(6): 3674–3679.

McKinsey D, Wheat J, Cloud G, et al (1999). Itraconazole is effective primary prophylaxis against systemic fungal infections in patients with advanced HIV infection: randomized, placebo-controlled, double-blind study. National Institute of Allergy and Infectious Diseases Mycoses Study Group. Clin Infect Dis 28: 1049–1056.

Monari C, Pericolini E, Bistoni G, et al (2005). Influence of indinavir on virulence and growth of *Cryptococcus neoformans*. J Infect Dis 191(2): 307–311.

Mussini C, Pezzotti P, Miro JM, et al (2004). Discontinuation of maintenance therapy for cryptococcal meningitis in patients with AIDS treated with highly active antiretroviral therapy: an international observation study. Clin Infect Dis 38(4): 565–571.

Mwaba P, Mwansa J, Chintu C, et al (2001). Clinical presentation, natural history and cumulative death rates of 230 adults with cryptococcal meningitis in Zambian AIDS patients treated under local conditions. Postgrad Med J 77 (914): 769–773.

Neuville S, Dromer F, Chretien F, et al (2002). Physiopathology of meningoencephalitis caused by *Cryptococcus neoformans*. Ann Med Interne (Paris) 153(5): 323–328.

Newton PH, Thai le H, Tip NQ, et al (2002). A randomised double-blind, placebo-controlled trial of acetazolamide for the treatment of elevated intracranial pressure in cryptococcal meningitis. Clin Infect Dis 35(6): 769–772.

Nielsen K, Cox GM, Wang P, et al (2003). Sexual cycles of *Cryptococcus neoformans* var. *grubbi* and virulence of congenic a and alpha isolates. Infect Immun 71(9): 4831–4841.

Pappas PG, Bustamante B, Ticona E, et al (2004). Recombinant interferon gamma 1b as adjunctive therapy for AIDS-related acute cryptococcal meningitis. J Infect Dis 189(12): 2185–2191.

Perfect JR, Granger DL, Durack DT (1987). Effects of antifungal agents and gamma interferon on macrophage cytotoxicity for fungi and tumor cells. J Infect Dis 156: 316–320.

Pfaller MA, Messer SA, Hollis RJ, et al (2001). In vitro activities of posaconazole (Sch 56592) compared with those of itraconazole and fluconazole against 3,685 clinical isolates of *Candida* spp. and *Cryptococcus neoformans*. Antimicrob Agents Chemother 45(10): 2862–2864.

Pitisuttithum P, Tansuphasawadikul S, Simpson AJ, et al (2001). A prospective study of AIDS-associated cryptococcal meningitis in Thailand treated with high-dose amphoterecin B. J Infect 43(4): 226–233.

Powderly WG (1993). Cryptococcal meningitis and AIDS. Clin Infect Dis 17: 837–841.

Powderly WG (1996). Recent advances in the management of cryptococcal meningitis in patients with AIDS. Clin Infect Dis 22(Suppl 2): S119–S123.

Powderly WG (2000). Current approach to the acute management of cryptococcal infections. J Infect 41: 18–22.

Powderly WG, Saag MS, Cloud GA, et al (1992). A controlled trial of fluconazole or amphotericin B to prevent relapse of cryptococcal meningitis in patients with the acquired immune deficiency syndrome. N Eng J Med 326: 793–798.

Powderly WG, Cloud GA, Dismukes WE, et al (1994). Measurement of cryptococcal antigen in serum and cerebrospinal fluid: value in the management of AIDS associated cryptococcal meningitis. Clin Infect Dis 18: 789.

Powderly WG, Finkelstein D, Feinberg J, et al (1995). A randomized trial comparing fluconazole with clotrimazole troches for the prevention of fungal infections in patients with advanced human immunodeficiency virus infection. NIAID AIDS Clinical Trials Group. N Engl J Med 332: 700–705.

Saag MS, Cloud GC, Graybill JR, et al (1999). A comparison of itraconazole versus fluconazole as maintenance therapy for AIDS-associated cryptococcal meningitis. Clin Infect Dis 28: 291–296.

Saag MS, Graybill RJ, Larsen RA, et al (2000). Practice guidelines for the management of cryptococcal disease. Infectious Diseases Society of America. Clin Infect Dis 30(4): 710–718.

Sar B, Monchy D, Vann M, et al (2004). Increasing in vitro resistance to fluconazole in *Cryptococcus neoformans* isolates: April 2000 to March 2002. J Antimicrob Chemother 54(2): 563–565.

Shelbourne S, Darcourt J, White C, et al (2005). The role of immune reconstitution inflammatory syndrome in AIDS-related Cryptococcus neoformans disease in the era of highly active antiretroviral therapy. Clin Infect Dis 40: 1049–1052.

Siddiqui AA, Brouwer AE, Wuthiekanun V, et al (2005). IFN-gamma at the site of infection determines rate of clearance of infection in cryptococcal meningitis. J Immunol 174(3): 1746–1750.

Singh N, Lortholary O, Alexander BD, et al (2005). An immune reconstitution syndrome-like illness associated with *Cryptococcus neoformans* infection in organ transplant recipients. Clin Infect Dis 40: 1756–1761.

Sirinavin S, Intusoma U, Tuntirungsee S (2004). Mother-to-time transmission of *Cryptococcus neoformans*. Pediatr Infect Dis J 23(3): 278–279.

Sorrel TC (2001). *Cryptococcus neoformans* variety *gattii*. Med Mycol 39(2): 155–168.

Steen BR, Lian T, Zuyderduyn S, et al (2002). Temperature-regulated transcription in the pathogenic fungus *Cryptococcus neoformans*. Genome Res 12(9): 1386–1400.

Steenbergen JN, Casadevall A (2000). Prevalence of *Cryptococcus neoformans* var. *neoformans* (serotype D) and *Cryptococcus neoformans* var. *grubii* (serotype A) isolates in New York City. J Clin Microbiol 38(5): 1974–1976.

Sungkanuparph S, Vibhagool A, Mootsikapun P, et al (2003). Efavirenz-based regimen as treatment of advanced AIDS with cryptococcal meningitis. J Acquir Immune Defic Syndr 33: 118–119.

Van der Horst CM, Saag MS, Cloud GA, et al (1997). Treatment of cryptococcal meningitis associated with the acquired immunodeficiency syndrome. N Engl J Med 337: 15–21.

van Duin D, Casadevall A, Nosanchuk JD (2002). Melanization of *Cryptococcus neoformans* and *Histoplasma capsulatum* reduces their susceptibilities to amphotericin B and caspofungin. Antimicrob Agents Chemother 46(11): 3394–3400.

van Duin D, Cleare W, Zaragoza O, et al (2004). Effects of voriconazole on *Cryptococcus neoformans*. Antimicrob Agents Chemother 48(6): 2014–2020.

Wang CY, Wu HD, Hsueh PR (2005). Nosocomial transmission of cryptococcosis. N Engl J Med 352: 1271–1272.

Handbook of Clinical Neurology, Vol. 85 (3rd series)
HIV/AIDS and the Nervous System
P. Portegies, J. R. Berger, Editors

Chapter 13

Progressive multifocal leukoencephalopathy

JOSEPH R. BERGER*

University of Kentucky College of Medicine, Lexington, KY, USA

13.1. Introduction

The recognition of three cases of progressive multifocal leukoencephalopathy (PML) attending the use of natalizumab, a monoclonal antibody directed against $\alpha4\beta1$ integrin for the treatment of multiple sclerosis and Crohn's disease, has heightened interest in this formerly rare disease (Berger and Koralnik, 2005; Kleinschmidt-DeMasters and Tyler, 2005; Langer-Gould et al., 2005; Van Assche et al., 2005). However, the illness ceased being a rare disorder with the advent of the HIV pandemic. First described as a distinct entity by Astrom and coworkers in 1958 (Astrom et al., 1958), the syndrome was identified chiefly on the basis of a triad of unique histopathological features which included demyelination, abnormal oligodendroglial nuclei and giant astrocytes. Each of the three patients they described had an underlying lymphoproliferative disorder. In the 26 years from 1958 to 1984, comprehensive review of PML by Brooks and Walker (1984) was able to identify 230 published and unpublished cases of which 69 cases were pathologically confirmed and 40 cases both virologically and pathologically confirmed. In this series, only five of the 230 patients had HIV/AIDS (Brooks and Walker, 1984), an illness first described in 1981, and only two of these AIDS-associated PML cases had been previously reported (Miller et al., 1982; Bedri et al., 1983). HIV/AIDS accounted for only 2.1% of the underlying illnesses in that series; however, by the end of the twentieth century, the spectrum of predisposing illnesses for PML changed dramatically. In some locales, HIV/AIDS accounted for more than 90% of the predisposing disorders associated with PML (Berger et al., 1998a, b). HIV/AIDS has resulted in a remarkable increase in the frequency of PML.

13.2. Molecular biology of JC virus

Following its initial description, the etiology of PML remained a mystery. However, in 1959, the year following this seminal description, Cavanaugh and Greenbaum suggested the possibility of a viral etiology based on the electron microscopic appearance of inclusion bodies in the enlarged oligodendroglial nuclei (Cavanaugh et al., 1959). The specific nature of these inclusion bodies on electron microscopic study appeared to be consistent with papovavirus (Silverman and Rubinstein, 1965; ZuRhein, 1965), a virus not previously known to result in central nervous system (CNS) disorders. Viral isolation in glial cell cultures by Padgett at the University of Wisconsin from the brain of a patient with PML confirmed that a papova virus was responsible (Padgett et al., 1971). The virus was labeled the JC virus using the initials of the person from whom it was isolated. JC virus is in the genus *Polyomavirus* in the family Papovavirus. Previous descriptions of PML occurring in association with other polyomaviruses, e.g. BK virus and SV40, are now believed to have been incorrect. The cases attributed to SV40 have been poorly characterized and, in some instances, re-examination of these brain tissues by in situ DNA hybridization has revealed JC virus, not SV40 (Stoner and Ryschkewitsch, 1991). BK virus, another human polyomavirus and an important cause of renal transplant rejection (Gardner et al., 1971; Tooze, 1980), not surprisingly, can be isolated from the urines of some patients with PML, but has not been proven to be neuropathogenic. BK virus shares more than 70% nucleotide homology with JCV (Frisque et al., 1979).

Our understanding of PML has increased greatly over the past 25 years. First, there has been an increased incidence of PML due to AIDS and, consequently, more

*Correspondence to: Joseph R. Berger, MD, Professor and Chairman, Department of Neurology, Kentucky Clinic L-445, 740 South Limestone St, University of Kentucky College of Medicine, Lexington, KY 40536-0284, USA. E-mail: jrbneuro@email.uky.edu; Tel: +1-859-323-1279, Fax: +1-859-323-5943.

opportunity to study the disease. Second, there has been development of highly sensitive molecular techniques that allow detection of very few copies of a viral genome including advances in in situ hybridization and amplification of viral genomes using polymerase chain reaction (PCR) (Arthur et al., 1989; Weber et al., 1990). The application of these techniques to tissues available from PML patients has, among other studies, enabled investigations of the mechanisms of viral multiplication (Lynch and Frisque, 1990, 1991; Major et al., 1990), the cellular control over viral gene expression (Amemiya et al., 1989, 1992; Tada et al., 1989) and the delivery of virus to the CNS (Houff et al., 1988; Weber et al., 1990).

JCV has a simple DNA genome of 5.1 kilobases in a double-stranded, supercoiled form, encapsidated in an icosahedral protein structure measuring 40 nm in diameter. The DNA codes for three nonstructural proteins and three capsid proteins (VP1, VP2, VP3) (Fig. 13.1). The T protein, a nonstructural but multifunctional protein, is a DNA binding protein responsible for initiation of viral DNA replication and transcription of the capsid proteins which are transcribed from opposite strands of the DNA genome. In certain rodent and nonhuman primate cells, JCV T protein expression is consistent with a malignant transformation or tumor induction particularly of astroglial cells into astrocytomas (ZuRhein, 1967; Walker et al., 1973; London et al., 1978). In these cells only the T protein, named for its tumor-promoting function, is expressed. The role of JC virus and its protein products in the pathogenesis of human glial and other tumors remains controversial. A smaller t protein has not been considered important for pathogenicity. There are approximately 200 nucleotide base pairs of noncoding sequence referred to as the regulatory region. It is located between the two coding sequence areas. This region of the genome contains the signals for DNA replication as well as for promotion and enhancement of transcription (ZuRhein, 1967; Walker et al., 1973; London et al., 1978; Frisque et al., 1979, 1984). This region is responsible for the cellular tropism of JCV (Khalili et al., 1988). This region of the viral genome demonstrates the most sequence variability in the brains of patients with PML as a consequence of deletions and rearrangements perhaps acquired during propagation in brain or in extraneural host tissues (Dorries, 1984; Martin et al., 1985).

Initial studies of JC virus in cultures of human fetal brain cells indicated an exclusive neurotropism for glial cells (ZuRhein, 1965; Padgett et al., 1971; Dorries, 1984; Martin et al., 1985). While not infecting neurons, JC virus does infect both oligodendrocytes and astrocytes (Wroblewska et al., 1980; Aksamit and Proper, 1988). Recent studies prompted by the appearance of a unique disorder of the cerebellum in AIDS patients suggest that JC virus may also infect the granular cells of the cerebellum (Koralnik et al., 2005).

Atwood and colleagues have demonstrated that JC virus uses serotonin receptor 5HT2a for binding to the cell surface (Elphick et al., 2004). It is quite likely that other receptors which remain yet to be identified can also permit JC virus binding. Following binding,

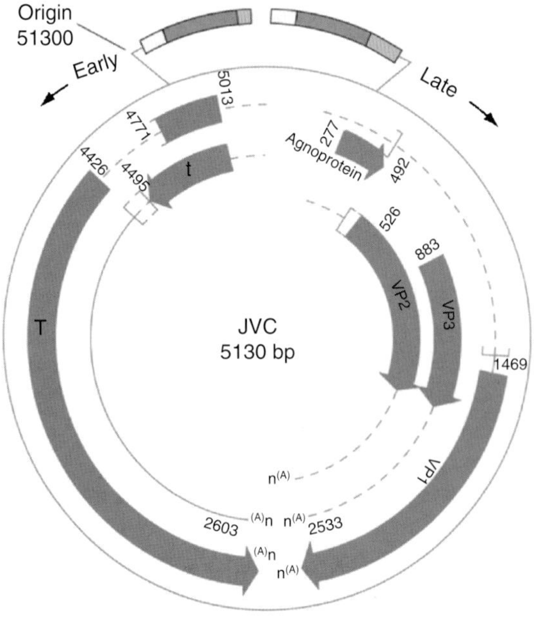

Fig. 13.1. For full color figure see plate section. The JC virus genome.

the virus enters the cell through clathrin and eps15-dependent pathways (Pho et al., 2000; Querbes et al., 2004) following which it is transported to the endoplasmic reticulum through caveosomes. From there, it enters the nucleus (Fig. 13.2). Nuclear DNA binding proteins that selectively interact with the regulatory region of the genome are critical to the tropism of the virus. In particular, binding of nuclear factor 1 (NF-1), a protein that functions in both transcriptional control and replication of DNA, is important to JC virus replication (Tamura et al., 1988; Amemiya et al., 1989, 1992; Ahmed et al., 1990). Cells that are not permissive to JCV infection probably do not have these same protein factors and/or have other proteins that bind the JCV regulatory sequences and block transcription (Tada et al., 1989; Amemiya et al., 1992).

13.3. Pathogenesis of PML and the role of HIV infection

For PML to develop, a number of steps must ensue: (1) infection with JC virus; (2) latency of JC virus in extraneural tissues; (3) rearrangement of JC virus into a neurotropic strain if not already; (4) re-activation of the neurotropic JC virus strain from sites of viral latency; (5) entry into the brain; (6) an ineffective immune system that prevents immunosurveillance from eliminating the infection; and (7) establishment of productive infection of oligodendrocytes. Following the initial infection of JC virus, the virus enters latency in selected tissues. These include the tonsils (Monaco et al., 1998), lung, spleen, bone marrow and kidney (Caldarelli-Stefano et al., 1999). Approximately 5–6% of the population excrete JC virus in urine as detected by either PCR or virus isolation (Coleman et al., 1980; Arthur et al., 1989; Myers et al., 1989; Flaegstad et al., 1991; Berger et al., 2005). These observations suggest that the kidney is a site of viral latency, but it is unlikely that this is the site of latency that leads to the development of PML as the DNA sequence of the regulatory region from kidney or urine in these individuals is markedly different from the sequence found in the brain of PML patients (Yogo et al., 1991). The JC virus isolated from the kidney is referred to as the archetype sequence (Yogo et al., 1990); its regulatory region contains 187 nucleotide pairs with no tandem repeats as is observed in the PML brain isolates (Martin et al., 1985; Henson et al., 1992). To convert the archetype sequence to that most often found in PML brain tissue would require deletions, substitutions and duplications (Yogo et al., 1991; Tominaga et al., 1992).

Current evidence implicates the importance of viral latency in lymphocytes in bone marrow or other lymphoid tissues that can be activated during immune suppression and enter the peripheral blood (Major et al., 1992). Importantly, several regulatory region sequences have been identified from JC virus DNA in peripheral blood of PML patients that are not related to the archetype but closely related to sequences found in PML brain (Tornatore et al., 1992a, b). Circulating infected lymphocytes may be able to cross the blood–brain barrier and pass infection to astrocytes at the border of vessels, which in turn augments infection through multiplication to eventually infect oligodendrocytes.

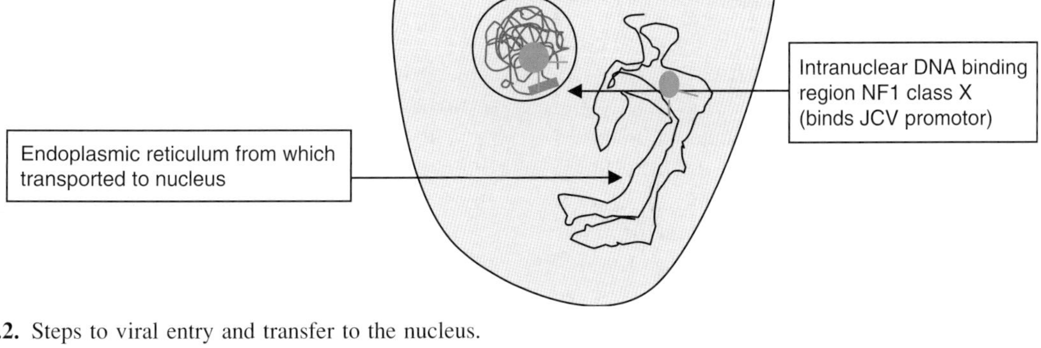

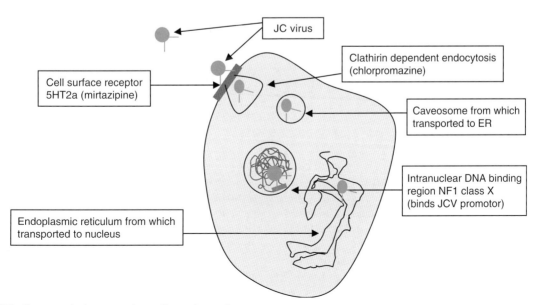

Fig. 13.2. Steps to viral entry and transfer to the nucleus.

Using in situ DNA hybridization, JC virus-infected cells are frequently found near blood vessels in the brain, in B lymphocytes in bone marrow (Houff et al., 1988) and in the brain (Major et al., 1990). In a report of 19 patients with biopsy-proven PML, over 90% had JC virus DNA in peripheral blood lymphocytes using the PCR technology (Tornatore et al., 1992a, b). Data derived from other groups of individuals without PML revealed that 60% of HIV-1-seropositive individuals, 30% of renal transplant recipients and approximately 5% of normal, healthy volunteers also had JC virus DNA in their peripheral circulation (Tornatore et al., 1992a, b). In relatively immunologically healthy HIV-infected persons on highly active antiretroviral therapy (HAART), the likelihood of finding JC virus DNA in circulating lymphocytes appears to parallel that of the normal population (Berger et al., 2005). However, any group whose immune systems are affected either by immunosuppressive regimens or by diseases would be considered to be at risk for the development of PML.

13.4. Epidemiology of JC virus

The ability of JC virus to cause hemagglutination of type O erythrocytes has enabled the performance of antibody studies to determine evidence of prior infection. To date, no disease has been convincingly associated with acute infection. The mechanism of viral spread remains speculative. Respiratory transmission has been postulated. The presence of JC virus in tonsillar tissues (Monaco et al., 1998) suggested that saliva and oropharyngeal secretions may be a means of transmission. However, recent studies of these fluids in HIV-infected persons and healthy controls indicate that the virus is rarely demonstrated in them and, when present, there is very low titer (Berger et al., 2005). Between the ages of 1 and 5 years, approximately 10% of children demonstrate antibody to JC virus, and by age 10 it can be observed in 40–60% of the population (Taguchi et al., 1982; Walker and Padgett, 1983a, b). By adulthood, this figure rises almost sevenfold (Walker and Padgett, 1983a, b) (Fig. 13.3). Seroconversion rates to JC virus have exceeded 90% in some urban areas (Walker and Padgett, 1983a, b).

The high prevalence of antibodies in the adult population and the rarity of the PML in children support the contention that PML is the consequence of reactivation of JC virus in individuals who become immunosuppressed. Most importantly in addressing whether PML is the consequence of a recently acquired infection or due to reactivation is whether high titers of IgM antibody specific for JC virus can be demonstrated in patients with PML. However, in one study addressing the nature of the immunoglobulin response, only 1 of

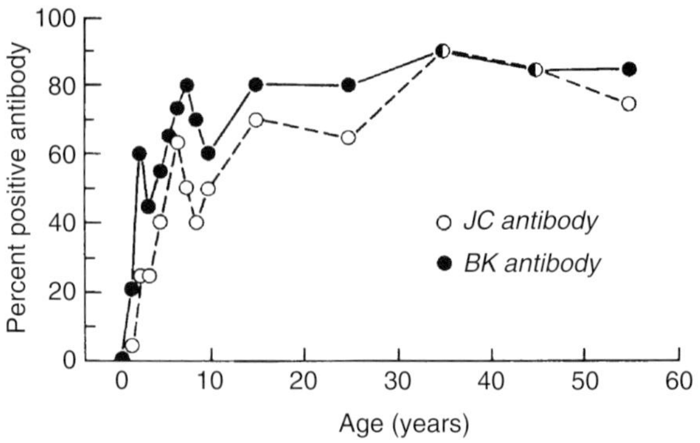

Fig. 13.3. Sero-epidemiology of JC virus (reproduced from Brooks and Walker (1984) with permission by Elsevier).

21 patients with PML had IgM specific for JC virus in their sera, whereas 20 of 21 had IgG antibody specific for JC virus (Padgett and Walker, 1983). Some investigators have argued that the latter study does not exclude the possibility of PML resulting from acute JC virus infection as many of these patients were studied late in the course of their disease.

13.5. Host factors and underlying diseases

In virtually all patients with PML, an underlying immunosuppressive condition has been recognized. Typically, the abnormality is one of cell-mediated immunity, more specifically, a general impairment of the Th1-type T-helper cell function (Weber et al., 2001). The first three patients described by Astrom et al. (1958) had either chronic lymphocytic leukemia or lymphoma as the underlying illness. In the large review of Brooks and Walker (1984), lymphoproliferative diseases were the most common underlying disorders accounting for 62.2% of the cases. Other predisposing illnesses in that series included myeloproliferative diseases in 6.5%, carcinoma in 2.2%, granulomatous and inflammatory diseases, such as tuberculosis and sarcoidosis, in 7.4% and other immune deficiency states in 16.1%. Although AIDS was included in the latter category, it accounted for only 5 of the 230 cases (2.1%) (Brooks and Walker, 1984).

Indeed, until the AIDS epidemic, PML remained a rare disease. To most practicing neurologists, it remained a medical curiosity about which one learned from the textbooks. However, the AIDS pandemic changed the incidence of this formerly rare illness quite remarkably. While only 230 cases were identified in a review from 1958 to 1984 (Brooks and Walker, 1984), 156 cases were identified from the University of Miami Medical Center and the Broward County Medical Examiner's Office in south Florida during the 14-year period 1980–1994 (Berger et al., 1998a, b). All but two of these cases were associated with HIV/AIDS, indicating how dramatically the AIDS pandemic had changed the epidemiology of PML. A 20-fold increase was observed in the disease frequency when comparing the five-year intervals 1980–1984 with 1990–1994 (Berger et al., 1998a, b). In the USA, AIDS has been estimated to be the underlying cause of immunosuppression in 55% to more than 85% of all cases of PML (Major et al., 1992). Data accumulated from death certificates reported to the Centers for Disease Control (CDC) indicated that 87% of all reported cases of PML in 1993 were associated with HIV infection (Selik et al., 1997).

13.6. PML in the era of AIDS

The first description of PML complicating AIDS was published in 1982, one year after the initial description of AIDS (Miller et al., 1982). In less than a decade, HIV/AIDS became the most common underlying disorder predisposing to the development of PML at institutions in New York (Krupp et al., 1985) and Miami (Berger et al., 1987). The epidemiology of PML changed dramatically with HIV/AIDS. Gillespie et al. (1991), studying the prevalence of AIDS-related illnesses in the San Francisco Bay area, estimated a prevalence for PML of 0.3% and opined that this figure may have significantly underestimated the true prevalence. Death certificate reporting of AIDS to the CDC between 1981 and June 1990 revealed that 971 (0.72%) of 135,644 individuals dying with AIDS also had PML (Holman et al., 1991). Due to the notorious inaccuracies in death certificate reporting (Messite and Stellman, 1996) and the requirement of pathologic confirmation for inclusion in this study, it, too, likely significantly underestimated the true prevalence of this disorder. Other studies have suggested that the prevalence of PML at this time in the AIDS pandemic was substantially higher with most estimates ranging between 1 and 5% in clinical studies and as high as 10% in pathological series (Krupp et al., 1985; Berger et al., 1987; Lang et al., 1989; Kure et al., 1991; Kuchelmeister et al., 1993). In 1987, a large, retrospective, hospital-based clinical study (Berger et al., 1987) found PML in approximately 4% of patients hospitalized with AIDS.

Some 4% of all patients dying with AIDS had PML in a combined series of seven separate neuropathological studies comprising a total of 926 patients with AIDS (Kure et al., 1991). Two other large neuropathologic series found PML in 7% (Lang et al., 1989) and 9.8% (Kuchelmeister et al., 1993) of autopsied AIDS patients. The authors of the latter study acknowledged that an unusually high estimate may have resulted from numerous referral cases from outside the study center and the inherent bias imposed (Kuchelmeister et al., 1993). However, a study of 548 consecutive, unselected autopsies between 1983 and 1991 performed on HIV seropositive individuals by the Broward County (Florida) Medical Examiner revealed that 29 (5.3%) had PML confirmed at autopsy (Whiteman et al., 1993). Similarly, the Multicenter AIDS Cohort Study (MACS) also identified a dramatic rise in the incidence of PML over a similar time period. Specifically, the MACS identified 22 cases of PML among the cohort of AIDS cases studied from 1985 to 1992: the average annual incidence of PML was 0.15 per 100 person-years with a yearly rate of increase of 24% between 1985 and 1992 (Bacellar et al., 1994). Although these estimates may be susceptible to selection and other biases, there is an indisputable markedly increased incidence and prevalence of PML since the inception of the AIDS pandemic. Indeed, it appears that the incidence of PML complicating HIV/AIDS is higher than that of any other immunosuppressive disorder relative to their frequency. A number of possible explanations have been proposed (Berger, 2003) including: (1) differences in the degree and duration of the cellular immunosuppression in HIV infection; (2) facilitation of the entry into the brain of JC virus-infected B-lymphocytes by alterations in the blood–brain barrier due to HIV; (3) facilitation of the entry of these infected cells by the upregulation of adhesion molecules on the brain vascular endothelium due to HIV infection; (4) transactivation of JC virus by the HIV proteins, particularly *tat*; and (5) transactivation of JC virus by the cytokines and chemokines elaborated by the microglial cells of HIV-infected brains.

Concomitant with the increase in PML in association with AIDS has been the not unexpected alteration in the demographics of the affected population. Prior to the AIDS epidemic, PML was chiefly a disease of the elderly with males and females virtually equally affected (Brooks and Walker, 1984). Currently, in the USA, PML occurs chiefly in men between the ages of 20 and 50 (Berger et al., 1998a, b), reflective of the demographics of the HIV-infected population. As the demographics of the HIV-infected population changes in the USA or in countries in which transmission of HIV is by means other than homosexual relationships, the demographics of PML will undoubtedly mirror that of the HIV-infected population at large. Children regardless of the cause of underlying immunosuppression rarely develop PML. As exposure to JC virus occurs sometime during childhood, a minority of young children is at risk for the disease. However, it has been observed in HIV-infected children (Berger et al., 1992b; Singer et al., 1993; Morriss et al., 1997). Curiously, there seems to be a higher degree of prevalence of PML in white males compared to African American males (Holman et al., 1998). Additionally, there may be some geographical differences in the prevalence of PML. For example, PML is considered rare in Africa and a neuropathological study from southern India suggests an incidence of 1% (Satishchandra et al., 2000). The population differences observed may be the consequence of the nature of medical care rendered as PML is typically observed in advanced HIV infection, and therefore patients succumbing to other AIDS-related disorders early in the course of their infection may not live sufficiently long to develop PML.

In 1996, there was an expansion of available antiretroviral therapies with the development of HAART. Opportunistic infections, e.g. cytomegalovirus (CMV) (Verbraak et al., 1999; Baril et al., 2000) and toxoplasmosis (Maschke et al., 2000), and primary CNS lymphomas (Sparano et al., 1999) have been reported to have declined significantly following their introduction. D'Arminio Monforte et al. (2000) detected a 95% risk reduction in all CNS AIDS-related conditions following the adoption of HAART in their cohort. There were 20 cases of PML identified in this study, but a specific analysis for PML was not undertaken. Others (Maschke et al., 2000) have similarly noted a decline in HIV-related CNS disorders, but have had too few cases to comment specifically about PML. Data from the MACS suggest that the incidence of PML has declined in the HAART era (Sacktor, 2002). However, the survival of patients with PML on HAART has increased substantially (Albrecht et al., 1998; Miralles et al., 1998; Clifford et al., 1999; De Luca et al., 2000; Antinori et al., 2003) and the prevalence of the disease may actually be relatively unaffected.

Table 13.1

Signs and symptoms of AIDS-associated progressive multifocal leukoencephalopathy

Symptoms
Weakness
Cognitive impairment
Speech abnormalities
Headache
Gait impairment
Visual abnormalities
Sensory loss
Signs
Hemiparesis
Gait disturbance
Cognitive impairment
Dysarthria
Dysphasia
Hemisensory loss
Visual field defect
Ocular palsy

Adapted from Berger et al. (1998b).

13.7. Clinical disease

PML heralds AIDS in approximately 1% of all HIV-infected persons (Berger et al., 1998a, b). Therefore, the occasional patient is seen with neurological features antedating knowledge of HIV infection. This may lead to significant diagnostic confusion in the otherwise healthy individual with unsuspected HIV infection. A high index of suspicion for underlying HIV infection is required when confronted with unusual neurological illnesses.

The most common symptoms (Table 13.1) reported by patients with PML or their caregivers are weakness and cognitive disturbances (Berger et al., 1998a, b). Other common symptoms included speech abnormalities, headaches, gait disorders, visual impairment and sensory loss. In a large series with more than 150 AIDS-related PML patients (Berger et al., 1998a, b), each of these symptoms was seen in more than 15% of patients. In general, these symptoms are similar to those identified in a series of non-HIV-associated PML cases. In comparison to the series of Brooks and Walker (1984), headaches were significantly more common in the HIV-infected population and visual disturbances were more common in the non-HIV infected. Seizures are seen in up to 10% of patients and are usually focal in nature although secondary generalization may occur. Seizures in AIDS-associated PML may reflect involvement of the cortical astrocytes by the JC virus (Sweeney et al., 1994) or may be secondary to some other process or HIV infection of the brain itself (Wong et al., 1990).

Limb weakness is the most common sign observed with HIV-associated PML (Table 13.1) (Berger et al., 1998b). It was observed in over 50% (Berger et al., 1998b). Cognitive disturbances and gait disorders are seen in approximately one-quarter to one-third of patients (Berger et al., 1998a, b). Diplopia, noted by 9% of patients, is usually the consequence of involvement of the third, fourth or sixth cranial nerves and is typically observed in association with other brainstem findings (Berger et al., 1998a, b). Visual field loss due to involvement of the retrochiasmal visual pathways is significantly more common than diplopia or other visual disturbances (Brooks and Walker, 1984; Ormerod et al., 1996; Berger et al., 1998a, b). Optic nerve disease does not occur with PML and, although the lesions of PML confirmed by JC virus antigen presence were detected in the spinal cord of one of 138 HIV-infected patients dying with AIDS (Henin et al., 1992), clinical myelopathy secondary to PML must be vanishingly rare. PML does not involve the peripheral nervous system.

13.8. Neuroimaging

In the appropriate clinical context, radiographic imaging may strongly support the diagnosis of PML. Computed tomography (CT) of the brain in PML reveals hypodense lesions of the affected white matter (Fig. 13.4). On CT scan, the lesions of PML exhibit no mass effect and infrequently contrast enhance. A "scalloped" appearance beneath the cortex is noted when there is involvement of the subcortical arcuate fibers (Whiteman et al., 1993). Cranial magnetic resonance imaging (MRI) is far more sensitive to the presence of the white matter lesions of PML than CT scan (Whiteman et al., 1993). MRI shows a hyperintense lesion on T2-weighted images in the affected regions (Fig. 13.5) and, as with CT scan, faint, typically peripheral, contrast enhancement may be observed in 5–10% of cases (Whiteman et al., 1993). The lesions of PML may occur virtually anywhere in the brain. The frontal lobes and parieto-occipital regions are commonly affected, presumably as a consequence of their volume. However, isolated or associated involvement of the basal ganglia, external capsule and posterior fossa structures (cerebellum and brainstem) may be seen as well (Whiteman et al., 1993). Other diseases may affect the white matter in a similar manner in association with HIV infection. Particularly notable in this regard are AIDS dementia and CMV infection. With respect to AIDS dementia, radiographic distinctions include a greater propensity of PML lesions to involve the subcortical white matter, hypointensity on T1W1 images and occasional contrast enhancement (Whiteman et al., 1993). CMV lesions are typically located in the periventricular white matter and centrum semiovale (Kalayjian et al., 1993; Miller et al., 1997). Subependymal enhancement is often observed as a consequence of CMV infection (Miller et al., 1997). Other potentially HIV-associated disorders that may result in hyperintense signal abnormalities of the white matter resembling PML include varicella-zoster leukoencephalitis (Gray et al., 1994), a multiple sclerosis-like illness (Berger et al., 1992b), acute disseminated encephalomyelitis (Chetty et al., 1997; Bhigjee et al., 1999), CNS vasculitis (Scaravilli et al., 1989), a reversible leukoencephalopathy associated with nucleoside analogue antiretrovirals (Church, 2002) and white matter edema associated with primary or metastatic brain tumors. Almost always, the clinical features, laboratory findings and associated radiographic features enable the correct diagnosis.

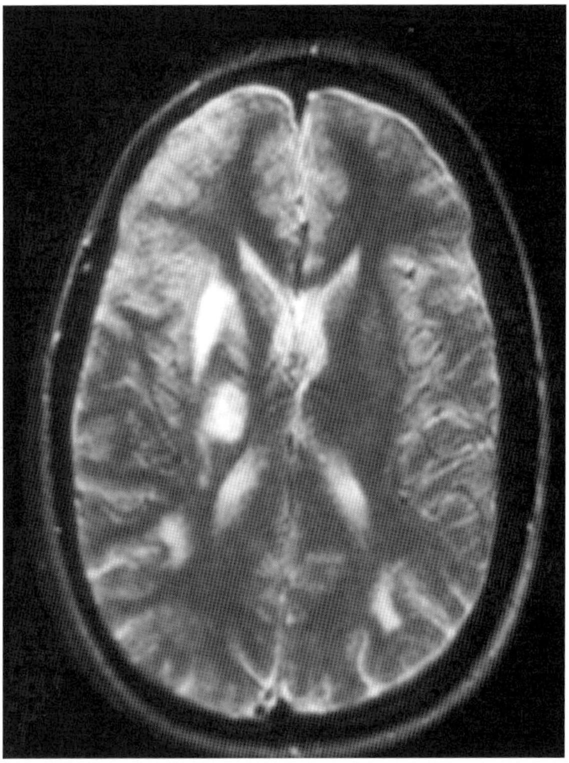

Fig. 13.5. MRI of PML.

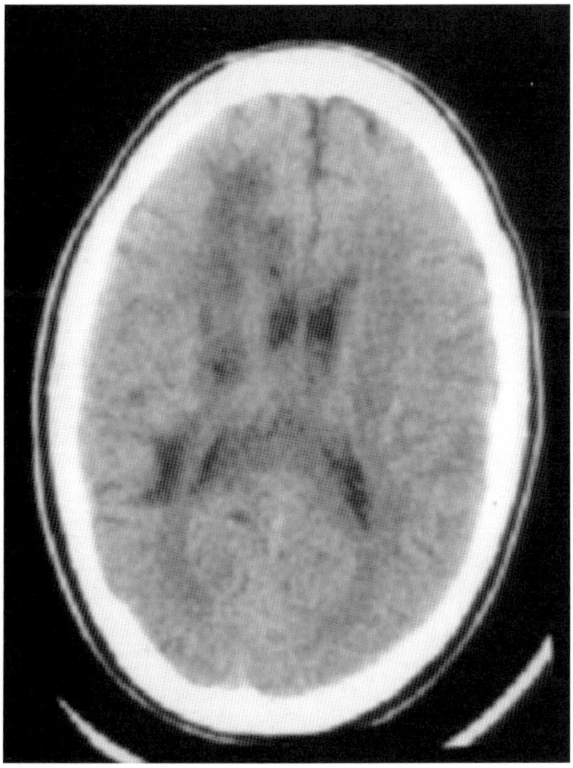

Fig. 13.4. CT scan of PML.

Magnetization transfer MRI studies have been suggested as an effective means of monitoring the degree of demyelination in PML (Brochet and Dousset, 1999).

Magnetic resonance spectroscopy reveals a decrease in N-acetylaspartate and creatine and increased choline products, myoinositol and lactate in the lesions of PML (Chang et al., 1997). These changes likely reflect neuronal loss and cell membrane and myelin breakdown consequent to PML (Chang et al., 1997). Cerebral angiography is not routinely performed, but exhibited arteriovenous shunting and a parenchymal blush in the absence of contrast enhancement on MRI in four of six patients in one study (Nelson et al., 1999). Pathological studies suggested that small vessel proliferation and perivascular inflammation were the explanation for these unexpected angiographic features (Nelson et al., 1999). Thallium-201 single photon emission computed tomography (^{201}Tl SPECT) generally reveals no uptake in the lesions of PML (Iranzo et al., 1999); however, a single case report of a contrast-enhancing lesion with a positive ^{201}Tl SPECT has been reported (Port et al., 1999).

13.9. Laboratory studies

In the overwhelming majority of HIV-infected patients with PML, severe cellular immunosuppression, as defined by CD4 lymphocyte counts <200 cells/mm^3, is observed. In three separate series of AIDS-related PML (von Einsiedel et al., 1993; Fong et al., 1995; Berger et al., 1998a, b), the mean CD^4 count study ranged from 84 to 104 cells/mm^3. However, in the largest series of HIV-associated PML (Berger et al., 1998a, b), 10% or more of patients had CD4 lymphocyte counts in excess of 200 cells/mm^3.

Cerebrospinal fluid (CSF) examination is very helpful in excluding other diagnoses. Cell counts are usually less than 20 cells/mm^3 (Berger et al., 1998a, b). In one large study, the median cell count was 2 cells/mm^3 and the mean was 7.7 cells/mm^3 (Berger et al., 1998a, b). In that same study, 55% had an abnormally elevated CSF protein (Berger et al., 1998a, b) with the highest recorded value being 208 mg/dl (2.08 g/l). Hypoglycorrhachia was observed in less than 15%. These abnormalities are not inconsistent with that previously reported to occur with HIV infection alone (Elovaara et al., 1987; Marshall et al., 1988; Katz et al., 1989). Several studies (Weber et al., 1994; Fong et al., 1995; McGuire et al., 1995) demonstrate a high sensitivity and specificity of CSF PCR for JC virus in PML. Many authorities have regarded the demonstration of JC viral DNA coupled with the appropriate clinical and radiological features sufficiently suggestive of PML to be diagnostic, thus obviating the need for brain biopsy. Quantitative PCR techniques for JC virus in biological fluids continues to be refined (Drews et al., 2000). The sensitivity of CSF PCR for JC virus in PML is of the order of 75% (Fong et al., 1995). Amplification of the virus from the CSF in the absence of PML is very unlikely.

13.10. Pathology

The cardinal feature of PML is demyelination which is apparent both macroscopically and microscopically. Demyelination may, on rare occasions, be monofocal, but it typically occurs as a multifocal process suggesting a hematological spread of the virus. These lesions may occur in any location in the white matter and range in size from 1 mm to several centimeters (Astrom et al., 1958; Richardson, 1970); larger lesions are not infrequently the result of coalescence of multiple smaller lesions. The histopathological hallmarks of PML are a triad (Astrom et al., 1958; Richardson, 1970) of multifocal demyelination (Fig. 13.3), hyperchromatic, enlarged oligodendroglial nuclei (Fig. 13.4) and enlarged bizarre astrocytes with lobulated hyperchromatic nuclei (Fig. 13.5). The latter may be seen to undergo mitosis and appear to be malignant. The latter has resulted in a mistaken diagnosis of glioma on occasion (Van Assche et al., 2005). Electron microscopic examination or immunohistochemistry will reveal the JC virus in the oligodendroglial cells. These virions measure 28–45 nm in diameter and appear singly or in dense crystalline arrays (Astrom et al., 1958; Richardson, 1970). Less frequently, the virions are detected in reactive astrocytes, and they are uncommonly observed in macrophages that are engaged in removing the affected oligodendrocytes (Mazlo and Herndon, 1977; Mazlo and Tariska, 1982). The virions are generally not seen in the large, bizarre astrocytes (Mazlo and Tariska, 1982). In situ hybridization and in situ PCR for JC virus antigen allows for detection of the virion in the infected cells even in formalin-fixed archival tissue (Samorei et al., 2000).

13.11. Prognosis

Prior to the HAART era, the median survival of PML complicating AIDS was 6 months and the mode was one month (Berger et al., 1998b). In the absence of effective antiretroviral therapy, the survival of AIDS patients with PML is not significantly different than PML occurring with any other immunosuppressive condition. However, even in the pre-HAART era, recovery of neurological function, improvement of PML lesions in radiographic imaging and survival exceeding 12 months had been observed in as many as 10% of patients with AIDS-associated PML (Berger et al., 1998b). Factors that appear to be associated with prolonged survival include PML as the presenting manifestation of AIDS (Berger et al., 1998a), higher CD^4 lymphocyte counts

(>300 cells/mm^3) (Berger et al., 1998a), presence of a perivascular inflammatory infiltrate in the PML lesions (Richardson, 1974) and contrast enhancement on radiographic imaging (Berger et al., 1998a; Arbusow et al., 2000), although another study failed to link any radiological features with prognosis (Post et al., 1999). Additionally, a correlation between low titers of JC viral DNA load in the CSF and prolonged survival has also been demonstrated (Taoufik et al., 1998; De Luca et al., 1999; Yiannoutsos et al., 1999). The longest reported survival has been 92 months from onset of illness with death due to another illness (Berger et al., 1998b).

The cellular immune response against JC virus appears to be tightly correlated with a favorable clinical outcome in PML (Du Pasquier et al., 2001; Koralnik et al., 2001; Koralnik, 2002). The presence of JC virus-specific cytotoxic T-lymphocytes (CTLs) in these patients is likely related to the presence of inflammatory infiltrates in the PML lesions which are responsible for the alterations of the blood–brain barrier and marginal contrast enhancement seen on imaging studies. These JC virus-specific CTLs are probably instrumental in destroying infected oligodendrocytes, preventing further disease progression and decreasing CSF JC viral load. The lack of recurrence of PML in some of the patients exhibiting long-term survival and recovery reflects clearance of the JC virus from the CSF (De Luca et al., 2000).

13.12. Treatment

To date there are no unequivocally successful therapeutic modalities for PML. Most of the extant literature consists of anecdotal reports. In the pre-HAART era, zidovudine (AZT) and other antiretrovirals had been proposed as adjunctive therapy for AIDS-associated PML. Anecdotal responses to antiretroviral therapy (Conway et al., 1990) led to the suggestion to administer zidovudine. However, in vitro assays failed to demonstrate an effect of zidovudine on JC virus replication (Hou and Major, 1998). The survival of PML in the era of HAART has changed quite considerably, however, with as many as 50% of patients demonstrating long-term survival (>12 months) (Albrecht et al., 1998; Miralles et al., 1998; Clifford et al., 1999; Inui et al., 1999; Tantisiriwat et al., 1999; De Luca et al., 2000; Antinori et al., 2001). The benefit of HAART in AIDS-associated PML has not been universally observed, however (De Luca et al., 1998), as the benefit seems to be chiefly confined to treatment-naive patients (Wyen et al., 2004). The remarkable success of HAART in the treatment of AIDS-related PML has had its downside as well. A syndrome referred to as the immune reconstitution inflammatory syndrome (IRIS) may acutely result in new or increased neurological deficits, an increased number or size of lesions observed by neuroimaging, contrast enhancement of these lesions and brain edema (Cinque et al., 2001; Thurnher et al., 2001; Safdar et al., 2002; Cinque et al., 2003; Du Pasquier and Koralnik, 2003; Hoffmann et al., 2003). Fatal outcomes have been reported (Safdar et al., 2002; Di Giambenedetto et al., 2004) and the development of this syndrome with infratentorial PML may be especially dangerous (Kastrup et al., 2005).

Nucleoside analogues have been employed because they impede the synthesis of DNA (Goodman et al., 1985). In vitro studies (Hou and Major, 1998) have clearly demonstrated the ability of cytosine arabinoside (Cytarabine, ARA-C), a cytosine analogue, to inhibit JC virus replication and anecdotal reports of intravenous and intrathecal administration suggested the value of this therapy in PML (Bauer et al., 1973; Conomy et al., 1974; Van Horn et al., 1978; Tashiro et al., 1987; O'Riordan et al., 1990; Lidman et al., 1991). However, a carefully conducted clinical trial of AIDS-related PML failed to show any value of either intravenous or intrathecal administration of ARA-C when compared to placebo (Hall et al., 1998). Theoretically, neither method of administration permitted adequate concentrations of the drug to reach the disease sites, and trials with novel intraparenchymal delivery systems have been suggested. Alternatively, higher doses of ARA-C than those employed in the randomized study may prove beneficial. Despite anecdotal reports of the value of other nucleoside analogues in PML, such as adenine arabinoside (Vidarabine, ARA-A) (Wolinsky et al., 1976; Rand et al., 1977; Tashiro et al., 1987), none has been convincingly demonstrated to ameliorate the disease course.

Interferons (IFNs) have also had occasional positive results both subcutaneously (Steiger et al., 1993) and intrathecally (Tashiro et al., 1987) when used in conjunction with ARA-C. The antiretroviral activity of the IFNs may be the consequence of their ability to stimulate natural killer (NK) cells. In a pilot study of 17 patients with AIDS and PML treated with alpha 2a IFN and zidovudine, two had long-term clinical stabilization, though none improved (Berger et al., 1992a). A retrospective study that compared patients with AIDS-associated PML receiving a minimum treatment of 3 weeks of 3 million units of alpha-IFN daily with untreated historical controls suggested that alpha-IFN treatment delayed the progression of the disease, palliated symptoms and significantly prolonged survival (Huang et al., 1998). However, re-examination of that data indicated that the improved survival could be explained by the concomitant administration of HAART (Geschwind et al., 2001).

The antineoplastic drug camptothecin, a DNA topoisomerase I inhibitor, has been demonstrated to block JC virus replication in vitro when administered in pulsed doses in amounts non-toxic to cells (Kerr et al., 1993). Its therapeutic usefulness in PML has been entirely anecdotal (O'Reilly, 1997; Vollmer-Haase et al., 1997). Another antineoplastic drug, topotecan, may also inhibit JC virus replication (Kerr et al., 1993). However, both these drugs display significant systemic toxicity and their value in the treatment of PML remains open to question.

Cidofovir (HPMPC; (S)-1-(3-hydroxy-2-phosphonylmethoxypropyl)cytosine) and its cyclic counterpart have demonstrated selective anti-polyomavirus activity (Andrei et al., 1997). The 50% inhibitory concentrations for HPMPC were in the range 4–7 μg/ml, and its selectivity index varied from 11 to 20 for mouse polyomavirus and from 23 to 33 for SV40 strains in confluent cell monolayers (Andrei et al., 1997). It has been proposed as an agent for the treatment of PML (Sadler et al., 1998) and there is anecdotal evidence to support its use (Blick et al., 1998; Brambilla et al., 1999; De Luca et al., 1999; Dodge, 1999; Meylan et al., 1999). However, an open-label AIDS Clinical Trials Group study addressing the value of cidofovir in the same fashion failed to show any benefit (Marra et al., 2002). Furthermore, the drug has serious side effects including ocular hypotony, bone marrow depression and renal disorders.

Increased understanding of the molecular biology of JC virus and new technologies will likely result in novel strategies. One strategy may be to attempt to block JC virus cell binding or entry with specific inhibitors. Another possibility is the use of antisense oligonucleotides. An antisense oligonucleotide that is properly designed with a specific complementary base sequence that binds selectively to a targeted region of mRNA can prevent the translation of the mRNA into protein. Antisense oligonucleotide directed to JC virus T antigen may reduce viral expression by 80% (K. Khalili, personal communication, 1994). Antisense oligonucleotides that target other sites of the viral genome, such as transcription sites, may prove to be effective therapeutic strategies. As a strong JC virus-specific cellular immunity has recently been associated with a favorable clinical outcome of PML (Du Pasquier et al., 2001; Koralnik et al., 2001; Koralnik, 2002), the enrichment of an autologous population of JC virus-specific CTL populations using tetrameric MHC class-I/JC virus peptide complexes may be demonstrated to be a therapeutic option. Koralnik and colleagues demonstrated the ability to boost immunity to JC virus with a vaccine based on a newly discovered JC virus-specific CTL epitope (Koralnik, 2002). Conceivably, this approach, too, may have some therapeutic merit.

References

Ahmed S, Rappaport J, Tada H, et al (1990). A nuclear protein derived from brain cells stimulates transcription of the human neurotropic virus promoter, JCVE, in vitro. J Biol Chem 265 (23): 13899–13905.

Aksamit A, Proper J (1988). JC virus replicates in primary adult astrocytes in culture. Ann Neurol 24: 471.

Albrecht H, Hoffmann C, Degen O, et al (1998). Highly active antiretroviral therapy significantly improves the prognosis of patients with HIV-associated progressive multifocal leukoencephalopathy. AIDS 12(10): 1149–1154.

Amemiya K, Traub R, Durham L, et al (1989). Interaction of a nuclear factor-1-like protein with the regulatory region of the human polyomavirus JC virus. J Biol Chem 264 (12): 7025–7032.

Amemiya K, Traub R, Durham L, et al (1992). Adjacent nuclear factor-1 and activator protein binding sites in the enhancer of the neurotropic JC virus. A common characteristic of many brain-specific genes. J Biol Chem 267 (20): 14204–14211.

Andrei G, Snoeck R, Vandeputte M, et al (1997). Activities of various compounds against murine and primate polyomaviruses. Antimicrob Agents Chemother 41(3): 587–593.

Antinori A, Ammassari A, Giancola ML, et al (2001). Epidemiology and prognosis of AIDS-associated progressive multifocal leukoencephalopathy in the HAART era. J Neurovirol 7(4): 323–328.

Antinori A, Cingolani A, Lorenzini P, et al (2003). Clinical epidemiology and survival of progressive multifocal leukoencephalopathy in the era of highly active antiretroviral therapy: data from the Italian Registry Investigative Neuro AIDS (IRINA). J Neurovirol 9(Suppl 1): 47–53.

Arbusow V, Strupp M, Pfister HW, et al (2000). Contrast enhancement in progressive multifocal leukoencephalopathy: a predictive factor for long-term survival? [letter]. J Neurol 247(4): 306–308.

Arthur RR, Dagostin S, Shah KV, et al (1989). Detection of BK virus and JC virus in urine and brain tissue by the polymerase chain reaction. J Clin Microbiol 27(6): 1174–1179.

Astrom K, Mancall E, Richardson EP, et al (1958). Progressive multifocal leukoencephalopathy. Brain 81(1): 93–127.

Bacellar H, Munoz A, Miller EN, et al (1994). Temporal trends in the incidence of HIV-1-related neurologic diseases: multicenter AIDS Cohort Study, 1985–1992. Neurology 44(10): 1892–1900.

Baril L, Jouan M, Agher R, et al (2000). Impact of highly active antiretroviral therapy on onset of *Mycobacterium avium* complex infection and cytomegalovirus disease in patients with AIDS [In Process Citation]. AIDS 14(16): 2593–2596.

Bauer WR, Turel AP Jr, Johnson KP (1973). Progressive multifocal leukoencephalopathy and cytarabine. Remission with treatment. JAMA 226(2): 174–176.

Bedri J, Weinstein W, De Gregorio P, et al (1983). Progressive multifocal leukoencephalopathy in acquired immunodeficiency syndrome. N Engl J Med 309(8): 492–493.

Berger JR (2003). Progressive multifocal leukoencephalopathy in acquired immunodeficiency syndrome: explaining the high incidence and disproportionate frequency of the illness relative to other immunosuppressive conditions. J Neurovirol 9(Suppl 1): 38–41.

Berger JR, Koralnik IJ (2005). Progressive multifocal leukoencephalopathy and natalizumab: unforeseen consequences. N Engl J Med 353(4): 414–416.

Berger JR, Kaszovitz B, Post MJ, et al (1987). Progressive multifocal leukoencephalopathy associated with human immunodeficiency virus infection. A review of the literature with a report of sixteen cases. Ann Intern Med 107 (1): 78–87.

Berger J, Pall L, McArthur J, et al (1992a). A pilot study of recombinant alpha 2a interferon in the treatment of AIDS-related progressive multifocal leukoencephalopathy [abstract]. Neurology 42(Suppl 3): 257.

Berger JR, Scott G, Albrecht J, et al (1992b). Progressive multifocal leukoencephalopathy in HIV-1-infected children. AIDS 6(8): 837–841.

Berger JR, Levy RM, Flomenhoft D, et al (1998a). Predictive factors for prolonged survival in acquired immunodeficiency syndrome-associated progressive multifocal leukoencephalopathy. Ann Neurol 44(3): 341–349.

Berger JR, Pall L, Lanska D, et al (1998b). Progressive multifocal leukoencephalopathy in patients with HIV infection. J Neurovirol 4(1): 59–68.

Berger JR, Miller C, et al (2005). Are saliva and oropharyngeal secretions a source of JCV infection? J Neurovirol 11 (Suppl 2): 83.

Bhigjee AI, Patel VB, Bhagwan B, et al (1999). HIV and acute disseminated encephalomyelitis [letter]. S Afr Med J 89(3): 283–284.

Blick G, Whiteside M, Griegor P, et al (1998). Successful resolution of progressive multifocal leukoencephalopathy after combination therapy with cidofovir and cytosine arabinoside [see comments]. Clin Infect Dis 26(1): 191–192.

Brambilla AM, Castagna A, Novati R, et al (1999). Remission of AIDS-associated progressive multifocal leukoencephalopathy after cidofovir therapy [letter]. J Neurol 246(8): 723–725.

Brochet B, Dousset V (1999). Pathological correlates of magnetization transfer imaging abnormalities in animal models and humans with multiple sclerosis. Neurology 53(5): S12–S17.

Brooks BR, Walker DL (1984). Progressive multifocal leukoencephalopathy. Neurol Clin 2(2): 299–313.

Caldarelli-Stefano R, Vago L, Omodea-Zorini E, et al (1999). Detection and typing of JC virus in autopsy brains and extraneural organs of AIDS patients and non-immunocompromised individuals. J Neurovirol 5(2): 125–133.

Cavanaugh J, Greenbaum D, Marshall A, et al (1959). Cerebral dymyelination associated with disorders of the reticuloendothelial system. Lancet 2: 524–529.

Chang L, Ernst T, Tornatore C, et al (1997). Metabolite abnormalities in progressive multifocal leukoencephalopathy by proton magnetic resonance spectroscopy. Neurology 48(4): 836–845.

Chetty KG, Kim RC, Mahutte CK (1997). Acute hemorrhagic leukoencephalitis during treatment for disseminated tuberculosis in a patient with AIDS. Int J Tuberc Lung Dis 1(6): 579–581.

Church JA (2002). Reversible leukoencephalopathy in a patient with nucleoside analogue-associated mitochondrial DNA depletion and metabolic disease. AIDS 16(17): 2366–2367.

Cinque P, Pierotti C, Vigano MG, et al (2001). The good and evil of HAART in HIV-related progressive multifocal leukoencephalopathy. J Neurovirol 7(4): 358–363.

Cinque P, Bossolasco S, Brambilla AM, et al (2003). The effect of highly active antiretroviral therapy-induced immune reconstitution on development and outcome of progressive multifocal leukoencephalopathy: study of 43 cases with review of the literature. J Neurovirol 9(Suppl 1): 73–80.

Clifford DB, Yiannoutsos C, Glicksman M, et al (1999). HAART improves prognosis in HIV-associated progressive multifocal leukoencephalopathy. Neurology 52(3): 623–625.

Coleman DV, Wolfendale MR, Daniel RA, et al (1980). A prospective study of human polyomavirus infection in pregnancy. J Infect Dis 142(1): 1–8.

Conomy J, Beard N, Matsumoto H, et al (1974). Cytarabine treatment of progressive multifocal leukoencephalopathy. JAMA 229: 1313–1316.

Conway B, Halliday WC, Brunhan RC, et al (1990). Human immunodeficiency virus-associated progressive multifocal leukoencephalopathy: apparent response to 3'-azido-3'-deoxythymidine. Rev Infect Dis 12(3): 479–482.

d'Arminio Monforte A, Duca PG, Vago L, et al (2000). Decreasing incidence of CNS AIDS-defining events associated with antiretroviral therapy. Neurology 54(9): 1856–1859.

De Luca A, Ammassari A, Cingolani A, et al (1998). Disease progression and poor survival of AIDS-associated progressive multifocal leukoencephalopathy despite highly active antiretroviral therapy [letter]. AIDS 12(14): 1937–1938.

De Luca A, Fantoni M, Tartaglione T, et al (1999). Response to cidofovir after failure of antiretroviral therapy alone in AIDS-associated progressive multifocal leukoencephalopathy. Neurology 52(4): 891–892.

De Luca A, Giancola ML, Ammassari A, et al (2000). The effect of potent antiretroviral therapy and JC virus load in cerebrospinal fluid on clinical outcome of patients with AIDS-associated progressive multifocal leukoencephalopathy. J Infect Dis 182(4): 1077–1083.

Di Giambenedetto S, Vago G, Pompucci A, et al (2004). Fatal inflammatory AIDS-associated PML with high CD4 counts on HAART: a new clinical entity? Neurology 63(12): 2452–2453.

Dodge RT (1999). A case study: the use of cidofovir for the management of progressive multifocal leukoencephalopathy. J Assoc Nurses AIDS Care 10(4): 70–74.

Dorries K (1984). Progressive multifocal leucoencephalopathy: analysis of JC virus DNA from brain and kidney tissue. Virus Res 1(1): 25–38.

Drews K, Bashir T, Dorries K (2000). Quantification of human polyomavirus JC in brain tissue and cerebrospinal fluid of patients with progressive multifocal leukoencephalopathy by competitive PCR. J Virol Methods 84(1): 23–36.

Du Pasquier RA, Koralnik IJ (2003). Inflammatory reaction in progressive multifocal leukoencephalopathy: harmful or beneficial? J Neurovirol 9(Suppl 1): 25–31.

Du Pasquier RA, Clark KW, Smith PS, et al (2001). JCV-specific cellular immune response correlates with a favorable clinical outcome in HIV-infected individuals with progressive multifocal leukoencephalopathy. J Neurovirol 7(4): 318–322.

Elovaara I, Iivanainen M, Valle SL, et al (1987). CSF protein and cellular profiles in various stages of HIV infection related to neurological manifestations. J Neurol Sci 78 (3): 331–342.

Elphick GF, Querbes W, Jordan JA, et al (2004). The human polyomavirus, JCV, uses serotonin receptors to infect cells. Science 306(5700): 1380–1383.

Flaegstad T, Sundsfjord A, Arthur RR, et al (1991). Amplification and sequencing of the control regions of BK and JC virus from human urine by polymerase chain reaction. Virology 180(2): 553–560.

Fong IW, Britton CB, Luinstra KE, et al (1995). Diagnostic value of detecting JC virus DNA in cerebrospinal fluid of patients with progressive multifocal leukoencephalopathy. J Clin Microbiol 33(2): 484–486.

Frisque RJ, Martin JD, Padgett BL, et al (1979). Infectivity of the DNA from four isolates of JC virus. J Virol 32(2): 476–482.

Frisque RJ, Bream GL, Cannella MT (1984). Human polyomavirus JC virus genome. J Virol 51(2): 458–469.

Gardner SD, Field AM, Coleman DV, et al (1971). New human papovavirus (B.K.) isolated from urine after renal transplantation. Lancet 1(7712): 1253–1257.

Geschwind MD, Skolasky RI, Royal WS, et al (2001). The relative contributions of HAART and alpha-interferon for therapy of progressive multifocal leukoencephalopathy in AIDS. J Neurovirol 7(4): 353–357.

Gillespie SM, Chang Y, Lemp G, et al (1991). Progressive multifocal leukoencephalopathy in persons infected with human immunodeficiency virus, San Francisco, 1981–1989. Ann Neurol 30(4): 597–604.

Goodman A, Goodman L, Rall T, et al (1985). The Pharmacological Basis of Therapeutics, Macmillan, New York.

Gray F, Belec L, Lescs MC, et al (1994). Varicella-zoster virus infection of the central nervous system in the acquired immune deficiency syndrome. Brain 117(Pt 5): 987–999.

Hall CD, Dafni U, Simpson D, et al (1998). Failure of cytarabine in progressive multifocal leukoencephalopathy associated with human immunodeficiency virus infection. AIDS Clinical Trials Group 243 Team [see comments]. N Engl J Med 338(19): 1345–1351.

Henin D, Smith TW, De Girolami U (1992). Neuropathology of the spinal cord in the acquired immunodeficiency syndrome. Hum Pathol 23(10): 1106–1114.

Henson J, Saffer J, Furneaux H (1992). The transcription factor Sp1 binds to the JC virus promoter and is selectively expressed in glial cells in human brain. Ann Neurol 32 (1): 72–77.

Hoffmann C, Horst HA, Albrecht H, et al (2003). Progressive multifocal leucoencephalopathy with unusual inflammatory response during antiretroviral treatment. J Neurol Neurosurg Psych 74(8): 1142–1144.

Holman RC, Janssen RS, Buehler JW, et al (1991). Epidemiology of progressive multifocal leukoencephalopathy in the United States: analysis of national mortality and AIDS surveillance data. Neurology 41(11): 1733–1736.

Holman RC, Torok TJ, Belay ED, et al (1998). Progressive multifocal leukoencephalopathy in the United States, 1979–1994: increased mortality associated with HIV infection. Neuroepidemiology 17(6): 303–309.

Hou J, Major EO (1998). The efficacy of nucleoside analogs against JC virus multiplication in a persistently infected human fetal brain cell line. J Neurovirol 4(4): 451–456.

Houff SA, Major EO, Katz DA, et al (1988). Involvement of JC virus-infected mononuclear cells from the bone marrow and spleen in the pathogenesis of progressive multifocal leukoencephalopathy. N Engl J Med 318(5): 301–305.

Huang SS, Skolasky RL, Dal Pan GJ, et al (1998). Survival prolongation in HIV-associated progressive multifocal leukoencephalopathy treated with alpha-interferon: an observational study. J Neurovirol 4(3): 324–332.

Inui K, Miyagawa H, Sashihara J, et al (1999). Remission of progressive multifocal leukoencephalopathy following highly active antiretroviral therapy in a patient with HIV infection. Brain Dev 21(6): 416–419.

Iranzo A, Marti-Fabregas J, Domingo P, et al (1999). Absence of thallium-201 brain uptake in progressive multifocal leukoencephalopathy in AIDS patients. Acta Neurol Scand 100(2): 102–105.

Kalayjian RC, Cohen ML, Bonomo RA, et al (1993). Cytomegalovirus ventriculoencephalitis in AIDS. A syndrome with distinct clinical and pathologic features. Medicine (Baltimore) 72(2): 67–77.

Kastrup O, Wanke I, Esser S, et al (2005). Evolution of purely infratentorial PML under HAART: negative outcome under rapid immune reconstitution. Clin Neurol Neurosurg 107(6): 509–513.

Katz RL, Alappattu C, Glass JP, et al (1989). Cerebrospinal fluid manifestations of the neurologic complications of human immunodeficiency virus infection. Acta Cytol 33 (2): 233–244.

Kerr DA, Chang CF, Gordon J, et al (1993). Inhibition of human neurotropic virus (JCV) DNA replication in glial cells by camptothecin. Virology 196(2): 612–618.

Khalili K, Rappaport J, Khoury G (1988). Nuclear factors in human brain cells bind specifically to the JCV regulatory region. EMBO J 7(4): 1205–1210.

Kleinschmidt-DeMasters BK, Tyler KL (2005). Progressive multifocal leukoencephalopathy complicating treatment with natalizumab and interferon beta-1a for multiple sclerosis. N Engl J Med 353(4): 369–374.

Koralnik IJ (2002). Overview of the cellular immunity against JC virus in progressive multifocal leukoencephalopathy. J Neurovirol 8(Suppl 2): 59–65.

Koralnik IJ, Du Pasquier RA, Letvin NL (2001). JC virus-specific cytotoxic T lymphocytes in individuals with progressive multifocal leukoencephalopathy. J Virol 75(7): 3483–3487.

Koralnik IJ, Wuthrich C, Dang X, et al (2005). JC virus granule cell neuronopathy: a novel clinical syndrome distinct from progressive multifocal leukoencephalopathy. Ann Neurol 57(4): 576–580.

Krupp LB, Lipton RB, Swerdlow ML, et al (1985). Progressive multifocal leukoencephalopathy: clinical and radiographic features. Ann Neurol 17(4): 344–349.

Kuchelmeister K, Gullotta F, Bergmann M, et al (1993). Progressive multifocal leukoencephalopathy (PML) in the acquired immunodeficiency syndrome (AIDS). A neuropathological autopsy study of 21 cases. Pathol Res Pract 189(2): 163–173.

Kure K, Llena JF, Lyman WD, et al (1991). Human immunodeficiency virus-1 infection of the nervous system: an autopsy study of 268 adult, pediatric, and fetal brains. Hum Pathol 22(7): 700–710.

Lang W, Miklossy J, Deruaz JP, et al (1989). Neuropathology of the acquired immune deficiency syndrome (AIDS): a report of 135 consecutive autopsy cases from Switzerland. Acta Neuropathol 77(4): 379–390.

Langer-Gould A, Atlas SW, Green AJ, et al (2005). Progressive multifocal leukoencephalopathy in a patient treated with natalizumab. N Engl J Med 353(4): 375–381.

Lidman C, Lindqvist L, Mathiesen T, et al (1991). Progressive multifocal leukoencephalopathy in AIDS. AIDS 5 (8): 1039–1041.

London WT, Houff SA, Madden DL, et al (1978). Brain tumors in owl monkeys inoculated with a human polyomavirus (JC virus). Science 201(4362): 1246–1249.

Lynch KJ, Frisque RJ (1990). Identification of critical elements within the JC virus DNA replication origin. J Virol 64(12): 5812–5822.

Lynch KJ, Frisque RJ (1991). Factors contributing to the restricted DNA replicating activity of JC virus. Virology 180(1): 306–317.

Major EO, Amemiya K, Elder G, et al (1990). Glial cells of the human developing brain and B cells of the immune system share a common DNA binding factor for recognition of the regulatory sequences of the human polyomavirus, JCV. J Neurosci Res 27(4): 461–471.

Major EO, Amemiya K, Tornatore CS, et al (1992). Pathogenesis and molecular biology of progressive multifocal leukoencephalopathy, the JC virus-induced demyelinating disease of the human brain. Clin Microbiol Rev 5(1): 49–73.

Marra CM, Rajicic N, Barner DE, et al (2002). A pilot study of cidofovir for progressive multifocal leukoencephalopathy in AIDS. AIDS 16(13): 1791–1797.

Marshall DW, Brey RL, Cahill WT, et al (1988). Spectrum of cerebrospinal fluid findings in various stages of human immunodeficiency virus infection. Arch Neurol 45(9): 954–958.

Martin JD, King DM, Slauch JM, et al (1985). Differences in regulatory sequences of naturally occurring JC virus variants. J Virol 53(1): 306–311.

Maschke M, Kastrup O, Esser S, et al (2000). Incidence and prevalence of neurological disorders associated with HIV since the introduction of highly active antiretroviral therapy (HAART). J Neurol Neurosurg Psych 69(3): 376–380.

Mazlo M, Herndon R (1977). Progressive multifocal leukoencephalopathy: ultrastructural findings in two brain biopsies. Neuropathol Appl Neurobiol 3: 323–339.

Mazlo M, Tariska I (1982). Are astrocytes infected in progressive multifocal leukoencephalopathy (PML)? Acta Neuropathol 56(1): 45–51.

McGuire D, Barhite S, Hollander H, et al (1995). JC virus DNA in cerebrospinal fluid of human immunodeficiency virus-infected patients: predictive value for progressive multifocal leukoencephalopathy [published erratum appears in Ann Neurol 1995; 37(5): 687]. Ann Neurol 37(3): 395–399.

Messite J, Stellman SD (1996). Accuracy of death certificate completion: the need for formalized physician training. JAMA 275(10): 794–796.

Meylan PR, Vuadens P, Maeder P, et al (1999). Monitoring the response of AIDS-related progressive multifocal leukoencephalopathy to HAART and cidofovir by PCR for JC virus DNA in the CSF. Eur Neurol 41(3): 172–174.

Miller JR, Barrett RE, Britton CB, et al (1982). Progressive multifocal leukoencephalopathy in a male homosexual with T-cell immune deficiency. N Engl J Med 307(23): 1436–1438.

Miller RF, Lucas SB, Hall-Craggs MA, et al (1997). Comparison of magnetic resonance imaging with neuropathological findings in the diagnosis of HIV and CMV associated CNS disease in AIDS. J Neurol Neurosurg Psych 62(4): 346–351.

Miralles P, Berenguer J, Garcia de Viedma D, et al (1998). Treatment of AIDS-associated progressive multifocal leukoencephalopathy with highly active antiretroviral therapy. AIDS 12(18): 2467–2472.

Monaco MC, Jensen PN, Hou J, et al (1998). Detection of JC virus DNA in human tonsil tissue: evidence for site of initial viral infection. J Virol 72(12): 9918–9923.

Morriss MC, Rutstein RM, Rudy B, et al (1997). Progressive multifocal leukoencephalopathy in an HIV-infected child. Neuroradiology 39(2): 142–144.

Myers C, Frisque RJ, Arthur RR (1989). Direct isolation and characterization of JC virus from urine samples of renal and bone marrow transplant patients. J Virol 63(10): 4445–4449.

Nelson PK, Masters LT, Zagzag D, et al (1999). Angiographic abnormalities in progressive multifocal leukoencephalopathy: an explanation based on neuropathologic findings. AJNR Am J Neuroradiol 20(3): 487–494.

O'Reilly S (1997). Efficacy of camptothecin in progressive multifocal leucoencephalopathy. Lancet 350(9073): 291.

O'Riordan T, Daly P, Hutchinson M, et al (1990). Progressive multifocal leukoencephalopathy: remission with cytarabine. J Infect Dis 20: 51–54.

Ormerod L, Rhodes R, Gross SA, et al (1996). Ophthalmologic manifestations of acquired immune deficiency syndrome-associated progressive multifocal leukoencephalopathy. Ophthalmology 103: 899–906.

Padgett BL, Walker DL (1983). Virologic and serologic studies of progressive multifocal leukoencephalopathy. Prog Clin Biol Res 105: 107–117.

Padgett BL, Walker DL, ZuRhein GM, et al (1971). Cultivation of papova-like virus from human brain with progressive multifocal leucoencephalopathy. Lancet 1(7712): 1257–1260.

Pho MT, Ashok A, Atwood WJ, et al (2000). JC virus enters human glial cells by clathrin-dependent receptor-mediated endocytosis. J Virol 74(5): 2288–2292.

Port JD, Miseljic S, Lee RR, et al (1999). Progressive multifocal leukoencephalopathy demonstrating contrast enhancement on MRI and uptake of thallium-201: a case report. Neuroradiology 41(12): 895–898.

Post MJ, Yiannoutsos C, Simpson D, et al (1999). Progressive multifocal leukoencephalopathy in AIDS: are there any MR findings useful to patient management and predictive of patient survival? AIDS Clinical Trials Group, 243 Team. AJNR Am J Neuroradiol 20(10): 1896–1906.

Querbes W, Benmerah A, Tosoni D, et al (2004). A JC virus-induced signal is required for infection of glial cells by a clathrin- and eps15-dependent pathway. J Virol 78(1): 250–256.

Rand KH, Johnson KP, Rubinstein LJ, et al (1977). Adenine arabinoside in the treatment of progressive multifocal leukoencephalopathy: use of virus-containing cells in the urine to assess response to therapy. Ann Neurol 1(5): 458–462.

Richardson EJ (1970). Progressive multifocal leukoencephalopathy. In: P Vinken, G Bruyn (Eds.), Handbook of Clinical Neurology, Vol. 9. Elsevier, New York, pp. 485–499.

Richardson EP Jr (1974). Our evolving understanding of progressive multifocal leukoencephalopatha. Ann N Y Acad Sci 230: 358–364.

Sacktor N (2002). The epidemiology of human immunodeficiency virus-associated neurological disease in the era of highly active antiretroviral therapy. J Neurovirol 8(Suppl 2): 115–121.

Sadler M, Chinn R, Healy J, et al (1998). New treatments for progressive multifocal leukoencephalopathy in HIV-1-infected patients. AIDS 12(5): 533–535.

Safdar A, Rubocki RJ, Horvath JA, et al (2002). Fatal immune restoration disease in human immunodeficiency virus type 1-infected patients with progressive multifocal leukoencephalopathy: impact of antiretroviral therapy-associated immune reconstitution. Clin Infect Dis 35(10): 1250–1257.

Samorei IW, Schmid M, Pawlita M, et al (2000). High sensitivity detection of JC-virus DNA in postmortem brain tissue by in situ PCR. J Neurovirol 6(1): 61–74.

Satishchandra P, Nalini A, Gourie-Devi M, et al (2000). Profile of neurologic disorders associated with HIV/AIDS from Bangalore, south India (1989–96). Indian J Med Res 111: 14–23.

Scaravilli F, Daniel SE, Harcourt-Webster N, et al (1989). Chronic basal meningitis and vasculitis in acquired immunodeficiency syndrome. A possible role for human immunodeficiency virus [see comments]. Arch Pathol Lab Med 113(2): 192–195.

Selik RM, Karon JM, Ward JW (1997). Effect of the human immunodeficiency virus epidemic on mortality from opportunistic infections in the United States in 1993. J Infect Dis 176(3): 632–636.

Silverman L, Rubinstein LJ (1965). Electron microscopic observations on a case of progressive multifocal leukoencephalopathy. Acta Neuropathol (Berl) 5(2): 215–224.

Singer C, Berger JR, Bowen BC, et al (1993). Akinetic-rigid syndrome in a 13-year-old girl with HIV-related progressive multifocal leukoencephalopathy. Mov Disord 8(1): 113–116.

Sparano JA, Anand K, Desai J, et al (1999). Effect of highly active antiretroviral therapy on the incidence of HIV-associated malignancies at an urban medical center. J Acquir Immune Defic Syndr 21(Suppl 1): S18–S22.

Steiger MJ, Tarnesby G, Gabe S, et al (1993). Successful outcome of progressive multifocal leukoencephalopathy with cytarabine and interferon. Ann Neurol 33(4): 407–411.

Stoner G, Ryschkewitsch C (1991). Evidence of JC virus in two progressive multifocal leukoencephalopathy (PML) brains previously reported to be infected with SC40. J Neuropathol Exp Neurol 50: 342.

Sweeney BJ, Manji H, Miller RF, et al (1994). Cortical and subcortical JC virus infection: two unusual cases of AIDS associated progressive multifocal leukoencephalopathy. J Neurol Neurosurg Psychiatry 57(8): 994–997.

Tada H, Lashgari M, Rappaport J, et al (1989). Cell type-specific expression of JC virus early promoter is determined by positive and negative regulation. J Virol 63(1): 463–466.

Taguchi F, Kajioka L, Miyamura T, et al (1982). Prevalence rate and age of acquisition of antibodies against JC virus and BK virus in human sera. Microbiol Immunol 26(11): 1057–1064.

Tamura T, Inoue T, Nagata K, et al (1988). Enhancer of human polyoma JC virus contains nuclear factor I-binding sequences; analysis using mouse brain nuclear extracts. Biochem Biophys Res Commun 157(2): 419–425.

Tantisiriwat W, Tebas P, Clifford DB, et al (1999). Progressive multifocal leukoencephalopathy in patients with AIDS receiving highly active antiretroviral therapy. Clin Infect Dis 28(5): 1152–1154.

Taoufik Y, Gasnault J, Karaterki A, et al (1998). Prognostic value of JC virus load in cerebrospinal fluid of patients with progressive multifocal leukoencephalopathy. J Infect Dis 178(6): 1816–1820.

Tashiro K, Doi S, Moriwaka F, et al (1987). Progressive multifocal leucoencephalopathy with magnetic resonance imaging verification and therapeutic trials with interferon. J Neurol 234(6): 427–429.

Thurnher MM, Post MJ, Rieger A, et al (2001). Initial and follow-up MR imaging findings in AIDS-related progressive multifocal leukoencephalopathy treated with highly active antiretroviral therapy. AJNR Am J Neuroradiol 22(5): 977–984.

Tominaga T, Yogo Y, Kitamura T, et al (1992). Persistence of archetypal JC virus DNA in normal renal tissue derived from tumor-bearing patients. Virology 186(2): 736–741.

Tooze J (1980). Human papovaviruses. In: J Tooze (Ed.), Molecular Biology of Tumor Viruses, Vol. 2. Cold Spring Harbor Laboratory, Cold Spring Harbor, NY.

Tornatore C, Berger J, Houft SA, et al (1992a). Detection of JC viral genome in the lymphocytes of non-PML HIV positive patients: association with B cell lymphopenia. Neurology 42(Suppl 3): 211.

Tornatore C, Berger JR, Houft JL, et al (1992b). Detection of JC virus DNA in peripheral lymphocytes from patients with and without progressive multifocal leukoencephalopathy. Ann Neurol 31(4): 454–462.

Van Assche G, Van Ranst M, Sciot R, et al (2005). Progressive multifocal leukoencephalopathy after natalizumab therapy for Crohn's disease. N Engl J Med 353(4): 362–368.

Van Horn G, Bastian F, Moake JL (1978). Progressive multifocal leukoencephalopathy: failure of response to transfer factor and cytarabine. Neurology 28: 794–797.

Verbraak FD, Boom R, Wertheim-van Dillen PM, et al (1999). Influence of highly active antiretroviral therapy on the development of CMV disease in HIV positive patients at high risk for CMV disease. Br J Ophthalmol 83(10): 1186–1189.

Vollmer-Haase J, Young P, Ringelstein EB (1997). Efficacy of camptothecin in progressive multifocal leucoencephalopathy. Lancet 349(9062): 1366.

Von Einsiedel RW, Fife TD, Aksamit AJ, et al (1993). Progressive multifocal leukoencephalopathy in AIDS: a clinicopathologic study and review of the literature. J Neurol 240 (7): 391–406.

Walker D, Padgett B (1983a). The epidemiology of human polyomaviruses. In: J Sever, , D Madden (Eds.), Polyomaviruses and Human Neurological Disease. Alan R. Liss, New York, pp. 99–106.

Walker D, Padgett B (1983b). Progressive multifocal leukoencephalopathy. In: H Fraenkel-Conrat, R Wagner (Eds.), Comprehensive Virology. Plenum, New York, pp. 161–193.

Walker DL, Padgett BL, ZuRhein GM, et al (1973). Human papovavirus (JC): induction of brain tumors in hamsters. Science 181(100): 674–676.

Weber F, Goldmann C, Kramer M, et al (2001). Cellular and humoral immune response in progressive multifocal leukoencephalopathy. Ann Neurol 49(5): 636–642.

Weber T, Turner R, Frye S, et al (1990). JC virus detected by polymerase chain reaction in cerebrospinal fluid of AIDS patients with progressive multifocal leukoencephalopathy. Neurological and neuropsychological complications of HIV infection. Proceedings of the Satellite Meeting of the International Conference on AIDS.

Weber T, Turner RW, Frye S, et al (1994). Progressive multifocal leukoencephalopathy diagnosed by amplification of JC virus-specific DNA from cerebrospinal fluid. AIDS 8 (1): 49–57.

Whiteman ML, Post MJ, Berger JR, et al (1993). Progressive multifocal leukoencephalopathy in 47 HIV-seropositive patients: neuroimaging with clinical and pathologic correlation. Radiology 187(1): 233–240.

Wolinsky JS, Johnson KP, Rand K, et al (1976). Progressive multifocal leukoencephalopathy: clinical pathological correlates and failure of a drug trial in two patients. Trans Am Neurol Assoc 101: 81–82.

Wong MC, Suite ND, Labar DR (1990). Seizures in human immunodeficiency virus infection. Arch Neurol 47(6): 640–642.

Wroblewska Z, Wellish M, Gilden D (1980). Growth of JC virus in adult human brain cell cultures. Arch Virol 65: 141–148.

Wyen C, Hoffmann C, Schmeisser N, et al (2004). Progressive multifocal leukoencephalopathy in patients on highly active antiretroviral therapy: survival and risk factors of death. J Acquir Immune Defic Syndr 37(2): 1263–1268.

Yiannoutsos CT, Major EO, Curfman B, et al (1999). Relation of JC virus DNA in the cerebrospinal fluid to survival in acquired immunodeficiency syndrome patients with biopsy-proven progressive multifocal leukoencephalopathy. Ann Neurol 45(6): 816–821.

Yogo T, Kitamura T, Sugimoto C, et al (1990). Isolation of a possible archetypal JC virus DNA sequence from nonimmunocompromised individuals. J Virol 64: 3139–3143.

Yogo Y, Kitamura T, Sugimoto C, et al (1991). Sequence rearrangement in JC virus DNAs molecularly cloned from immunosuppressed renal transplant patients. J Virol 65(5): 2422–2428.

ZuRhein G (1965). Particles resembling papovavirions in human cerebral demyelinating disease. Science 148: 1477–1479.

ZuRhein G (1967). Polyona-like virions in a human demyelinating disease. Acta Neuro Pathol (Berl) 8: 57–68.

Handbook of Clinical Neurology, Vol. 85 (3rd series)
HIV/AIDS and the Nervous System
P. Portegies, J. R. Berger, Editors

Chapter 14

Other opportunistic infections of the central nervous system in AIDS

BRUCE A. COHEN[1]* AND JOSEPH R. BERGER[2]

[1]*Department of Neurology, Feinberg School of Medicine, Northwestern University, Chicago, IL, USA*

[2]*Department of Neurology, University of Kentucky Medical Center, Lexington, KY, USA*

With the introduction of modern highly active antiretroviral therapy (HAART), and routine prophylaxis for common pathogens, the frequency of many HIV-associated neurological opportunistic infections (Table 14.1) has been found to decrease (Jellinger et al., 2000; Sactor et al., 2001; Vago et al., 2002). However, other series have not shown such widespread reductions (Masliah et al., 2000; Ives et al., 2001), and central nervous system (CNS) opportunistic infections continue to appear as presenting syndromes in previously unrecognized HIV-infected individuals, in those who fail to maintain a response to HAART because of the emergence of resistance and in those who for social and economic reasons lack access to modern therapeutics. In recent autopsy series from the HAART era, the brain has been found to be the second most commonly affected organ after the lung (Jellinger et al., 2000; Masliah et al., 2000). A French autopsy series suggests that the spectrum of CNS opportunistic infections may be changing with a shift toward agents such as varicella-zoster virus and herpes simplex virus (HSV) which can be seen in less severely immune-compromised patients, and re-emergence of other entities such as toxoplasmosis in new forms (Gray et al., 2003). Additionally, as attention turns to AIDS in under-developed countries with limited health systems, CNS opportunistic infections, some of which result from alternative pathogens endemic in these environments, are being recognized as important causes of HIV-associated morbidity and mortality. Consequently, discussion of CNS opportunistic infections remains relevant and timely in any comprehensive review of neuro-AIDS.

The most common neurologic opportunistic entities, toxoplasmosis, cryptococcus and progressive multifocal leukoencephalopathy are all reviewed in dedicated chapters elsewhere in this volume. This chapter addresses the broad category of other opportunistic processes in order to facilitate the recognition of these currently less common entities. The CNS has a limited repertoire of pathological expression, and the more severely immune-suppressed AIDS patient has a limited ability to mount an inflammatory response. Consequently the discussion of these entities will illustrate how different etiologies may yield similar clinical features as well as how CNS opportunistic infections may vary in disease features in AIDS patients when compared to immune-competent individuals.

14.1. Bacterial infections

14.1.1. Mycobacterium tuberculosis

Tuberculosis is the most common opportunistic infection and a leading cause of death in persons infected with HIV worldwide (Shafer and Edlin, 1996). CNS tuberculosis occurs in about 15% of HIV-infected patients with tuberculosis (Shafer et al., 1991). Early in the HIV pandemic, a study in south Florida found HIV-associated tuberculosis to be more common in those of Haitian and African-American ancestry with a 60% prevalence among the Haitian population (Pitchenik et al., 1984) and high rates of tuberculosis in HIV-infected populations can be found worldwide (Nunn and McAdam, 1988). Intravenous drug use appears to be a significant

*Correspondence to: Bruce A. Cohen, MD, Davee Department of Neurology, Feinberg School of Medicine, Northwestern University, 710 North Lake Shore Drive, Abbott Hall 1121, Chicago, IL 60611, USA. E-mail: bac106@northwestern.edu, Tel: +1-312-908-5631, Fax: +1-312-908-5073.

Table 14.1

Frequency of CNS opportunistic infection at the time of autopsy (Kure et al., 1991)

Cytomegalovirus	15.8%
Toxoplasmosis	13.6%
Cryptococcus	7.6%
Progressive multifocal leukoencephalopathy	4.0%
HSV encephalitis	1.6%
Candidiasis	1.1%
HZV encephalitis	0.6%
Histoplasmosis	0.4%
Tuberculosis	0.3%
Aspergillosis	0.3%

risk factor for HIV-associated tuberculosis (Sunderam et al., 1986; Lesprit et al., 1997).

14.1.1.1. Neuropathology

Primary infection with *M. tuberculosis* is usually acquired by inhalation which is followed by hematogenous dissemination to lymph nodes where a local inflammatory response results in the formation of granulomata and the bacteria are contained within membrane bound compartments in tissue macrophages. Spread to the CNS is thought to occur via initial hematogenous dissemination resulting in the formation of "Rich foci" which are small nodules within blood vessels containing inflammatory cells, and sometimes tuberculoma. Progression of these lesions extends from the subendothelial region and may ultimately result in vascular occlusion. When these vascular lesions rupture into the subarachnoid space, the bacteria are released resulting in meningitis. Growth of the tubercles without rupture may lead to mass lesions with or without abscess formation. Pathological examination of patients with tuberculous meningitis (TBM) reveals a gelatinous exudate composed of inflammatory cells and erythrocytes with focal areas of caseous necrosis, extending along perivascular channels into the parenchyma. The exudate is most abundant at the base of the brain and commonly obstructs cerebrospinal fluid (CSF) flow resulting in ventricular dilation and hydrocephalus. Findings may be different in some HIV-infected patients, reflecting their diminished ability to mount an effective inflammatory response. In a series from India, patients with HIV-associated TBM were found to have thinner serous exudates without tubercles. In the absence of a granulomatous response, organisms were found in large numbers in cerebral tissue. Despite the absence of occlusive exudates surrounding large vessels at the base of the brain, vasculitis with infarctions was common (Katrack et al., 2000). Similarly diminished granulomatous responses with endarteritis and infarctions were noted in a small South African series (Schutte, 2001).

The neurological complications of tuberculosis are summarized in Table 14.2.

14.1.1.2. Clinical manifestations of HIV-associated CNS tuberculosis

The most common CNS manifestation of infection with *M. tuberculosis* irrespective of associated HIV infection is meningitis (Berenguer et al., 1992). Co-infection with HIV has been associated with increased risk of meningitis compared to patients with non-HIV-associated tuberculosis (Berenguer et al., 1992; Rana et al., 2000) although not all series have found this to be the case (Shafer et al., 1991; Taelman et al., 1992). The clinical features of TBM appear to be similar irrespective of HIV co-infection and consist of fever, headache and altered mentation evolving over a period ranging from days to months, although most commonly progressing over weeks. Cranial nerves are commonly involved; most frequently the sixth followed by the third, fourth, seventh, second, eighth, tenth, eleventh and twelfth (Lincoln et al., 1960; Traub et al., 1984). Meningismus is seen in one-half to two-thirds of patients. Focal signs are found in about 20% and seizures in 3–18%. About half of patients with TBM will have preceding prodromal symptoms potentially attributable to tuberculosis, including malaise, anorexia, weight loss, myalgias

Table 14.2

The neurological complications of tuberculosis

- Meningeal
  - Purulent tuberculous meningitis
    1. Disseminated miliary tuberculosis
    2. Focal caseating plaques
    3. Inflammatory caseous meningitis
    4. Proliferative meningitis
  - Serous tuberculous meningitis
- Cerebral
  - Tuberculous encephalopathy
  - Tuberculoma
  - Tuberculous brain abscess
  - Cerebral infarction
  - Tuberculous of the skull
- Spinal
  - Spinal tuberculoma
  - Tuberculous spinal abscess
  - Tuberculous spinal osteomyelitis
  - Necrotizing myelopathy
  - Tuberculous radiculomyelitis

and fatigue. Nausea and emesis occur in one-fourth to one-half of cases (Dube et al., 1992; Berenguer et al., 1992; Kent et al., 1993; Sanchez-Portocarrero et al., 1999; Katrack et al., 2000). A polyradiculitis resulting in subacute paraparesis, radicular pain and bladder disturbances has also been described with tuberculous meningitis (Hernandez-Albujar et al., 2000). A staging system for the severity of tuberculous meningitis has been devised (Table 14.3).

Mass lesions seen in HIV-associated CNS tuberculosis include both granulomatous tuberculomas and abscesses (Bishburg et al., 1986; Velasco-Martinez et al., 1995; Lesprit et al., 1997) and it has been suggested that intracerebral lesions may be more common in HIV-associated CNS tuberculosis (Bishburg et al., 1986; Dube et al., 1992). Presenting features do not differ from other space occupying CNS lesions and may include headache and fever, seizures, limb weakness, congnitive impairment, impaired sensorium or cognition, and diplopia (Bishburg et al., 1986; Lesprit et al., 1997; Minagar et al., 2000; Thonell et al., 2000). Intramedullary tuberculous lesions resulting in myelopathy and paraparesis (Doll et al., 1987; Mohit et al., 2004), and obstructive myelopathic symptoms resulting from TBM (Vlcek et al., 1984), are less common presentations. Tuberculous epidural abscesses have been reported (Gupta et al., 1989; Gettler and El-Sadr, 1993).

14.1.1.3. Diagnosis of HIV-associated CNS tuberculosis

14.1.1.3.1. Imaging studies

A triad of findings including basal meningeal enhancement, hydrocephalus and infarctions involving both cerebral and brainstem parenchyma are suggestive of TBM (Bernaerts et al., 2003). The most common abnormality is ventricular enlargement and hydrocephalus which is usually of the communicating type and is due to obstruction of CSF resorption by the basal meningeal exudates. Occasionally, obstructive hydrocephalus due to narrowing of the aqueduct of Sylvius or resulting from a mass lesion obstructing ventricular outflow may be seen. Enlarged ventricles may be found in 5–50% of patients and hydrocephalus is more common in children with TBM. The basal cisterns may be obliterated on non-contrast CT; however, such scans may be normal in 20–33% of patients with TBM (Berenguer et al., 1992; Sanchez-Portocarrero et al., 1999; Katrak et al., 2000; Schutte, 2001; Bernaerts et al., 2003).

Meningeal enhancement is seen in 15–29% by CT (Berenguer et al., 1992; Sanchez-Portocarrero et al., 1999; Katrak et al., 2000; Schutte, 2001) but up to 60% with MRI. Early in the course of TBM, meningeal enhancement may be the only abnormality on MRI. Some studies have suggested that meningeal enhancement is less likely in HIV-coinfected patients due to diminished immune response (Katrack et al., 2000) while others have found no significant imaging differences between HIV-coinfected and non-HIV-infected patients with TBM (Villoria et al., 1992). Enhancement may be prominant in the basal cisterns where sites of predilection include the interpeduncular fossa, the ambient cistern and the chiasmatic region (Whiteman, 1997).

Cerebral infarctions may be seen by imaging studies in about one-fourth to one-third of cases (Offenbacher et al., 1991; Berenguer et al., 1992; Villoria et al., 1992; Whiteman, 1997). A panarteritis with secondary thrombosis and occlusion is responsible for most of the infarctions which, due to encasement of penetrating arteries by the basal meningitic process, occur most commonly in the basal ganglia and internal capsule. Larger arteries are involved less commonly and may result in hemorrhagic infarction. Conventional angiography reveals narrowing of arteries at the base of the brain, narrowed and occluded small and medium sized vessels and early draining veins (Whiteman, 1997; Bernaerts et al., 2003).

Mass lesions associated with HIV-associated cerebral tuberculosis may occur in 15–44% of patients (Whiteman, 1997). They may be ring or nodular enhancing lesions, or show mixed patterns. Both multiple and single lesions occur. Multiloculated abscesses may be seen in up to one-fourth of patients. Edema and associated meningeal enhancement may also be present (Lesprit et al., 1997). Tuberculomas have a predilection to involve corticomedullary and periventricular regions reflecting hematagenous dissemination, but may involve cerebrum, cerebellum, subarachnoid, subdural and epidural spaces (Villoria et al., 1995; Whiteman,

Table 14.3

British Medical Research Council clinical staging of tuberculous meningitis (Thwaites and Hien, 2005)

Stage I (early):
Nonspecific symptoms and signs
Consciousness undisturbed
No neurological signs
Stage II (intermediate):
Consciousness disturbed but not comatose nor delirious
Signs of meningeal irritation
Minor focal neurological signs, e.g. cranial nerve palsies
Stage III (advanced):
Seizures
Abnormal movements
Stupor or coma
Severe neurological deficits, e.g. paresis

1997). The radiologic appearance is related to the pathological patterns. A noncaseating granuloma is isodense or slightly dense on non-contrast CT and enhances homogenously after contrast infusion. On MRI, the lesion is hypointense on T1 and hyperintense on T2 with homogenous enhancement after contrast infusion on T1 sequences. Caseating granulomas with solid centers enhance heterogeneously in the center with uniformly thick ring enhancement of the capsule. These lesions are hypointense or isointense on T1 sequences and isointense to hypointense on T2. Lesions with central liquefaction demonstrate a hypodense core on CT or T1 MRI surrounded by enhancement of the rim which may be indistinguishable from pyogenic abscesses (Whiteman, 1997; Bernaerts et al., 2003). Radiographic clues suggested to favor CNS tuberculosis over primary CNS lymphoma or toxoplasmosis include a multiloculated abscess, cisternal enhancement, basal ganglia infarction and communicating hydrocephalus (Whiteman et al., 1995). A target sign described as a central nidus of calcification surrounded by a ring of enhancement has been suggested to be pathognomonic of a CNS tuberculoma (Bargallo et al., 1996).

Involvement of the spinal axis is common in TBM and may result from exudates filling the spinal subarachnoid space causing adhesive radiculitis, infarctions of the spinal cord or nerve roots due to inflammation of spinal arteries, intramedullary tuberculomas or abscesses, or epidural abscesses extending contiguously from vertebral or perispinal infection. Findings on MRI include loss of the definition between spinal cord and subarchnoid space, clumping of nerve roots, loculated collections of spinal fluid and enhancement of nerve roots, spinal cord or epidural inflammation after contrast infusion (Whiteman, 1997; Bernaerts et al., 2003).

Chest X-rays will be abnormal in up to 60% of patients with TBM (Berenguer et al., 1992; Sanchez-Portocarrero et al., 1999; Schutte, 2001) but may be of limited value in areas where a high degree of prevalence for previously acquired infection exists. The presence of military tuberculosis, however, is highly suggestive of multiorgan involvement (Thwaites and Hien, 2005).

14.1.1.3.2. Cerebrospinal fluid

CSF analysis is the most useful test in the diagnosis of TBM and will be abnormal in the majority of cases. Typical features include an elevated opening pressure, and a pleocytosis numbering in the hundreds which may be predominantly lymphocytic, neutrophilic or mixed. Cases of acellular CSF are noted in most large HIV-associated series (Dube et al., 1992; Berenguer et al., 1992; Sanchez-Portocarrero et al., 1999) and may obscure the diagnosis in life (Laguna et al., 1992). Protein levels are typically elevated and may exceed 1000 mg/dl with xanthochromia; however, levels were normal in 43% of one large series (Berenguer et al., 1992). CSF glucose levels are classically diminished but may be normal.

Definitive diagnosis requires demonstration of either acid-fast bacteria on CSF smears or cultures, or demonstration of antigenic or genetic markers unique to *M. tuberculosis*.

While CSF smears are rarely positive in most clinical settings, repeated sampling may improve the yield (Kennedy and Fallon, 1979) and meticulous laboratory technique with large volumes of CSF has recently been reported to yield a bacteriologic diagnosis in 81% of cases (Thwaites et al., 2004a). Series of HIV patients with TBM selected by the presence of positive CSF cultures have reported positive smears in 22–27% of cases (Berenguer et al., 1992; Dube et al., 1992). Because CSF cultures take weeks to grow and have a sensitivity of only about 70% (Thwaites et al., 2004a), efforts to develop more rapid assays have focused on immunologic and molecular diagnostic techniques. Polymerase chain reaction amplification (PCR) offers the potential for highly specific rapid diagnosis (Folgueira et al., 1994; Scarpellini et al., 1995); however, a recent meta-analysis of commercial tests found a sensitivity of only 56% (Pai et al., 2003). Sensitivity appears to be higher in patients with culture proven TBM (Chedore and Jamieson, 2002). Testing of CSF for antibodies to a panel of specific *M. tuberculosis* antigens yielded a sensitivity of 87% for clinically diagnosed TBM and 72% for culture-positive TBM in HIV-uninfected individuals, but a lower sensitivity of 70% for clinically diagnosed TBM in HIV-coinfected patients (Chandramuki et al., 2002). Testing for tubulostearic acid, a component of the mycobacterial cell wall, has been reported to have a sensitivity of over 90% (French et al., 1987). CSF adenosine deaminase levels have been reported to be highly specific for TBM (Ribera et al., 1987; Lopez-Cortes et al., 1995); however, in a series of HIV-coinfected patients, false positive results with neurological cytomegalovirus, cryptococcal and lymphomatous disease reduced specificity to 87% with a sensitivity of 57% (Corral et al., 2004).

14.1.1.4. Treatment of HIV-associated CNS tuberculosis

Current recommendations regarding treatment of TBM employ four agents for initial therapy including isoniazid (INH), rifampin, pyrazinamide and streptomycin, ethambutol or ethionamide. Two or three agents are typically continued for 9–12 months while the others are discontinued after 2 months. INH is dosed at 5 mg/kg body weight up to a maximum of 300 mg/day, and rifampin at 10 mg/kg to a maximum of 600 mg/day. The dose of pyrazinamide is 1000 mg/day for patients with body weight of 55 kg or less, 1500 mg/day for those with weights of 56–75 kg and 2000 mg/day for those weighing more than 75 kg. For ethambutol, the doses corresponding to these weights are 800, 1200 and 1600 mg/day, respectively. Streptomycin is dosed at 15 mg/kg up to 1000 mg/day (Benson et al., 2004; Portegies et al., 2004; Thwaites and Hien, 2005). Pyridoxine must be given in addition to INH to prevent toxic peripheral neuropathy. Hepatotoxicity may complicate any of these agents and patients on ethambutol must be monitored for a toxic optic neuropathy. Streptomycin can cause otologic and vestibular neuropathy. Unfortunately, strains of *M. tuberculosis* resistant to multiple antibiotics have emerged complicating treatment (Horn et al., 1994; Patel et al., 2004). This is a particular problem for treatment of HIV-associated CNS tuberculosis since many affected patients have previously been treated for pulmonary tuberculosis (Silber et al., 1999). A recent report of recurrent tuberculosis with the same strain in a child with HIV and TBM, almost two years following treatment under observation, may indicate a need for longer courses of therapy in this setting (Schaaf et al., 2000).

Adjunctive steroid therapy for patients with TBM is controversial. A recent controlled study of dexamethasone in Vietnamese adults showed benefit on survival but not on the outcome of severe disability or the combined endpoint of death and severe disability. In a subgroup analysis, patients coinfected with HIV failed to benefit on either endpoint (Thwaites et al., 2004b).

Individuals with cerebral tuberculomas or abscesses may also respond to medical therapy alone (Thonell et al., 2000). Diagnosis can often be made with low risk by stereotactic biopsy (Lesprit et al., 1997); however, empiric therapy has been advocated in settings where *M. tuberculosis* is endemic, particularly when serology to *Toxoplasma gondii* is negative. When toxoplasmic serology is positive in such a setting, empiric therapy for both agents has been recommended (Modi et al., 2004). Radiologic response to therapy may be seen within 4–6 weeks (Bernaerts et al., 2003).

14.1.2. Atypical mycobacteria

Non-tuberculous mycobacterial infections are common in HIV-infected individuals typically involving the lungs, but often disseminating to other organs. The most common agent seen is *M. avium-intracellulare* (MAI), but a variety of other strains have also been found (Nunn and McAdam, 1988). MAI has been reported to occur in 15–20% of AIDS patients in life and up to 50% postmortem (Hawkins et al., 1986). Constitutional symptoms include fever, malaise, night sweats, generalized weakness and weight loss (Hawkins et al., 1986; Young et al., 1986) and bacteremia is common (Whimbey et al., 1986). Animal studies have demonstrated CNS infection following intravenous inoculation of MAI (Wu et al., 2000).

The CNS is uncommonly involved, but when it is, the most common manifestations are brain abscess and septic meningitis. Clinical features are not etiologically distinctive and most reports consist of isolated cases or small series. A variety of species have been associated with CNS infection. Brain abscess and meningitis have been described with MAI (Jacob et al., 1993; Dwork et al., 1994; Dickerman et al., 2003; Lascaux et al., 2003), *M. kansaai* (Haas et al., 1977; Gordon and Blumberg, 1992; Bergen et al., 1993), *M. triplex* (Zeller et al., 2003), *M. genavense* (Berman and Kim, 1994) and *M. fortuitum* (Jacob et al., 1993). On occasion, brain abscess due to atypical mycobacteria have been reported in the absence of apparent systemic manifestations (Berman and Kim, 1994; Dickerman et al., 2003). A patient successfully treated for systemic MAI for three years presented with a late cerebral relapse after stopping effective antiretroviral therapy with subsequent recurrence of immune suppression. Genotyping of the isolates from both episodes confirmed that they were identical. The patient responded to re-initiation of antimicrobial therapy with rifabutin, ethambutol and clarithromycin which was followed by resumption of HAART (Lascaux et al., 2003).

Initial treatment of mycobacterium avium complex includes clarithromycin 500 mg twice daily and ethambutol as for mycobacterium tuberculosis. Rifabutin 300 mg/day may be added for paitents with CD^4 counts less than 50, but the dosage must be adjusted when used with HAART depending on the particular agents used (see Benson et al., 2004). Alternative agents for severe infections include ciprofloxacin 500–750 mg/day, levofloxacin 500 mg/day or amikacin 10–15 mg/kg/day. Azithromycin 500–600 mg/day may be substituted for clarithromycin (Benson et al., 2004).

14.1.3. Syphilis

Syphilis is caused by the bacterium *Treponema pallidum*, a long, slender, coil-shaped organism that measures 6–15 μm in length, but only 0.15 μm in width, a dimension below the resolution of light microscopy. The organism has regular spirals numbering 5 to 20 and is actively motile using a rotational screw-like activity, flexion and back-and-forth motion. Syphilis is a sexually transmitted disease. Shortly after infection, a spirochetemia results with dissemination of *T. pallidum* to virtually any organ, including the CNS. Both humoral and cellular immunity play a role in the ensuing infection. Antibodies to *T. pallidum* are detectable within 10 to 21 days of infection. The humoral response is of little consequence in containing the infection; however, cellular immunity appears to be effective in controlling the infection as evidenced by immunity during rechallenge with the degree of protection directly proportional to the extent of the response. Impairment of cellular immunity due to drugs, pregnancy, AIDS, etc., results in a more aggressive syphilitic infection than otherwise anticipated.

14.1.3.1. Clinical manifestations of syphilis

Infection with *T. pallidum* is divided into several stages. Primary syphilis, the initial manifestation of infection occurring approximately 3 weeks (3–90 days) after inoculation, is generally heralded by an ulcerated, painless lesion with firm borders referred to as a chancre. This lesion develops at the site of epidermal or mucous membrane inoculation and is accompanied by regional adenopathy. Secondary syphilis, attributable to a bacteremic phase, typically appears within 2 to 8 weeks of the appearance of the chancre. It is characterized by a macular, maculopapular or pustular rash that often involves the palms and soles; mucous patches; alopecia; constitutional signs; diffuse adenopathy; iridocyclitis; hepatitis; periostitis; and arthritis. Symptomatic aseptic meningitis may occur in up to 5% of immunologically normal persons and is more common with HIV infection. Although the majority of patients with CSF abnormalities occurring in association with secondary syphilis are neurologically asymptomatic, approximately 5% of all patients with secondary syphilis will have associated meningitis. Headaches, meningismus, impaired vision, cranial nerve palsies (chiefly, in descending order of frequency, VII, VIII, VI and II), hearing loss, tinnitus and vertigo may be observed in isolation or combination in upwards of 40% of patients with secondary syphilis. The symptoms of syphilitic meningitis (Fig. 14.1) include headache, photophobia and a stiff neck. Encephalopathic features resulting from vascular compromise or increased intracranial pressure may be observed. Acute sensorineural hearing loss and acute optic neuritis may occur in association with syphilitic meningitis or independently.

Latent syphilis is a quiescent phase that is divided into early (within 2 years of infection) when contagious syphilis is more likely to recrudesce and late (longer than 2 years). It may be followed by tertiary syphilis which is characterized chiefly by skin, osseous, cardiovascular and neurologic complications. Clinically apparent neurologic complications of tertiary syphilis affect less than

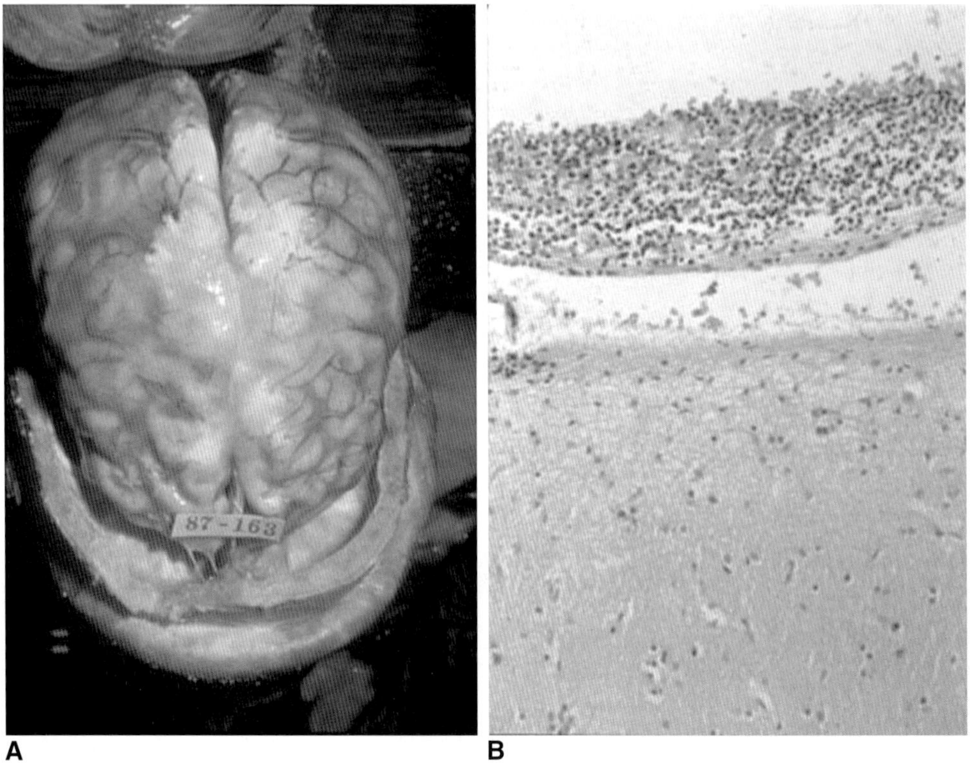

Fig. 14.1. For full color figure, see plate section. Syphilitic meningitis. A. Gross pathology of syphilitic meningitis with opacification of the meninges. B. Histopathological examination showing lymphocytic infiltration of the meninges.

10% of untreated patients. The common neurological complications of tertiary syphilis include meningovascular syphilis, general paresis (a progressive dementing disorder) and spinal cord syphilis, although a large variety of abnormalities may be seen resulting from effects of the organism on the osseous structures, meninges, brain and spinal cord parenchyma, cranial nerves, dorsal roots or cerebral blood vessels.

14.1.3.2. Effect of concomitant HIV infection

Concomitant human immunodeficiency virus infection may significantly alter the natural history of neurosyphilis (Johns et al., 1987; Katz and Berger, 1989; Katz et al., 1993). Syphilis appears to be not only more aggressive but also more difficult to treat when it occurs in association with HIV infection (Berry et al., 1987; Musher et al., 1990; Katz et al., 1993). These observations suggest that the host's immune response is critical in controlling this infection. The inability of the HIV-infected patient to establish delayed hypersensitivity to *T. pallidum* may prevent secondary syphilis from evolving to latency or may cause a spontaneous relapse from a latent state. This impairment of delayed hypersensitivity may account for a more rapid progression of neurosyphilis in HIV-infected individuals than would otherwise be expected. *T. pallidum* can be isolated from the CSF of HIV-seropositive patients with primary, secondary and latent syphilis following currently Centers for Disease Control-recommended penicillin therapy (Lukehart et al., 1988). Despite the associated immunosuppression, serum nontreponemal titers at the time of presentation of neurosyphilis in the HIV-infected individual are typically high, averaging 1:128 (Flood et al., 1998).

In HIV infection, acute, symptomatic, syphilitic meningitis during the course of secondary syphilis is not uncommon. A decrease in the latent period to the development of some neurosyphilitic manifestations, such as meningovascular syphilis and general paresis, has been suggested. The development of meningovascular syphilis within 4 months of primary infection despite the administration of accepted penicillin regimens (Johns et al., 1987), as well as the neurologic relapse of syphilis in HIV-infected individuals after appropriate doses of benzathine penicillin for secondary syphilis (Berry et al., 1987), has been reported. Other unusual manifestations of syphilis that have been reported in association with HIV infection include unexplained fever (Chung et al., 1983), bilateral optic neuritis with blindness (Zambrano et al., 1987), Bell palsy, severe bilateral sensorineural hearing loss (Fernandez-Guerrero et al., 1988), syphilitic meningomyelitis (Berger, 1992), syphilitic polyradiculopathy (Lanska et al., 1988) and syphilitic cerebral gumma presenting as a mass lesion (Berger et al., 1992).

14.1.3.3. Diagnosis of neurosyphilis

There are two categories of serological study: (1) nontreponemal tests that are flocculation tests using cardiolipin, lecithin and cholesterol as antigen; and (2) treponemal tests, which rely on specific treponemal cellular components as antigens. Nontreponemal tests include the Venereal Disease Research Laboratory test, rapid plasma reagin, Wasserman and Kolmer. The treponemal tests include the fluorescent treponemal antibody absorption test, microhemagglutination assay, hemagglutination treponemal test for syphilis and the treponemal immobilization test.

Unfortunately, there is no readily applicable "gold standard" for the diagnosis of neurosyphilis. Culturing the organism from the CSF is cumbersome and available in few laboratories. Furthermore, the fragility of *T. pallidum* may result in a low sensitivity of the test. The presence of a reactive Venereal Disease Research Laboratory test in the CSF is specific, with rare reports of false positives, but the test is not sufficiently sensitive to exclude the diagnosis of neurosyphilis on the basis of a negative study. The serum Venereal Disease Research Laboratory test is positive in 72% of patients with primary syphilis, nearly 100% of patients with secondary syphilis, 73% of patients with latent syphilis and 77% of patients with tertiary syphilis. Therefore, as many as one-quarter of patients with neurosyphilis are anticipated to have a negative serum Venereal Disease Research Laboratory test. Its frequency of reactivity appears to vary with the clinical form of neurosyphilis and its presence in asymptomatic neurosyphilis may be substantially lower than in symptomatic disease. The CSF Venereal Disease Research Laboratory test is too insensitive to be relied on to exclude the diagnosis of neurosyphilis. Therefore, other measures must be used to supplement the establishment of the diagnosis. The frequency with which the CSF Venereal Disease Research Laboratory test is negative in the presence of neurosyphilis is not known, but has been estimated to exceed 25%. In many respects, neurosyphilis is a diagnosis established on clinical grounds. To date, there is no consensus regarding diagnostic criteria and the physician should probably refrain from rigid adherence to narrow guidelines in making the diagnosis.

A cardinal requirement for the diagnosis of neurosyphilis is a reactive serum treponemal test. Neurosyphilis should be diagnosed in anyone with serologies reactive for a treponemal test occurring in association with a reactive CSF Venereal Disease Research Laboratory test. A diagnosis of neurosyphilis should be considered in patients with serological evidence of syphilis and one or more of the following abnormalities in their CSF: a mononuclear pleocytosis, an elevated protein, increased immunoglobulin G or the presence of oligoclonal bands. Undoubtedly, neurosyphilis is overdiagnosed using these criteria. The CSF fluorescent treponemal antibody absorption test has been suggested as a sensitive screening test. Unlike the CSF Venereal Disease Research Laboratory test, which requires gross blood contamination of the CSF to be rendered falsely positive, small amounts of blood contamination of the CSF may give false positive tests with the fluorescent treponemal antibody absorption test. Furthermore, the fluorescent treponemal antibody absorption test is dependent on immunoglobulin G antibody that may cross the blood–brain barrier to result in a false-positive test for neurosyphilis. The CSF fluorescent treponemal antibody-immunoglobulin G test has been suggested as an alternative to avoid the latter possibility. Coinfection with HIV considerably complicates the interpretation of CSF abnormalities as a mononuclear pleocytosis, increased protein, increased immunoglobulin G and the presence of oligoclonal bands may all attend HIV infection in the absence of neurosyphilis (Hollander, 1988). A schema has been proposed for diagnosing neurosyphilis in the face of HIV infection:

Definite neurosyphilis:

1. + blood treponemal serology, e.g. FTA-ABS, MHA-TP, etc.
2. + CSF VDRL

Probable neurosyphilis:

1. + blood treponemal serology
2. − CSF VDRL
3. CSF mononuclear pleocytosis (>20 cells/mm^3). Or + CSF protein (>60 mg/dL). Neurologic complications compatible with neurosyphilis, such as cranial nerve palsies, stroke, etc., or evidence of ophthalmological syphilis

Possible neurosyphilis:

1. + blood treponemal serology
2. − CSF VDRL
3. CSF mononuclear pleocytosis (>20 cells/mm^3). Or + CSF protein (>60 mg/dL). No neurologic or ophthalmological complications compatible with syphilis

Although not diagnostic of neurosyphilis, radiologic studies may be suggestive and are certainly helpful in excluding other pathologies. Radiologic manifestations of neurosyphilis include meningeal enhancement, nonspecific white matter abnormalities, hydrocephalus, gummas, periostitis, generalized cerebral atrophy and infarction (Brightbill et al., 1995; Good and Jager, 2000).

14.1.3.4. Treatment

The treatment regimen for neurosyphilis should be 12 million units to 24 million units of crystalline aqueous penicillin administered intravenously daily (2 million units to 4 million units every 4 hours) for a period of 10 to 14 days. This regimen generally requires hospitalization, but a prolonged hospitalization may be avoided in some reliable, well-motivated patients by placement of an indwelling catheter and home administration of penicillin after the first 24 to 48 hours of therapy. Penicillin should be administered at 4-hour intervals to maintain the penicillin levels consistently at or above treponemicidal values and to avoid the subtherapeutic troughs that occur when it is administered at less frequent intervals. An alternative approach to the use of parenteral penicillin is the daily oral administration of amoxicillin (3.0 g) and probenecid (0.5 g) administered twice daily for 15 days (Faber et al., 1983). This regimen achieves treponemicidal levels of amoxicillin in the CSF (Faber et al., 1983; Morrison et al., 1985). The Center for Disease Control has recommended that the initial therapy of intravenous aqueous penicillin be followed in HIV-infected individuals by weekly intramuscular injections of 2.4 million units of benzathine penicillin for 3 weeks. However, in light of lack of treponemicidal levels of penicillin with the latter and evidence that high-dose penicillin regimens are not consistently effective in patients infected with HIV (Gordon et al., 1994), a more logical course may be the administration of a 30-day course of doxycycline, 200 mg twice daily, following the completion of intravenous therapy. Although secondary prophylaxis is extensively employed, further studies are warranted before secondary prophylaxis (or some permutation of it) can be broadly recommended. HIV-seropositive patients should be carefully monitored for relapse of neurosyphilis for 2 or more years following initial treatment.

14.1.4. Bartonella

Bartonella henselae (formerly *Rochalimaea henselae*) is a Gram-negative bacillus known to cause disease following bites or scratches from cats. Domestic cats are a reservoir for this agent (Regnery et al., 1995). Common symptoms include painful lymphadenitis and less frequently retinitis (Wong et al., 1995). Neurologic manifestations described in immune-competent patients include retinitis, cerebral arteritis, myelitis, encephalitis, aseptic meningitis and dementia (Wong et al., 1995; George et al., 1998). In HIV-coinfected patients, bacillary angiomatosis (Spach et al., 1992), aseptic meningitis (Wong et al., 1995) and meningoencephalitis with multifocal enhancing mass lesions (George et al., 1998) have been described. Antibiotic therapy with erythromycin (Spach et al., 1992) and a combination of doxycycline and rifampin (George et al., 1998) was successful in resolving neurologic deficits.

In a series of HIV-infected patients from Los Angeles, Schwartzman found a seroprevalence rate for *B. henselae* of 32% in those with neurologic manifestations compared to 4% in HIV-infected subjects without neurologic disease. CSF IgG antibodies to *B. henselae* were found in 26% of this group. The neurologic manifestations included acute and subacute dementia, encephalitis, optic neuritis, chronic headache and myelitis (Schwartzman et al., 1994). In a sample of 369 HIV-seropositive individuals participating in a prospective longitudinal HIV natural history study, IgM antibodies to *B. henselae* were associated with a 1.7-fold increased risk of neuropsychological decline or dementia over a five-year follow-up period and was also associated with cat ownership (Schwartzman et al., 1995). In contrast, a more recent report from another longitudinal natural history cohort found similar seropositivity rates for *B. henselae* among both HIV-infected and uninfected subjects. Similar rates of neurocognitive decline occurred in *B. henselae* seropositive subjects regardless of HIV status, and also in a handful of patients in whom seroconversion to *B. henselae* was documented, thereby failing to demonstrate an association of cognitive decline to HIV, and arguing against a causative role for the agent in HIV-dementia (Wallace et al., 2001). Nonetheless, the possibility of *B. henselae* as a co-morbid or opportunistic cause of reversible neurological morbidity in HIV-infected patients should be considered in relevant clinical settings.

14.1.5. Nocardia

Nocardia is a Gram-positive bacterium found in soil which is most often acquired by inhalation or less frequently by percutaneous inoculation. The lung is most commonly involved; however, CNS involvement has been described in about 20% of HIV-associated cases. Most of these cases result from *N. asteroides* and result in cerebral abscesses with or without accompanying meningitis (Sharer and Kapila, 1985; Adair et al., 1987; Kim et al., 1991; Javaly et al., 1992; Uttamchandani et al., 1994). On occasion, meningitis may be seen without abscess formation (Khorrami and Heffeman, 1993). An apparent association between nocardial infection and co-morbid mycobacterial infection has been noted in both HIV-infected and non-HIV-infected patients (Kim et al., 1991; Javaly et al., 1992). A number of reported HIV-infected individuals with nocardia infections have been intravenous drug users (Kim et al., 1991; Javaly et al., 1992; Pintado et al., 2003) suggesting inadvertent self-inoculation as a potential route of entry.

Presenting features in most cases are nonspecific. A large review of nocardial brain abscesses in both immune suppressed and non-immune suppressed patients noted focal deficits in 42%, seizures in 30% and nonfocal findings in 27% (Mamelak et al., 1994). Nonfocal presentations may include fever, headache, changes in sensorium, malaise and generalized weakness (Kim et al., 1991; Javaly et al., 1992; Khorrami and Heffeman, 1993). When meningitis is associated, the CSF often yields a mixed pleocytosis which may show a neutrophilic predominance, elevated protein and normal or depressed glucose levels (Javaly et al., 1992; Khorrami and Heffeman, 1993; Minamoto and Sordillo, 1998). The presence of filamentous, beaded, branching, Gram-positive rods on Gram-stained CSF or other body fluid should suggest the diagnosis. Cultures require prolonged growth and isolation may be difficult when other bacteria are present. On selective Lowenstein–Jensen media, colonies resemble atypical mycobacteria except that the former organisms are more strongly acid-fast and do not demonstrate branching on smears (Javaly et al., 1992). Blood or sputum cultures may be positive (Ogg et al., 1997; Minamoto and Sordillo, 1998). Treatment options include sulfonamides, minocycline, amikacin and third-generation cephalosporins. In vitro studies demonstrate efficacy for imipenem and amikacin, and synergistic suppression with imipenem and trimethoprim-sulfamethoxazole, imipenem and cefotaxime, and amikacin and trimethoprim-sulfamethoxazole (Javaly et al., 1992; Khorrami and Heffeman, 1993; Ogg et al., 1997).

Although excision and drainage of cerebral nocardial abscesses is often recommended (Mamelak et al., 1994), resolution with only antibiotic therapy using amikacin 15 mg/kg/day, imipenem 500 mg three times daily and ceftriaxone 4000 mg/day (Ogg et al., 1997) or sulfadiazine 1500 mg every six hours, minocycline 200 mg twice daily and imipenem 500 mg every six hours (Kim et al., 1991) has been reported. Recurrence following successful antibiotic therapy has also been described, possibly in relation to short treatment courses (Kim et al., 1991) or discontinuation of both antibiotic and antiretroviral therapy (Ogg et al., 1997).

14.1.6. Listeria

Listeria monocytogenes is known to cause meningitis in immune-suppressed individuals and has been noted to occur with increased frequency in HIV-infected populations. In a study from Atlanta, 19% of all cases of invasive listeriosis over a 2-year period occurred in HIV-infected individuals, resulting in rates of 65–145 times higher than the general population for HIV-infected and AIDS patients respectively. All of the HIV-associated cases met the current definition of AIDS (defining illness or CD^4 count <200/μl) and two of the patients were infants (Jurado et al., 1993). Similarly, an increased incidence of listeriosis in three New York medical centers was attributed to increased incidence of HIV infection (Kales and Holzman, 1990). In addition to meningitis, brain abscess due to *L. monocytogenes* has also been reported (Harris et al., 1980; Patey et al., 1989; Berenguer et al., 1991; Jurado et al., 1993). Neurologic presentations include fever, headache, nausea, emesis, neck stiffness, weakness and cognitive changes and seizures (Mascola et al., 1988; Berenguer et al., 1991; Decker et al., 1991). CSF typically reveals a neutrophilic predominant pleocytosis with low sugar and elevated protein (Benninger et al., 1988; Decker et al., 1991); however, initial CSF studies may be normal (Decker et al., 1991). Treatment with ampicillin with or without added gentamycin may be successful (Gould et al., 1986; Berenguer et al., 1991; Decker et al., 1991). Trimethoprim-sulfamethoxazole has also been reported to be effective (Spitzer et al., 1986; Berenguer et al., 1991).

14.2. Viral opportunistic infections

14.2.1. Cytomegalovirus

Serologic evidence of infection with human cytomegalovirus (CMV) has been found in 60–80% of adults in the USA and more than 90% of HIV-infected patients (Drew et al., 1981; Bale, 1984; Drew, 1988). In the pre-HAART era, CMV was a common opportunistic infection in AIDS patients; however, its frequency since the introduction of modern combination therapy has dramatically declined. In a large long-running natural history study of HIV-infected individuals, the Multicenter AIDS Cohort Study (MACS), the risk of opportunistic CMV infection was reduced by 90% in the HAART era compared to prior years of monotherapy (Detels et al., 2001).

14.2.1.1. Neuropathology

In neuropathological series, pathology due to CMV has ranged from isolated cytomegalic cells to severe necrotizing encephalitis, myelitis and neuritis. The most common pattern is a diffuse microglial nodular encephalitis disseminated in deep gray and, less frequently, white matter sites (Morgello et al., 1987; Vinters et al., 1989). With the increased sensitivity of molecular diagnostic techniques, CMV antigens or DNA has been identified in more than one-third of brains from AIDS patients, localized in microglial nodules, macrophages, astrocytes, oligodendrocytes, neurons, ependymal cells, endothelia and meninges (Wiley and Nelson, 1988; Schmidbauer et al., 1989; Belec et al., 1990). A necrotizing ventriculitis occurs in up to 10% of cases with inflammatory infiltrates and cytomegalic cells and may be associated with vasculitis, meningitis involving the cranial roots and choroid plexus and centrifugal extension of inflammation (Morgello et al., 1987; Vinters et al., 1988; Kalayjian et al., 1993). Multifocal necrotizing encephalitis or myelitis is associated with discrete areas of parenchymal inflammatory infiltration with macrophages and cytomegalic cells, and infarction and hemorrhages due to vascular involvement. Disease is typically more extensive than revealed by imaging studies in life (Morgello et al., 1997; Vinters et al., 1988; Gungor et al., 1993).

A necrotizing radiculitis with associated vasculitis and segmental thrombosis accompanied by a polymorphonuclear or mixed inflammatory infiltrate is typical of CMV polyradiculitis (Eidelberg et al., 1986a; Mahieux et al., 1989; Miller et al., 1996). Extension into the adjacent spinal cord may result in a necrotizing myelitis (Jacobsen et al., 1988; Miller et al., 1990). Demyelination may be seen in conjunction with necrotizing radiculitis (Moskowitz et al., 1984; Bishopric et al., 1985).

A multifocal necrotizing neuritis and arteritis with polymorphonuclear infiltration and variable presence of cytomegalic cells may involve peripheral (Said et al., 1991; Roullet et al., 1994) or cranial nerves (Small et al., 1989). Necrotizing optic neuritis associated with CMV retinitis has been described (Grossniklaus et al., 1987) and dorsal root ganglionitis may be associated with necrotizing myelitis (Tucker et al., 1985). Demyelinating neuropathy has been attributed to CMV (Robert et al., 1989; Cornford et al., 1992a; Morgello and Simpson, 1994), and cytomegalocytes containing CMV antigens have been found in Schwann cells (Bishopric et al., 1985; Grafe and Wiley, 1989). In a series of 115 adult autopsies, 27% had evidence of CMV in perineurial, epineurial, perimysial or epimysial sites by light microscopy. The frequency of CMV was related to the duration of survival following the diagnosis of AIDS, increasing from 19% in those dying within three months to 46% in those surviving two years or more (Cornford et al., 1992b).

The neuropathologic features suggest hematagenous dissemination to the nervous system, although subsequent extension of infection may involve CSF seeding or contiguous extension (Morgello et al., 1987; Wiley and Nelson, 1988). Virtually all patients with CMV encephalitis have evidence of systemic CMV infection (Kalayjian et al., 1993; Holland et al., 1994).

14.2.1.2. Clinical features

A number of clinical syndromes have come to be associated with CMV infections in AIDS, although each of them can be mimicked by other pathologic entities seen in this population.

CMV encephalitis is characterized by a subacute diffuse encephalopathy evolving over weeks with impairments of cognition, sensorium and mood. Neurologic examination reveals deficits in memory, attention and cognitive processing, accompanied by variable motor features and ataxia. The short time course and delirium distinguish this entity from HIV-associated dementias which evolve more gradually and do not cause impairment of sensorium (Holland et al., 1994). When CMV ventriculitis is present, cranial neuropathies, nystagmus and progressive ventricular enlargement are characteristic (Kalayjian et al., 1993). Progressive CMV encephalitis following in utero infection has been reported in an HIV-infected infant (Curless et al., 1987). A patient presenting with acute HIV seroconversion and concurrent CMV encephalitis has been described (Berger et al., 1996).

Other presentations include headaches with fever and focal neurological findings (Edwards et al., 1985; Laskin et al., 1987a; Masdeu et al., 1988), hypopituitarism due to hypothalamic infection (Sullivan et al., 1992) and cerebellar and brainstem encephalitis presenting with ataxia, oculomotor impairments, quadraparesis and bulbar dysfunction (Fuller et al., 1989; Pierelli et al., 1997). Cerebral infarctions resulting from CMV vasculitis may progress in a stepwise fashion (Kieburtz et al., 1993) and both acute subarachnoid hemorrhage (Hawley et al., 1983) and intracerebral hemorrhage (Hawley et al., 1983; Dyer et al., 1995) due to CMV have been reported.

Much less common are CMV mass lesions which have been reported in both brain and spinal cord. Symptoms include weakness, aphasia, sensory loss and pain. Imaging studies reveal ring-enhancing lesions with or without associated mass effect and edema (Dyer et al., 1995; Moulignier et al., 1996; Huang et al., 1997). In one case, thallium-201 scanning demonstrated increased uptake suggestive of tumor (Gorniak et al., 1997). Diagnosis has required biopsy.

Cerebral imaging studies are of limited sensitivity in patients with diffuse CMV encephalitis. Often, they are unremarkable or show nonspecific atrophic changes (Clifford et al., 1996). Ependymal enhancement suggests ventriculitis and meningeal enhancement may be seen (Grafe et al., 1990; Kalayjian et al., 1993; Holland et al., 1994). Areas of focal infarction or necrosis (Masdeu et al., 1988; Grafe et al., 1990) and both patchy and diffuse leukoencephalopathy (Miller et al., 1997) have been described.

CSF findings are also variable. Leukocytes may be absent, or range from a modest lymphocytosis (Holland et al., 1994) to a prominent polymorphonuclear pleocytosis (Kalayjian et al., 1993). Rarely the virus can be cultured from the CSF of patients with encephalitis (Dix et al., 1985a) but diagnosis today is commonly made by demonstrating CMV DNA in CSF using PCR techniques (see below).

A CMV necrotizing mylitis in AIDS patients has also been described, usually in association with polyradiculitis (Moskowitz et al., 1984; Tucker et al., 1985; Jacobsen et al., 1988). Necrotizing myelitis without polyradiculitis may present with acute or progressive paraparesis and urinary and rectal sphincter disturbances. Unless there is a concurrent polyneuropathy, reflexes are enhanced and a sensory level may be demonstrated (Said et al., 1991; Gungor et al., 1993). The necrosis involves both gray and white matter structures within the spinal cord (Vinters et al., 1989). A patient with a non-necrotizing demyelinating myelopathy associated with astrocytic gliosis, cytomegalic cells and spongy changes in the posterior columns has been described (Moskowitz et al., 1984).

CMV polyradiculitis typically presents with paresthesias or pain in the legs and perineum followed by a progressive paraparesis with hypotonia and hyporeflexia. Urinary sphincter disturbance, particularly retention, is characteristic. Examination may disclose features of myelitis including a sensory level and Babinski signs. The process typically evolves over days to weeks and may ascend to the upper extremities and involve cranial nerves (Tucker et al., 1985; Eidelberg et al., 1986a; Behar et al., 1987; Miller et al., 1990; Cohen et al., 1993; So and Olney, 1994; Anders and Goebel, 1998).

MRI may be normal or reveal enhancement of nerve roots, the conus medullaris, cauda equine and meninges in about one-third of cases (Bazan et al., 1991; Talpos et al., 1991; Whiteman et al., 1994; Anders et al., 1998). Electrophysiologic studies reveal features of an axonal polyneuropathy with denervation changes. Variable conduction slowing may also be present (Behar et al., 1987; Miller et al., 1990).

CSF is usually characterized by a polymorphonuclear pleocytosis, increased protein and hypoglycorrhachia; however, any or all of these features may be absent (Miller et al., 1996). Although cauda equina syndrome in a patient with AIDS is often due to CMV, other entities can mimic the presentation, and even in some cases the CSF findings. Differential considerations include other herpes viruses, particularly varicella-zoster, as well as syphilis, bacterial meningitis, toxoplasmic meningitis and lymphomatous meningitis. CMV may be cultured from CSF although demonstration of growth may require several weeks (Tucker et al., 1985; Cohen et al., 1993; So and Olney, 1994). CMV DNA can be demonstrated in most cases using PCR techniques.

14.2.1.3. Diagnosis

Until the common availability of molecular diagnostic techniques, diagnosis of opportunistic CNS CMV infections was difficult in life and treatment was often started empirically in the face of active CMV infection in the retina or systemically, or a compatible neurologic syndrome. Currently, diagnosis is facilitated by the availability of PCR which has been found to be highly sensitive and specific in both retrospective (Cinque et al., 1992; Gozlan et al., 1992; Wolf and Spector, 1992; Clifford et al., 1993) and prospective (Gozlan et al., 1995) studies. Quantification of CMV DNA levels has been correlated with histopathologic evidence of infection and found to be highest in patients with ventriculitis (Arribas et al., 1995). Another study challenged the specificity of CSF PCR for CMV encephalitis but did not examine spinal cord or nerve roots (Achim et al., 1994). Quantification of CMV genomes using PCR appears to be useful in discriminating active from latent infection (Wildemann et al., 1998).

In patients with polymorphonuclear pleocytosis, detection of the CMV matrix phosphoprotein pp65 within the nuclei can provide a rapid diagnostic test (Revello et al., 1994). Detection of CMV DNA in peripheral blood neutrophils can be seen in patients with CMV encephalitis indicative of hematangenous infection (Musiani et al., 1994).

14.2.1.4. Therapy

There are currently three antiviral agents used in the treatment of CMV opportunistic infections in AIDS. All act by inhibiting the CMV DNA polymerase, thereby preventing viral replication. Ganciclovir is an acyclic nucleoside which requires phosphorylation initiated by a CMV phophotransferase within the infected cell for its effect. The triphosphate form is incorporated into the viral DNA preventing replication (Matthews and Boehme, 1988). Efficacy has been established in CMV retinitis, and gastroenteritis. Ganciclovir is given in dosages of 5 mg/kg intravenously every 12 hours initially in patients with normal renal function. Hematologic toxicity is a common adverse effect (Collaborative DHPG Treatment Study Group, 1986; Laskin et al., 1987b; Dieterich et al., 1988). Resistance mutations in which the initial CMV mediated phosphorylation does not occur can emerge with prolonged exposure to the drug and have been found to occur most commonly in the UL97 gene (Erice et al., 1989; Baldanti et al., 2004).

Foscarnet is a pyrophosphate analogue which exerts its effect by directly inhibiting the pyrophosphate binding site on the CMV DNA polymerase. It also has been shown to be effective in CMV retinitis (Studies of Ocular Complications of AIDS Research Group et al., 1992). Renal toxicity is a significant adverse effect. Seizures and paresthesias may occur with foscarnet and anemia, electrolyte disturbances and myelosuppression may also be seen (Palestine et al., 1991). Foscarnet is active against CMV strains resistant to ganciclovir; however, CMV strains resistant to foscarnet may also emerge and are typically related to mutations in the UL54 gene (Baldanti et al., 2004). Foscarnet is initially dosed at 60 mg/kg every 8 hours, or 90 mg/kg every 12 hours intravenously in patients with normal renal function. Concurrent administration of ganciclovir and foscarnet has been reported to provide suppression of active CMV retinitis or gastrointestinal disease after each has failed as monotherapy (Dieterich et al., 1993; Studies of Ocular Complications of AIDS Research Group et al., 1996).

Cidofovir is a nucleoside analogue which is phosphorylated by host cell enzymes and selectively inhibits CMV DNA synthesis. It has been shown to be effective in AIDS patients with CMV retinitis progressing despite therapy with ganciclovir and foscarnet (Lalezari et al., 1998). Potential adverse effects include nephrotoxicity which is mitigated by hydration and probenecid, and ocular hypotony which requires regular ocular pressure monitoring to prevent blindness. In patients with normal renal function, cidofovir is initially dosed at 5 mg/kg intravenously weekly for two weeks then every other week, preceded by probenecid, 2000 mg orally 3 hours before the dose, 1000 mg 2 hours after the dose and 1000 mg 8 hours after the dose (Benson et al., 2004). Cidofovir resistance may also emerge from mutations in the UL54 gene (Baldanti et al., 2004).

No controlled studies of optimal therapy for CMV CNS opportunistic infections have been carried out to date. Anecdotal reports have documented responses of CMV encephalitis to ganciclovir (Sullivan et al., 1992; Cohen, 1996). Patients with acute ventriculitis have responded poorly to treatment, although many had prior exposure to ganciclovir (Price et al., 1992; Kalayjian et al., 1993). The occurrence of CMV encephalitis in patients on maintenance ganciclovir for CMV retinitis has responded to foscarnet transiently (Enting et al., 1992), and to cidofovir (Blick et al., 1997; Sadler et al., 1997). Combination therapy with ganciclovir and foscarnet has been effective in some reported cases with prolonged survivals (Peters et al., 1992; Cohen, 1996). A recently published prospective pilot trial demonstrated that initial combination therapy could be given safely in 31 patients with CMV encephalitis or myelitis. Clinical improvement or stabilization occurred in 74% at the end of the induction phase; however, the median time to first relapse on maintenance therapy was only 126 days (Anduze-Faris et al., 2000).

Similarly, anecdotal reports in patients with CMV polyradiculomyelitis have indicated response to prompt initiation of ganciclovir therapy, although clinical recovery may be slow taking place over several months (Graveleau et al., 1989; Miller et al., 1990; Cohen et al., 1993; Kim and Hollander, 1993; Anders et al., 1998). Not all patients have responded (Jacobsen et al., 1988; de Gans et al., 1990) and progression of polyradiculomyelitis despite ganciclovir has been linked to resistant strains (Cohen et al., 1993). Addition of foscarnet may be beneficial (Decker et al., 1994) and combination therapy may also be beneficial despite failure of both agents as monotherapy (Karmochkine et al., 1994). Due to the slowness of the clinical responses, serial CSF studies may be useful for therapeutic monitoring. Persistence of polymorphonuclear pleocytosis and hypoglycorrhachia are indicative of a poor response and persistence of CMV DNA in CSF may also be a useful marker (Cinque et al., 1995; Cohen, 1996).

14.2.2. Varicella-zoster

Varicella-zoster virus (VZV), the cause of chicken pox, resides in the dorsal root ganglia following primary infection. It typically causes segmental rash and dermatomal pain when it reactivates in adults. Studies using restriction enzyme analysis have shown no difference between viral isolates obtained during episodes of chicken pox and subsequent zoster from the same individual (Ilyid et al., 1977). Reactivation of zoster occurs more commonly in individuals with immune suppression, and the age adjusted risk of VZV radiculitis in HIV-infected persons has been reported to be 17 times that of a non-infected homosexual control population (Buchbinder et al., 1992). The frequency of CNS VZV in AIDS autopsy series has been reported to range from 2 to 4.4% (Petito et al., 1986; Gray et al., 1994; Jellinger et al., 2000). In most reports, however, a recent series of autopsies from the HAART era found evidence of CNS zoster in 4 of 23 cases prompting the suggestion that the increased frequency might be due to milder degrees of immune suppression in HAART treated patients (Gray et al., 2003).

14.2.2.1. Clinical syndromes of VZV infection in AIDS

Manifestations of VZV radiculitis are similar in HIV- and non-HIV-infected individuals. A painful cutaneous eruption characterized by vesicles on an erythematous base in a dermatomal distribution evolves over days. The trunk is most commonly affected although limbs and face may also be involved. Pain may occur without rash, or cutaneous dissemination suggestive of viremia may occur (Gelb, 1993). VZV DNA can be detected in circulating mononuclear cells during clinical eruptions (Gilden et al., 1989). In HIV-infected individuals, dermatomal VZV may be multisegmental and recurrent. Herpes zoster oticus (Ramsay Hunt Syndrome) has been reported (Mishell and Applebaum, 1989) and segmental myoclonus preceding or occurring concurrently with the dermatomal eruption has been reported (Koppel and Daras, 1992). A recent report noted a high frequency of otherwise unremarkable zoster dermatitis in HIV-infected individuals shortly after the addition of a protease inhibitor to their antiretroviral treatment regimen. Most episodes occurred within one to four months following initiation of the new regimen and risk was unrelated to CD^4 count, viral load, age, or specific agent added. However, a CD^8 level of $> 66\%$ at baseline or a CD^8 increase of $>5\%$ at one month was strongly associated with the subsequent risk of zoster radiculitis (Martinez et al., 1998).

VZV involvement of the CNS causes a number of syndromes which often occur concurrently in AIDS patients. Gray has classified these as multifocal encephalitis, ventriculitis, acute meningomyeloradiculitis with necrotizing vasculitis, focal necrotizing myelitis and vasculopathy resulting in cerebral infarction (Gray et al., 1994). Kleinschmidt-DeMasters has classified the vascular patterns into a large vessel vasculopathy associated with bland or hemorrhagic infarctions, a small vessel vasculitis affecting arterioles and producing demarcated ischemic-demyelinative lesions characteristically found at the cortical gray–white junctions associated with inflammation and viral inclusions in adjacent glial cells, and ventriculitis-ependymitis which may be a result of the small vessel vasculitis extending to the ependyma (Kleinschmidt-DeMasters et al., 1996). Isolated aseptic meningitis (de la Blanchardiere et al., 2000) and focal brainstem encephalitis (Rosenblum, 1989; Moulignier et al., 1995) have also been reported. Retrobulbar optic neuritis may precede the development of zoster retinitis (Rousseau et al., 1993; Friedlander et al., 1996; Lee et al., 1998; Meenken et al., 1998) and accompany aseptic meningitis (Franco-Paredes et al., 2002).

Individuals with HIV-associated CNS VZV may have only moderate immune suppression with CD^4 counts >200 cells/μl (Gray et al., 1994). Cutaneous zoster is absent in up to one-third of cases (Morgello et al., 1988; Chretien et al., 1993; Gray et al., 1994; Moulignier et al., 1995; de la Blanchardiere et al., 2000). In a retrospective series of 34 cases, only 3 had neither history of previous zoster dermatitis nor evidence of concurrent rash or acute retinitis at the time of CNS zoster (de la Blanchardiere et al., 2000).

VZV multifocal leukoencephalitis may follow a subacute course with clinical symptoms of progressive confusion, focal weakness and sensory loss, ataxia or visual impairments. Seizures may be a presenting or concomitant neurological manifestation. Symptoms may evolve over weeks to months with a clinical course resembling progressive multifocal leukoencephalopathy (PML), although cases of indolent progression more suggestive of primary HIV-associated dementia (Gilden et al., 1988) and of acute encephalitis (Gray et al., 1992; Aygun et al., 1998) have also been described.

14.2.2.2. Neuropathology

Pathological studies of HIV-infected patients with CNS zoster have demonstrated discrete and confluent white matter lesions with a predilection for cortical gray–white junctions and periventricular regions. Central cavitation necrosis with surrounding myelin pallor and edema, reactive astrocytosis and microglial proliferation occur in demyelinated lesions. Intranuclear and cytoplasmic inclusions can be found in astrocytes, oligodendroglia, endothelial cells and macrophages. A necrotizing vasculitis may involve leptomeningeal vessels as well as parencymal arterioles and result in ventriculitis with ependymal necrosis. Both ischemic and hemorrhagic infarctions may be found (Morgello et al., 1988; Gray et al., 1994; Amlie-Lefond et al., 1995; Weaver et al., 1999). A recent autopsy on a patient responding to HAART and subsequently succumbing to pneumonia demonstrated necrotic foci without evidence of inflammation or active zoster. Lesions had the target-like appearance but the rim consisted only of macrophages and necrosis. It was suggested that the immune recovery resulting from effective therapy (demonstrated by increase in CD^4 and reduction of HIV plasma viral load) resulted in a "burned out" VZV encephalitis (de la Grandmaison et al., 2005).

The source of VZV causing zoster encephalitis may be hematogenous spread as suggested by the predilection for involvement of the cortical gray–white junctions and deep white matter regions (Morgello et al., 1988; Gray et al., 1994; Amlie-Lefond et al., 1995; Kleinschmidt-DeMasters et al., 1996). Transaxonal spread has been demonstrated to result in focal brainstem encephalitis (Rosenblum, 1989) and along visual pathways (Rostad et al., 1989).

14.2.2.3. Diagnosis

Imaging features felt to be suggestive of VZV include well-defined discrete ovoid and confluent lesions with a predilection to involve the cortical gray–white junctions as well as deep white matter including periventricular regions. The former feature is felt to be highly suggestive of zoster (Kleinschmidt-DeMasters et al., 1996) as are target-like lesions which may show ring-enhancing patterns (Aygun et al., 1998) and subsequently coalesce (Aygun et al., 1998; Weaver et al., 1999). Imaging studies may, however, be unremarkable (Ryder et al., 1986; Morgello et al., 1988; Poscher, 1994; de la Blanchardiere et al., 2000). In a series of patients with VZV retinal necrosis and optic neuritis, imaging studies demonstrated extension along the central optic pathways suggestive of transynaptic neuronal migration (Bert et al., 2004).

CSF most often reveals a lymphocytic meningitis with elevated protein and normal glucose levels (Dix et al., 1985a; Ryder et al., 1986; Morgello et al., 1988; Gilden et al., 1988; Gray et al., 1992; Poscher, 1994; Kleinschmidt-DeMasters et al., 1998; de la Blanchardiere et al., 2000). The cellular and protein levels may be profoundly increased, reflecting widespread necrotizing vasculitis and resulting in Froin's syndrome (Kleinschmidt-DeMasters et al., 1998). Alternatively, the CSF may be acellular (Gilden et al., 1988; Amlie-Lefond et al., 1995) and even normal (Moulignier et al., 1995). VZV can uncommonly be cultured from CSF (Dix et al., 1985a; de la Blanchardiere et al., 2000). VZV-specific IgG antibodies may be detected in CSF by immunofluorescent assays (Poscher, 1994; de Silva et al., 1996); however, wide availability of PCR amplification for detection of specific viral DNA has now made this technique the most common means of diagnosis (de la Blanchardiere et al., 2000; Quereda et al., 2000). The availability of PCR techniques has led to recovery of zoster DNA in individuals with neurologic symptoms judged due to other concurrent pathologies raising the question of whether other CNS pathologies may activate resident zoster at a subclinical level leading to CSF detection (Cinque et al., 1997). Clearance of the zoster DNA on subsequent sampling appears to correspond to virological response (Cinque et al., 1997; de la Blanchardiere et al., 2000).

Large vessel vasculopathy due to VZV may also occur in HIV-infected patients and typically results in bland or hemorrhagic infarction. Herpes zoster ophthalmicus with subsequent infarction resulting in contralateral hemiparesis has been described and may follow the radiculitis by many months (Eidelberg et al., 1986b; Pillai et al., 1989; Amlie-Lefond et al., 1995). VZV has been documented in larger blood vessels (Kleinschmidt-DeMasters et al., 1996).

14.2.2.4. Zoster myelitis

Myelitis is a less common opportunistic complication of VZV in AIDS and may occur in association with a concurrent rash or retinitis, a remote history of zoster radiculitis or in the absence of previously known eruption. The course is typically acute or subacute with ascending sensory and motor deficits accompanied by urinary sphincter dysfunction typical of myelitis, and may precede the cutaneous eruption (Vinters et al., 1988; Devinsky et al., 1991; Chretien et al., 1993; Gilden et al., 1994; Gomez-Tortosa et al., 1994; Gray et al., 1994; de Silva et al., 1996). Pathologic features are those of a necrotizing myelitis with associated vasculitis involving nerve roots and dorsal ganglia in addition to the spinal cord. Cowdry type A inclusion bodies may be seen in oligodendrocytes or mononuclear cells, and VZV genetic material may be demonstrated by in situ hybridization techniques (Vinters et al., 1988; Devinsky et al., 1991; Chretien et al., 1993). Spinal cord infarction has been reported in the absence of active inflammation at the spinal level but in association with VZV ventriculo-encephalitis (Kenyon et al., 1996). A polyradiculomyelitis similar to that seen with CMV may also be seen with VZV (Vinters et al., 1988; Chretien et al., 1993; Gray et al., 1994).

MRI usually reveals intramedullary pathology which may be enhancing with prominent swelling of the cord (Chretien et al., 1993; de Silva et al., 1996) or non-enhancing (Gilden et al., 1994; de Silva et al., 1996). Rarely, spinal cord MRI may be normal early in the course (Gomez-Tortosa et al., 1994).

CSF typically reveals lymphocytic pleocytosis with increased protein (Devinsky et al., 1991; Gilden et al., 1994; de Silva et al., 1996); however, a mixed pleocytosis with lymphocytes and neutrophils may be seen in more fulminant cases (Vinters et al., 1988; Chretien et al., 1993; Gray et al., 1994; Snoeck et al., 1994; Kenyon et al., 1996). Virus can rarely be recovered from CSF (Snoeck et al., 1994); however, diagnosis can now usually be made by the detection of VZV DNA in CSF using PCR techniques (Gilden et al., 1994).

Response to therapy of CNS VZV in AIDS has been variable. Progression of disease has occurred in some individuals despite use of aciclovir 10 mg/kg intravenously every eight hours or ganciclovir (Chretien et al., 1993; Gilden et al., 1994) while others have responded to aciclovir or famciclovir (de Silva et al., 1996; Otero et al., 1998). Aciclovir-resistant VZV has been shown to cause CNS disease (Snoek et al., 1994) and foscarnet offers an effective alternative (Safrin et al., 1991; de la Blanchardiere et al., 2000). VZV strains resistant to foscarnet have also been described, however (Visse et al., 1998), and some might advocate combination therapy in patients with CNS zoster, although no prospective treatment trial has been undertaken. Dosages of ganciclovir and foscarnet used for patients with opportunistic varicella-zoster infections of the brain or spinal cord are the same as those used for CMV.

14.2.3. Herpes simplex virus

Herpes simplex virus (HSV) type 1 is typically acquired early in life while evidence of HSV type 2 infection appears later in life and increases with numbers of sexual partners. HSV-2 is acquired through contact with infected mucosa, and as many as 70% of heterosexual and 90% of homosexual sexually active men may have serologic evidence of infection (Nahmias et al., 1990). Retrograde axonal transport leads to infection of the sacral ganglia and latency (Baringer, 1974). HSV-1 latency in the CNS has been demonstrated in trigeminal, superior cervical and vagal ganglia (Baringer and Swoveland, 1973; Warren et al., 1978). Despite the frequency of HSV in the adult population, reports of HSV encephalitis in HIV-infected patients have been surprisingly infrequent. Classic necrotizing HSV encephalitis has been described on occasion (Tan et al., 1993). In a large cohort of 918 HIV-infected patients with neurological symptoms, however, only 2% were found to have HSV 1 or 2 DNA in their CSF. Of these, three-fourths also had CMV DNA in their CSF. This appears to be more than a casual association as the group as CMV DNA was found in only 16% of the entire cohort (Cinque et al., 1998). Other authors have also noted the concurrence of HSV and CMV encephalitis in AIDS patients. Cases of concurrent necrotizing retinitis and ventriculitis (Pepose et al., 1984), CMV ventriculitis and anatomically separated HSV focal encephalitis (Vital et al., 1995; Chretien et al., 1996) and necrotizing encephalitis with evidence of both CMV and HSV in areas of necrosis (Laskin et al., 1987a) have been described. In an autopsy series of 82 cases of CMV encephalitis, concomitant HSV infection was found in 16% (Vago et al., 1996). Clinical features in these patients with combined viral infection have usually been subacute in time course, with cognitive impairments, ataxia, weakness and lethargy described. Seizures and acute delirium, headache with fever, and more indolent temporal courses have also been reported (Pepose et al., 1984; Laskin et al., 1987a; Chretien et al., 1996; Cinque et al., 1998; Grover et al., 2004).

In a subset of 258 subjects from the French cohort, on which autopsy data were available, HSV 1 or 2 infections were documented in 7 (3%). One of these had a necrotizing frontal temporal encephalitis typical of HSV-1 while the other 6 had mixed HSV and CMV encephalitis (Cinque et al., 1998). In another retrospective study of 486 cases of patients with HIV infection and neurological symptoms, only nine (1.85%)

individuals had HSV detected either by CSF PCR or by pathological examination. One of these patients had a focal temporal-frontal lesion consistent with classic HSV encephalitis. Another had concurrent CMV and three others had concurrent Epstein–Barr virus (EBV) DNA in their CSF (Grover et al., 2004).

It has been suggested that the absence of vigorous immune responses may modify the clinical and pathological features of HSV encephalitis (Price et al., 1973; Sage et al., 1985) and that the incidence of HSV encephalitis could be underestimated in HIV-infected patients as a result (Schiff and Rosenblum, 1998). In the 258 autopsied patients from the French series noted above, inflammatory responses were noted to be mild to moderate in six of the seven patients with HSV (Cinque et al., 1998).

Acute HSV rhombencephalitis presenting with ataxia and fever has been described, with virus demonstrated by immunostaining in oligodendrocytes. In this case, aciclovir in usual doses failed to suppress the infection and at autopsy an HSV isolate with markedly increased virulence was identified (Hamilton et al., 1995). In another individual with microglial nodular foci, HSV DNA was identified in neuronal nuclei by in situ hybridization (Schmidbauer et al., 1989).

A diffuse meningoencephalitis is most commonly seen in HIV-infected patients with neurologically symptomatic HSV-2. The clinical presentation is nonspecific with fever, headache, lethargy, delirium, seizures and tremors in varying combinations. Imaging studies are usually normal. Two cases of individuals with focal temporal encephalitis have been reported in whom HSV-2 was successfully grown from cultures of biopsied brain tissue (Dix et al., 1985b; Madhoun et al., 1991).

Myelitis and radiculitis may also occur in HIV-infected individuals with concurrent HSV infection. Herpes genitalis may be associated with perineal paresthesiae, neuralgia and urninary retention (Caplan et al., 1977) and this pattern has been described in HIV-infected individuals (Yoritaka et al., 2005). Meningitis with ascending myelitis may occur with (Craig and Nahmias, 1973; Britton et al., 1985; Wiley et al., 1987) or in the absence of (Klastesky et al., 1972; Bergstrom et al., 1990) vesicular lesions. Progressive back pain, paresthesias and neuralgia were followed by weakness and urinary retention. CSF was unrevealing and diagnosis was ultimately made at autopsy (Britton et al., 1985). Dual CMV and HSV myelitis has also been described in a patient with subacute progressive perineal and sacral hypesthesia, followed by back and leg pain and then acute weakness and urinary retention evolving over two weeks. Anal HSV-2 vesicular lesions appeared as further neurological progression led to quadriplegia and cranial nerve dysfunction. At autopsy, disseminated CMV infection and anal HSV were present, with both viruses recovered from sites in the spinal cord (Tucker et al., 1985).

CSF findings have been variable in these cases, ranging from normal to remarkable lymphocytic pleocytosis and elevated protein. Glucose levels are usually normal or mildly low. Currently, diagnosis usually results from the demonstration of HSV-specific DNA in the CSF using PCR techniques.

Treatment of neurological HSV infections in HIV-infected patients is typically initiated with aciclovir 10 mg/kg every eight hours intravenously for 14–21 days (Benson et al., 2004). In some instances, progression of HSV encephalitis has been seen despite therapy with usual doses (Tan et al., 1993; Hamilton et al., 1995). Treatment of anal HSV-2 lesions in another individual failed to prevent ascending myelitis (Britton et al., 1985), and encephalitis improved but symptoms suggestive of radiculitis failed to respond in another (Madhoun et al., 1991). Other individuals with HSV-2 infections do respond, however, with either stabilization (Dix et al., 1985b) or recovery (Grover et al., 2004). Aciclovir-resistant strains of HSV which lack or have altered activity of the thymidine kinase required to initiate intracellular phosphorylation of aciclovir to its active form occur (Erlich et al., 1989; Englund et al., 1990) and have been demonstrated to cause neurological infections in AIDS (Gateley et al., 1990). Persistent CSF HSV DNA on serial sampling suggests the possibility of resistance and should lead to alternate therapy. Foscarnet has been shown to be effective in treating aciclovir-resistant mucocutaneous lesions in HIV-infected individuals (Chatis et al., 1989; Safrin et al., 1991; Safrin, 1992) and is now recommended at doses of 120–200 mg/kg/day divided in two to three doses for patients with aciclovir-resistant HSV (Benson et al., 2004). HSV encephalitis has also occurred in a patient on chronic foscarnet therapy, however, presumably due to viral isolates subsequently shown to be resistant to this agent (Read et al., 1998).

14.2.4. Hepatitis C virus

Neurological symptoms involving both the peripheral and central nervous systems have been identified in individuals with HCV infection. Peripheral neuropathy with pain, hyspesthesia, sensory ataxia and dysesthesias has been associated with HCV and cryoglobulinemia. Both distal sensory polyneuropathy and sensory-motor polyneuropathy have been reported, and superimposed multifocal neuropathy has also been reported. One case of anterior optic neuropathy with visual loss was described. Sural nerve biopsies have revealed features of vasculitis with or without necrotizing changes in epineural vessels (Heckmann et al., 1999; Tembl et al., 1999).

Syndromes involving the CNS include a leukoencephalopathy presenting with subacute impairments of cognition and sensorium or dementia and ataxia, spasticity and myoclonus, headache, vertigo and sensory impairments. Cryoglobulins were variably present in these cases (Origgi et al., 1998; Heckmann et al., 1999; Tembl et al., 1999). In some instances, treatment with steroids or cyclophosphamide has been associated with resolution of the CNS symptoms (Origgi et al., 1998; Tembl et al., 1999). A patient with subacute partial myelitis has also been described (Nolte et al., 2002).

A cross sectional study comparing HIV-infected individuals coinfected with HCV to HIV subjects matched on HIV disease characteristics found the HCV-coinfected subjects had an increased likelihood of depression and demonstrated increased perseveration on tests of executive function. Results were independent of hepatic disease severity; however, the presence of HCV coinfection was also associated with increased history of substance abuse potentially confounding attribution of the differences (Ryan et al., 2004). Another study failed to show significant differences on tests of attention, memory or intelligence (von Giesen et al., 2004). A recent review of current data notes the complexity of confounding drug and alcohol use in the HCV-coinfected population but suggests that preliminary data may favor an impact on neurocognitive function in HIV-infected individuals coinfected with HCV (Hilsabeck et al., 2005).

14.3. Fungal infections

The most common fungal infection associated with HIV infection is *Cryptococcus neoformans* which is covered elsewhere in this volume. The entities that follow are far less commonly seen; however, for that reason, they may escape early detection.

14.3.1. Coccidioides immitis

Coccidioides immitis is found as a mold in soil of semiarid regions of the southwestern USA, northern Mexico and Central and South America. Its growth is enhanced by bat and rodent droppings and human infection results most commonly from inhalation of aerosolized arthroconidia. At body temperature, the arthroconidia rupture releasing endospores. Normally contained by a granulomatous cell-mediated immune response, the organisms survive and may produce productive infection in immune compromised hosts. In HIV-infected patients, CNS coccidiomycosis has resulted in meningitis (Fish et al., 1990; Galgiani et al., 1993), meningoencephalitis, myelitis and radiculitis (Mischel and Vinters, 1995) and cerebral abscesses (Jarvik et al., 1988; Levy et al., 1988).

Autopsy reports of AIDS patients describe a necrotizing granulomatous, and at times suppurative, meningitis extending along the Virchow-Robin spaces into the parenchyma. Adjacent vessels are inflamed with organisms evident in the adventitia, and an endarteritis obliterans may be found. With cavitary necrosis, abscess formation occurs (Jarvik et al., 1988; Mischel and Vinters, 1995). A retrospective evaluation of 91 cases of HIV-associated coccidiomycosis identified meningeal involvement in 15% (Singh et al., 1996).

In patients with meningitis, CSF may show prominent pleocytosis, increased protein and hypoglycorrhachia. Organisms may be cultured from the CSF or complement fixation antibodies may be detected. Some reported patients have had CD^4 levels >200 cells/mm^3 (Fish et al., 1990). With cerebral abscess, headache and fever with or without focal neurological findings may be seen. However, coccidioides serology may be negative in almost one-third of patients (Singh et al., 1996).

Imaging studies may reveal single or multiple enhancing lesions which may be small and without significant mass effect (Jarvik et al., 1988; Levy et al., 1988). A patient with dureal and cerebral venous thrombosis due to *C. immitis* has been described (Kleinschmidt-DeMasters et al., 2000). Some success in HIV patients with CNS coccidiomycosis has followed treatment with amphotericin B, ketoconazole or fluconazole, and mortality appears to relate to CD^4 level at the time of infection (Fish et al., 1990; Galgiani et al., 1993). Current recommendations are to treat acute meningeal infections with fluconazole in doses of 400–800 mg daily either intravenously or orally. Intrathecal amphotericin is considered an alternative option. Fluconazole 400 mg/day or itraconazole 200 mg twice daily are recommended for chronic suppressive maintenance therapy (Benson et al., 2004).

14.3.2. Histoplasma capsulatum

Histoplasma capsulatum is an ascomycete which is found worldwide. The highest prevalence is in temperate and tropical environments, including the north and south central portions of the USA. Enrichment of the soil by bird or bat guano favors growth of the organism and aerosolization is promoted by disruption of the soil due to natural factors or construction activity. Tuberculate microconidia and infectious macroconidia are acquired by inhalation (Wheat, 1995). Following inhalation, *H. capsulatum* transforms into a yeast at body temperature and is contained by a granulomatous response. The organism parasitizes in macrophages and is disseminated in immune-compromised AIDS patients either as part of the primary infection or reactivation when CD^4 counts fall (Johnson et al., 1988; Wheat et al., 1990a, b).

In the early years of the AIDS pandemic, histoplasmosis was noted in as many as 2–5% of patients in endemic areas of the USA. In Indianapolis, histoplasmic infection in as many as 25% of AIDS patients was noted, and was part of the first AIDS presentation in 75% of those affected (Johnson et al., 1988; Wheat et al., 1990b).

The CNS is affected by *H. capsulatum* in up to 20% of HIV-associated cases. A mononuclear meningitis involves adjacent blood vessels resulting in endothelial proliferation, granulomatous vasculitis and fibrinoid necrosis (Anaissie et al., 1988; Wheat et al., 1990a, b; Weidenheim et al., 1992). Clinical features include headache and fever, changes in cognition and sensorium, and cranial neuropathies. Meningismus is uncommon. Focal neurologic features are noted in about 10% and seizures in 10–30%. Presentation with stroke due to meningovasculitis, thrombosis of basal or meningeal vessels or septic emboli from infected cardiac valves may also be encountered (Wheat et al., 1990a; Wheat, 1995). Optic neuritis with pathologically confirmed *H. capsulatum* invasion of the optic nerve sheath has been reported (Yau et al., 1996).

CSF may reveal a lymphocytic pleocytosis with elevated protein and hypoglycorrhachia, or be entirely normal. Histoplasma antigen can be detected in CSF in 40% and antibodies to histoplasma in 60%. Culture of the organism requires weeks (Wheat et al., 1990a, b). CNS imaging may be normal or show enhancing mass lesions, infarctions or meningeal enhancement (Anaissie et al., 1988; Wheat et al., 1990a; Weidenheim et al., 1992).

H. capsulatum may be recovered from the blood, bone marrow or respiratory secretions in about 85% of AIDS patients with disseminated disease. Serologic tests are of limited value in patients from endemic regions, although high titers will be seen in about 60% of patients with CNS histoplasmosis irrespective of concurrent HIV infection (Wheat et al., 1990a, b; Wheat, 1995). Neuropathological specimens yield organisms on culture or by histological staining with methenamine silver in about 80% of cases (Wheat et al., 1990a, b; Weidenheim et al., 1992).

Prognosis for AIDS patients with CNS histoplasmosis has been reported to be poor with mortalities of greater than 60% (Wheat et al., 1990b). Liposomal amphotericin B appears to offer more rapid clearance of fungus than itraconazole (Wheat et al., 2001a) which was found to be more effective than fluconazole (Wheat et al., 2001b). However, CNS relapse has been described despite treatment with amphotericin B (Weidenheim et al., 1992) or amphotericin and itraconazole in combination (Vullo et al., 1997). Amphotericin B in doses of 0.7–1.0 mg/kg/day for 10–12 weeks has recently been recommended for initial therapy (Medical Letter, 2002).

14.3.3. Blastomyces dermatitides

Blastomyces dermatitides is an ascomycete with pear-shaped conidia found in moist acidic soil worldwide. At body temperature, it transforms to a broad-based budding yeast. A polymorphonuclear and lymphocytic inflammatory response results in pyogranuloma formation in normal hosts; however, disseminated infection may occur in those who are immune suppressed (Pappas et al., 1992; Wheat, 1995). In a review of 24 reported cases, CNS disease was seen in 46%, five to ten times the rate expected from non-HIV-associated series. Mortality was 54%, also five times the rate expected. Most of the patients had CD^4 counts less than $200/mm^3$ (Witzig et al., 1994).

Neuropathologic features include basilar meningitis, necrotizing arteritis and encephalitis with abscess formation. Organisms may be demonstrated in meninges and around blood vessels (Fraser et al., 1991; Harding, 1991; Pappas et al., 1992; Tan et al., 1992). Clinical features are nonspecific, resulting from meningitis or mass lesions (Fig. 14.2), and include fever, headache, changes in mentation and sensorium, seizures and variable focal signs (Harding, 1991; Pappas et al., 1992; Witzig et al., 1994). Isolated CNS infection without apparent pulmonary involvement may be a presenting manifestation (Pappas et al., 1992) or a site of recurrent disease (Witzig et al., 1994).

Diagnosis is made from pulmonary specimens in most individuals with accompanying lung disease; however, culture of cutaneous lesions, blood, CSF or abscess fluid may be required in others. Serologic studies have not been helpful in most cases (Pappas et al., 1992; Witzig et al., 1994). Response to therapy with amphotericin B, ketoconazole or fluconazole has not been encouraging in most reported cases (Fraser et al., 1991; Pappas et al., 1992; Tan et al., 1993; Witzig et al., 1994); however, success has been reported with sequential amphotericin B followed by ketoconazole in one patient (Pappas et al., 1992). Itraconazole has been useful in the treatment of non-HIV-infected patients with blastomycosis (Wheat, 1995). Amphotericin B in doses of 0.7–1.0 mg/kg/day for up to 12 weeks has recently been recommended as initial therapy (Medical Letter, 2002).

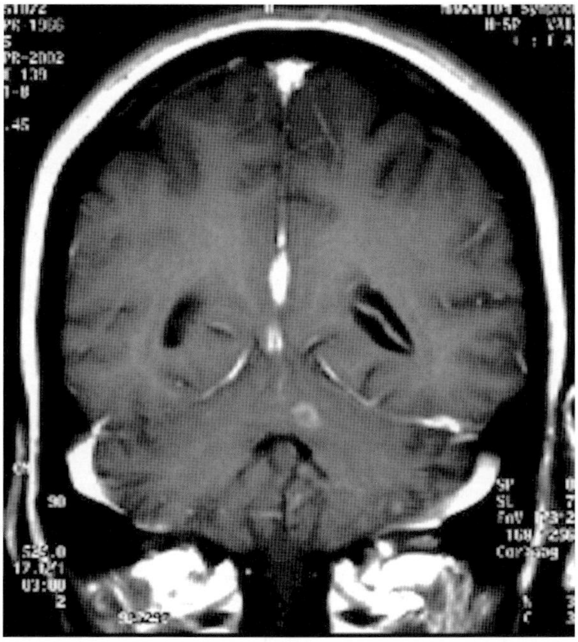

Fig. 14.2. Blastomycosis abscess of cerebellum. Contrast-enhanced T1-weighted MR image: coronal section showing ring-enhancing lesion of right cerebellar hemisphere.

14.3.4. Aspergillus

Aspergillus species are found in most environments and acquired by inhalation. A polymorphonuclear response is important in containing the mycelial form while macrophages kill the conidia (Minamoto et al., 1992). Invasive aspergillosus occurs in individuals with impaired neutrophil and macrophage function and may be associated with neutropenia, corticosteroid therapy or both. Other conditions associated with increased risk for aspergillosus include alcoholism, hyperglycemia or diabetes mellitus, intravenous drug use and cavitary chronic pulmonary disease (Singh et al., 1991; Minamoto et al., 1992; Pursell et al., 1992; Lortholary et al., 1993).

In a large HIV-infected population database, the incidence of aspergillosis has been estimated at 3.5 cases per 1000 patient-years, with increased risk related to CD^4 counts less than 100/mm^3, age over 35 years, prior opportunistic infections or medications associated with neutropenia. Median survival was three months after the diagnosis of aspergillosis; however, 26% survived over a year (Holding et al., 2000).

Aspergillus may invade the CNS by contiguous extension from sites in cranial or dural sinuses or spinal structures (Woods and Goldsmith, 1990; Strauss and Fine, 1991; Hall and Farrior, 1993). Vascular invasion may lead to hematogenous dissemination with necrotizing vasculitis and bland or hemorrhagic infarction. CNS involvement is found in 10–25% of immunosuppressed patients with aspergillus and may be higher in those with AIDS. Most CNS cases are caused by *A. fumigatus* with *A. flavus* seen less frequently (Singh et al., 1991; Minamoto et al., 1992; Pursell et al., 1992; Kleinschmidt-DeMasters, 2002). Septic embolization from cardiac valvular vegetations may result in infarction, cerebral abscess, or mycotic aneurysm formation (Henochowicz et al., 1985; Cox et al., 1990).

Neuropathological examination reveals a necrotizing vasculitis with hemorrhagic or bland infarctions, hemorrhages, microscopic or large abscess formation or purulent meningitis. The fungi, with characteristic acute or right-angle branching septate hyphae, may be demonstrated in vessels or parenchyma (Asnis et al., 1988; Vinters and Anders, 1990; Woods and Goldsmith, 1990; Carrazana et al., 1991; Kleinschmidt-DeMasters et al., 2002). Extension from cranial sinuses may result in dural sinus thrombosis (Strauss and Fine, 1991; Hall and Farrior, 1993) while extension from the lungs may lead to vertebral osteomyelitis and subsequent meningomyelitis (Woods and Goldsmith, 1990) or epidural abscess formation with compressive myelopathy (Go et al., 1993).

Diagnosis of CNS aspergillosis is difficult and requires a high index of suspicion, and pathological sampling for culture or demonstration of the organisms. Presenting symptoms may be nonspecific with fever, headache, infarction, seizures, weakness, changes in mentation or cranial neuropathies (Woods and Goldsmith, 1990; Carrazana et al., 1991; Singh et al., 1991; Pursell et al., 1992; Mylonakis et al., 2000). Acute embolic infarction from a septic embolus may result in fever and headache (Henochowicz et al., 1985). Fever with chronic back pain and progressive myelopathy may be the presentation of meningomyelitis (Woods and Goldsmith, 1990) or epidural abscess with compressive myelopathy (Go et al., 1993). Facial neuropathy preceded by otalgia and otorhea may precede lethargy and headache from CNS extension of an otologic focus (Strauss and Fine, 1991; Hall and Farrior, 1993; Lyos et al., 1993). The most common sites of extraneural involvement are the lungs, sinuses and orbit (Mylonakis et al., 2000).

CSF sampling may show a lymphocytic or neutrophilic pleocytosis (Woods and Goldsmith, 1990; Carrazana et al., 1991) or be unremarkable despite meningitis (Asnis et al., 1988; Woods and Goldsmith, 1990). Cultures are usually unrevealing and diagnosis in HIV-infected patients has been made by examination or culture of tissue specimens in about half the cases, and at autopsy in the other half (Mylonakis et al., 2000).

CNS imaging may demonstrate enhancing mass lesions or non-enhancing infarctions, hemorrhages or multifocal white matter lesions (Woods and Goldsmith, 1990; Singh et al., 1991; Pursell et al., 1992). MRI failed to reveal a necrotizing myelitis in one reported case (Woods and Goldsmith, 1990).

Therapy of HIV-associated CNS aspergillosis has been unrewarding with short-term mortality regardless of therapy in most instances (Mylonakis et al., 2000). Acute treatment with amphotericin B at doses of 1–1.5 mg/kg/day has been recommended (Medical Letter, 2002). Voriconazole, dosed at 6 mg/kg intravenously every 12 hours for the first day, followed by 4 mg/kg every 12 hours for at least seven days, was shown to be more effective than amphotericin B in patients with invasive aspergillosis, most of whom were receiving immune suppressive therapy following transplantation procedures or for treatment of malignancy. An oral dose of 200 mg twice daily can be used once the infection is adequately suppressed (Herbrecht et al., 2002). However, the success with AIDS-related cerebral aspergillosis has been modest (Denning et al., 2002). Caspofungin, an inhibitor of fungal β-1–3-glucan synthesis, in dosages of 70 mg the first day followed by 50 mg/day, has been used in combination with voriconazole resulting in improved three-month survivals in non-AIDS-immune-suppressed patients with invasive aspergillosis failing initial treatment with amphotericin B (Marr et al., 2004), and has recently been reported to be successful in combination with liposomal amphotericin B in a patient with non-AIDS-associated cerebral aspergillosis which failed to respond to voriconazole and liposomal amphotericin B (Ehrmann et al., 2005). Another patient with non-AIDS-related cerebral aspergillosis who failed voriconazole monotherapy responded to a combination of liposomal amphotericin B and voriconazole in combination (Ehrmann et al., 2005).

14.3.5. Candida albicans

Candida albicans is a common mucocutaneous infection of the oropharynx and esophagus in HIV-infected individuals. In the CNS, *C. albicans* causes meningitis, microabscess formation, cereberal infarctions, subarachoid hemorrhage or large abscesses behaving as cerebral mass lesions (Sanchez-Portocarrero et al., 2000). Dissemination of *C. albicans* is presumed to be hematogenous, and both meningitis and abscesses have been reported in HIV-infected individuals, sometimes discovered only at autopsy (Snider et al., 1983; Levy et al., 1985; Petito et al., 1986; Kure et al., 1991). Risk factors for CNS candida infection in addition to immune suppression include intravenous drug use and prior therapy with antibiotics or corticosteroids, yet the frequency of candidal infection of the CNS is a rare event (Casado et al., 1998; Sanchez-Portocarrero et al., 2000).

Clinical features are nonspecific. Microabscesses commonly cause a diffuse encephalopathy which may only be discovered postmortem. Meningitis is typically subacute in onset with fever and headache. Meningismus may or may not be seen in HIV-infected individuals and CD^4 count is usually low with a median of 135 cells/mm^3. CSF may show a moderate pleocytosis of up to 500 cells/mm^3, which is either neutrophilic (Ehni and Ellison, 1987) or mononuclear (Bruinsma-Adams, 1991). Protein is typically elevated and hypoglycorrhachia is seen in the majority of cases. Adenosine deaminase levels may be elevated. Since cultures and smears are usually unrevealing, differentiation of other causes of chronic meningitis such as tuberculosis is difficult. Repeated CSF sampling with large volumes of CSF sent for culture may be required to successfully culture the organism (Casado et al., 1998).

Abscesses can be seen on contrast MRI as ring-enhancing lesions. Large abscesses may be associated with focal findings (Sanchez-Portocarrero et al., 2000). Colonization of cerebral vasculature with infarction has been reported (Kieburtz et al., 1993). Culture of *C. albicans* from blood samples is considered evidence of invasive candidiasis; however, blood cultures may be negative even when invasive candidiasis is demonstrated pathologically.

Treatment of CNS candidiasis in AIDS is most often undertaken with amphotericin B and flucytosine. In a small series of patients with candidal meningitis this combination was associated with 70% survival (Casado et al., 1998). Treatment may need to be initiated empirically in the setting of notable risk factors such as intravenous drug use (Sanchez-Portocarrero et al., 2000). Fluconazole is effective against *C. albicans* and may be used for maintenance regimens; however, fluconazole-resistant strains of *C. albicans* have been recovered from CSF (Heinic et al., 1993; Berenguer et al., 1996). In patients with cerebral abscess, surgical drainage and amphotericin without (Pitlik et al., 1983) or with flucytosine (Levy et al., 1985) has been associated with response. Both voriconazole and caspofungin have activity against *Candida* species and could be considered in CNS candidiasis which fails to respond to amphotericin (see Section 14.3.4).

14.4. Parasitic infections

The most common of the parasitic opportunistic infections is *Toxoplasma gondii* which is covered elsewhere in this volume. Less common pathogens are addressed below.

14.4.1. Trypanosoma cruzi (Chagas' disease)

Trypanosoma cruzi is a flagellated parasite endemic in South America which is typically acquired in childhood and has been estimated to infect 16–18 million people in the western hemisphere. Immune-competent individuals are often asymptomatic; however, with immune deficiency, reactivation of latent infection may result in significant symptoms. Reactivation of Chagas' disease has been reported as the initial manifestation leading to a diagnosis of AIDS (Cohen et al., 1998; de Olivera et al., 2002). *T. cruzi* may cause cardiomyopathy with congestive heart failure or arrhythmia, impairment of esophageal or intestinal motility due to involvement of autonomic cells or meningoencephalitis and mass lesions which resemble those of toxoplasmosis in the CNS (Rosemberg et al., 1992; Rocha et al., 1994; Di Lorenzo et al., 1996; Ferreira et al., 1997; Silva et al., 1999).

Histopathological features include multifocal necrosis with *T. cruzi* amastigotes demonstrable within astrocytes, macrophages, endothelial cells and less frequently neurons using methenamine silver stains. Free amastigotes may be seen in perivascular and intercellular spaces (Del Castillo et al., 1990; Gluckstein et al., 1992; Solari et al., 1993; Rocha et al., 1994). Purulent meningitis and vasculitis has been described (Rosemberg et al., 1992; Solari et al., 1993). In comparison to non-immune-suppressed patients, HIV-infected individuals with *T. cruzi* infestation are more likely to have necrotizing hemorrhagic lesions of larger dimension, with dense exudates of macrophages, lymphocytes and plasma cells within the perivascular spaces. In contrast, the majority of patients with chronic Chagas' disease have mild encephalitis with few inflammatory foci and glial scars (Lazo et al., 1998).

Clinical features include meningoencephalitis with headache, fever, variable focal findings and seizures (Del Castillo et al., 1990; Ferreira et al., 1991; Gluckstein et al., 1992; Metze and Maciel, 1993; Solari et al., 1993; Ferreira et al., 1997). Progressive hydrocephalus with ataxia and emesis has been described (Rosemberg et al., 1992).

Multifocal CNS parenchymal disease is common and cerebral imaging often reveals ring or irregular enhancing lesions as well as non-enhancing areas of diminished signal. Mass effect may be seen. In many cases the lesions resemble those found in CNS toxoplasmosis and primary CNS lymphoma (Rocha et al., 1994).

CSF often reveals a lymphocytic pleocytosis with elevated protein and sometimes hypoglycorrhachia. *T. cruzi* trypomastigotes can be identified in giemsa-stained smears of CSF and peripheral blood. Motile parasites can sometimes be seen in fresh unstained CSF specimens (Rocha et al., 1994; Ferreira et al., 1997; Silva et al., 1999). Identification of the organism in blood smears and positive blood cultures may be found in individuals with chronic Chagas' disease irrespective of CNS involvement (Ferreira et al., 1997). Antibodies to *T. cruzi* may or may not be present in patients with CNS trypanosomiasis (Gluckstein et al., 1992; Metze and Maciel, 1993; Silva et al., 1999). A recent report used PCR techniques to detect and monitor *T. cruzi* in CSF of a patient with Chagasic meningoencephalitis undergoing therapy (Lages-Silva et al., 2002).

Therapy with nifurtimox (Del Castillo et al., 1990) and benznidazole with or without the addition of itraconazole and fluconazole has also been associated with clinical response (Solari et al., 1993; Silva et al., 1999). One review recommends induction therapy for two months with benznidazole 5 mg/kg daily, split into two doses, or nifurtimox 8–10 mg/kg daily, split into three doses. Subsequent maintenance therapy is recommended at lower dose levels (Ferreira et al., 1997). A more recent recommendation is benznidazole 5–8 mg/kg/day in two divided doses (Benson et al., 2004).

Since many cases appear to be identified only after clinical worsening occurs despite empiric therapy for presumed toxoplasmosis, the possibility of CNS trypanosomiasis should be considered early in HIV-infected individuals from endemic areas or with relevant travel exposures who are being empirically treated for presentations suspected to be CNS toxoplasmosis. A recent case report describes a patient with concurrent CNS toxoplasmosis and trypanosomiasis in whom initial therapy led to initial resolution of CNS lesions; however, a new lesion and related focal signs prompted further investigation with demonstration of *T. cruzi* in CSF (Yoo et al., 2004).

14.4.2. Granulomatous amebic encephalitis

Free-living acanthameba and *Balmuthia mandillaris* (leptomyxid ameba) have rarely been reported to produce granulomatous infection in HIV-infected individuals. These agents are commonly found in moist soil in warm regions, and are thought to be acquired by inhalation or cutaneous contact. In contrast naegleria is acquired via the olfactory epithelium following exposure by swimming in contaminated water and results in a fulminant fatal necrotizing encephalitis in previously normal individuals. Both a humoral and cellular immune response occurs (Bottone, 1993; Sison et al., 1995). Neuropathological examination reveals a necrotizing arteritis with fibrinoid necrosis (Wiley et al., 1987; Di Gregorio et al., 1992) and thrombo-occlusive angiitis (Gardner et al., 1991). Suppurative meningitis may be extensive and amebic trophozoites can be demonstrated invading perivascular spaces and vessel walls using periodic acid Schiff or methenamine silver stains (Wiley et al., 1987; Gardner et al., 1991; Di Gregorio et al., 1992; Gordon et al., 1992; Zagardo et al., 1997).

Clinical features of meningoencephalitis due to acanthameba or leptomyxid amebic infestation include nonspecific fever and headache with or without focal neurological findings. Meningismus, seizures and signs of increased intracranial pressure may be present (Wiley et al., 1987; Anzil et al., 1991; Gardner et al., 1991; Di Gregorio et al., 1992; Gordon et al., 1992; Tan et al., 1993; Zagardo et al., 1997). Cutaneous nodules, papules or pustules may precede amebic encephalitis in as many as half the cases (Tan et al., 1993; Sison et al., 1995).

CNS imaging studies may show nodular homogenous or ringlike enhancing lesions of variable size. Mass effect and surrounding edema may be present. Non-enhancing lesions resembling infarctions are also seen (Wiley et al., 1987; Gardner et al., 1991; Gordon et al., 1992; Zagardo et al., 1997; Martinez et al., 2000).

CSF may show a neutrophilic or lymphocytic pleocytosis and elevated protein (Wiley et al., 1987; Zagardo et al., 1997) or be acellular (Gardner et al., 1991; Gordon et al., 1992). Although CSF is not likely to yield the organisms, cutaneous lesions may be revealing (Tan et al., 1993; Sison et al., 1995). Diagnosis has most often resulted from tissue examination.

Although most described cases have been fatal, a patient with a solitary abscess due to *Acanthameba castellanii* responded to treatment with fluconazole and sulfadiazine following surgical excision (Martinez et al., 2000).

14.4.3. Taenia solium (cysticercosis)

Neurocysticercosis, although a common parasitic infection of the CNS in Mexico, Latin America, Asia and Africa, has not apparently been found with increased frequency in HIV-infected patients as the number of reported cases are few. The parasite is the most common helminthic infection of the CNS and is typically acquired by ingestion of food contaminated with *Taenia solium* eggs. The parasite may subsequently be asymptomatic, or cause meningitis or parenchymal lesions with seizures and focal signs. A series of 107 HIV-infected patients from Zimbabwe revealed 13 with intracranial mass lesions, four of whom had cysticercosis. These individuals presented with headaches and seizures or focal deficits and were found to have multiple intracranial lesions with cerebral imaging. The frequency of 30% of cerebral mass lesions due to cysticercosis was compared with 6% in a series of 51 intracranial mass lesions in non-HIV-infected individuals, and the authors suggested an increased frequency due to reactivation in the setting of immune suppression (Thornton et al., 1992). In contrast, two series from Mexico did not reveal an increased frequency of cysticercosis in HIV-infected individuals. In one autopsy series of 97 patients, only one case was identified compared to three cases in 197 controls. The authors suggested that any association was incidental (Barron-Rodriguez et al., 1990). In a series of 91 patients from a referral center in Mexico City, two cases were identified, one presenting with a large rapidly progressive cyst and increased intracranial pressure and another as an incidental concurrent finding in a patient with cerebral toxoplasmosis whose symptoms responded to treatment of the latter infection. In their review of several other reported cases, the authors noted the relative frequency of giant cysts compared to larger series of non-HIV-infected individuals with cysticercosis (Soto-Hernandez et al., 1996). Recently, an HIV-infected patient with cerebral and epidural spinal cysticercosis producing a cauda equine syndrome has been reported (Delobel et al., 2004). Currently, it seems unlikely that *T. solium* is a prominent opportunistic pathogen in AIDS, although there may be an effect of HIV or related immune suppression on the frequency of large cystic lesions. A recent treatment recommendation for *T. solium* is albendazole 400 mg twice daily for 8–30 days or paraziquantal 50–100 mg/kg/day in three divided doses for 30 days (Medical Letter, 2002).

14.4.4. Cerebral microsporidiosis

Microsporidia are obligate intracellular spore-forming protozoa which are seen most often as opportunistic causes of gastrointestinal disease in AIDS patients. Disseminated *Encephalitozoon cuniculi* has been reported to cause CNS symptoms in HIV-infected patients. Autopsy studies have demonstrated areas of central necrosis with both free spores and intracellular spores in surrounding macrophages and astrocytes (Yachnis et al., 1996).

Clinical presentations include headache, alterations of cognition and sensorium, seizures and increased intracranial pressure. Imaging studies reveal multiple small ring-enhancing and nodular lesions. CSF may be normal or show a neutrophilic pleocytosis with intracellular and extracellular microsporidial spores. Spores may also be detected in urinary sediment, stools and sputum (Yachnis et al., 1996; Weber et al., 1997).

Treatment with albendazole has been associated with clinical and imaging response in one patient; however, recrudescence with recurrence of the CNS lesions occurred despite continued therapy (Weber et al., 1997). The recommended dose of albendazole is 400 mg twice daily (Medical Letter, 2002).

References

Achim CL, Nagra RM, Wang R, et al (1994). Detection of cytomegalovirus in cerebrospinal fluid autopsy specimens from AIDS patients. J Infect Dis 169: 623–627.

Adair JC, Beck AC, Apfelbaum AC, et al (1987). Nocardial cerebal abscess in the acquired immunodeficiency syndrome. Arch Neurol 449: 548–550.

Amlie-Lefond C, Kleinschmidt DeMasters BK, Mahalingam R, et al (1995). The vasculopathy of varicella-zoster virus encephalitis. Ann Neurol 37: 784–790.

Anaissie E, Fainstein V, Samo T, et al (1988). Central nervous system histoplasmosis. An unappreciated complication of the acquired immunodeficiency syndrome. Am J Med 84: 215–217.

Anders H-J, Goebel FD (1998). Cytomegalovirus polyradiculopathy in patients with AIDS. Clin Infect Dis 27: 345–352.

Anders HJ, Weiss N, Bogner JR, et al (1998). Ganciclovir and foscarnet efficacy in AIDS-related CMV polyradiculopathy. J Infect 36: 29–33.

Anduze-Faris BM, Fillet A-M, Gozlan J, et al (2000). Induction and maintenance therapy of cytomegalovirus central nervous system infection in HIV-infected patients. AIDS 14: 517–524.

Anzil AP, Rao C, Wrzole MA, et al (1991). Amebic meningoencephalitis in a patient with AIDS caused by a newly recognized opportunistic pathogen: leptomyxid ameba. Arch Pathol Lab Med 115: 21–25.

Arribas JR, Clifford DB, Fichtenbaun CJ, et al (1995). Level of cytomegalovirus (CMV) DNA in cerebrospinal fluid of subjects with AIDS and CMV infection of the central nervous system. J Infect Dis 172: 527–531.

Asnis DS, Chitkara RK, Jacobsen M, et al (1988). Invasive aspergillosis: an unusual manifestation of AIDS. N Y State J Med 88: 653–655.

Aygun N, Finelli DA, Rodgers MS, et al (1998). Multifocal varicella-zoster virus leukoencephalitis in a patient with AIDS: MR findings. Am J Neuroradiol 19: 1897–1899.

Baldanti F, Lurain N, Gerna G (2004). Clinical and biologic aspects of human cytomegalovirus resistance to antiviral drugs. Hum Immunol 65: 403–409.

Bale JF Jr (1984). Human cytomegalovirus infection and disorders of the nervous system. Arch Neurol 41: 310–320.

Bargallo J, Berenguer J, Garcia-Barrionuevo J, et al (1996). The "target sign": is it a specific sign of CNS tuberculoma? Neuroradiology 38: 547–550.

Baringer JR (1974). Recovery of herpes simplex virus from human sacral ganglions. N Engl J Med 291: 828–830.

Baringer JR, Swoveland P (1973). Recovery of herpes simplex virus from human trigeminal ganglions. N Engl J Med 288: 648–650.

Barron-Rodriguez LP, Jessurun J, Hernandez-Avila M (1990). The prevalence of invasive amebiasis and cysticercosis is not increased in Mexican patients dying of AIDS. International Conference on AIDS, 6. San Francisco, p. 253 (abstr Th B 524).

Bazan C III, Jackson C, Jinkins JR, et al (1991). Gadolinium-enhanced MRI in a case of cytomegalovirus polyradiculopathy. Neurology 41: 1522–1523.

Behar R, Wiley C, McCutchan JA (1987). Cytomegalovirus polyradiculoneuropathy in acquired immune deficiency syndrome. Neurology 37: 557–561.

Belec L, Gray F, Mikoll J, et al (1990). Cytomegalovirus (CMV) encephalomyeloradiculitis and human immune deficiency virus (HIV) encephalitis: presence of HIV and CMV co-infected multinucleated giant cells. Acta Neuropathol 81: 99–104.

Benninger PR, Savoia MC, Davis CE (1988). *Listeria monocytogenes* meningitis in a patient wth AIDS-related complex [letter]. J Infect Dis 158: 1396–1397.

Benson CA, Kaplan JE, Masur H, et al (2004). Treating opportunistic infections among HIV-infected adults and adolescents. MMWR 53 (RR 15): 1–112.

Berenguer J, Solera J, Diaz MD, et al (1991). Listeriosis in patients infected with human immunodeficiency virus. Rev Infect Dis 13: 115–119.

Berenguer J, Moreno S, Laguna F, et al (1992). Tuberculous meningitis in patients infected with the human immunodeficiency virus. N Engl J Med 326: 668–672.

Berenguer J, Diaz-Guerra TM, Ruiz-Diez B, et al (1996). Genetic dissimilarity of two fluconazole-resistant candida albicans strains causing meningitis and oral candidiasis in the same AIDS patient. J Clin Microbiol 34: 1542–1545.

Bergen GA, Yangco BG, Adelman HM (1993). Central nervous system infection with *Mycobacterium kansasii*. Ann Intern Med 118: 396.

Berger JR (1992). Spinal cord syphilis associated with human immunodeficiency virus infection: a treatable myelopathy. Am J Med 92: 101–103.

Berger JR, Waskin H, Pall L, et al (1992). Syphilitic cerebral gumma with HIV infection. Neurology 42: 1282–1287.

Berger DS, Bucher G, Nowak JA, et al (1996). Acute primary human immunodeficiency virus type 1 infection in a patient with concomitant cytomegalovirus encephalitis. Clin Infect Dis 23: 66–70.

Bergstrom J, Vahlne A, Alestig K (1990). Primary and recurrent herpes simplex virus type 2 induced meningitis. J Infect Dis 162: 322–330.

Berman SM, Kim RC (1994). The development of cytomegalovirus encephalitis in AIDS patients receiving ganciclovir. Am J Med 96: 415–419.

Bernaerts A, Vanhoenacker FM, Parizel PM, et al (2003). Tuberculosis of the central nervous system: overview of neuroradiological findings. Eur Radiol 13: 1876–1890.

Berry CD, Hooten TM, Collier AC, et al (1987). Neurologic relapse after benzathine penicillin therapy for secondary syphilis in a patient with HIV infection. N Engl J Med 316: 1587–1589.

Bert RJ, Samawareerwa R, Melhem ER (2004). CNS MR and CT findings associated with a clinical presentation of herpetic acute retinal necrosis and herpetic retrobulbar optic neuritis: five HIV-infected and one non-infected patients. Am J Neuroradiol 25: 1722–1729.

Bishburg E, Sunderam G, Reichman LB, et al (1986). Central nervous system tuberculosis with the acquired immunodeficiency syndrome and its related complex. Ann Intern Med 105: 210–213.

Bishopric G, Bruner J, Butler J (1985). Guillain-Barre syndrome with cytomegalovirus infection of peripheral nerves. Arch Pathol Lab Med 109: 1106–1108.

Blick G, Garton T, Hopkins U, et al (1997). Successful use of cidofovir in treating AIDS-related cytomegalovirus retinitis, encephalitis, and esophagitis. J Acquir Immune Defic Syndr 15: 84–85.

Bottone EJ (1993). Free-living amebas of the genera acanthamoeba and naegleria: an overview and basic microbiologic correlates. Mt Sinai J Med 60: 260–270.

Brightbill TC, Ihmedian IH, Post MJ, et al (1995). Neurosyphilis in HIV-positive and HIV-negative patients: neuroimaging findings. AJNR Am J Neuroradiol 16: 703–711.

Britton CB, Mesa-Tejada R, Fenoglio CM, et al (1985). A new complication of AIDS: thoracic myelitis caused by herpes simplex virus. Neurology 35: 1071–1074.

Bruinsma-Adams IK (1991). AIDS presenting as *Candida albicans* meningitis: a case report [letter]. AIDS 5: 1268.

Buchbinder S, Katz MH, Hessol N, et al (1992). Herpes zoster and human immunodeficiency virus infection. J Infect Dis 166: 1153–1156.

Caplan LR, Kleeman FL, Berg S (1977). Urinary retention probably secondary to herpes genitalis. N Engl J Med 297: 920–921.

Carrazana EJ, Rossitch E Jr, Morris J (1991). Isolated central nervous system aspergillosis in the acquired immunodeficiency syndrome. Clin Neurol Neurosurg 93: 227–230.

Casado JL, Quereda C, Corral I (1998). Candidal meningitis in HIV-infected patients. AIDS Patient Care STDs 12: 681–686.

Chandramuki A, Lyashchenko K, Kumari HBV, et al (2002). Detection of antibody to Mycobacterium tuberculosis protein antigens in the cerebrospinal fluid of patients with tuberculous meningitis. J Infect Dis 186: 678–683.

Chatis PA, Miller CH, Schrager LE, et al (1989). Successful treatment with foscarnet of an aciclovir resistant mucocutaneous infection with herpes simplex in a patient with acquired immunodeficiency syndrome. N Engl J Med 320: 297–300.

Chedore P, Jamieson FB (2002). Rapid molecular diagnosis of tuberculous meningitis using the Gen-probe amplified mycobacterium tuberculosis direct test in a large Canadian public health laboratory. Int J Tubercul Lung Dis 6: 913–919.

Chretien F, Gray F, Lescs MC, et al (1993). Acute varicella-zoster virus ventriculitis and meningo-myelo-radiculitis in acquired immunodeficiency syndrome. Acta Neuropathol 86: 659–665.

Chretien F, Belec L, Hilton DA, et al (1996). Herpes simplex virus type 1 encephalitis in acquired immunodeficiency syndrome. Neuropathol Appl Neurobiol 22: 394–404.

Chung WM, Pien FD, Grekin JL (1983). Syphilis: a cause of fever of unknown origin. Cutis 31: 537–540.

Cinque P, Vago L, Brytting M, et al (1992). Cytomegalovirus infection of the central nervous system in patients with AIDS: diagnosis by DNA amplification from cerebrospinal fluid. J Infect Dis 166: 1408–1411.

Cinque P, Baldanti F, Vago L, et al (1995). Ganciclovir therapy for cytomegalovirus (CMV) infection of the central nervous system in AIDS patients: monitoring by CMV DNA detection in cerebrospinal fluid. J Infect Dis 171: 1603–1606.

Cinque P, Bossolasco S, Vago L, et al (1997). Varicella-zoster virus (VZV) DNA in cerebrospinal fluid of patients infected with human immunodeficiency virus: VZV disease of the central nervous system or subclinical reactivation of VZV infection. Clin Infect Dis 25: 634–639.

Cinque P, Vago L, Marenzi R, et al (1998). Herpes simplex virus infections of the central nervous system in human immunodeficiency virus-infected patients: clinical management by polymerase chain reaction assay of cerebrospinal fluid. Clin Infect Dis 27: 303–309.

Clifford DB, Buller RS, Mohammed S, et al (1993). Use of polymerase chain reaction to demonstrate cytomegalovirus DNA in CSF of patients with human immunodeficiency virus infection. Neurology 43: 75–79.

Clifford DB, Arribas JR, Storch GA, et al (1996). Magnetic resonance brain imaging lacks sensitivity for AIDS associated cytomegalovirus encephalitis. J Neurovirol 2: 397–403.

Cohen BA (1996). Prognosis and response to therapy of CMV encephalitis and meningomyelitis in AIDS. Neurology 46: 444–450.

Cohen BA, McArthur JC, Grohman S, et al (1993). Neurologic prognosis of cytomegalovirus polyradiculomyelopathy in AIDS. Neurology 43: 493–499.

Cohen JE, Tsai EC, Ginsberg HJ, et al (1998). Pseudotumoral chagasic meningoencephalitis as the first manifestation of acquired immunodeficiency syndrome. Surg Neurol 49: 324–327.

Collaborative DHPG Treatment Study Group (1986). Treatment of serious cytomegalovirus infections with 9-(1,3-dihydroxy-2-propoxymethyl) guanine in patients with AIDS and other immunodeficiencies. N Engl J Med 314: 801–805.

Cornford ME, Ho HW, Vinters HV (1992a). Correlation of neuromuscular pathology in acquired immunodeficiency syndrome patients with cytomegalovirus infection and zidovudine treatment. Acta Neuropathol 84: 516–529.

Cornford ME, Holden JK, Boyd MC, et al (1992b). Neuropathology of the acquired immunodeficiency syndrome (AIDS): report of 39 autopsies from Vancouver British Columbia. Canad J Neurol Sci 19: 442–452.

Corral I, Quereda C, Navas E, et al (2004). Adenosine deaminase activity in cerebrospinal fluid of HIV-infected patients: limited value for diagnosis of tuberculous meningitis. Eur J Clin Microbiol Infect Dis 23: 471–476.

Cox JN, Di Dió F, Pizzolato GP, et al (1990). Aspergillus endocarditis and myocarditis in a patient with the acquired immunodeficiency syndrome (AIDS). Virchows Archives [A] 417: 255–259.

Craig CP, Nahmias AJ (1973). Different patterns of neurologic involvement with herpes simplex virus types 1 and 2: isolation of herpes simplex virus type 2 from the buffy coat of two adults with meningitis. J Infect Dis 127: 365–372.

Curless RG, Scott GB, Post MJ, et al (1987). Progressive cytomegalovirus encephalopathy following congenital infection in an infant with acquired immunodeficiency syndrome. Childs Nerv Syst 3: 255–257.

Decker CF, Simon GL, Digioia RA, et al (1991). *Listeria monocytogenes* infections in patients with AIDS: report of five cases and review. Rev Infect Dis 13: 413–417.

Decker CF, Tarver III JH, Murray DF, et al (1994). Prolonged concurrent use of ganciclovir and foscarnet in the treatment of polyradiculopathy due to cytomegalovirus in a patient with AIDS [letter]. Clin Infect Dis 19: 548–549.

De Gans J, Portegies P, Tiessens G, et al (1990). Therapy for cytomegalovirus polyradiculomyelitis in patients with AIDS: treatment with ganciclovir. AIDS 4: 421–425.

De Olivera SE, Dos Reis CJ, Moncao HCG, et al (2002). Reactivation of Chagas' disease leading to the diagnosis of acquired immune deficiency syndrome. Brazilian J Infect Dis 6: 317–321.

De Silva SM, Mark AS, Gilden DH, et al (1996). Zoster myelitis: improvement with antiviral therapy in two cases. Neurology 47: 929–931.

De la Blanchardiere A, Rozenberg F, Caumes E, et al (2000). Neurological complications of varicella-zoster virus infection in adults with human immunodeficiency virus infection. Scand J Infect Dis 32: 263–269.

De la Grandmaison GL, Carlier R, Chretien F, et al (2005). Burnt out varicella-zoster virus encephalitis in an AIDS patient following treatment by highly active antiretroviral therapy. Clin Radiol 60: 613–617.

Del Castillo M, Mendoza G, Oviedo J, et al (1990). AIDS and Chagas' disease with central nervous system tumorlike lesion. Am J Med 88: 693–694.

Delobel P, Signate A, Guedj ME, et al (2004). Unusual form of neurocysticercosis associated with HIV infection. Eur J Neurol 11: 55–58.

Denning DW, Ribaud P, Milpied N, et al (2002). Efficacy and safety of voriconazole in the treatment of acute invasive aspergillosis. Clin Infect Dis 34: 563–571.

Detels R, Tarwater P, Phair JP, et al (2001). Multicenter AIDS Cohort Study. AIDS 15: 347–355.

Devinsky O, Cho E-S, Petito CK, et al (1991). Herpes zoster myelitis. Brain 114: 1181–1196.

Dickerman RD, Stevens QE, Rak R, et al (2003). Isolated intracranial infection with *Mycobacterium avium* complex. J Neurol Sci 47: 101–105.

Dieterich DT, Chachoua A, Lafleur F, et al (1988). Ganciclovir treatment of gastrointestinal infections caused by cytomegalovirus in patients with AIDS. Rev Infect Dis 10 (Suppl 3): S532–S537.

Dieterich DT, Poles MA, Lew EA, et al (1993). Concurrent use of ganciclovir and foscarnet to treat cytomegalovirus infection in AIDS patients. J Infect Dis 167: 1184–1188.

Di Gregorio C, Rivasi F, Mongiardo N, et al (1992). Acanthamoeba meningoencephalitis in a patient with acquired immunodeficiency syndrome. Arch Pathol Lab Med 116: 1363–1365.

Di Lorenzo GA, Pagano MA, Taratuto AL, et al (1996). Chagasic granulomatous encephalitis in immunosuppressed patients. Computed tomography and magnetic resonance imaging findings. J Neuroimaging 6: 94–97.

Dix RD, Bredsen DE, Erlich KS, et al (1985a). Recovery of herpesviruses from cerebrospinal fluid of immunodeficient homosexual men. Ann Neurol 18: 611–614.

Dix RD, Waitzman DM, Follansbee S, et al (1985b). Herpes simplex virus type 2 encephalitis in two homosexual men with persistent lymphadenopathy. Ann Neurol 17: 203–206.

Doll DC, Yarbro JW, Phillip SK, et al (1987). Mycobacterial spinal cord abscess with an ascending polyneuropathy [letter]. Ann Intern Med 106: 333–334.

Drew WL (1988). Cytomegalovirus infection in patients with AIDS. J Infect Dis 158: 449–456.

Drew WL, Mintz L, Miner RC, et al (1981). Prevalence of cytomegalovirus infection in homosexual men. J Infect Dis 143: 188–192.

Dube MP, Holtom PD, Larsen RA (1992). Tuberculous meningitis in patients with and without human immunodeficiency virus infection. Am J Med 93: 520–524.

Dwork AJ, Chin S, Boyce LA (1994). Intracerebral *Mycobacterium avium*-intracellulare in a child with acquired immunodeficiency syndrome. Pediatric Infect Dis J 13: 1149–1151.

Dyer JR, French MAH, Mallal SA (1995). Cerebral mass lesions due to cytomegalovirus in patients with AIDS: report of two cases. J Infect 30: 147–151.

Edwards RH, Messing R, Mckendall RR (1985). Cytomegalovirus meningoencephalitis in a homosexual man with Kaposi's sarcoma: isolation of CMV from CSF cells. Neurology 35: 560–562.

Ehni WF, Ellison RT (1987). Spontaneous *Candida albicans* meningitis in a patient with the acquired immune deficiency syndrome [letter]. Am J Med 83: 806–807.

Ehrmann S, Bastides F, Gissot V, et al (2005). Cerebral aspergillosis in the critically ill: two cases of successful medical treatment. Intensive Care Med 31: 738–742.

Eidelberg D, Sotrel A, Vogel H, et al (1986a). Progressive polyradiculopathy in acquired immune deficiency syndrome. Neurology 36: 912–916.

Eidelberg D, Sotrel A, Horoupian DS, et al (1986b). Thrombotic cerebral vasculopathy associated with herpes zoster. Ann Neurol 19: 7–14.

Englund JA, Zimmerman ME, Swierkosz EM, et al (1990). Herpes simplex virus resistance to aciclovir. Ann Intern Med 112: 416–422.

Enting R, De Gans J, Reiss P, et al (1992). Ganciclovir/foscarnet for cytomegalovirus meningoencephalitis in AIDS [letter]. Lancet 340: 559–560.

Erice A, Chou S, Biron KK, et al (1989). Progressive disease due to ganciclovir resistant cytomegalovirus in immunocompromised patients. N Engl J Med 320: 289–293.

Erlich KS, Mills J, Chatis P, et al (1989). Aciclovir-resistant herpes simplex virus infections in patients with the acquired immunodeficiency syndrome. N Engl J Med 320: 293–296.

Faber WR, Bos JD, Rietra PJ, et al (1983). Treponemicidal levels of amoxicillin in cerebrospinal fluid after oral administration. Sex Transm Dis 10: 148–150.

Fernandez-Guerrero ML, Miranda C, Cenjor C, et al (1988). The treatment of neurosyphilis in patients with HIV infection [letter]. JAMA 259: 1495–1496.

Ferreira MS, Nishioka SDA, Rocha A, et al (1991). Acute fatal *Trypanosoma cruzi* meningoencephalitis in a human immunodeficiency virus-positive hemophiliac patient. Am J Trop Med Hyg 45: 723–727.

Ferreira MS, Mishioka SA, Silvestre MTA, et al (1997). Reactivation of Chagas' disease in patients with AIDS: report of three new cases and review of the literature. Clin Infect Dis 25: 1397–1400.

Fish DG, Ampel NM, Galgiani JN, et al (1990). Coccidioidomycosis during human immunodeficiency virus infection, a review of 77 patients. Medicine 69: 384–391.

Flood JM, Weinstock HS, Guroy ME, et al (1998). Neurosyphilis during the AIDS epidemic: San Francisco, 1985–1992. J Infect Dis 177: 931–940.

Folgueira L, Delgado R, Palenque E, et al (1994). Polymerase chain reaction for rapid diagnosis of tuberculous meningitis in AIDS patients. Neurology 44: 1336–1338.

Franco-Paredes C, Bellehemeur T, Merchant A, et al (2002). Aseptic meningitis and optic neuritis preceding varicella-zoster progressive outer retinal necrosis in a patient with AIDS. AIDS 16: 1045–1049.

Fraser VJ, Keath EJ, Powderly WG (1991). Two cases of blastomycosis from a common source: use of DNA restriction analysis to identify strains. J Infect Dis 163: 1378–1381.

French GL, Teoh R, Chan CY, et al (1987). Diagnosis of tuberculous meningitis by detection of tuberculostearic acid in cerebrospinal fluid. Lancet 2: 117–119.

Friedlander SM, Rahhal FM, Ericson L, et al (1996). Optic neuropathy preceding acute retinal necrosis in acquired immunodeficiency syndrome. Arch Opthalmol 114: 1481–1485.

Fuller GN, Guiloff RJ, Scaravilli F, et al (1989). Combined HIV-CMV encephalitis presenting with brainstem signs. J Neurol Neurosurg Psych 52: 975–979.

Galgiani JN, Catanzaro A, Cloud GA, et al (1993). Fluconazole therapy for coccidioidal meningitis. Ann Intern Med 119: 28–35.

Gardner HAR, Martinez AJ, Visvesvara GS, et al (1991). Granulomatous amebic encephalitis in an AIDS patient. Neurology 41: 1993–1995.

Gateley A, Gander RM, Johnson PC, et al (1990). Herpes simplex virus type 2 meningoencephalitis resistant to aciclovir in a patient with AIDS. J Infect Dis 161: 711–715.

Gelb LD (1993). Varicella zoster virus: clinical aspects. In: B Roizman, RJ Whitley, C Lopez (Eds.), The Human Herpesviruses. New York, Raven Press, pp. 281–308.

George TI, Manley G, Koehler JE, et al (1998). Detection of Bartonella henselae by polymerase chain reaction in brain tissue of an immunocompromised patient with multiple enhancing lesions. J Neurosurg 89: 640–644.

Gettler JF, El-Sadr W (1993). Cranial epidural abscess due to Mycobacterium tuberculosis in a patient infected with the human immunodeficiency virus [letter]. Clin Infect Dis 17: 289–290.

Gilden DH, Murray RS, Wellish M, et al (1988). Chronic progressive varicella-zoster virus encephalitis in an AIDS patient. Neurology 38: 1150–1153.

Gilden DH, Devlin H, Wellish M, et al (1989). Persistence of varicella-zoster virus DNA in blood mononuclear cells of patients with varicella or zoster. Virus Genes 2: 299–305.

Gilden DH, Befinlich BR, Rubinstien EM, et al (1994). Varicella-zoster virus myelitis: an expanding spectrum. Neurology 44: 1818–1823.

Gluckstein D, Ciferri F, Ruskin J (1992). Chagas's disease: another cause of cerebral mass in the acquired immunodeficiency syndrome. Am J Med 92: 429–432.

Go BM, Ziring DJ, Kountz DS (1993). Spinal epidural abscess due to aspergillus sp in a patient with acquired immunodeficiency syndrome. South Med J 86: 957–960.

Gomez-Tortosa E, Gadea I, Gegundez MI, et al (1994). Development of myelopathy before herpes zoster rash in a patient with AIDS. Clin Infect Dis 18: 810–812.

Good CD, Jager HR (2000). Contrast enhancement of the cerebrospinal fluid on MRI in two cases of spirochaetal meningitis. Neuroradiology 42: 448–450.

Gordon SM, Blumberg HM (1992). *Mycobacterium kansasii* brain abscess in a patient with AIDS. Clin Infect Dis 14: 789–790.

Gordon SM, Steinberg JP, Du Puis MH, et al (1992). Culture isolation of acanthamoeba species and leptomyxid amebas from patients with amebic meningoencephalitis, including two patients with AIDS. Clin Infect Dis 15: 1024–1030.

Gordon SM, Eaton ME, George R, et al (1994). The response of symptomatic neurosyphilis to high-dose intravenous penicillin G in patients with human immunodeficiency virus infection. N Engl J Med 331: 1469–1473.

Gorniak RJT, Kramer EL, McMeeking AA, et al (1997). Thallium-201 uptake in cytomegalovirus encephalitis. J Nucl Med 38: 1386–1388.

Gould IA, Belok LC, Handwerger S (1986). *Listeria monocytogenes*: a rare cause of opportunistic infection in the acquired immunodeficiency syndrome (AIDS) and a new cause of meningitis in AIDS. A case report. AIDS Res 2: 231–234.

Gozlan J, Salord J-M, Roullet E, et al (1992). Rapid detection of cytomegalovirus DNA in cerebrospinal fluid of AIDS patients with neurologic disorders. J Infect Dis 166: 1416–1421.

Gozlan J, Amrani ME, Baudrimont M, et al (1995). A prospective evaluation of clinical criteria and polymerase chain reaction assay of cerebrospinal fluid for the diagnosis of cytomegalovirus related neurologic diseases during AIDS. AIDS 9: 253–260.

Grafe MR, Wiley CA (1989). Spinal cord and peripheral nerve pathology in AIDS: the roles of cytomegalovirus and human immunodeficiency virus. Ann Neurol 25: 561–566.

Grafe MR, Press GA, Berthoty DP, et al (1990). Abnormalities of the brain in AIDS patients: correlation of post mortem MR findings with neuropathology. Am J Roentgenol 11: 905–911.

Graveleau P, Perol R, Chapman A (1989). Regression of cauda equina syndrome in AIDS patient being treated with ganciclovir [letter]. Lancet 2: 511–512.

Gray F, Mohr M, Rozenberg F, et al (1992). Varicella-zoster virus encephalitis in acquired immunodeficiency syndrome: report of four cases. Neuropathol Appl Neurobiol 18: 502–514.

Gray F, Belec L, Lescs MC et al (1994). Varicella-zoster virus infection of the central nervous system in the acquired immune deficiency syndrome. Brain 117: 987–999.

Gray F, Chretien F, Vallat-Decouvelaere AV, et al (2003). The changing pattern of HIV neuropathology in the HAART era. J Neuropathol Exp Neurol 62: 429–440.

Grossniklaus HE, Frank E, Tomsak RL (1987). Cytomegalovirus retinitis and optic neuritis in acquired immune deficiency syndrome. Opthalmology 94: 1601–1604.

Grover D, Hewsholme W, Brink N, et al (2004). Herpes simplex virus infection of the central nervous system in human immunodeficiency virus-type 1-infected patients. Int J STD AIDS 15: 597–600.

Gungor T, Funk M, Linde R, et al (1993). Cytomegalovirus myelitis in perinatally acquired HIV. Arch Dis Child 68: 399–401.

Gupta RK, Jena A, Sharma A (1989). Sellar abscesses associated with tuberculous osteomyelitis of the skull: MR findings. Am J Neuroradiol 10: 448.

Haas EJ, Madhavan T, Quinn EL, et al (1977). Tuberculous meningitis in an urban general hospital. Arch Intern Med 137: 1518–1521.

Hall PJ, Farrior JB (1993). Aspergillus mastoiditis. Otolaryngol Head Neck Surg 108: 167–170.

Hamilton RL, Achim C, Grafe MR, et al (1995). Herpes simplex virus brainstem encephalitis in an AIDS patient. Clin Neuropathol 14: 45–50.

Harding CV (1991). Blastomycosis and opportunistic infections in patients with acquired immunodeficiency syndrome: an autopsy study. Arch Pathol Lab Med 115: 1133–1136.

Harris JO, Marquez J, Swerdloff MA, et al (1980). Listeria brain abscess in the acquired immunodeficiency syndrome [letter]. Arch Neurol 46: 250.

Hawkins CH, Gold JWM, Whimbey E, et al (1986). *Mycobacterium avium* complex infections in patients with the acquired immunodeficiency syndrome. Ann Intern Med 105: 184.

Hawley DA, Schaefer JF, Schulz DM, et al (1983). Cytomegalovirus encephalitis in acquired immunodeficiency syndrome. Am J Clin Pathol 80: 874–877.

Heckmann JG, Kayser C, Heuss D, et al (1999). Neurological manifestations of chronic hepatitis C. J Neurol 246: 486–491.

Heinic GS, Stevens DA, Greenspan D, et al (1993). Fluconazole-resistant *Candida* in AIDS patients. Oral Surg Oral Med Oral Pathol 76: 711–715.

Henochowicz S, Mustafa M, Lawrinson WE, et al (1985). Cardiac aspergillosis in acquired immune deficiency syndrome. Am J Cardiol 55: 1239–1240.

Herbrecht R, Denning DW, Patterson TF, et al (2002). Voriconazole versus amphotericin B for primary therapy of invasive aspergillosis. N Engl J Med 347: 408–415.

Hernandez-Albujar S, Arribas JR, Royo A, et al (2000). Tuberculous radiculomyelitis complicating tuberculous meningitis: case report and review. Clin Infect Dis 30: 915–921.

Hilsabeck RC, Castelion SA, Hinkin CH (2005). Neuropsychological aspects of coinfection with HIV and hepatitis C virus. Clin Infect Dis 41 (Suppl 1): S38–S44.

Holding KJ, Dworkin MS, Wan PC, et al (2000). Aspergillosis among people infected with human immunodeficiency virus: incidence and survival. Clin Infect Dis 31: 1253–1257.

Holland NR, Power C, Matthews VP, et al (1994). Cytomegalovirus encephalitis in acquired immunodeficiency syndrome (AIDS). Neurology 44: 507–514.

Hollander H (1988). Cerebrospinal fluid normalities and abnormalities in individuals infected with human immunodeficiency virus. J Infect Dis 158: 855–858.

Horn DL, Hewlett D Jr, Haas WH, et al (1994). Superinfection with rifampin-isoniazid-streptomycin-tuberculosis in three patients with AIDS: confirmation by polymerase chain reaction fingerprinting. Ann Intern Med 121: 115–116.

Huang PP, McMeeking AA, Stempien MJ, et al (1997). Cytomegalovirus disease presenting as a focal brain mass: report of two cases. Neurosurgery 40: 1074–1079.

Ilyid JP, Oakes JE, Hyman RW, et al (1977). Comparison of the DNAs of varicella-zoster viruses isolated from clinical cases of varicella and herpes zoster. Virology 82: 345–352.

Ives NJ, Gazzard G, Easterbrook PJ (2001). The changing pattern of AIDS-defining illnesses with the introduction of highly active antiretroviral therapy (HAART) in a London clinic. J Infect 42: 134–139.

Jacob CN, Henein SS, Heurich AI, et al (1993). Nontuberculous mycobacterial infection of the central nervous system in patients with AIDS. South Med J 86: 638–640.

Jacobsen MA, Mills J, Rush J, et al (1988). Failure of antiviral therapy for acquired immunodeficiency syndrome related cytomegalovirus myelitis. Arch Neurol 45: 1090–1092.

Jarvik JG, Hesselink JR, Wiley C, et al (1988). Coccidioidomycotic brain abscess in an HIV infected man. West J Med 149: 83–86.

Javaly K, Horowitz HW, Wormser GP (1992). Nocardiosis in patients with human immunodeficiency virus infection. Medicine 71: 128–138.

Jellinger KA, Setinek U, Drlicek M, et al (2000). Neuropathology and general autopsy findings in AIDS during the last 15 years. Acta Neuropathol 100: 213–220.

Johns DR, Teirney M, Felsenstein D (1987). Alteration in the natural history of neurosyphilis by concurrent infection with the human immunodeficiency virus. N Engl J Med 316: 1569–72.

Johnson PC, Khardori N, Najjar AF, et al (1988). Progressive histoplasmosis in patients with acquired immunodeficiency syndrome. Am J Med 85: 152–158.

Jurado RL, Farley MM, Pereira E, et al (1993). Increased risk of meningitis and bacteremia due to *Listeria monocytogenes* in patients with human immunodeficiency virus isolation. Clin Infect Dis 17: 224–227.

Kalayjian RC, Cohen ML, Bonomo R, et al (1993). Cytomegalovirus ventriculoencephalitis in AIDS. Medicine 72: 67–77.

Kales CP, Holzman RS (1990). Listeriosis in patients with HIV infection: clincal manifestations and response to therapy. J Acquir Immune Defic Syndr 3: 139–143.

Karmochkine M, Molina J-M, Scieux C, et al (1994). Combined therapy with ganciclovir and foscarnet for cytomegalovirus polyradiculitis in patients with AIDS. Am J Med 97: 196–197.

Katrak SM, Shembalkar PK, Bijwe SR, et al (2000). The clinical, radiological, and pathological profile of tuberculous meningitis in patients with and without human immunodeficiency virus infection. J Neurol Sci 181: 18–126.

Katz DA, Berger JR (1989). Neurosyphilis in acquired immunodeficiency syndrome. Arch Neurol 46: 895–8.

Katz DA, Berger JR, Duncan RC (1993). Neurosyphilis. A comparative study of the effects of concomitant HIV-1 infection. Arch Neurol 50: 243–9.

Kennedy DH, Fallon RJ (1979). Tuberculous meningitis. JAMA 241: 264–268.

Kent SJ, Crowe SM, Yung A, et al (1993). Tuberculous meningitis: a 30 year review. Lancet 7: 987–994.

Kenyon LC, Dulaney E, Montone KT, et al (1996). Varicella-zoster ventriculo-encephalitis and spinal cord infarction in a patient with AIDS. Acta Neuropathol 92: 202–205.

Khorrami P, Heffeman EJ (1993). Pneumonia and meningitis due to *Nocardia asteroides* in a patient with AIDS. (letter). Clin Infect Dis 17: 1084–1085.

Kieburtz KD, Eskin TA, Ketonen L, et al (1993). Opportunistic cerebral vasculopathy and stroke in patients with the acquired immunodeficiency syndrome. Arch Neurol 50: 430–432.

Kim YS, Hollander H (1993). Polyradiculopathy due to cytomegalovirus: report of two cases in which improvement occurred after prolonged therapy and review of the literature. Clin Infect Dis 17: 32–37.

Kim J, Minamoto GY, Grieco MH (1991). Nocardial infection as a complication of AIDS: report of six cases and review. Rev Infect Dis 13: 624–629.

Klastersky J, Carpel R, Snoeck JM, et al (1972). Ascending myelitis in association with herpes simplex virus. N Engl J Med 287: 182–184.

Kleinschmidt-DeMasters BK (2002). Central nervous system aspergillosis: a 20 year retrospective series. Hum Pathol 33: 116–124.

Kleinschmidt-DeMasters BK, Amlie-Lefond C, Gilden DH (1996). The patterns of varicella zoster virus encephalitis. Hum Pathol 27: 927–938.

Kleinschmidt-DeMasters BK, Mahalingam R, Shimek C, et al (1998). Profound cerebrospinal fluid peeocytosis and Froin's syndrome secondary to widespread necrotizing vasculitis in an HIV-positive patient with varicella zoster encephalomyelitis. J Neurol Sci 159: 213–218.

Kleinschmidt-DeMasters BK, Mazowiecki M, Bonds LA, et al (2000). Coccidiomycosis meningitis with massive dural and cerebral venous thrombosis and tissue arthroconidia. Arch Pathol Lab Med 2124: 310–314.

Koppel BS, Daras M (1992). Segmental myoclonus preceding herpes zoster radiculitis. Eur Neurol 32: 264–266.

Kure K, Llena JF, Lyman WD, et al (1991). Human immunodeficiency virus-1 infection of the nervous system: an autopsy study of 268 adult pediatric and fetal brains. Hum Pathol 22: 700–710.

Lages-Silva E, Ramirez LE, Silva-Vergara ML, et al (2002). Chagasic meningoencephalitis in a patient with acquired immunodeficiency syndrome: diagnosis, follow-up and genetic characterization of *Trypanosoma cruzi*. Clin Infect Dis 34: 118–123.

Laguna F, Andrados M, Ortega A, et al (1992). Tuberculous meningitis with acellular cerebrospinal fluid in AIDS patients. AIDS 6: 1165–1167.

Lalezari JP, Holland GN, Kramer F, et al (1998). Randomized, controlled study of the safety and efficacy of intravenous cidofovir for the treatment of relapsing cytomegalovirus retinitis in patients with AIDS. J Acquir Immune Defic Syndr 17: 339–34.

Lanska MJ, Lanska DJ, Schmidley JW (1988). Syphilitic polyradiculopathy in an HIV-positive man. Neurology 38: 1297–301.

Lascaux A-S, Lesprit P, Deforges L, et al (2003). Late cerebral relapse of a mycobacterium avium complex disseminated infection in an HIV-infected patient after cessation of antiretroviral therapy. AIDS 17: 1410–1411.

Laskin OL, Stahl-Bayliss CM, Morgello S (1987a). Concomitant herpes simplex virus type 1 and cytomegalovirus ventriculoencephalitis in acquired immunodeficiency syndrome. Arch Neurol 44: 843–847.

Laskin OL, Cederberg DM, Mills J for the Ganciclovir Study Group (1987b). Ganciclovir for the treatment and suppression of serious infections caused by cytomegalovirus. Am J Med 83: 201–207.

Lazo J, Meneses ACO, Rocha A, et al (1998). Chagasic meningoencephalitis in the immunodeficient. Arq Neuropsiquiatr 56: 93–97.

Lee MS, Cooney EL, Stoessel KM, et al (1998). Varicella zoster virus retrobulbar optic neuritis preceding retinitis in patients with acquired immune deficiency syndrome. Ophthalmology 105: 467–471.

Lesprit P, Zagdanski A-M, Blanchardiere ADL, et al (1997). Cerebral tuberculosis in patients with the acquired immunodeficiency syndrome (AIDS). Medicine 76: 423–431.

Levy RM, Bredesen DE, Rosenblum ML (1985). Neurological manifestations of the acquired immunodficiency syndrome (AIDS): Experience at UCSF and review of the literature. J Neurosurg 62: 475–495.

Levy RM, Bredesen DE, Rosenblum ML (1988). Opportunistic central nervous system pathology in patients with AIDS. Ann Neurol 23 (suppl): S7–S12.

Lincoln EM, Sordillo SVR, Davies PA (1960). Tuberculous meningitis in children: a retrospecitive review of 167 untreated and 74 treated patients with special reference to early diagnosis. J Pediatr 57: 807–823.

Lopez-Cortes LF, Cruz-Ruiz M, Gomez-Mateos J, et al (1995). Adenosine deaminase activity in the CSF of patients with aseptic meningitis: utility in the diagnosis of tuberculous meningitis or neurobrucellosis. Clin Infect Dis 20: 525–530.

Lortholary O, Meyohas M-C, Dupont B, et al (1993). Invasive aspergillosis in patients with the acquired immunodeficiency syndrome: report of 33 cases. Am J Med 95: 177–187.

Lukehart SA, Hook EW 3rd, Baker-Zander SA, et al (1988). Invasion of the central nervous system by *Treponema pallidum:* implications for diagnosis and treatment. Ann Intern Med 109: 855–62.

Lyos AT, Malpica A, Estrada R, et al (1993). Invasive aspergillosis of the temporal bone: an unusual manifestation of acquired immunodeficiency syndrome. Am J Otolaryngol 14: 444–448.

Madhoun ZT, Du Bois DB, Rosenthal J, et al (1991). Central diabetes insipidus: a complication of herpes simplex type 2 encephalitis in a patient with AIDS. Am J Med 90: 658–659.

Mahieux F, Gray F, Fenelon G, et al (1989). Acute myeloradiculitis due to cytomegalovirus as the initial manifestition of AIDS. J Neurol Neurosurg Psychiatry 52: 270–274.

Mamelak AN, Obana WG, Flaherty HF, et al (1994). Nocardial brain abscess: treatment strategies and factors influencing outcome. Neurosurgery 35: 622–631.

Marr KA, Boeckh M, Carter RA, et al (2004). Combination antifungal therapy for invasive aspergillosis. Clin Infect Dis 39: 797–802.

Martinez E, Gatell J, Moran Y, et al (1998). High incidence of Herpes zoster in patients with AIDS soon after therapy with protease inhibitors. Clin Infect Dis 27: 1510–1513.

Martinez MS, Gonzalez-Mediero G, Santiago P, et al (2000). Granulomatous amebic encephalitis in a patient with AIDS: isolation of Acanthamoeba sp. group II from brain tissue and successful treatment with sulfadiazine and fluconazole. J Clin Microbiol 38: 3892–3895.

Mascola L, Lieb L, Chiu J, et al (1988). Listeriosis: an uncommon opportunistic infection in patients with acquired immunodeficiency syndrome. Am J Med 84: 162–164.

Masdeu JC, Small CB, Weiss L, et al (1988). Multifocal cytomegalovirus encephalitis in AIDS. Ann Neurol 23: 97–99.

Masliah E, Deteresa RM, Mallory ME, et al (2000). Changes in pathological findings at autopsy in AIDS cases for the last 15 years. AIDS 14: 69–74.

Matthews T, Boehme R (1988). Antiviral activity and mechanism of action of ganciclovir. Rev Infect Dis 10 (suppl 3): S490–S494.

Medical Letter on Drugs and Therapeutics Handbook of Antimicrobial Therapy, 16th Edition., 2002. The Medical Letter Inc., New Rochelle NY.

Meenken C, van den Horn GJ, de Smet MD, et al (1998). Optic neuritis heralding varicella zoster virus retinitis in a patient with acquired immunodeficiency syndrome. Ann Neurol 43: 534–536.

Metze K, Maciel JAJR (1993). AIDS and Chagas's disease. Neurology 43: 447–448(letter).

Miller RG, Storey JR, Greco CM (1990). Ganciclovir in the treatment of progressive AIDS related polyradiculopathy. Neurology 40: 569–574.

Miller RF, Fox JD, Thomas P, et al (1996). Acute lumbosacral polyradiculopathy due to cytomegalovirus in advanced HIV disease: CSF findings in 17 patients. J Neurol Neurosurg Psychiatry 61: 456–460.

Miller RF, Lucas SB, Hall-Craggs MA, et al (1997). Comparison of magnetic resonance imaging with neuropatholical findings in the diagnosis of HIV and CMV associated CNS disease in AIDS. J Neurol Neurosurg Psychiatry 62: 346–351.

Minamoto GY, Sordillo EM (1998). Disseminated nocardiosis in a patient with AIDS: diagnosis by blood and cerebrospinal fluid cultures. Clin Infect Dis 26: 242–243.

Minamoto GY, Barlam TF, Vander NJ (1992). ELS: Invasive aspergillosis in patients with AIDS. Clini Infect Dis 14: 66–74.

Minagar A, Schatz NJ, Glaser JS (2000). Case report: one and a half syndrome and tuberculosis of the pons in a patient with AIDS. AIDS Patient Care STDs 14: 461–463.

Mischel PS, Vinters HV (1995). Coccidioidomycosis of the central nervous system: neuropathological and vasculopathic manifestations and clinical correlates. Clin Infect Dis 20: 400–405.

Mishell JH, Applebaum EL (1990). Ramsay-Hunt syndrome in a patient with HIV infection. Otolaryngol Head Neck Surg 102: 177–179.

Modi M, Mochan A, Modi G (2004). Management of HIV-associated focal brain lesions in developing countries. Q J Med 97: 413–421.

Mohit AA, Santiago P, Rostomily R (2004). Intramedullary tuberculoma mimicking primary CNS lymphoma. J Neurol Neurosurg Psychiatry 75: 1636–1638.

Morgello S, Simpson DM (1994). Multifocal cytomegalovirus demyelinative polyneuropathy associated with AIDS. Muscle Nerve 17: 176–182.

Morgello S, Cho ES, Nielsen S (1987). Cytomegalovirus encephalitis in patients with acquired immunodeficiency syndrome. Hum Pathol 18: 289–297.

Morgello S, Block GA, Price RW, et al (1988). Varicella-zoster virus leukoencephalitis and cerebral vasculopathy. Arch Pathol Lab Med 112: 173–177.

Morrison RE, Harrison S, Tramont EC (1985). Oral amoxicillin, an alternative treatment for neurosyphilis. Genitourin Med 61: 359–62.

Moskowitz LB, Gregorios JB, Hensley GT, et al (1984). Cytomegalovirus induced demyelination associated with acquired immuno deficiency syndrome. Arch Pathol Lab Med 108: 873–877.

Moulignier A, Pialoux G, Dega H, et al (1995). Brain stem encephalitis due to varicella-zoster virus in a patient with AIDS. Clin Infect Dis 20: 1378–1380.

Moulignier A, Mikol J, Gonzalez-Canali G, et al (1996). AIDS-associated cytomegalovirus infection mimicking central nervous system tumors: a diagnostic challenge. Clin Infect Dis 22: 626–631.

Musher DM, Hamill RJ, Baughn RE (1990). Effect of human immunodeficiency virus (HIV) infection on the course of syphilis and on the response to treatment. Ann Intern Med 113: 872–81.

Musiani M, Zerbini M, Venturoli S, et al (1994). Rapid diagnosis of cytomegalovirus encephalitis in patients with AIDS using in situ hybridization. J Clin Pathol 47: 886–891.

Mylonakis E, Paliou M, Sax PE, et al (2000). Central nervous system aspergillosis in patients with human immunodeficiency virus infection: report of 6 cases and review. Medicine 79: 269–280.

Nahmias AJ, Lee FK, Beckman-Nahmias S (1990). Seroepidemiologic and sociological patterns of herpes simplex virus infection in the world. Scand J Infect Dis 69: 19–36.

Nolte CH, Endres AS, Meisel H (2002). Sensory ataxia in myelopathy with chronic hepatitis C infection. Neurology 59: 958–0.

Nunn PP, McAdam KPWJ (1988). Mycobacterial infections and AIDS. Br Med Bull 44: 801–813.

Offenbacher H, Fazekas F, Schmidt R, et al (1991). MRI in tuberculous meningoencephalitis: report of four cases and review of the neuroimaging literature. J Neurol 238: 340–344.

Ogg G, Lynn WA, Peters M, et al (1997). Cerebral nocardial abscesses in a patient with AIDS: correlation of magnetic resonance and white celll scanning images with neuropathological findings. J Infect 35: 311–313.

Origgi L, Massimo V, Carbone A, et al (1998). Central nervous system involvement in patients with HCV-related cryoglobulinemia. Am J Med Sci 315: 208–210.

Otero J, Ribera E, Gavalda J, et al (1998). Response to aciclovir in two cases of herpes zoster leukoencephalitis and review of the literature. Eur J Clin Microbiol Infect Dis 17: 286–289.

Pai M, Flores LI, Pai N, et al (2003). Diagnostic accuracy of nucleic acid amplification tests for tuberculous meningitis: a systematic review and meta-analysis. Lancet Infect Dis 3: 633–643.

Palestine AG, Polis MA, De Smet MD, et al (1991). A randomized, controlled trial of foscarnet in the treatment of cytomegalovirus retinitis in patients with AIDS. Ann Intern Med 115: 665–673.

Pappas PG, Pottage JC, Powderly WG, et al (1992). Blastomycosis in patients with acquired immunodeficiency syndrome. Ann Intern Med 116: 847–853.

Patel VB, Padayatchi N, Bhigjee AI, et al (2004). Multidrug-resistant tuberculous meningitis in Kwa Zulu-Natal, South Aftrica. Clin Infect Dis 38: 851–856.

Patey O, Nedelec C, Emond JP, et al (1989). *Listeria moncytogenes* septicemia in an AIDS patient with a brain abscess. Eur J Clin Microbiol Infect Dis 8: 746–748.

Pepose JS, Hilborne LH, Cancilla PA, et al (1984). Concurrent herpes simplex and cytomegalovirus retinitis and encephalitis in the acquired immune deficiency syndrome (AIDS). Ophthalmology 91: 1669–1677.

Peters M, Timm U, Schurmann D, et al (1992). Combined and alternating ganciclovir and foscarnet in acute and maintenance therapy of human immuno deficiency virus-related cytomegalovirus encephalitis refractory to ganciclovir alone. Clin Investig 70: 456–458.

Petito CK, Cho E-S, Lemann W, et al (1986). Neuropathology of acquired immunodeficiency syndrome (AIDS): an autopsy review. J Neuropathol Exp Neurol 45: 635–646.

Pierelli F, Tilia G, Damiani A, et al (1997). Brainstem CMV encephalitis in AIDS: clinical case and MRI features. Neurology 48: 529–530.

Pillai S, Mahmood MA, Limaye SR (1989). Herpes-zoster ophthalmicus, contralateral hemiplegia and recurrent ocular toxoplasmosis in a patient with acquired immune deficiency syndrome-related complex. J Clin Neuroophthalmol 9: 229–233.

Pintado V, Gomez-Mampaso E, Cobo J, et al (2003). Nocardia infection in patients infected with the human immunodeficiency virus. Clin Microbiol Infect 9: 716–720.

Pitchenik AE, Cole C, Russell BW, et al (1984). Tuberculosis, atypical mycobacteriosis, and the acquired immunodeficiency syndrome among Haitian and non-Haitian patients in South Florida. Ann Intern Med 101: 641–645.

Pitlik SD, Fainstein V, Bolivar R, et al (1983). Spectrum of central nervous system complications in homosexual men with acquired immune deficiency syndrome. J Infect Dis 148: 771–772(letter).

Portegies P, Solod L, Cinque P, et al (2004). Guidelines for the diagnosis and management of neurological complications of HIV infection. Eur J Neurol 11: 297–304.

Poscher ME (1994). Successful treatment of varicella zoster virus meningoencephalitis in patients with AIDS: report of four cases and review. AIDS 8: 1115–1117.

Price R, Chernik NL, Horta-Barbosa L, et al (1973). Herpes simplex encephalitis in an anergic patient. Am J Med 54: 222–227.

Price TA, Digioia RA, Simon GL (1992). Ganciclovir treatment of cytomegalovirus ventriculitis in a patient infected with human immunodeficiency virus. Clin Infect Dis 15: 606–608.

Pursell KJ, Telzak EE, Armstrong D (1992). Aspergillus species colonization and invasive disease in patients with AIDS. Clin Infect Dis 14: 141–148.

Quereda C, Corral I, Laguna F, et al (2000). Diagnostic utility of a multiplex herpesvirus pcr assay performed with cerebrospinal fluid from human immunodeficiency virus-infected patients with neurological disorders. J Clin Microbiol 38: 3061–3067.

Rana FS, Hawken MP, Mwachari C, et al (2000). Autopsy study of HIV-1-postitive and HIV-1-negative adult medical patients in Nairobi, Kenya. J Acquir Immune Defic Syndr 24: 23–29.

Read RC, Vilar FJ, Smith TL (1998). AIDS-related herpes simplex virus encephalitis during maintenance foscarnet therapy. Clin Infect Dis 26: 513–514.

Regnery RL, Childs JE, Koehler JE (1995). Infections associated with *Bartonella* species in persons infected with human immunodeficiency virus. Clin Infect Dis 21: S94–S98.

Revello MG, Percivalle E, Sarasini A, et al (1994). Diagnosis of human cytomegalovirus infection of the nervous system by pp 65 detection in polymorphonuclear leukocytes of cerebrospinal fluid from AIDS patients. J Infect Dis 170: 1275–1279.

Ribera E, Martinez-Vazquez JM, Ocana I, et al (1987). Activity of adenosine deaminase in cerebrospinal fluid for the diagnosis and follow-up of tuberculous meningitis in adults. J Infect Dis 155: 603–607.

Robert ME, Geraghty JJ III, Miles SA, et al (1989). Severe neuropathy in a patient with acquired immune deficiency syndrome (AIDS). Evidence for widespread cytomegalovirus infection of peripheral nerve and human immunodeficiency virus like immunoreactivity of anterior horn cells. Acta Neuropathol 79: 255–261.

Rocha A, De Menses ACO, Da Silva AM, et al (1994). Pathology of patients with Chagas' desease and acquired immunodeficiency syndrome. Am J Trop Med Hyg 50: 261–268.

Rosemberg S, Chaves CJ, Higuchi ML, et al (1992). Fatal meningoencephalitis caused by reactivation of *Trypanosoma cruzi* infection in a patient with AIDS. Neurology 42: 640–642.

Rosenblum MK (1989). Bulbar encephalitis complicating trigeminal zoster in the acquired immune deficiency syndrome. Hum Pathol 20: 292–295.

Rostad SW, Olson K, McDougall J, et al (1989). Transsymaptic spread of varicella-zoster virus through the visual system: a mechanism of viral dissemination in the central nevous system. Hum Pathol 20: 174–179.

Roullet E, Assuerus V, Gozlan J, et al (1994). Cytomegalovirus multifocal neuropathy in AIDS: analysis of 15 consecutive cases. Neurology 44: 2174–2182.

Rousseau F, Perronne C, Raguin G, et al (1993). Necrotizing retinitis and cerebral vasculitis due to varicella-zoster virus in patients infected with the human immunodeficiency virus. Clin Infect Dis 17: 943–944 (letter).

Ryan EL, Morgello S, Isaacs K, et al (2004). Neuropsychiatric impact of hepatitis C on advanced HIV. Neurology 62: 957–962.

Ryder JW, Croen K, Kleinschmidt-Demasters BK, et al (1986). Progressive encephalitis three months after resolution of cutaneous zoster in a patient with AIDS. Ann Neurol 19: 182–188.

Sactor N, Lyles RH, Skolasky R, et al (2001). HIV-associated neurologic disease incidence changes: Multicenter AIDS Cohort Study 1990–1998. Neurology 56: 257–260.

Sadler M, Morris-Jones S, Nelson M, et al (1997). Successful treatment of cytomegalovirus encephalitis in an AIDS patient using cidofovir. AIDS 11: 1293–1294.

Safrin S (1992). Treatment of aciclovir-resistant herpes simplex virus infections in patients with AIDS. J Acquir Immune Defic Syndr 5 (suppl 1): S29–S32.

Safrin S, Crumpacker C, Chatis P (1991). A controlled trial comparing foscarnet with vidarabine for aciclovir-resistant muco-cutaneous herpes simplex in the acquired immunodeficiency syndrome. N Engl J Med 325: 551–555.

Sage JI, Weinstein MP, Miller DC (1985). Chronic encephalitis possibly due to herpes simplex virus: two cases. Neurology 35: 1470–1472.

Said G, Lacroix C, Chemovilli P (1991). Cytomegalovirus neuropathy in acquired immunodeficiency syndrome: a clinical and pathological study. Ann Neurol 29: 139–146.

Sanchez-Portocarrero J, Perez-Cecilia E, Romero-Vivas J (1999). Infection of the central nervous system by *Mycobacterium tuberculosis* in patients infected with human immunodeficiency virus (the new neurotuberculosis). Infection 27: 313–317.

Sanchez-Portocarrero J, Perez-Cecilia E, Corral O, et al (2000). The central nervous system and infection by candida species. Diagn Mircrobiol Infect Dis 37: 169–179.

Scarpellini P, Racca S, Cinque P, et al (1995). Nested polymerase chain reaction for diagnosis and monitoring treatment response in AIDS patients with tuberculous meningitis. AIDS 9: 895–900.

Schaaf HS, Gie RP, Van Rie A, et al (2000). Second episode of tuberculosis in an HIV-infected child: relapse or reinfection? J Infect 41: 100–103.

Schiff D, Rosenblum MK (1998). Herpes simplex encephalitis (HSE) and the immunocompromised: a clinical and autopsy study of HSE in the settings of cancer and human immunodeficiency virus-type 1 infection. Hum Pathol 29: 215–222.

Schmidbauer M, Budka H, Ulrich W, et al (1989). Cytomegalovirus (CMV) disease of the brain in AIDS and connatal infection: a comparative study by histology, immunocytochmistry and in situ hybridization. Acta Neuropathol 79: 286–293.

Schmidbauer M, Budka H, Ambros P (1989). Herpes simplex virus (HSV) DNA in microglial nodular brainstem encephalitis. J Neuropathol Exp Neurol 48: 645–652.

Schutte C-M (2001). Clinical, cerebrospinal fluid and pathological findings and outcomes in HIV-positive and HIV-negative patients with tuberculous meningitis. Infection 29: 213–217.

Schwartzman WA, Patnaik M, Barka NE, et al (1994). *Rochalemaea* antibodies in HIV-associated neurologic disease. Neurology 44: 1312–1316.

Schwartzman WA, Patnaik M, Angulo FJ, et al (1995). *Bartonella* (*Rochalimaea*) antibodies, dementia, and cat ownership among men infected with human immunodeficiency virus. Clin Infect Dis 21: 954–959.

Shafer RW, Edlin BR (1996). Tuberculosis in patients infected with human immunodeficiency virus: perspective on the past decade. Clin Infect Dis 22: 683–704.

Shafer RW, Kim DS, Weiss JP, et al (1991). Extrapulmonary tuberculosis in patients with human immunodeficiency syndrome. Medicine 70: 384–397.

Sharer LR, Kapila R (1985). Neuropatholgic observations in acquired immunodeficiency syndrome (AIDS). Acta Neuropathol 66: 188–198.

Silber E, Sonnenberg P, Ho KC, et al (1999). Meningitis in a community with a high prevalence of tuberculosis and HIV infection. J Neurol Sci 162: 20–26.

Silva N, O'Bryan L, Medeiros E, et al (1999). *Trypanosoma cruzi* meningoencephalitis in HIV-infected patients. AIDS Res Hum Retroviruses 20: 342–349.

Singh N, Yu VL, Rihs JD (1991). Invasive aspergillosis in AIDS. South Med J 84: 822–827.

Singh VR, Smith DK, Lawrence J, et al (1996). Coccidiomycosis in patitents infected with human immunodeficiency virus: review of 91 cases at a single institution. Clin Infect Dis 23: 563–568.

Sison JP, Kemper CA, Loveless M, et al (1995). Disseminated acanthamoeba infection in patients with AIDS: case reports and review. Clin Infect Dis 20: 1207–1216.

Small P, McPhaul LW, Sooy CD, et al (1989). Cytomegalovirus infection of the laryngeal nerve presenting as hoarseness in patients with acquired immunodeficiency syndrome. Am J Med 86: 108–110.

Snider WD, Simpson DM, Nielsen S, et al (1983). Neurological complications of acquired immune deficiency syndrome: Analysis of 50 patients. Ann Neurol 14: 403–418.

Snoeck R, Gerard M, Sadzot-Delvaux C, et al (1994). Meningoradiculoneuritis due to aciclovir-resistant varicella zoster virus in an acquired immune deficiency syndrome patient. J Med Virol 42: 338–347.

So YT, Olney RK (1994). Acute lumbosacral polyradiculopathy in acquired immunodeficiency syndrome: experience in 23 patients. Ann Neurol 35: 53–58.

Solari A, Saavedra H, Sepulveda C, et al (1993). Successful treatment of *Trypanosoma cruzi* encephalitis in a patient with hemophilia and AIDS. Clin Infect Dis 16: 255–259.

Soto-Hernandez JL, Ostrosky-Zeichner L, Tavera G, et al (1996). Neurocysticercosis and HIV infection: report of two cases and review. Surg Neurol 45: 57–61.

Spach DH, Panther LA, Thorning DR, et al (1992). Intracerebral bacillary angiomatosis in a patient infected with human immunodeficiency virus. Ann Intern Med 116: 740–742.

Spitzer PG, Hammer SM, Karchmer AW (1986). Treatment of *Listeria monocytogenes* infection with trimethoprim-sulfamethoxazole: case report and review of the literature. Rev Infect Dis 8: 427–430.

Strauss M, Fine E (1991). *Aspergillus otomastoiditis* in acquired immunodeficiency syndrome. Am J Otolaryngol 12: 49–53.

Studies of Ocular Complications of Aids Research Group In Collaboration with the Aids Clinical Trials Group (1992). Mortality in patients with acquired immunodeficiency syndrome treated with either foscarnet or ganciclovir for cytomegalovirus retinitis. N Engl J Med 326: 213–220.

Studies Of Ocular Complications Of Aids Research Group In Collaboration With The Aids Clincal Trials Group (1996). Combination foscarnet and ganciclovir therapy vs monotherapy for the treatment of relapsed cytomegalovirus retinitis in patients with AIDS. Arch Ophthalmol 114: 23–33.

Sullivan WM, Kelley GG, O'Connor P, et al (1992). Hypopituitarism associated with a hypothalamic CMV infection in a patient with AIDS. Am J Med 92: 221–223.

Sunderam G, McDonald RJ, Maniatis T, et al (1986). Tuberculosis as a manifestation of the acquired immunodeficiency syndrome (AIDS). JAMA 256: 362–366.

Taelman H, Batungwanayo J, Clerinx J, et al (1992). (letter). N Engl J Med 327: 1171.

Talpos D, Tien RD, Hesselink JR (1991). Magnetic resonance imaging of AIDS related polyradiculopathy. Neurology 41: 1996–1997.

Tan G, Kaufman L, Peterson EM, et al (1992). Disseminated atypical blastomycosis in two patients with AIDS. Clin Infect Dis 16: 107–111.

Tan B, Weldon-Linne M, Rhone DP, et al (1993). Acanthamoeba infection presenting as skin lesions in patients with the acquired immunodeficiency syndrome. Arch Pathol Lab Med 117: 1043–1046.

Tan SV, Guiloff RJ, Scaravilli F, et al (1993). Herpes simplex type 1 encephalitis in acquired immunodeficiency syndrome. Ann Neurol 34: 619–622.

Tembl JI, Ferrer JM, Sevilla MT, et al (1999). Neurologic complications associated with hepatitis C virus infection. Neurology 53: 861–864.

Thonell L, Pendle S, Sacks L (2000). Clinical and radiological features of South Aftrican with tuberculomas of the brain. Clin Infect Dis 31: 619–620.

Thornton CA, Houston S, Latif S (1992). Neurocysticercosis and human immunodeficiency virus infection, a possible association. Arch Neurol 49: 963–965.

Thwaites GE, Hien TT (2005). Tuberculous meningitis: many questions, too few answers. Lancet Neurol 4: 160–170.

Thwaites GE, Caws M, Chau TTH, et al (2004a). Comparison of conventional bacteriology with nucleic acid amplification (amplified mycobacterium direct test) for diagnosis of tuberculous meningitis before and after inception of antituberculosis chemotherapy. J Clin Microbiol 42: 996–1002.

Thwaites GE, Bang HD, Dung NH, et al (2004b). Dexamethasone for the treatment of tuberculous meningitis in adolescents and adults. N Engl J Med 351: 1741–1751.

Traub M, Colchester ACF, Kingsley DPE, et al (1984). Tuberculosis of the central nervous system. Q J Med 53: 81–100.

Tucker T, Dix RD, Katzen C, et al (1985). Cytomegalovirus and herpes simplex ascending myelitis in a patient with acquired immune deficiency syndrome. Ann Neurol 18: 74–79.

Uttamchandani RB, Daikos GL, Reyes RR, et al (1994). Nocardiosis in 30 patients with advanced human immunodeficiency virus infection: clinical features and outcome. Clin Infect Dis 18: 348–353.

Vago L, Nebuloni M, Sala E, et al (1996). Coinfection of the central nervous system by cytomegalovirus and herpes simplex virus type 1 or 2 in AIDS patients: autopsy study on 82 cases by immunohistochemistry and polymerase chain reaction. Acta Neuropathol 92: 404–408.

Vago L, Bonetto S, Nebuloni M, et al (2002). Arminio-Monforte, A. Pathological findings in the central nervous system of AIDS patients on assumed antiretroviral therapeutic regimens: retrospective study of 1597 autopsies. AIDS 16: 1925–1928.

Velasco-Martinez JJ, Guerrero-Espejo A, Gomez-Mampaso E, et al (1995). Tuberculous brain abscess should be considered in HIV/AIDS patients. AIDS 9: 1197–1199.

Villoria MF, De La Torre J, Fortea F, et al (1992). Intracranial tuberculosis in AIDS: CT and MRI findings. Neuroradiology 34: 11–14.

Villoria MF, Fortea F, Moreno S, et al (1995). MR imaging and CT of central nervous system tuberculosis in the patient with AIDS. Radiol Clin North Am 33: 805–820.

Vinters HV, Anders KH (1990). Neuropathology of AIDS. CRC Press, Boca Raton, pp. 61–65, 69.

Vinters HV, Guerra WF, Eppolito L, et al (1988). Necrotizing vasculitis of the nervous system in a patient with AIDS-related complex. Neuropathol Appl Neurobiol 14: 417–424.

Vinters HV, Kwok MK, Ho HW, et al (1989). Cytomegalovirus in the nervous system of patients with the acquired immune deficiency syndrome. Brain 112: 245–268.

Visse B, Dumont B, Huraux J-M, et al (1998). Single amino acid change in DNA polymerase is associated with foscarnet resistance in a varicella-zoster virus strain recovered from a patient with AIDS. J Infect Dis 178 (suppl 1): S55–57.

Vital C, Monlum E, Vital A, et al (1995). Concurrent herpes simplex type 1 necrotizing encephalitis, cytomegalovirus ventriculoencephalitis and cerebral lymphoma in an AIDS patient. Acta Neuropathol 89: 105–108.

Vlcek B, Burchiel KJ, Gordon T (1984). Tuberculous meningitis presenting as an obstructive myelopathy: case report. J Neurosurg 60: 196–199.

Von Giesen H-J, Heintges T, Abbasi-Boroudjeni N, et al (2004). Psychomotor slowing in hepatitis C and HIV infection. J Acquir Immune Defic Syndr 35: 131–137.

Vullo V, Mastroianni M, Ferone U, et al (1997). Nervous system involvement as a relapse of disseminated histoplasmosis in an Italian AIDS patient. J Infect 35: 83–84.

Wallace MR, Persing DH, McCutchan JA, et al (2001). *Bartonella henselae* serostatus is not correlated with neurocognitive decline in HIV infection. Scand J Infect Dis 33: 593–595.

Warren KG, Brown SM, Wroblewska Z, et al (1978). Isolation of latent herpes simplex virus from superior cervical and vagus ganglions of human beings. New Engl J Med 298: 1068–1069.

Weaver S, Rosenblum MK, Deangelis LM (1999). Herpes varicella zoster encephalitis in immunocompromised patients. Neurology 52: 193–195.

Weber R, Deplazes P, Flepp M, et al (1997). Cerebral microsporidiosis due to *Encephalitozoon cuniculi* in a patient with human immunodeficiency virus infection. New Engl J Med 336: 474–478.

Weidenheim KM, Nelson SJ, Kure K, et al (1992). Unusual patterns of histoplasma capsulatuum in a patient with the acquired immunodeficiency virus. Hum Pathol 23: 581–586.

Wheat J (1995). Endemic mycoses in AIDS: a clinical review. Clin Microbiol Rev 8: 146–159.

Wheat LJ, Batteiger BE, Sathapatayavongs B (1990a). Histoplasma capsulatum infections of the nervous system: a clinical review. Medicine 69: 244–260.

Wheat LJ, Connolly-Stringfield PA, Baker RL, et al (1990b). Disseminated histoplasmosis in the acquired immune deficiency syndrome: clinical findings, diagnosis and treatment, and review of the literature. Medicine 69: 361–374.

Wheat LJ, Cloud G, Johnson PC, et al (2001a). Clearance of fungal burden during treatment of disseminated histoplasmosis with liposomal amphotericin B versus itraconazole. Antimicrob Agents Chemother 45: 2354–2357.

Wheat LJ, Connolly P, Smedema M, et al (2001b). Emergence of resistance t fluconazole as a cause of failure during treatment of histoplasmosis in patients with acquired immunodeficiency syndrome. Clin Infect Dis 33: 1910–1913.

Whimbey E, Gold JWM, Polsky B, et al (1986). Bacteremia and fungemia in patients with the acquired immunodeficiency syndrome. Ann Intern Med 10: 511–514.

Whiteman ML (1997). Neuroimaging of central nervous system tuberculosis in HIV-infected patients. Neuroimaging Clin North Am 7: 199–214.

Whiteman MLH, Dandapani BK, Shebert RT, et al (1994). MRI of AIDS-related polyradiculomyelitis. J Comput Assist Tomogr 18: 7–11.

Whiteman M, Espinoza L, Post MJ, et al (1995). Central nervous system tuberculosis in HIV-infected patients: clinical and radiographic findings. Am J Neuroradiol 16: 1319–1327.

Wildemann B, Haas J, Lynen N, et al (1998). Diagnosis of cytomegalovirus encephalitis in patients with AIDS by quantitation of cytomegalovirus genomes in cells of cerebrospinal fluid. Neurology 50: 693–697.

Wiley CA, Nelson JA (1988). Role of human immunodeficiency virus and cytomegalovirus in AIDS encephalitis. Am J Pathol 133: 73–81.

Wiley CA, Van Patten PD, Carpenter PM, et al (1987). Acute ascending necrotizing myelopathy caused by herpes simplex virus type 2. Neurology 37: 1791–1794.

Wiley CA, Sofrin RE, Davis CE, et al (1987). Acanthomoeba meningoencephalitis in an AIDS patient. J Infect Dis 155: 130–133.

Witzig RS, Hoadley DJ, Greer DL, et al (1994). Blastomycosis and human immunodeficiency virus: three new cases and review. South Med J 87: 715–719.

Wolf DG, Spector SA (1992). Diagnosis of human cytomegalovirus central nervous system disease in AIDS patients by DNA amplification from cerebrospinal fluid. J Infect Dis 166: 1412–1415.

Wong MT, Dolan MJ, Lattuada CP Jr, et al (1995). Neuroretinitis, aseptic meningtis, and lymphadenitis associated with *Bartonella* (*Rochalimaea*) *henselae* infection in immunocompetent patients and patients infected with human immunodeficiency virus type 1. Clin Infect Dis 21: 352–360.

Woods GL, Goldsmith JC (1990). Aspergillus infection of the central nervous system in patients with acquired immunodeficiency syndrome. Arch Neurol 47: 181–184.

Wu HS, Kolonoski P, Chang YY, et al (2000). Invasion of the brain anc chronic central nervous system infection aftger systemic mycobacterium avium complex infection in mice. Infect Immun 68: 2979–2984.

Yachnis AT, Berg J, Martinez-Salazar A, et al (1996). Disseminated microsporidiosis especially infecting the brain, heart, and kidneys. Report of a newly recognized pansporoblastic species in two symptomatic AIDS patients. Am J Clin Pathol 106: 535–543.

Yau TH, Rivera-Velazquez PM, Mark AS, et al (1996). Unilateral optic neuritis caused by Histoplasma capsulatum in a patient with the acquired immune deficiency syndrome. Am J Ophthalmol 121: 324.

Yoo TW, Mikotic A, Cornford ME, et al (2004). Concurrent cerebral American trypanosomiasis and toxoplasmosis in a patient with AIDS. Clin Infect Dis 39: e30–e34.

Yoritaka A, Ohta K, Kishida S (2005). Herpetic lumbosacral radiculoneuropathy in patients with human immunodeficiency virus infection. Eur Neurol 53: 179–181.

Young LS, Inderlied CV, Berlin OG, et al (1986). Mycobacterial infection in AIDS patients, with an emphasis on the *Mycobacterium avium* complex. Rev Infect Dis 8: 1024–1033.

Zagardo MT, Castellani RJ, Zoarski GH, et al (1997). Granulomatous amebic encephalitis caused by leptomyxid amebae in an HIV-infected patient. Am J Neuroradiol 18: 903–908.

Zambrano W, Perez GM, Smith JL (1987). Acute syphilitis blindness in AIDS. J Clin Neuro-ophthalmol 7: 1–5.

Zeller V, Nardi A-L, Truffot-Pernot C, et al (2003). Disseminated infection with a mycobacterium related to *Mycobacterium triplex* with central nervous system involvement associated with AIDS. J Clin Microbiol 41: 2785–2787.

Further Reading

Afghani B, Lieberman JM (1994). Paradoxical enlargement or development of intracranial tuberculomas during therapy: case report and review of the literature. Clin Infect Dis 19: 1092–1099.

Alvarez S, McCabe WR (1984). Extrapulmonary tuberculosis revisited: a review of experience at Boston City and other hospitals. Medicine 63: 25–54.

Ampel NM, Dols CL, Galgiani JN (1992). Coccidioidomycosis during human immunodeficiency virus infection: results of a prospective study in a coccidiodal endemic area. Am J Med 94: 235–239.

Anders H-J, Goebel FD (1999). Neurological manifestations of cytomegalovirus infection in the acquired immunodeficiency syndrome. Int J STD AIDS 10: 151–161.

Arribas JR, Storch GA, Clifford DB, et al (1996). Cytomegalovirus encephalitis. Ann Intern Med 125: 577–587.

Bacellar H, Muñoz A, Miller EN, et al (1994). Temporal trends in the incidence of HIV-1-related neurolgic diseases: Multicenter AIDS Cohort Study, 1985–1992. Neurology 44: 1892–1900.

Balfour HH, Benson C, Braun J (1994). Management of aciclovir-resistant herpes simplex and varicella-zoster virus infections. J Acquir Immune Defic Syndr 7: 254–260.

Berger JR, Moskowitz L, Fischl M, et al (1987). Neurologic disease as the presenting manifestation of acquired immunodeficiency syndrome. South Med J 80: 683–686.

Bishburg E, Eng RH, Slim J, et al (1989). Brain lesions in patients with acquired immunodeficiency syndrome. Arch Intern Med 149: 941–943.

British Medical Research Council (1948). Streptomycin treatment of tuberculous meningitis. Lancet 1: 582–596.

Bronnimann DA, Adam RD, Galgiani JN, et al (1987). Coccidioidomycosis in the acquired immunodeficiency syndrome. Ann Intern Med 106: 372–379.

Chaisson RE, Schecter GF, Theuer CP, et al (1987). Tuberculosis in patients with the acquired immunodeficiency syndrome. Clinical features, response to therapy, and survival. Am Rev Respir Dis 136: 570–574.

Cinque P, Cleator GM, Weber T, et al (1998a). Diagnosis and clinical management of neurological disorders caused by cytomegalovirus in AIDS patients. J Neurovirol 4: 120–132.

Cleland PG, Lawande RV, Onyemelukwe G, et al (1982b). Chronic amoebic meningo encephalitis. Arch Neurol 39: 56–57.

Cole EL, Meisler DM, Calabiese LH, et al (1984). Herpes zoster ophthalmicus and acquired immune deficiency syndrome. Arch Ophthalmol 102: 1027–1029.

Dahan P, Haettich B, Le Parc JM, et al (1988). Meningoradiculitis due to herpes simplex virus disclosing HIV infection [letter]. Ann Rheum Dis 47: 440.

De Angelis L, Weaver S, Rosenblum M (1994). Herpes zoster (HVZ) encephalitis in immunocompromised patients [abstract]. Neurology 44 (Suppl 2): A332.

Donald PR, Victor TC, Jordaan AM, et al (1993). Polymerase chain reaction in the diagnosis of tuberculous meningitis. Scand J Infect Dis 25: 613–617.

Dutcher JP, Marcus SL, Tanowitz HB, et al (1990). Disseminated strongyloidiasis with central nervous system involvement diagnosed antemortem in a patient with acquired immunodeficiency syndrome and Burkitts lymphoma. Cancer 66: 2417–2420.

Eide FF, Gean AD, So YT (1993). Clinical and radiographic findings in disseminated tuberculosis of the brain. Neurology 43: 1427–1429.

Gray F, Gherardi R, Keohane C, et al (1988). Pathology of the central nervous system in 40 cases of acquired immune deficiency syndrome (AIDS). Neuropathol Appl Microbiol 14: 365–380.

Harvey RL, Chandrasekar PH (1988). Chronic meningitis caused by Listeria in a patient infected with human immunodeficiency virus [letter]. J Infect Dis 157: 1091–1092.

Holtz HA, Lavery DP, Kapila R (1985). Actinomycetales infection in patients with acquired immunodeficiency syndrome. Ann Intern Med 102: 203–205.

Idemyor V, Cherubin CE (1992). Pleurocerebral nocardia in a patient with human immunodeficiency virus. Ann Pharmacother 26: 188–189.

Jacobs JL, Murray HW (1986). Why is *L. monocytogenes* not a pathogen in the acquired immunodeficiency sydrome? Arch Intern Med 146: 1299–1300.

Jennings TS, Hardin TC (1993). Treatment of aspergillosis with itraconazole. Ann Pharmacother 27: 1206–1211.

Kaneko K, Onodera O, Miyatake T, et al (1990). Rapid diagnosis of tuberculous meningitis by polymerase chain reaction (PCR). Neurology 40: 1617–1618.

Laissy JP, Soyer P, Parlier C, et al (1994). Persistent enhancement after treatment for cerebral toxoplasmosis in patients with AIDS: predictive value for subsequent recurrence. Am J Neuroradiol 15: 1773–1778.

Mazlo M, Tariska I (1982). Are astrocytes infected in progressive multifocal leukoencephalopathy. Acta Neuropathol 56: 45–51.

McCutchan JA (1995). Cytomegalovirus infections of the nervous system in patients with AIDS. Clin Infect Dis 20: 747–754.

Mullin GE, Sheppell AL (1987). *Listeria monocytogenes* and the acquired immunodeficiency syndrome. Arch Intern Med 147: 176.

Norden CW, Ruben FL, Selker R (1983). Nonsurgical treatment of cerebral nocardiosis. Arch Neurol 40: 594–595.

Palestine AG, Rodrigues MM, Macher AM, et al (1984). Ophthalmic involvement in acquired immunodeficiency syndrome. Ophthalmology 91: 1092–1099.

Wheat LJ, Kohler RB, Tewari RP, et al (1989). Significance of histoplasma antigen in the cerebrospinal fluid of patients with meningitis. Arch Intern Med 149: 302–304.

Wheat LJ, Connolly-Stringfield P, Blair R, et al (1991). Histoplasma relapse in patients with AIDS: detection using histoplasma capsulatum variety capsulatum antigen levels. Ann Intern Med 115: 936–941.

White AC, Dakik H, Diaz P (1995). Asymptomatic neurocysticercosis in a patient with AIDS and cryptococcal meningitis. Am J Med 99: 101–102.

Handbook of Clinical Neurology, Vol. 85 (3rd series)
HIV/AIDS and the Nervous System
P. Portegies, J. R. Berger, Editors

Chapter 15

Primary central nervous system lymphoma

EDO RICHARD,[1] MATTHIJS C. BROUWER,[1] AND PETER PORTEGIES[2]*

[1]*Department of Neurology, Academic Medical Center, University of Amsterdam, Amsterdam, The Netherlands*
[2]*Department of Neurology, OLVG Hospital, Amsterdam, The Netherlands*

15.1. Introduction

Before the advent of the human immunodeficiency virus (HIV) epidemic, primary central nervous system lymphoma (primary CNS lymphoma) used to be a rare malignancy, accounting for less than 2% of central nervous system (CNS) tumors and 0.5–2% of all non-Hodgkin's lymphomas (NHLs) (Freeman et al., 1972; Jellinger et al., 1975; Zimmerman, 1975). In the 1980s a significant increase in incidence was found among HIV-infected patients which initially led to the belief that primary CNS lymphoma was going to be the most frequent of CNS tumors (Eby et al., 1988; Hochberg and Miller, 1988; Fine and Mayer, 1993). The Center for Disease Control and Prevention (USA) recognized primary CNS lymphoma as an AIDS-defining illness in 1993 (Center for Disease Control, 1993). At that point, the risk of developing primary CNS lymphoma in HIV patients was 3600 times increased compared to the normal population (Coté et al., 1996). Following the introduction of highly active antiretroviral therapy (HAART), the incidence of primary CNS lymphoma decreased dramatically (Ammassari et al., 2000; Kadan-Lottick et al., 2001; Newell et al., 2004; Robotin et al., 2004). Despite this decrease, primary CNS lymphoma remains the second most common cause of focal brain lesions in AIDS after toxoplasmosis with limited therapeutic options and a dismal prognosis (Ammassari et al., 2000).

15.2. Epidemiology

Until the 1980s primary CNS lymphoma was a rare malignancy, mainly occurring in immunosuppressed patients in the sixth decade with an incidence in 1973 of 0.27 per 1,000,000 in the USA. A steady increase was seen before the HIV epidemic up to 0.75 per 1,000,000 (Eby et al., 1988; Schabet, 1999). This was followed by a dramatic increase in primary CNS lymphoma, soon recognized to be largely attributable to HIV (Levy et al., 1985; Gray et al., 1988). Peak incidence was reached in 1995 at 10.2 per 1,000,000, two years after the peak incidence of AIDS (Kadan-Lottick et al., 2002). A meta-analysis of 47,936 patients showed a risk of developing primary CNS lymphoma in AIDS patients of 1.7 cases per 1000 per year from 1992 to 1995 (International Collaboration on HIV and Cancer, 2000). The same meta-analysis showed a 58% decrease in incidence to 0.7 cases per 1000 per year after the introduction of HAART. Other studies confirmed the observed decline in incidence in the late 1990s (Ammassari et al., 2000; Kirk et al., 2001; Bonnet et al., 2004; Panageas et al., 2005).

Final diagnosis in HIV patients presenting with a focal brain lesion was primary CNS lymphoma in 27%, making it the second most prevalent cause after toxoplasmosis encephalitis (Ammassari et al., 2000). In a study of 111 patients concurrent cerebral toxoplasmosis was found in 10% of primary CNS lymphoma patients (Newell et al., 2004). The prevalence of primary CNS lymphoma in AIDS patients found in autopsy studies was 9–14%, making it the third most seen AIDS CNS complication after HIV encephalopathy and cytomegalovirus encephalitis (Davies et al., 1997; Masliah et al., 2000). The patients in these studies, however, were mostly studied before HAART. Before 1997 primary CNS lymphoma was the AIDS-defining illness in 17–30%; with HAART this number rose to 56% in an Australian study (Robotin et al., 2004). In known AIDS patients, average time from AIDS diagnosis to primary CNS lymphoma is approximately 10 months (Newell et al., 2004). In 2000 primary CNS lymphoma was

*Correspondence to: Peter Portegies, MD, PhD, Department of Neurology, OLVG Hospital, PO Box 95500, 1090 HM Amsterdam, The Netherlands. E-mail: p.portegies@olvg.nl, Tel: +31-20-599-3044, Fax: +31-20-599-3845.

reported to be responsible for 10% of HIV malignancy-related deaths (Bonnet et al., 2004).

Mean age of AIDS patients with primary CNS lymphoma is 35–37 years in contrast to non-AIDS patients who are mostly over 60 (Coté et al., 1996; Newell et al., 2004). Some 96% of all primary CNS lymphoma patients are male, comparable with the male predominance in HIV patients of 91% (Coté et al., 1996). On presentation, average CD^4 count is 10 and opportunistic infections are regularly seen. This did not significantly change in the HAART era (Newell et al., 2004; Robotin et al., 2004). Higher CD^4 counts are associated with prolonged survival (Newell et al., 2004). The route of HIV transmission did not affect the chance of developing primary CNS lymphoma (Kirk et al., 2001; Gendelman et al., 2005).

15.3. Pathogenesis

The pathogenesis of primary CNS lymphoma in AIDS is complicated and only partly understood and is different from primary CNS lymphoma in immunocompetent patients (Paulus, 1999). The Epstein–Barr virus (EBV) plays a very prominent role in the pathogenesis of primary CNS lymphoma in AIDS, as opposed to immunocompetent patients, in which EBV infection is less commonly observed. Primary CNS lymphoma in AIDS are B-cell lymphomas which display large-cell and immunoblastic histologies and usually contain large areas of necrosis. EBV is present in virtually all primary CNS lymphoma in AIDS. Several EBV proteins, such as latent membrane protein-1 (LMP-1) and Epstein–Barr nuclear antigen-2 (EBNA-2), are expressed in these tumors. These proteins might contribute to the malignant transformation in several ways. They may activate B-cells, upregulate cell adhesion, enable their entry into the CNS and interfere with apoptosis. They might also be able to inactivate specific tumor suppressor genes (Schlegel et al., 2000). Several other genetic factors are involved as well, of which translocations involving different oncogenes, e.g. bcl-6 and c-myc, are involved. Considering the consistent monoclonality of this tumor, other genetic factors are probably involved which have yet to be discovered (Carbone, 2002; Schlegel et al., 2000). In addition to translocations, gross chromosomal aberrations occur, although whether these are pathogenetically important is not yet clear.

Several other viruses, like cytomegalovirus, other herpes viruses (especially HHV-6) and polyoma viruses have been implicated in the pathogenesis of primary CNS lymphoma in AIDS, but no solid evidence of involvement of any of these has been found.

15.4. Pathology

Primary CNS lymphomas in HIV patients are 98% B-cell NHLs, usually of high or intermediate malignancy grade (Table 15.1) (Coté et al., 1996; Bindal et al., 1997; Paulus, 1999; Gendelman et al., 2005). Further classification of primary CNS lymphoma has proven to be difficult using histopathologic classification schemes for systemic lymphomas like the Kiel Classification and the Working Formulation. Using the more recently developed REAL or WHO classifications, primary CNS lymphomas are usually classified as high-grade B-cell lymphoma, Burkitt-type and diffuse large B-cell lymphoma (Harris et al., 1997; Paulus, 1999). It remains unclear what clinical relevance these classifications have in primary CNS lymphoma in HIV patients.

The macroscopic appearance of primary CNS lymphoma is that of a quite well demarcated multifocal tumor. Affected brain tissue is softened and discolored. If primary CNS lymphoma had spread to the subarachnoid space (lymphomatose meningitis) the meninges look thickened (Brew, 2001; Ellison et al., 2004). Microscopically primary CNS lymphoma shows sheets of lymphoma separated by areas of necrosis. The cells invade the endothelium of small cerebral blood vessels and accumulate in perivascular spaces. At the edge primary CNS lymphoma diffusely infiltrates CNS tissue mimicking the growth pattern of gliomas. Primary CNS lymphoma can be distinguished from gliomas or metastases by immunolabeling with CD^{45} antibodies. Differentiation between B- and T-cell lymphoma can be done with CD^{20}, CD^{79a} (B-cell markers) and CD^3 and CD^{45}RO antibodies (T-cell markers) (Ellison et al., 2004). Histopathologically, primary CNS lymphoma is indistinguishable from systemic NHL (Jellinger and Paulus, 1995).

Table 15.1

Histological Working Formulation grade in HIV-related primary CNS lymphoma (%)

Study	Coté et al. (1996)	Newell et al. (2004)
Number of patients	269	41
High	21	46
Medium	27	51
Low	2	2
Miscellaneous[a]	0.3	–
NOS[b]	51	–

[a]Histology is microglia, mycosis, T-cell or others.
[b]Not otherwise specified.

15.5. Clinical presentation

Since primary CNS lymphoma can be localized at any intracranial localization, including the meninges, a wide variety of clinical presentations is possible. Both generalized neurological dysfunction and focal deficits are common. Depending on the localization, the most common initial clinical features are mental changes and behavioral disturbances (51–54%), focal deficits like limb paresis (51–61%), headache and nausea (14–43%) and seizures (22–28%) (Fine and Mayer, 1993; Chamberlain and Kormanik, 1999; Skiest, 2002; Newell et al., 2004) (Table 15.2). Other symptoms and signs are cranial nerve dysfunction (usually indicating meningeal involvement), ataxia and visual disturbances which are either caused by raised intracranial pressure or ocular involvement. Presentation with an isolated spinal cord syndrome is extremely unusual, as is occurrence of hypothalamic dysfunction with secondary pituitary dysfunction (Boadle and Tattersall, 1989; Patrick et al., 1989; Jellinger and Paulus, 1992; Balmaceda et al., 1994; Herrlinger et al., 1998; Herrlinger et al., 1999; Panageas et al., 2005). There seem to be no major differences in symptoms and signs at presentation between immunocompetent and immunodeficient patients. Symptoms are usually rapidly progressive: time from first symptoms to first diagnostic imaging is 13–27 days. Time from symptoms to diagnosis takes an average of 22–54 days (Skiest, 2002). This period is often prolonged because it is common to treat focal CNS lesions in AIDS patients with anti-toxoplasmosis therapy for 2–4 weeks and await the result. Only when no result is observed, additional diagnostic procedures like cerebrospinal fluid (CSF) examination or biopsy are usually performed.

Extracranial localization of the primary CNS lymphoma is extremely uncommon, except for the eye, in which involvement of retina, uvea and vitreous are all possible, and slit-lamp examination by an ophthalmologist is therefore recommended in all primary CNS lymphoma patients with visual complaints like decreased visual acuity or floaters (Ferreri et al., 2002).

Table 15.2

Presenting symptoms in primary CNS lymphoma (%)

	Chamberlain	Skiest	Fine	Newell
Focal neurologic deficits	52	54	51	61
Mental status change	52	54	53	51
Seizures	28	28	27	22
Headache	30	25	14	43
Ataxia	nr	nr	nr	nr

nr = not reported.
Data from Fine and Mayer (1993), Chamberlain and Kormanik (1999), Skiest (2002), Newell et al. (2004).

15.6. Investigations

In HIV-positive patients, imaging of the brain will usually be performed when they present with neurological symptoms or signs. The differential diagnosis of a focal intracerebral lesion in the absence of a normal cellular immunity is different from the immunocompetent patient. The main diagnoses to consider are toxoplasmosis, primary CNS lymphoma, progressive multifocal leucoencephalopathy (PML) and abscesses, frequently mycotic. Other diagnoses to be considered are tuberculosis, cryptococcoma, aspergillosis and gliomas. Neuroimaging, CSF examination and brain biopsy are fundamental diagnostic procedures of value.

15.6.1. CT and MRI

Both CT and MRI are used for brain imaging in AIDS patients, the latter being superior in differentiating between focal CNS lesions. At presentation the lesions of primary CNS lymphoma can be solitary (in 20–30% of immunocompromised patients as opposed to around 65% of immunocompetent patients) or multiple. On a non-contrast CT primary CNS lymphoma can be hypodense, isodense or sometimes slightly hyperdense (Dina, 1991; Thurner et al., 2001). On post-contrast CT variable enhancement can be seen. Often this is ring-shaped in immunocompromised patients, in contrast to immunocompetent patients, in whom more diffuse enhancement is usually seen. In two series the contrast enhancement was often irregular and sometimes homogenous or rim-enhancement (Dina, 1991; Thurner et al., 2001). Most MRI series are in immunocompetent, although some smaller series of HIV-positive patients are reported as well (Dina, 1991; Buhring et al., 2001; Thurner et al., 2001; Kuker et al., 2005). On MRI lymphoma is usually iso- or hypointense from white matter on T2WI. On T1WI the lesions are iso- or hypointense. On T1WI with gadolinium diffuse or ring enhancement can be seen, although in one series of AIDS patients enhancement was absent in up to 27% (Fig. 15.1) (Thurner et al., 2001). On FLAIR images primary CNS lymphoma has a hyperintense aspect. On DWI lymphoma might show a slightly hyperintese signal, but often the signal is indistinguishable from normal cortex (Kuker et al., 2005). MR spectroscopy shows a high choline and lipid peak and a low N-acetyl aspartate (NAA) peak (Kuker et al., 2005;

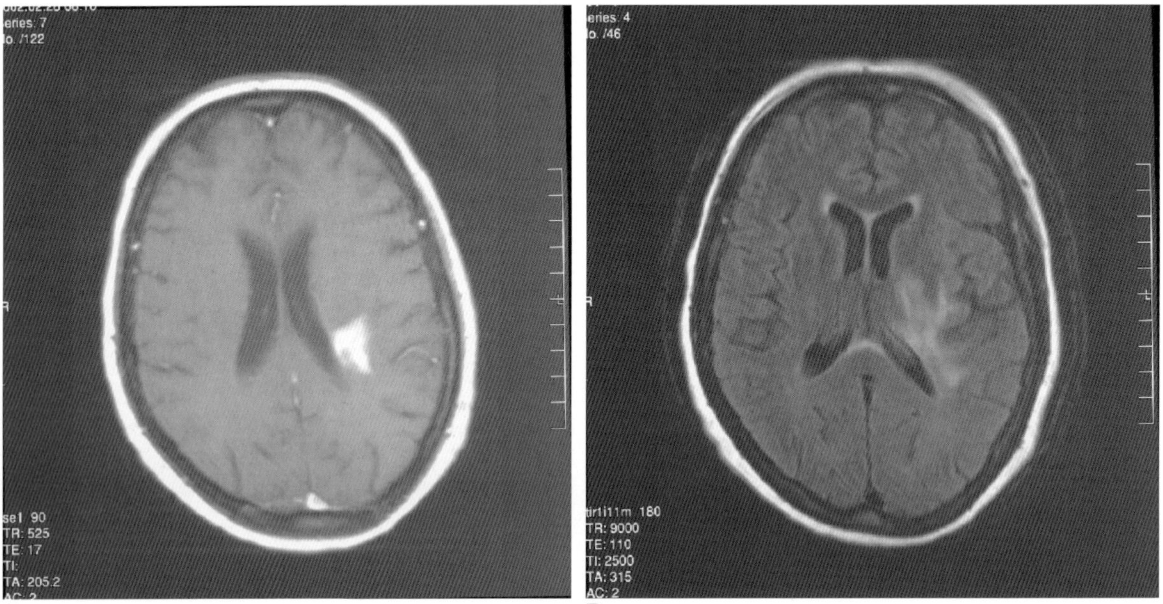

A B

Fig. 15.1. MRI of primary CNS lymphoma with periventricular localization.

Raizer et al., 2005). The MR spectrum is not homogenous throughout the tumor, and transition zones occur on the border between tumor and normal brain tissue.

The most important differential diagnosis is toxoplasmosis, and distinction between the two cannot always be made, even using all the above mentioned techniques. Both primary CNS lymphoma and toxoplasmosis tend to be multiple, although toxoplasmosis lesions are usually more numerous and smaller than primary CNS lymphoma. The location can help one differentiate between the two: primary CNS lymphoma is often periventricular and subependymal spread is quite common, as is ventricular encasement. Spread across the corpus callosum is highly suspicious for primary CNS lymphoma. Leptomeningeal involvement may occur, visible on MRI as leptomeningeal enhancement after gadolineum (Coulon et al., 2002). Toxoplasmosis, in contrast, tends to be located in the basal ganglia and does not spread in a periventricular pattern. Both primary CNS lymphoma and toxoplasmosis can occur supra- and infratentorial. Hemorrhage can occur in toxoplasmosis lesions, but is uncommon in primary CNS lymphoma. The enhancement pattern cannot always distinguish the two, since small lesions of both primary CNS lymphoma and toxoplasmosis can show diffuse enhancement and larger lesions of both can show either ring or diffuse enhancement.

In conclusion the radiological presentation of primary CNS lymphoma is very diverse and is hard to distinguish from other focal brain lesions, especially toxoplasmosis. There are no radiological features specific for primary CNS lymphoma.

15.6.2. CSF

Often CSF analysis cannot be performed because of present cerebral edema and midline shift. CSF cytology can reveal tumor cells in 18–35% of patients, and multiple lumbar punctures increase the chance of positive cytology (Hayakawa et al., 1994; Balmaceda et al., 1995; Herrlinger et al., 1998, 1999). Cell count, total protein, IgG and lactate are raised in 40–85% of patients (Herrlinger et al., 1999), but these are not very helpful in distinguishing from other CNS disorders. Polymerase chain reaction (PCR) can reveal the presence of EBV DNA in the CSF. Some authors report a very high accuracy of this test with a sensitivity of up to 80% and a specificity of up to 100% (Cingolani et al., 1998; Bossolasco et al., 2002). Whether these high sensitivity and specificity reflect the accuracy of using this test in daily practice is doubted by others (Ivers et al., 2004). A negative EBV PCR does not exclude the possibility of primary CNS lymphoma and a positive EBV PCR in addition to imaging studies compatible with the diagnosis of primary CNS lymphoma makes the diagnosis likely. Possibly the combination of single photon emission computed tomography (SPECT) and EBV PCR yields a very high sensitivity and specificity for the diagnosis of primary CNS lymphoma (Antinori et al., 1999).

15.6.3. SPECT

Imaging with thallium-201 SPECT can be used to differentiate primary CNS lymphoma from an infectious cause of focal brain lesions in AIDS. The additional value of this technique in the diagnostic process has not been clearly established. Focal increased thallium uptake can be seen in primary CNS lymphoma, as opposed to infections, most importantly, of course, toxoplasmosis, in which there is low thallium uptake. Some authors report a very high sensitivity and specificity for discrimination between primary CNS lymphoma and toxoplasmosis, especially in combination with either toxoplasmosis serology or PCR for EBV in CSF (Lorberboym, 1998; Antinori et al., 1999; Skiest et al., 2000). Other authors report worse sensitivity and specificity, suggesting the technique is not accurate enough for regular use in clinical practice (Licho et al., 2002). Interestingly enough, patients taking HAART seem to have a higher thallium uptake whether having primary CNS lymphoma or toxoplasmosis, suggesting that this technique might be less useful in this group of patients, which is by far the largest group of AIDS patients in which SPECT scanning would be considered (Giancola et al., 2004). In summary, thallium-201 SPECT can be used to discriminate between primary CNS lymphoma and toxoplasmosis, but its accuracy and additional value after CT, MRI, toxoplasma serology and EBV PCR in CSF is unclear.

15.6.4. Brain biopsy

With the advancing possibilities in neuroimaging and CSF analysis as described above, the indication for brain biopsy in AIDS patients with focal brain lesions is becoming less obvious. Probably biopsy should only be performed in patients with focal brain lesions where after imaging and CSF examination (if feasible, therefore not precluded for intracranial mass effect or other contraindication), there is still doubt about the diagnosis (Antinori et al., 1999). Since the main differential diagnosis is toxoplasmosis, and this infection usually shows rapid response (within two weeks) to therapy either clinically or radiologically, it is recommended to start this therapy first and await the result before considering a brain biopsy (Skolasky et al., 1999; Antinori et al., 2000). Obviously this approach depends also on the clinical situation the patient is in. If this treatment fails, biopsy can be considered. A stereotactic approach is usually chosen. This procedure is relatively safe, if performed by an experienced surgeon in a selected group of patients without profound coagulopathies and a Karnofsky performance scale of more than 50–60. The morbidity (mainly caused by hemorrhage) and mortality are 7–11.5% and 0–5.3%, respectively, depending on center and probably patient selection (Levy et al., 1992; Luzzati et al., 1996; Skolasky et al., 1999; Antinori et al., 2000). The diagnostic yield of the procedure is high, with a definite diagnosis in 88–96% of the biopsies. The yield is highest if multiple samples are taken, both from the core and the edge of the lesion (Skolasky et al., 1999; Antinori et al., 2000). If steroids have been administered to the patient, the diagnostic value of biopsy can drop dramatically in primary CNS lymphoma (Weller, 1999).

15.7. Treatment

In general, median survival of HIV-related primary CNS lymphoma is poor. Without treatment average survival is approximately 1 month. In most review articles comfort care is chosen for up to 60% of patients, because of a poor performance status or other HIV-related illnesses (Baumgartner et al., 1990; Fine and Mayer 1993; Newell et al., 2004; Robotin et al., 2004).

15.7.1. Steroids

Glucocorticoids have a strong lympholytic effect which can lead to a dramatic improvement in the clinical condition of the patient (Fine and Mayer, 1993; Weller, 1999). Glucocorticoids induce apoptosis in lymphoid cells leading to a quick decrease in tumor mass. They should be withheld if biopsy is considered since it can obscure the results, decreasing the yield of a biopsy (Geppert et al., 1990). If steroids are given prior to biopsy, rapid clinical improvement combined with reduction or disappearance of lesions on CT or MRI will make primary CNS lymphoma the most probable diagnosis. This effect is usually short-lived and glucocorticoid-resistant lymphoma cells will recur. There are a few case reports on long-term remissions in primary CNS lymphoma, but this was not in HIV-related primary CNS lymphoma (Herrlinger et al., 1996; Pirotte et al., 1997).

HIV-related primary CNS lymphoma is less suited for glucocorticoid treatment because of the immunosuppressive effects. Long-term treatment can lead to electrolyte disturbance, adrenal insufficiency, steroid diabetes, osteoporosis, cataract, myopathy and cognitive dysfunction (Weller, 1999). Glucocorticoids can have a role in treatment of HIV-related primary CNS lymphoma in patients in a reasonable condition who are likely to endure the side effects, e.g. patients with a higher CD^4.

15.7.2. Radiotherapy

So far radiotherapy is the only well-documented effective therapy for prolonging survival. Multiple studies show an increase in median survival from 1 to 3–4 months (Baumgartner et al., 1990; Fine and Mayer, 1993; Chamberlain and Kormanik, 1999; Sparano, 2001; Robotin et al., 2004). This increase in survival found could partly be based on a selection bias. Mode of radiation is whole brain radiotherapy with fractions of 1.8–3.0 Gy. Newell et al. (2004) report only improved survival in patients who received a total of 30 Gy or more. Others report total doses of 20–60 Gy (Sparano, 2001; Skiest, 2002). Serious side effects of radiation therapy are noted in patients surviving more than a year (Skiest and Crosby, 2003). Patients who finished radiotherapy usually die of other AIDS-related illnesses, whereas patients who did not receive radiotherapy die of primary CNS lymphoma (Baumgartner et al., 1990).

15.7.3. Chemotherapy

Since the 1990s, encouraging results have been achieved with chemotherapy in non-HIV-related primary CNS lymphoma (Panageas et al., 2005). HIV-related primary CNS lymphoma patients, however, tolerate cytotoxic chemotherapy poorly and develop a high rate of opportunistic infections and bone marrow toxicity (Sparano, 2001). Only in selected patients (Karnofsky performance score ≥ 60, $CD^4 > 200$, no opportunistic infections) does chemotherapy seem to improve survival to 11–16 months (Chamberlain, 1994; Chamberlain and Kormanik, 1999). Since the introduction of antiretroviral agents, more patients seem to be eligible for chemotherapy since recovery of the immune system improves the tolerability of chemotherapy (Sparano, 2003). A small series showed complete remission with intravenous methotrexate in combination with glucocorticoids in 47% of patients (Jacomet et al., 1997). Although not many data are available on chemotherapy in primary CNS lymphoma since the introduction of HAART, it seems that chemotherapy will play a greater role in primary CNS lymphoma treatment in the future.

15.7.4. Neurosurgery

Neurosurgical removal of primary CNS lymphoma does not improve survival since macroscopic gross total resection always leaves microscopic tumor residue. This is caused by diffuse invasion of CNS tissue of primary CNS lymphoma (Ellison et al., 2004). Data on the effect of biopsy, subtotal resection and gross total resection show there is no difference in survival in these three groups (Bellinzona et al., 2005). Therefore the role of neurosurgery is limited to a diagnostic one and has no part in therapy.

15.7.5. HAART

Considering the decline in incidence in primary CNS lymphoma since the introduction of HAART it appeared that primary CNS lymphoma is a manifestation of the advanced immunosuppression phase of AIDS and can be regarded as an opportunistic infection (Ammassari et al., 2000; Kadan-Lottick et al., 2002). This sparked hopes for treatment of primary CNS lymphoma with HAART, but to date results have been ambiguous. Conti et al. (2000) reported no prolonged survival in 32 patients and Levine et al. (2000) in 17 patients on HAART. However, Hoffmann et al. (2001) found an increase in survival in four out of six patients to 1.7–4.8 years and Skiest and Crosby (2003) reported more than two-year survival for six out of seven patients treated with HAART. A recent study of 26 patients of Newell et al. (2004) showed a more modest but significantly better prognosis in 25 patients with a median survival of 59 days versus 49 days if two or more antiretroviral agents were taken. No systematic review of these data has been performed so no definitive conclusion can be drawn about the effect of HAART on primary CNS lymphoma. Until then starting HAART can be advocated to prevent other HIV-related complications since patients with primary CNS lymphoma present with very low CD^4 counts (<50). The true effect of HAART on survival in primary CNS lymphoma patients will become clear in the next few years.

References

Ammassari A, Cingolani A, Pezzotti P, et al (2000). AIDS-related focal brain lesions in the era of highly active antiretroviral therapy. Neurology 55: 1194–1200.

Antinori A, De Rossi G, Ammassari A, et al (1999). Value of combined approach with thallium-201 single photon emission computed tomography and Epstein-Barr virus DNA polymerase chain in CSF for the diagnosis of AIDS-related primary CNS lymphoma. J Clin Oncol 17: 554–560.

Antinori A, Ammassari A, Luzzati R, et al (2000). Role of brain biopsy in the management of focal brain lesions in HIV-infected patients. Neurology 54: 993–997.

Balmaceda CM, Fetell MR, Setman JE, et al (1994). Diabetes insipidus as first manifestation of primary central nervous system lymphoma. Neurology 44: 358–359.

Balmaceda C, Gaynor JJ, Sun M, et al (1995). Leptomeningeal tumor in primary central nervous system lymphoma: recognition, significance and implication. Ann Neurol 38: 202–209.

Baumgartner JE, Rachlin JR, Beckstead JH, et al (1990). Primary central nervous system lymphomas: natural history and response to radiation therapy in 55 patients with acquired immunodeficiency syndrome. J Neurosurg 73: 206–211.

Bellinzona M, Rosera F, Ostertagc H, et al (2005). Surgical removal of primary central nervous system lymphomas (PCNSL) presenting as space occupying lesions: a series of 33 cases. Eur J Surg Oncol 31: 100–105.

Bindal AK, Blisard KS, Melin-Aldama H, et al (1997). Primary T-cell lymphoma of the brain in acquired immunodeficiency syndrome: case report. J Neurooncol 31: 267–271.

Boadle DJ, Tattersall MHN (1989). Primary cerebral lymphoma: presentation of eight cases and review of the literature. Aust N Z J Med 19: 682–686.

Bonnet F, Lewden C, May T, et al (2004). Malignancy-related causes of death in human immunodeficiency virus-infected patients in the era of highly active antiretroviral therapy. Cancer 101: 317–324.

Bossolasco S, Cinque P, Ponzoni M, et al (2002). Epstein–Barr virus DNA load in cerebrospinal fluid and plasma of patients with AIDS-related lymphoma. J Neurovirol 8(5): 432–438.

Brew JB (2001). HIV Neurology. Oxford University Press, Oxford.

Buhring U, Herrlinger U, Krings T, et al (2001). MRI features of primary central nervous system lymphomas at presentation. Neurology 57: 393–396.

Carbone A (2002). AIDS-related non-Hodgkin lymphomas: from pathology and molecular pathogenesis to treatment. Hum Pathol 33(4): 392–404.

Center for Disease Control: Castro KG, Ward JW, Slutsker L, et al (1993). Revised classification system for HIV infection and expanded surveillance case definition for AIDS among adolescents and adults. MMWR Recomm Rep 39: 39–40.

Chamberlain MC (1994). Long survival in patients with acquired immune deficiency syndrome-related primary central nervous system lymphoma. Cancer 73: 1728–1730.

Chamberlain MC, Kormanik PA (1999). AIDS-related central nervous system lymphomas. J Neurooncol 43: 269–276.

Cingolani A, DeLuca A, Larocca LM, et al (1998). Minimally invasive diagnosis of acquired immunodeficiency syndrome-related primary central nervous system lymphoma. J Natl Cancer Inst 90(5): 364–369.

Conti S, Masocco M, Pezzotti P, et al (2000). Differential impact of combined antiretroviral therapy on the survival of Italian patients with specific AIDS-defining illnesses. J Acquir Immun Defic Syndr 25: 451–458.

Coté TR, Manns A, Hardy CRM, et al (1996). Epidemiology of brain lymphoma among people with or without acquired immunodeficiency syndrome. J Natl Cancer Inst 10: 675–679.

Coulon A, Lafitte F, Hoang-Xuan K, et al (2002). Radiographic findings in 37 cases of primary CNS lymphoma in immunocompetent patients. Eur Neurol 12(2): 329–340.

Davies J, Everall IP, Weich S, et al (1997). HIV-associated brain pathology in the United Kingdom: an epidemiological study. AIDS 11: 1145–1150.

Dina TS (1991). Primary central nervous system lymphoma versus toxoplasmosis in AIDS. Radiology 179: 823–828.

Eby NL, Grufferman S, Flannelly CM, et al (1988). Increasing incidence of primary brain lymphoma in the US. Cancer 62: 2461–2465.

Ellison D, Love S, Chimelli L, et al (2004). A Reference Text to CNS Pathology. Elsevier, London.

Ferreri AJM, Blay JY, Reni M, et al (2002). Relevance of intraocular involvement in the management of primary central nervous system lymphomas. Ann Oncol 13: 531–538.

Fine HA, Mayer RJ (1993). Primary central nervous system lymphoma. Ann Intern Med 119: 1093–1104.

Freeman C, Berg JW, Cutler SJ (1972). Occurrence and prognosis of extranodal lymphomas. Cancer 29: 252–260.

Gendelman HE, Grant I, Everall IP, et al (2005). The Neurology of AIDS. Oxford University Press, Oxford.

Geppert M, Ostertag CB, Seitz G, et al (1990). Glucocorticoid treatment therapy obscures the diagnosis of cerebral lymphoma. Acta Neuropathol 80: 629–634.

Giancola ML, Rizzi EB, Schiavo R, et al (2004). Reduced value of thallium-201 single-photon emission computed tomography in the management of HIV-related focal brain lesions in the era of highly active antiretroviral therapy. AIDS Res Hum Retrovir 20(6): 584–588.

Gray F, Gherardi R, Scaravilli F (1988). The neuropathology of the acquired immunodeficiency syndrome (AIDS). A review. Brain 111: 245–266.

Harris NL, Jaffe ES, Diebold J, et al (1997). The World Health Organization classification of neoplastic diseases of the hematopoietic and lymphoid tissues. Report of the Clinical Advisory Committee meeting, Airlie House, Virginia, November. Ann Oncol 10: 1419–1432.

Hayakawa T, Takakura K, Abe H, et al (1994). Primary central nervous system in Japan: a retrospective, cooperative study by the CNS-lymphoma study group in Japan. J Neurooncol 19: 197–215.

Herrlinger U, Schabet M, Eichhorn M, et al (1996). Prolonged corticosteroid-induced remission in primary central nervous system lymphoma: report of a case and review of the literature. Eur Neurol 36: 241–243.

Herrlinger U, Schabet M, Clemens M, et al (1998). Clinical presentation and therapeutic outcome in 26 patients with primary CNS lymphoma. Acta Neurol Scand 97(4): 257–264.

Herrlinger U, Schabet M, Bitzer M, et al (1999). Primary central nervous system lymphoma: from clinical presentation to diagnosis. J Neurooncol 43(3): 219–226.

Hochberg FH, Miller DC (1988). Primary central nervous system lymphoma. J Neurosurg 68: 835–853.

Hoffmann C, Tabriziah S, Wolf E, et al (2001). Survival of AIDS patients with primary central nervous system lymphoma is dramatically improved by HAART induced immune recovery. AIDS 15: 2119–2127.

International Collaboration on HIV and Cancer (2000). Highly active antiretroviral therapy and incidence of cancer in human immunodeficiency virus infected adults. J Natl Cancer Inst 92: 1823–1830.

Ivers LC, Kim AY, Sax PE (2004). Predictive value of polymerase chain reaction of cerebrospinal fluid for detection of Epstein–Barr virus to establish the diagnosis of HIV-related primary central nervous system lymphoma. Clin Infect Dis 38: 1629–1632.

Jacomet C, Girard PM, Lebrette MG, et al (1997). Intravenous methotrexate for primary central nervous system non-Hodgkin's lymphoma in AIDS. AIDS 11: 1725–1730.

Jellinger KA, Paulus W (1992). Primary central nervous system lymphomas: an update. J Cancer Res Clin Oncol 119: 7–27.

Jellinger KA, Paulus W (1995). Primary central nervous system lymphomas: new pathological development. J Neurooncol 24: 33–36.

Jellinger K, Radaskiewicz TH, Slowik F (1975). Primary malignant lymphomas of the central nervous system in man. Acta Neuropathol 6: 95–102.

Kadan-Lottick NS, Skluzacek MC, Gurney JG (2001). Decreasing incidence rates of primary central nervous system lymphoma. Cancer 95: 193–202.

Kirk O, Pedersen C, Cozzi-Lepri A, et al (2001). Non-Hodgkin lymphoma in HIV-infected patients in the era of highly active antiretroviral therapy. Blood 98: 3406–3412.

Kuker W, Nagele T, Korfel A, et al (2005). Primary central nervous system lymphoma (PCNSL): MRI features at presentation in 100 patients. J Neurooncol 72: 169–177.

Levine AM, Seneviratne L, Espina BM, et al (2000). Evolving characteristics of AIDS-related lymphoma. Blood 96: 4084–4090.

Levy RM, Bredesen DE, Rosenblum ML (1985). Neurological manifestations of the acquired immunodeficiency syndrome (AIDS): experience at UCSF and review of the literature. J Neurosurg 62: 475–479.

Levy RM, Russell E, Yungbluth M, et al (1992). The efficacy of image-guided stereotactic brain biopsy in neurologically symptomatic acquired immunodeficiency syndrome patients. Neurosurgery 30(2): 189–190.

Licho R, Scott Litofsky N, Senitko M, et al (2002). Inaccuracy of Tl-201 brain SPECT in distinguishing cerebral infections from lymphoma in patients with AIDS. Clin Nucl Med 27(2): 81–86.

Lorberboym M, Wallach F, Estok L, et al (1998). Thallium-201 retention in focal intracranial lesions for differential diagnosis of primary lymphoma and nonmalignant lesions in AIDS patients. J Nucl Med 39(8): 1366–1369.

Luzzati R, Ferrari S, Nicolato A, et al (1996). Stereotactic brain biopsy in human immunodeficiency virus-infected patients. Arch Intern Med 156(5): 565–568.

Masliah E, DeTeresa RM, Mallory ME, et al (2000). Changes in pathological findings at autopsy in AIDS cases for the last 15 years. AIDS 14: 69–74.

Newell MA, Hoy JF, Cooper SG, et al (2004). Human immunodeficiency virus-related primary central nervous system lymphoma. Cancer 100: 2627–2636.

Panageas KC, Elkin EB, DeAngelis LM, et al (2005). Trends in survival from primary central nervous system lymphoma. Cancer 104: 2466–2472.

Patrick AW, Campbell IW, Ashworth B, et al (1989). Primary cerebral lymphoma presenting with cranial diabetes insipidus. Postgrad Med J 65: 771–772.

Paulus W (1999). Classification, pathogenesis and molecular pathology of primary CNS lymphomas. J Neurooncol 43: 203–208.

Pirotte B, Levivier M, Goldman S, et al (1997). Glucocorticoid-induced longterm remission in primary central nervous system lymphoma: case report and review of the literature. J Neurooncol 32: 63–69.

Raizer JJ, Koutcher JA, Abrey LE, et al (2005). Proton magnetic resonance spectroscopy in immunocompetent patients with primary central nervous system lymphoma. J Neurooncol 71: 173–180.

Robotin MC, Law MG, Milliken S, et al (2004). Clinical features and predictors of survival of AIDS-related non-Hodgkin's lymphoma in a population-based case series in Sydney, Australia. HIV Med 5: 377–384.

Schabet M (1999). Epidemiology of primary CNS lymphoma. J Neurooncol 43: 199–201.

Skiest DJ (2002). Focal neurologic disease in patients with acquired immunodeficiency syndrome. Clin Infect Dis 34: 103–115.

Skiest DJ, Crosby C (2003). Survival is prolonged by highly active antiretroviral therapy in AIDS patients with primary central nervous system lymphoma. AIDS 17: 1787–1793.

Skiest DJ, Erdman W, Chang WE, et al (2000). SPECT thallium-201 combined with toxoplasma serology for the presumptive diagnosis of focal central nervous system mass lesions in patients with AIDS. J Infect 40: 274–281.

Skolasky RL, Dal Pan GJ, Olivi A, et al (1999). HIV-associated primary CNS lymorbidity and utility of brain biopsy. J Neurol Sci 163: 32–38.

Sparano JA (2001). Clinical aspects and management of AIDS-related lymphoma. Eur J Cancer 37: 1296–1305.

Sparano JA (2003). Human immunodeficiency virus associated lymphoma. Curr Opin Oncol 15: 372–378.

Thurner MM, Rieger A, Kleibl-Popov C, et al (2001). Primary central nervous system lymphoma in AIDS: a wider spectrum of CT and MRI findings. Neuroradiology 43: 29–35.

Weller M (1999). Glucocorticoid treatment of primary CNS lymphoma. J Neurooncol 43: 237–239.

Zimmerman HM (1975). Malignant lymphomas of the nervous system. Acta Neuropathol 6: 69–74.

Further Reading

Bower M, Fife K, Sullivan A, et al (1999). Treatment outcome in presumed and confirmed AIDS-related primary cerebral lymphoma. Eur J Cancer 4: 601–604.

DeAngelis LM (1999). Primary CNS lymphoma: treatment with combined chemotherapy and radiotherapy. J Neurooncol 42: 249–257.

Freeman CR, Shustik CR, Brisson ML, et al (1986). Primary malignant lymphoma of the central nervous system. Cancer 58: 1106–1111.

Goplen AK, Dunlop O, Liestol K, et al (1997). The impact of primary central nervous system lymphoma in AIDS patients: a population-based autopsy study from Oslo. J Acq Immune Defic Syndr Hum Retrovirol 14: 351–354.

Herrlinger U (1999). Primary CNS lymphoma: findings outside the brain. J Neurooncol 43: 227–230.

Nelson DF (1999). Radiotherapy in treatment of primary central nervous system lymphoma (PCNSL). J Neurooncol 43: 241–247.

Salvesen Haldorsen I, Aarseth JH, Hollender A, et al (2004). Incidence, clinical features, treatment and outcome of primary central nervous system lymphoma in Norway. Acta Oncol 43(6): 520–529.

Schlegel U, Schmidt-Wolf GH, Deckert M (2000). Primary CNS lymphoma: clinical presentation, pathological classification, molecular pathogenesis and treatment. J Neurol Sci 181: 1–12.

Handbook of Clinical Neurology, Vol. 85 (3rd series)
HIV/AIDS and the Nervous System
P. Portegies, J. R. Berger, Editors

Chapter 16

Neuroimaging of the HIV/AIDS patient

CURTIS A. GIVEN II*

University of Kentucky Chandler Medical Center, Lexington, KY, USA

16.1. Introduction

Historically, up to 10% of all patients with the acquired immune deficiency syndrome (AIDS) initially present with symptoms related to central nervous system (CNS) pathology, and ultimately up to 60% of the patients will develop neurological abnormalities at some point in the course of the illness (Snider et al., 1983; Levy et al., 1985). Despite progress in the treatment and prevention of the life-threatening effects of the human immunodeficiency virus (HIV) on the immune system in the highly active antiretroviral therapy (HAART) era, the incidence of CNS disease remains high. Imaging plays an important role in the evaluation of HIV-positive patients with neurologic signs and symptoms. In general, the CNS disease processes common to the HIV patient can be divided into three main categories: HIV-associated diseases, opportunistic infections and malignancies. There is often overlap in the appearance of various pathologic processes affecting the CNS in the HIV-positive patient, prohibiting definitive diagnosis based on imaging alone. However, there are several imaging characteristics which are suggestive of a specific diagnosis or at least useful in limiting differential considerations. Often the imaging appearance can be used in conjunction with serology to reasonably establish a diagnosis and permit institution of empiric therapy. Establishing a diagnosis is often complicated by the not infrequent coexistence of multiple CNS pathologic processes in HIV patients. Imaging can be utilized for surveillance of disease in AIDS patients, assessing lesion response to therapy and guiding biopsy of equivocal lesions.

16.2. HIV-associated diseases

16.2.1. HIV encephalitis

HIV encephalitis is the most common CNS manifestation of AIDS. HIV encephalitis represents a subacute encephalomyelitis directly related to the HIV infection. Patients tend to present with varying symptoms including headache, movement and sensory disorders and declining mental status and function. Symptoms may progress on to dementia (HIV dementia complex or progressive dementia complex). Prior to the advent of HAART, HIV encephalitis was diagnosed in 10–30% of HIV-positive patients (Post et al., 1992; Thurnher et al., 1997b; Kandanearatchi et al., 2003). With the advent of HAART, there has been a profound reduction in the incidence of CNS opportunistic infections and malignancies, but with a somewhat less impressive effect observed on HIV encephalitis and dementia (Neuenburg et al., 2002; Sacktor, 2002; Sperber and Shao, 2003).

Imaging of HIV encephalitis most commonly demonstrates diffuse cerebral atrophy with a central predominance, out of proportion to age-related changes (Fig. 16.1). Atrophy is nonspecific in the HIV patient and may be attributable to other infectious etiologies (e.g. cytomegalovirus, neurosyphilis). There will often be progression of atrophy with development of patchy and/or confluent (nearly symmetrical) abnormalities within the deep white matter structures. The white matter abnormalities demonstrate a predilection for the frontal lobes and are shown as areas of low attenuation on unenhanced CT examinations, or areas of hyperintensity on T2-weighted or fluid-attenuated inversion recovery (FLAIR) MR sequences (Fig. 16.2). The changes of

*Correspondence to: Curtis A. Given II, MD, Department of Radiology, University of Kentucky Chandler Medical Center, Room HX-311C, 800 Rose Street, Lexington, KY 40536, USA. E-mail: cagive2@uky.edu.

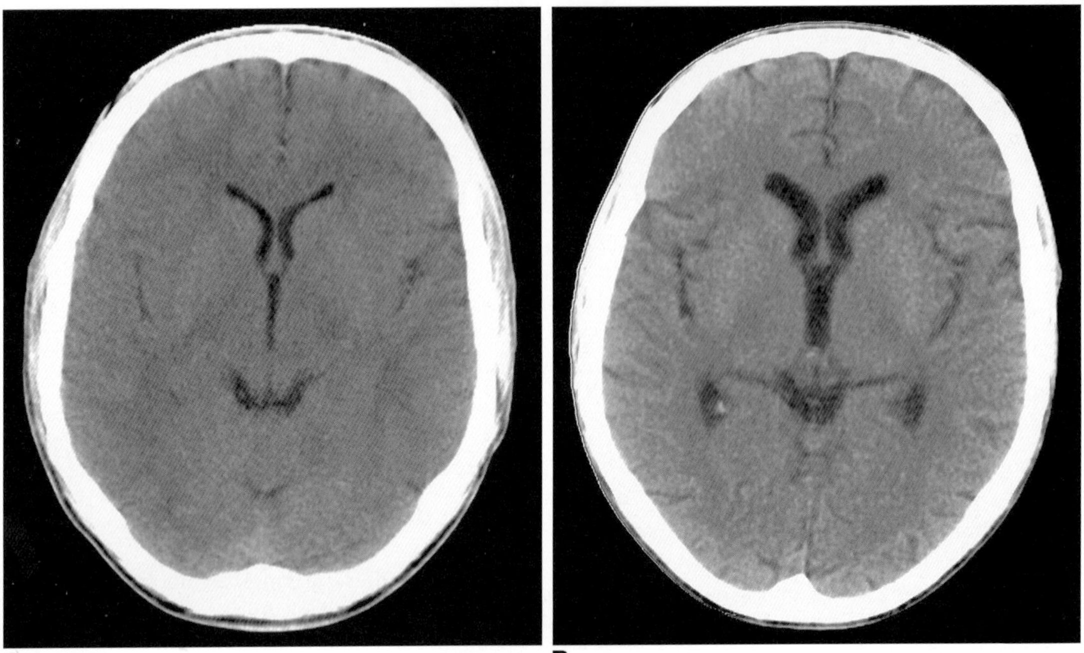

Fig. 16.1. HIV encephalitis with brain atrophy. A. Un-enhanced axial CT image of the brain is unremarkable. B. Un-enhanced axial CT image of the brain at a similar slice obtained 17 months later demonstrates rapid development of atrophy with widening of the sulci and ventricles.

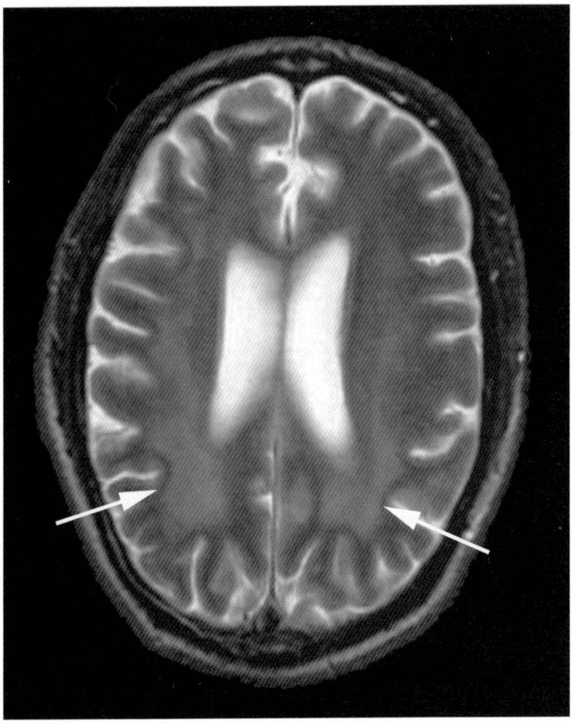

Fig. 16.2. HIV encephalitis. Axial T2-weighted MR image reveals atrophy and confluent, symmetrical areas of hyperintensity within the cerebral white matter (arrows).

HIV encephalitis do not enhance following intravenous contrast administration and are not associated with mass effect. HIV encephalitis can also display asymmetric white matter abnormalities on imaging studies, a finding that warrants inclusion of progressive multifocal leukoencephalopathy (PML) into differential consideration. Not infrequently, HIV encephalitis may coexist with other opportunistic infections and malignancies (Levy et al., 1985). The presence of mass effect or contrast enhancement associated with any lesion should prompt investigation of possible opportunistic infection or neoplasm.

16.2.2. AIDS dementia complex

The AIDS dementia complex is a complication occurring late in the course of the disease. The diagnosis is generally made based on clinical findings comprising a classic triad of cognitive changes (ranging from decreased mentation to global dementia), impaired motor skills and behavioral alterations (Heald and Hicks, 1997). Imaging will typically show the aforementioned changes of HIV encephalitis with atrophy and confluent T2 white matter signal abnormalities. MR spectroscopy has shown promise in the evaluation and monitoring of the AIDS dementia complex, with increased concentrations of myoinositol and choline, with relatively suppressed concentrations of N-acetyl aspartate (NAA) in the frontal lobes compared with their HIV-positive counterparts without cognitive impairment (Lee et al., 2003; Sacktor et al., 2005). Perfusion studies, with either MR or nuclear medicine imaging, display decreased relative cerebral blood flow within the frontal lobes of patients with the AIDS dementia complex (Chang et al., 2000). Further advances in perfusion imaging and spectroscopy are likely to yield increasing utility in the diagnosis and surveillance of the AIDS dementia complex.

16.3. Opportunistic infections

16.3.1. Toxoplasmosis

Toxoplasmosis is the most common intracranial mass lesion in the HIV-positive patient (Provenzale and Jinkins, 1997; Ramsey and Gean, 1997), with 28% of AIDS patients eventually developing CNS toxoplasmosis (Grant et al., 1990). There has been a significant decrease in the incidence of CNS lesions (including toxoplasmosis) since the development of HAART (Abgrall et al., 2001; Jones et al., 2002). Toxoplasmosis was uncommonly encountered prior to the AIDS epidemic. Toxoplasmosis is the result of active infection with obligate intracellular protozoan *Toxoplasma gondii*. *T. gondii* is typically acquired through ingestion of infected cat feces, either through contamination of food products or via cat litter. Transmission may also occur to the fetus from an infected mother. *T. gondii* infection results in only a subclinical infection or a mild flu-like illness in the immunocompetent host. In the immunocompromised state, such as AIDS, active infection may progress to toxoplasmosis. CNS manifestations of toxoplasmosis are variable, with the most significant symptoms related to a focal mass lesion (e.g. focal neurological deficit, seizure).

Imaging will typically display multiple lesions throughout the brain, with a propensity for involvement of the supratentorial corticomedullary junction (likely secondary to hematogenous dissemination), the basal ganglia and the thalami (Fig. 16.3). Solitary lesions may be observed in up to one-third of the cases (Porter and Sande, 1992; Ramsey and Gean, 1997). These lesions are typically isodense to gray matter on unenhanced CT, but are usually discernible secondary to accompanying vasogenic edema (Fig. 16.4). Lesions will demonstrate enhancement following intravenous contrast administration, and enhancement may be either solid or ring-like in nature. The eccentric target sign, a ring-enhancing lesion with an eccentric mural nodule along the ring wall, is reported as being highly specific for toxoplasmosis lesions (Ramsey and Geremia, 1998) (Figs. 16.5 and 16.7). Not surprisingly MR imaging is more sensitive for lesion detection and characterization than CT. As with CT, the lesions may display either solid or ring-like enhancement on MR imaging. The lesions may be of varying signal on the T2-weighted images.

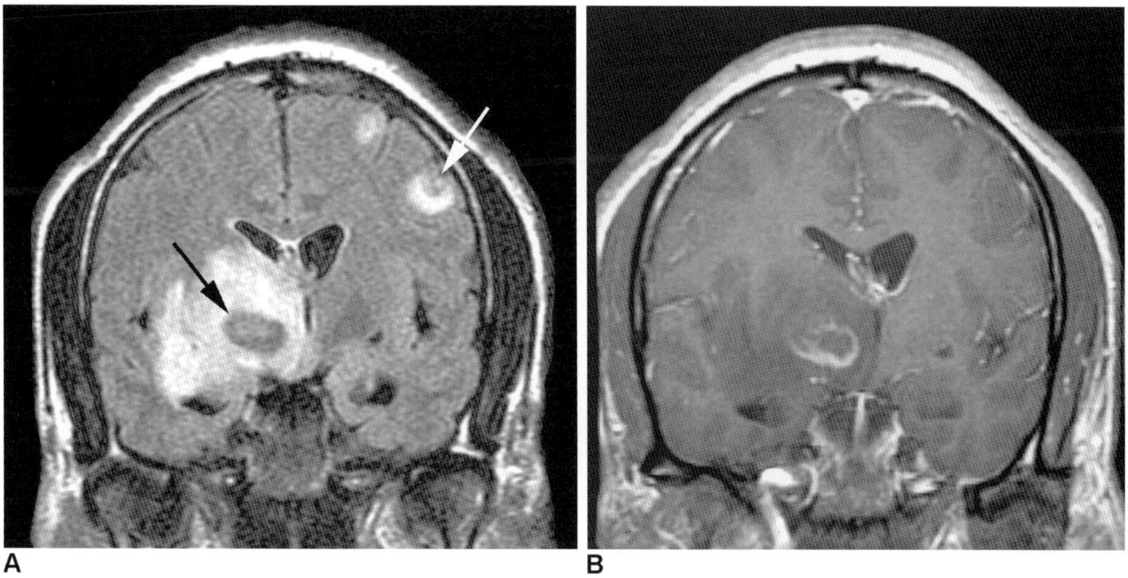

Fig. 16.3. Toxoplasmosis. A. Coronal FLAIR MR image reveals lesions within the right basal ganglia and the left frontal corticomedullary junction with surrounding vasogenic edema. B. Contrast-enhanced coronal T1-weighted MR image demonstrates ring enhancement of the basal ganglia lesion.

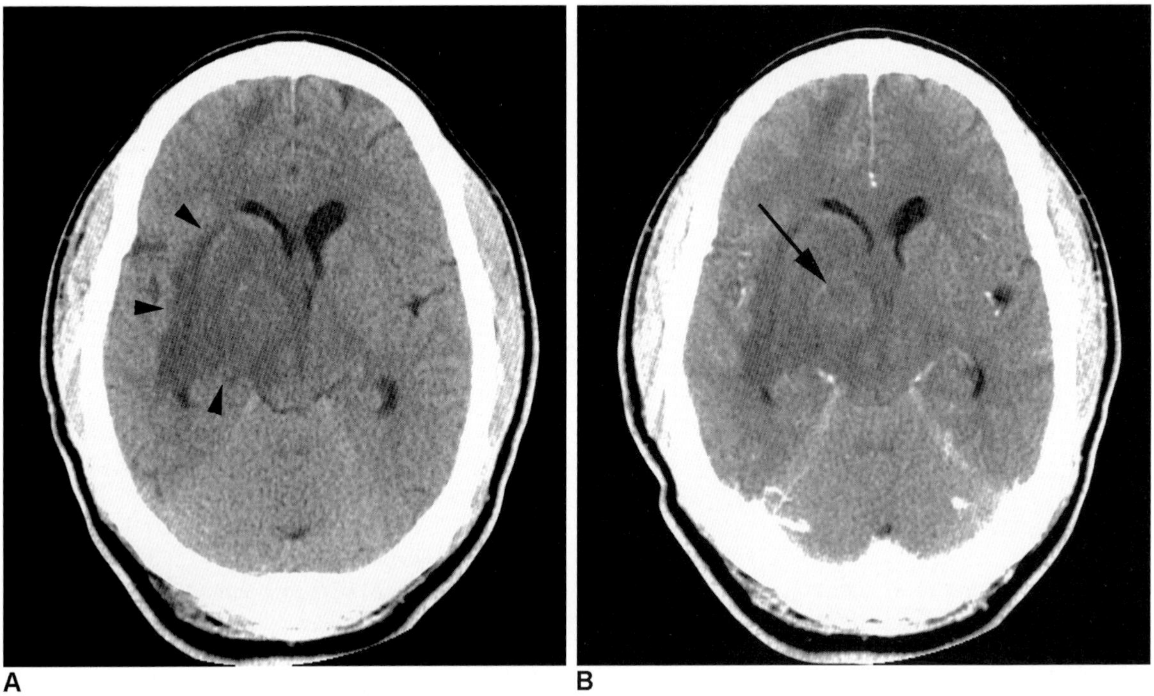

Fig. 16.4. Toxoplasmosis. A. Un-enhanced axial CT image of the brain reveals edema tracking through the right basal ganglia (arrowheads). B. Contrast-enhanced axial CT image of the brain reveals a ring-enhancing toxoplasmosis lesion within the right basal ganglia (arrow).

The presence of multiple enhancing brain lesions in an HIV-positive patient may be attributable to a number of diagnostic entities, with the main differential considerations including toxoplasmosis and primary CNS lymphoma (PCNSL). Both toxoplasmosis and PCNSL may present with multiple lesions, with ring-like or nodular enhancement. While definitive diagnosis may not be possible, there are several imaging findings that may favor the diagnosis of toxoplasmosis. Basal ganglia and thalamic involvement is frequently seen with both toxoplasmosis and PCNSL, but toxoplasmosis lesions tend to be scattered throughout the parenchyma and at the corticomedullary junction (Fig. 16.3). The presence of leptomeningeal and/or ependymal involvement of a lesion is highly suggestive of PCNSL, as this finding is exceedingly rare with toxoplasmosis (Sell et al., 2005). It is rare for toxoplasmosis lesions to involve the corpus callosum (Supiot et al., 1997), present as a ventriculitis or hydrocephalus (Sell et al., 2005) or present as an intracranial hemorrhage (Trenkwalder et al., 1992; Berlit et al., 1996).

Advanced MR imaging with diffusion, perfusion and spectroscopy techniques has shown mixed results with regards to their usefulness in differentiating toxoplasmosis from PCNSL. Diffusion-weighted imaging may be helpful in differentiating lesions greater than 1 cm in size, with neither the core nor rim of toxoplasmosis lesions having restricted diffusion (Fig. 16.6). Restricted diffusion within a lesion is more suggestive of PCNSL (Camacho et al., 2003; Chong-Han et al., 2003), but this is a nonspecific finding also observed within the core of pyogenic abscesses. Perfusion MR imaging reveals decreased perfusion throughout the toxoplasmosis lesions, with PCNSLs showing areas of increased perfusion within the solid portions of the tumor (Ernst et al., 1998). Proton MR spectroscopy has proven to be of limited utility in differentiating these two entities, with a large overlap in spectral appearances (Chinn et al., 1995; Simone et al., 1998; Pomper et al., 2002). Both toxoplasmosis and PCNSL lesions possess elevated levels of choline, suppressed levels of NAA and the presence of lipid materials (Fig. 16.7). The presence of lipids within these lesions may reflect proliferation or breakdown of cell membranes, or the presence of inflammation and infiltrating macrophages associated with the lesions (Simone et al., 1998).

Radionuclide imaging with either thallium-201 single photon emission computed tomography (SPECT) or fluoro-deoxyglucose-18 positron emission tomography (FDG-PET) has proven useful in differentiating larger toxoplasmosis lesions from PCNSL (Hoffman et al., 1993; Ruiz et al., 1994; Villringer et al., 1995; Heald et al., 1996; Ramsey and Gean, 1997; Skiest

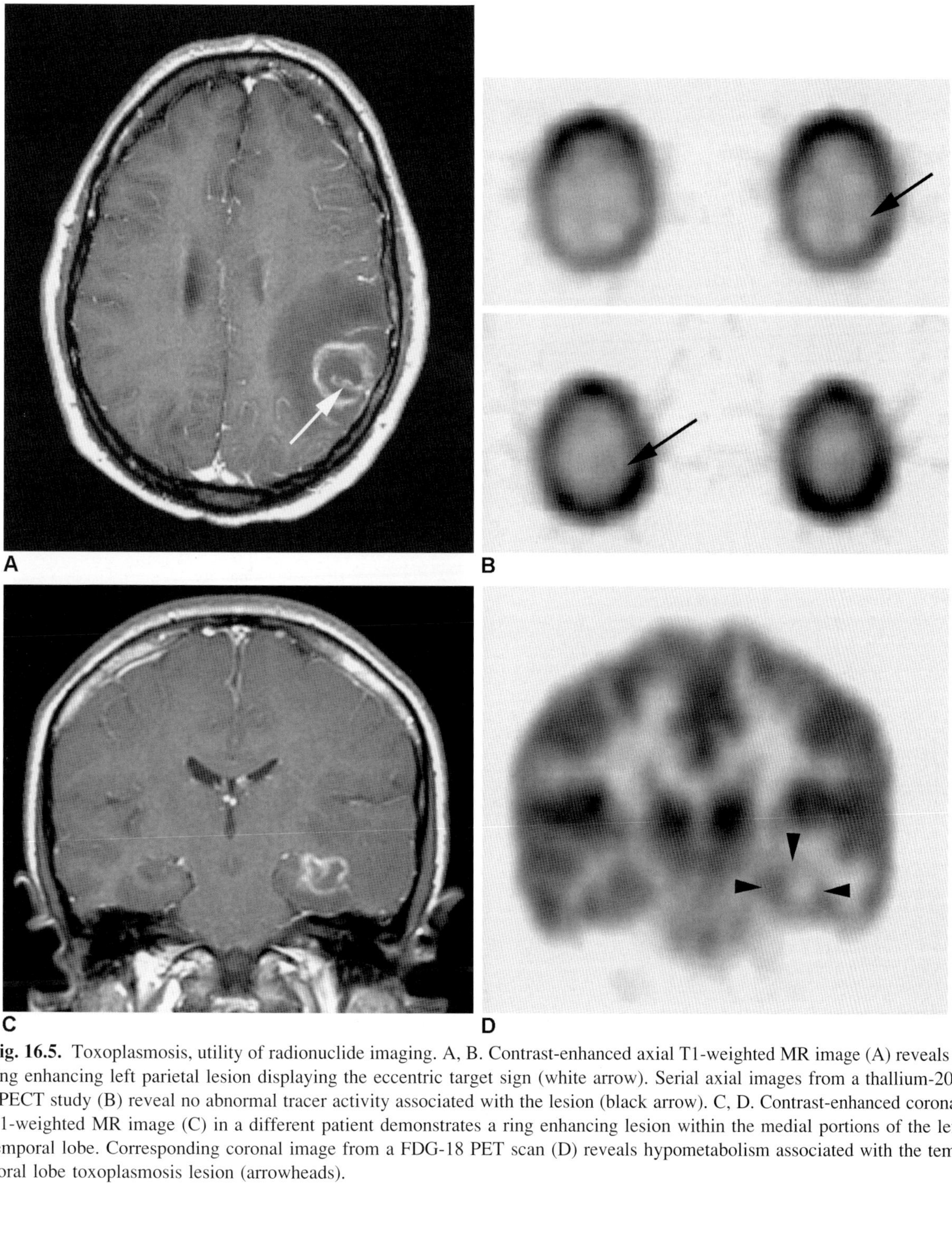

Fig. 16.5. Toxoplasmosis, utility of radionuclide imaging. A, B. Contrast-enhanced axial T1-weighted MR image (A) reveals a ring enhancing left parietal lesion displaying the eccentric target sign (white arrow). Serial axial images from a thallium-201 SPECT study (B) reveal no abnormal tracer activity associated with the lesion (black arrow). C, D. Contrast-enhanced coronal T1-weighted MR image (C) in a different patient demonstrates a ring enhancing lesion within the medial portions of the left temporal lobe. Corresponding coronal image from a FDG-18 PET scan (D) reveals hypometabolism associated with the temporal lobe toxoplasmosis lesion (arrowheads).

et al., 2000). PCNSL lesions display increased tracer activity related to increased metabolic activity on both techniques, whereas infectious lesions (such as toxoplasmosis) show little or absent activity (Figs. 16.5 and 16.7).

Accurately diagnosing toxoplasmosis may not be possible on imaging, and AIDS patients may present multiple lesions of various pathologies (Post et al., 1992). An algorithm constructed by the Quality Standards Subcommittee of the American Academy of Neurology (1998)

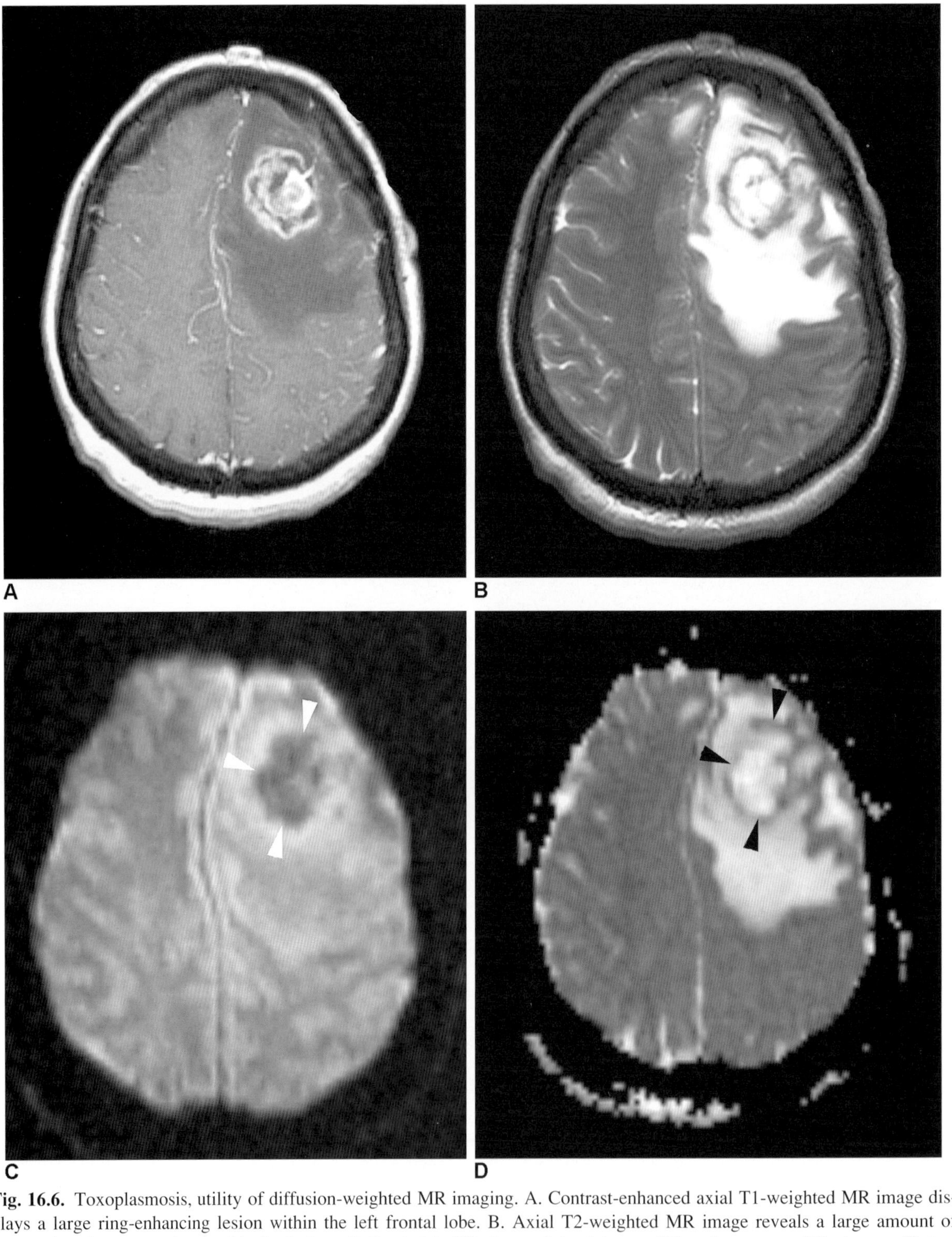

Fig. 16.6. Toxoplasmosis, utility of diffusion-weighted MR imaging. A. Contrast-enhanced axial T1-weighted MR image displays a large ring-enhancing lesion within the left frontal lobe. B. Axial T2-weighted MR image reveals a large amount of vasogenic edema associated with the lesion. C, D. Axial diffusion weighted image (C) and apparent diffusion coefficient (ADC) map (D) demonstrate unrestricted diffusion within the toxoplasmosis lesion (arrowheads).

suggests that the presence of two or more lesions with positive toxoplasmosis serology supports the diagnosis of toxoplasmosis, allowing patients to be placed empirically on an antitoxoplamosis regimen. Toxoplasmosis lesions tend to respond quickly to treatment, and thus if the lesions decrease in size on short interval follow-up imaging studies, then a presumptive diagnosis of toxoplasmosis can be made (Fig. 16.8). Care must be taken to ensure that all lesions are responding to antitoxoplasmosis therapy, as not infrequently HIV-positive patients will possess more than one active disease process (Post et al., 1992). A solitary lesion in the presence of negative serology argues against toxoplasmosis and may warrant biopsy or additional imaging techniques.

16.3.2. Cryptococcus

Cryptococcus is the most common fungal infection in the central nervous system of AIDS patients, resulting from inhalation of the *Cryptococcus neoformans* organism from soil contaminated with bird excrement. Immunocompetent individuals rarely develop symptoms related to cryptococcus, but approximately 5% of the AIDS population develops CNS infection with the fungus (Zuger et al., 1988; Andreula et al., 1993; Harris and Enterline, 1997). The incidence of cryptococcus in HIV patients has declined due to newer treatments (e.g. HAART) and the use of antifungal treatments. Where HAART therapy is not freely available, prophylaxis with either itraconazole or fluconazole will prevent most cryptococcus infections.

Patients with cryptococcus infection generally present with a meningoencephalitis, with early symptomatology including headache, nausea, confusion, blurred vision and cranial nerve deficits. Meningeal signs of fever and nuchal rigidity are frequently absent (Harris and Enterline, 1997). The diagnosis can usually be made with lumbar puncture, with India ink stains demonstrating the encapsulated organism or with positive serum cryptococcal antigen assays. At the time of obtaining cerebrospinal fluid (CSF), opening pressures should be recorded as there can be markedly elevated intracranial pressure in patients with cryptococcus meningoencephalitis. In such instances, high-volume CSF taps can be therapeutic. Some patients will require ventriculoperitoneal shunting to alleviate the elevated intracranial pressure (Woodworth et al., 2005).

CNS involvement with cryptococcus infection may present with characteristic findings on imaging studies. After accessing the CSF, the fungus tends to spread along the perivascular (Virchow–Robin) subarachnoid spaces. The organism produces mucoid material that expands the perivascular spaces, forming gelatinous pseudocysts (Garcia et al., 1985; Popovich et al., 1990; Harris and Enterline, 1997). The pseudocysts favor the basal ganglia, thalami and periventricular regions as they extend along the perivascular spaces. On imaging, the pseudocysts are typically seen as non-enhancing lesions that are hypodense on CT. But nearly half of CT scans in patients with cryptococcus will be normal (Popovich et al., 1990; Tien et al., 1991). MR imaging is more sensitive to lesion detection and characterization, with the lesions being hypointense on T1-weighted and hyperintense on T2-weighted imaging (Fig. 16.9). Cryptococcal lesions typically illicit only a mild host immune response, and therefore edema and contrast enhancement are usually faint or absent (When et al., 1989; Arnder et al., 1996). However, if the disease is not treated, progression may occur with parenchymal involvement and the development of several nodular areas of enhancement within the parenchyma (Harris and Enterline, 1997). Once the parenchyma is invaded, a cryptococcoma may develop from a collection of organisms, inflammatory cells and mucoid debris (Harris and Enterline, 1997). The cryptococcomas may display surrounding vasogenic edema and contrast enhancement (Harris and Enterline, 1997). After successful treatment, the lesions and enhancement often resolve,

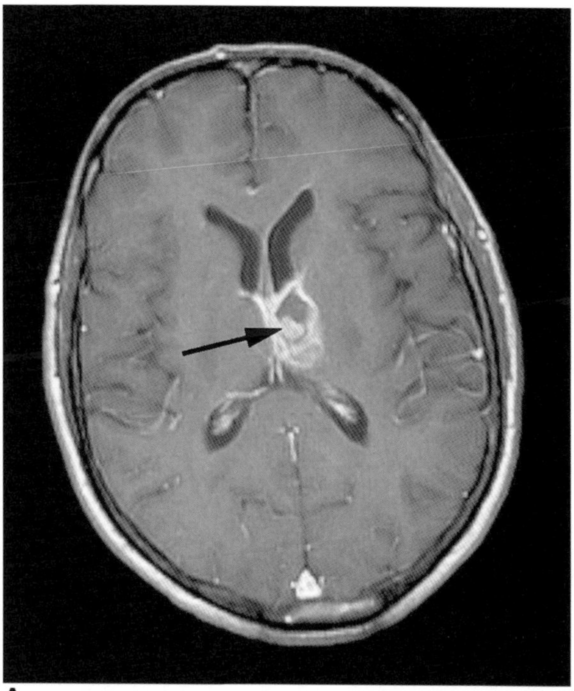

A

Fig. 16.7. Toxoplasmosis, spectroscopy and thallium imaging. A. Contrast-enhanced axial T1-weighted image reveals a ring-enhancing lesion within the left basal ganglia with an eccentric mural nodule (arrow) representing the eccentric target sign.

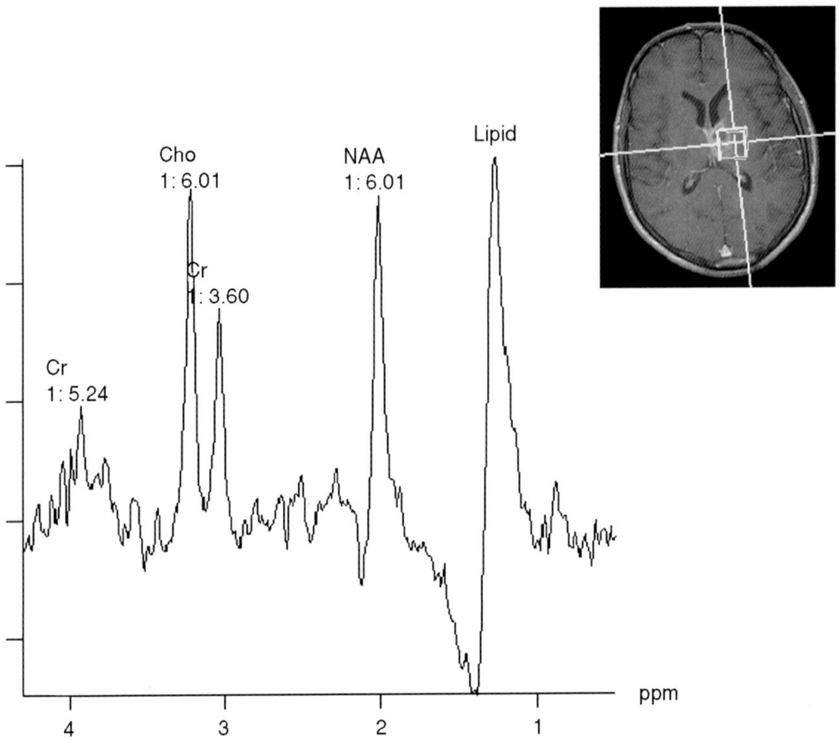

B

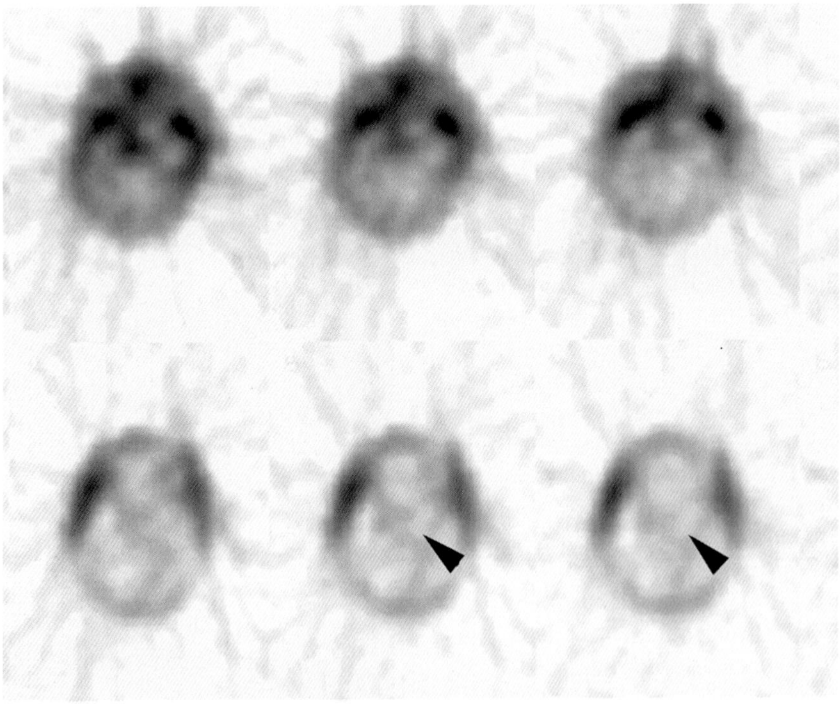

C

Fig. 16.7. (Continued) Toxoplasmosis, spectroscopy and thallium imaging. B. Single voxel MR spectroscopy (TE = 270) displays nonspecific elevation of choline, suppression of NAA, and the presence of lipids. C. Serial axial images from a thallium-201 SPECT scan demonstrate no abnormal tracer activity in the region of the left basal ganglia lesion (arrowheads).

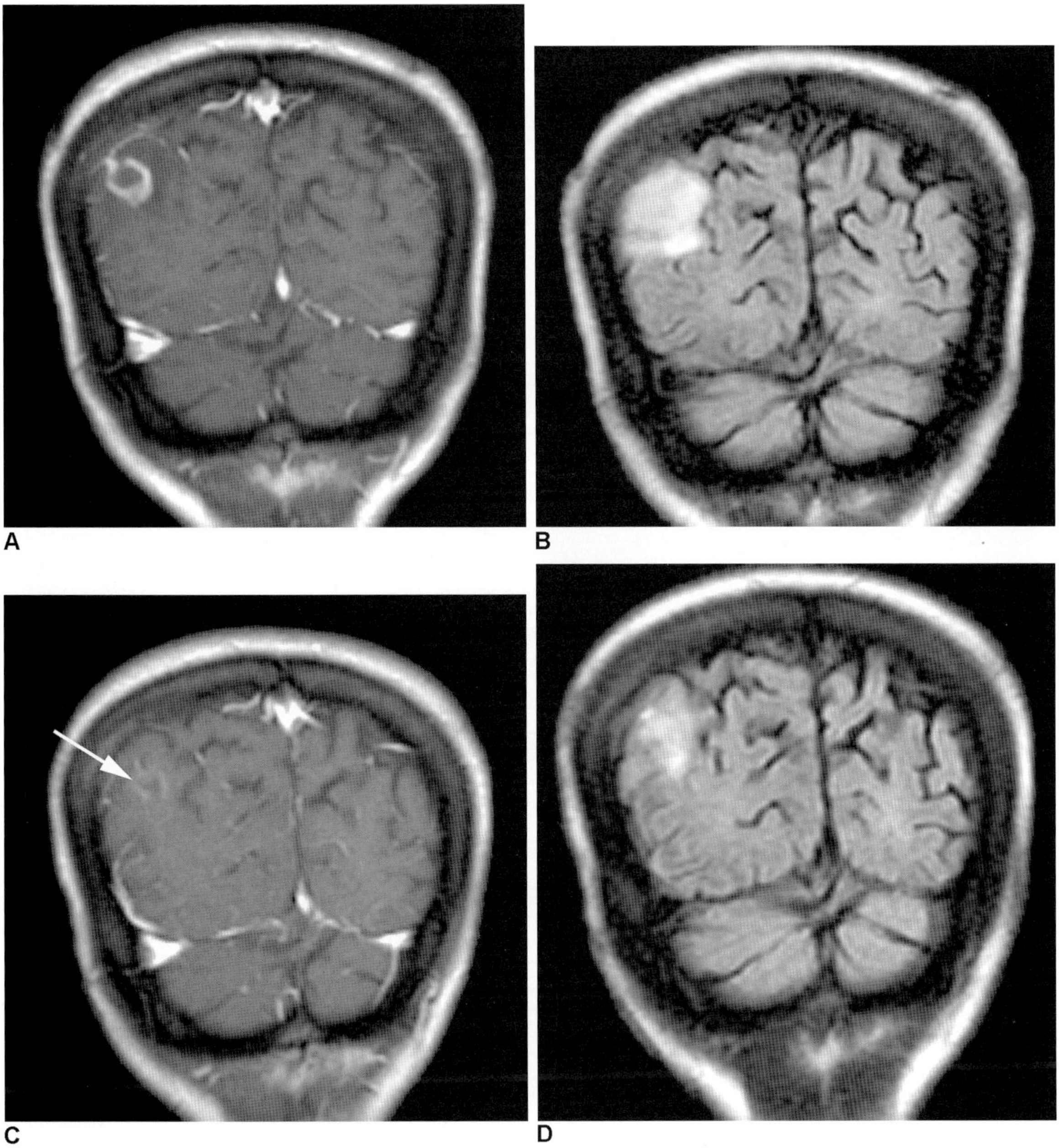

Fig. 16.8. Toxoplasmosis, response to empiric therapy. A, B. Contrast-enhanced coronal T1-weighted (A) and coronal FLAIR (B) MR images reveal a ring-enhancing right posterior parietal lesion with associated edema. C, D. Contrast-enhanced coronal T1-weighted (C) and coronal FLAIR (D) MR images following 3 weeks of empiric anti-toxoplasmosis therapy demonstrate marked interval reduction in lesion size (arrow) and accompanying edema.

but residual cystic or calcified lesions can persist for years.

The lack of enhancement of the basal cisterns and leptomeningeal surfaces with typical cryptococcus infections is useful in distinguishing it from typical bacterial meningitides (Mathews et al., 1992; Arnder et al., 1996; Harris and Enterline, 1997; Provenzale and Jinkins, 1997; Berkefeld et al., 1999). One should consider cryptococcus infection when presented with hydrocephalus in an AIDS patient, as there is a propensity for involvement of the choroids plexus and obstructive hydrocephalus (Andreula et al., 1993; Patronas and Makariou, 1993; Malessa et al., 1994; Provenzale and Jinkins, 1997). Cryptococcus can rarely produce a vasculopathy with resultant cerebral infarction (Leite et al., 2004).

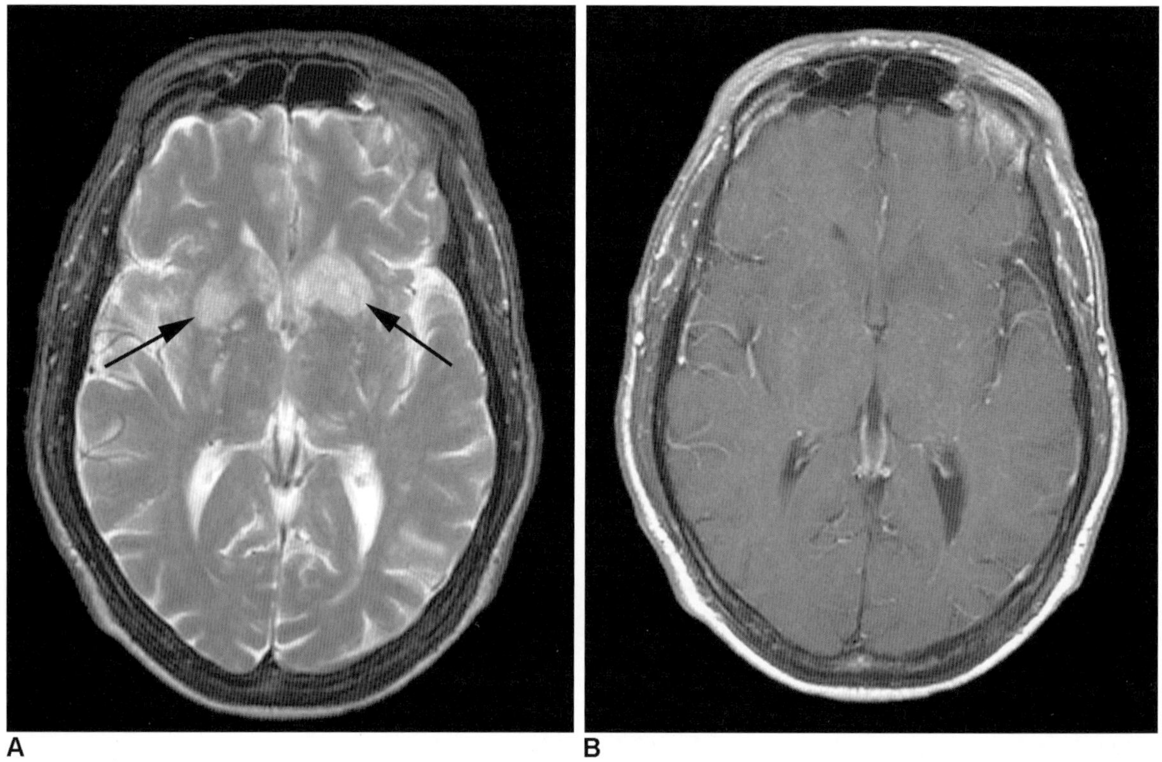

Fig. 16.9. Cryptococcosis. A. Axial T2-weighted image reveals abnormally increased signal within the basal ganglia (arrows), representing gelatinous pseudocysts along the perivascular spaces. B. Axial contrast-enhanced T1-weighted MR image reveals no appreciable contrast enhancement associated with the lesions.

The characteristic appearance on routine MR imaging in combination with CSF analysis allows for confident diagnosis in the majority of cases, and thus advanced imaging techniques are rarely required for diagnosis. In equivocal cases, radionuclide imaging is useful in differentiating infectious processes (such as a cryptococcoma) from neoplastic processes in the AIDS patient. Cryptococcus lesions typically display decreased activity on thallium-201 SPECT imaging, as opposed to their malignant counterparts. The addition of gallium-67 SPECT scans may further increase sensitivity, with cryptococcal lesions demonstrating increased activity (Lee et al., 1999; Skiest et al., 2000).

16.3.3. Cytomegalovirus

Cytomegalovirus (CMV) is a form of herpes virus that is generally asymptomatic and latent in immunocompetent individuals, but may become active in the AIDS patient resulting in encephalitis. CNS involvement with CMV may be seen in up to 10% of the AIDS population (Post et al., 1985b). The virus tends to spread to the CNS from a systemic infection, with preferential involvement of the meninges and cranial nerves. CMV infection is confirmed with either CSF cultures positive for CMV DNA or histological by

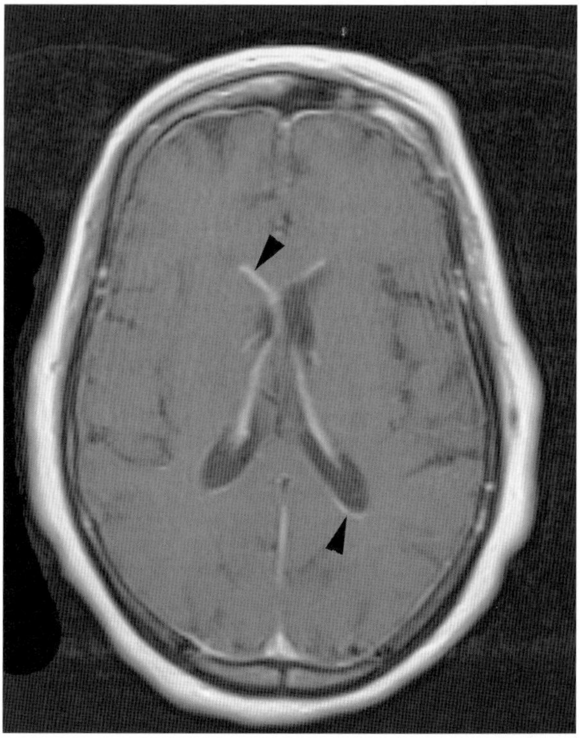

Fig. 16.10. Cytomegalovirus (CMV) ventriculitis. Contrast-enhanced axial T1-weighted MR image reveals abnormal enhancement of the ependymal surface of the ventricles (arrowheads). Case courtesy of Zoran Rumboldt, MD, Columbia, SC.

the presence of microglial nodules and viral inclusions producing a characteristic owl's eye appearance (Post et al., 1986; Salazar et al., 1995; Clifford et al., 1996).

CMV meningoencephalitis may produce abnormalities within the periventricular region on imaging studies. CT will demonstrate confluent hypodense periventricular lesions on un-enhanced scans, with subependymal enhancement following contrast administration (Post et al., 1986; Lizerbram and Hesselink, 1997). Unfortunately, CT scans are not sensitive for the detection of CMV involvement, with abnormalities demonstrated in only 30% of cases (Post et al., 1985a, 1986). MR imaging is more sensitive in detecting CMV involvement, demonstrating hypointense T1-weighted and hyperintense T2-weighted confluent abnormalities within the periventricular regions with subependymal enhancement following contrast administration (Fig. 16.10). Rarely the lesions can become large and exhibit mass effect, mimicking a neoplasm (Dyer et al., 1995; Moulignier et al., 1996).

CMV may also involve the spinal cord and peripheral nerves, presenting as a polyradiculopathy. With MR imaging, diffuse enhancement of the pial lining the conus medullaris and lumbosacral nerve roots can be seen with CMV polyradiculopathy (Bazan et al., 1991; Talpos et al., 1991; Whiteman et al., 1994; Lizerbram and Hesselink, 1997) (Fig. 16.11). The lumbosacral nerve roots may also appear thickened on non-contrast images, but generally contrast administration is required for the diagnosis.

16.3.4. Progressive multifocal leukoencephalopathy

Reactivation of a latent JC papovavirus can result in a fulminate opportunistic infection known as PML in the immunocompromised host. PML has been reported to occur in up to 5% of AIDS patients (Berger et al., 1998; Berger and Major, 1999), with 85% of all PML cases occurring in the HIV population (Manji and Miller, 2000). The organism targets and infects the oligodendroglial cells resulting in altered function and subsequent demyelination (Petito, 1988; Berger and Major, 1999). PML typically presents with a focal neurologic deficit, weakness, gait disturbance, headache or visual impairment. The neurologic deficits will progress to death, typically in less than six months without therapy.

Diagnosis can generally be made with reasonable confidence based on a combination of epidemiological data, typical features of PML on imaging and CSF analysis positive for JC virus DNA (Von Giesen et al., 1997; Berger and Major, 1999; Happe et al., 2000).

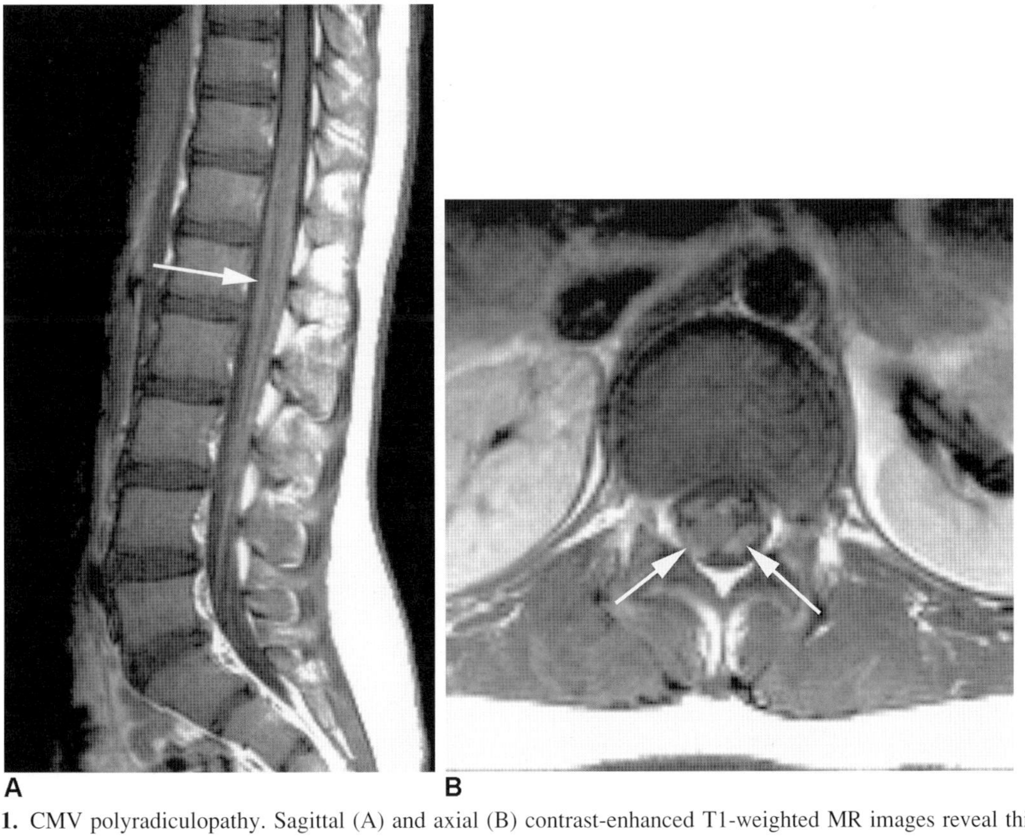

Fig. 16.11. CMV polyradiculopathy. Sagittal (A) and axial (B) contrast-enhanced T1-weighted MR images reveal thickening and abnormal enhancement of the lumbosacral nerve roots (arrows).

Ultimately, stereotactic biopsy of the lesions may be necessary for definitive diagnosis. The prognosis of patients with PML has generally been regarded as dismal, although clinical and radiologic improvement with increased survival may be seen with HAART (Berger and Major, 1999; Thurnher et al., 2001; Nuttall et al., 2004).

Imaging studies will generally display involvement of the subcortical white matter in the initial stages. CT scans classically reveal asymmetric hypodense lesions within the subcortical white matter of the parieto-occipital region (Fig. 16.12). The lesions typically do not enhance with contrast administration and lack appreciable mass effect. While enhancement within a PML lesion is thought to be a rare occurrence (Woo et al., 1996; Berger et al., 1998; Post et al., 1999), inflammation after induction of HAART may result in enhancement of the lesions. MR better delineates the lesions than CT, revealing lesions that are hypointense on T1-weighted and hyperintense on T2-weighted sequences within the subcortical white matter (Fig. 16.12). As the lesions progress, there is involvement of the deep white matter and corpus callosum (Fig. 16.13), and lesions may even extend into the deep gray matter structures. Posterior fossa involvement is common with PML, and has been shown to exist in nearly 50% of cases by imaging (Berger et al., 1998; Post et al., 1999) (Fig. 16.12).

The main differential consideration in cases of PML is HIV encephalitis. PML tends to be a more asymmetrical process with multiple or confluent lesions, with HIV encephalitis typically displaying symmetric diffuse areas of abnormality on T2-weighted sequences (Figs. 16.2 and 16.13). Clinical features such as dementia may help in establishing a diagnosis in equivocal cases, as dementia is not a prominent clinical feature of PML. Advanced imaging techniques may also prove useful in distinguishing these two entities. Magnetization transfer MR imaging has been shown useful in differentiating white matter lesions of PML from those of HIV encephalitis, with PML showing a significantly greater reduction in magnetization transfer ratios than those of HIV encephalitis (Ernst et al., 1999; Hurley et al., 2003). PML lesions may present with a variable appearance on diffusion-weighted imaging depending on the stage of disease; areas of restricted diffusion are thought to correspond to areas of active infection (Fig. 16.14). Early trials suggest proton MR spectroscopy is a useful adjunct in differentiating PML from HIV encephalitis, with PML lesions typically displaying significantly reduced levels of NAA, elevation of choline and the presence of lactate (Fig. 16.14) (Simone et al., 1998; Iranzo et al., 1999b; Hurley et al., 2003). These findings will need to be validated with future trials. The PML lesions may demonstrate varying degrees of metabolic activity with either thallium-201 or FDG-18 PET radionuclide imaging (Fig. 16.13) (Heald et al., 1996; Iranzo et al., 1999a; Lee et al., 1999; Port et al., 1999).

16.3.5. Tuberculosis

The incidence of tuberculosis is nearly 500 times greater in the AIDS population than in their immunocompetent counterparts (Whiteman, 1997). Approximately 10% of AIDS patients with tuberculosis have CNS manifestations (Bishburg et al., 1986; Berenguer et al., 1992). Infection with the *Mycobacterium tuberculosis* organism is generally from reactivation of latent pulmonary disease, and typically occurs early in the course of immunodeficiency (Chaisson and Slutkin, 1989; Whiteman, 1997). Clinical manifestations of CNS tuberculosis are rather nonspecific, with headache, fever and altered mental status being the primary symptoms in over half the cases (Villoria et al., 1992; Whiteman et al., 1995). Diagnosis is typically confirmed with CSF analysis revealing acid-fast bacilli.

CNS tuberculosis may manifest in a variety of forms, including meningitis, encephalitis with tuberculoma or abscess, and infarction. Leptomeningitis is the most frequent presentation of tuberculosis in the AIDS patient, with a thick exudate filling the basal cisterns (Villoria et al., 1992). While the basal cisterns are a favored location, leptomeningeal enhancement can be observed anywhere, including over the cerebral convexities and the ventricular ependyma (Villoria et al., 1992). Because CNS tuberculosis tends to occur early in the course of immunosuppression, a vigorous host inflammatory response is generated, resulting often in striking enhancement of the basal cisterns that is evident on either CT or MR imaging (Figs. 16.15 and 16.16). Communicating hydrocephalus will frequently complicate cases of tuberculous meningitis, seen in over 50% of the cases (Chang et al., 1990; Villoria et al., 1992). The marked inflammatory response within the leptomeninges and accompanying hydrocephalus will often allow differentiation from other forms of meningitis affecting the AIDS patient (e.g. cryptococcus, CMV).

Parenchymal involvement with CNS tuberculosis is estimated to occur in 15–44% of cases (Whiteman et al., 1995). Tuberculous granulomas, or tuberculomas, can occur anywhere in the brain or extra-axial spaces, albeit most lesions occur within the supratentorial region (Villoria et al., 1992). Tuberculomas may result from hematogenous seeding or direct extension from the leptomeningeal disease, and as such may be seen in the presence or absence of meningeal

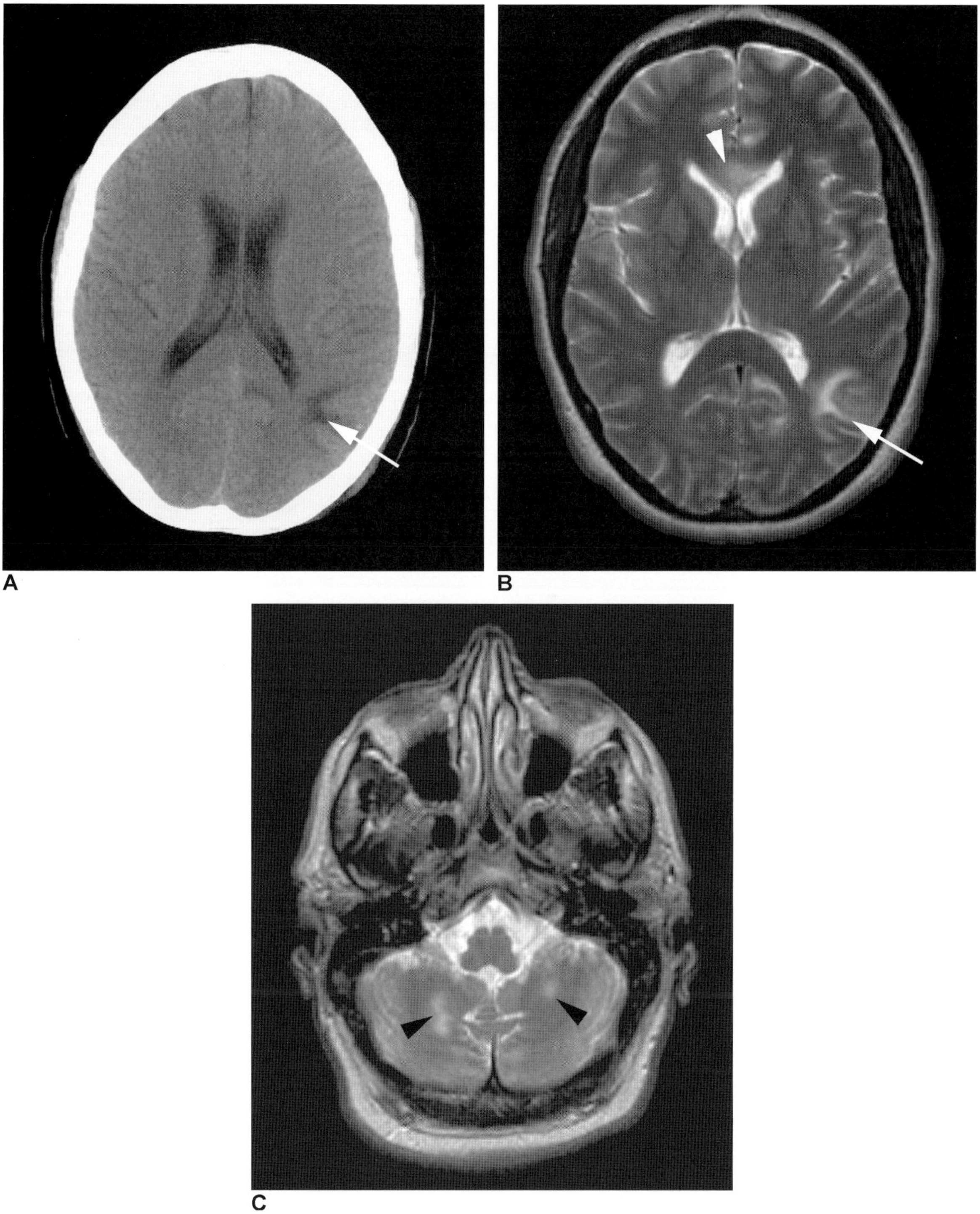

Fig. 16.12. Progressive multifocal leukoencephalopathy (PML). A. Un-enhanced axial CT image of the brain displays hypointense lesion within the subcortical white matter of the left parieto-occipital region (arrow). B, C. Axial T2-weighted MR images better delineate the subcortical involvement (arrow), as well as involvement of the corpus callosum (white arrowhead) (B) and cerebellum (black arrowheads) (C).

involvement (Whiteman, 1997). On imaging studies, tuberculomas appear as ring-enhancing lesions with surrounding edema. Central calcification or enhancement within the hypodense center of a ring-enhancing lesion has been called the target sign; a nonspecific finding that can be seen with tuberculomas on CT scans (Van Dyk, 1988; Bargallo et al., 1996). On MR imaging, tuberculomas are generally isointense on

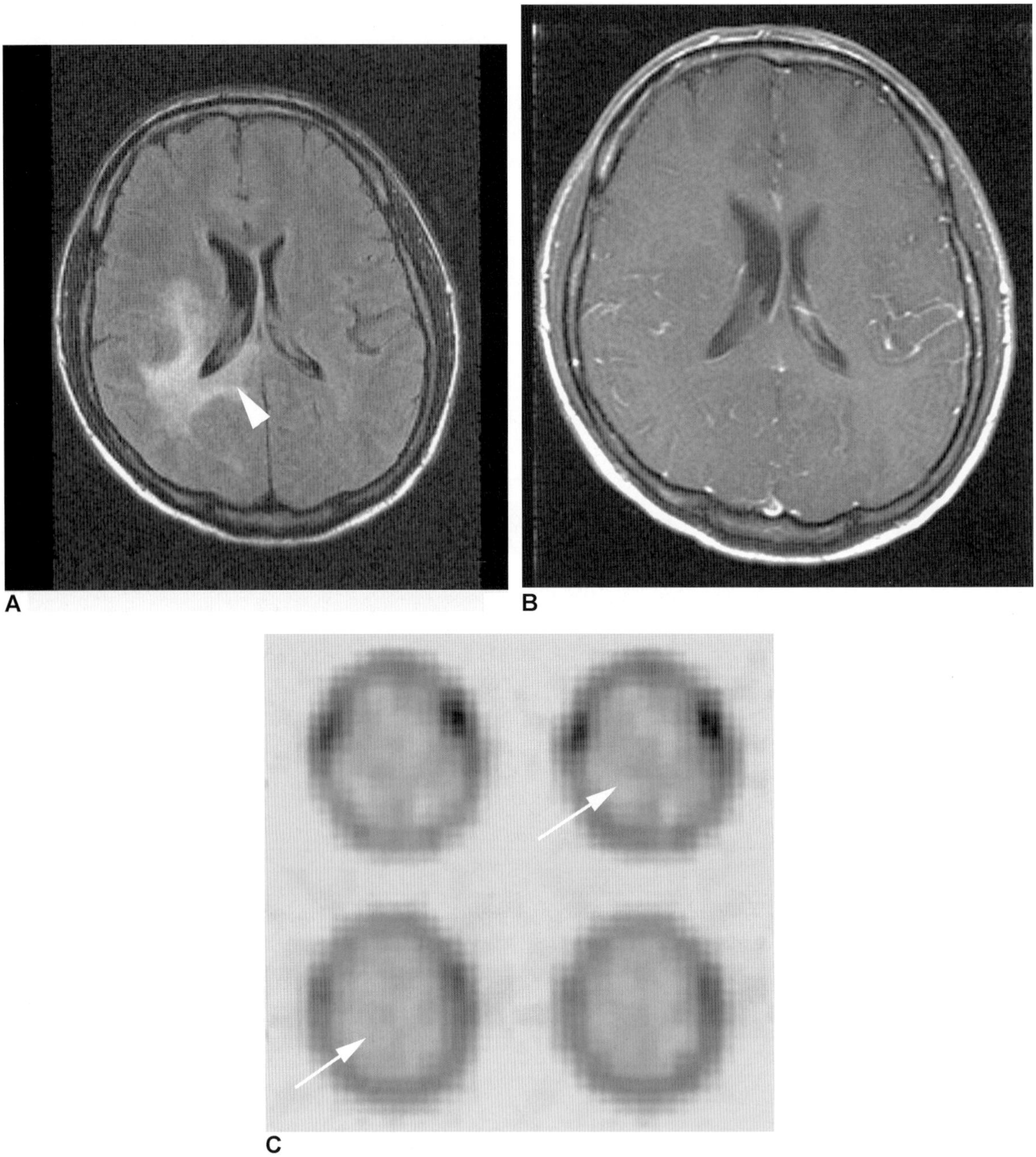

Fig. 16.13. Progressive multifocal leukoencephalopathy (PML), utility of radionuclide imaging. A, B. Axial FLAIR (A) and contrast-enhanced T1-weighted (B) MR images reveal asymmetric periventricular white matter hyperintensity on the FLAIR image without associated contrast enhancement or mass effect. Extension into the corpus callosum is present (arrow). C. Serial axial images from a thallium-201 SPECT study demonstrate no abnormal tracer activity associated with the PML lesion (arrows).

T1-weighted imaging. The appearance on T2-weighted sequences is variable depending on the stage of the tuberculoma. Noncaseating tuberculomas are generally hyperintense on T2-weighted imaging with solid, nodular contrast enhancement. Caseating granulomas are isointense to markedly hypointense on T2-weighted images, with ring-like enhancement following contrast administration (Chang et al., 1990; Gupta et al., 1991) (Fig. 16.15). The unusual hypointense appearance on the T2 sequences results from paramagnetic effects of the free radicals released by infiltrating macrophages (Sze and Zimmerman, 1988). The granulomas may also exhibit central T2 hyperintensity with peripheral contrast enhancement. Tuberculous abscesses are generally multiloculated, larger lesions (often greater than 3 cm) with greater degrees of associated edema and mass

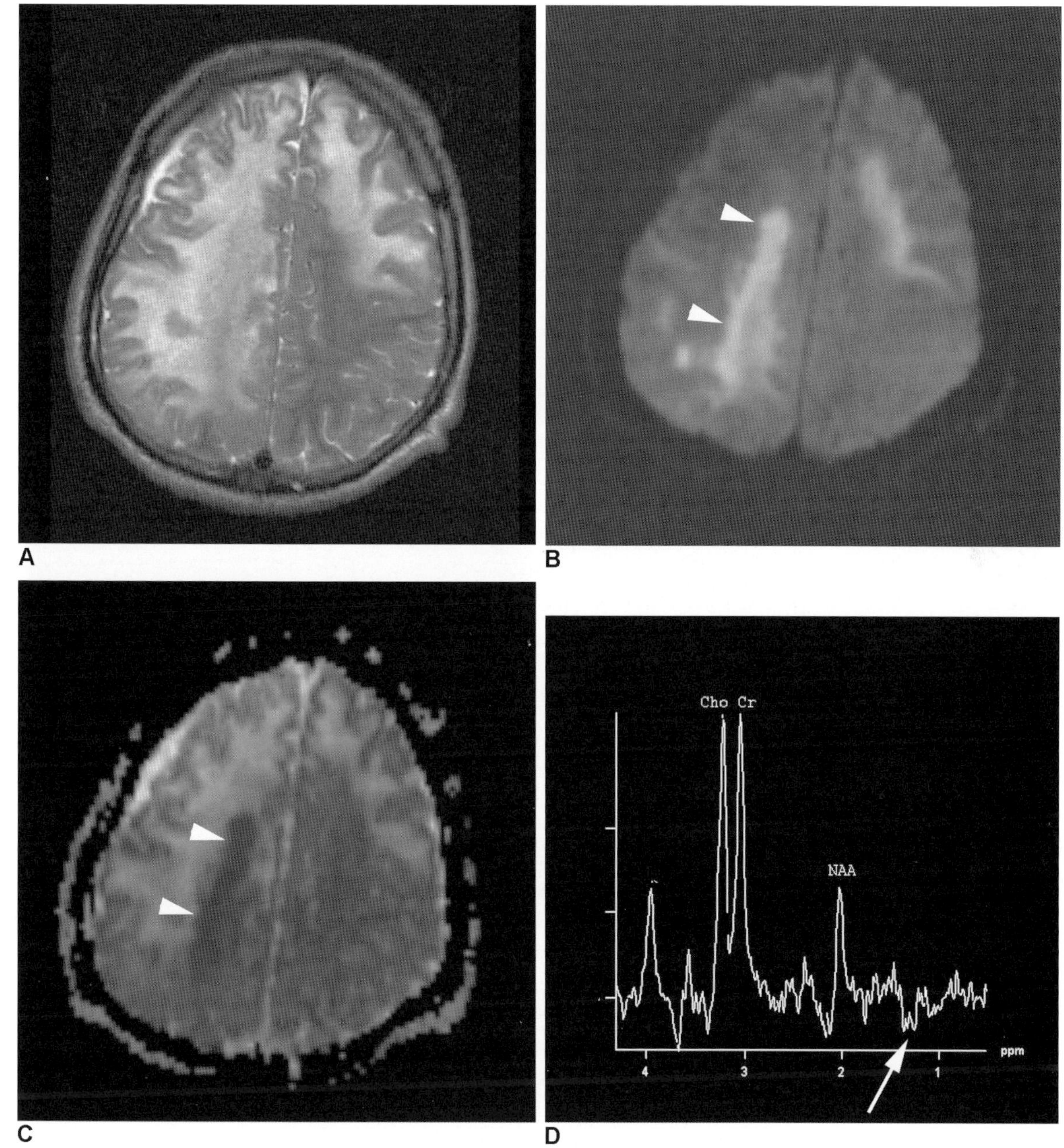

Fig. 16.14. PML, utility of MR diffusion imaging and spectroscopy. A. Axial T2-weighted MR image reveals extensive, asymmetric abnormal hyperintensity throughout the cerebral white matter. B, C. Axial diffusion-weighted image (B) and apparent diffusion coefficient (ADC) map (C) reveal heterogeneous signal within the PML lesion, with areas of restricted diffusion along the medial margin (arrowheads). D. Single voxel MR spectroscopy (TE = 270) of the PML lesion reveals mild elevation of choline, suppression of NAA and the presence of lactate (arrow).

effect than tuberculomas (Yang et al., 1987; Sze and Zimmerman, 1988; Whiteman et al., 1995). Tuberculous abscesses cannot reliably be distinguished from their bacterial counterparts with imaging; as such the majority of these lesions will undergo aspiration for diagnosis and treatment.

The arteries in contact with the inflammatory exudate associated with tuberculosis meningitis may undergo spasm resulting in thrombosis and infarction (Whiteman, 1997). Cerebral infarction is reported to occur in 18–34% of AIDS patients (Gillams et al., 1997). Infarction is frequently seen with infections such as CMV, tuberculosis, cryptococcus, toxoplasmosis and syphilis (Gillams et al., 1997). Infarction is reported to occur in greater than 23% of cases of AIDS-related CNS tuberculosis (Villoria et al., 1992;

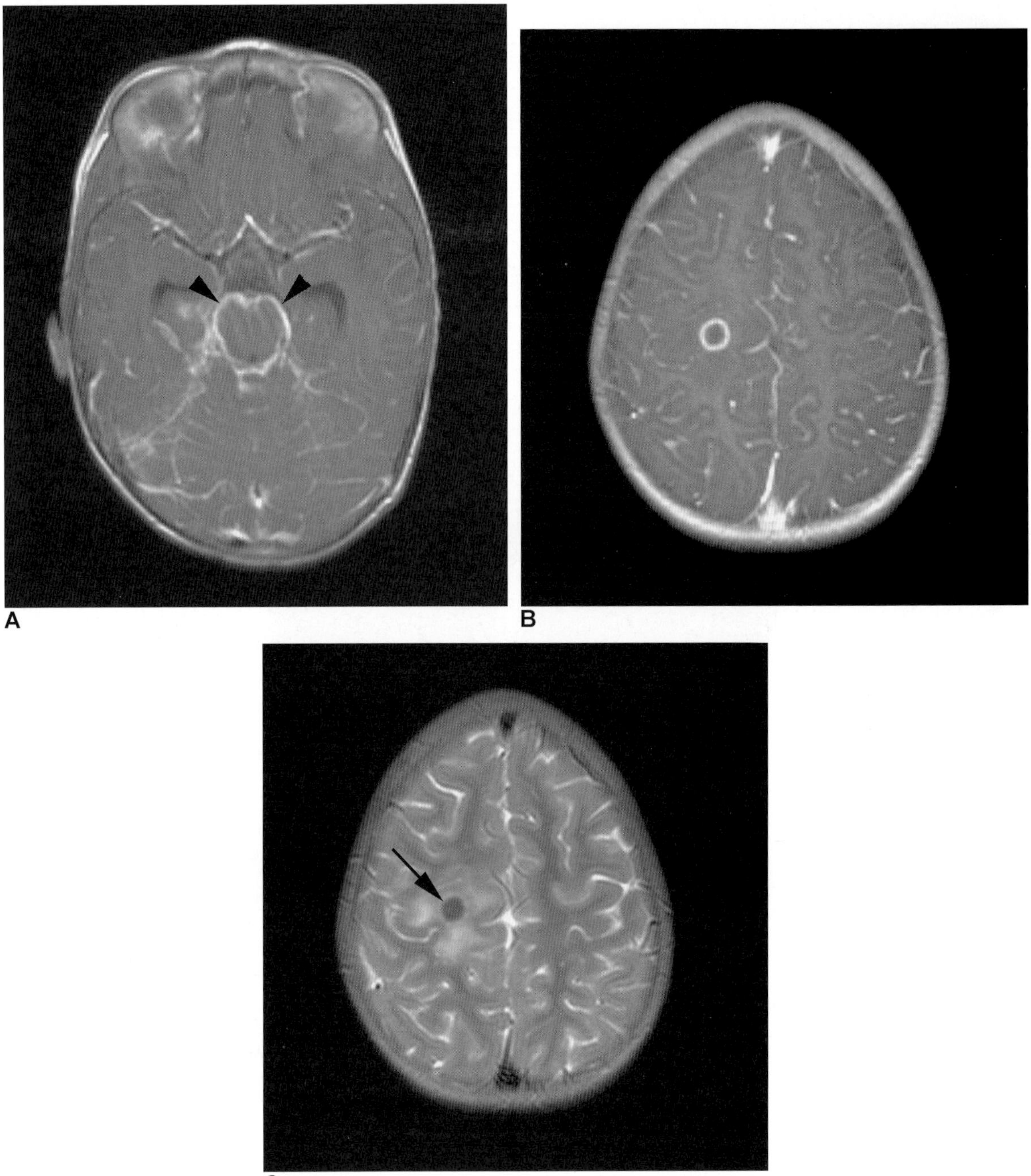

Fig. 16.15. Tuberculous meningitis and tuberculoma. A. Axial contrast-enhanced T1-weighted MR image reveals intense enhancement of the inflammatory exudates within the basal cisterns (arrowheads). B. Axial contrast-enhanced T1-weighted MR image reveals a smooth walled ring-enhancing lesion within the right frontal lobe. C. Axial T2-weighted MR image shows marked hypointensity within the central portions of the lesion (arrow).

Whiteman et al., 1995). The middle cerebral artery and lenticulostriate perforating vessels are most commonly affected, resulting in basal ganglia infarcts that are readily apparent with MR imaging (Sheller and Des Prez, 1986) (Fig. 16.16).

16.3.6. Neurosyphilis

The occurrence of syphilis has risen with the AIDS epidemic, with a threefold increase in the incidence of neurosyphilis since 1983 (Katz and Berger, 1989).

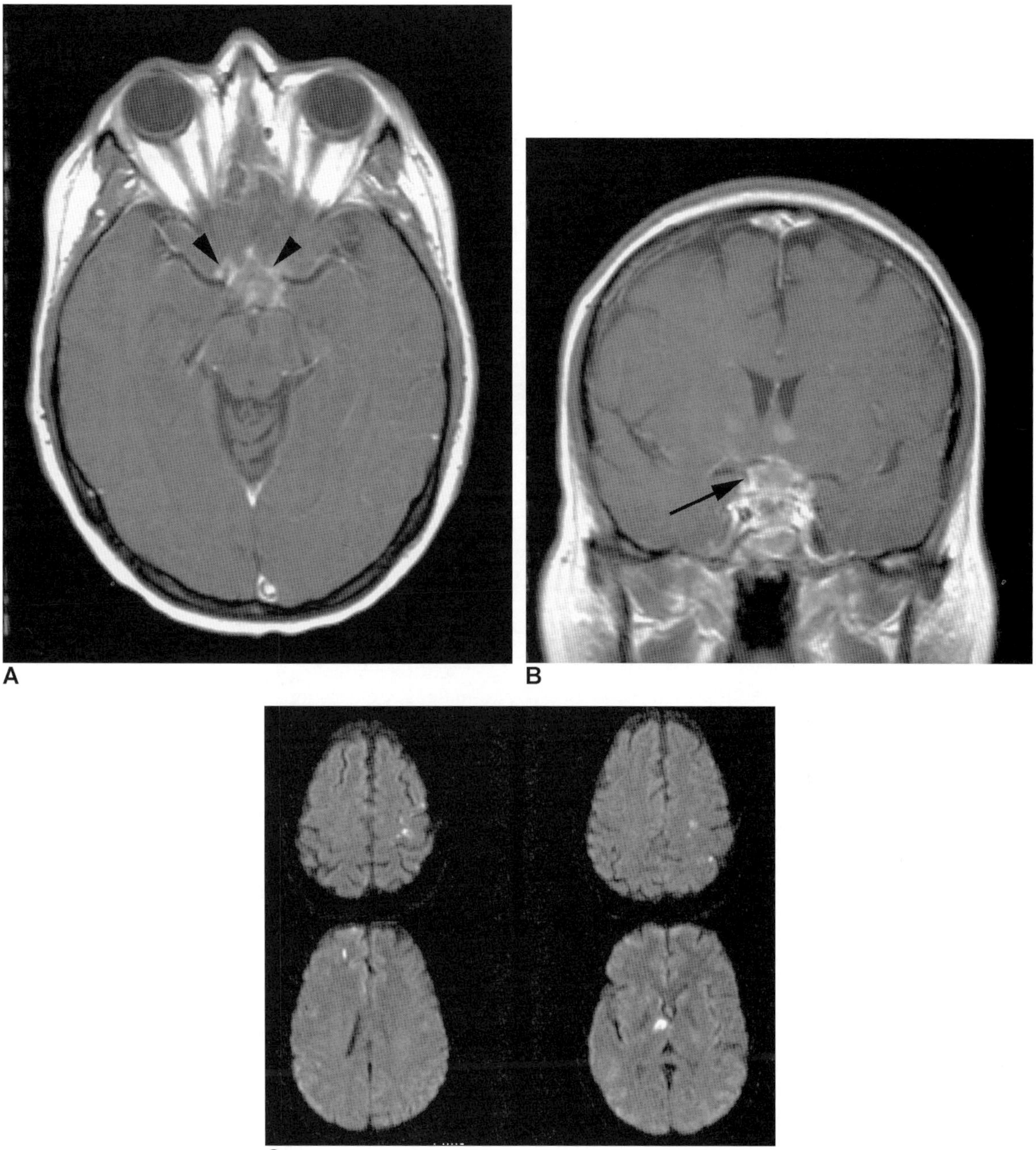

Fig. 16.16. Tuberculous meningitis and vasculitis. A, B. Axial (A) and coronal (B) contrast-enhanced T1-weighted MR images display intense enhancement exudates filling the basal cisterns (arrowheads), intimately associated with the internal carotid arteries (arrow). C. Serial diffusion-weighted MR images reveal multiple small cortical and deep infarcts, not confined to a single vascular distribution, presumably related to complicating vasculitis.

Syphilis is a sexually transmitted disease resulting from infection with the spirochete *Treponema pallidum*. The infectious course of syphilis was classically divided into three stages in the pre-AIDS era. Primary syphilis is represented by a chancre (ulcer or sore) at the site of initial infection with regional lymphadenopathy. Secondary syphilis, resulting from hematogenous dissemination, can occur several weeks after the initial inoculation if appropriate therapy is not instituted. Aseptic meningitis may be seen in a small number of patients with secondary syphilis. One-third to one-half of patients with secondary syphilis will progress to the tertiary stage of infection, with one-third of these patients displaying CNS involvement and neurosyphilis

(Knox et al., 1976; Harris et al., 1997). Neurosyphilis usually takes greater than 5 years to develop following initial infection in the immunocompetent host, and the majority of patients remain asymptomatic (Clark and Danboldt, 1964). In the immune deficient state, notably AIDS, the natural progression of the disease process is altered resulting in a more rapid involvement of the CNS (Johns et al., 1987; Katz et al., 1993; Harris et al., 1997). The prevalence of neurosyphilis approaches 10% in untreated patients (Holland et al., 1986; Brightbill et al., 1995). Diagnosis is generally achieved with a positive CSF-VDRL sample, and treatment encompasses a penicillin antibiotic regimen.

Neurosyphilis most commonly presents with meningeal and vascular involvement (meningovascular), with tabes dorsalis and paresis occurring less commonly (Harris et al., 1997; Provenzale and Jinkins, 1997). The meningeal manifestations include symptoms of acute meningitis, hydrocephalus, cranial neuropathy and the formation of leptomeningeal granulomas (gummas). The vascular manifestations are that of an endarteritis that may involve any size vessel resulting in areas of infarction. The combined meningovascular findings on imaging studies parallel those seen with traditional infarcts, low density on CT, hypointense on T1-weighted imaging, hyperintense in T2-weighted imaging, with restricted diffusion in the acute phase on diffusion weighted imaging. The infarcts may involve any portion of the brain, with infarcts occurring in the cortex, deep white matter, gray matter nuclei and brainstem. Nonspecific hyperintensities on T2-weighted imaging in the deep white matter may also be present (Harris et al., 1997). Arteriography is nonspecific with segmental narrowing of intracranial vessels (Fig. 16.17) (Kaplan et al., 1981). The abnormalities are more evident within the larger vessels on arteriography. Parenchymal involvement with the gummas is uncommon, but on imaging these lesions appear as homogenously or ring-enhancing nodules (Brightbill et al., 1995; Harris et al., 1997). Neurosyphilis is also an important cause of sensorineural hearing loss in the AIDS patient with abnormal enhancement of the labyrinthian structures with MR imaging (Harris et al., 1997).

16.4. Neoplasms

16.4.1. Primary CNS lymphoma

PCNSL is the second most common intracranial mass lesion in the AIDS patient, exceeded in frequency by only toxoplasmosis. PCNSL remains the most common intracranial neoplasm affecting the AIDS patient, occurring in at least 3% of patients (Ruiz et al., 1997; Berger, 2003). While the overall incidence of PCNSL has markedly increased over the past several decades (paralleling the increase in incidence of AIDS), there has been a more recent decline in PCNSL secondary to HAART (Kirk et al., 2001). A causal relationship between PCNSL and viral infections, notably the Epstein–Barr virus (EBV), has gained widespread acceptance. Proliferation of a lymphocyte population is thought to occur in response to a chronic viral infection in the setting of an inadequate immune surveillance system (McMahon et al., 1991; Koeller et al., 1997; Ruiz et al., 1997; Antinori et al., 1999). PCNSL is almost exclusively a B-cell, non-Hodgkin's form of lymphoma (Koeller et al., 1997; Provenzale and Jinkins, 1997; Ruiz et al., 1997). PCNSL most commonly develops in AIDS patients during the third and fourth decades, with men being affected more commonly than women (Ruiz et al., 1997). The presenting clinical signs and symptoms of PCNSL are diverse and relate to the regions affected by the tumor. Patients may represent with symptoms relating to a rapidly enlarging intracranial mass with a focal neurologic deficit, headache, encephalitis, personality change or cranial nerve palsy (Koeller et al., 1997; Ruiz et al., 1997).

PCNSL can display a variety of appearances on neuroimaging examinations. PCNSL may occur at virtually any site within the brain, and commonly presents as multiple lesions (Poon et al., 1989; Dina, 1991; Koeller et al., 1997; Ruiz et al., 1997). Favored locations include the periventricular white matter, corpus callosum, cerebellum and corticomedullary junction. The classically described "deep gray matter" involvement with PCNSL is present in only 33% of cases (Koeller et al., 1997). The cerebral white matter of the frontal lobe is the most common site for PCNSL (Poon et al., 1989; Koeller et al., 1997). PCNSL lesions tend to abut and spread along the ependyma and meninges (Dina, 1991).

PCNSL characteristically produces isodense or hyperdense lesions on un-enhanced CT examinations, related to the compact nature of the tumor cells with a relatively high nuclear to cytoplasmic ratio within the tumor (Lee et al., 1986; Provenzale and Jinkins, 1997; Ruiz et al., 1997) (Fig. 16.18). PCNSL lesions may be well circumscribed with surrounding vasogenic edema, or infiltrative in nature with poorly defined margins. Virtually all PCNSL lesions display marked enhancement following contrast administration. Ring enhancement with central necrosis is commonly seen in AIDS patients, as opposed to the uniform enhancement expected in the immunocompetent state (Fig. 16.19). Rarely a gyriform pattern of enhancement has been reported, mimicking cerebral infarction (Lee et al., 1986). Calcification is not seen in pretreated PCNSL lesions (Koeller et al., 1997). Areas of hemorrhage are rarely seen within PCNSL, albeit accompanying hemorrhage is more commonly seen in

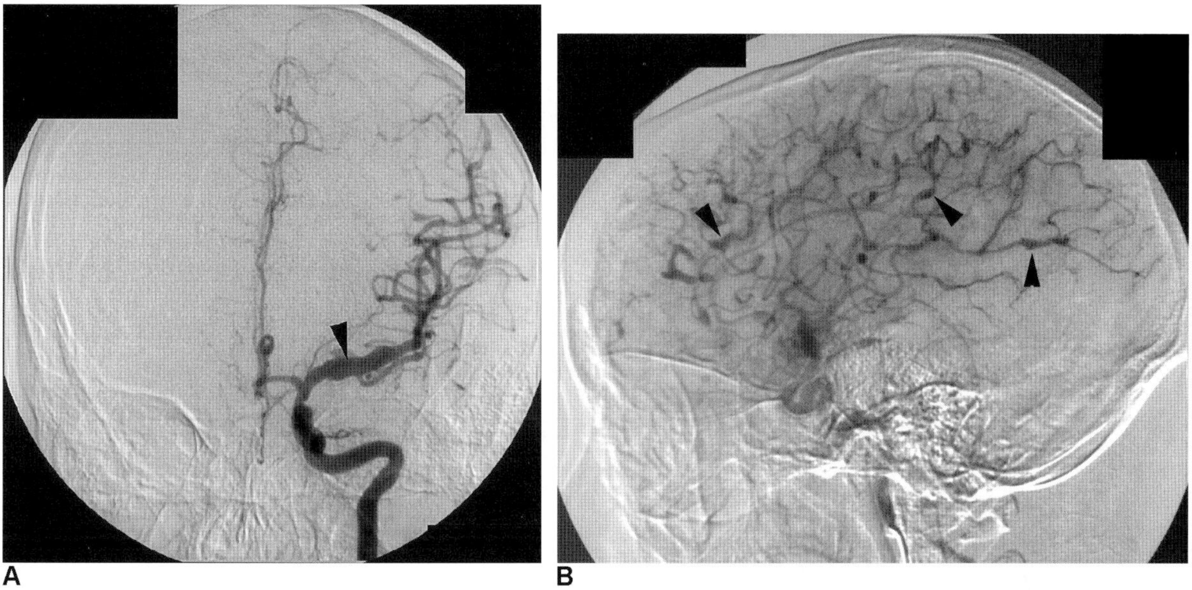

Fig. 16.17. Vasculitis, presumed secondary to neurosyphilis. Frontal (A) and lateral (B) views from a left internal carotid digital subtraction angiogram reveal multiple areas of narrowing and dilatation within the proximal and distal vasculature (arrowheads).

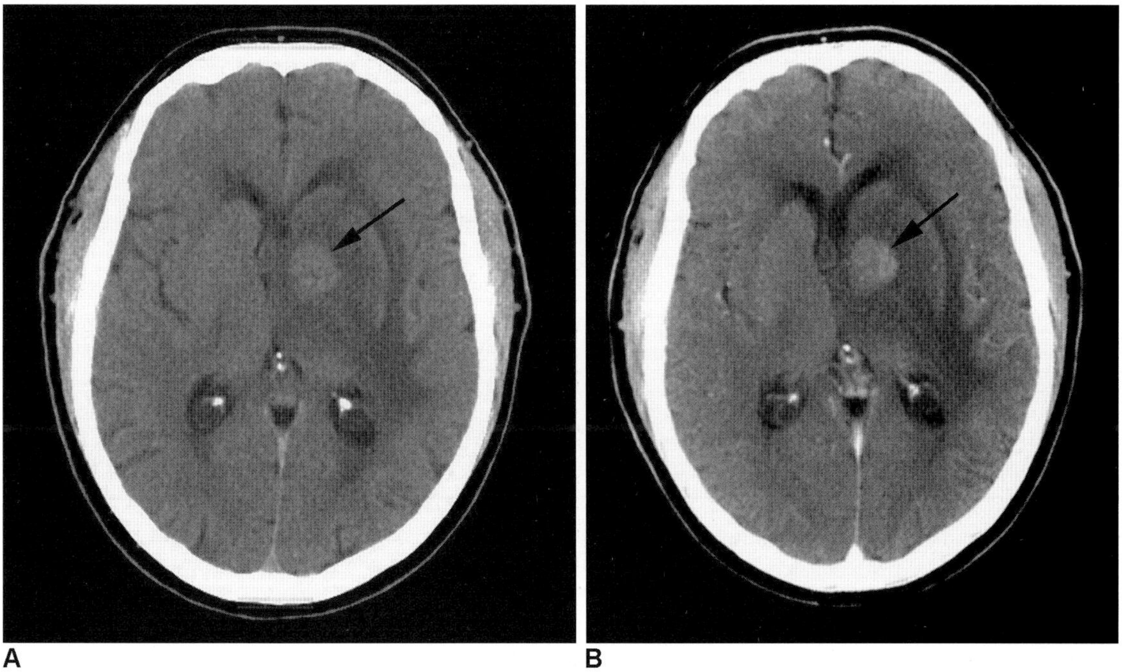

Fig. 16.18. Primary CNS lymphoma. A. Un-enhanced axial CT image reveals a hyperdense mass within the left basal ganglia (arrow) with surrounding vasogenic edema. B. Contrast-enhanced axial CT image reveals intense enhancement of the mass (arrow).

the immunocompromised patient (Koeller et al., 1997; Ruiz et al., 1997).

MR imaging is more sensitive than CT for the detection and characterization of the PCNSL lesions. PCNSL lesions appear iso- to hypointense on T2-weighted sequences (Fig. 16.20). In lesions with central necrosis, the rim appears iso- to hypointense on T2-weighted sequences, with the central areas displaying T2 hyperintensity. As with CT the lesions demonstrate intense contrast enhancement, with the solid lesions displaying

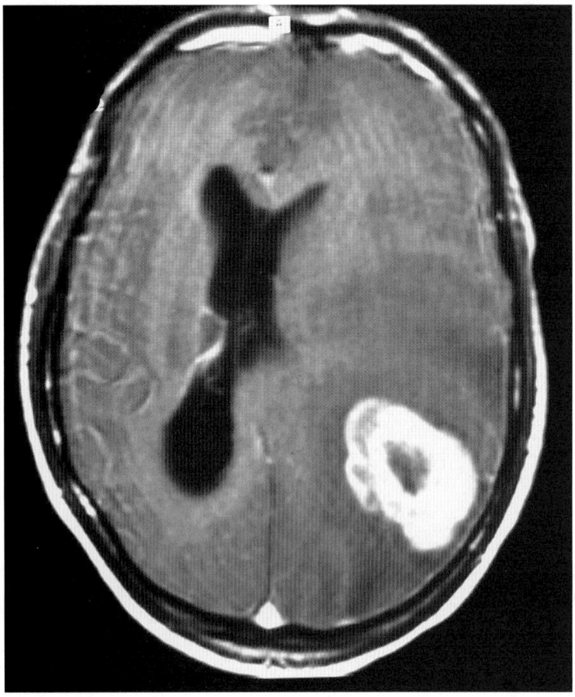

Fig. 16.19. Primary CNS lymphoma. Contrast-enhanced axial T1-weighted MR image reveals a solitary, large lesion with ring enhancement and central necrosis within the left parietal lobe with mass effect and surrounding edema.

uniform enhancement and the necrotic lesions having a ring-like pattern of enhancement. MR imaging is superior to CT in detecting ependymal or meningeal involvement with PCNSL (Fig. 16.21).

Statistically, a large solitary enhancing lesion in the AIDS patient is most commonly PCNSL (Ruiz et al., 1997). Distinguishing PCNSL from toxoplasmosis (and other infectious etiologies) when presented with multiple ring-enhancing lesions can be challenging, if not impossible, based on routine imaging. Several imaging characteristics favor PCNSL over toxoplasmosis, including ependymal or meningeal involvement (Dina, 1991), extension into the corpus callosum, lack of hemorrhagic component (Provenzale and Jinkins, 1997) and rapid growth despite adequate anti-toxoplasmosis therapy. Thallium-201 SPECT or FDG-18 PET imaging appears to be most useful in distinguishing PCNSL from toxoplasmosis. PCNSL typically shows intense tracer uptake within solid lesions and within the rim of necrotic lesions, corresponding to areas of increased metabolic activity (Ruiz et al., 1994; Heald et al., 1996; Antinori et al., 1999; Lee et al., 1999) (Figs. 16.21 and 16.22). Some authors advocate the use of ^{99}Tc-sestamibi SPECT imaging in evaluation of PCNSL, and report similar sensitivities and specificities to that of thallium-201 SPECT (De La Pena et al., 1998; Naddaf et al., 1998). Increased tracer uptake within a lesion on thallium-201 SPECT imaging, particularly when combined with positive CSF for the EBV DNA, has proven reliable in diagnosing PCNSL in the AIDS patient (Miller et al., 1998; Antinori et al., 1999). Diffusion-weighted MR imaging proves less helpful than the nuclear medicine examinations with overlap in appearances, but typically PCNSL lesions will display areas of restricted diffusion not commonly encountered with toxoplasmosis (Berger, 2003; Camacho et al., 2003) (Fig. 16.20). The problem in differentiating toxoplasmosis from PCNSL is compounded by the patient who is administered empiric therapy for presumed toxoplasmosis, but also receives steroids for associated cerebral edema. The PCNSL lesions may show striking response to steroid therapy on imaging, dramatically shrinking the size and enhancement of the lesions, as well as associated edema (Koeller et al., 1997). One may misinterpret lesion improvement on imaging as a response to the anti-toxoplasmosis regimen (supporting a diagnosis of toxoplasmosis), when in reality it represented a transient response of PCNSL to steroid therapy.

16.5. Immune reconstitution inflammatory syndrome

The development of effective therapy, namely HAART, for HIV infection has improved the prognosis of patients with AIDS, decreasing both the development of opportunistic infection and mortality rates. An acute inflammatory leukoencephalopathy has recently been reported in HIV-positive patients after institution of HAART therapy and reconstitution of the immune system, known as immune reconstitution inflammatory syndrome (IRIS). Alternatively IRIS has been referred to as "inflammatory" PML or immune restoration disease. IRIS has been described as a "paradoxical deterioration in clinical status attributable to the recovery of the immune system during HAART" (Shelburne et al., 2002; Vendrely et al., 2005). IRIS has been associated with a variety of infectious pathogens, and the complications of the syndrome vary depending on the offending organism. IRIS has been shown to occur with tuberculosis, cryptococcus, CMV, PML and other opportunistic infections (DeSimmone et al., 2000; Miralles et al., 2001; Shelburne et al., 2002; Du Pasquier and Koralnik, 2003; Cattelan et al., 2004; Nuttall et al., 2004; Vendrely et al., 2005).

Patients generally have a low CD^4 count, followed by a marked increase in the CD^4 count and marked reduction in HIV viral load after institution of HAART (DeSimmone et al., 2000; Miralles et al., 2001; Nuttall et al., 2004). With restoration of the immune system, there is a resultant intense inflammatory response

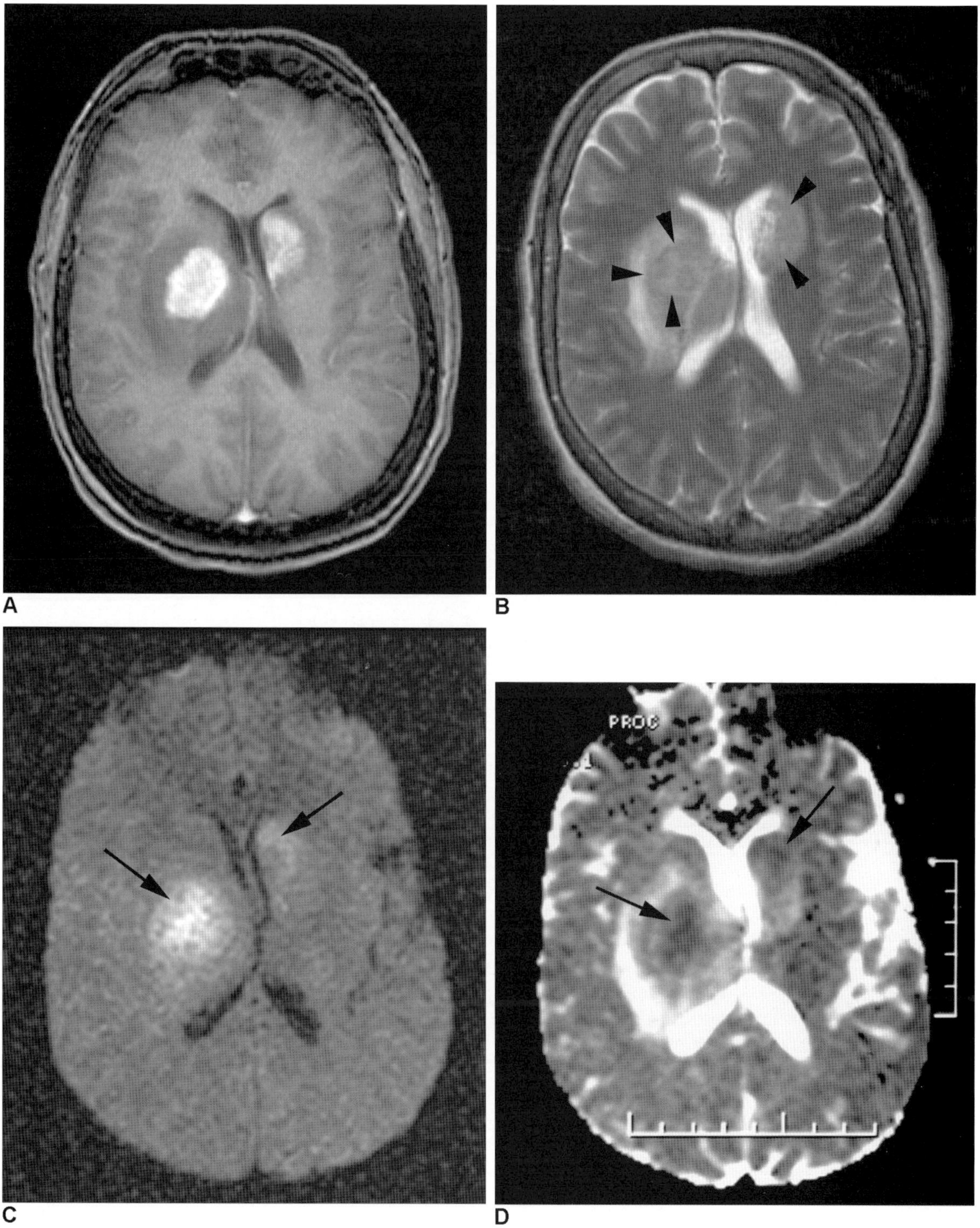

Fig. 16.20. Primary CNS lymphoma, utility of MR diffusion-weighted imaging. A. Contrast-enhanced axial T1-weighted MR image reveals homogeneously enhancing masses within the bilateral basal ganglia. B. Axial T2-weighted MR image reveals isointense masses (arrowheads) with surrounding vasogenic edema. C, D. Axial diffusion-weighted image (C) and apparent diffusion coefficient (ADC) map (D) demonstrate restricted diffusion within the lesions (arrows).

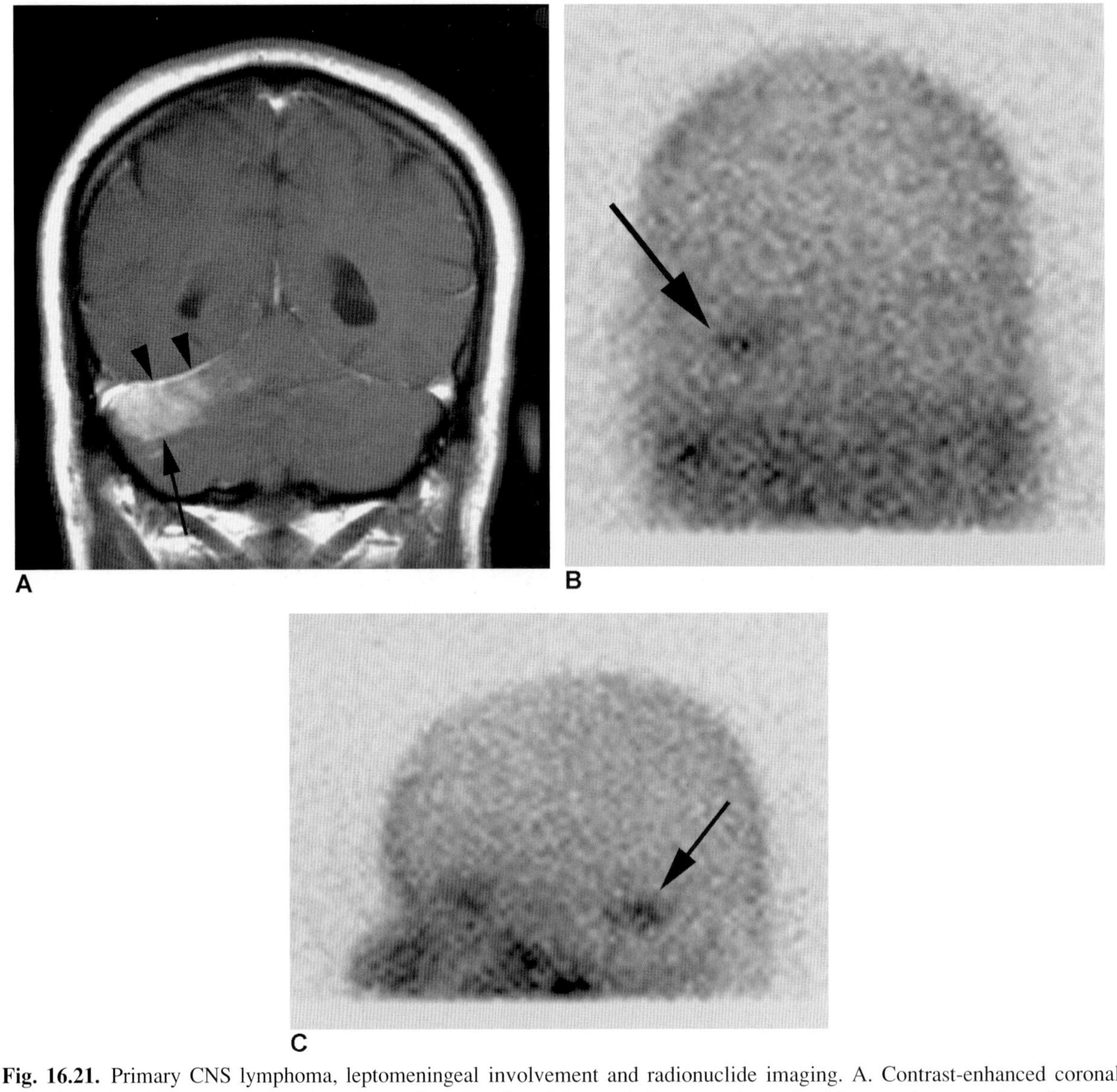

Fig. 16.21. Primary CNS lymphoma, leptomeningeal involvement and radionuclide imaging. A. Contrast-enhanced coronal T1-weighted MR image reveals a homogenously enhancing right cerebellar mass (arrow) with extension to and involvement of the leptomeningeal surface (arrowheads). B, C. Posterior (B) and lateral (C) projection planar images from a thallium-201 nuclear medicine brain scan reveal intense abnormal metabolic activity associated with the cerebellar mass (arrow).

towards offending pathogens. Cryptococcal and mycobacterial infections appear particularly sensitive to the development of IRIS, with reported 30% occurrence rates following institution of HAART (Shelburne et al., 2005a, b). IRIS is manifested with both a clinical and radiological worsening shortly after institution of HAART. Biopsy of the IRIS lesions demonstrates a marked lymphocytic infiltration. Imaging often demonstrates mass effect and enhancement, findings which may have not been present on initial imaging studies and characteristically not associated with initial pathogen (Fig. 16.23) (Miralles et al., 2001; Hoffman et al., 2003; Nuttall et al., 2004). Care must thus be taken not to mistake IRIS for disease progression. Preliminary investigations have suggested that the mass effect and enhancement may represent a positive predictive factor in patient survival (Thurnher et al., 2001; Du Pasquier and Koralnik, 2003). The majority of patients with IRIS will show improvement with continuation of HAART, primary therapy against the offending pathogen, and symptomatic treatment with anti-inflammatory agents (DeSimmone et al., 2000; Miralles et al., 2001; Du Pasquier and Koralnik, 2003; Shelburne and Hamill, 2003).

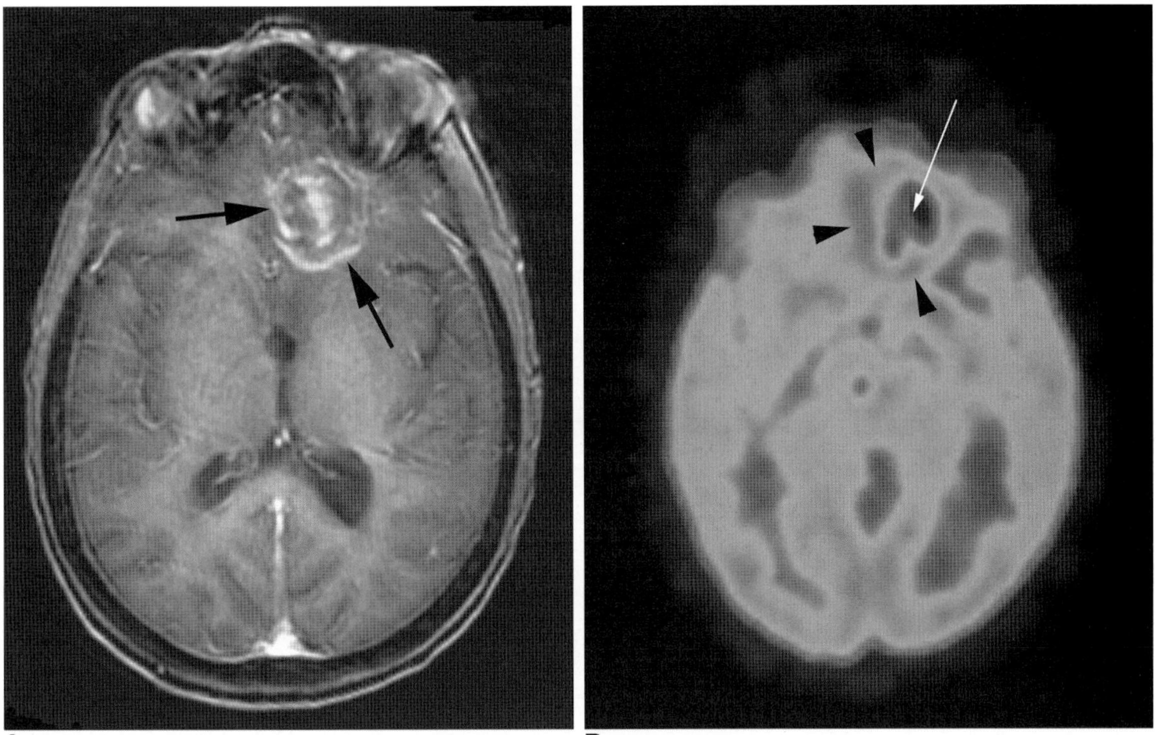

A B

Fig. 16.22. For full color figure, see plate section. Primary CNS lymphoma, utility of PET imaging. A. Contrast-enhanced axial T1-weighted MR image reveals a ring-enhancing lesion (black arrows) within the inferior portions of the left frontal lobe with central areas of necrosis. B. Axial image from a FDG-18 PET scan reveals hypermetabolic activity associated with the rim of the lesion (arrowheads). There is hypometabolic activity within center of the lesion corresponding to areas of necrosis (white arrow).

16.6. Vertebral and spinal cord lesions

16.6.1. Vertebral infections and neoplasms

Spondylodiscitis in the AIDS patient is generally the result of *Staphylococcal aureus*, the most common causative agent in both the immunocompetent and immunosuppressed populations (Heary et al., 1994). The imaging findings of spondylodiscitis parallel those found in the immunocompetent patient with involvement/destruction of the vertebral body end-plates, increased T2-weighted signal within the disc, enhancement of the disc and end-plates following IV contrast administration and inflammation of the surrounding paraspinal and epidural soft tissues. In addition to the common pathogens seen in immunocompetent state, infection with *Salmonella* and *Mycobacterium tuberculosis* species frequently may present with spondylodiscitis and epidural abscess in the AIDS patient. Tuberculosis will typically begin with involvement of the anterior half of the vertebral body with extension into the paraspinal and epidural soft tissues, producing large abscesses (Thurnher et al., 1997a, 2000). Adjacent vertebral bodies may also be affected by tuberculosis via subligamentous spread, with a relative sparing of the intervertebral disc compared to that of pyogenic spondylodiscitis (Fig. 16.24). Vertebral body collapse is also commonly encountered in cases of tuberculous spondylodiscitis (Shanley, 1995; Schinina et al., 2001).

Lymphomatous involvement of the vertebral bodies has been reported in nearly 22% of AIDS patients (Safai et al., 1992). Radiographs and CT examinations may reveal lytic lesions throughout the spinal axis. MR imaging is similarly nonspecific with focal hypointense T1-weighted and hyperintense T2-weighted lesions (Fig. 16.25), the majority revealing patchy contrast enhancement (Thurnher et al., 1997a). Lymphomatous involvement of the vertebral body may extend into the epidural space, simulating an epidural abscess.

Spinal MR examinations in AIDS patients may also demonstrate loss of the normal hyperintense appearance of the vertebral bodies on T1-weighted imaging, felt to represent increased iron storage within the marrow related to chronic anemia (Geremia et al., 1990; Thurnher et al., 2000). The uniform and diffuse involvement of

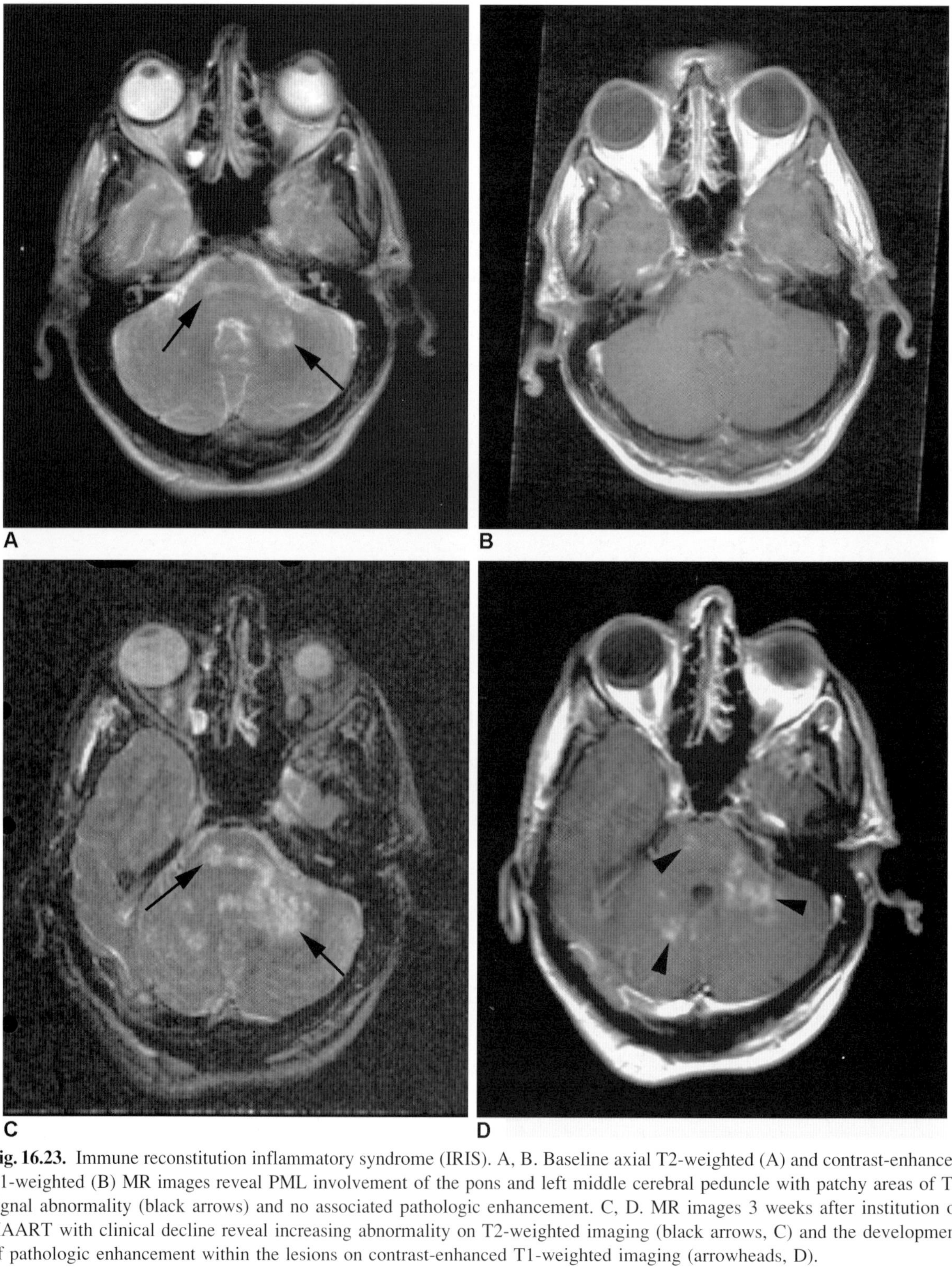

Fig. 16.23. Immune reconstitution inflammatory syndrome (IRIS). A, B. Baseline axial T2-weighted (A) and contrast-enhanced T1-weighted (B) MR images reveal PML involvement of the pons and left middle cerebral peduncle with patchy areas of T2 signal abnormality (black arrows) and no associated pathologic enhancement. C, D. MR images 3 weeks after institution of HAART with clinical decline reveal increasing abnormality on T2-weighted imaging (black arrows, C) and the development of pathologic enhancement within the lesions on contrast-enhanced T1-weighted imaging (arrowheads, D).

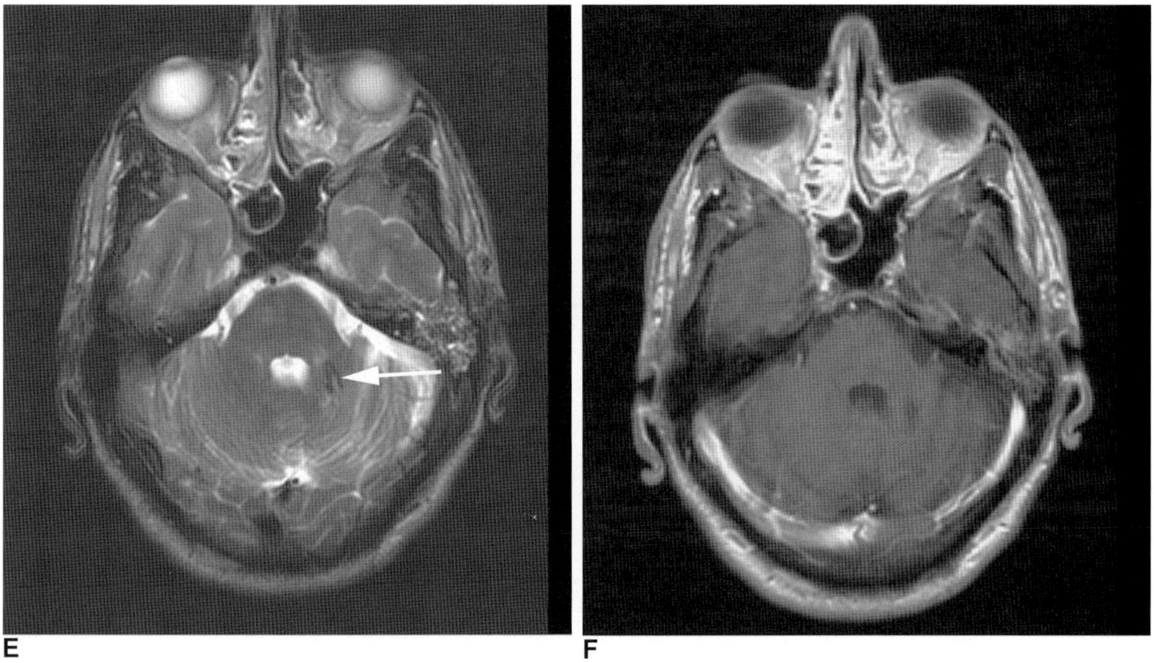

Fig. 16.23. (Continued) Immune reconstitution inflammatory syndrome. E, F. The patient subsequently underwent stereotactic biopsy confirming IRIS complicating PML. With continuation of HAART, the patient showed both clinical and radiological improvement with resolving T2 abnormality (E) and absence of pathologic contrast enhancement (F) on follow-up MR imaging 3 months later. Note site of prior biopsy with small area of hemosiderin (white arrow, E).

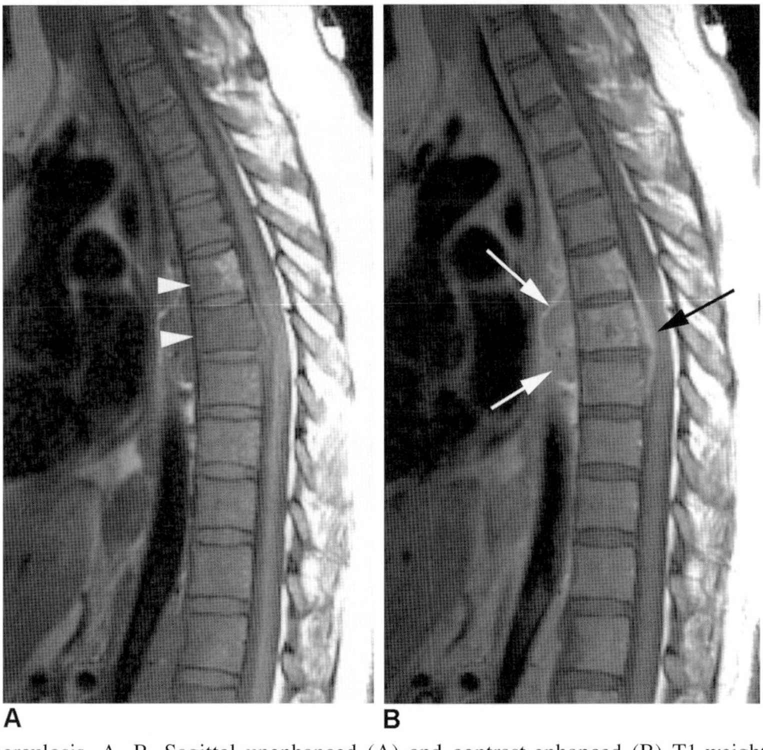

Fig. 16.24. Spinal tuberculosis. A, B. Sagittal unenhanced (A) and contrast-enhanced (B) T1-weighted MR images reveal abnormal signal and inflammation within several mid-thoracic vertebral bodies (arrowheads) with relatively sparing of the disc space. There is sub-ligamentous spread (black arrow) of the inflammatory process over several vertebral body levels and extension into the paravertebral soft tissues (white arrows).

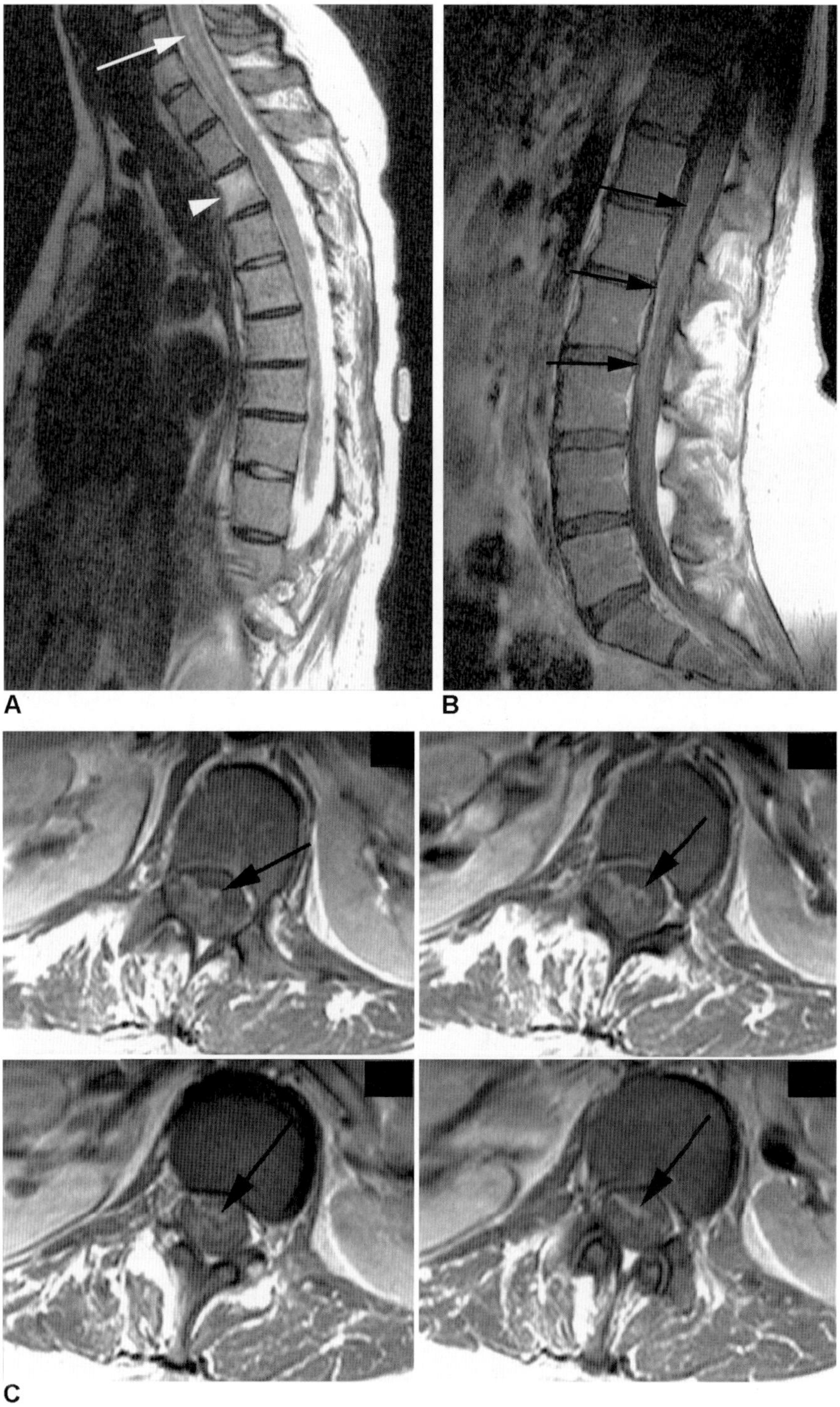

Fig. 16.25. Spinal lymphoma. A. Sagittal T2-weighted MR image of the cervicothoracic junction reveals abnormal T2 signal and enlargement of the lower cervical spinal cord (white arrow), and involvement of an upper thoracic vertebral body (white arrowhead). B, C. Sagittal (B) and axial (C) contrast-enhanced T1-weighted MR images of the lumbar spine reveal abnormal contrast enhancement on the surface of the spinal cord and lumbosacral nerve roots (black arrows).

the vertebral bodies seen with anemia of chronic disease generally permits distinction from the patchy involvement typically seen with lymphomatous infiltration.

16.6.2. Spinal cord lesions

The spinal cord is not isolated from pathologic processes in the AIDS patient, and CNS lesions frequent to the brain are also those most commonly implicated in cases of myelopathy. AIDS patients have been reported to have spinal cord involvement with HIV myelitis in 5–8% of cases, vacuolar myelopathy in 14–54% of cases, opportunistic infections in 8–15% of cases and lymphomatous involvement in 2–8% of cases (Petito, 1997). In all, spinal cord pathology has been reported to occur in up to 50% of AIDS patients (Levy et al., 1985; Henin et al., 1992; Petito, 1997). Despite this high prevalence, imaging of spinal cord pathology has received little attention in the literature. Reasons for this are multifactorial, but likely masking of the various myelopathies by coexisting brain lesions and the lack of pathologic correlation with either biopsy or autopsy specimens makes imaging and diagnosis of spinal cord pathology less than ideal (Quencer and Post, 1997).

The imaging appearance of the various AIDS-related myelitides is frequently nonspecific, with MR imaging the modality of choice. Quencer and Post (1997) characterized spinal cord pathology in the AIDS patient by describing three common imaging presentations: intramedullary lesion with mass effect, intramedullary lesion without mass effect and intramedullary lesion with accompanying meningeal disease.

16.6.2.1. Intramedullary lesion with mass effect

Given the frequency of cerebral involvement, it is not surprising that the two most common spinal cord lesions to present with mass effect are toxoplasmosis and lymphoma, albeit their occurrence is rare compared to their intracranial counterparts. The imaging appearance of both lesions includes contrast-enhancing lesions with surrounding edema on T2-weighted imaging (Fig. 16.25) (Harris et al., 1990; Poon et al., 1992; Quencer and Post, 1997; Thurnher et al., 2000). Differentiating toxoplasmosis from lymphoma in the spinal cord is no less challenging than in the brain. Often adequate characterization of brain lesions will allow a presumptive diagnosis of the spinal lesion, but unfortunately multiple, concurrent pathologic processes are frequently encountered in the AIDS patient (Quencer and Post, 1997). Most patients with spinal toxoplasma myelitis will have concurrent intracranial involvement (Thurnher et al., 2000). Fungal and viral lesions, such as CMV and herpes, may less commonly present with cord lesions producing this imaging appearance.

16.6.2.2. Intramedullary without mass effect

The finding of a normal-sized spinal cord with T2 signal abnormality on MR imaging poses a differential diagnosis consisting of vacuolar myelopathy, tract pallor, PML and HIV myelitis. Similar to the mass-producing lesion of the spinal cord, correlation with imaging studies of the brain will often permit a presumptive diagnosis. Establishing the process as focal or diffuse proves useful when evaluating the spinal cord abnormality. Vacuolar myelopathy and tract (myelin) pallor are poorly understood processes in the AIDS patient, with uncertain etiologies. Microscopically, vacuolar myelopathy represents vacuolation of the spinal white matter and the presence of lipid-laden macrophages (Santosh et al., 1995; Thurnher et al., 2000). Tract pallor represents loss of myelin staining without accompanying vacuolation or macrophages (Santosh et al., 1995). Vacuolar myelopathy is strongly associated with HIV myelitis, and there is evidence implicating the myelin destruction by HIV-infected macrophages (Grafe and Wiley, 1989; Quencer and Post, 1997). Nonetheless, the diagnosis of vacuolar myelopathy remains one of exclusion (Thurnher et al., 2000). MR imaging is reported to display diffuse and symmetrical T2-weighted signal abnormalities, without contrast enhancement, within a normal-sized spinal cord (Fig. 16.26) (Santosh et al., 1995; Quencer and Post, 1997; Thurnher et al., 2000). The distribution of vacuolar myelopathy favors the lateral and posterior columns of the cervical and thoracic spinal cord (Maier et al., 1989; Artigas et al., 1990; Quencer and Post, 1997; Thurnher et al., 2000). HIV myelitis tends to involve both the gray and white matter portions of the spinal cord and is typically associated with HIV encephalitis (Petito, 1997). MR imaging of HIV myelitis is reported to demonstrate more focal and asymmetrical T2-weighted abnormalities than those of vacuolar myelopathy, helping to distinguish these two entities (Santosh et al., 1995; Quencer and Post, 1997; Thurnher et al., 2000). Contrast enhancement is thought to be with either vacuolar myelopathy or HIV encephalitis.

16.6.2.3. Intramedullary with accompanying meningeal disease

The identification of meningeal disease proves useful in evaluating spinal cord pathology in the AIDS patient. CMV, tuberculosis, neurosyphilis, toxoplasmosis and lymphoma can all present with cord abnormalities accompanied by meningeal enhancement on MR

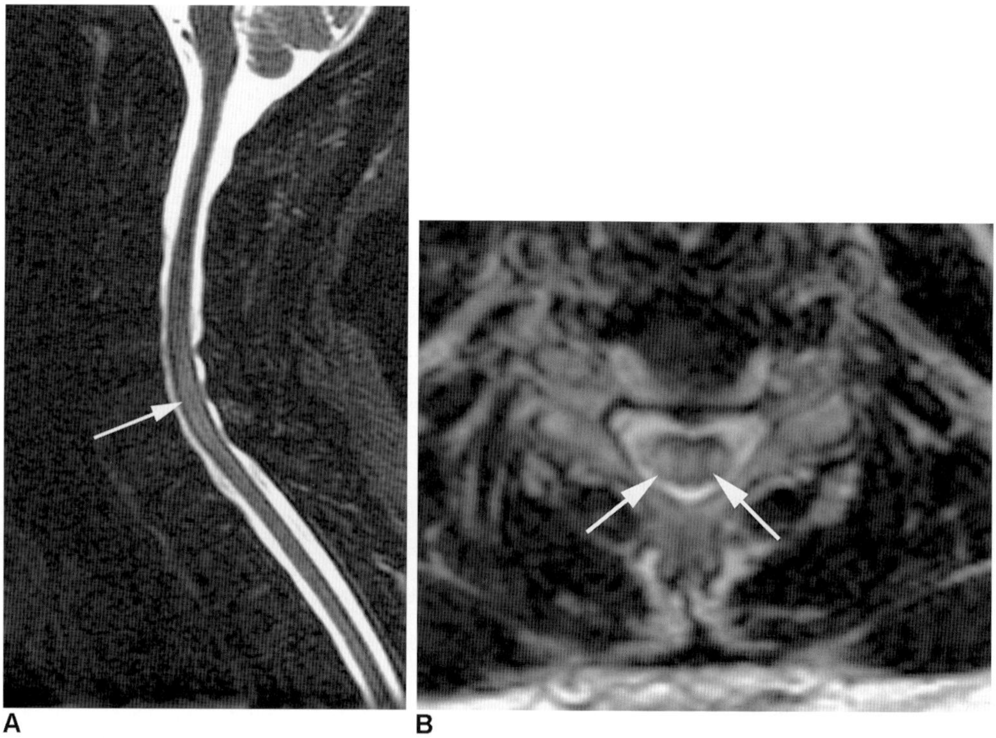

Fig. 16.26. Presumed vacuolar myelopathy. A, B. Sagittal short tau inversion recovery (STIR, A) and axial T2 (B) MR images reveal abnormal T2 signal (arrows) diffusely within the substance of the lower cervical spinal cord. Contrast-enhanced images (not shown) revealed no associated pathologic enhancement.

imaging (Quencer and Post, 1997). Cryptococcal infection will typically not produce meningeal findings. CMV ventriculitis has a propensity for development of spinal involvement, either with subpial involvement of the cord or in the form of a polyradiculopathy (Fig. 16.11). With MR imaging, one may observe diffuse enhancement lining the conus medullaris and lumbosacral nerve roots in cases of CMV polyradiculomyelitis (Bazan et al., 1991; Talpos et al., 1991; Whiteman et al., 1994; Lizerbram and Hesselink, 1997). The pattern of lumbosacral enhancement with CMV may be linear or nodular in nature, and involvement is not typically appreciable without intravenous contrast administration. Tuberculosis and lymphoma can also present with involvement of the lumbosacral nerve roots, with clumping and enhancement of the lumbosacral nerve roots on MR imaging (Fig. 16.25) (Quencer and Post, 1997).

16.7. Conclusion

Despite increasing effectiveness of HIV therapies, namely HAART, CNS involvement by various pathologic processes in the AIDS population continues to be a significant cause of morbidity and mortality. Imaging techniques, particularly MR imaging, are vital to the detection and characterization of CNS pathology. While a definitive diagnosis cannot often be established based on conventional imaging alone, advanced imaging techniques (e.g. thallium-201 SPECT) in combination with serum and CSF analysis can often establish a presumptive diagnosis and permit institution of empiric therapies. Imaging also plays an important role in disease surveillance and in assessing treatment response.

References

Abgrall S, Rabaud C, Costagliola D, et al (2001). Incidence and risk factors for toxoplasmic encephalitis in human immunodeficiency virus-infected patients before and during the highly active antiretroviral therapy era. Clin Infect Dis 33(10): 1747–1755.

Andreula CF, Burdi N, Carella A (1993). CNS cryptococcosis in AIDS: spectrum of MR findings. J Comput Assist Tomogr 17(3): 438–441.

Antinori A, DeRossi G, Ammassari A, et al (1999). Value of combined approach with thallium-201 single-photon emission computed tomography and Epstein-Barr virus DNA polymerase chain reaction in CSF for the diagnosis of AIDS-related primary CNS lymphoma. J Clin Oncol 17(2): 554–560.

Arnder L, Castillo M, Heinz ER, et al (1996). Unusual pattern of enhancement in cryptococcal meningitis: in vivo findings with postmortem correlation. J Comput Assist Tomogr 20(6): 1023–1026.

Artigas J, Grosse G, Niedobitek F (1990). Vacuolar myelopathy in AIDS. A morphological analysis. Pathol Res Pract 186: 228–237.

Bargallo J, Berenguer J, Garcia-Barrionuevo J, et al (1996). The "target sign": is it a specific sign of CNS tuberculoma? Neuroradiology 38(6): 547–550.

Bazan III C , Jackson C, Jinkins JR, et al (1991). Gadolinium-enhanced MRI in a case of cytomegalovirus polyradiculopathy. Neurology 41: 1522–1523.

Berenguer J, Moreno S, Laguna F, et al (1992). Tuberculous meningitis in patients infected with the human immunodeficiency virus. N Engl J Med 326: 668–672.

Berger JR (2003). Mass lesions of the brain in AIDS: the dilemmas of distinguishing toxoplasmosis from primary CNS lymphoma. AJNR Am J Neuroradiol 24(4): 554–555.

Berger JR, Major EO (1999). Progressive multifocal leukoencephalopathy. Semin Neurol 19(2): 193–200.

Berger JR, Pall L, Lanska D, et al (1998). Progressive multifocal leukoencephalopathy in patients with HIV infection. J Neurovirol 4(1): 59–68.

Berkefeld J, Enzensberger W, Lanfermann H (1999). Cryptococcus meningoencephalitis in AIDS: parenchymal and meningeal forms. Neuroradiology 41(2): 129–133.

Berlit P, Popescu O, Weng Y, et al (1996). Disseminated cerebral hemorrhages as unusual manifestation of toxoplasmic encephalitis in AIDS. J Neurol Sci 143(1–2): 187–189.

Bishburg E, Sunderam G, Reichman LB, et al (1986). Central nervous system tuberculosis with the acquired immunodeficiency syndrome and its related complex. Ann Intern Med 105: 210–213.

Brightbill TC, Ihmeidan IH, Post MJ, et al (1995). Neurosyphilis in HIV-positive and HIV-negative patients: neuroimaging findings. AJNR Am J Neuroradiol 16(4): 703–711.

Camacho DL, Smith JK, Castillo M (2003). Differentiation of toxoplasmosis and lymphoma in AIDS patients by using apparent diffusion coefficients. AJNR Am J Neuroradiol 24: 633–637.

Cattelan AM, Trevenzoli M, Sasset L, et al (2004). Multiple cerebral cryptococcomas associated with immune reconstitution in HIV-1 infection. AIDS 18(2): 349–351.

Chaisson RE, Slutkin G (1989). Tuberculosis and human immunodeficiency virus infection. J Infect Dis 159: 96–100.

Chang K-H, Han M-H, Roh JK, et al (1990). Gd-DTPA enhanced MR imaging in intracranial tuberculosis. Neuroradiology 32: 19–25.

Chang L, Ernst T, Leonido-Yee M, et al (2000). Perfusion MRI detects rCBF abnormalities in early stages of HIV-cognitive motor complex. Neurology 54(2): 389–396.

Chinn RJ, Wilkinson RD, Hall-Craggs MA, et al (1995). Toxoplasmosis and primary central nervous system lymphoma in HIV infection: diagnosis with MR spectroscopy. Radiology 197(3): 649–654.

Chong-Han CH, Cortex SC, Tung GA (2003). Diffusion-weighted MRI of cerebral toxoplasma abscess. AJR Am J Roentgenol 181(6): 1711.

Clark EG, Danboldt N (1964). The Oslo study of the natural course of untreated syphilis. Med Clin North Am 48: 613.

Clifford DB, Arribas JR, Storch GA, et al (1996). Magnetic resonance brain imaging lacks sensitivity for AIDS associated cytomegalovirus encephalitis. J Neurovirol 2(6): 397–403.

De La Pena RC, Ketonen L, Villanueva-Meyer J (1998). Imaging of brain tumors in AIDS patients by means of dual-isotope thallium-201 and technetium-99m single photon emission tomography. Eur J Nucl Med 25(10): 1404–1411.

DeSimmone JA, Pomerantz RJ, Babinchak TJ (2000). Inflammatory reactions in HIV-1-infected persons after initiation of highly active antiretroviral therapy. Ann Intern Med 133(6): 447–454.

Dina TS (1991). Primary central nervous system lymphoma vs. toxoplasmosis in AIDS. Radiology 179: 823–828.

Du Pasquier RA, Koralnik IJ (2003). Inflammatory reaction in progressive multifocal leukoencephalopathy: harmful or beneficial? J Neurovirol 9(Suppl 1): 25–31.

Dyer JR, French MA, Mallal SA (1995). Cerebral mass lesions due to cytomegalovirus in patients with AIDS: report of two cases. J Infect 30(2): 147–151.

Ernst TM, Chang L, Witt MD, et al (1998). Cerebral toxoplasmosis and lymphoma in AIDS: perfusion MR imaging experience in 13 patients. Radiology 208(3): 663–669.

Ernst T, Chang L, Witt M, et al (1999). Progressive multifocal leukoencephalopathy and human immunodeficiency virus-associated white matter lesion in AIDS: magnetization transfer MR imaging. Radiology 210: 539–543.

Garcia CA, Weisberg LA, Lacorte WS (1985). Cryptococcal intracerebral mass lesions: CT-pathologic considerations. Neurology 35(5): 731–734.

Geremia GK, McCluney KW, Adler SS, et al (1990). The magnetic resonance hypointense spine of AIDS. J Comput Assist Tomogr 14(5): 785–789.

Gillams AR, Allen E, Hrieb K, et al (1997). Cerebral infarction in patients with AIDS. AJNR Am J Neuroradiol 18(8): 1581–1585.

Grafe MR, Wiley CA (1989). Spinal cord and peripheral nerve pathology in AIDS: the roles of cytomegalovirus and human immunodeficiency virus. Ann Neurol 25(6): 561–566.

Grant IH, Gold JWM, Rosenblum M, et al (1990). Toxoplasma gondii serology in HIV-infected patients: the development of central nervous system toxoplasmosis in AIDS. AIDS 4: 519–521.

Gupta RK, Pandey R, Khan EM, et al (1991). Intracranial tuberculomas: MRI signal intensity correlation with histopathology and localized proton spectroscopy. Magn Reson Imaging 11: 443–449.

Happe S, Lunenborg N, Ricker CH, et al (2000). Progressive multifocal leukoencephalopathy in AIDS. Overview and retrospective analysis of 17 patients. Nervenarzt 71(2): 96–104.

Harris DE, Enterline DS (1997). Fungal infections of the central nervous system. Neuroimaging Clin North Am 7(2): 187–198.

Harris TM, Smith RR, Bognanno JR, et al (1990). Toxoplasmic myelitis in AIDS: gadolinium-enhanced MR. J Comput Assist Tomogr 14(5): 809–811.

Harris DE, Enterline DS, Tien RD (1997). Neurosyphilis in patients with AIDS. Neuroimaging Clin North Am 7(2): 215–221.

Heald AE, Hicks CB (1997). Clinical complications of HIV infection. Radiol Clin North Am 35(5): 1007–1027.

Heald AE, Hoffman JM, Bartlett JA, et al (1996). Differentiation of central nervous system lesions in AIDS using positron emission tomography (PET). Int J STD AIDS 7(5): 337–346.

Heary RF, Hunt CD, Kreiger AJ, et al (1994). HIV status does not affect microbiologic spectrum or neurologic outcome in spinal infections. Surg Neurol 42: 417–423.

Henin D, Smith TW, De Girolami U, et al (1992). Neuropathology of the spinal cord in the acquired immunodeficiency syndrome. Hum Pathol 23: 1106–1114.

Hoffman C, Horst HA, Albrecht H, et al (2003). Progressive multifocal leukoencephalopathy with unusual inflammatory response during antiretroviral treatment. J Neurol Neurosurg Psych 74: 1142–1144.

Hoffman JM, Waskin HA, Schifter T, et al (1993). FDG-PET in differentiating lymphoma from nonmalignant central nervous system lesions in patients with AIDS. J Nucl Med 34(4): 567–575.

Holland BA, Perrett LV, Mills CM (1986). Meningovascular syphilis: CT and MR findings. Radiology 158: 439–442.

Hurley RA, Ernst T, Khalili K, et al (2003). Identification of HIV-associated progressive multifocal leukoencephalopathy: magnetic resonance imaging and spectroscopy. J Neuropsych Clin Neurosci 15(1): 1–6.

Iranzo A, Marti-Fabregas J, Domingo P, et al (1999a). Absence of thallium-201 brain uptake in progressive multifocal leukoencephalopathy in AIDS patients. Acta Neurol Scand 100(2): 102–105.

Iranzo A, Moreno A, Pujol J, et al (1999b). Proton magnetic resonance spectroscopy pattern of progressive multifocal leukoencephalopathy in AIDS. J Neurol Neurosurg Psych 66(4): 520–523.

Johns DR, Tierney M, Felsenstein D (1987). Alteration in the natural course of neurosyphilis by concurrent infection with the human immunodeficiency virus. N Engl J Med 316: 1569–1572.

Jones JL, Sehgal M, Maguire JH (2002). Toxoplasmosis-associated deaths among human immunodeficiency virus-infected persons in the United States, 1992–1998. Clin Infect Dis 34(8): 1161.

Kandanearatchi A, Williams B, Everall IP (2003). Assessing the efficacy of highly active antiretroviral therapy in the brain. Brain Pathol 13: 104–110.

Kaplan JG, Sterman AB, Horoupian D, et al (1981). Leutic meningitis with gumma: clinical, radiographic, and neuropathologic features. Neurology 31: 464–467.

Katz DA, Berger JR (1989). Neurosyphilis in acquired immunodeficiency syndrome. Arch Neurol 46: 895–898.

Katz DA, Berger JR, Duncan RC (1993). Neurosyphilis. A comparative study of the effects of infection with human immunodeficiency virus. Arch Neurol 50(3): 243–249.

Kirk O, Pederson C, Cozzi-Lepri A, et al (2001). Non-Hodgkin lymphoma in HIV-infected patients in the era of highly active antiretroviral therapy. Blood 98(12): 406–412.

Knox JM, Musher D, Guzick ND (1976). The pathogenesis of syphilis and related treponematoses. In: RC Johnson, (Ed.), The Biology of Parasitic Spirochetes. Academic Press, San Diego, CA, pp. 249–259.

Koeller KK, Smirniotopoulos JG, Jones RV (1997). Primary central nervous system lymphoma: radiologic–pathologic correlation. Radiographics 17: 1497–1526.

Lee VW, Antonacci V, Tilak S, et al (1999). Intracranial mass lesions: sequential thallium and gallium scintigraphy in patients with AIDS. Radiology 211(2): 507–512.

Lee YY, Bruner JM, Van Tassel P, et al (1986). Primary central nervous system lymphoma: CT and pathologic correlation. AJR Am J Roentgenol 147(4): 747–752.

Lee PL, Yiannoutsos CT, Ernst T (2003). A multi-center 1H MRS study of the AIDS dementia-complex: validation and preliminary analysis. J Magn Reson Imaging 17(6): 625–633.

Leite AG, Vidal JE, Bonasser Filho F, et al (2004). Cerebral infarction related to cryptococcal meningitis in an HIV-infected patient: case report and literature review. Braz J Infect Dis 8(2): 175–179.

Levy R, Bredesen D, Rosenblum M (1985). Neurological manifestations of the acquired immunodeficiency syndrome (AIDS): experience at the UCSF and review of the literature. J Neurosurg 62: 475–495.

Lizerbram EK, Hesselink JR (1997). Neuroimaging of AIDS: I. Viral infections. Neuroimaging Clin North Am 7(2): 261–280.

Maier H, Budka H, Lassmann H, et al (1989). Vacuolar myelopathy with multinucleated giant cells in the acquired immune deficiency syndrome (AIDS). Light and electron microscopic distribution of human immunodeficiency virus (HIV) antigens. Acta Neuropathol (Berl) 78(5): 497–503.

Malessa R, Krams M, Hengge U, et al (1994). Elevation of intracranial pressure in acute AIDS-related cryptococcal meningitis. Clin Invest 72(12): 1020–1026.

Manji H, Miller RF (2000). Progressive multifocal leukoencephalopathy: progress in the AIDS era. J Neurol Neurosurg Psych 69: 569–571.

Mathews VP, Alo PL, Glass JD, et al (1992). AIDS-related CNS cryptococcosis: radiologic–pathologic correlation. AJNR Am J Neuroradiol 13(5): 1477–1486.

McMahon MEM, Glass JD, Hayward SD, et al (1991). Epstein-Barr virus detection in AIDS-related primary central nervous system lymphoma. Lancet 338: 969–973.

Miller RF, Hall-Craggs MA, Costa DC, et al (1998). Magnetic resonance imaging, thallium-201 SPECT scanning, and laboratory analyses for discrimination of cerebral lymphoma and toxoplasmosis in AIDS. Sex Transm Infect 74 (4): 258–264.

Miralles P, Berenguer J, Lacruz C, et al (2001). Inflammatory reactions in progressive multifocal leukoencephalopathy after highly active antiretroviral therapy. AIDS 15(14): 1900–1902.

Moulignier A, Mikol J, Gonzalez-Canali G, et al (1996). AIDS-associated cytomegalovirus infection mimicking central nervous system tumors: a diagnostic challenge. Clin Infect Dis 22: 626–631.

Naddaf SY, Akisik MF, Aziz M, et al (1998). Comparison between 201-Tl-chloride and 99Tc(m)-sestamibi SPET brain imaging for differentiating intracranial lymphoma from non-malignant lesions in AIDS patients. Nucl Med Commun 19(1): 47–53.

Neuenburg JK, Brodt HR, Herndier BG, et al (2002). HIV-related neuropathy, 1985 to 1999: rising prevalance of HIV encephalopathy in the era of highly active antiretroviral therapy. J Acquir Immune Defic Syndr 31: 171–177.

Nuttall JJ, Wilmshurst JM, Ndondo AP, et al (2004). Progressive multifocal leukoencephalopathy after initiation of highly active antiretroviral therapy in a child with advanced human immunodeficiency virus infection: a case of immune reconstitution inflammatory syndrome. Pediatr Infect Dis J 23(7): 683–685.

Patronas NJ, Makariou EV (1993). MRI of choroidal plexus involvement in intracranial cryptococcosis. J Comput Assist Tomogr 17: 547–550.

Petito CK (1988). Review of central nervous system pathology in human immunodeficiency virus infection. Ann Neurol 23: S54–S57.

Petito CK (1997). The neuropathology of human immunodeficiency virus infection of the spinal cord. In: JR Berger, RM Levy (Eds.), AIDS and the Nervous System, 2nd edn. Lippincott-Raven, Philadelphia, PA, pp. 451–459.

Pomper MG, Constantinides CD, Barker PB, et al (2002). Quantitative MR spectroscopic imaging of brain lesion in patients with AIDS: correlation with [11C-methyl]thymidine PET and thallium-201 SPECT. Acad Radiol 9(4): 398–409.

Poon T, Matoso I, Tchertkoff V, et al (1989). CT features of primary cerebral lymphoma in AIDS and non-AIDS patients. J Comput Assist Tomogr 13(1): 6–9.

Poon TP, Tchertkoff V, Pares GF, et al (1992). Spinal cord toxoplasma lesion in AIDS: MR findings. J Comput Assist Tomogr 16(5): 817–819.

Popovich MJ, Arthur RH, Helmes E (1990). CT of intracranial cryptococcosis. AJNR Am J Neuroradiol 11: 139–142.

Port JD, Miseljic S, Lee RR, et al (1999). Progressive multifocal leukoencephalopathy demonstrating contrast enhancement on MRI and uptake of thallium-201: a case report. Neuroradiology 41(12): 895–898.

Porter SB, Sande MA (1992). Toxoplasmosis of the central nervous system in the acquired immunodeficiency syndrome. N Engl J Med 327: 1643–1648.

Post MJ, Kursunoglu SJ, Hensley GT, et al (1985a). Cranial CT in acquired immunodeficiency syndrome: spectrum of diseases and optimal contrast enhancement technique. AJR Am J Roentgenol 145(5): 929–940.

Post MJD, Kursunoglu SJ, Hensley GT, et al (1985b). Cranial CT in acquired immunodeficiency syndrome: spectrum of diseases and optimal contrast enhancement technique. AJNR Am J Neuroradiol 6: 743–754.

Post MJD, Hensley CT, Moskowitz LB, et al (1986). Cytomegalic inclusion virus encephalitis in patients with AIDS: CT, clinical and pathologic correlation. AJR Am J Roentgenol 146(6): 1229–1234.

Post MJ, Levin BE, Berger JR, et al (1992). Sequential cranial MR findings of asymptomatic and neurologically symptomatic HIV+ subjects. AJNR Am J Neuroradiol 13(1): 359–370.

Post MJ, Yiannoutsos C, Simpson D, et al (1999). Progressive multifocal leukoencephalopathy in AIDS: are there any MR findings useful to patient management and predictive of patient survival? AJNR Am J Neuroradiol 20: 1896–1906.

Provenzale JM, Jinkins JR (1997). Brain and spine imaging findings in AIDS patients. Radiol Clin North Am 35(5): 1127–1166.

Quality Standards Subcommittee of the American Academy of Neurology (1998). Evaluation and management of intracranial mass lesions in AIDS. Report of the Quality Standards Subcommittee of the American Academy of Neurology. Neurology 50: 21–26.

Quencer RM, Post MJ (1997). Spinal cord lesions in patients with AIDS. Neuroimaging Clin North Am 7(2): 359–373.

Ramsey RG, Gean AD (1997). Neuroimaging of AIDS: I. Central nervous system toxoplasmosis. Neuroimaging Clin North Am 7(2): 171–186.

Ramsey R, Geremia GK (1998). CNS complications of AIDS: CT and MR findings. Am J Roentgenol 151: 449.

Ruiz A, Ganz WI, Post MJ, et al (1994). Use of thallium-201 brain SPECT to differentiate cerebral lymphoma from toxoplasma encephalitis in AIDS patients. AJNR Am J Neuroradiol 15: 1885–1894.

Ruiz A, Post MJ, Bundschu C, et al (1997). Primary central nervous system lymphoma in patients with AIDS. Neuroimaging Clin North Am 7(2): 281–296.

Sacktor N (2002). The epidemiology of human immunodeficiency virus-associated neurological disease in the era of highly active antiretroviral therapy. J Neurovirol 8 (Supp2): 115–121.

Sacktor N, Skolasky RL, Ernst T, et al (2005). A multicenter study of two magnetic resonance spectroscopy techniques in individuals with HIV dementia. J Magn Reson Imaging 21(4): 325–333.

Safai B, Diaz B, Schwartz J (1992). Malignant neoplasms associated with human immunodeficiency virus infection. Cancer J Clin 42: 74–95.

Salazar A, Padzamczer D, Rene R, et al (1995). Cytomegalovirus ventriculoencephalitis in AIDS patients. Scand J Infect Dis 27(2): 165–169.

Santosh CG, Bell JE, Best JJ (1995). Spinal tract pathology in AIDS: postmortem MRI correlation with neuropathology. Neuroradiology 37: 134–138.

Schinina V, Rizzi EB, Rovighi L, et al (2001). Infectious spondylodiscitis: magnetic resonance imaging in HIV-infected and HIV-uninfected patients. Clin Imaging 25(5): 362–367.

Sell M, Sander B, Klingebiel B (2005). Ventriculitis and hydrocephalus as the primary manifestations of cerebral toxoplasmosis associated with AIDS. J Neurol 252: 234–236.

Shanley DJ (1995). Tuberculosis of the spine: imaging features. AJR Am J Roentgenol 164(3): 659–664.

Shelburne SA 3rd, Hamill RJ (2003). The immune resconstitution inflammatory syndrome. AIDS Rev 5(2): 67–79.

Shelburne SA 3rd, Hamill RJ, Rodriguez-Barradas MC (2002). Immune reconstitution inflammatory syndrome: emergence of a unique syndrome during highly active antiretroviral therapy. Medicine (Baltimore) 81(3): 213–227.

Shelburne SA 3rd, Darcourt J, White AC Jr, et al (2005a). The role of immune reconstitution inflammatory syndrome in AIDS-related Cryptococcus neoformans disease in the era of highly active antiretroviral therapy. Clin Infect Dis 40(7): 1049–1052.

Shelburne SA 3rd, Visnegarwala F, Darcourt J, et al (2005b). Incidence and risk factors for immune reconstitution inflammatory syndrome during highly active antiretroviral therapy. AIDS 19(4): 399–406.

Sheller JR, Des Prez RM (1986). CNS tuberculosis. Neurol Clin 4: 143–158.

Simone IL, Federico F, Tortorella C, et al (1998). Localized 1H-MR spectroscopy for metabolic characterization of diffuse and focal brain lesions in patients infected with HIV. J Neurol Neurosurg Psych 64(4): 516–523.

Skiest DJ, Erdman W, Chang WE, et al (2000). SPECT thallium-201 combined with toxoplasma serology for the presumptive diagnosis of focal central nervous system mass lesions in patients with AIDS. J Infect 41: 274–281.

Snider WD, Simpson DM, Nielson S, et al (1983). Neurological complications of acquired immunodeficiency syndrome: analysis of 50 patients. Ann Neurol 14: 403–418.

Sperber K, Shao L (2003). Neurologic consequences of HIV infection in the era of HAART. AIDS Patient Care STDs 17(10): 509–518.

Supiot F, Guillaume MP, Hermanus N, et al (1997). Case report: toxoplasma encephalitis in a HIV patient: unusual involvement of the corpus callosum. Clin Neurol Neurosurg 99: 287–290.

Sze G, Zimmerman RD (1988). The magnetic resonance imaging of infections and inflammatory diseases. Radiol Clin North Am 26: 839–859.

Talpos D, Tien RD, Hesselink JR (1991). Magnetic resonance imaging of AIDS-related polyradiculopathy. Neurology 41: 1996–1997.

Thurnher MM, Jinkins JR, Post MJ (1997a). Diagnostic imaging of infections and neoplasms affecting the spine in patients with AIDS. Neuroimaging Clin North Am 7(2): 341–357.

Thurnher MM, Thurnher SA, Schindler E (1997b). Pictorial review: CNS involvement in AIDS: spectrum of CT and MR findings. Neuroradiology 7: 1091–1097.

Thurnher MM, Post MJ, Jinkins JR (2000). MRI of infections and neoplasms of the spine and spinal cord in 55 patients with AIDS. Neuroradiology 42(8): 551–563.

Thurnher MM, Post MJ, Rieger A, et al (2001). Initial and follow-up MR imaging findings in AIDS-related progressive multifocal leukoencephalopathy treated with highly active antiretroviral therapy. AJNR Am J Neuroradiol 22: 977–984.

Tien RD, Chu PK, Hesselink JR, et al (1991). Intracranial cryptococcosis in immunocompromised patients: CT and MR findings in 29 cases. AJNR Am J Neuroradiol 12(2): 283–289.

Trenkwalder P, Trenkwalder C, Feiden W, et al (1992). Toxoplasmosis with early intracerebral hemorrhage in a patient with the acquired immunodeficiency syndrome. Neurology 42(2): 436–438.

Van Dyk A (1988). CT of intracranial tuberculosis with specific reference to the "target sign". Neuroradiology 30: 329–336.

Vendrely A, Bienvenu B, Gasnault J, et al (2005). Case report. Fulminant inflammatory leukoencephalopathy associated with HAART-induced immune restoration in AIDS-related progressive multifocal leukoencephalopathy. Acta Neuropathol 109: 449–455.

Villoria MF, de la Torre J, Fortea F, et al (1992). Intracranial tuberculosis in AIDS: CT and MRI findings. Neuroradiology 34: 11–14.

Villringer K, Jager H, Dichgans M, et al (1995). Differential diagnosis of CNS lesions in AIDS patients by FDG-PET. J Comput Assist Tomogr 19(4): 532–536.

Von Giesen HJ, Neuen-Jacob E, Dorries K, et al (1997). Diagnostic criteria and clinical procedures in HIV-1 associated progressive multifocal leukoencephalopathy. J Neurol Sci 147(1): 63–72.

When SM, Heinz ER, Burger PC, et al (1989). Dilated Virchow-Robin spaces in cryptococcal meningitis associated with AIDS: CT and MRI findings. J Comput Assist Tomogr 13(5): 756–762.

Whiteman ML (1997). Neuroimaging of central nervous system tuberculosis in HIV-infected patients. Neuroimaging Clin North Am 7(2): 199–214.

Whiteman ML, Dandapani BK, Shebert RT, et al (1994). MRI of AIDS-related polyradiculomyelitis. J Comput Assist Tomogr 18(1): 7–11.

Whiteman M, Espinoza L, Post MJ, et al (1995). Central nervous system tuberculosis in HIV-infected patients: clinical and radiologic findings. AJNR Am J Neuroradiol 16: 1319–1327.

Woo HH, Rezai AR, Knopp EA, et al (1996). Contrast-enhancing progressive multifocal leukoencephalopathy: radiological and pathological correlations: case report. Neurosurgery 39(5): 1031–1034.

Woodworth GF, McGirt MJ, Williams MA, et al (2005). The use of ventriculoperitoneal shunts for uncontrollable intracranial hypertensions without ventriculomegally secondary to HIV-associated cryptococcal meningitis. Surg Neurol 63(6): 529–531.

Yang PJ, Reger KM, Seeger JF, et al (1987). Brain abscess: an atypical CT appearance of CNS tuberculosis. AJNR Am J Neuroradiol 8: 919–920.

Zuger A, Louie E, Hozman RS, et al (1988). Cryptococcal disease in patients with the acquired immunodeficiency syndrome: diagnostic features and outcome of treatment. Ann Intern Med 104: 234–240.

Handbook of Clinical Neurology, Vol. 85 (3rd series)
HIV/AIDS and the Nervous System
P. Portegies, J. R. Berger, Editors

Chapter 17

Cerebrospinal fluid markers in central nervous system HIV infection and AIDS dementia complex

PAOLA CINQUE,[1]* BRUCE J. BREW,[2] MAGNUS GISSLEN,[3] LARS HAGBERG[3] AND RICHARD W. PRICE[4]

[1]*Clinic of Infectious Diseases, San Raffaele Scientific Institute, Milan, Italy*

[2]*Departments of Neurology and HIV Medicine, St Vincent's Hospital and University of New South Wales, Sydney, Australia*

[3]*Department of Infectious Diseases, Göteborg University, Sahlgrenska University Hospital, Göteborg, Sweden*

[4]*Department of Neurology, University of California San Francisco, San Francisco General Hospital, San Francisco, CA, USA*

17.1. Introduction

The analysis of cerebrospinal fluid (CSF) remains among the most important procedures in the diagnosis of central nervous system (CNS) infections (Fishman, 1992). CSF analysis can directly identify the underlying etiological agent and it can also characterize specific local immune responses that establish etiology and predict prognosis (Cinque and Linde, 2003). While of less immediate practical consequence, CSF assessment of biological and biochemical changes can provide important information contributing to the understanding of the underlying pathogenic process.

In the context of HIV-1 (in the following, referred to as simply HIV) infection, CSF analysis has proved useful in the diagnosis of opportunistic infections and tumors. Both traditional microbiological methods (e.g. CSF culture or antigen detection for diagnosis of CNS cryptococcosis), and more recent nucleic acid amplification techniques (primarily the polymerase chain reaction, PCR), enable a rapid and certain etiological diagnosis of most of these complications. Notable examples are the detection of cytomegalovirus (CMV) DNA in patients with CMV encephalitis, of JC virus DNA in patients with progressive multifocal leukoencephalopathy (PML) and Epstein–Barr virus (EBV) DNA in primary CNS lymphoma. These methods have been validated in large case-control and prospective studies against standard diagnostic procedures (Cinque et al., 1997).

The study of CSF has been less useful in the diagnosis of the AIDS dementia complex (ADC). This is because the relationship between HIV infection and brain dysfunction is complex, requiring more than simply the presence of the virus within the CNS. HIV reaches the CNS, or at least the meninges and perivascular spaces, at the time of primary infection (Pilcher et al., 2001). Subsequently, the virus is almost always detected in the CSF during the neurologically asymptomatic stages of infection, both in early and late disease. In the absence of neurological symptoms, virus detection is likely related to either ongoing local replication and/or continuous seeding by trafficking blood cells entering the CSF, enhanced by systemic and local immune activation (Price, 2000). Thus, the presence of the etiological agent of ADC, HIV, in the CSF does not serve as a specific marker of the disease. Indeed, ADC remains a syndromic, clinical diagnosis based upon the setting of HIV infection, a constellation of symptoms and signs indicating a progressive "subcortical" type dementia with characteristic cognitive, motor and behavioral disturbances, and, importantly, exclusion of alternative etiologies (American Academy of Neurology, 1991).

One of the greatest needs in HIV neurology is a more objective approach to diagnosis of ADC and underlying HIV encephalitis. This is similarly important as a basis of targeted therapy. One needs a marker that is sensitive in the early detection of ADC and specific despite this background of asymptomatic CNS HIV infection or of other CNS disorders that can occur in this setting. Unfortunately, despite a number of studies of viral, inflammatory and neural CSF markers, no single analyte has emerged as useful for the diagnosis of ADC. This is in

*Correspondence to: Paola Cinque, MD, PhD, Division of Infectious Diseases, San Raffaele Scientific Institute, Via Stamira d'Ancona 20, 20127 Milano, Italy. E-mail: cinque.paola@hsr.it, Tel: +39-022-643-7985, Fax: +39-022-643-7989.

part because no single marker fulfills the objectives of diagnostic sensitivity and specificity, but also in part because clinical application of known markers and efforts to assess additional markers have often focused more on their pathogenetic implications than direct clinical utility. While understanding pathogenesis is clearly a useful step in approaching discovery of markers, it is also important to thoroughly test clinical utility of single markers and combinations of markers if this approach to diagnosis and management is going to be integrated into patient care.

When considering CSF markers of ADC, it is also important to emphasize the impact of highly active antiretroviral therapies (HAART) on the frequency, clinical presentation and course of this condition. These treatments have dramatically reduced the incidence of ADC and modified its natural history, including perhaps its clinical and biological features. Because HAART is effective in improving neurological and CSF abnormalities in patients with ADC, it has been suggested that this disease might now manifest with altered phenotypes, for instance with less clear "subcortical" features and a more indolent and chronic course. On the other hand, inflammatory and hyperacute presentations may be found in association with immune reconstitution following HAART initiation or interruption (Langford et al., 2002; Gray et al., 2003). In addition, a number of confounding neurological or psychiatric conditions might develop in those who are long-term survivors of HIV infection due to HAART, making diagnosis of ADC even more difficult. These may include CNS diseases typical of the elderly, such as Alzheimer's disease, cerebrovascular diseases and other forms of dementia. Antiretroviral drugs themselves might be an adjunctive risk factor for development of CNS disease, either directly or through their effects on metabolic functions. Hence, the longevity and susceptibility to other conditions greatly increases the need for objective diagnositic approaches to ADC.

This review will examine published studies of individual CSF markers, with attention to both their implications for ADC pathogenesis and their present or potential future use for clinical management, including diagnosis, assessment of disease activity, prognosis and response to anti-HIV treatments. We have separated CSF markers into three groups related to the three major components of the process ultimately leading to ADC, i.e. viral markers, host response markers and indicators of CNS tissue damage (see Table 17.1). The host response markers have further been divided according to their main role in the pathogenesis of ADC, i.e. immune-activation markers, neurotoxic molecules and markers of apoptosis. This classification is clearly imperfect because some of the markers could be placed in more than one of these categories, e.g. tumor necrosis factor-α (TNF-α) and quinolinic acid are markers of inflammation but may also exert toxic effects on neuronal cells. Nonetheless, this classification has the advantage of simplicity for the purposes of this review.

17.2. Markers of HIV-1 infection

Unlike most other viruses responsible for brain infections, both infectious HIV and its components (proteins, nucleic acids) can be easily identified and quantified in the CSF. In addition, HIV sequences retrieved in the CSF can be analyzed, and the intrathecal antibody response can be quantified. As discussed before, however, none of these approaches is sufficiently specific and sensitive for diagnosis of ADC. Nonetheless, their characterization has been important in understanding CNS HIV infection, and some of the details may hint at future application to clinical practice. In this section we review some of the main findings using various methods to identify and measure components of HIV, characterize the local virus by sequence analysis and define local specific immune responses to CSF.

17.2.1. Detection and quantification of HIV-1

A number of techniques are available for HIV detection in CSF, including virus culture, antigen detection and nucleic acid measurement. HIV quantitation has historically been accomplished by the use of p24 antigen detection and, more recently, by nucleic acid amplification techniques that measure HIV RNA concentration (the *viral load*). Measuring the viral load in tissues or body fluids is important because it provides direct information on HIV replication, in both untreated and treated patients.

17.2.1.1. Viral culture

HIV can be cultured from the CSF of many infected individuals, regardless of the presence of neurological symptoms (Hollander and Levy, 1987). The virus is usually recovered in CSF during acute infection, whereas the rate of recovery in asymptomatic seropositive patients is more variable across studies: between 0 and 50% (Resnick et al., 1988). At later stages of systemic infection, and whether or not neurological symptoms and/or dementia is present, HIV is again recovered in about two-thirds of patients (Hollander and Levy, 1987). CSF HIV culture had a sensitivity of 30% and a specificity of 80% for diagnosing ADC and only 25% of patients with stages 3 or 4 ADC had a positive culture (Brew et al., 1994). In general, infectious virus is less frequently isolated

Table 17.1

Cerebrospinal fluid markers studied in HIV-infected patients

Viral markers	
Viral culture	
p24 antigen	
HIV-1 RNA	
HIV-1 DNA	
HIV-1 genetic markers	*Pol*
	Env
Detection of anti-HIV Ab	
Host response markers	
Markers of inflammation	
White blood cells	
Adhesion molecules	Soluble intercellular adhesion molecule-1 (ICAM-1)
	Soluble vascular cell adhesion molecule-1 (VCAM-1)
Cytokines	Tumor necrosis factor-α (TNF-α)
	Interleukin-1 (IL-1)
	Interleukin-2 (IL-2)
	IL-2 receptor (IL-2R)
	Interleukin-6 (IL-6)
	Interleukin-10 (IL-10)
	Interferon-α (IFN-α)
	Interferon-γ (IFN-γ)
	Transforming growth factor-β (TGF-β)
Chemokines	Monocyte chemotactic protein-1 (MCP-1) / CCL2
	Interferon-γ inducible protein 10 (IP-10) / CXCL10
	Macrophage inflammatory protein-α (MIP1-α) / CCL3
	Macrophage inflammatory protein-β (MIP1-β) / CCL4
	RANTES / CCL5
	Fractalkine / CX3CL1
Neopterin	
Beta-2 microglobulin (β2M)	
Matrix metalloproteases (MMPs)	MMP-2
	MMP-7
	MMP-9
Urokinase plasminogen activator system	Urokinase plasminogen activator (uPA)
	Soluble uPA receptor (suPAR)
	Plasminogen activator inhibitor-1 (PAI-1)
Neurotoxic host factors	
Quinolinic acid and tryptophan	
Prostaglandins	
Platelet activating factor (PAF)	
Glutamate	
Nitric oxide (NO)	
Markers of apoptosis	
Fas and Fas ligand (FasL)	
Sphingomyelin and ceramide	
Markers of CNS cell damage	
Microglia markers	GD3
Astrocyte markers	S-100β
	Glial fibrillary acidic protein (gFAP)
Neuronal markers	Neurofilament protein light chain (NFL)
	Tau protein (tau)
	14-3-3 protein

from CSF than from blood (Andersson et al., 2000) and from cell-depleted than cell-associated CSF fractions. As in blood, infectious HIV can also be quantified in the CSF, but, given the low sensitivity of this method, this is of limited diagnostic utility, particularly with the advent of quantitative nucleic acid hybridization methods (Peeters et al., 1995).

High variability of virus isolation rate between studies likely depends on several factors, including technical variables (e.g. CSF volume, cell content and time to culture) and the viral burden in the CSF. It has been shown that culture-positive CSF samples yield higher HIV RNA levels than culture-negative samples. A positive predictive value for positive virus isolation from CSF samples equal to or above 90% was achieved with a cut off level of >5000 HIV RNA copies/ml (Andersson et al., 2000). Though no formal studies have been performed in patients before and after antiretroviral treatment, the rate of recovery has been noted to be substantially lower in treated than untreated patients, with the recovery of virus from CSF affected more than in plasma (Andersson et al., 2000). Because of low sensitivity, reproducibility and relative technical complexity, virus culture from CSF has not found an application for patient management. Indeed, this test is now infrequently used even in the research setting (e.g. for characterizing CSF isolates in functional studies) and has been largely replaced by direct nucleic acid amplification of genetic regions of interest followed by insertion into recombinant viruses.

17.2.1.2. p24 antigen

The p24 antigen (p24-Ag) is the major core component of HIV. Immunoassays based on the detection of this protein in serum were developed at the end of the 1980s and were widely used to assess virus replication in serum/plasma and other body fluids, including CSF. However, this method has now been replaced by the more sensitive nucleic acid amplification-based techniques, though it is still used to assess viral replication in cell culture studies.

p24-Ag is detectable in the serum of most patients at the advanced stage of disease, but rarely in symptomless seropositive patients. It appears early in primary HIV infection (PHI), before antibody production, and usually disappears afterwards. Similarly p24-Ag can be present transiently in CSF before the development of CSF antibody (Goudsmit et al., 1986). CSF p24-Ag levels are higher in subjects with HIV-associated neurologic disease than in neurologically asymptomatic persons, while reports have suggested no differences in these levels in relation to systemic disease stage (Singer et al., 1994). Higher levels, however, have been noted in association with low CD^4-positive blood T lymphocyte counts and high CSF β2-microglobulin (β2M) concentration (Brew et al., 1994). The sensitivity of p24-Ag detection for diagnosis of ADC was only between 21 and 47%. In contrast, specificity of the assay was higher, of between 95 and 98% (Brew et al., 1994; Royal et al., 1994). When histopathologic diagnoses were considered, and control patients those with CNS opportunistic infections rather than asymptomatic patients, this test was shown to yield a sensitivity of 53% for diagnosis of HIV encephalitis (HIV-E) and a specificity of 100% using a cut off value of 70 pg/mL (Pierotti et al., 2005).

In the pre-HAART era p24-Ag was detected less frequently in CSF from demented subjects on antiretroviral drugs than untreated demented individuals (Royal et al., 1994). Early studies documented the clearance of p24-Ag from CSF following two months of zidovudine treatment (de Gans et al., 1988). Since the advent of HAART temporarily coincided with the availability of molecular assay for quantitation of HIV RNA, the effect of HAART on p24-Ag in CSF has not been evaluated in longitudinal studies.

Although it was shown to be highly specific for the diagnosis of ADC and HIV encephalitis, the use of this marker has now lost favor as a potential diagnostic marker. The main limitations are represented by low sensitivity, narrow dynamic range and the confounding effect of local antibody binding. During recent years, the sensitivity of p24-Ag detection has been improved by measures that include heat or acid denaturation of interfering antibodies, signal amplification and use of virus disruption buffers (Schupbach, 2002), but none of these approaches has been formally evaluated in CSF analysis. Nonetheless, this test may still be useful as a low cost marker of HIV replication (Respess et al., 2005; Stevens et al., 2005). Since its diagnostic specificity for HIV-E is similar to that of HIV RNA assays (see below), p24-Ag should probably be evaluated in settings where amplification-based assays are too costly.

17.2.1.3. Nucleic acid amplification

17.2.1.3.1. HIV-1 RNA (viral load)

HIV RNA genomes can be measured in virtually all body fluids, including plasma and CSF. A number of methods are in use, including reverse transcriptase (RT)-PCR, nucleic acid sequence-based amplification assay (NASBA), ligase chain reaction and those based on signal amplification, e.g. branched DNA. Several in-house protocols were also developed in earlier times, although commercial assays are currently used in most cases. While more expensive, the latter have the advantage of standardization over time and between laboratories. The most recent, commercially available assays have improved their sensitivity, enabling quantitation of levels as low as 2.5 genomes per milliliter (Havlir et al., 2001). Furthermore, the spectrum of viral subtypes that can be detected has been expanded. Today, most assays can reliably be used to quantify both subtype B virus — largely prevalent in the Americas, Europe and Australia — and non-B viruses — widely represented in Africa and Asia, and becoming increasingly frequent also in the western world.

The assessment of HIV RNA load in plasma has led to revolutionary advances in our understanding of the viral dynamics within the human host and provided a unique means for clinical monitoring of HIV infection (Mellors et al., 1996; Perelson et al., 1996). In untreated patients, plasma HIV RNA levels are an important prognostic marker, as they can predict subsequent disease evolution (Mellors et al., 1996). In patients receiving anti-HIV drugs, plasma viral load measurement is used to assess whether treatment is virologically effective.

In contrast to viral load measurement in plasma, the diagnostic and prognostic significance of HIV RNA levels in CSF is less clear. In general, viral RNA is detected in the CSF from the vast majority of untreated patients, with levels varying widely from just above the limit of detection to over one million genomes/ml. In patients with primary HIV infection (PHI), CSF RNA levels may be high in parallel with the peak of plasma viremia (Tambussi et al., 2000; Pilcher et al., 2001). Higher levels are observed in PHI patients with neurological symptoms, most commonly those with aseptic meningitis (Tambussi et al., 2000). As observed in plasma, CSF HIV RNA levels usually decline spontaneously after the early weeks of infection (Boufassa et al., 1995). During the subsequent course of HIV infection, CSF HIV RNA levels tend to remain steady in clinically stable patients over several years (Ellis et al., 2003). However, levels increase in patients showing clinical disease progression and loss of CD^4 cells (McArthur et al., 1997; Gisslen et al., 1998; Ellis et al., 2003). Irrespective of the CD^4 counts, CSF HIV RNA levels are higher in ADC than in neurologically asymptomatic patients and in patients with severe dementia compared to those with moderate or mild neuropsychological impairment, though there is wide individual variation (Brew et al., 1997; Ellis et al., 1997; McArthur et al., 1997). HIV RNA levels are also highly correlated with the presence and severity of HIV-E at neuropathological examination (McArthur et al., 1997; Cinque et al., 1998a). Using a cut-off value of 32,000 copies/ml, CSF HIV RNA was 93% specific and 59% sensitive for predicting HIV-E in a population of neurologically impaired patients with postmortem examination (Cinque et al., 1998a).

Among patients with neurological symptoms, high CSF HIV RNA levels can also be found in patients with opportunistic CNS infections characterized by increased inflammatory meningeal infiltrates, e.g. cryptococcosis or tuberculous meningitis (Brew et al., 1997; Morris et al., 1998). There is a correlation in these diseases between the number of CSF lymphocytes and HIV burden, supporting the idea that, in these cases, HIV RNA is produced by infected CSF cells. However, the reasons behind elevated CSF viral load in opportunistic CNS diseases might be more complex, since in vitro studies have shown that some pathogens may enhance HIV replication (Pettoello-Mantovani et al., 1992; Lederman et al., 1994).

The association between high CSF levels of HIV RNA and HIV-associated neurological disease supports the view that in ADC and HIV-E CSF virus is mainly derived from parenchymal brain infection. In favor of this hypothesis is also the frequent dissociation, in these cases, between CSF and plasma viral loads (Brew et al., 1997; Ellis et al., 1997; McArthur et al., 1997; Cinque et al., 1998a). It can thus be hypothesized that, in ADC, CNS macrophages and microglial cells produce viral particles that are released into the extracellular space and flow into the CSF. In patients without evidence of HIV-induced neurological complications, extraparenchymal sources of HIV, primarily blood cells, have been hypothesized. This idea has recently been supported by studies in simian immune deficiency virus (SIV)-infected macaques, which showed HIV RNA in CSF at a time when it could not be retrieved in the brain (Clements et al., 2002). Indeed, aseptic meningitis caused by HIV is frequently observed in patients with acute infection, and it occurs asymptomatically in HIV-infected patients throughout the course of infection. Lymphocytic meningeal infiltration might be present in these cases, and CSF pleocytosis is a frequent finding in association with high CSF levels of immune activation markers (Hagberg et al., 1988; Brew et al., 1997; Martin et al., 1998; Morris et al., 1998; Spudich et al., 2005b). On the other hand, CSF virus might originate from blood cells trafficking through the blood–CSF barrier. Activated lymphocytes are facilitated to cross the brain barriers either as HIV-infected or uninfected cells. Based on the decay kinetics of viral variants in CSF and plasma upon HAART, it has recently been suggested that blood lymphocytes might become infected once they reach the CSF, through the virus released by CNS resident or other trafficking infected cells (see also below — virus compartmentalization) (Harrington et al., 2005).

A number of studies have measured the changes of CSF HIV RNA levels in response to antiretroviral treatment. Earlier studies of neurologically asymptomatic patients receiving monotherapy regimens had already shown reductions of CSF viral load following a few months of treatment with zidovudine or lamivudine (Lewis et al., 1996; Gisslen et al., 1997b). Similarly, the association of two nucleoside reverse transcriptase (RT) inhibitors (NRTIs) — lamivudine plus zidovudine or stavudine — in neurologically asymptomatic patients led to significant decline of HIV levels in the CSF, which became undetectable in all treated patients after three months of therapy (Foudraine et al., 1998). The efficacy of HAART in reducing CSF viral loads is now extensively documented in patients at different disease stage. These included patients with PHI (Enting et al., 2001), neurologically asymptomatic patients, patients with mild or severe HIV-related neurological symptoms, as well as patients with opportunistic CNS complications (Stellbrink et al., 1997; Staprans et al., 1999; Ellis et al., 2000; Garcia et al., 2000; Gisslen et al., 2000; Polis et al., 2003; Spudich et al., 2005b). In patients with HIV-induced neurological or neuropsychological impairment, the decline of HIV RNA level was observed in parallel with an improvement of neurological and neuropsychological test performances (Eggers et al., 2003; Marra et al., 2003; Robertson et al., 2004).

Although the virus is eventually cleared from the CSF in most of the patients receiving HAART, different dynamics of response between CSF and plasma can be observed. Following 4–12 weeks of HAART, HIV-RNA load decreases in parallel in plasma and CSF in some patients, whereas slower declines in the CSF than in plasma are observed in others (Staprans et al., 1999; Ellis et al., 2000; Cinque et al., 2001; Eggers et al., 2003; Spudich et al., 2005b). This dual response pattern is observed with virtually any HAART regimen, including lopinavir/ritonavir monotherapy and four-drug combinations of all drug classes (Polis et al., 2003; van den Brande et al., 2005). These different modalities of response to HAART are likely to reflect different origin of the virus and a variable degree of compartmentalization of the infection between CSF and plasma. The slower response in CSF may mirror the response to treatment in cells with slow virus turnover, such as brain macrophages, and it has been observed in patients with HIV encephalopathy and/or low CD^4 cell counts. Thus, in these cases, CSF infection would be principally sustained by productive brain infection. In contrast, the parallel pattern of response is observed mainly in patients with high CD^4 cell count and CSF pleocytosis and appears consistent with the presence in CSF or brain of "infiltrating" lymphocytes (Staprans et al., 1999). Indeed, ultraintensive CSF sampling of asymptomatic HIV-infected subjects during the first hours to days of HAART showed early rapid viral decay in CSF and plasma, consistent with the hypothesis that CD^4 T lymphocytes are the primary source of CSF virus in this setting (Haas et al., 2003).

Despite the differences between classes of drugs and individual drugs with respect to penetration into the CSF and metabolism in brain macrophages, virtually any HAART regimen that is effective in reducing the plasma viral load has a similar effect on the CSF viral load (Mellgren et al., 2005). It is uncertain to what extent this effect depends on a reduction of viral load in blood, which, in turn, continuously replenishes CSF virus, or on direct action against intrathecal viral replication. In this regard, it has been shown that the combination of saquinavir and ritonavir, two protease inhibitors which achieve very low CSF concentrations, is less effective on CSF viral load than the combination of the same two protease inhibitors plus stavudine (a NRTI known to achieve CSF concentrations above those required to inhibit HIV replication) despite similar responses in plasma (Gisolf et al., 2000a). This observation suggests that a degree of local replication contributes to viral load in CSF and that at least one "neuroactive" drug is necessary to suppress viral load in this compartment. Several cross-sectional studies have also evaluated the contribution of individual compounds contained in HAART regimens to viral load suppression in the CSF. While some have shown that the number of "CNS penetrating" drugs is associated with better virological response in CSF (Antinori et al., 2002; De Luca et al., 2002; Letendre et al., 2004), others have not confirmed these findings (Evers et al., 2004). These discrepant results likely reflect different study designs, and in particular the difficulties of accurately defining the penetrating potential of individual drugs.

Although the CSF viral load decreases in most patients treated with HAART, concern has been raised that virus may rebound in the long term with the onset of drug-resistant virus as is noted in the plasma. Because of the limited penetration of certain drugs into the CSF, this might theoretically occur more easily and earlier in CSF compared to blood. However, there is evidence that HIV RNA remains suppressed in CSF despite virological failure in plasma and presence of resistant virus. HIV may actually rebound in the CSF concomitant with plasma failure, but only rarely is a "CSF escape" observed during long-term treatments (Staprans et al., 1999; Gunthard et al., 2001; Bestetti et al., 2004; Spudich et al., 2005b).

Because of the dramatic and sustained effect of HAART on CSF viral load, the correlation between this marker and HIV-induced brain diseases needs to be reexamined in treated patients. High CSF replication levels have recently been shown not to predict development of ADC in this population (Sevigny et al., 2004). Additionally, a relationship between CSF HIV RNA and neurological status was no longer observed in persons treated with HAART (McArthur et al., 2004), and low or undetectable virus levels were frequently observed in neurologically impaired but treated patients (Cysique et al., 2005). It is not clear whether neurological impairment in these cases is caused by other, non-HIV-related conditions, by an active brain process that occurs or progresses despite undetectable virus in the CSF or is simply a consequence of previously active HIV disease. The use of ultrasensitive molecular techniques that may detect virus at very low copy numbers may prove useful to assess these hypotheses. Actually, CSF viral loads $\geq$200 copies/ml in treated patients have been shown to predict progression to neuropsychological impairment (Ellis et al., 2002). In addition, neurologically asymptomatic patients successfully treated with HAART may show mildly elevated levels of CSF immune activation markers despite complete viral suppression in the same compartment (Gisolf et al., 2000b; Abdulle et al., 2002). Thus, the use of ultrasensitive HIV RNA measurements in combination with other CSF markers might prove useful to understand the biology of HIV brain infection in patients treated with HAART and possibly also for management purposes.

17.2.1.3.2. HIV-1 DNA

Besides the HIV RNA load, HIV cell-associated, or proviral, DNA can also be measured by PCR. The test is usually performed on peripheral blood mononuclear cells (PBMC), but a number of studies have examined HIV DNA quantitation in CSF white blood cells (WBC) or in tissues. Since viral DNA may remain latent in the host cells despite successful inhibition of viral replication, detection and quantitation of proviral DNA may be useful to monitor more in depth the efficacy of antiretroviral treatments.

Before commercial assays for HIV RNA assessment became available, a few studies started investigating the presence and number of HIV-infected cells in CSF by measuring proviral DNA. Overall, the reported frequency of HIV DNA detection in CSF in untreated patients varied largely between studies, up to values of 90% (Shaunak et al., 1990; Buffet et al., 1991; Shaunak et al., 1991; Sonnerborg et al., 1991; Steuler et al., 1992a; Schmid et al., 1994). The DNA copy number recovered in CSF was also variable, usually higher than in blood, and with values up to 1 copy per 20 CSF cells (Steuler et al., 1992a; Schmid et al., 1994). HIV provirus was initially shown to be recovered more frequently in CSF from patients with neurological symptoms than in neurologically asymptomatic patients (Shaunak et al., 1990), and CSF HIV DNA levels were found to be higher in AIDS than in asymptomatic patients. However, these findings were not confirmed by subsequent studies, which reported similar HIV DNA levels irrespective of neurological status and disease stage (Sonnerborg et al., 1991; Steuler et al., 1992a). These latter observations are consistent with recent findings in SIV-infected macaques, showing the persistence of steady-state levels of viral DNA in the brain tissue throughout the infection (Clements et al., 2002). Based on this evidence, it seems unlikely that measuring HIV DNA levels in CSF could prove useful for diagnosis of HIV infection of the CNS.

Nevertheless, virological characterization of CSF proviral DNA may provide additional information related to the analysis of RNA sequences (see below). For instance, longitudinal studies of treated patients show that RNA *pol* sequences from plasma change earlier than DNA sequences in PBMC, and archived sequences in PBMC may reexpand following interruptions or resumption of treatment (Imamichi et al., 2001; Siliciano and Siliciano, 2004). Similarly, studying proviral DNA in CSF might provide a measure of the evolution of viral variants in this compartment.

Finally, proviral DNA is considered a useful marker for exploring cellular reservoirs in treated patients, particularly when HIV RNA is undetectable. Indeed, quantification of HIV DNA in PBMC is used to assess residual viral burden in HAART-treated patients with suppressed HIV replication in plasma. A decline in DNA reflects the long-term efficacy of HAART on the reservoirs (Ibanez et al., 1999). The same principle may theoretically apply also to the study of CSF.

17.2.2. HIV-1 genetic studies

In common with other RNA viruses, HIV exhibits considerable sequence diversity among variants that are retrieved in the same patient at different times or in different body tissues. Several factors allow for the continuous generation of genetic HIV variants and influence the speed with which these viruses evolve. These include the high rate of HIV replication, the poor fidelity of the RNA polymerase and the rapid selection for viruses of distinct fitness due to immune pressure, coreceptor selection and antiviral therapy (Coffin, 1995). The analysis of genetic variation in CSF compared to plasma or other body sites provides information related to the origin of CSF virus, i.e. intrathecal versus systemic, as well as to quasispecies selection and evolution within the CNS.

Viral variants can be studied by different techniques based on analysis of HIV genome amplification products (Arens, 1999). Among these, population or clonal analysis with genome sequencing, the hybridization-based assays and the heteroduplex tracking assay (HTA) have been informative when applied to paired CSF and plasma samples. Viral genome sequencing is now relatively easy to perform using automated procedures. Sequence analysis can be applied to amplification products that have been directly generated from CSF or plasma ("direct sequencing" or "population sequencing"). In this case the sequences generated will represent the predominant viral variants present in the clinical specimen. Direct sequencing finds several practical applications, primarily in detection of HIV-resistance mutations. Alternatively, genome amplification may be followed by molecular cloning, directed at separating all or most of the principal virus quasispecies that are present in the specimen. By molecular cloning, compartmentalization will be evident if the distances between CSF and plasma derived sequences are greater than those between sequences derived within each compartment. In addition, phylogenetic analyses may add information on CSF and plasma sequence evolution over time. While this method is more precise and informative, it is cumbersome and may fail to characterize the whole spectrum of variants present in a sample. Hybridization-based methods include modern high-stringency hybridization or point-mutation assays, which can identify minimal variations in the genome composition, such as single mutations. An example is the reverse hybridization technique, incorporated into the commercial Line Probe Assay, that has been used to detect HIV resistance mutations in CSF and plasma pairs (Cunningham et al., 2000). By this test, low proportions of resistant variants, up to 10%, can be detected (Stuyver et al., 1997). These assays, however, are designed only to detect selected mutations and may occasionally generate false negative results due to mutations near the site of hybridization.

The HTA assay has been used to compare sequences obtained from different sites or at different times from same subjects (Barlow et al., 2000). By HTA, a labeled probe anneals to the amplification products from either CSF or plasma, to generate probe-PCR product heteroduplexes. Point mutations, insertions or deletions will modify migration of heteroduplexes on a polyacrilamide gel, thus revealing distinct genomic virus variants in each compartment. Comparison between CSF and plasma patterns will provide information on both presence and relative quantity of viral subpopulations in the two compartments. HTA has recently been used to compare CSF with plasma HIV *env* sequences between patients at different stage of HIV infection and at various time points during HAART (see below) (Ritola et al., 2004; Harrington et al., 2005; Ritola et al., 2005).

Different HIV genome regions have been investigated in genotypic studies on HIV CNS compartmentalization. The greatest attention has been paid to *pol* and *env* sequences, the first in relation to antiretroviral drug resistance, the latter because of its high variability and implications with cell tropism. However, other regions, for example *gag*, *nef* and *tat*, have shown differences between CNS and plasma isolates and may also be studied in the CSF (Hughes et al., 1997; Bratanich et al., 1998; Churchill et al., 2004).

17.2.2.1. Studies of *pol*

Pol is approximately 3000 nucleotides long and encodes for the viral enzymes protease, RT and integrase. The RT and protease regions have been widely studied because of their implications for antiretroviral drug resistance, including NRTIs, non-nucleoside RT inhibitors (NNRTIs) and protease inhibitors (PIs). For resistance to an antiretroviral agent to occur, the target enzyme, for instance RT or protease, will undergo genetic change, yet preserve its function in the presence of the drug. In the case of incomplete suppression of viral replication, the selective pressure of antiretroviral therapy gives drug-resistant mutants a competitive advantage and they eventually come to represent the dominant variant. Resistance mutations of the RT and protease genes are well known for the principal available drugs and are associated with reduced drug susceptibility in cell cultures (Hirsch et al., 2000) (hivdb.stanford.edu, hiv-web.lanl.gov/seq-db.html). In clinical practice, genotypic assays based on direct sequencing have largely supplanted phenotypic tests, which measure drug susceptibility of patient-derived strains or recombinant viruses with patient sequences inserted (Richman, 2000). Recently, new approaches have been introduced for interpretation of genotypic patterns. These include software programs, which run algorithms based on current knowledge from both literature data and expert opinion (Vandamme et al., 2001), and virtual phenotype, which is determined by matching the patient's genotype with genotypes with known phenotype profiles in a large database (Harrigan et al., 2001).

The analysis of *pol* sequences derived from the CSF has mainly focused on the distribution of resistance mutations between CSF and plasma. The presence in the CNS of anatomical and functional barriers provides a tissue "reservoir" restricting drug entry and potentially establishing an environment that promotes development of distinct resistant mutants. Early studies to evaluate this hypothesis were performed by sequencing the RT gene following amplification of proviral DNA directly from CSF and blood or from culture isolates derived from these fluids. However, the number of patients studied was low, and results inconsistent, with discordant resistant mutations between the two compartments demonstrated by some authors, but not by others (Wildemann et al., 1993; Di Stefano et al., 1995). More recent work has focused on viral sequences obtained by direct amplification of RNA from clinical specimens and in larger patient groups. Overall, up to over one-half of patients with advanced HIV disease show significantly different resistance mutation profiles between CSF and blood (Chien et al., 1999; Cunningham et al., 2000; Venturi et al., 2000; Lanier et al., 2001; Stingele et al., 2001; Tashima et al., 2002; Strain et al., 2005). Resistance mutations may be present in the blood and not in the CSF as well as vice versa, although, in general, these are more frequently recognized in blood. Recently, CSF and plasma mutation profiles have also been analyzed by virtual phenotype, disclosing differences in fold-change of resistance between CSF and plasma virus for at least one drug in approximately half of the patients. These figures are consistent with findings of genotypic studies, and suggest that different mutation patterns between CSF and plasma might have practical relevance (Antinori et al., 2005).

Some of the above studies have also addressed the question of whether resistance mutations in CSF could affect the virological response to HAART in this compartment. As expected, the presence of mutations to NRTIs in both CSF and plasma was associated with reduced virological response to treatment in both compartments (Lanier et al., 2001). However, patients exhibiting a slower CSF than plasma virological response, or failing to respond in CSF only, did not show resistance mutations restricted to this compartment (Cinque et al., 2001). Also, despite the high frequency of cases with discordant resistance mutations between CSF and plasma, it remains to be demonstrated that mutations that develop in CSF spread to the systemic circulation with any significant frequency (Lanier et al., 2001). Studying the kinetics of mutations onset, however, is limited by the requirement of frequent sampling of CSF, and only a few studies have assessed drug resistance mutations in serial samples drawn days to weeks apart. In general these showed that differences between CSF and plasma sequences at baseline are maintained during treatment as long as viral RNA is detectable (Tang et al., 2000; Cinque et al., 2001). Similarly, the analysis of serial CSF and plasma samples taken following structured HAART interruption showed simultaneous changes of both phenotypic and genotypic profiles, i.e. from drug resistant to drug susceptible. This occurred in parallel with viral load increases in both compartments, suggesting interchange between blood and CSF variants (Price et al., 2001).

Beyond the analysis of resistance codons, whole *pol* sequences have been analyzed to investigate the distribution of viral quasispecies in CSF and plasma. The analysis of RT sequences from proviral DNA or viral RNA obtained by molecular cloning showed that brain and CSF-derived sequences were distinct from spleen and lymph node-derived sequences either considering or excluding resistance codons in the analysis (Wong et al., 1997). Population sequencing of RT and protease disclosed increasing discordance between CSF and plasma sequences with progression of disease, reflecting increased segregation of infection. Furthermore, CSF plasma diversity was higher in patients with ADC than in neurologically asymptomatic patients or patients with other CNS complications (Sala et al., 2005). *Pol* mutations that accumulate selectively in the CNS or CSF are likely to result from different replication kinetics in this compartment compared to plasma, and this might depend on different drug levels achieved intrathecally. Recently, however, a significant accumulation of nonsynonymous substitutions in RT positions corresponding to areas of known structural and functional significance have been observed in CNS-derived sequences from untreated patients, suggesting that RT may be under positive evolutionary selection within the CNS independent of treatment (Huang et al., 2002).

17.2.2.2. Studies of *env*

Env is approximately 2600 nucleotides in length and encodes the external glycoprotein gp120 and the transmembrane glycoprotein gp41. Gp120 contains the binding site for the CD^4 receptor and the seven-transmembrane domain chemokine receptor that serves as coreceptor for HIV entry in target cells. The gp120 subunit consists of five variable regions, V1 to V5, interspersed with conserved regions (C1–C5). Overall, *env* shows up to 25% inter-patient variation when comparing HIV strains from geographically separated locations. For this reason it has historically been the target of studies on virus transmission and evolution. Progression of HIV infection is associated with an increasing diversity of *env* sequences within individual patients, as a result of environmental pressure, including immune and coreceptor usage selection (Coffin, 1995). A large number of studies have been carried out to evaluate the distribution of viral quasispecies in different body compartments according to *env*, including the brain and the CSF. Different *env* regions have been targeted, more frequently the V3 loop and other gp120 hypervariable regions. Both DNA and RNA have been analyzed, by direct sequencing, analysis of cloned sequences or HTA. These studies have assessed compartmentalization of infection, and the related issues of neurotropism and neurovirulence.

Initial studies of cloned *env* sequences already showed that viral variants in the brain and CSF differed from those present in circulating lymphocytes or other tissues within the same individuals (Steuler et al., 1992b; Kuiken et al., 1995). Also, the intra-patient genetic distance between paired CSF and plasma sequences seemed to increase with advancing disease stage, being more prominent in AIDS than in asymptomatic patients (Keys et al., 1993) and in patients suffering from AIDS encephalopathy (Steuler et al., 1992b). These findings, indicating a different evolution between CSF and plasma subpopulations with disease progression and development of productive brain infection, have more recently been corroborated by HTA studies. By HTA, paired CSF and plasma sequences from patients with PHI were virtually identical, consistent with rapid and efficient penetration of the primary infecting variant into the intrathecal compartment (Ritola et al., 2004). CSF variants appeared to become compartmentalized later during the course of infection, either as "enriched" variants (detected in both CSF and plasma but differing in two compartments with respect to their relative abundance) or "unique" variants (detected in CSF or in plasma only). In addition, patients with ADC displayed a significant increase of more unique variants in CSF compared to plasma (Ritola et al., 2005). HTA analysis also showed that CSF variants declined rapidly during the first days of HAART in asymptomatic patients, with a half-life similar to that of viruses infecting blood CD^4 T cells. This observation is consistent with the hypothesis that, during asymptomatic infection, CD^4 T cells trafficking in the CNS are the cells supporting HIV replication in this compartment. Because these variants are unique to CSF but differ from blood variants, it was suggested that "trafficking" CD^4 cells might become infected once they enter the intrathecal compartment, with virus produced by resident infected macrophages or lymphocytes (Harrington et al., 2005).

It is not known whether divergent *env* variants arise in the intrathecal compartment following genetic drift of a segregated population or if they are more actively selected as a result of the peculiar biological and anatomical characteristics of CNS itself. In the latter case, selection of CSF or CNS *env* variants might result from differences in local immune selection pressure and also in virus entry and replication rate, due to different target cells (van Marle and Power, 2005). In this regard, several studies have addressed the hypothesis that HIV infection of the CNS is sustained by CNS-specialized viruses, i.e. neurotropic and/or "neurovirulent" strains.

Early analyses of the V3 region identified "brain signature" patterns, which had also been recognized in macrophage-tropic isolates. It was thus suggested that macrophage tropism could be the biological constraint determining viral brain signature pattern and neurotropism (Korber et al., 1994). Although subsequent studies failed to confirm the presence of brain-specific signature sequences (Ohagen et al., 2003), a small number of positions in the C2–V3 region that appear to discriminate CSF from plasma virus have recently been reported (Strain et al., 2005). Of note, some of these corresponded to those initially indicated (Korber et al., 1994) or to positions possibly implicated in neurovirulence (see below) (Power et al., 1994). Coreceptor usage is theoretically a determinant of neurotropism. It can be assessed by in vitro functional entry assays either using patient isolates or recombinant viruses containing amplified patient *env* sequences, and also deduced by genetic analysis of V3 sequences. A number of studies suggest that brain-derived viruses are macrophage-tropic, and that principally use CC chemokine receptor 5 (CCR5) for virus entry ("R5 viruses", as opposed to "X4 viruses", i.e. those that principally use CXC chemokine receptor 4 (CXCR4) for virus entry) (He et al., 1997; Ghorpade et al., 1998; Albright et al., 1999). Accordingly, R5 virus was recently shown to predominate in CSF by both functional assays and V3 sequencing (Ritola et al., 2004; Spudich et al., 2005a). Also, flow cytometry studies of CSF cells showed that a high percentage of monocytes express CCR5 and that most T cells express both CXCR3 and CCR5 (Shacklett et al., 2004; Neuenburg et al., 2005b). However, and more generally, determinants of neurotropism are likely more complex than predictions based on *env* sequences or virus coreceptor usage. Indeed, mixtures of R5 and X4 viruses as well as viruses able to utilize either coreceptor are also found in CSF (Spudich et al., 2005a), and, on the other hand, not all R5 viruses seem to be able to infect macrophage cells in vitro, suggesting that CCR5 usage alone may not be sufficient for macrophage-tropism (Gorry et al., 2001).

Finally, variants specialized for replication in the CNS, or "neurovirulent", might play a role in the development of ADC. Indeed, associations have been reported between presence of ADC and the amino acid composition of given positions within the V3 region of viral strains amplified from the brain or CSF (Power et al., 1994; Kuiken et al., 1995). However, these associations have been disputed (Di Stefano et al., 1996; Reddy et al., 1996; Strain et al., 2005), and thus whether particular sequence motifs are relevant in determining development of HIV-related lesions or neurological abnormalities is still unresolved. Whether coreceptor usage may be a determinant of neurovirulence also remains to be assessed. According to preliminary observations, brain isolates from patients with ADC and HIV encephalitis were not restricted to the use of CCR5 for virus entry (Gorry et al., 2001), but, on the other hand, R5 tropism in CSF appeared to be associated with ADC (Spudich et al., 2005a).

In addition to their relevance in understanding HIV dynamics in the CNS, CSF *env* studies of viral compartmentalization, as well as those aiming at identifying neurotropic or neurovirulent variants, may eventually have more practical applications in patient management. If defined with more certainty, genetic markers of disease might be used to predict, diagnose and manage HIV infection of the CNS. These may include quantitative measures of CSF compartmentalization as well as genetic signatures. In addition, the analysis of coreceptor usage in the CSF may also have implications for monitoring antiretroviral therapy containing CCR5 and CXCR4 inhibitors — the former is currently being evaluated in clinical trials (Coakley et al., 2005).

17.2.3. Detection of anti-HIV antibodies

Since early studies, it was clear that HIV-infected patients have virus-specific antibodies in their CSF beginning at the early stages of infection. Indeed, detection of unique oligoclonal immunoglobulin (Ig) G bands and high indexes of intrathecal IgG synthesis were among the earliest observations suggesting that HIV infection of the CNS occurs in the majority of infected patients with or without neurologic disease (Goudsmit et al., 1986; Grimaldi et al., 1988; Resnick et al., 1988). CSF lymphocytes from HIV-positive subjects produced specific IgG in vitro, providing direct evidence that CSF HIV-specific antibody resulted from intrathecal synthesis (Amadori et al., 1988). Intrathecal IgG production seems to increase during early HIV infection, while specific HIV antibodies are less frequently detected during advanced infection, indicating declining B cell response with disease progression (Andersson et al., 1988; Elovaara et al., 1988; Marshall et al., 1988; Van Wielink et al., 1990). Indeed, fewer anti-HIV antibody bands in CSF by Western blot correlated with poor outcome (Desai et al., 2000), and anti-*pol* and anti-*env* antibody titers declined with severity of HIV disease (Elovaara et al., 1993). When compared to symptom-free patients, however, neurologically symptomatic patients had increased intrathecal HIV-specific and total IgG synthesis (Singer et al., 1994). Likewise, increased CSF anti-*env* antibody responses were more common in patients with ADC than in HIV-infected asymptomatic subjects (Trujillo et al., 1996).

Intrathecal production of antibodies to HIV appears to be strongly associated with virus isolation from CSF and p24-Ag CSF levels, suggesting that intrathecal synthesis of antibodies to HIV is related to a persistent HIV antigenic stimulation in the CNS (Sonnerborg et al., 1989b). Indeed, intrathecal HIV antibody synthesis could no longer be detected in half of the patients after one year of zidovudine therapy (Elovaara et al., 1994). More recent studies of neurologically asymptomatic patients receiving HAART for up to two years, however, showed only a modest reduction of intrathecal IgG production, which is consistent with persisting intrathecal immune activation despite complete viral load suppression (Abdulle et al., 2005).

17.3. Host response markers

17.3.1. Markers of inflammation

HIV encephalitis is associated with inflammation, which largely exceeds the extent of viral infection in the brain. HIV-induced immune activation of brain macrophages and microglial cells induces these cells to release soluble factors leading to activation of uninfected cells. In addition, these products lead to migration of HIV-infected cells into the CNS, thus contributing to the amplification of the inflammatory process. Parenchymal HIV brain infection eventually leads to neuronal injury and death, through a panoply of direct or indirect neurotoxic pathways (Gonzalez-Scarano and Martin-Garcia, 2005; Kaul et al., 2005; Jones and Power, 2006). A number of these soluble factors, either secreted by brain cells or resulting from cleavage of structural cell components, can be measured in the CSF, thereby providing indicators of the underlying biological processes. As discussed, however, most of these markers are not specific for ADC or HIV encephalitis, because they may also be found in the CSF of neurologically asymptomatic patients or those with opportunistic CNS infections. Nonetheless, some of these markers remain of proven or potential relevance for studying in vivo the dynamics of ADC.

17.3.1.1. White blood cells

In normal adults, CSF is acellular containing only up to 4–5 WBCs/μl, predominantly lymphocytes. Higher cell counts, or CSF pleocytosis, is frequent in persons with HIV infection, with values usually of 5–15 cells/μl, but occasionally higher (Appleman et al., 1988; Hagberg et al., 1988; Marshall et al., 1988; McArthur et al., 1988; Gisslen et al., 1998; Martin et al., 1998). Pleocytosis is more common in patients with early, asymptomatic stages than in those with advanced infection and low CD^4 cell counts, e.g., below 50 cells/μl (Elovaara et al., 1988; Spudich et al., 2005b). Patients with ADC may show either normal or elevated cell numbers (Navia et al., 1986). In the early stages of infection, higher CSF cell counts are associated with elevated CSF HIV RNA (Martin et al., 1998; Price, 2000; Spudich et al., 2005b) and with smaller differences between plasma and CSF HIV RNA levels (Spudich et al., 2005b), supporting the idea that CSF mononuclear cells are the principal source of CSF virus.

Pleocytosis usually resolves with anti-HIV treatment. CSF cell numbers declined in CSF after 6 weeks of zidovudine therapy (Elovaara et al., 1994), as well as in patients receiving HAART (Polis et al., 2003). In HAART-treated subjects showing slower initial viral decay in CSF than in plasma, pleocytosis resolves when HIV RNA eventually approaches or reaches the limit of detection in CSF (Spudich et al., 2005b). Of note, pleocytosis recurs promptly in patients undergoing structured treatment interruption when treatment is stopped (Staprans et al., 1999; Price et al., 2001; Price and Deeks, 2004).

Recent flow cytometry studies have begun to more fully characterize pleocytosis in HIV-infected persons, especially in terms of degree of activation and status of infection of CSF cells. It is known that T cells are the most common cell type in normal CSF (Svenningsson et al., 1993). While this was also observed in HIV-infected patients without opportunistic infections, the proportion of activated CD^4 and CD^8 T cells (CD^{38}) was higher in these subjects than in uninfected controls. Also, T cell activation was reduced by HAART (Neuenburg et al., 2005a). By intracellular HIV p24-Ag detection, most CSF T cells appeared to be uninfected, even in CSF samples with high HIV RNA levels. Most of the virus-producing cells were shown to be activated CD^4 T cells (CD^{38}), although most of the activated T cells were not producing HIV (Neuenburg et al., 2004). To study the role of CD^8 T lymphocytes in CSF trafficking in greater depth, recent studies also investigated the expression of adhesion molecules by these cells. CSF CD^8 T cells were found to express high level of the adhesion molecules very late antigen (VLA)-4 and leukocyte function antigen (LFA)-1, which thus likely play a role in lymphocyte trafficking to CSF during HIV infection. These findings, together with the demonstration of pleocytosis and enhanced CCR5 expression by these cells compared to blood, also support a model by which lymphocyte extravasation is driven by cell activation, expression of adhesion molecules, and increased vascular permeability, and is coupled with chemokine-mediated trafficking to inflammatory sites in the CNS (Shacklett et al., 2004).

Monocytes also represent a significant proportion of CSF cells, and their presence and activation in the CSF may be particularly important because cells from the monocyte-macrophage lineage sustain HIV replication in the brain. In untreated subjects, elevated CSF viral load was associated with higher CSF monocyte activation (CD^{16}/CD^{14}). Surprisingly, the number of activated CSF monocytes was higher in HAART-treated than untreated patients, and in HAART successes than in HAART failures (defined by plasma viral load >500 copy number/ml) in association with lower viral loads in CSF (Neuenburg et al., 2005b). The highest monocyte activation level appeared to associate with use of PIs. Several hypotheses have been suggested to explain this observation, including the possible effects of PIs on lipid levels, antigen presentation, sterol metabolism and apoptosis, which might all affect monocytes regulation (Neuenburg et al., 2005b), or, possibly, continued low level viral replication because of relatively poor CNS penetration by PIs.

17.3.1.2. Adhesion molecules

Adhesion molecules are a large family of molecules that enable contact between cells. Adhesion molecules fall into two broad groups, the integrins (e.g., LFA-1, VLA-4), which principally bind molecules in the intercellular matrix, and a second group that mainly binds to cells. The latter group comprises three categories, the cadherins, the Ig superfamily (e.g. intercellular adhesion molecule-1, ICAM-1, and vascular cell adhesion molecule-1, VCAM-1), and the selectins (e.g. E-selectin). Adhesion molecules relevant to CNS infections are expressed by brain endothelial cells and leukocytes and are essential mediators of leukocyte migration across the vascular brain endothelium (Engelhardt and Ransohoff, 2005).

Several experimental findings suggest that HIV entry into the CNS is, in part, a consequence of the ability of virus-infected and immune activated monocytes to induce adhesion molecules on brain endothelium (Nottet and Gendelman, 1995). High levels of E-selectin and VCAM-1 were found to be expressed by endothelial cells cocultured with activated HIV-infected monocytes (Nottet et al., 1996). ICAM-1 was induced in vascular endothelial cells following stimulation by gp120 (Stins et al., 2003), while both ICAM-1 and VCAM-1 were expressed by astrocytes following *tat* stimulation (Woodman et al., 1999). In brain tissue of HIV-infected subjects with encephalitis, expression of E-selectin and, to a lesser degree, VCAM-1 by endothelial cells was associated with expression of both HIV antigen and proinflammatory cytokines and with macrophage infiltration (Nottet et al., 1996). Abundant VCAM-1 protein was also expressed on endothelial brain cells in macaques with SIV encephalitis (Sasseville et al., 1992), whereas other endothelial-related adhesion molecules, including ICAM-1, E-selectin and P-selectin, were not uniformly associated with encephalitis in this model (Sasseville et al., 1992).

Adhesion molecules expressed by CSF cells can be detected and measured by flow cytometric analyses, whereas some adhesion molecules expressed by these cells as well as by the brain endothelium can be detected in their soluble forms. In patients with HIV infection, soluble ICAM was detected at low levels in the CSF of patients with neurological symptoms, not differing from those of neurologically asymptomatic patients (Rieckmann et al., 1993; Heidenreich et al., 1994). Soluble VCAM-1 concentrations in CSF were also increased approximately 20-fold in animals with encephalitis above those without encephalitis. More recently, flow cytometry studies of CSF cells showed that CD^8 T cells trafficking to CSF express high levels of VLA-4 and LFA-1 (Shacklett et al., 2004).

17.3.1.3. Cytokines and chemokines

Cytokines are a group of small secreted proteins that mediate and regulate a number of physiological processes, including immunity and inflammation. Cytokines are variously classified, mainly according to their principal or initially discovered functions. We will separately review here "classic" cytokines and chemokines.

17.3.1.3.1. Cytokines

In CNS infections, cytokines are produced by resident cells (neurons, astrocytes, oligodendrocytes and microglia) and cells from the immune system that traffick through the CNS (CD^4 and CD^8 T lymphocytes, B lymphocytes, monocyte/macrophages and natural killer cells). Their activation causes a myriad of changes related to local inflammation, cell proliferation and structural and functional damage of brain cells and tissues.

In HIV infection, brain cells have been shown to abundantly express both proinflammatory cytokines, e.g. tumor necrosis factor (TNF)-α, interleukin (IL)-1β and IL-6, and immunoregulatory cytokines, e.g. IL-2, interferon (IFN)-γ and IL-10, during the entire course of infection. This reflects continued immune activation and points to the role for the immune system in the control of HIV infection but also in the pathogenesis of HIV encephalopathy. Certain cytokines, including IL-1β, TNF-α and IFN-α, have been found to be overexpressed in brain tissues from HIV-infected compared to HIV-negative subjects, primarily by perivascular macrophages and microglial cells, and astrocytes (Tyor et al., 1992; Rho et al., 1995). These cytokines have been found upregulated in the CNS irrespective of neurological dysfunction (Tyor et al., 1992) and, in patients with HIV-induced brain damage, without correlation with severity of encephalitis (Achim et al., 1993). A semiquantitative study of mRNA expression, however, showed selectively elevated TNF-α mRNA levels in the brains of subjects with ADC (Griffin, 1997). In contrast, IL-1β was decreased and IL-4 mRNA was undetectable in ADC patients, whereas no differences were seen with IL-6, transforming growth factor (TGF)-β 1 and IFN-γ between ADC and nondemented patients (Wesselingh et al., 1993).

A number of cytokines have been measured in the CSF of HIV-infected patients, and some were found at increased concentration. These include TNF-α (see below), IL-1 (Gallo et al., 1989; Rimaniol et al., 1997), IL-2 and its receptor (Sethi and Naher, 1986; Ciardi et al., 1993), IL-6 (Gallo et al., 1989; Laurenzi et al., 1990), IL-10 (Gallo et al., 1994), IFN-γ (Griffin et al., 1991), IFN-α (Rho et al., 1995) and TGF-β1 (Perrella et al., 2001). In general, none of these cytokines emerged as a sensitive and specific marker for clinical management of ADC. Also, no association was found between CSF abnormalities and degree of HIV encephalitis, likely reflecting the concomitant presence of opportunistic infections or the intense immune activation in the absence of CNS lesions (Wiley et al., 1992). However, elevations in these CSF cytokines have provided a useful pathogenetic view of pathways involved in ADC and HIV-E and, likely, extension of such studies will continue to expand this understanding. We selectively review here the most relevant findings relating to cytokine detection in CSF.

17.3.1.3.1.1. Tumor necrosis factor-α

TNF-α is a proinflammatory cytokine produced by activated macrophages and microglial cells. It exerts its effects upon binding with the receptors tumor necrosis factor receptor (TNFR)-I and TNFR-II, both expressed by activated phagocytes. This cytokine and its receptors have been implicated in the pathogenesis of a number of CNS inflammatory diseases, including ADC. In vitro, TNF-α and HIV are mutually upregulated in microglial cells (Peterson et al., 1992; Wilt et al., 1995). TNF-α was found to be overexpressed in the brain of patients with ADC (Achim et al., 1993; Glass et al., 1993; Wesselingh et al., 1993) and the profile of TNF-α mRNA correlated with the onset, progression and severity of dementia (Griffin, 1997). In more severe cases of encephalitis, the white and cortical gray matter showed the highest TNF-α concentrations, whereas only deep gray matter appeared to be involved in mild cases (Achim et al., 1993). Also its soluble receptors TNFR-I (p55) and TNFR-II (p75) were found to be overexpressed in the brain of patients with AIDS and particularly in those with HIV-E (Sippy et al., 1995). Besides being a proinflammatory cytokine, in vitro studies suggest that TNF-α might itself be involved in neurodegeneration, either directly, by inducing apoptosis of neurons (Garden et al., 2002), or less directly by boosting the effects of other neurotoxic mediators, including glutamate, nitric oxide, viral proteins and other cytokines (Saha and Pahan, 2003).

Measures of TNF-α in the CSF have produced discrepant results. While this cytokine was detected in the majority of cases in some studies, others failed to show detectable concentrations (Gallo et al., 1989; Shaskan et al., 1992b). These differences have been ascribed to technical discrepancies in measuring TNF-α and to the lability of this protein in biological fluids. Increased CSF expression of TNF-α was found in patients with HIV infection versus HIV-negative subjects (Tyor et al., 1992), and in patients with ADC or opportunistic CNS disorders compared to HIV asymptomatic subjects (Grimaldi et al., 1991; Franciotta et al., 1992; Mastroianni et al., 1992). TNF-α levels correlated with CSF HIV RNA levels (Lafeuillade et al., 1996) and CSF level of other cytokines, such as IL-2 and its soluble receptor (sIL-2R) (Ciardi et al., 1994). Since TNF-α is rapidly cleared from CSF as well as from other body fluids, measuring its soluble receptors TNFR-I and TNFR-II (sTNFR-I, sTNFR-II) has been suggested to be more reliable than assessing TNF-α itself (Gallo et al., 1989). Since TNF-α receptor levels correlate with TNF-α levels, finding elevated receptor concentrations may reflect activation of the whole TNF-α system (Rimaniol et al., 1997). Indeed, increased levels of TNF-α receptors were found in patients with ADC and CNS opportunistic infections compared to HIV-infected asymptomatic controls (Rimaniol et al., 1997).

Receptor levels have been measured longitudinally in CSF of patients receiving anti-HIV treatment. Following treatment with two NRTIs, sTNFR-I concentrations did not change from baseline (Enting et al., 2000). Among patients receiving treatment with two PIs with or without NRTIs, CSF sTNFR-II levels increased in the patients who did not suppress their CSF HIV-RNA and also in some patients with good CSF HIV-RNA responses (Gisolf et al., 2000b). Recently, TNF-α levels, but not HIV RNA, were weakly associated with development of cognitive symptoms and dementia in a cohort of patients receiving HAART (Sevigny et al., 2004).

17.3.1.3.1.2. Other cytokines

Initial studies reported elevated CSF levels of IL-1β and IL-6 in approximately half of patients with HIV infection, although their association with CNS involvement by HIV was not clear (Gallo et al., 1989; Perrella et al., 1992). A more recent study, however, failed to detect both IL-1β and IL-1, but found high CSF concentrations of the IL-1 receptor antagonist (Rimaniol et al., 1997). IL-6 was detected more frequently and at higher concentrations in the advanced stages than in the early stages of HIV infection (Laurenzi et al., 1990), including patients with ADC or opportunistic infections (Torre et al., 1992).

IFN-α was found to be elevated in the CSF of HIV-positive patients with ADC compared to those without dementia and controls (Rho et al., 1995). IFN-α levels were also higher in severe than in mild ADC and correlated positively with CSF HIV RNA (Krivine et al., 1999; Perrella et al., 2001). CSF concentrations of IFN-γ were only slightly higher in CSF of HIV-infected subjects than HIV seronegative controls, and without differences between patients with or without neurological disease (Griffin et al., 1991). In the absence of neurologic disease and treatment, CSF IFN-γ remained at stable levels for up to 4 years after seroconversion (Griffin et al., 1991). Finally, TGF-β1 was undetectable in the CSF of patients with severe ADC. In subjects with mild dementia, there was a significant inverse correlation between CSF levels of this cytokine and both CSF HIV RNA and IFN-α (Perrella et al., 2001).

17.3.1.3.2. Chemokines

Chemokines are chemotactic cytokines that mediate the migration of leukocytes into tissues (Baggiolini, 2001). They are small proteins, approximately 70–130 aminoacids, and, at present, include more than 50 different molecules, sharing 20–70% aminoacid homology. Based on the position of the amino-terminal cysteines, chemokines are classified into four subfamilies (CC, C, CXC, CXXXC). Original names have recently been replaced by a newer nomenclature based on their family classification and their cell receptors (Murphy et al., 2000). In the CNS, chemokines are produced by a variety of cell types, including both circulating blood and CNS resident cells. Although some are produced under physiological conditions, they are usually released in response to infections or infiltrating inflammatory cells. Released chemokines interact specifically with chemokine receptors on the surface of different inflammatory and native CNS cell types. Individual chemokines may bind to more than one chemokine receptor and many chemokine receptors bind more than one chemokine. Chemokine receptors belong to the seven-transmembrane domain, G-protein-coupled receptor superfamily. Upon binding, chemokine receptors activate G protein for signal transduction leading to a variety of cellular responses, including regulation of leukocyte migration into and within the CNS, production of cytokines and regulation of neuropoiesis (Martin-Garcia et al., 2002).

In HIV infection of the CNS, as in other CNS diseases, chemokines are primarily mediators of inflammatory processes. Their most prominent effect is on the selective migration of HIV-infected or uninfected leukocytes across the brain–blood barrier to areas of infection. CXC chemokines recruit predominantly T lymphocytes, whereas CC chemokines exert their effect on both T lymphocytes and monocytic cells. Infiltrating cells may contribute to productive HIV brain infection by carrying new virus into the CNS and by recruiting a pool of fresh, susceptible cells. Furthermore, these cells will secrete additional chemokines, thus amplifying the whole process. In addition to their chemotactic role, it has been suggested that chemokines may also be implicated in HIV-induced neurodegeneration. In vitro, chemokines can induce calcium mobilization in neurons (Bolin et al., 1998), promote neuronal apoptosis (Pulliam et al., 1996; Hesselgesser et al., 1998; Sanders et al., 1998; Zheng et al., 1999) and stimulate the release of glia-derived neurotoxins (Kaul et al., 2001). Besides these functions, the chemokines importantly influence HIV infection by virtue of the role of their receptors in HIV binding to target cells, in conjunction with CD^4. The most important chemokine receptors for HIV are CCR5 and CXCR4, but several others have been identified as minor coreceptors (Martin-Garcia et al., 2002). Chemokines such as macrophage inflammatory protein (MIP)-1α, MIP-1β and RANTES may, in fact, inhibit HIV replication in vitro by blocking the use of these coreceptors by HIV (Cocchi et al., 1995). It is thus possible, at least theoretically, that some chemokines contribute to suppression of viral replication in the brain. Of note, coreceptor antagonists are currently in preclinical and clinical development stages, including CCR5 and CXCR4 inhibitors (Shaheen and Collman, 2004).

CSF studies have contributed to the understanding of the role of chemokines in the pathogenesis of CNS disease. Because chemokines released by CNS cells act locally rather than systemically, their levels in CSF, unlike plasma levels, may reflect both the type and severity of the underlying CNS disease. For this reason, chemokine levels in CSF have been suggested as surrogate markers for inflammatory events occurring within the CNS. Their levels can easily be measured by immunosorbent procedures. By cytofluorimetry it is also possible to study chemokine receptor expression on CSF cells. As with other inflammatory markers, the limitation thus far has related to the nonspecificity of their elevation in the CSF.

17.3.1.3.2.1. Monocyte chemotactic protein-1 (MCP-1)/CCL2

Monocyte chemotactic protein (MCP)-1 is a CC or β chemokine (CCL2) that plays a central role in recruitment and activation of monocytes and macrophages (Lu et al., 1998; Gu et al., 1999). In HIV infection, mononuclear cells are crucial for virus entry and dissemination to the CNS (Koenig et al., 1986), and their presence and number in the CNS is an important correlate of both HIV-E and ADC (Glass et al., 1995; Gonzalez-Scarano and Baltuch, 1999). Based on such observations, a central role of MCP-1 in productive CNS HIV infection has been suggested, and this hypothesis has been supported during recent years by a number of observations. These include the demonstration that MCP-1 is upregulated in vitro by HIV (Conant et al., 1998; Mengozzi et al., 1999), the finding of MCP-1 expression in brain macrophages in and around multinucleated HIV-infected giant cells (Sanders et al., 1998), the increased MCP-1 levels in CSF of patients with ADC, and the correlation between presence of a mutant MCP-1 allele (2578G) and increased risk of developing ADC and of ADC progression (Gonzalez et al., 2002). Additionally, astrocytes and endothelial cells likely contribute to MCP-1 production. MCP-1 expression was shown to be upregulated in astrocytes by *tat* (Conant et al., 1998), quinolinic acid (Guillemin et al., 2003), as well by MCP-1 binding with its receptor CCR2 (Cota et al., 2000; Dorf et al., 2000). Astrocytes and endothelial cells have been shown to express MCP-1 in HIV-E (Sanders et al., 1998). Following SIV infection of macaques, CSF expression of glial fibrillary acid protein (GFAp), an astrocyte marker of astrogliosis (see below), paralleled the increase of MCP-1 levels and preceded the development of brain inflammation (Zink et al., 2001). Taken together, all these observations support a model in which HIV infection of mononuclear phagocytes induces MCP-1 release by these and other cell types, especially astrocytes. These initial events are followed by further recruitment of mononuclear cells to the site of infection, thus providing a mechanism of amplifying the CNS inflammatory responses.

A number of CSF studies have suggested that MCP-1 is a reliable marker of activation of macrophages and microglial cells in the CNS. CSF MCP-1 levels are usually low in HIV-positive, neurologically asymptomatic patients, with levels similar to those found in HIV-negative controls (Enting et al., 2000). In contrast, high CSF MCP-1 levels are found in patients with ADC (Conant et al., 1998; Kelder et al., 1998) and HIV-E (Cinque et al., 1998b). In these patients, CSF MCP-1 correlates with CSF HIV RNA levels, but not with MCP-1 or HIV RNA levels in plasma, strongly arguing for its intrathecal origin (Cinque et al., 1998b; Kelder et al., 1998). Elevated MCP-1 CSF levels were also found in patients with CNS opportunistic infections, especially in CMV encephalitis which is associated with an intense macrophage infiltrate (Bernasconi et al., 1996). MCP-1 is also a potential predictive marker of ADC, as shown by the observation that an increase of its CSF levels seems to precede the development of neurological disease (Kelder et al., 1998). This was confirmed by studies on SIV-infected macaques, which showed a progressive increase of CSF MCP-1 levels throughout the course of HIV infection in the animals that develop encephalitis (Zink et al., 2001). Recent studies showed a correlation between elevated CSF MCP-1 and contrast enhancement with standard magnetic resonance imaging (MRI) (Avison et al., 2004) and neuronal dysfunction by proton MR spectroscopy (Chang et al., 2004). Of interest, higher MCP-1 CSF levels were also found in patients carrying the allele 2578G, which was previously associated with increased risk for ADC (Gonzalez et al., 2002; Letendre et al., 2004).

Following HAART, CSF MCP-1 levels did not vary significantly among asymptomatic patients, who already showed low levels of this chemokine at baseline (Gisolf et al., 2000b). In contrast, MCP-1 levels decreased dramatically in ADC patients, who showed higher baseline concentrations, and in parallel with CSF HIV RNA (our unpublished observation). It was also observed that CSF MCP-1 may rebound slightly after a few months of HAART, despite continued undetectable CSF HIV RNA (Gisolf et al., 2000b). In patients undergoing treatment interruption, increases in CSF MCP-1 immediately preceded or coincided with a rebound of CSF VL, confirming the strict interrelationship between expression of this chemokine and viral replication (Monteiro de Almeida et al., 2005). Overall, these observations indicate that, by reducing viral replication in the brain, HAART interferes locally with the inflammatory process. However, the presence of significant CSF MCP-1 levels despite virologically effective HAART might reflect the persistence of macrophage activation, possibly due to ongoing low-grade viral replication in brain tissue or to slow downregulation of immune activation.

In summary, MCP-1 has provided a useful tool for monitoring HIV infection of the CNS. This chemokine has largely been evaluated in clinical series that have included virtually all patient groups as well as untreated and treated patients. The body of information collected so far indicates that MCP-1 is a useful marker of activation of macrophages and microglial cells.

17.3.1.3.2.2. Interferon-γ-inducible protein 10(IP-10)/ CXCL10

Interferon-γ-inducible protein 10 (IP-10, also CXCL10) is a CXC or α-chemokine that was first identified as an early response gene to IFN-γ. It can be produced by a variety of cells, including T cells, macrophages, endothelial cells and astrocytes (Taub et al., 1993; Bajetto et al., 2002; Melchjorsen et al., 2003). Upon binding with its receptor, CXCR3, it attracts activated T cells, natural killer cells and blood monocytes, and appears to be an important participant in a variety of inflammatory conditions (Taub et al., 1993; Loetscher et al., 1996, 1998; Qin et al., 1998; Bajetto et al., 2002; Melchjorsen et al., 2003).

IP-10 has been suggested to enhance HIV infection in the brain either directly or by attracting leukocytes to the CNS. In vitro, IP-10 may indeed promote HIV replication directly (Lane et al., 2003) or through recruitment of activated target cells (Reinhart, 2003). In turn, HIV may induce IP-10 expression through its gene products, including gp120 and *tat* (Wetzel et al., 2002; Izmailova et al., 2003). CXCR3 was shown to be upregulated on CSF T cells (Shacklett et al., 2004), which is consistent with IP-10 serving as one of the major chemotactic signals attracting these cells into the CSF (Kivisakk et al., 2002; Trebst et al., 2003; Shacklett et al., 2004). Since expression of both IP-10 message and CXCR3 has been identified on astrocytes and neurons, this chemokine has also been suggested to be a mediator of neuronal injury (Xia and Hyman, 1999; Xia et al., 2000; Liu and Lane, 2001; Sui et al., 2004). The receptor has actually been shown to serve as an intermediary in the neurotoxicity of the HIV *Nef* gene product (van Marle et al., 2004). In addition, IP-10 treatment of brain cell cultures induced neuronal apoptosis in vitro, and both IP-10 and its receptor were detected in neurons of retrovirus-infected macaques with encephalitis in association with markers of apoptosis (Xia et al., 2000; Sui et al., 2004).

In addition to supporting a role for IP-10 in amplification of local infection, CSF studies indicate that this chemokine is also associated with pleocytosis. High CSF levels of this chemokine correlated with increased chemotactic activity of CSF for activated T cells, while chemotactic activity of CSF was neutralized by anti-IP-10 antibody (Kolb et al., 1999). Also in vivo, CSF levels of IP-10 were found to associate closely with both mononuclear pleocytosis and CSF HIV concentrations (Lahrtz et al., 1998; Kolb et al., 1999; Cinque et al., 2005). In contrast, CSF IP-10 concentrations did not correlate with plasma IP-10 or HIV RNA, or CD^4 counts. CSF levels of IP-10 were increased in subjects with acute or asymptomatic HIV infections or ADC, although less frequently in those with more advanced infection, with or without CNS opportunistic diseases, except in patients with CMV encephalitis (Cinque et al., 2005).

In patients receiving HAART, the CSF levels of this chemokine decreased significantly in parallel with the virological response and decreases in CSF cell counts. The decrease was more marked in ADC than in neurologically asymptomatic patients, who had lower baseline concentrations (Cinque et al., 2005). However, as with MCP-1, IP-10 concentrations were found to increase in the CSF after one year of successful HAART (Gisolf et al., 2000b). Increasing values of IP-10 were also detected in subjects stopping HAART, in parallel with increasing HIV RNA levels and CSF cell counts (Cinque et al., 2005).

In summary, measuring IP-10 in CSF is not a specific marker of ADC and of little immediate utility for clinical management. Nevertheless, this marker proved to be useful to study in vivo the events regulating traffic of CSF cells (Shacklett et al., 2004).

17.3.1.3.2.3. Macrophage inflammatory protein-1α (MIP-1α)/CCL3, MIP-1β/CCL4, RANTES/CCL5

MIP-1α, MIP-1β and RANTES are CC chemokines that can attract mononuclear cells into the CNS. They are expressed by microglial and other resident CNS cells, including astrocytes and neurons. Their main receptors, CCR3 (for MIP-1α and MIP-1β) and CCR5 (for RANTES), are expressed by neurons and microglial cells (Sanders et al., 1998). CCR5 and, to a lesser extent, CCR3 are coreceptors used by HIV to enter target cells in conjunction to CD^4, and MIP-1α and MIP-1β have actually been shown to inhibit infection in microglial cell cultures (Kitai et al., 2000). In fact, these chemokines are induced in human monocyte cultures in response to HIV infection (Schmidtmayerova et al., 1996) or *tat* (D'Aversa et al., 2004), and RANTES is upregulated in astrocytic cells by HIV infection (Cota et al., 2000). Notably, these chemokines have also been shown to protect neurons from gp120-induced cell death (Meucci et al., 1998). Since MIP-1α and MIP-1β are potent chemoattractants for both monocytes and lymphocytes, their increased expression might influence the trafficking of leukocytes during HIV infection.

These chemokines were expressed at high levels in brains with HIV or SIV encephalitis (Sasseville et al., 1996; Schmidtmayerova et al., 1996), but only minimally in tissues from HIV-infected patients without encephalitis (Sanders et al., 1998). MIP-1α and RANTES were abundantly expressed in normal microglia and in microglial cells surrounding multinucleated giant cells, while MIP-1β was expressed to a lesser extent, predominantly in astrocytes (Sanders et al., 1998; Vago et al., 2001).

Unlike MCP-1, MIP chemokines and RANTES were infrequently found in the CSF from HIV-infected patients. Discordant results were found between studies with respect to the proportion of HIV-positive patients with detectable CSF levels, so that reported values have ranged from less than 20 to 80% for RANTES, and from undetectable to 38% for MIP-1α or to 66% for MIP-1β (Bernasconi et al., 1996; Kelder et al., 1998; Kolb et al., 1999; Letendre et al., 1999). Also the association between MIP chemokine levels and HIV infection of CNS was not consistent between studies (Bernasconi et al., 1996; Kelder et al., 1998; Kolb et al., 1999). The rate of patients with detectable RANTES, as well as the CSF levels of this chemokine, seemed to be higher in patients with ADC, compared to those with no or mild neuropsychological impairment (Kelder et al., 1998; Letendre et al., 1999). In one study, however, non-demented patients had the highest CSF levels of MIP-1α, and these correlated inversely with CSF HIV RNA levels (Letendre et al., 1999).

Overall, the role for these chemokines in HIV infection of the CNS is less clear than that of other chemokines. While in vitro and tissue studies point to increased expression of these chemokines in response to HIV infection, it is unclear whether this response is sustained and whether it may contribute to the brain insult. The uncertainty of their role, together with difficulties in detecting these chemokines in the CSF, do not support their use of monitoring HIV CNS infection in the clinic.

17.3.1.3.2.4. Fractalkine/CX3CL1

Fractalkine (CX3CL1) is a recently identified CX3CL chemokine constitutively expressed in the CNS, principally by neurons and to a lesser degree by astrocytes. Its receptor, CX3CR1, is unique for fractalkine and is expressed on neurons, astrocytes and monocyte-macrophage cells, allowing for different types of interactions. This chemokine is involved in regulation of cellular injury and survival. It exists in two forms, as a membrane protein, which has adhesion properties, and a cleaved soluble protein. Both forms induce chemotaxis and activation of leukocytes, macrophages and microglial cells (Harrison et al., 1998; Cotter et al., 2002).

Fractalkine production is augmented in astrocyte, macrophage and neuronal cell cultures upon treatment with HIV, its viral products or pro-inflammatory cytokines (Pereira et al., 2001; Erichsen et al., 2003). In neurons, this increase was induced by cell treatment with TNF-α and neurotoxic molecules, such as glutamate and quinolinic acid (Guillemin et al., 2003), and seemed to result from membrane cleavage rather than upregulation of fractalkine expression. In addition, fractalkine potently induced migration of monocytes across a model of brain barrier (Tong et al., 2000). Thus, brain infection and neuronal injury might induce fractalkine production, which might further attract phagocytic cells to the site of injury. Neurons have been proposed as the primary source of fractalkine, at least in the initial phases following neuronal injury. In this view, neurons may themselves be regarded as effector cells producing chemokines and regulating the activity of macrophage and microglial cells (Erichsen et al., 2003). On the other hand, this chemokine seems also to promote neuron survival, as it was shown to protect cultured neurons from the neurotoxicity induced by gp120, *tat* and platelet activating factor (PAF) (Meucci et al., 2000; Tong et al., 2000). Consistent with in vitro findings, fractalkine expression was increased in the brain of AIDS patients with ADC or HIV-E, with immunoreactivity detected in astrocytes (Pereira et al., 2001) and neurons (Tong et al., 2000). Conversely, fractalkine receptor,

CX3CR1, localized to both neurons and microglia in CNS tissue from patients with ADC (Tong et al., 2000).

CSF levels of this chemokine were found to be increased in cognitively impaired patients compared to neuroasymptomatic patients and HIV negative controls (Erichsen et al., 2003; Sporer et al., 2003). CX3CL1 levels neither correlated with CSF or serum HIV load nor with other CSF parameters (Sporer et al., 2003), which is consistent with the production of this chemokine by neurons, rather than infected macrophage and microglial cells.

Although this chemokine has not yet been extensively studied in HIV infection of the CNS, preliminary observations indicate that it may be considered a neuroimmune modulator, because of both chemoattracting and neuroprotective functions. The possible use of this chemokine as CSF marker of ADC requires additional studies.

17.3.1.4. Neopterin

Neopterin is a low molecular product of the guanosine triphosphate metabolism, mainly produced by macrophages and other mononuclear phagocytes, including microglial and dendritic cells. Its production is triggered by IFN-γ, costimulated by other cytokines. Neopterin concentrations increase significantly in body fluids during activation of cell-mediated immunity, reflecting disease extent and activity. Thus, increased concentrations of this marker can be found in a variety of conditions, including infections, autoimmune disorders, tumors and allograft rejection (Wirleitner et al., 2005).

In HIV infection, neopterin concentrations measured in urine or serum is a well-established marker of systemic disease progression and useful to monitor the efficacy of anti-HIV therapy. Increased formation of neopterin occurs during acute HIV infection, before antibody seroconversion. After this stage, serum and urine neopterin closely correlate with systemic virus load (Tsoukas and Bernard, 1994; Wirleitner et al., 2005). In addition to the role of neopterin as a biomarker of HIV disease, neopterin derivatives might be involved in HIV immunopathogenesis, by activating redox-sensitive transcription factors, inducing pre-apoptotic signaling and promoting the release of cytokines and adhesion molecules (Wirleitner et al., 2005).

Increased CSF levels of neopterin have been found in several CNS diseases, including bacterial and viral infections, multiple sclerosis and several neurodegenerative diseases (Fredrikson et al., 1987; Furukawa et al., 1992; Hagberg et al., 1993). Elevated levels of neopterin were found in patients with primary HIV infection and in subjects in the early stages of chronic HIV infection (Sonnerborg et al., 1989a). Although CSF concentrations of neopterin increased in parallel with progression of infection (Sonnerborg et al., 1989a), the highest CSF levels were found in patients with ADC and opportunistic CNS infections (Griffin et al., 1991; Hagberg, 1995). These findings mirror neopterin concentrations in brain tissue, where HIV encephalitis was associated with elevated subcortical levels of neopterin and CMV encephalitis with enhanced levels in the frontal cortex (Shaskan et al., 1992a). Among patients with ADC, CSF levels of neopterin correlated with ADC stage (Fuchs et al., 1989; Brew et al., 1990). High neopterin levels were associated with a higher risk of developing ADC (Brew et al., 1996).

CSF neopterin decreased promptly following initiation of antiretroviral therapy, including zidovudine monotherapy (Gisslen et al., 1997b) and HAART (Abdulle et al., 2002). However, it was observed that levels fell to normal in only half of the patients following two years of treatment with CSF virological suppression (Abdulle et al., 2002). These findings, in line with the observation of persistent MCP-1 levels in long-term treated patients, suggest that immune activation persists in the brain despite effective anti-HIV treatment, which might reflect residual low level replication of HIV in the brain.

17.3.1.5. β-2-Microglobulin (β2M)

β2M is the small subunit of the major histocompatibility class I molecule. It is constitutively expressed on the surface of all nucleated cells, with the exception of neurons, and with the highest density on lymphocytes. β2M is required for the transport of class I heavy chains from the endoplasmic reticulum to the cell surface. It is present in small amounts in serum, CSF, and urine of healthy people. As a result of increased cell turnover, its levels are increased in body fluids of patients with inflammatory conditions and lymphoproliferative malignancies.

Serum or plasma β2M is a reliable marker of early systemic HIV infection and disease progression. Most HIV-infected subjects have increased levels within the first months of seroconversion, which tend to remain stable through the chronic phases of infection. Levels of β2M correlate inversely with the CD^4 cell counts and, before the introduction of viral load measurements, elevated concentrations of this marker was the most powerful predictor of AIDS in several clinical cohort studies (Tsoukas and Bernard, 1994).

CSF β2M levels are above normal values in patients with acute or asymptomatic HIV infection (Enting et al., 2000; Enting et al., 2001). Higher concentrations are found in patients with ADC and these increase with severity of dementia (Brew et al., 1992; McArthur et al., 1992). Comparing CSF β2M values between ADC patients and HIV-asymptomatic controls, a cutoff value of 3.8 mg/l provided a 44% sensitivity for ADC diagnosis, with specificity of 90%, and positive predictive value of 88% (McArthur et al., 1992). Indeed, until measurement of CSF HIV viral load became available, β2M levels were regarded as the best CSF marker of this condition (Brew et al., 1992). However, high CSF β2M concentrations can also be found in patients with opportunistic CNS infections or lymphoma (Lazzarin et al., 1992). Like neopterin, β2M seems to be a predictive marker of ADC, at least in HIV-infected neurologically asymptomatic subjects with less than 200 CD^4 cells in whom high CSF concentrations were associated with an increased risk of ADC development within the following six months (Brew et al., 1996).

After 6 weeks of zidovudine therapy, levels of β2M declined in both CSF and serum (Elovaara et al., 1994). CSF β2M levels fell after antiretroviral treatment with two NNRTIs (Enting et al., 2000) and HAART (Abdulle et al., 2002) in parallel with CSF VL. Unlike MCP-1 and neopterin, β2M concentrations in CSF remained normal in all treated patients up to two years of treatment (Enting et al., 2000; Abdulle et al., 2002). The predictive value of β2M was recently reassessed in patients receiving HAART, and this marker actually failed to predict neuropsychological worsening. Also, neuropsychological abnormalities were not associated with altered β2M CSF levels, suggesting that neuropsychological impairment in HAART-treated patients might be associated with biologically inactive disease (Cysique et al., 2005).

17.3.1.6. Proteases

The human protease systems include the families of matrix metalloprotease (MMPs) and serine proteases. These systems are classically and primarily involved in extracellular proteolysis and exert important functions in maintaining homeostasis during development and adulthood. Recently, human proteases have been shown to be involved in a number of inflammatory, degenerative and neoplastic diseases of the CNS. Molecules belonging to the protease systems have been proposed as CSF markers of neurological disease.

17.3.1.6.1. Matrix metalloproteases

MMPs consist of a family of over 20 enzymes identified to date. These proteins originate from the sequential proteolysis and consequent activation of latent pro-MMPs. Several MMPs, together with their endogenous inhibitors, the tissue inhibitors of metalloproteinase (TIMPs 1–4), are expressed within the CNS by both resident glial and infiltrating inflammatory cells. MMPs and TIMPs are important in maintaining CNS homeostasis through remodeling of the extracellular matrix. An imbalance between TIMPs and MMPs is associated with many pathologic conditions (Lukes et al., 1999; Gardner and Ghorpade, 2003).

MMPs can be activated by several cytokines and chemokines involved in the pathogenesis of ADC (Stuve et al., 1996; Conant et al., 1999) and were also induced in macrophages following stimulation with brain-derived *tat* sequences from ADC patients (Johnston et al., 2001). MMPs secreted by HIV-infected monocytes have been shown to degrade components of the blood–brain barrier (Dhawan et al., 1992; Lafrenie et al., 1996). Recently, a novel mechanism of HIV-induced neurodegeneration was proposed, which involves MMP-2 cleavage of the stromal cell-derived factor 1-α (SDF-1), a chemokine overexpressed by astrocytes during HIV infection, resulting in the formation of a highly neurotoxic protein (Zhang et al., 2003). In encephalitic brain tissue, MMPs were expressed within perivascular and parenchymal macrophages and multinucleated giant cells (Ghorpade et al., 2001). Reduced levels of TIMP-1 were also found in brain tissues of ADC patients in association with increased MMP-2 (Suryadevara et al., 2003).

MMP-2, MMP-7 and MMP-9, which have been shown to exert a specific proteolitic action on blood–brain barrier proteins, including laminin, entactin and collagen type IV, have been found at increased concentrations in the CSF of patients with ADC and CNS opportunistic infections (Sporer et al., 1998; Conant et al., 1999; Liuzzi et al., 2000). Elevated CSF levels of pro-MMP-2 and pro-MMP-7 were found in patients with ADC compared to neurologically asymptomatic or seronegative controls. By zymography, a sensitive functional assay, MMP-2 or pro-MMP-9 activity was more frequently detectable in the CSF of patients with ADC than in the same control groups (Sporer et al., 1998; Conant et al., 1999). MMP levels also correlated with CSF markers of barrier leakage, such as elevated mononuclear cell counts and CSF/serum albumin ratios (Conant et al., 1999; Liuzzi et al., 2000).

17.3.1.6.2. Urokinase plasminogen activation system

The urokinase plasminogen activator (uPA), or urokinase, belongs to the family of serine proteases. The uPA system is composed of uPA, its receptor uPAR, its inhibitors plasminogen activator inhibitor (PAI-1) and protease nexin-1 (PN-1) and various binding molecules. The classic and more defined physiological function of the uPA system is extracellular proteolysis, involving tissue fibrinolysis and tissue remodeling. Activated uPA, which originates from the conversion of its inactive form (pro-uPA) following binding to the receptor uPAR, converts plasminogen into plasmin. Plasmin is able to break down components of the extracellular matrix, including proteins of the blood–brain barrier, either directly or through the activation of MMPs. More recently, the uPA system has also been shown to be involved in the regulation of cell migration and intracellular signaling. Through dysregulation of these functions, the uPA system has been implicated in the pathogenesis of a number of pathological conditions, including human cancers, neurodegenerative diseases and infections (Mondino and Blasi, 2004).

HIV was initially shown to modulate uPAR expression on monocytes and T-cells (Nykjaer et al., 1994; Speth et al., 1998). More recently, uPA, inactive pro-uPA or its aminoterminal have been demonstrated to inhibit virus production in both acutely and chronically HIV-infected cells (Alfano et al., 2002). Retrospective clinical studies have shown that high serum levels of suPAR are a major negative prognostic factor in HIV infection independently of the clinical stage, viremia levels and CD^4 cell counts (Sidenius et al., 2000). uPAR was found to be overexpressed in the CNS of patients with HIV-E compared to controls. Immunochemical studies showed a prevalent expression of uPAR in macrophages and microglial cells, with predominantly membrane positivity, and a partial co-localization with HIV p24 antigen (Cinque et al., 2004). In contrast, uPA was detected only on rare cells in most of the cases (Sidenius et al., 2004).

Elevated CSF levels of uPA, soluble uPAR (suPAR) or PAI-1 have been found in patients with different CNS diseases, including bacterial meningitis, viral infections and various neoplastic and neurodegenerative diseases (Garcia-Monco et al., 2002; Winkler et al., 2002). Among patients with HIV infection, CSF levels of suPAR were higher in ADC and CNS opportunistic infections than in neurologically asymptomatic patients, with the highest levels found in patients with ADC or CMV encephalitis (Cinque et al., 2004; Sporer et al., 2005). In neurologically impaired patients, CSF suPAR levels correlated with CSF HIV RNA, but not with plasma suPAR, supporting the intrathecal origin of this molecule. Also CSF uPA levels were higher in ADC than HIV negative and asymptomatic HIV-infected subjects (Sidenius et al., 2004). Although uPA and uPAR concentration were intercorrelated, the differences of uPA levels between patients with or without ADC appeared less marked than those observed for suPAR. Levels of the uPA inihibitor, PAI-1, were also measured in the CSF of HIV-infected patients, but these were not higher in patients with ADC compared to the others (Sporer et al., 2005).

CSF suPAR levels decrease significantly following 6–12 weeks of HAART in parallel with decreasing HIV RNA, both in ADC and neuroasymptomatic patients (Cinque et al., 2004).

17.3.2. Neurotoxic host factors

HIV-associated neuronal damage is thought to occur via an indirect pathway in which virus-infected macrophages and other uninfected brain cells release substances with neurotoxic properties. Candidate substances include proteins encoded by the viral genome (such as gp120, gp41, *tat* and *vpr*) and endogenous products including cytokines, e.g., TNF-α, platelet activating factor (PAF), quinolinic acid and glutamate (Kaul et al., 2001). Measuring neurotoxins in the CSF provides support to experimental findings of neurotoxic activity in culture systems. Furthermore, in a few cases, neurotoxins are potential biomarkers for clinical use.

17.3.2.1. Quinolinic acid and tryptophan

Quinolinic acid is a product in the kynurenine pathway, by which the aminoacid tryptophan is catabolized through the enzyme indoleamine-2,3-dioxygenase to L-kynurenine and then to quinolinic acid. Tryptophan is, in turn, involved in the synthesis of proteins and of monoamine neurotransmitters.

Quinolinic acid is synthesized within normal brain, but its formation is increased during immune activation, produced in large quantities by activated macrophages and microglia (Smith et al., 2001). It is an endogenous brain excitotoxin, i.e. a substance that damages neurons through its overactivity and upon binding to certain cell receptors, and thus a putative mediator of brain injury in ADC (Heyes et al., 1991). In vitro, chronic exposure of human neurons to quinolinic acid results in neuronal changes (Kerr et al., 1998), and inhibition of quinolinic acid production reduces neurotoxicity of HIV-infected macrophage supernatants (Kerr et al., 1997). Quinolinic acid likely contributes to neuronal injury through its activity as a N-methyl-D-aspartate (NMDA) receptor agonist resulting in alterations of intracellular calcium (Stone et al., 2003). In additon to its neurotoxic properties, quinolinic acid has been shown to induce apoptosis of astrocytes (Guillemin et al., 2005). Also, it was shown to upregulate astrocyte production of MCP-1 and other chemokines, and to increase expression of CXCR4 and CCR5, thus theoretically also promoting brain inflammation (Guillemin et al., 2003). Concentrations of quinolinic acid in brain tissues were substantially increased in HIV-infected patients compared to controls, although the correlation with severity of HIV encephalitis was not clear (Achim et al., 1993).

Elevations in quinolinic acid in the brain and CSF have been detected during primary HIV infection (Heyes et al., 1992a; Smith et al., 1995; Lane et al., 1996), and neurologically asymptomatic HIV seropositive patients showed higher CSF concentrations compared to non-HIV-infected controls (Heyes et al., 1992a). CSF quinolinic acid concentrations were higher in patients with ADC than asymptomatic HIV infection (Heyes et al., 1991) and correlated with neuropsychological deficits and ADC severity (Heyes et al., 1992a). By comparing CSF to postmortem findings, CSF quinolinic acid levels have been shown to be highly representative of parenchymal values (Heyes et al., 1998). Increased CSF quinolinic acid concentrations also correlated with higher CSF HIV RNA levels (Heyes et al., 2001; Valle et al., 2004), pleocytosis and increased CSF protein concentrations (Valle et al., 2004). Higher levels were also associated with brain volume loss, as quantified by MRI, in regions selectively vulnerable to excitotoxic injury, supporting the neurotoxic role of this marker in ADC (Heyes et al., 2001). Like the other inflammatory markers, however, CSF quinolinic acid was also found to be elevated in patients with opportunistic CNS diseases (Heyes et al., 1992a).

Of note, increases in CSF quinolinic acid in HIV infection and other inflammatory CNS disease were shown to be accompanied by proportional increases in CSF kynurenine and reductions in CSF tryptophan (Heyes et al., 1992b; Gisslen et al., 1994). Also, tryptophan concentrations correlated inversely with markers of immune activation, such as CSF neopterin and cell counts (Fuchs et al., 1990; Heyes et al., 1992a). These responses are consistent with induction of indoleamine-2,3-dioxygenase, the first enzyme of the kynurenine pathway which converts tryptophan to kynurenine and quinolinic acid. Indeed, studies in retrovirus-infected macaques have shown increased activity of this enzyme and increased concentrations of quinolinic acid in the brain (Heyes et al., 1992b). On the other hand, no significant change was observed in CSF or blood levels of monoamine neurotransmitters or their metabolites, including the serotonin metabolite 5-hydroxyindoleacetic acid (5-HIAA) and the dopamine metabolite homovanillic acid (HVA) (Larsson et al., 1989; Gisslen et al., 1994).

Following HAART, CSF quinolinic acid levels decreased with a time course similar to the kinetics of CSF HIV decay (Valle et al., 2004). Conversely, CSF tryptophan concentrations increased during treatment with anti-HIV drugs, as noted in patients receiving zidovudine (Fuchs et al., 1996). These observations indicate that anti-HIV treatment contributes to a gradual normalization of tryptophan metabolism in patients with HIV infection, likely through decreasing immune activation.

17.3.2.2. Prostaglandins

Prostaglandins are lipid substances that are found in virtually all tissues and organs. They exert a wide variety of actions, including mediating inflammation. They are synthesized in the cells from arachidonic acid and the other essential fatty acids, gamma-linolenic acid and eicosapentaenoic acid, by either the cyclooxygenase or the lipoxygenase pathway. The cyclooxygenase pathway produces the biological active compounds thromboxane, prostacyclin and prostaglandin D, E and F. The lipoxygenase pathway synthesizes leukotrienes.

Disturbances in the arachidonic acid cascade have been proposed to play a role in HIV pathogenesis. The in vitro interaction of HIV-infected monocytes with astroglia leads to release of high levels of leukotrienes, in addition to cytokines and PAF. These molecules lead to neuronotoxicity, suggesting that they may mediate neuronal dysfunction in ADC (Genis et al., 1992). In additon, cyclooxygenase-1 was found to be expressed in the brains of patients with ADC, at slightly higher levels than in non-demented patients (Griffin et al., 1994).

Increased CSF levels of prostaglandin E2, F2-α and thromboxane B2, all products of the cyclooxygenase pathway of arachidonic acid metabolism, have been found in patients with ADC. In contrast, levels similar to controls were found for leukotriene C4, a product of the lipoxygenase pathway (Griffin et al., 1994). The increase of CSF prostaglandin E2 levels was associated with severity of dementia and correlated with CSF neopterin and β2M concentrations. These initial observations, however, were not followed up by further evidence supporting the use of prostaglandins for monitoring ADC. In fact, these markers have not found widespread use in other CNS diseases. Recently, measuring molecules related to prostaglandins, such as F2-isoprostanes (isomers of prostaglandin F2 produced exclusively from free-radical-catalyzed peroxidation of arachidonic acid), have been promoted as new potential biomarkers of oxidative brain damage (Montine et al., 1999).

17.3.2.3. Glutamate

L-glutamate is the major excitatory neurotransmitter in the CNS. It binds to multiple receptors including NMDA receptors. Glutamate can also act as a potent neurotoxin leading to increased cell calcium conductance and cell death, so-called excitotoxicity (Greenamyre and Porter, 1994).

Both increased release from glutamatergic neurons and impaired reuptake of glutamate have been shown in astrocytic cultures treated with HIV gp120 (Vesce et al., 1997). Also, increased glutamate levels were produced by HIV-infected macrophages (Jiang et al., 2001). Increased glutamate formation might be mediated by the activation of phosphate-activated mitochondrial glutaminase, the primary enzyme for the production of glutamate that catalyzes the conversion of glutamine to glutamate (Zhao et al., 2004).

Whether CSF levels of glutamate are associated with ADC and can thus be used as a biomarker for disease is not resolved. A fivefold increase in glutamate CSF levels was initially reported in ADC patients compared to HIV-infected patients without dementia and healthy controls. Glutamate levels also correlated with cognitive decline and brain atrophy as assessed by computed tomography scan. In addition, normal CSF glutamate was found in patients with Alzheimer's disease, pointing to the specificity of this marker for HIV-related dementia (Ferrarese et al., 1997, 2001). However, these findings were not confirmed by subsequent studies, which found no difference in glutamate concentrations between these same patient groups (Espey et al., 2002). It is possible that these discrepant findings relate to technical differences in both processing of samples before storing — it has been suggested that glutamate activity in CSF may be degraded due to enzymatic activity in unprocessed samples — and methods used for high performance liquid cromatography analysis (Ferrarese et al., 2001; Espey et al., 2002).

17.3.2.4. Platelet activating factor

PAF is a phospholipid inflammatory agent with a range of biological activities; its original naming relates to its main effects on platelet aggregation. PAF both stimulates and is produced by microglia (Jaranowska et al., 1995). It contributes to inflammatory responses in the brain and is upregulated in various CNS diseases (Lindsberg et al., 1990). PAF has also been shown to mediate NMDA excitotoxicity and can kill neurons at high concentrations (Ogden et al., 1998).

PAF was shown to be overexpressed in HIV-infected monocytes upon stimulation (Nottet et al., 1995) and produced neuronal death in cell culture. Because its effect is partially blocked by NMDA receptor antagonists, PAF has been considered another HIV-induced endogenous excitotoxin (Gelbard et al., 1994). The neurotoxic effect of PAF could also be mediated by quinolinic acid, since this molecule is produced by HIV-infected macrophage on PAF stimulation (Smith et al., 2001).

PAF has been detected at high levels in CSF of HIV-infected patients with neurological impairment. However, it has also been found in CSF of patients with other CNS or systemic diseases accompanied by neurological dysfunction (Gelbard et al., 1994). Although CSF studies are limited, the wide range of physiological functions of PAF and its involvement in an array of pathological conditions suggests that it may not prove useful in clinical management of ADC.

17.3.2.5. Nitric oxide

Nitric oxide (NO) is a free radical synthesized by the enzyme nitric oxide synthase (NOS) in a number of different cell types and for a variety of biological functions. Two major isoforms of NOS have been identified: a constitutive isoform found in neurons and endothelial cells and an inducible isoform (iNOS), found in macrophages, fibroblasts and hepatocytes. The latter produces NO in relatively large amounts in response to inflammation and acts in a host defensive role through its oxidative toxicity. Because of its very short half-life ($T_{1/2} = 4$ seconds), NO is immediately converted in vivo to either nitrate (NO_3^-) or nitrite (NO_2^-). Because the ratio of nitrite to nitrate is variable and unpredictable, the best index of NO production is the total of both nitrate and nitrite (NO metabolites).

There is evidence that NO formation is involved in neurodegenerative disorders (Sarchielli et al., 2003), and it has also been proposed to contribute to HIV-induced neurologic disease by mediating neurotoxicity. In vitro, iNOS upregulation and NO formation are induced in response to stimulation with IL-1β or HIV gp41 (Adamson et al., 1996; Liu et al., 1996). In addition, cocultures of HIV-infected macrophages and astrocytes released NO differently from uninfected cocultures (Bukrinsky et al., 1995), suggesting that interactions between HIV-infected macrophages/microglia and astrocytes are critical in the induction of factors that lead to neuronal injury. Direct neurotoxic effects of NO were enhanced by reacting with superoxide anion to form peroxynitrite, and increased peroxynitrite activity was found in ADC compared with non-demented patients (Boven et al., 1999). In turn, the synthesis of NO resulted in a significant increase in HIV replication (Blond et al., 2000). Also, the severity and rate of ADC progression correlated with levels of iNOS in brain tissues (Adamson et al., 1999). Altogether these findings are in agreement with a putative model by which HIV influences the expression of iNOS and NO formation, leading to neuronal dysfunction and further upregulation of HIV replication in the brain.

CSF studies only partially corroborated in vitro and tissue findings. In SIV-infected macaques, increased levels of NO metabolites in the CSF correlate with iNOS detection in postmortem brain tissue (Lane et al., 1996). However, the association between CSF levels of NO metabolites, HIV infection and ADC was less clear in human studies. No significant increases in the CSF levels of NO metabolites were initially observed in patients with HIV infection, nor in other neurodegenerative diseases, including Huntington's, Alzheimer's disease and amyotrophic lateral sclerosis (Milstien et al., 1994). Another study reported elevated concentrations of NO metabolites in the CSF and serum of patients with AIDS and CNS complications, compared to non-HIV-infected patients with inflammatory and non-inflammatory neurological disorders (Giovannoni et al., 1998). A progressive decrease of NO metabolites in the CSF was also reported in ADC patients following HAART in parallel with CSF viral load suppression and decline in other CSF immune activation markers (Gendelman et al., 1998). In summary, it is not yet documented that this measure is consistently elevated in ADC and thus provide a clinically useful disease biomarker.

17.3.3. Markers of apoptosis

Evidence from several sources suggests that apoptosis, or programmed cell death, is an important mechanism of CNS injury in HIV infection. Thus, HIV infection induces apoptosis of astrocytes and neurons in vitro, and apoptosis of astrocytes, neurons and endothelial cells is frequently detected in brains of HIV-infected patients (Sabri et al., 2003; Li et al., 2005). Moreover, the proportion of apoptotic astrocytes is greater in subjects with rapidly progressive dementia than in slow progressors or non-demented HIV patients (Thompson et al., 2001). Apoptosis is characterized by changes in an array of intracellular and extracellular signaling pathways, which include induction of certain soluble factors. Important molecules for the initiation and execution of the apoptotic program are the receptor Fas and caspases, a group of intracellular cysteine proteases.

17.3.3.1. Fas and Fas ligand (FasL)

Fas, also named CD^{95} and Apo-1, is a member of the TNF/nerve growth factor superfamily. It is located on the cell surface, and can regulate inflammatory responses (Nagata and Golstein, 1995). Fas is not detected in the brain of healthy subjects, but it has been shown to be expressed in various CNS diseases, including cancer, ischemia and Parkinson's disease (Matsuyama et al., 1994; Mogi et al., 1996). Cells expressing Fas can undergo apoptosis upon binding to its endogenous ligand, FasL (Ju et al., 1995). Both Fas and FasL exist in membrane-bound and soluble forms. The soluble form of the Fas receptor, sFas, is generated by activated cells and regulates cell death by inhibiting the binding between Fas and FasL on the cell surface (Cheng et al., 1994). FasL is expressed on activated T cells and its soluble form, sFasL, is released by cleavage through metalloproteinases. It serves as a death-inducing element, but under experimental conditions is less potent than FasL itself (Tanaka et al., 1998).

In vitro, HIV-infected macrophages were shown to release soluble FasL, which triggered apoptosis of uninfected astrocytes (Aquaro et al., 2000), whereas Fas expression was found to be upregulated in post-mortem brain tissues (Elovaara et al., 1999). It is thus hypothesized that in ADC HIV-infected macrophages and other inflammatory cells express FasL, which binds to reactive astrocytes, leading to apoptosis of these cells.

Both sFas and sFasL were found to be elevated in the CSF of patients with ADC compared to those without this condition (Sabri et al., 2001; Towfighi et al., 2004). Subjects with moderate/severe dementia showed higher CSF sFas than patients with mild forms (Towfighi et al., 2004). Soluble FasL, but not sFas, correlated with CSF HIV RNA levels (Sabri et al., 2001), whereas the correlation with blood CD^{4+} cell counts was less clear (Sporer et al., 2000; Sabri et al., 2001; Towfighi et al., 2004). These findings support the concept that formation of sFas and sFasL is strictly associated with local CNS infection. A decrease in the CSF level of sFas was observed following a few months of HAART in parallel with decreasing CSF HIV load. Also sFasL levels tended to decrease over time, although less markedly than sFas (Sabri et al., 2001). Based on these preliminary experiences, sFas and sFasL in CSF appear promising biomarkers of ADC. Extension of initial findings to larger populations, including patients with opportunistic CNS infections, different disease stages and during HAART, appears worth pursuing.

17.3.3.2. Sphingomyelin and ceramide

Sphingomyelin and ceramide are lipidic components of the cell membrane. Sphingomyelin belongs to the class of sphingolipids, and is the most abundant of these in the brain. Ceramide is produced from the degradation of sphingomyelin by sphingomyelinases. In addition to their structural role, sphingolipids and the products of their metabolism exert important signaling functions. Ceramide regulates a variety of events, including cell differentiation and survival, and is an important inducer of apoptosis in various cell types, including glial and neuronal cells (Brugg et al., 1996).

Sphingomyelin and ceramide were found to be increased in the brain of ADC patients, compared to neurologically asymptomatic HIV-positive and uninfected control patients (Haughey et al., 2004). Increased cellular levels of sphingomyelin and ceramide were also generated in cultured neurons following exposure to gp120 and *Tat* (Haughey et al., 2004), in association with sphingomyelinases activation (Jana and Pahan, 2004). Thus, experimental findings suggest that HIV may promote a lipid imbalance in neural cells, resulting in an overproduction of ceramide and consequent cellular dysfunction and death. Interestingly, reactive oxygen species were found to be involved in the activation of sphingomyelinases, through gp120 coupling with CXCR4, induction of NADPH oxidase and production of superoxide radicals (Jana and Pahan, 2004). Cellular survival is likely to be affected by the redox status of lipids in the cell membrane (Power and Patel, 2004), and there is additional experimental evidence that reactive oxygen species may cause neurodegeneration in ADC (Viviani et al., 2001; Brooke et al., 2002). Indeed, sphingomyelin and ceramide have recently been proposed as markers of oxidative stress in ADC (Sacktor et al., 2004).

Recently, the CSF levels of sphingomyelin and ceramide have been measured in HIV-infected patients, together with concentrations of derivative products of 4-hydroxinonenal (HNE), a highly reactive compound generated from lipid peroxidation that exerts toxic effects on cell components. CSF levels of sphingomyelins, ceramides and HNE were higher in ADC patients than in HIV-infected or HIV-negative control subjects. Levels observed in neurologically asymptomatic patients did not differ from HIV-negative subjects. Also, patients with moderate to severe dementia had higher sphingomyelin and ceramide levels than patients with mild disease (Haughey et al., 2004). Higher concentrations of HNE and ceramide were found in HAART-treated patients regarded to have active dementia, compared to those with inactive disease (Sacktor et al., 2004). Further studies are required to assess the association of these promising markers with ongoing damage of neuronal cells and other biomarkers of ADC activity.

17.4. Markers of CNS cell damage

The markers discussed above all relate to HIV and host immune and inflammatory responses, and, hence, might be perturbed by CNS and meningeal infection, but do not necessarily indicate brain injury. Thus, it is also important to assess markers that signal an impact of infection on this critical target organ. Indeed, it would be both pathogenetically and clinically useful to have brain-specific injury markers that would predict ADC, aid in its diagnosis and follow the effects of treatment.

Common to other neuroinflammatory and neurodegenerative diseases, three steps in the pathology of HIV disease of the CNS can be segregated: first, initial microglial and astrocytic activation; second, neuro-axonal degeneration; and third, astrogliosis (Charcot, 1868). Based on this general model, biomarkers of macrophage/microglia, astrocyte and neuron damage/activation will be reviewed here.

17.4.1. Microglial markers

Brain mononuclear phagocytes — monocytes, macrophages and microglial cells — play a central role in HIV-E and ADC. These cells serve as a reservoir for persistent viral infection, a vehicle for viral dissemination throughout the brain and a source of neurotoxic products (Gonzalez-Scarano and Baltuch, 1999). The dynamics of HIV infection result in more protracted release of progeny virus than from lymphocytes and likely underlies the chronicity of compartmentalized CNS infection. Moreover, infected and uninfected cells of this lineage release important signal and toxic molecules into the extracellular fluid and into the CSF.

17.4.1.2. Gangliosides: GD3

Gangliosides are sialic acid-containing glycosphingolipids located in the outer surface of plasma membranes. Together with sulfatide, cerebroside, sphingomyelin and other lipidic fractions, gangliosides constitute the lipidic fraction of the myelin sheath and membranes of CNS cells. The family of gangliosides consists of a number of molecules, differentiated on the basis of their composition and cell specificity. These include, among the others, GD3, preferentially expressed by macrophages and microglial cells, GM1, on astrocytes and neuronal cells, GD1b, on astrocytes, GD1a and GD2, on oligodendrocyte precursors or mature oligodendrocytes, GD2 and A2B5, on neuronal cells (Marconi et al., 2005). In general, increased CSF concentrations of these molecules reflect increased turnover of cell membranes, as a consequence of cell destruction or activation. Of additional broad interest to neurology are immune responses to these molecules that may result in autoimmune CNS injury and with the detection of specific CSF antibodies.

The ganglioside GD3 is expressed at low levels in normal adult brains. Increased amounts of GD3 have been documented in a variety of neurodegenerative disorders, including multiple sclerosis, Creutzfeldt–Jakob disease and subacute sclerosing panencephalitis (Ando et al., 1984). Although GD3 expression indicates microglia activation, it is usually observed in association with reactive gliosis. Therefore, GD3 concentrations in CSF need to be measured in combination with a specific astroglial marker, for example GFAp.

Increased CSF GD3 concentration combined with normal GFAp is likely to reflect microglial activation. This association was found in approximately one-third of asymptomatic HIV seropositive individuals (Andersson et al., 1998). It is not known whether GD3 levels are selectively increased in patients with ADC. Of note, the ganglioside GM1, which is preferentially expressed on astrocytes and neurons, was found at abnormally high concentrations in the CSF of HIV-positive individuals, irrespective of disease stage (Gisslen et al., 1997a). The other lipid constituent, sulfatide, which is mainly expressed in oligodendrocytes, was found at abnormal levels in patients with AIDS more frequently than in asymptomatic HIV-positive patients. However, there was no difference between patients with or without ADC (Gisslen et al., 1996). The potential role of the other two cell membrane lipidic fractions, ceramide and sphingomyelin, as markers of HIV brain infection is discussed above together with markers of apoptosis.

17.4.2. Astrocyte markers

17.4.2.1. S-100β

S-100 is a calcium-binding protein synthesized in astroglial cells and found both intracellularly and extracellularly in the brain. It is present in the body in different subchains, of which the β form predominates in the brain. Its physiologic function is not entirely understood (Kapural et al., 2002).

CSF S-100β has been used for many years as a marker for astrocytic proliferation. Increased S-100β concentrations in CSF are regarded as an index of the active phase of cell injury in patients with neurologic diseases, including acute multiple sclerosis exacerbations, acute encephalomyelitis, and Alzheimer's disease, presumably related to chronic astroglial activation (Massaro and Tonali, 1998; Petzold et al., 2003). In addition, CSF S-100β concentrations are also increased in the presence of tumors involving astroglial cells or their precursors, including gliomas and highly differentiated neuroblastomas (Heizmann et al., 2002). Besides serving as a marker of astrocyte damage, S-100β might also be involved in causing cerebral damage. CSF concentrations of S-100β have been detected in the micromolar range concentrations that have been shown to induce neuronal apoptosis both in vitro and in animal models (Rothermundt et al., 2003).

In HIV infection, CSF S-100β concentrations were higher in patients with ADC compared to neurologically asymptomatic patients, and among ADC patients, in those with moderate or severe disease compared to those with mild symptoms (Pemberton and Brew, 2001). Moreover, CSF S-100β levels were more elevated in individuals with rapid ADC progression compared with slow progressors (Pemberton and Brew, 2001). However, as with other markers, CSF S-100β was also elevated in opportunistic CNS complications of HIV (Green et al., 1999).

17.4.2.2. Glial fibrillary acidic protein (GFAp)

GFAp is the main constituent of the glial intermediate filament, a major structural component of astrocytes. GFAp was first isolated from multiple sclerosis plaques and subsequently found in normal astrocytes (Lucas et al., 1980). Whereas astrocytic and microglial activation describe the immediate cellular response to any CNS challenge (Streit et al., 1988; Eng and Ghirnikar, 1994; Barron, 1995), astrogliosis is defined as the fibrinoid scar that replaces lost tissue (Charcot, 1868). In multiple sclerosis and other neurodegenerative diseases, GFAp may serve as a biomarker for disease progression, probably reflecting the increasing rate of astrogliosis (Malmestrom et al., 2003).

Because astrogliosis is also one of the neuropathological hallmarks of ADC and HIV-E, measuring GFAp levels in the CSF might find an application in HIV infection. Body fluid levels of GFAp are indeed an important tool for estimating astrogliosis and astrocytic activation in vivo (Rosengren et al., 1995). However, in a preliminary report, neither CSF GFAp levels nor the frequency of increased GFAp concentrations differed between patients with ADC, opportunistic CNS infections or neurologically asymptomatic infection (Sporer et al., 2004). Although this marker did not appear sensitive for astrogliosis in HIV-infected patients with active neurological disease, it might, at least theoretically, retain a role in monitoring chronic infection, possibly indicating progression of CNS infection and astrogliosis. In this regard, and similar to what has been observed in multiple sclerosis, GFAp might also represent a marker of disability (Petzold et al., 2002).

17.4.3. Neuronal markers

17.4.3.1. Neurofilament

The neurofilament is a major structural component of axons, which consists of three components differing in their molecular size: a light chain (NFL), which forms the backbone of the protein, an intermediate (NFM) and a heavy chain (NFH). Its main function is to stabilize the axonal compartment, which is crucial to maintain morphological integrity and conduction velocity of nerve impulses. The amount of phosphorylation determines the axonal caliber, and, therefore, neurofilament and its state of phosphorylation are potential markers for neurodegeneration. Measuring NFL in CSF has provided useful in studies of neurodegeneration, particularly axonal degeneration (Rosengren et al., 1996). Healthy individuals show very low NFL concentrations in CSF, whereas markedly elevated levels have been observed following acute CNS damage, such as in viral encephalitis or acute stroke, in multiple sclerosis and in chronic neurodegenerative diseases, e.g. Alzheimer's disease and amyotrophic lateral sclerosis (Studahl et al., 2000; Blennow, 2004; Teunissen et al., 2005).

Preliminary studies in HIV-infected patients show that the highest levels are found in patients with ADC or opportunistic CNS infections. However, increased CSF concentrations of NFL were also found in some neurologically asymptomatic patients, suggesting that subclinical axonal injury may occur. In asymptomatic patients the highest levels of NFL were observed in those with lower CD^4 cell counts and/or more advanced systemic disease stage (Gisslen et al., 2006). Furthermore, NFL concentrations were found to correlate with CSF neopterin (Hagberg et al., 2000).

CSF NFL levels decreased following HAART in parallel with the decrease of HIV RNA (Gisslen et al., 2006). Interestingly, NFL increased in some patients after treatment interruption, probably reflecting a subclinical axonal injury following onset of viral replication (Gisslen et al., 2005).

17.4.3.2. Tau protein

Tau is a microtubule-associated protein expressed by CNS axons that plays an important role in microtubule assembly and stability. The hyperphosphorylated form (p-tau) is the main component of the neurofibrillary tangles in Alzheimer's disease (Buee et al., 2000). Tau proteins accumulate preferentially in neurons, but also in astrocytes and oligodendrocytes in several neurodegenerative diseases. However, different forms of phosphorylated tau are identified in neurons and glial cells (Ikeda et al., 1998). Both total tau (t-tau) and p-tau can be measured in the CSF, and raised levels indicate neuronal damage or degeneration. T-tau and p-tau, together with amyloid-β_{1-42} ($A\beta_{1-42}$), are considered highly specific markers of Alzheimer's disease (Blennow, 2004), although elevated CSF tau is also found in a number of diseases associated with brain injury, including frontotemporal dementia, Creutzfeldt–Jakob disease and stroke (Andreasen et al., 2003). Tau is also detectable in serum, but its concentration is about ten times lower than in CSF (Reiber, 2001).

A few studies have addressed the significance of CSF tau protein level in HIV infection (Ellis et al., 1998; Andersson et al., 1999; Brew et al., 2005). Although these reached somewhat different conclusions, most agreed that both t-tau and p-tau are elevated in patients with ADC compared to neurologically asymptomatic patients, and that the latter have levels similar to HIV negative healthy controls (Andersson et al., 1999; Brew et al., 2005). Of note, t-tau and p-tau concentrations in ADC were similar to those observed in patients with Alzheimer's disease, whereas CSF $A\beta_{1-42}$ was decreased in both conditions. In addition to indicating these markers as potentially useful for diagnosis of ADC, these observations also suggest that ADC may be associated with an Alzheimer's disease-like process.

17.4.3.3. 14-3-3 protein

14-3-3 proteins are major cytosolic proteins expressed in abundance in the whole body as well as in all neuronal cells. In neurons, they exist in five different isoforms: β, γ, ζ (zeta), ϵ (epsilon) and η (eta). These proteins have different regulatory functions, including signal transduction, neurotransmitter synthesis and apoptosis. Expression of 14-3-3 proteins is increased in several neurodegenerative diseases, and high CSF levels serve as a marker of Creutzfeldt–Jakob disease (Berg et al., 2003).

Three isoforms, 14-3-3 ϵ, γ and ζ, have been found to be selectively elevated in the CSF of HIV-positive patients with ADC or CMV encephalitis, compared to AIDS patients without neurological symptoms and HIV-negative patients. Also, increased frequency of 14-3-3 detection in CSF correlated with the advanced stage of systemic disease and low CD^4 cell counts (Wakabayashi et al., 2001). Of note, the isoform patterns in HIV-infected patients were different from those reported in Creutzfeldt–Jakob disease and herpes simplex encephalitis, suggesting the existence of disease-related specificity in the neuronal synthesis of these proteins. Expression of 14-3-3 proteins in CSF was also studied in the SIV-macaque model, where their presence was associated with high CSF viral loads after acute infection and high levels of viral RNA and protein in brain tissues, but not with CNS microglial/macrophage activation (Helke et al., 2005). Together these findings indicate that CSF levels of 14-3-3 protein may be a real-time marker of neuronal damage associated with viral replication in CNS, though application of this in the clinic requires further study.

17.5. Conclusions

In the research setting associations between CSF markers and HIV-related neurological conditions, primarily ADC, may provide important insight into pathogenesis. In the clinical setting, useful markers of ADC need to be sufficiently sensitive and specific to be used for diagnosis, to predict disease development and outcome and to assess the response to treatments.

A large number of the CSF markers discussed above have indeed met the requirements for research utility, either confirming a hypothesized mechanism or pointing to new avenues. CSF has often served to support hypotheses generated by in vitro findings. In several instances CSF studies have triggered additional experimental studies. On the practical side, however, no single marker has been shown to be sufficiently sensitive and specific for ADC diagnosis. Some markers have indeed shown acceptable sensitivity and specificity values in studies that used neurologically asymptomatic HIV-infected patients as controls. However, this may not be a realistic approach, because the differential diagnosis of patients with suspected ADC will include other neurological conditions in clinical practice. A few markers have also been useful to predict ADC development or progression, and most of those tested also seem reliable during follow-up of antiretroviral treatment. However, no CSF marker is currently used for these purposes in the clinic.

While ADC is now less common than formerly in the developed world, diagnosis has, if anything, become more difficult. This is because there is an increased prevalence of confounding conditions as individuals survive infection for decades and grow older. In this setting, more severe ADC now occurs most commonly in those with background co-morbidities including head injury, alcohol and substance abuse and psychiatric disease — all of which can impair treatment adherence. Moreover, a number of newer disease phenotypes have been reported that may result from the interactions of HIV and treatment. This makes clinical diagnosis more problematic and increases the need for objective markers of disease processes, whether due to HIV or not. Clinical diagnosis, even when supplemented by detailed neuropsychological testing, is now more difficult. While modern neuroimaging is of enormous help in differential diagnosis, it is less helpful in subjects without macroscopic focal lesions. It is in this setting that reliable disease markers in CSF might be invaluable.

Hence, work needs be done to improve diagnosis and clinical management of ADC using CSF biomarkers. First, some potential relevant markers may need to be re-evaluated more accurately and in extended patient series. Indeed only a few CSF markers have been assessed in thoroughly controlled studies with the aim of testing their clinical utility. Assessing the diagnostic value of a biomarker of disease requires the use of methodological standards to define study populations, the tests employed and analysis of results (Reid et al., 1995; Linde et al., 1997). Second, it is possible that a single marker will never prove to be the gold standard of ADC diagnosis. Therefore, a new approach that uses different markers in combination might be more reliable. A hypothetical CSF marker combination including a viral marker, a marker of immune activation and a marker of cell damage might provide information on HIV replication and both activity and status of CNS disease. Third, new biomarkers of ADC are needed in any event. In this regard, new pathogenic pathways are being explored as research in neuro-AIDS and in neuroscience progresses. These will likely suggest new possible factors to be tested in CSF. The opposite, though not alternative, approach is the direct search in CSF of substances that are specifically and sensitively expressed in ADC. The advent of new technologies, primarily proteomics, enables one to sieve CSF for proteins, including previously unknown proteins or protein fragments, which may be specifically present or found at abnormal concentrations in ADC (Berger et al., 2005; Davidsson and Sjogren, 2006; Irani et al., 2006).

There is still the need to recognize and, possibly, to predict the neurological complications resulting from HIV infection of the CNS. The need for reliable markers of ADC remains essential in the countries where HAART is available and HIV infection of the CNS is evolving towards a chronic, persistent infection. The need for biomarkers of ADC is even more important in the resource-limited countries, where the access to antiretroviral drugs is still limited and the impact of neurological diseases, including ADC, is at least as relevant as it was in the developed world before the advent of HAART. CSF markers, especially if they use simple and inexpensive methodologies, would be particularly relevant in these settings because of the higher costs of other diagnostic approaches, including neuroimaging.

Acknowledgment

Our own work and preparation of this review was supported by NIH grant R01 NS043103.

References

Abdulle S, Hagberg L, Svennerholm B, et al (2002). Continuing intrathecal immunoactivation despite two years of effective antiretroviral therapy against HIV-1 infection. AIDS 16: 2145–2149.

Abdulle S, Hagberg L, Gisslen M (2005). Effects of antiretroviral treatment on blood-brain barrier integrity and intrathecal immunoglobulin production in neuroasymptomatic HIV-1-infected patients. HIV Med 6: 164–169.

Achim CL, Heyes MP, Wiley CA (1993). Quantitation of human immunodeficiency virus, immune activation factors, and quinolinic acid in AIDS brains. J Clin Invest 91: 2769–2775.

Adamson DC, Wildemann B, Sasaki M, et al (1996). Immunologic NO synthase: elevation in severe AIDS dementia and induction by HIV-1 gp41. Science 274: 1917–1921.

Adamson DC, McArthur JC, Dawson TM, et al (1999). Rate and severity of HIV-associated dementia (HAD): correlations with Gp41 and iNOS. Mol Med 5: 98–109.

Albright AV, Shieh JT, Itoh T, et al (1999). Microglia express CCR5, CXCR4, and CCR3, but of these, CCR5 is the principal coreceptor for human immunodeficiency virus type 1 dementia isolates. J Virol 73: 205–213.

Alfano M, Sidenius N, Panzeri B, et al (2002). Urokinase-urokinase receptor interaction mediates an inhibitory signal for HIV-1 replication. Proc Natl Acad Sci USA 99: 8862–8867.

Amadori A, De Rossi A, Gallo P, et al (1988). Cerebrospinal fluid lymphocytes from HIV-infected patients synthesize HIV-specific antibody in vitro. J Neuroimmunol 18: 181–186.

American Academy of Neurology (1991). Nomenclature and research case definitions for neurologic manifestations of human immunodeficiency virus-type 1 (HIV-1) infection. Report of a Working Group of the American Academy of Neurology AIDS Task Force. Neurology 41: 778–785.

Andersson L, Blennow K, Fuchs D, et al (1999). Increased cerebrospinal fluid protein tau concentration in neuro-AIDS. J Neurol Sci 171: 92–96.

Andersson LM, Fredman P, Lekman A, et al (1998). Increased cerebrospinal fluid ganglioside GD3 concentrations as a marker of microglial activation in HIV type 1 infection. AIDS Res Hum Retrovir 14: 1065–1069.

Andersson LM, Svennerholm B, Hagberg L, et al (2000). Higher HIV-1 RNA cutoff level required in cerebrospinal fluid than in blood to predict positive HIV-1 isolation. J Med Virol 62: 9–13.

Andersson MA, Bergstrom TB, Blomstrand C, et al (1988). Increasing intrathecal lymphocytosis and immunoglobulin G production in neurologically asymptomatic HIV-1 infection. J Neuroimmunol 19: 291–304.

Ando S, Toyoda Y, Nagai Y, et al (1984). Alterations in brain gangliosides and other lipids of patients with Creutzfeldt-Jakob disease and subacute sclerosing panencephalitis (SSPE). Jpn J Exp Med 54: 229–234.

Andreasen N, Sjogren M, Blennow K (2003). CSF markers for Alzheimer's disease: total tau, phospho-tau and Abeta42. World J Biol Psychiatry 4: 147–155.

Antinori A, Giancola ML, Grisetti S, et al (2002). Factors influencing virological response to antiretroviral drugs in cerebrospinal fluid of advanced HIV-1-infected patients. AIDS 16: 1867–1876.

Antinori A, Perno CF, Giancola ML, et al (2005). Efficacy of cerebrospinal fluid (CSF)-penetrating antiretroviral drugs against HIV in the neurological compartment: different patterns of phenotypic resistance in CSF and plasma. Clin Infect Dis 41: 1787–1793.

Appleman ME, Marshall DW, Brey RL, et al (1988). Cerebrospinal fluid abnormalities in patients without AIDS who are seropositive for the human immunodeficiency virus. J Infect Dis 158: 193–199.

Aquaro S, Panti S, Caroleo MC, et al (2000). Primary macrophages infected by human immunodeficiency virus trigger CD^{95}-mediated apoptosis of uninfected astrocytes. J Leukoc Biol 68: 429–435.

Arens M (1999). Methods for subtyping and molecular comparison of human viral genomes. Clin Microbiol Rev 12: 612–626.

Avison MJ, Nath A, Greene-Avison R, et al (2004). Inflammatory changes and breakdown of microvascular integrity in early human immunodeficiency virus dementia. J Neurovirol 10: 223–232.

Baggiolini M (2001). Chemokines in pathology and medicine. J Intern Med 250: 91–104.

Bajetto A, Bonavia R, Barbero S, et al (2002). Characterization of chemokines and their receptors in the central nervous system: physiopathological implications. J Neurochem 82: 1311–1329.

Barlow KL, Green J, Clewley JP (2000). Viral genome characterisation by the heteroduplex mobility and heteroduplex tracking assays. Rev Med Virol 10: 321–335.

Barron KD (1995). The microglial cell. A historical review. J Neurol Sci 134(Suppl): 57–68.

Berg D, Holzmann C, Riess O (2003). 14-3-3 proteins in the nervous system. Nat Rev Neurosci 4: 752–762.

Berger JR, Avison M, Mootoor Y, et al (2005). Cerebrospinal fluid proteomics and human immunodeficiency virus dementia: preliminary observations. J Neurovirol 11: 557–562.

Bernasconi S, Cinque P, Peri G, et al (1996). Selective elevation of monocyte chemotactic protein-1 in the cerebrospinal fluid of AIDS patients with cytomegalovirus encephalitis. J Infect Dis 174: 1098–1101.

Bestetti A, Presi S, Pierotti C, et al (2004). Long-term virological effect of highly active antiretroviral therapy on cerebrospinal fluid and relationship with genotypic resistance. J Neurovirol 10(Suppl 1): 52–57.

Blennow K (2004). Cerebrospinal fluid protein biomarkers for Alzheimer's disease. NeuroRx 1: 213–225.

Blond D, Raoul H, Le Grand R, et al (2000). Nitric oxide synthesis enhances human immunodeficiency virus replication in primary human macrophages. J Virol 74: 8904–8912.

Bolin LM, Murray R, Lukacs NW, et al (1998). Primary sensory neurons migrate in response to the chemokine RANTES. J Neuroimmunol 81: 49–57.

Boufassa F, Bachmeyer C, Carre N, et al (1995). Influence of neurologic manifestations of primary human immunodeficiency virus infection on disease progression. SEROCO Study Group. J Infect Dis 171: 1190–1195.

Boven LA, Gomes L, Hery C, et al (1999). Increased peroxynitrite activity in AIDS dementia complex: implications for the neuropathogenesis of HIV-1 infection. J Immunol 162: 4319–4327.

Bratanich AC, Liu C, McArthur JC, et al (1998). Brain-derived HIV-1 tat sequences from AIDS patients with dementia show increased molecular heterogeneity. J Neurovirol 4: 387–393.

Brew BJ, Bhalla RB, Paul M, et al (1990). Cerebrospinal fluid neopterin in human immunodeficiency virus type 1 infection. Ann Neurol 28: 556–560.

Brew BJ, Bhalla RB, Paul M, et al (1992). Cerebrospinal fluid beta 2-microglobulin in patients with AIDS dementia complex: an expanded series including response to zidovudine treatment. AIDS 6: 461–465.

Brew BJ, Paul MO, Nakajima G, et al (1994). Cerebrospinal fluid HIV-1 p24 antigen and culture: sensitivity and specificity for AIDS-dementia complex. J Neurol Neurosurg Psychiatry 57: 784–789.

Brew BJ, Dunbar N, Pemberton L, et al (1996). Predictive markers of AIDS dementia complex: CD^4 cell count and cerebrospinal fluid concentrations of beta 2-microglobulin and neopterin. J Infect Dis 174: 294–298.

Brew BJ, Pemberton L, Cunningham P, et al (1997). Levels of human immunodeficiency virus type 1 RNA in cerebrospinal fluid correlate with AIDS dementia stage. J Infect Dis 175: 963–966.

Brew BJ, Pemberton L, Blennow K, et al (2005). CSF amyloid beta42 and tau levels correlate with AIDS dementia complex. Neurology 65: 1490–1492.

Brooke SM, McLaughlin JR, Cortopassi KM, et al (2002). Effect of GP120 on glutathione peroxidase activity in cortical cultures and the interaction with steroid hormones. J Neurochem 81: 277–284.

Brugg B, Michel PP, Agid Y, et al (1996). Ceramide induces apoptosis in cultured mesencephalic neurons. J Neurochem 66: 733–739.

Buee L, Bussiere T, Buee-Scherrer V, et al (2000). Tau protein isoforms, phosphorylation and role in neurodegenerative disorders. Brain Res Brain Res Rev 33: 95–130.

Buffet R, Agut H, Chieze F, et al (1991). Virological markers in the cerebrospinal fluid from HIV-1-infected individuals. AIDS 5: 1419–1424.

Bukrinsky MI, Nottet HS, Schmidtmayerova H, et al (1995). Regulation of nitric oxide synthase activity in human immunodeficiency virus type 1 (HIV-1)-infected monocytes: implications for HIV-associated neurological disease. J Exp Med 181: 735–745.

Chang L, Ernst T, St Hillaire C, et al (2004). Antiretroviral treatment alters relationship between MCP-1 and neurometabolites in HIV patients. Antivir Ther 9: 431–440.

Charcot J (1868). Histologie de la sclérose en plaques. Gaz Hop (Paris) 41: 405–406, 554, 557, 566.

Cheng J, Zhou T, Liu C, et al (1994). Protection from Fas-mediated apoptosis by a soluble form of the Fas molecule. Science 263: 1759–1762.

Chien JW, Valdez H, McComsey G, et al (1999). Presence of mutation conferring resistance to lamivudine in plasma and cerebrospinal fluid of HIV-1-infected patients. J Acquir Immune Defic Syndr 21: 277–280.

Churchill M, Sterjovski J, Gray L, et al (2004). Longitudinal analysis of nef/long terminal repeat-deleted HIV-1 in blood and cerebrospinal fluid of a long-term survivor who developed HIV-associated dementia. J Infect Dis 190: 2181–2186.

Ciardi M, Sharief MK, Noori MA, et al (1993). Intrathecal synthesis of interleukin-2 and soluble IL-2 receptor in asymptomatic HIV-1 seropositive individuals. Correlation with local production of specific IgM and IgG antibodies. J Neurol Sci 115: 117–122.

Ciardi M, Sharief MK, Thompson EJ, et al (1994). High cerebrospinal fluid and serum levels of tumor necrosis factor-alpha in asymptomatic HIV-1 seropositive individuals. Correlation with interleukin-2 and soluble IL-2 receptor. J Neurol Sci 125: 175–179.

Cinque P, Linde A (2003). CSF analysis in diagnosis of viral encephalitis and meningitis. In: A Nath, JR Berger (Eds.), Clinical Neurovirology. Marcel Dekker, New York.

Cinque P, Scarpellini P, Vago L, et al (1997). Diagnosis of central nervous system complications in HIV-infected patients: cerebrospinal fluid analysis by the polymerase chain reaction. AIDS 11: 1–17.

Cinque P, Vago L, Ceresa D, et al (1998a). Cerebrospinal fluid HIV-1 RNA levels: correlation with HIV encephalitis. AIDS 12: 389–394.

Cinque P, Vago L, Mengozzi M, et al (1998b). Elevated cerebrospinal fluid levels of monocyte chemotactic protein-1 correlate with HIV-1 encephalitis and local viral replication. AIDS 12: 1327–1332.

Cinque P, Presi S, Bestetti A, et al (2001). Effect of genotypic resistance on the virological response to highly active antiretroviral therapy in cerebrospinal fluid. AIDS Res Hum Retrovir 17: 377–383.

Cinque P, Nebuloni M, Santovito ML, et al (2004). The urokinase receptor is overexpressed in the AIDS dementia complex and other neurological manifestations. Ann Neurol 55: 687–694.

Cinque P, Bestetti A, Marenzi R, et al (2005). Cerebrospinal fluid interferon-gamma-inducible protein 10 (IP-10, CXCL10) in HIV-1 infection. J Neuroimmunol 168: 154–163.

Clements JE, Babas T, Mankowski JL, et al (2002). The central nervous system as a reservoir for simian immunodeficiency virus (SIV): steady-state levels of SIV DNA in brain from acute through asymptomatic infection. J Infect Dis 186: 905–913.

Coakley E, Petropoulos CJ, Whitcomb JM (2005). Assessing chemokine co-receptor usage in HIV. Curr Opin Infect Dis 18: 9–15.

Cocchi F, DeVico AL, Garzino-Demo A, et al (1995). Identification of RANTES, MIP-1 alpha, and MIP-1 beta as the major HIV-suppressive factors produced by CD^{8+} T cells. Science 270: 1811–1815.

Coffin JM (1995). HIV population dynamics in vivo: implications for genetic variation, pathogenesis, and therapy. Science 267: 483–489.

Conant K, Garzino-Demo A, Nath A, et al (1998). Induction of monocyte chemoattractant protein-1 in HIV-1 Tat-stimulated astrocytes and elevation in AIDS dementia. Proc Natl Acad Sci USA 95: 3117–3121.

Conant K, McArthur JC, Griffin DE, et al (1999). Cerebrospinal fluid levels of MMP-2, 7, and 9 are elevated in association with human immunodeficiency virus dementia. Ann Neurol 46: 391–398.

Cota M, Kleinschmidt A, Ceccherini-Silberstein F, et al (2000). Upregulated expression of interleukin-8, RANTES and chemokine receptors in human astrocytic cells infected with HIV-1. J Neurovirol 6: 75–83.

Cotter R, Williams C, Ryan L, et al (2002). Fractalkine (CX3CL1) and brain inflammation: implications for HIV-1-associated dementia. J Neurovirol 8: 585–598.

Cunningham PH, Smith DG, Satchell C, et al (2000). Evidence for independent development of resistance to HIV-1 reverse transcriptase inhibitors in the cerebrospinal fluid. AIDS 14: 1949–1954.

Cysique LA, Brew BJ, Halman M, et al (2005). Undetectable cerebrospinal fluid HIV RNA and beta-2 microglobulin do not indicate inactive AIDS dementia complex in highly active antiretroviral therapy-treated patients. J Acquir Immune Defic Syndr 39: 426–429.

D'Aversa TG, Yu KO, Berman JW (2004). Expression of chemokines by human fetal microglia after treatment with the human immunodeficiency virus type 1 protein Tat. J Neurovirol 10: 86–97.

Davidsson P, Sjogren M (2006). Proteome studies of CSF in AD patients. Mech Ageing Dev 127: 133–137.

de Gans J, Lange JM, Derix MM, et al (1988). Decline of HIV antigen levels in cerebrospinal fluid during treatment with low-dose zidovudine. AIDS 2: 37–40.

De Luca A, Ciancio BC, Larussa D, et al (2002). Correlates of independent HIV-1 replication in the CNS and of its control by antiretrovirals. Neurology 59: 342–347.

Desai A, Ravi V, Satishchandra P, et al (2000). HIV antibody profiles in serum and CSF of patients with neurological disease can serve as predictors of outcome. J Neurol Neurosurg Psychiatry 68: 86–88.

Dhawan S, Vargo M, Meltzer MS (1992). Interactions between HIV-infected monocytes and the extracellular matrix: increased capacity of HIV-infected monocytes to adhere to and spread on extracellular matrix associated with changes in extent of virus replication and cytopathic effects in infected cells. J Leukoc Biol 52: 62–69.

Di Stefano M, Sabri F, Leitner T, et al (1995). Reverse transcriptase sequence of paired isolates of cerebrospinal fluid and blood from patients infected with human immunodeficiency virus type 1 during zidovudine treatment. J Clin Microbiol 33: 352–355.

Di Stefano M, Wilt S, Gray F, et al (1996). HIV type 1 V3 sequences and the development of dementia during AIDS. AIDS Res Hum Retrovir 12: 471–476.

Dorf ME, Berman MA, Tanabe S, et al (2000). Astrocytes express functional chemokine receptors. J Neuroimmunol 111: 109–121.

Eggers C, Hertogs K, Sturenburg HJ, et al (2003). Delayed central nervous system virus suppression during highly active antiretroviral therapy is associated with HIV encephalopathy, but not with viral drug resistance or poor central nervous system drug penetration. AIDS 17: 1897–1906.

Ellis RJ, Hsia K, Spector SA, et al (1997). Cerebrospinal fluid human immunodeficiency virus type 1 RNA levels are elevated in neurocognitively impaired individuals with acquired immunodeficiency syndrome. HIV Neurobehavioral Research Center Group. Ann Neurol 42: 679–688.

Ellis RJ, Seubert P, Motter R, et al (1998). Cerebrospinal fluid tau protein is not elevated in HIV-associated neurologic disease in humans. HIV Neurobehavioral Research Center Group (HNRC). Neurosci Lett 254: 1–4.

Ellis RJ, Gamst AC, Capparelli E, et al (2000). Cerebrospinal fluid HIV RNA originates from both local CNS and systemic sources. Neurology 54: 927–936.

Ellis RJ, Moore DJ, Childers ME, et al (2002). Progression to neuropsychological impairment in human immunodeficiency virus infection predicted by elevated cerebrospinal fluid levels of human immunodeficiency virus RNA. Arch Neurol 59: 923–928.

Ellis RJ, Childers ME, Zimmerman JD, et al (2003). Human immunodeficiency virus-1 RNA levels in cerebrospinal fluid exhibit a set point in clinically stable patients not receiving antiretroviral therapy. J Infect Dis 187: 1818–1821.

Elovaara I, Seppala I, Poutiainen E, et al (1988). Intrathecal humoral immunologic response in neurologically symptomatic and asymptomatic patients with human immunodeficiency virus infection. Neurology 38: 1451–1456.

Elovaara I, Albert PS, Ranki A, et al (1993). HIV-1 specificity of cerebrospinal fluid and serum IgG, IgM, and IgG1-G4 antibodies in relation to clinical disease. J Neurol Sci 117: 111–119.

Elovaara I, Poutiainen E, Lahdevirta J, et al (1994). Zidovudine reduces intrathecal immunoactivation in patients with early human immunodeficiency virus type 1 infection. Arch Neurol 51: 943–950.

Elovaara I, Sabri F, Gray F, et al (1999). Upregulated expression of Fas and Fas ligand in brain through the spectrum of HIV-1 infection. Acta Neuropathol (Berl) 98: 355–362.

Eng LF, Ghirnikar RS (1994). GFAP and astrogliosis. Brain Pathol 4: 229–237.

Engelhardt B, Ransohoff RM (2005). The ins and outs of T-lymphocyte trafficking to the CNS: anatomical sites and molecular mechanisms. Trends Immunol 26: 485–495.

Enting RH, Foudraine NA, Lange JM, et al (2000). Cerebrospinal fluid beta2-microglobulin, monocyte chemotactic protein-1, and soluble tumour necrosis factor alpha receptors before and after treatment with lamivudine plus zidovudine or stavudine. J Neuroimmunol 102: 216–221.

Enting RH, Prins JM, Jurriaans S, et al (2001). Concentrations of human immunodeficiency virus type 1 (HIV-1) RNA in cerebrospinal fluid after antiretroviral treatment initiated during primary HIV-1 infection. Clin Infect Dis 32: 1095–1099.

Erichsen D, Lopez AL, Peng H, et al (2003). Neuronal injury regulates fractalkine: relevance for HIV-1 associated dementia. J Neuroimmunol 138: 144–155.

Espey MG, Basile AS, Heaton RK, et al (2002). Increased glutamate in CSF and plasma of patients with HIV dementia. Neurology 58: 1439; author reply 1439–1440.

Evers S, Rahmann A, Schwaag S, et al (2004). Prevention of AIDS dementia by HAART does not depend on cerebrospinal fluid drug penetrance. AIDS Res Hum Retrovir 20: 483–491.

Ferrarese C, Riva R, Dolara A, et al (1997). Elevated glutamate in the cerebrospinal fluid of patients with HIV dementia. JAMA 277: 630.

Ferrarese C, Aliprandi A, Tremolizzo L, et al (2001). Increased glutamate in CSF and plasma of patients with HIV dementia. Neurology 57: 671–675.

Fishman RA (1992). Cerebrospinal Fluid in Diseases of the Nervous System. WB Saunders, Philadelphia, PA.

Foudraine NA, Hoetelmans RM, Lange JM, et al (1998). Cerebrospinal-fluid HIV-1 RNA and drug concentrations after treatment with lamivudine plus zidovudine or stavudine. Lancet 351: 1547–1551.

Franciotta DM, Melzi d'Eril GL, Bono G, et al (1992). Tumor necrosis factor alpha levels in serum and cerebrospinal fluid of patients with AIDS. Funct Neurol 7: 35–38.

Fredrikson S, Eneroth P, Link H (1987). Intrathecal production of neopterin in aseptic meningo-encephalitis and multiple sclerosis. Clin Exp Immunol 67: 76–81.

Fuchs D, Chiodi F, Albert J, et al (1989). Neopterin concentrations in cerebrospinal fluid and serum of individuals infected with HIV-1. AIDS 3: 285–288.

Fuchs D, Forsman A, Hagberg L, et al (1990). Immune activation and decreased tryptophan in patients with HIV-1 infection. J Interferon Res 10: 599–603.

Fuchs D, Gisslen M, Larsson M, et al (1996). Increase of tryptophan in serum and in cerebrospinal fluid of patients with HIV infection during zidovudine therapy. Adv Exp Med Biol 398: 131–134.

Furukawa Y, Nishi K, Kondo T, et al (1992). Significance of CSF total neopterin and biopterin in inflammatory neurological diseases. J Neurol Sci 111: 65–72.

Gallo P, Frei K, Rordorf C, et al (1989). Human immunodeficiency virus type 1 (HIV-1) infection of the central nervous system: an evaluation of cytokines in cerebrospinal fluid. J Neuroimmunol 23: 109–116.

Gallo P, Sivieri S, Rinaldi L, et al (1994). Intrathecal synthesis of interleukin-10 (IL-10) in viral and inflammatory diseases of the central nervous system. J Neurol Sci 126: 49–53.

Garcia F, Alonso MM, Romeu J, et al (2000). Comparison of immunologic restoration and virologic response in plasma, tonsillar tissue, and cerebrospinal fluid in HIV-1-infected patients treated with double versus triple antiretroviral therapy in very early stages: The Spanish EARTH-2 Study. Early Anti-Retroviral Therapy Study. J Acquir Immune Defic Syndr 25: 26–35.

Garcia-Monco JC, Coleman JL, Benach JL (2002). Soluble urokinase receptor (uPAR, CD^{87}) is present in serum and cerebrospinal fluid in patients with neurologic diseases. J Neuroimmunol 129: 216–223.

Garden GA, Budd SL, Tsai E, et al (2002). Caspase cascades in human immunodeficiency virus-associated neurodegeneration. J Neurosci 22: 4015–4024.

Gardner J, Ghorpade A (2003). Tissue inhibitor of metalloproteinase (TIMP)-1: the TIMPed balance of matrix metalloproteinases in the central nervous system. J Neurosci Res 74: 801–806.

Gelbard HA, Nottet HS, Swindells S, et al (1994). Platelet-activating factor: a candidate human immunodeficiency virus type 1-induced neurotoxin. J Virol 68: 4628–4635.

Gendelman HE, Zheng J, Coulter CL, et al (1998). Suppression of inflammatory neurotoxins by highly active antiretroviral therapy in human immunodeficiency virus-associated dementia. J Infect Dis 178: 1000–1007.

Genis P, Jett M, Bernton EW, et al (1992). Cytokines and arachidonic metabolites produced during human immunodeficiency virus (HIV)-infected macrophage–astroglia interactions: implications for the neuropathogenesis of HIV disease. J Exp Med 176: 1703–1718.

Ghorpade A, Nukuna A, Che M, et al (1998). Human immunodeficiency virus neurotropism: an analysis of viral replication and cytopathicity for divergent strains in monocytes and microglia. J Virol 72: 3340–3350.

Ghorpade A, Persidskaia R, Suryadevara R, et al (2001). Mononuclear phagocyte differentiation, activation, and viral infection regulate matrix metalloproteinase expression: implications for human immunodeficiency virus type 1-associated dementia. J Virol 75: 6572–6583.

Giovannoni G, Miller RF, Heales SJ, et al (1998). Elevated cerebrospinal fluid and serum nitrate and nitrite levels in patients with central nervous system complications of HIV-1 infection: a correlation with blood-brain-barrier dysfunction. J Neurol Sci 156: 53–58.

Gisolf EH, Enting RH, Jurriaans S, et al (2000a). Cerebrospinalfluid HIV-1 RNA during treatment with ritonavir/saquinavir or ritonavir/saquinavir/stavudine. AIDS 14: 1583–1589.

Gisolf EH, van Praag RM, Jurriaans S, et al (2000b). Increasing cerebrospinal fluid chemokine concentrations despite undetectable cerebrospinal fluid HIV RNA in HIV-1-infected patients receiving antiretroviral therapy. J Acquir Immune Defic Syndr 25: 426–433.

Gisslen M, Larsson M, Norkrans G, et al (1994). Tryptophan concentrations increase in cerebrospinal fluid and blood after zidovudine treatment in patients with HIV type 1 infection. AIDS Res Hum Retrovir 10: 947–951.

Gisslen M, Fredman P, Norkrans G, et al (1996). Elevated cerebrospinal fluid sulfatide concentrations as a sign of increased metabolic turnover of myelin in HIV type I infection. AIDS Res Hum Retrovir 12: 149–155.

Gisslen M, Hagberg L, Norkrans G, et al (1997a). Increased cerebrospinal fluid ganglioside GM1 concentrations indicating neuronal involvement in all stages of HIV-1 infection. J Neurovirol 3: 148–152.

Gisslen M, Norkrans G, Svennerholm B, et al (1997b). The effect on human immunodeficiency virus type 1 RNA levels in cerebrospinal fluid after initiation of zidovudine or didanosine. J Infect Dis 175: 434–437.

Gisslen M, Hagberg L, Fuchs D, et al (1998). Cerebrospinal fluid viral load in HIV-1-infected patients without antiretroviral treatment: a longitudinal study. J Acquir Immune Defic Syndr Hum Retrovirol 17: 291–295.

Gisslen M, Svennerholm B, Norkrans G, et al (2000). Cerebrospinal fluid and plasma viral load in HIV-1-infected patients with various anti-retroviral treatment regimens. Scand J Infect Dis 32: 365–369.

Gisslen M, Rosengren L, Hagberg L, et al (2005). Cerebrospinal fluid signs of neuronal damage after antiretroviral treatment interruption in HIV-1 infection. AIDS Res Ther 2: 6.

Gisslen M, Cinque P, Hagberg L, et al (2006). High cerebrospinal fluid neurofilament protein levels in AIDS dementia complex. 13th Conference of Retrovirus and Opportunistic Infections. Denver, CO.

Glass JD, Wesselingh SL, Selnes OA, et al (1993). Clinical-neuropathologic correlation in HIV-associated dementia. Neurology 43: 2230–2237.

Glass JD, Fedor H, Wesselingh SL, et al (1995). Immunocytochemical quantitation of human immunodeficiency virus in the brain: correlations with dementia. Ann Neurol 38: 755–762.

Gonzalez E, Rovin BH, Sen L, et al (2002). HIV-1 infection and AIDS dementia are influenced by a mutant MCP-1 allele linked to increased monocyte infiltration of tissues and MCP-1 levels. Proc Natl Acad Sci USA 99: 13795–13800.

Gonzalez-Scarano F, Baltuch G (1999). Microglia as mediators of inflammatory and degenerative diseases. Annu Rev Neurosci 22: 219–240.

Gonzalez-Scarano F, Martin-Garcia J (2005). The neuropathogenesis of AIDS. Nat Rev Immunol 5: 69–81.

Gorry PR, Bristol G, Zack JA, et al (2001). Macrophage tropism of human immunodeficiency virus type 1 isolates from brain and lymphoid tissues predicts neurotropism independent of coreceptor specificity. J Virol 75: 10073–10089.

Goudsmit J, de Wolf F, Paul DA, et al (1986). Expression of human immunodeficiency virus antigen (HIV-Ag) in serum and cerebrospinal fluid during acute and chronic infection. Lancet 2: 177–180.

Gray F, Chretien F, Vallat-Decouvelaere AV, et al (2003). The changing pattern of HIV neuropathology in the HAART era. J Neuropathol Exp Neurol 62: 429–440.

Green AJ, Giovannoni G, Miller RF, et al (1999). Cerebrospinal fluid S-100b concentrations in patients with HIV infection. AIDS 13: 139–140.

Greenamyre JT, Porter RH (1994). Anatomy and physiology of glutamate in the CNS. Neurology 44: S7–13.

Griffin DE (1997). Cytokines in the brain during viral infection: clues to HIV-associated dementia. J Clin Invest 100: 2948–2951.

Griffin DE, McArthur JC, Cornblath DR (1991). Neopterin and interferon-gamma in serum and cerebrospinal fluid of patients with HIV-associated neurologic disease. Neurology 41: 69–74.

Griffin DE, Wesselingh SL, McArthur JC (1994). Elevated central nervous system prostaglandins in human immunodeficiency virus-associated dementia. Ann Neurol 35: 592–597.

Grimaldi LM, Castagna A, Lazzarin A, et al (1988). Oligoclonal IgG bands in cerebrospinal fluid and serum during asymptomatic human immunodeficiency virus infection. Ann Neurol 24: 277–279.

Grimaldi LM, Martino GV, Franciotta DM, et al (1991). Elevated alpha-tumor necrosis factor levels in spinal fluid from HIV-1-infected patients with central nervous system involvement. Ann Neurol 29: 21–25.

Gu L, Tseng SC, Rollins BJ (1999). Monocyte chemoattractant protein-1. Chem Immunol 72: 7–29.

Guillemin GJ, Croitoru-Lamoury J, Dormont D, et al (2003). Quinolinic acid upregulates chemokine production and chemokine receptor expression in astrocytes. Glia 41: 371–381.

Guillemin GJ, Wang L, Brew BJ (2005). Quinolinic acid selectively induces apoptosis of human astrocytes: potential role in AIDS dementia complex. J Neuroinflammation 2: 16.

Gunthard HF, Havlir DV, Fiscus S, et al (2001). Residual human immunodeficiency virus (HIV) Type 1 RNA and DNA in lymph nodes and HIV RNA in genital secretions and in cerebrospinal fluid after suppression of viremia for 2 years. J Infect Dis 183: 1318–1327.

Haas DW, Johnson BW, Spearman P, et al (2003). Two phases of HIV RNA decay in CSF during initial days of multidrug therapy. Neurology 61: 1391–1396.

Hagberg L (1995). The clinical use of cerebrospinal fluid neopterin concentrations in central nervous system infections. Pteridins 6: 147–152.

Hagberg L, Forsman A, Norkrans G, et al (1988). Cytological and immunoglobulin findings in cerebrospinal fluid of symptomatic and asymptomatic human immunodeficiency virus (HIV) seropositive patients. Infection 16: 13–18.

Hagberg L, Dotevall L, Norkrans G, et al (1993). Cerebrospinal fluid neopterin concentrations in central nervous system infection. J Infect Dis 168: 1285–1288.

Hagberg L, Fuchs D, Rosengren L, et al (2000). Intrathecal immune activation is associated with cerebrospinal fluid markers of neuronal destruction in AIDS patients. J Neuroimmunol 102: 51–55.

Harrigan PR, Montaner JS, Wegner SA, et al (2001). Worldwide variation in HIV-1 phenotypic susceptibility in untreated individuals: biologically relevant values for resistance testing. AIDS 15: 1671–1677.

Harrington PR, Haas DW, Ritola K, et al (2005). Compartmentalized human immunodeficiency virus type 1 present in cerebrospinal fluid is produced by short-lived cells. J Virol 79: 7959–7966.

Harrison JK, Jiang Y, Chen S, et al (1998). Role for neuronally derived fractalkine in mediating interactions between neurons and CX3CR1-expressing microglia. Proc Natl Acad Sci USA 95: 10896–10901.

Haughey NJ, Cutler RG, Tamara A, et al (2004). Perturbation of sphingolipid metabolism and ceramide production in HIV-dementia. Ann Neurol 55: 257–267.

Havlir DV, Bassett R, Levitan D, et al (2001). Prevalence and predictive value of intermittent viremia with combination HIV therapy. JAMA 286: 171–179.

He J, Chen Y, Farzan M, et al (1997). CCR3 and CCR5 are co-receptors for HIV-1 infection of microglia. Nature 385: 645–649.

Heidenreich F, Arendt G, Jander S, et al (1994). Serum and cerebrospinal fluid levels of soluble intercellular adhesion molecule 1 (sICAM-1) in patients with HIV-1 associated neurological diseases. J Neuroimmunol 52: 117–126.

Heizmann CW, Fritz G, Schafer BW (2002). S100 proteins: structure, functions and pathology. Front Biosci 7: d1356–1368.

Helke KL, Queen SE, Tarwater PM, et al (2005). 14-3-3 protein in CSF: an early predictor of SIV CNS disease. J Neuropathol Exp Neurol 64: 202–208.

Hesselgesser J, Taub D, Baskar P, et al (1998). Neuronal apoptosis induced by HIV-1 gp120 and the chemokine SDF-1 alpha is mediated by the chemokine receptor CXCR4. Curr Biol 8: 595–598.

Heyes MP, Brew BJ, Martin A, et al (1991). Quinolinic acid in cerebrospinal fluid and serum in HIV-1 infection: relationship to clinical and neurological status. Ann Neurol 29: 202–209.

Heyes MP, Brew BJ, Saito K, et al (1992a). Inter-relationships between quinolinic acid, neuroactive kynurenines, neopterin and beta 2-microglobulin in cerebrospinal fluid and serum of HIV-1-infected patients. J Neuroimmunol 40: 71–80.

Heyes MP, Saito K, Crowley JS, et al (1992b). Quinolinic acid and kynurenine pathway metabolism in inflammatory and non-inflammatory neurological disease. Brain 115 (Pt 5): 1249–1273.

Heyes MP, Saito K, Lackner A, et al (1998). Sources of the neurotoxin quinolinic acid in the brain of HIV-1-infected patients and retrovirus-infected macaques. FASEB J 12: 881–896.

Heyes MP, Ellis RJ, Ryan L, et al (2001). Elevated cerebrospinal fluid quinolinic acid levels are associated with region-specific cerebral volume loss in HIV infection. Brain 124: 1033–1042.

Hirsch MS, Brun-Vezinet F, D'Aquila RT, et al (2000). Antiretroviral drug resistance testing in adult HIV-1 infection: recommendations of an International AIDS Society-USA Panel. JAMA 283: 2417–2426.

Hollander H, Levy JA (1987). Neurologic abnormalities and recovery of human immunodeficiency virus from cerebrospinal fluid. Ann Intern Med 106: 692–695.

Huang KJ, Alter GM, Wooley DP (2002). The reverse transcriptase sequence of human immunodeficiency virus type 1 is under positive evolutionary selection within the central nervous system. J Neurovirol 8: 281–294.

Hughes ES, Bell JE, Simmonds P (1997). Investigation of population diversity of human immunodeficiency virus type 1 in vivo by nucleotide sequencing and length polymorphism analysis of the V1/V2 hypervariable region of env. J Gen Virol 78(Pt 11): 2871–2882.

Ibanez A, Puig T, Elias J, et al (1999). Quantification of integrated and total HIV-1 DNA after long-term highly active antiretroviral therapy in HIV-1-infected patients. AIDS 13: 1045–1049.

Ikeda K, Akiyama H, Arai T, et al (1998). Glial tau pathology in neurodegenerative diseases: their nature and comparison with neuronal tangles. Neurobiol Aging 19: S85–91.

Imamichi H, Crandall KA, Natarajan V, et al (2001). Human immunodeficiency virus type 1 quasi species that rebound after discontinuation of highly active antiretroviral therapy are similar to the viral quasi species present before initiation of therapy. J Infect Dis 183: 36–50.

Irani DN, Anderson C, Gundry R, et al (2006). Cleavage of cystatin C in the cerebrospinal fluid of patients with multiple sclerosis. Ann Neurol 59: 237–247.

Izmailova E, Bertley FM, Huang Q, et al (2003). HIV-1 Tat reprograms immature dendritic cells to express chemoattractants for activated T cells and macrophages. Nat Med 9: 191–197.

Jana A, Pahan K (2004). Human immunodeficiency virus type 1 gp120 induces apoptosis in human primary neurons through redox-regulated activation of neutral sphingomyelinase. J Neurosci 24: 9531–9540.

Jaranowska A, Bussolino F, Sogos V, et al (1995). Platelet-activating factor production by human fetal microglia. Effect of lipopolysaccharides and tumor necrosis factor-alpha. Mol Chem Neuropathol 24: 95–106.

Jiang ZG, Piggee C, Heyes MP, et al (2001). Glutamate is a mediator of neurotoxicity in secretions of activated HIV-1-infected macrophages. J Neuroimmunol 117: 97–107.

Johnston JB, Zhang K, Silva C, et al (2001). HIV-1 Tat neurotoxicity is prevented by matrix metalloproteinase inhibitors. Ann Neurol 49: 230–241.

Jones G, Power C (2006). Regulation of neural cell survival by HIV-1 infection. Neurobiol Dis 21: 1–17.

Ju ST, Panka DJ, Cui H, et al (1995). Fas(CD^{95})/FasL interactions required for programmed cell death after T-cell activation. Nature 373: 444–448.

Kapural M, Krizanac-Bengez L, Barnett G, et al (2002). Serum S-100beta as a possible marker of blood-brain barrier disruption. Brain Res 940: 102–104.

Kaul M, Garden GA, Lipton SA (2001). Pathways to neuronal injury and apoptosis in HIV-associated dementia. Nature 410: 988–994.

Kaul M, Zheng J, Okamoto S, et al (2005). HIV-1 infection and AIDS: consequences for the central nervous system. Cell Death Differ 12(Suppl 1): 878–892.

Kelder W, McArthur JC, Nance-Sproson T, et al (1998). Beta-chemokines MCP-1 and RANTES are selectively increased in cerebrospinal fluid of patients with human immunodeficiency virus-associated dementia. Ann Neurol 44: 831–835.

Kerr SJ, Armati PJ, Pemberton LA, et al (1997). Kynurenine pathway inhibition reduces neurotoxicity of HIV-1-infected macrophages. Neurology 49: 1671–1681.

Kerr SJ, Armati PJ, Guillemin GJ, et al (1998). Chronic exposure of human neurons to quinolinic acid results in neuronal changes consistent with AIDS dementia complex. AIDS 12: 355–363.

Keys B, Karis J, Fadeel B, et al (1993). V3 sequences of paired HIV-1 isolates from blood and cerebrospinal fluid cluster according to host and show variation related to the clinical stage of disease. Virology 196: 475–483.

Kitai R, Zhao ML, Zhang N, et al (2000). Role of MIP-1beta and RANTES in HIV-1 infection of microglia: inhibition of infection and induction by IFNbeta. J Neuroimmunol 110: 230–239.

Kivisakk P, Trebst C, Liu Z, et al (2002). T-cells in the cerebrospinal fluid express a similar repertoire of inflammatory chemokine receptors in the absence or presence of CNS inflammation: implications for CNS trafficking. Clin Exp Immunol 129: 510–518.

Koenig S, Gendelman HE, Orenstein JM, et al (1986). Detection of AIDS virus in macrophages in brain tissue from AIDS patients with encephalopathy. Science 233: 1089–1093.

Kolb SA, Sporer B, Lahrtz F, et al (1999). Identification of a T cell chemotactic factor in the cerebrospinal fluid of

HIV-1-infected individuals as interferon-gamma inducible protein 10. J Neuroimmunol 93: 172–181.

Korber BT, Kunstman KJ, Patterson BK, et al (1994). Genetic differences between blood- and brain-derived viral sequences from human immunodeficiency virus type 1-infected patients: evidence of conserved elements in the V3 region of the envelope protein of brain-derived sequences. J Virol 68: 7467–7481.

Krivine A, Force G, Servan J, et al (1999). Measuring HIV-1 RNA and interferon-alpha in the cerebrospinal fluid of AIDS patients: insights into the pathogenesis of AIDS dementia complex. J Neurovirol 5: 500–506.

Kuiken CL, Goudsmit J, Weiller GF, et al (1995). Differences in human immunodeficiency virus type 1 V3 sequences from patients with and without AIDS dementia complex. J Gen Virol 76(Pt 1): 175–180.

Lafeuillade A, Poggi C, Pellegrino P, et al (1996). HIV-1 replication in the plasma and cerebrospinal fluid. Infection 24: 367–371.

Lafrenie RM, Wahl LM, Epstein JS, et al (1996). HIV-1-Tat protein promotes chemotaxis and invasive behavior by monocytes. J Immunol 157: 974–977.

Lahrtz F, Piali L, Spanaus KS, et al (1998). Chemokines and chemotaxis of leukocytes in infectious meningitis. J Neuroimmunol 85: 33–43.

Lane BR, King SR, Bock PJ, et al (2003). The C-X-C chemokine IP-10 stimulates HIV-1 replication. Virology 307: 122–134.

Lane TE, Buchmeier MJ, Watry DD, et al (1996). Expression of inflammatory cytokines and inducible nitric oxide synthase in brains of SIV-infected rhesus monkeys: applications to HIV-induced central nervous system disease. Mol Med 2: 27–37.

Langford TD, Letendre SL, Marcotte TD, et al (2002). Severe, demyelinating leukoencephalopathy in AIDS patients on antiretroviral therapy. AIDS 16: 1019–1029.

Lanier ER, Sturge G, McClernon D, et al (2001). HIV-1 reverse transcriptase sequence in plasma and cerebrospinal fluid of patients with AIDS dementia complex treated with Abacavir. AIDS 15: 747–751.

Larsson M, Hagberg L, Norkrans G, et al (1989). Indole amine deficiency in blood and cerebrospinal fluid from patients with human immunodeficiency virus infection. J Neurosci Res 23: 441–446.

Laurenzi MA, Siden A, Persson MA, et al (1990). Cerebrospinal fluid interleukin-6 activity in HIV infection and inflammatory and noninflammatory diseases of the nervous system. Clin Immunol Immunopathol 57: 233–241.

Lazzarin A, Castagna A, Cavalli G, et al (1992). Cerebrospinal fluid beta 2-microglobulin in AIDS related central nervous system involvement. J Clin Lab Immunol 38: 175–186.

Lederman MM, Georges DL, Kusner DJ, et al (1994). Mycobacterium tuberculosis and its purified protein derivative activate expression of the human immunodeficiency virus. J Acquir Immune Defic Syndr 7: 727–733.

Letendre S, Marquie-Beck J, Singh KK, et al (2004). The monocyte chemotactic protein-1-2578G allele is associated with elevated MCP-1 concentrations in cerebrospinal fluid. J Neuroimmunol 157: 193–196.

Letendre SL, Lanier ER, McCutchan JA (1999). Cerebrospinal fluid beta chemokine concentrations in neurocognitively impaired individuals infected with human immunodeficiency virus type 1. J Infect Dis 180: 310–319.

Letendre SL, McCutchan JA, Childers ME, et al (2004). Enhancing antiretroviral therapy for human immunodeficiency virus cognitive disorders. Ann Neurol 56: 416–423.

Lewis LL, Venzon D, Church J, et al (1996). Lamivudine in children with human immunodeficiency virus infection: a phase I/II study. The National Cancer Institute Pediatric Branch-Human Immunodeficiency Virus Working Group. J Infect Dis 174: 16–25.

Li W, Galey D, Mattson MP, et al (2005). Molecular and cellular mechanisms of neuronal cell death in HIV dementia. Neurotox Res 8: 119–134.

Linde A, Klapper PE, Monteyne P, et al (1997). Specific diagnostic methods for herpesvirus infections of the central nervous system: a consensus review by the European Union Concerted Action on Virus Meningitis and Encephalitis. Clin Diagn Virol 8: 83–104.

Lindsberg PJ, Yue TL, Frerichs KU, et al (1990). Evidence for platelet-activating factor as a novel mediator in experimental stroke in rabbits. Stroke 21: 1452–1457.

Liu J, Zhao ML, Brosnan CF, et al (1996). Expression of type II nitric oxide synthase in primary human astrocytes and microglia: role of IL-1beta and IL-1 receptor antagonist. J Immunol 157: 3569–3576.

Liu MT, Lane TE (2001). Chemokine expression and viral infection of the central nervous system: regulation of host defense and neuropathology. Immunol Res 24: 111–119.

Liuzzi GM, Mastroianni CM, Santacroce MP, et al (2000). Increased activity of matrix metalloproteinases in the cerebrospinal fluid of patients with HIV-associated neurological diseases. J Neurovirol 6: 156–163.

Loetscher M, Gerber B, Loetscher P, et al (1996). Chemokine receptor specific for IP10 and mig: structure, function, and expression in activated T-lymphocytes. J Exp Med 184: 963–969.

Loetscher M, Loetscher P, Brass N, et al (1998). Lymphocyte-specific chemokine receptor CXCR3: regulation, chemokine binding and gene localization. Eur J Immunol 28: 3696–3705.

Lu B, Rutledge BJ, Gu L, et al (1998). Abnormalities in monocyte recruitment and cytokine expression in monocyte chemoattractant protein 1-deficient mice. J Exp Med 187: 601–608.

Lucas CV, Reaven EP, Bensch KG, et al (1980). Immunoperoxidase staining of glial fibrillary acidic (GFA) protein polymerized in vitro: an ultramicroscopic study. Neurochem Res 5: 1199–1209.

Lukes A, Mun-Bryce S, Lukes M, et al (1999). Extracellular matrix degradation by metalloproteinases and central nervous system diseases. Mol Neurobiol 19: 267–284.

Malmestrom C, Haghighi S, Rosengren L, et al (2003). Neurofilament light protein and glial fibrillary acidic protein as biological markers in MS. Neurology 61: 1720–1725.

Marconi S, De Toni L, Lovato L, et al (2005). Expression of gangliosides on glial and neuronal cells in normal and pathological adult human brain. J Neuroimmunol 170: 115–121.

Marra CM, Lockhart D, Zunt JR, et al (2003). Changes in CSF and plasma HIV-1 RNA and cognition after starting potent antiretroviral therapy. Neurology 60: 1388–1390.

Marshall DW, Brey RL, Cahill WT, et al (1988). Spectrum of cerebrospinal fluid findings in various stages of human immunodeficiency virus infection. Arch Neurol 45: 954–958.

Martin C, Albert J, Hansson P, et al (1998). Cerebrospinal fluid mononuclear cell counts influence CSF HIV-1 RNA levels. J Acquir Immune Defic Syndr Hum Retrovirol 17: 214–219.

Martin-Garcia J, Kolson DL, Gonzalez-Scarano F (2002). Chemokine receptors in the brain: their role in HIV infection and pathogenesis. AIDS 16: 1709–1730.

Massaro AR, Tonali P (1998). Cerebrospinal fluid markers in multiple sclerosis: an overview. Mult Scler 4: 1–4.

Mastroianni CM, Paoletti F, Valenti C, et al (1992). Tumour necrosis factor (TNF-alpha) and neurological disorders in HIV infection. J Neurol Neurosurg Psychiatry 55: 219–221.

Matsuyama T, Hata R, Tagaya M, et al (1994). Fas antigen mRNA induction in postischemic murine brain. Brain Res 657: 342–346.

McArthur JC, Cohen BA, Farzedegan H, et al (1988). Cerebrospinal fluid abnormalities in homosexual men with and without neuropsychiatric findings. Ann Neurol 23 (Suppl): S34–S37.

McArthur JC, Nance-Sproson TE, Griffin DE, et al (1992). The diagnostic utility of elevation in cerebrospinal fluid beta 2-microglobulin in HIV-1 dementia. Multicenter AIDS Cohort Study. Neurology 42: 1707–1712.

McArthur JC, McClernon DR, Cronin MF, et al (1997). Relationship between human immunodeficiency virus-associated dementia and viral load in cerebrospinal fluid and brain. Ann Neurol 42: 689–698.

McArthur JC, McDermott MP, McClernon D, et al (2004). Attenuated central nervous system infection in advanced HIV/AIDS with combination antiretroviral therapy. Arch Neurol 61: 1687–1696.

Melchjorsen J, Sorensen LN, Paludan SR (2003). Expression and function of chemokines during viral infections: from molecular mechanisms to in vivo function. J Leukoc Biol 74: 331–343.

Mellgren A, Antinori A, Cinque P, et al (2005). Cerebrospinal fluid HIV-1 infection usually responds well to antiretroviral treatment. Antivir Ther 10: 701–707.

Mellors JW, Rinaldo CR Jr, Gupta P, et al (1996). Prognosis in HIV-1 infection predicted by the quantity of virus in plasma. Science 272: 1167–1170.

Mengozzi M, De Filippi C, Transidico P, et al (1999). Human immunodeficiency virus replication induces monocyte chemotactic protein-1 in human macrophages and U937 promonocytic cells. Blood 93: 1851–1857.

Meucci O, Fatatis A, Simen AA, et al (1998). Chemokines regulate hippocampal neuronal signaling and gp120 neurotoxicity. Proc Natl Acad Sci USA 95: 14500–14505.

Meucci O, Fatatis A, Simen AA, et al (2000). Expression of CX3CR1 chemokine receptors on neurons and their role in neuronal survival. Proc Natl Acad Sci USA 97: 8075–8080.

Milstien S, Sakai N, Brew BJ, et al (1994). Cerebrospinal fluid nitrite/nitrate levels in neurologic diseases. J Neurochem 63: 1178–1180.

Mogi M, Harada M, Kondo T, et al (1996). The soluble form of Fas molecule is elevated in parkinsonian brain tissues. Neurosci Lett 220: 195–198.

Mondino A, Blasi F (2004). uPA and uPAR in fibrinolysis, immunity and pathology. Trends Immunol 25: 450–455.

Monteiro de Almeida S, Letendre S, Zimmerman J, et al (2005). Dynamics of monocyte chemoattractant protein type one (MCP-1) and HIV viral load in human cerebrospinal fluid and plasma. J Neuroimmunol 169: 144–152.

Montine TJ, Beal MF, Cudkowicz ME, et al (1999). Increased CSF F2-isoprostane concentration in probable AD. Neurology 52: 562–565.

Morris L, Silber E, Sonnenberg P, et al (1998). High human immunodeficiency virus type 1 RNA load in the cerebrospinal fluid from patients with lymphocytic meningitis. J Infect Dis 177: 473–476.

Murphy PM, Baggiolini M, Charo IF, et al (2000). International union of pharmacology. XXII. Nomenclature for chemokine receptors. Pharmacol Rev 52: 145–176.

Nagata S, Golstein P (1995). The Fas death factor. Science 267: 1449–1456.

Navia BA, Jordan BD, Price RW (1986). The AIDS dementia complex: I. Clinical features. Ann Neurol 19: 517–524.

Neuenburg JK, Sinclair E, Nilsson A, et al (2004). HIV-producing T cells in cerebrospinal fluid. J Acquir Immune Defic Syndr 37: 1237–1244.

Neuenburg JK, Cho TA, Nilsson A, et al (2005a). T-cell activation and memory phenotypes in cerebrospinal fluid during HIV infection. J Acquir Immune Defic Syndr 39: 16–22.

Neuenburg JK, Furlan S, Bacchetti P, et al (2005b). Enrichment of activated monocytes in cerebrospinal fluid during antiretroviral therapy. AIDS 19: 1351–1359.

Nottet HS, Gendelman HE (1995). Unraveling the neuroimmune mechanisms for the HIV-1-associated cognitive/motor complex. Immunol Today 16: 441–448.

Nottet HS, Jett M, Flanagan CR, et al (1995). A regulatory role for astrocytes in HIV-1 encephalitis. An overexpression of eicosanoids, platelet-activating factor, and tumor necrosis factor-alpha by activated HIV-1-infected monocytes is attenuated by primary human astrocytes. J Immunol 154: 3567–3581.

Nottet HS, Persidsky Y, Sasseville VG, et al (1996). Mechanisms for the transendothelial migration of HIV-1-infected monocytes into brain. J Immunol 156: 1284–1295.

Nykjaer A, Moller B, Todd RF 3rd, et al (1994). Urokinase receptor. An activation antigen in human T lymphocytes. J Immunol 152: 505–516.

Ogden F, DeCoster MA, Bazan NG (1998). Recombinant plasma-type platelet-activating factor acetylhydrolase attenuates NMDA-induced hippocampal neuronal apoptosis. J Neurosci Res 53: 677–684.

Ohagen A, Devitt A, Kunstman KJ, et al (2003). Genetic and functional analysis of full-length human immunodeficiency virus type 1 env genes derived from brain and blood of patients with AIDS. J Virol 77: 12336–12345.

Peeters MF, Colebunders RL, Van den Abbeele K, et al (1995). Comparison of human immunodeficiency virus biological phenotypes isolated from cerebrospinal fluid and peripheral blood. J Med Virol 47: 92–96.

Pemberton LA, Brew BJ (2001). Cerebrospinal fluid S-100beta and its relationship with AIDS dementia complex. J Clin Virol 22: 249–253.

Pereira CF, Middel J, Jansen G, et al (2001). Enhanced expression of fractalkine in HIV-1 associated dementia. J Neuroimmunol 115: 168–175.

Perelson AS, Neumann AU, Markowitz M, et al (1996). HIV-1 dynamics in vivo: virion clearance rate, infected cell life-span, and viral generation time. Science 271: 1582–1586.

Perrella O, Guerriero M, Izzo E, et al (1992). Interleukin-6 and granulocyte macrophage-CSF in the cerebrospinal fluid from HIV infected subjects with involvement of the central nervous system. Arq Neuropsiquiatr 50: 180–182.

Perrella O, Carreiri PB, Perrella A, et al (2001). Transforming growth factor beta-1 and interferon-alpha in the AIDS dementia complex (ADC): possible relationship with cerebral viral load? Eur Cytokine Netw 12: 51–55.

Peterson PK, Gekker G, Hu S, et al (1992). Microglial cell upregulation of HIV-1 expression in the chronically infected promonocytic cell line U1: the role of tumor necrosis factor-alpha. J Neuroimmunol 41: 81–87.

Pettoello-Mantovani M, Casadevall A, Kollmann TR, et al (1992). Enhancement of HIV-1 infection by the capsular polysaccharide of Cryptococcus neoformans. Lancet 339: 21–23.

Petzold A, Eikelenboom MJ, Gveric D, et al (2002). Markers for different glial cell responses in multiple sclerosis: clinical and pathological correlations. Brain 125: 1462–1473.

Petzold A, Jenkins R, Watt HC, et al (2003). Cerebrospinal fluid S100B correlates with brain atrophy in Alzheimer's disease. Neurosci Lett 336: 167–170.

Pierotti C, Lillo F, Vago L, et al (2005). Cerebrospinal Fluid HIV-1 p24 Antigen versus HIV-1 RNA as Low Cost Marker Supporting the Diagnosis of HIV-Associated CNS Disease. HIV Infection and the Central Nervous System, Developed and Resource Limited Settings, Frascati, Italy.

Pilcher CD, Shugars DC, Fiscus SA, et al (2001). HIV in body fluids during primary HIV infection: implications for pathogenesis, treatment and public health. AIDS 15: 837–845.

Polis MA, Suzman DL, Yoder CP, et al (2003). Suppression of cerebrospinal fluid HIV burden in antiretroviral naive patients on a potent four-drug antiretroviral regimen. AIDS 17: 1167–1172.

Power C, Patel KD (2004). Neurolipidomics: an inflammatory perspective on fat in the brain. Neurology 63: 608–609.

Power C, McArthur JC, Johnson RT, et al (1994). Demented and nondemented patients with AIDS differ in brain-derived human immunodeficiency virus type 1 envelope sequences. J Virol 68: 4643–4649.

Price RW (2000). The two faces of HIV infection of cerebrospinal fluid. Trends Microbiol 8: 387–391.

Price RW, Deeks SG (2004). Antiretroviral drug treatment interruption in human immunodeficiency virus-infected adults: clinical and pathogenetic implications for the central nervous system. J Neurovirol 10(Suppl 1): 44–51.

Price RW, Paxinos EE, Grant RM, et al (2001). Cerebrospinal fluid response to structured treatment interruption after virological failure. AIDS 15: 1251–1259.

Pulliam L, Clarke JA, McGrath MS, et al (1996). Monokine products as predictors of AIDS dementia. AIDS 10: 1495–1500.

Qin S, Rottman JB, Myers P, et al (1998). The chemokine receptors CXCR3 and CCR5 mark subsets of T cells associated with certain inflammatory reactions. J Clin Invest 101: 746–754.

Reddy RT, Achim CL, Sirko DA, et al (1996). Sequence analysis of the V3 loop in brain and spleen of patients with HIV encephalitis. AIDS Res Hum Retrovir 12: 477–482.

Reiber H (2001). Dynamics of brain-derived proteins in cerebrospinal fluid. Clin Chim Acta 310: 173–186.

Reid MC, Lachs MS, Feinstein AR (1995). Use of methodological standards in diagnostic test research. Getting better but still not good. JAMA 274: 645–651.

Reinhart TA (2003). Chemokine induction by HIV-1: recruitment to the cause. Trends Immunol 24: 351–353.

Resnick L, Berger JR, Shapshak P, et al (1988). Early penetration of the blood–brain barrier by HIV. Neurology 38: 9–14.

Respess RA, Cachafeiro A, Withum D, et al (2005). Evaluation of an ultrasensitive p24 antigen assay as a potential alternative to human immunodeficiency virus type 1 RNA viral load assay in resource-limited settings. J Clin Microbiol 43: 506–508.

Rho MB, Wesselingh S, Glass JD, et al (1995). A potential role for interferon-alpha in the pathogenesis of HIV-associated dementia. Brain Behav Immun 9: 366–377.

Richman DD (2000). Principles of HIV resistance testing and overview of assay performance characteristics. Antivir Ther 5: 27–31.

Rieckmann P, Nunke K, Burchhardt M, et al (1993). Soluble intercellular adhesion molecule-1 in cerebrospinal fluid: an indicator for the inflammatory impairment of the blood-cerebrospinal fluid barrier. J Neuroimmunol 47: 133–140.

Rimaniol AC, Zylberberg H, Rabian C, et al (1997). Imbalance between IL-1 and IL-1 receptor antagonist in the cerebrospinal fluid of HIV-infected patients. J Acquir Immune Defic Syndr Hum Retrovirol 16: 340–342.

Ritola K, Pilcher CD, Fiscus SA, et al (2004). Multiple V1/V2 env variants are frequently present during primary infection with human immunodeficiency virus type 1. J Virol 78: 11208–11218.

Ritola K, Robertson K, Fiscus SA, et al (2005). Increased human immunodeficiency virus type 1 (HIV-1) env compartmentalization in the presence of HIV-1-associated dementia. J Virol 79: 10830–10834.

Robertson KR, Robertson WT, Ford S, et al (2004). Highly active antiretroviral therapy improves neurocognitive functioning. J Acquir Immune Defic Syndr 36: 562–566.

Rosengren LE, Lycke J, Andersen O (1995). Glial fibrillary acidic protein in CSF of multiple sclerosis patients: relation to neurological deficit. J Neurol Sci 133: 61–65.

Rosengren LE, Karlsson JE, Karlsson JO, et al (1996). Patients with amyotrophic lateral sclerosis and other neurodegenerative diseases have increased levels of neurofilament protein in CSF. J Neurochem 67: 2013–2018.

Rothermundt M, Peters M, Prehn JH, et al (2003). S100B in brain damage and neurodegeneration. Microsc Res Tech 60: 614–632.

Royal W 3rd, Selnes OA, Concha M, et al (1994). Cerebrospinal fluid human immunodeficiency virus type 1 (HIV-1) p24 antigen levels in HIV-1-related dementia. Ann Neurol 36: 32–39.

Sabri F, De Milito A, Pirskanen R, et al (2001). Elevated levels of soluble Fas and Fas ligand in cerebrospinal fluid of patients with AIDS dementia complex. J Neuroimmunol 114: 197–206.

Sabri F, Titanji K, De Milito A, et al (2003). Astrocyte activation and apoptosis: their roles in the neuropathology of HIV infection. Brain Pathol 13: 84–94.

Sacktor N, Haughey N, Cutler R, et al (2004). Novel markers of oxidative stress in actively progressive HIV dementia. J Neuroimmunol 157: 176–184.

Saha RN, Pahan K (2003). Tumor necrosis factor-alpha at the crossroads of neuronal life and death during HIV-associated dementia. J Neurochem 86: 1057–1071.

Sala S, Spudich S, Presi S, et al (2005). Comparison of HIV-1 Pol Gene Sequences between CSF and Plasma shows Increasing Compartmentalization of Replication with Disease Stage and AIDS Dementia Complex. HIV Infection and the Central Nervous System, Developed and Resource Limited Settings, Frascati, Italy.

Sanders VJ, Pittman CA, White MG, et al (1998). Chemokines and receptors in HIV encephalitis. AIDS 12: 1021–1026.

Sarchielli P, Galli F, Floridi A, et al (2003). Relevance of protein nitration in brain injury: a key pathophysiological mechanism in neurodegenerative, autoimmune, or inflammatory CNS diseases and stroke. Amino Acids 25: 427–436.

Sasseville VG, Newman WA, Lackner AA, et al (1992). Elevated vascular cell adhesion molecule-1 in AIDS encephalitis induced by simian immunodeficiency virus. Am J Pathol 141: 1021–1030.

Sasseville VG, Smith MM, Mackay CR, et al (1996). Chemokine expression in simian immunodeficiency virus-induced AIDS encephalitis. Am J Pathol 149: 1459–1467.

Schmid P, Conrad A, Syndulko K, et al (1994). Quantifying HIV-1 proviral DNA using the polymerase chain reaction on cerebrospinal fluid and blood of seropositive individuals with and without neurologic abnormalities. J Acquir Immune Defic Syndr 7: 777–788.

Schmidtmayerova H, Nottet HS, Nuovo G, et al (1996). Human immunodeficiency virus type 1 infection alters chemokine beta peptide expression in human monocytes: implications for recruitment of leukocytes into brain and lymph nodes. Proc Natl Acad Sci USA 93: 700–704.

Schupbach J (2002). Measurement of HIV-1 p24 antigen by signal-amplification-boosted ELISA of heat-denatured plasma is a simple and inexpensive alternative to tests for viral RNA. AIDS Rev 4: 83–92.

Sethi KK, Naher H (1986). Elevated titers of cell-free interleukin-2 receptor in serum and cerebrospinal fluid specimens of patients with acquired immunodeficiency syndrome. Immunol Lett 13: 179–184.

Sevigny JJ, Albert SM, McDermott MP, et al (2004). Evaluation of HIV RNA and markers of immune activation as predictors of HIV-associated dementia. Neurology 63: 2084–2090.

Shacklett BL, Cox CA, Wilkens DT, et al (2004). Increased adhesion molecule and chemokine receptor expression on CD^{8+} T cells trafficking to cerebrospinal fluid in HIV-1 infection. J Infect Dis 189: 2202–2212.

Shaheen F, Collman RG (2004). Co-receptor antagonists as HIV-1 entry inhibitors. Curr Opin Infect Dis 17: 7–16.

Shaskan EG, Brew BJ, Rosenblum M, et al (1992a). Increased neopterin levels in brains of patients with human immunodeficiency virus type 1 infection. J Neurochem 59: 1541–1546.

Shaskan EG, Thompson RM, Price RW (1992b). Undetectable tumor necrosis factor-alpha in spinal fluid from HIV-1-infected patients. Ann Neurol 31: 687–689.

Shaunak S, Albright RE, Klotman ME, et al (1990). Amplification of HIV-1 provirus from cerebrospinal fluid and its correlation with neurologic disease. J Infect Dis 161: 1068–1072.

Shaunak S, Albright RE, Bartlett JA (1991). Amplification of human immunodeficiency virus provirus from cerebrospinal fluid: results of long-term clinical follow-up. J Infect Dis 164: 818.

Sidenius N, Sier CF, Ullum H, et al (2000). Serum level of soluble urokinase-type plasminogen activator receptor is a strong and independent predictor of survival in human immunodeficiency virus infection. Blood 96: 4091–4095.

Sidenius N, Nebuloni M, Sala S, et al (2004). Expression of the urokinase plasminogen activator and its receptor in HIV-1-associated central nervous system disease. J Neuroimmunol 157: 133–139.

Siliciano JD, Siliciano RF (2004). A long-term latent reservoir for HIV-1: discovery and clinical implications. J Antimicrob Chemother 54: 6–9.

Singer EJ, Syndulko K, Fahy-Chandon BN, et al (1994). Cerebrospinal fluid p24 antigen levels and intrathecal immunoglobulin G synthesis are associated with cognitive disease severity in HIV-1. AIDS 8: 197–204.

Sippy BD, Hofman FM, Wallach D, et al (1995). Increased expression of tumor necrosis factor-alpha receptors in the brains of patients with AIDS. J Acquir Immune Defic Syndr Hum Retrovirol 10: 511–521.

Smith DG, Guillemin GJ, Pemberton L, et al (2001). Quinolinic acid is produced by macrophages stimulated by platelet activating factor, Nef and Tat. J Neurovirol 7: 56–60.

Smith MO, Heyes MP, Lackner AA (1995). Early intrathecal events in rhesus macaques (Macaca mulatta) infected with pathogenic or nonpathogenic molecular clones of simian immunodeficiency virus. Lab Invest 72: 547–558.

Sonnerborg AB, von Stedingk LV, Hansson LO, et al (1989a). Elevated neopterin and beta 2-microglobulin levels in blood and cerebrospinal fluid occur early in HIV-1 infection. AIDS 3: 277–283.

Sonnerborg AB, von Sydow MA, Forsgren M, et al (1989b). Association between intrathecal anti-HIV-1 immunoglobulin G synthesis and occurrence of HIV-1 in cerebrospinal fluid. AIDS 3: 701–705.

Sonnerborg A, Johansson B, Strannegard O (1991). Detection of HIV-1 DNA and infectious virus in cerebrospinal fluid. AIDS Res Hum Retrovir 7: 369–373.

Speth C, Pichler I, Stockl G, et al (1998). Urokinase plasminogen activator receptor (uPAR; CD^{87}) expression on monocytic cells and T cells is modulated by HIV-1 infection. Immunobiology 199: 152–162.

Sporer B, Paul R, Koedel U, et al (1998). Presence of matrix metalloproteinase-9 activity in the cerebrospinal fluid of human immunodeficiency virus-infected patients. J Infect Dis 178: 854–857.

Sporer B, Koedel U, Goebel FD, et al (2000). Increased levels of soluble Fas receptor and Fas ligand in the cerebrospinal fluid of HIV-infected patients. AIDS Res Hum Retrovir 16: 221–226.

Sporer B, Kastenbauer S, Koedel U, et al (2003). Increased intrathecal release of soluble fractalkine in HIV-infected patients. AIDS Res Hum Retrovir 19: 111–116.

Sporer B, Missler U, Magerkurth O, et al (2004). Evaluation of CSF glial fibrillary acidic protein (GFAP) as a putative marker for HIV-associated dementia. Infection 32: 20–23.

Sporer B, Koedel U, Popp B, et al (2005). Evaluation of cerebrospinal fluid uPA, PAI-1, and soluble uPAR levels in HIV-infected patients. J Neuroimmunol 163: 190–194.

Spudich SS, Huang W, Nilsson AC, et al (2005a). HIV-1 chemokine coreceptor utilization in paired cerebrospinal fluid and plasma samples: a survey of subjects with viremia. J Infect Dis 191: 890–898.

Spudich SS, Nilsson AC, Lollo ND, et al (2005b). Cerebrospinal fluid HIV infection and pleocytosis: relation to systemic infection and antiretroviral treatment. BMC Infect Dis 5: 98.

Staprans S, Marlowe N, Glidden D, et al (1999). Time course of cerebrospinal fluid responses to antiretroviral therapy: evidence for variable compartmentalization of infection. AIDS 13: 1051–1061.

Stellbrink HJ, Eggers C, van Lunzen J, et al (1997). Rapid decay of HIV RNA in the cerebrospinal fluid during antiretroviral combination therapy. AIDS 11: 1655–1657.

Steuler H, Munzinger S, Wildemann B, et al (1992a). Quantitation of HIV-1 proviral DNA in cells from cerebrospinal fluid. J Acquir Immune Defic Syndr 5: 405–408.

Steuler H, Storch-Hagenlocher B, Wildemann B (1992b). Distinct populations of human immunodeficiency virus type 1 in blood and cerebrospinal fluid. AIDS Res Hum Retrovir 8: 53–59.

Stevens G, Rekhviashvili N, Scott LE, et al (2005). Evaluation of two commercially available, inexpensive alternative assays used for assessing viral load in a cohort of human immunodeficiency virus type 1 subtype C-infected patients from South Africa. J Clin Microbiol 43: 857–861.

Stingele K, Haas J, Zimmermann T, et al (2001). Independent HIV replication in paired CSF and blood viral isolates during antiretroviral therapy. Neurology 56: 355–361.

Stins MF, Pearce D, Di Cello F, et al (2003). Induction of intercellular adhesion molecule-1 on human brain endothelial cells by HIV-1 gp120: role of CD^4 and chemokine coreceptors. Lab Invest 83: 1787–1798.

Stone TW, Mackay GM, Forrest CM, et al (2003). Tryptophan metabolites and brain disorders. Clin Chem Lab Med 41: 852–859.

Strain MC, Letendre S, Pillai SK, et al (2005). Genetic composition of human immunodeficiency virus type 1 in cerebrospinal fluid and blood without treatment and during failing antiretroviral therapy. J Virol 79: 1772–1788.

Streit WJ, Graeber MB, Kreutzberg GW (1988). Functional plasticity of microglia: a review. Glia 1: 301–307.

Studahl M, Rosengren L, Gunther G, et al (2000). Difference in pathogenesis between herpes simplex virus type 1 encephalitis and tick-borne encephalitis demonstrated by means of cerebrospinal fluid markers of glial and neuronal destruction. J Neurol 247: 636–642.

Stuve O, Dooley NP, Uhm JH, et al (1996). Interferon beta-1b decreases the migration of T lymphocytes in vitro: effects on matrix metalloproteinase-9. Ann Neurol 40: 853–863.

Stuyver L, Wyseur A, Rombout A, et al (1997). Line probe assay for rapid detection of drug-selected mutations in the human immunodeficiency virus type 1 reverse transcriptase gene. Antimicrob Agents Chemother 41: 284–291.

Sui Y, Potula R, Dhillon N, et al (2004). Neuronal apoptosis is mediated by CXCL10 overexpression in simian human immunodeficiency virus encephalitis. Am J Pathol 164: 1557–1566.

Suryadevara R, Holter S, Borgmann K, et al (2003). Regulation of tissue inhibitor of metalloproteinase-1 by astrocytes: links to HIV-1 dementia. Glia 44: 47–56.

Svenningsson A, Hansson GK, Andersen O, et al (1993). Adhesion molecule expression on cerebrospinal fluid T lymphocytes: evidence for common recruitment mechanisms in multiple sclerosis, aseptic meningitis, and normal controls. Ann Neurol 34: 155–161.

Tambussi G, Gori A, Capiluppi B, et al (2000). Neurological symptoms during primary human immunodeficiency virus (HIV) infection correlate with high levels of HIV RNA in cerebrospinal fluid. Clin Infect Dis 30: 962–965.

Tanaka M, Itai T, Adachi M, et al (1998). Downregulation of Fas ligand by shedding. Nat Med 4: 31–36.

Tang YW, Huong JT, Lloyd RM Jr, et al (2000). Comparison of human immunodeficiency virus type 1 RNA sequence heterogeneity in cerebrospinal fluid and plasma. J Clin Microbiol 38: 4637–4639.

Tashima KT, Flanigan TP, Kurpewski J, et al (2002). Discordant human immunodeficiency virus type 1 drug resistance mutations, including K103N, observed in cerebrospinal fluid and plasma. Clin Infect Dis 35: 82–83.

Taub DD, Lloyd AR, Conlon K, et al (1993). Recombinant human interferon-inducible protein 10 is a chemoattractant for human monocytes and T lymphocytes and promotes T cell adhesion to endothelial cells. J Exp Med 177: 1809–1814.

Teunissen CE, Dijkstra C, Polman C (2005). Biological markers in CSF and blood for axonal degeneration in multiple sclerosis. Lancet Neurol 4: 32–41.

Thompson KA, McArthur JC, Wesselingh SL (2001). Correlation between neurological progression and astrocyte apoptosis in HIV-associated dementia. Ann Neurol 49: 745–752.

Tong N, Perry SW, Zhang Q, et al (2000). Neuronal fractalkine expression in HIV-1 encephalitis: roles for macrophage recruitment and neuroprotection in the central nervous system. J Immunol 164: 1333–1339.

Torre D, Zeroli C, Ferraro G, et al (1992). Cerebrospinal fluid levels of IL-6 in patients with acute infections of the central nervous system. Scand J Infect Dis 24: 787–791.

Towfighi A, Skolasky RL, St Hillaire C, et al (2004). CSF soluble Fas correlates with the severity of HIV-associated dementia. Neurology 62: 654–656.

Trebst C, Staugaitis SM, Tucky B, et al (2003). Chemokine receptors on infiltrating leucocytes in inflammatory pathologies of the central nervous system (CNS). Neuropathol Appl Neurobiol 29: 584–595.

Trujillo JR, Navia BA, Worth J, et al (1996). High levels of anti-HIV-1 envelope antibodies in cerebrospinal fluid as compared to serum from patients with AIDS dementia complex. J Acquir Immune Defic Syndr Hum Retrovirol 12: 19–25.

Tsoukas CM, Bernard NF (1994). Markers predicting progression of human immunodeficiency virus-related disease. Clin Microbiol Rev 7: 14–28.

Tyor WR, Glass JD, Griffin JW, et al (1992). Cytokine expression in the brain during the acquired immunodeficiency syndrome. Ann Neurol 31: 349–360.

Vago L, Nebuloni M, Bonetto S, et al (2001). Rantes distribution and cellular localization in the brain of HIV-infected patients. Clin Neuropathol 20: 139–145.

Valle M, Price RW, Nilsson A, et al (2004). CSF quinolinic acid levels are determined by local HIV infection: cross-sectional analysis and modelling of dynamics following antiretroviral therapy. Brain 127: 1047–1060.

van den Brande G, Marquie-Beck J, Capparelli E, et al (2005). Kaletra independently reduces HIV replication in cerebrospinal fluid. 12th Conference on Retrovirus and Opportunistic Infections, Boston, MA.

van Marle G, Power C (2005). Human immunodeficiency virus type 1 genetic diversity in the nervous system: evolutionary epiphenomenon or disease determinant? J Neurovirol 11: 107–128.

van Marle G, Henry S, Todoruk T, et al (2004). Human immunodeficiency virus type 1 Nef protein mediates neural cell death: a neurotoxic role for IP-10. Virology 329: 302–318.

Van Wielink G, McArthur JC, Moench T, et al (1990). Intrathecal synthesis of anti-HIV IgG: correlation with increasing duration of HIV-1 infection. Neurology 40: 816–819.

Vandamme AM, Houyez F, Banhegyi D, et al (2001). Laboratory guidelines for the practical use of HIV drug resistance tests in patient follow-up. Antivir Ther 6: 21–39.

Venturi G, Catucci M, Romano L, et al (2000). Antiretroviral resistance mutations in human immunodeficiency virus type 1 reverse transcriptase and protease from paired cerebrospinal fluid and plasma samples. J Infect Dis 181: 740–745.

Vesce S, Bezzi P, Rossi D, et al (1997). HIV-1 gp120 glycoprotein affects the astrocyte control of extracellular glutamate by both inhibiting the uptake and stimulating the release of the amino acid. FEBS Lett 411: 107–109.

Viviani B, Corsini E, Binaglia M, et al (2001). Reactive oxygen species generated by glia are responsible for neuron death induced by human immunodeficiency virus-glycoprotein 120 in vitro. Neuroscience 107: 51–58.

Wakabayashi H, Yano M, Tachikawa N, et al (2001). Increased concentrations of 14-3-3 epsilon, gamma and zeta isoforms in cerebrospinal fluid of AIDS patients with neuronal destruction. Clin Chim Acta 312: 97–105.

Wesselingh SL, Power C, Glass JD, et al (1993). Intracerebral cytokine messenger RNA expression in acquired immunodeficiency syndrome dementia. Ann Neurol 33: 576–582.

Wetzel MA, Steele AD, Henderson EE, et al (2002). The effect of X4 and R5 HIV-1 on C, C-C, and C-X-C chemokines during the early stages of infection in human PBMCs. Virology 292: 6–15.

Wildemann B, Haas J, Ehrhart K, et al (1993). In vivo comparison of zidovudine resistance mutations in blood and CSF of HIV-1-infected patients. Neurology 43: 2659–2663.

Wiley CA, Achim CL, Schrier RD, et al (1992). Relationship of cerebrospinal fluid immune activation associated factors to HIV encephalitis. AIDS 6: 1299–1307.

Wilt SG, Milward E, Zhou JM, et al (1995). In vitro evidence for a dual role of tumor necrosis factor-alpha in human immunodeficiency virus type 1 encephalopathy. Ann Neurol 37: 381–394.

Winkler F, Kastenbauer S, Koedel U, et al (2002). Role of the urokinase plasminogen activator system in patients with bacterial meningitis. Neurology 59: 1350–1355.

Wirleitner B, Schroecksnadel K, Winkler C, et al (2005). Neopterin in HIV-1 infection. Mol Immunol 42: 183–194.

Wong JK, Ignacio CC, Torriani F, et al (1997). In vivo compartmentalization of human immunodeficiency virus: evidence from the examination of pol sequences from autopsy tissues. J Virol 71: 2059–2071.

Woodman SE, Benveniste EN, Nath A, et al (1999). Human immunodeficiency virus type 1 TAT protein induces adhesion molecule expression in astrocytes. J Neurovirol 5: 678–684.

Xia MQ, Hyman BT (1999). Chemokines/chemokine receptors in the central nervous system and Alzheimer's disease. J Neurovirol 5: 32–41.

Xia MQ, Bacskai BJ, Knowles RB, et al (2000). Expression of the chemokine receptor CXCR3 on neurons and the elevated expression of its ligand IP-10 in reactive astrocytes: in vitro ERK1/2 activation and role in Alzheimer's disease. J Neuroimmunol 108: 227–235.

Zhang K, McQuibban GA, Silva C, et al (2003). HIV-induced metalloproteinase processing of the chemokine stromal cell derived factor-1 causes neurodegeneration. Nat Neurosci 6: 1064–1071.

Zhao J, Lopez AL, Erichsen D, et al (2004). Mitochondrial glutaminase enhances extracellular glutamate production in HIV-1-infected macrophages: linkage to HIV-1 associated dementia. J Neurochem 88: 169–180.

Zheng J, Thylin MR, Ghorpade A, et al (1999). Intracellular CXCR4 signaling, neuronal apoptosis and neuropathogenic mechanisms of HIV-1-associated dementia. J Neuroimmunol 98: 185–200.

Zink MC, Coleman GD, Mankowski JL, et al (2001). Increased macrophage chemoattractant protein-1 in cerebrospinal fluid precedes and predicts simian immunodeficiency virus encephalitis. J Infect Dis 184: 1015–1021.

Handbook of Clinical Neurology, Vol. 85 (3rd series)
HIV/AIDS and the Nervous System
P. Portegies, J. R. Berger, Editors

Chapter 18

The neuropathology of HIV

BENJAMIN B. GELMAN*

Texas NeuroAIDS Research Center, University of Texas Medical Branch, Galveston, TX, USA

18.1. Historical perspective

The neuropathology of human immunodeficiency virus type 1 (HIV-1) infection has evolved in parallel with changing clinical patterns (McArthur, 2004). In the first decade of the HIV-1 pandemic, virus infection was relentless and progressed eventually to the acquired immunodeficiency syndrome (AIDS). AIDS was usually fatal within one or two years, and most decedents were relatively young adults with profound immunodeficiency. Disease progression was not interrupted substantially and hardly any patients survived to old age. It was recognized early in the epidemic that pathological changes were common in the central nervous system (CNS), and that CNS pathology was associated clinically with problems related to subcortical dementia, opportunistic infections and primary CNS lymphoma (Petito, 1993; Budka, 1998). The cause of death in people with AIDS was related to underlying CNS pathology in about 35% of the autopsies, second only to pulmonary failure (Klatt, 1988). Peripheral neuropathy was highly prevalent clinically and pathologically, and it produced painful disability, although it did not contribute substantially to the high mortality rates (Cornblath and McArthur, 1988). PNS pathology usually consisted primarily of selective degeneration of small fiber axons, leading to a painful distal sensory peripheral neuropathy (Griffin et al., 1998).

The scenario changed favorably after highly active antiretroviral therapy (HAART) was introduced. HAART produced clinical suppression of HIV-1 replication that, in turn, led to decreased AIDS morbidity and mortality in treated populations. In most people HAART transformed HIV-1 infection from a potentially lethal progressive disease into a persistent infection amenable to medical management and prolonged survival. The lack of disease progression and prolonged survival produced a parallel decline in CNS-related mortality. Neurological status can show some improvement after HAART (Sacktor et al., 1999; Brew, 2004; McArthur et al., 2005), but it remains unclear how these changing clinical patterns are reflected in the underlying neuropathology. The availability of postmortem information in the HAART era is not as plentiful because fewer autopsies are being performed. When autopsies are done nowadays one is likely to observe the end result of a convoluted disease history that is intertwined with multiple co-morbid conditions. The more advanced age of HIV-1-infected decedents has introduced a new spectrum of potential neuropathological change that was not evident in younger autopsy cohorts before HAART. Older decedents have increasingly prolonged biological "histories" of HIV-1 infection, and they usually undergo widely varying exposures to HAART medicines. Prolonged and less stereotypical battles with HIV-1 infection could be synergistic with co-morbid factors that can influence neurocognitive function. Some important comorbidities that might influence cognition or the underlying neuropathology are (1) biological and pathological aging of the brain, (2) side effects of HAART drugs, (3) increased cumulative exposure to CNS HIV infection, (4) addiction to drugs of abuse, (5) effects of treating psychiatric disturbances and (6) co-infection with hepatitis C virus (Bell, 2004). A key logistical advancement in the HAART era was the establishment of the National NeuroAIDS Tissue Consortium, which made it more feasible to perform research on the cliniconeuropathological correlation (Morgello et al., 2001). To avoid repetition, many important opportunistic infections of the CNS that occur in people with HIV/AIDS are not described yet again. The reader is referred to outstanding contributions made by previous writers (Petito, 1993; Budka, 1998) and to Chapter 19 in this volume.

*Correspondence to: Benjamin B. Gelman, MD, PhD, Professor of Pathology and Neuroscience and Cell Biology, Director, Texas NeuroAIDS Research Center, Department of Pathology, Route 0785, University of Texas Medical Branch, Galveston, TX 77555-0785, USA. Tel: +1-409-772-5316, Fax: +1-409-772-5220.

18.2. HIV encephalitis

Interest in HIV-associated neuropathology continues to be focused on the mechanism of HIV-associated dementia (Heaton et al., 1995; McArthur et al., 2005). Unlike the senile dementias, the cliniconeuropathological correlation in HIV-associated dementia remains to be clearly elucidated. The prime neuropathological candidate for causation of HIV dementia, and less severe neurocognitive changes, is HIV encephalitis (Fig. 18.1). HIV encephalitis is a subacute inflammatory response in the brain that occurs in association with an overt infection of brain cells by HIV-1. When other infectious CNS pathogens are present in the brain, such as cytomegalovirus (CMV) or *Toxoplasma gondii*, the diagnosis can be difficult to make. When no other pathogen is present that can provoke a subacute inflammatory picture, the diagnosis of HIV encephalitis is made much more readily. HIV-1 infects cells of the CNS in all infected people (Bell, 2004), but in HIV encephalitis, virus replication may be accelerated and proinflammatory changes are produced. The rate of HIV-1 replication and inflammatory responses in the brain vary substantially between infected people for reasons that remain unclear. Before HAART, HIV encephalitis occurred in about 15 to 20% of decedents, a prevalence that seemed to echo the prevalence of dementia in clinical cohorts (McArthur et al., 2005). The incidence of HIV dementia declined in the HAART era, but HIV encephalitis prevalence may not have undergone a parallel decrease in HAART-era autopsy cohorts (Jellinger et al., 2000; Masliah et al., 2000; Morgello et al., 2002; Vago et al., 2002; Neuenburg et al., 2002; Gray et al., 2003). The difference between autopsy and clinical surveys probably relates to the fact that autopsy cohorts are enriched with people with end-stage AIDS. These patients are the most vulnerable to HIV encephalitis and other AIDS-associated neurological problems, whereas clinical cohorts tend to contain fewer of these types of patients. Whatever the reason is, HIV encephalitis at autopsy remains prevalent.

Gross brain anatomy remains of interest due to the expanding technology of clinical brain imaging. The gross brain in people with HIV encephalitis does not contain a characteristic anomaly. Mild brain atrophy is a highly prevalent change in people with AIDS generally and is more prevalent in people with dementia (Gelman et al., 1996; McArthur et al., 2005). Cerebral atrophy is not, however, specific to HIV-associated dementia or HIV encephalitis. As well, atrophy is not linked solidly with any single neuropathological change that is used to diagnose HIV encephalitis (Gelman and Guinto, 1992). HIV encephalitis is diagnosed histologically based upon the presence of a reasonably characteristic constellation of changes. The most unique change is the presence of multinucleated giant cells, which is considered a hallmark lesion (Budka, 1986; Budka et al., 1991). Multinucleated giant cells are composed of fused mononuclear phagocytes actively infected with HIV-1. In the setting of increased inflammation, in a decedent known to have HIV-1 infection, the multinucleated giant cell is a specific sign of HIV encephalitis provided that other diseases that form giant cells are excluded (such as extrapulmonary tuberculosis, fungal infection, sarcoidosis or other granulomatous disease). HIV-1 glycoprotein 41 (gp41), a subunit of the virus envelope, is the putative fusion protein that leads to giant cell formation. Multinucleated giant cells most often appear in perivascular sectors of brain white matter, but they can appear in any area. Any doubt concerning specificity for active HIV replication in a particular case can be resolved by demonstrating positive immunostaining for HIV-1 envelope protein within them (i.e. HIV-1 gp41 and p24 antigens). HIV-1 antigen staining is not usually needed to recognize them or to make the diagnosis of HIV encephalitis. Indeed, some cases do not display multinucleated giant cells, and the diagnosis is sometimes made based upon the presence of other changes (see below). Many neuropathologists prefer to report cases that contain multinucleated giant cells as "HIV-associated multinucleated cell encephalitis", so as to make a clear distinction from cases of HIV encephalitis that do not contain them.

Another histological change usually present in HIV encephalitis is multiple microglial nodules. Microglial nodules are loosely aggregated clusters of activated microglial cells. These cells may, but do not always, contain HIV antigenicity (e.g. gp41 and p24) when immunostaining is done. Microglial nodules are not specific to HIV encephalitis and can be produced by infection with other CNS pathogens, the most relevant examples being CMV encephalitis and *T. gondii*. Nevertheless, when microglial nodules are present in combination with other characteristic features (to be described below), and no evidence of other CNS pathogen is detected, HIV encephalitis can be diagnosed even when multinucleated cells are not observed. When there is residual doubt concerning the specificity of the reaction for HIV-1, a diagnosis of microglial nodule encephalitis is made. These types of cases continue to cause substantial confusion because they raise the issue of whether microglial nodule encephalitis can be a subtype of HIV encephalitis that lacks the hallmark multinucleated giant cell formation, or represents other low-level infection, or is a combination of infections. At present, it is fair to say that some cases with microglial nodules lack enough adjunctive pathological evidence to justify a firm diagnosis of HIV encephalitis. We still do not know whether the neurobiology of those cases is substantially different from HIV encephalitis

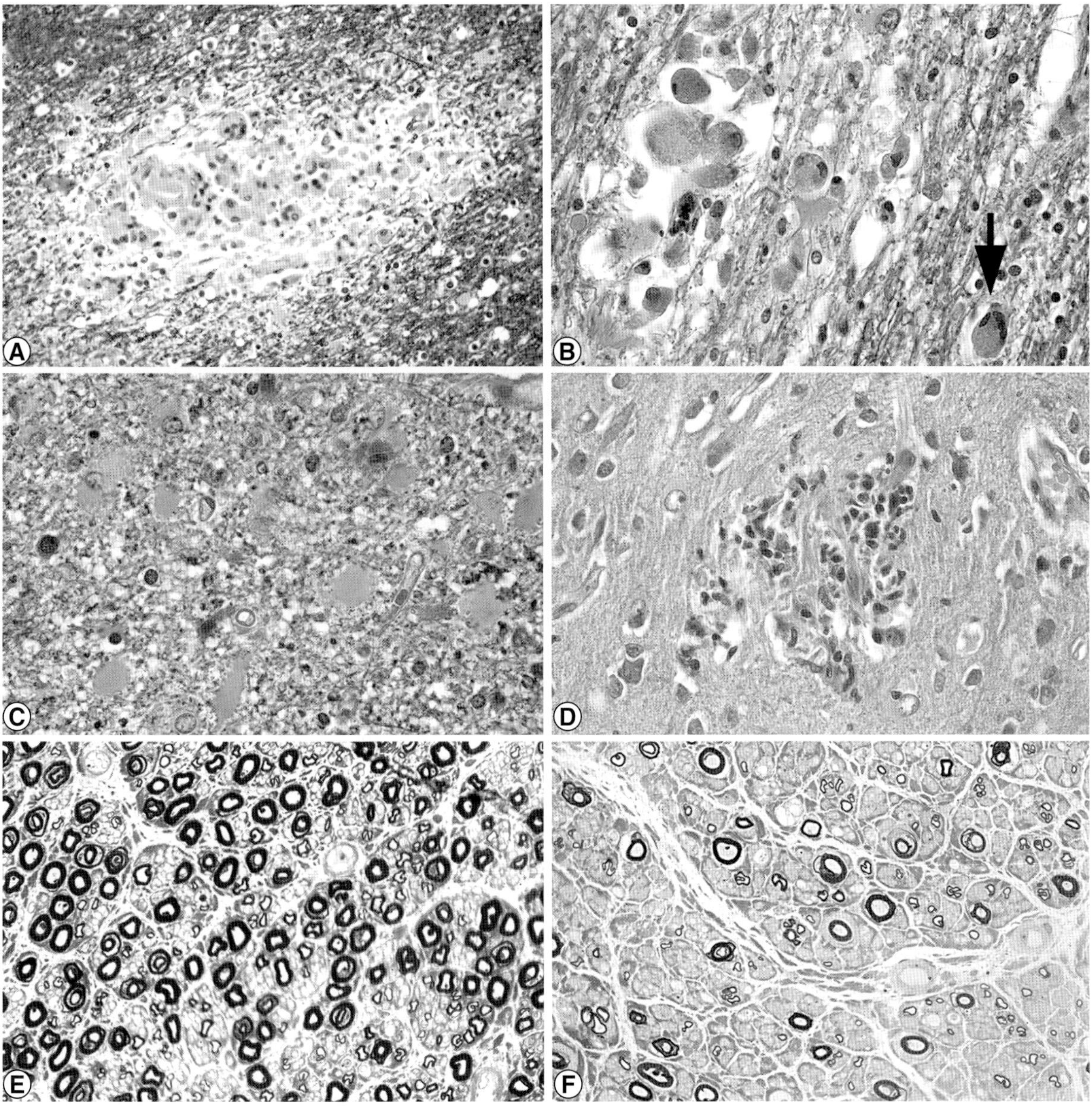

Fig. 18.1. For full color figure, see plate section. Neuropathological changes in HIV-associated encephalitis and peripheral neuropathy. A. A white matter lesion in HIV encephalitis contains numerous macrophages and microglial cells. Note the loss of blue-stained myelin in this lesion. B. A lesion in HIV encephalitis contains abundant plump macrophages; some contain more than one nucleus. A typical HIV-infected multinucleated cell is indicated by the arrow. C. Blue-stained myelin contains many abundant plump reactive astrocytes (pink areas). Astrocytes can support a nonproductive HIV infection and might participate in the pathophysiology of HIV encephalitis. D. A large microglial nodule is shown in brain gray matter. These clusters of microglial cells are usually present in HIV encephalitis, and often are immunostained with antibodies against HIV antigens. E, F. Cross-sections of the sural nerve from two HIV-infected subjects. E shows a subject that did not have neuropathy; F is from a subject with distal sensory peripheral neuropathy (DSP) and reveals a lack of myelinated nerve fibers. DSP is not in general correlated with the amount of HIV replication in the nerve tissue. (A–D: Luxol Fast Blue/hematoxylin/eosin stain; E, F: Semithin sections stained with toluidine blue.)

cases that do contain multinucleated giant cells. "Lumping" together all these types of HIV encephalitis cases is not completely justified; the rationale for "splitting" them into separate categories also has weakness.

Another characteristic pattern of change in HIV encephalitis is the appearance of rarefaction of myelin staining centered primarily about white matter blood vessels. The loss of white matter is usually angiocentric, and it often is associated with increased numbers of macrophages, microglial cells, multinucleated giant cells, reactive astrocytes or lymphocytes within the perivascular sectors. Macrophages in these numerous small sectors sometimes contain myelin debris, which demonstrates that degeneration of myelinated nerve fibers occurs to some extent. Sometimes, the microvasculature exhibits sclerotic changes of endothelial damage, which suggests that microvascular damage and secondary ischemia play a role (Smith et al., 1990). A more diffuse decrease of myelin staining of white matter also occurs that is not restricted to the perivascular spaces. This loss of white matter is usually accompanied by astroglial hypertrophy. It is possible that perivascular patterns of nerve fiber injury evolve into these more diffuse white matter lesions. There is little or no indication that the changes in white matter represent "primary" demyelination as in multiple sclerosis plaques (i.e. loss of myelin with relative preservation of axons). Indeed, the evidence suggests that myelinated axons do undergo degeneration in HIV encephalitis (Giometto et al., 1997; Raja et al., 1997). "Myelin pallor" is yet another pattern of diffuse white matter change that may be present in HIV encephalitis. Myelinated nerve fibers exhibit pale staining, but this change is not likely to reflect outright loss of myelinated nerve fibers. The lack of staining reflects leakage of plasma protein through the blood–brain barrier, which interferes with white matter staining technique (Petito and Cash, 1992; Power et al., 1993). Astroglial hypertrophy is present to some extent in the majority of HIV encephalitis cases. The intensity and regional pattern of astroglial hypertrophy are highly variable; it is more often present in white matter. The astrocytic changes could represent a nonspecific reaction to injury, or could be a participant in a multicellular inflammatory cascade. It also is possible that astrocytes become hypertrophic and/or apoptotic in response to a nonproductive and limited type of HIV-1 infection (Sabri et al., 2003; Thompson et al., 2004). Other variably seen abnormalities in HIV encephalitis include a diffuse microglial cell activation (not in clusters) (Gelman, 1993; Bell, 2004), and perivascular lymphocytic infiltration (Petito et al., 2003). All these cellular changes fit with the overall proinflammatory concept that is suggested to drive the pathophysiology of HIV-associated dementia (Williams and Hickey, 2002). In severe cases of HIV encephalitis there can be structural disorganization and vacuolization of the neuropil (poliodystrophy) as viewed in routine histological tissue sections (Budka et al., 1991). The technique of Golgi impregnation has been used to demonstrate that subtle disorganization of dendritic morphology can occur (Masliah et al., 1992a), and stereological technique has suggested that there may be dropout of neuronal cell bodies (Ketzler et al., 1990; Everall et al., 1993a; Fischer et al., 1999). Subtle synaptodendritic damage was quantified by one group of investigators and was significantly correlated with HIV-associated dementia (Masliah et al., 1997). Another very unusual change that is recognized to occur in HIV encephalitis is a severe vacuolar degeneration of cerebral white matter, known as vacuolar leukoencephalopathy (Schmidbauer et al., 1990).

Some cases of HIV encephalitis have changes that are merely suggestive of the diagnosis because histological anomalies either are very weakly present, or the changes are not specific enough after excluding other potential pathogens. This kind of autopsy case material seems to be more common in the HAART era (personal observation of the author). For example, a well-sampled brain can contain two or three multinucleated giant cells, or diffuse microglial cell activation, without any other substantial histopathological change. Measuring HIV p24 or gp41 antigenicity using immunostaining, or semiquantitative HIV-1 gene expression using in situ hybridization, can provide useful adjunctive information to establish whether or not replicating HIV-1 is present in the lesions (Budka, 1998). When HIV-1 antigens are present, a diagnosis of HIV encephalitis, albeit "mild", is usually justified. Some case material contains abundant evidence of HIV-1 antigenicity and/or gene expression in microglial cells, but has little obvious inflammatory reaction and no multinucleated giant cells. These challenging cases are usually categorized as HIV encephalitis when there is evidence that clinical dementia was present (Cherner et al., 2002). The fact that some demented people do not have HIV encephalitis and exhibit no neuropathological change has been recognized for many years (Glass et al., 1993; Wiley and Achim, 1994). That fact implies that the onset of cognitive impairment might actually precede the appearance of neuropathological changes instead of following them.

18.3. HIV encephalitis and HAART

Clinically, the incidence of dementia has declined, but the overall prevalence in the population of less severe neurocognitive disability is increasing because people survive longer (McArthur, 2004). As the clinical picture undergoes evolution with time, a qualitative change in the spectrum of neuropathology of HIV infection may be underway. However, autopsy reports suggest that the prevalence of HIV encephalitis has remained at about the same level in the HAART era, and indicate that people at risk of dying from HIV/AIDS still remain vulnerable to HIV encephalitis (Jellinger et al., 2000; Masliah et al., 2000; Morgello et al., 2002; Neuenberg et al., 2002; Vago et al., 2002; Gray et al., 2003). The *intensity* of HIV encephalitis might have undergone overall reduction in the HAART era in parallel with the decreased severity of impairment, but that is difficult to validate since the diagnosis remains essentially "all or none". Thus, the correlation of milder dementia with the presence of multinucleated cells is not strong (Cherner et al., 2002). The general impression is that the proportion of autopsies that contain anomalies lacking sufficient specificity or intensity to make the diagnosis of HIV encephalitis has increased (personal observation). One can speculate that these cases might represent the earliest and most formative stages of HIV encephalitis, or might reflect aborted infection and "burnt out" HIV encephalitis after effective virus suppression with HAART (Gray et al., 2003).

Several examples have been reported in which the intensity of white matter damage in HAART-treated decedents was increased relative to what was described in the era prior to HAART (Langford et al., 2002). White matter necrosis severe enough to be termed leukoencephalopathy was present, and it was linked to signs of HIV encephalitis and HIV-1 replication in most instances. Increased perivascular mononuclear inflammation also was observed, and it was suggested that immune reconstitution induced by HAART was the cause in some cases. This trend also was suggested to be present in a European autopsy cohort (Gray et al., 2003). Leukoencephalopathy was recognized generally as being fairly common prior to HAART (Budka, 1998). When a person is treated with HAART successfully, one may observe only nonspecific astrogliosis and damage to white or gray matter, but no multinucleated giant cells, microglial nodule encephalitis, inflammation or other infectious agent. As mentioned, a diagnosis of "burnt out HIV encephalitis" is suggested to apply to these cases (Gray et al., 2003).

Immune reconstitution inflammatory syndrome is an emerging adverse response to HAART that occurs in some people shortly after HIV-1 replication is suppressed (Shelburne et al., 2005). Decreasing virus load produces a restoration of the immune system as reflected by an increase in the number of circulating CD^4 lymphocytes, which then triggers an autoimmune-like inflammatory reaction. Immune reconstitution inflammatory syndrome is usually an over-response to a pre-existing infection, so the clinical picture resembles disease progression. This paradoxical deterioration in immune reconstitution inflammatory syndrome can produce serious and potentially fatal pathological effects in the CNS. Immune reconstitution inflammatory syndrome in the CNS can be a response to pre-existing infection with *Cryptococcus neoformans* (Wood et al., 1998; King et al., 2002), CMV (Jacobson et al., 1997; Deayton et al., 2000) or papovavirus (the cause of progressive multifocal leukoencephalopathy, PML) (Tantisiriwat et al., 1998; Du Pasquier and Koralnik, 2003; Corral et al., 2004), and possibly HIV encephalitis (Langford et al., 2002; Miller et al., 2004). CNS inflammatory responses usually were blunted in people with immunodeficiency prior to HAART, whereas in immune reconstitution inflammatory syndrome, the inflammatory response is heightened pathologically. For example, a fatal inflammatory variant of PML was reported in which some white matter lesions contained abundant inflammatory cells with little stained papovavirus (Miralles et al., 2001; Corral et al., 2004; Vendrely et al., 2005). These examples give the impression that "inflammatory PML" is becoming increasingly prevalent in the HAART era (Du Pasquier and Koralnik, 2003). Immune reconstitution inflammatory syndrome-related relapsing meningitis due to *Cryptococcus neoformans* infection and CMV retinitis (Wood et al., 1998; Deayton et al., 2000; King et al., 2002) also are documented. Concerning the impact of immune reconstitution inflammatory syndrome on HIV encephalitis, some examples of heightened inflammatory changes in necrotic white matter were suggested to be a form of immune reconstitution inflammatory syndrome, in which pre-existing HIV encephalitis is exacerbated (Langford et al., 2002). Two examples of panencephalitis (gray and white matter inflammation) with increased infiltration of T lymphocytes after initiation of HAART also were associated with an immune reconstitution inflammatory syndrome-like worsening brain HIV infection (Miller et al., 2004). These case reports seem to imply that an unusual but dangerous "inflammatory variant" of HIV encephalitis could have emerged in the HAART era due to immune reconstitution inflammatory syndrome.

Another qualitative change suggested to be possible in HAART-era brain specimens pertains to biological and pathological brain aging. These two pathologies could be independent and produce additive effects on cognitive impairment. They also might enter into mutual interaction and synergy. Either way, elderly HIV-infected people will experience escalating risk of neurocognitive disability in old age, as everyone does. If the biological interaction should prove to be synergistic, then the added burden of disability in infected people could prove to be clinically significant. Studies on pathological aging in HIV-infected people are few in number thus far, and they have focused on the pathological changes associated with Alzheimer's disease (Nath and Hersh, 2005). Specifically, extracellular deposition of fibrillary aggregations of the beta fragment of amyloid precursor protein (Aβ plaque or diffuse plaque) is linked with pathological brain aging. They are the probable precursor of neuritic ("senile") plaques, which are a neuropathological hallmark of Alzheimer's disease. One report suggested that plaques that contain argyrophilic deposits of Aβ (possibly senile plaques) might be greater than expected in people with HIV/AIDS (Esiri et al., 1998). Two reports suggested that Aβ plaque is observed more commonly in people with HIV/AIDS than expected (Green et al., 2005; Rempel and Pulliam, 2005). However, a study that controlled for the influence of chronological age found that hippocampi from elderly people with end-stage AIDS had diffuse Aβ plaque counts that were not different quantitatively from an age-matched comparison cohort (Gelman and Schuenke, 2004). Fibrillary Aβ begins to accumulate in most humans after the age of about 50 years and reflects biological brain age in part (Yamaguchi et al., 2001). There is no doubt that fibrillary Aβ will accumulate eventually in people with HIV/AIDS who survive to advanced age. The important question remains as to whether Aβ accumulation is significantly out of proportion to the chronological age of the decedent. Specifically, the age of onset and the rate of accumulation of Aβ must be carefully assessed. The kinetics of "neuritic" changes (i.e. pathological senile plaques) also need to be evaluated in parallel, because these are the lesions that produce dementia. These cross-sectional neuropathological studies are tedious and difficult to perform. One fact that can limit the meaning of senile plaque counting is that the correlation between them and the severity of senile dementia is fairly weak (Terry et al., 1991). Thus, even if senile plaques are eventually found to be prevalent in HIV/AIDS, their clinical impact on the incidence or severity of HIV-associated dementia would remain uncertain. It remains unclear whether HIV-infected people are destined to experience worsening of age-adjusted risk for the accumulation of senile plaque or other age-related neuropathologies. Numerous other age-associated changes in the brain will need to be considered to obtain a clear picture. For example, senile dementia and HIV-associated dementia both are correlated with damage to the neuropil (Terry et al., 1991; Masliah et al., 1997). Synaptodendritic changes might, therefore, prove to be a clinically relevant neuropathogical focus in aging populations. As well, the age-associated accumulation of ubiquitinylated "aggresomes" in brain tissue is more prevalent in HIV encephalitis, and might exacerbate biological brain aging (Gelman and Schuenke, 2004). Still another potential focus of synergy is the effect of microglial cell activation, which occurs in pathological brain aging and HIV infection both (Gelman, 1993; Streit, 2005). An "elderly variant" of HIV encephalitis could emerge eventually in which pathological processes that drive senile dementias, and diseases of aging generally, are exacerbated. That scenario remains a speculation since studies that carefully control for chronological age are required.

18.4. The cliniconeuropathological correlation in HIV-associated dementia

Many histochemical, morphological or biochemical anomalies occur in human brain specimens, cerebrospinal fluid (CSF) samples and animal models of HIV encephalitis. Pertaining to the cliniconeuropathological correlation, there are some persistent difficulties with this body of literature. First, this literature tends to focus on the neuropathologic diagnosis of HIV encephalitis instead of clinical dementia. Many studies did not have access to the resources needed to measure the neurocognitive status of a decedent before death. As a result the clinical relevance of many pathological reports remains vague. Second, co-morbid clinical conditions that can produce neurocognitive changes are very difficult to factor out of retrospective autopsy studies. Third, a standardized research paradigm to define dementia and co-morbid conditions in HIV-1 infected people has not been canonized. This makes cross-comparing the various reports quite difficult to do. And finally, in vitro systems often produce highly interesting results that are of unknown relevance to diseased brain tissue. It is difficult to compare all these kinds of studies to one another. Very few significant clinical correlations have undergone independent replication in more than one cohort. One limitation of the neuro-AIDS field before HAART was that the resources needed to address the clinical–pathological correlation were not available widely. That problem was addressed beginning in 1998, when the National NeuroAIDS Tissue Consortium was established (Morgello et al., 2001). A consortium of "brain banks" was formed to accumulate and distribute human brain specimens to the worldwide research community. Investigators can obtain brain tissue specimens from patients who underwent formal evaluation of neurocognitive impairment before undergoing an autopsy (www.hivbrainbanks.org). Other co-morbid factors are taken account of in the National NeuroAIDS Tissue Consortium cohort. Specimens and data are available from four different geographical cohorts in the USA, so attempts to "replicate" findings across these four cohorts is feasible. Expertise worldwide can be harnessed to pursue the pathophysiology of HIV-associated dementia.

Table 18.1 summarizes some of the key experimental findings that have been made using brain specimens from people with HIV/AIDS. A wide variety of brain

Table 18.1

Cliniconeuropathologic correlation in HIV-associated dementia

Neuropathological abnormality	Dementia measured antemortem?	Significant correlation suggested?	In more than a single cohort?
Inflammation, macrophages and HIV			
HIV encephalitis	Yes	Yes	Yes
HIV antigen, or RNA, or DNA	Yes	Sometimes (esp. when HIV replication not suppressed)	Yes
Increased nitric oxide synthesis	Yes	Yes	Yes
Brain density of macrophages	Yes	Yes	No
Tumor necrosis factor alpha synthesis	Yes	Yes	No
Increased matrix metalloproteinase activity	Yes	Yes	No
Increased monocyte chemoattractant protein-1	Yes	Yes	No
White matter damage, mixed glial cell changes, subcortical damage			
Blood–brain barrier change	Yes	Yes	No
Apoptotic death of astrocytes	Yes	Yes	Yes
White matter lysosome expansion	Yes	Yes	No
Increased chemokine receptor	No	No	No
Increased matrix metalloproteinase activity	No	No	No
Increased immunophilin synthesis	No	No	No
Neurodegeneration and neuropil damage			
Synaptodendritic damage	Yes	Yes	No
Dropout of neuron cell bodies	Yes	Inconclusive	Yes
Loss of neuropil matrix components	No	No	Yes
Active apoptotic death of neurons	Yes	No	Yes
Axon damage	Yes	No	No
Neuronal channelopathies	Yes	Possibly	No
Brain aging			
Fibrillary beta amyloid, diffuse	No	No	No
Fibrillary beta amyloid, argyrophilic	No	No	No
Ubiquitinylated aggresomes	No	No	No

abnormalities have been characterized. Few have been compared to the presence of neurocognitive performance, and still fewer to the severity of impairment. Of those in which clinical correlation was searched for, only a handful have exhibited a significant correlation after rigorous study so far, and fewer still were linked with dementia in more than one cohort.

Abnormalities grouped according to macrophage dysfunction and/or HIV-1 infection of macrophages remain the best documented in terms of the cliniconeuropathological correlation (Williams and Hickey, 2002). HIV encephalitis remains a bellwether change in the diagnostic realm of autopsy neuropathology. It has been correlated with dementia in multiple cohorts worldwide. HAART-era data have continued to support that tenet, and, indeed, the diagnosis of HIV encephalitis remains perhaps the only neuropathological category that has withstood the test of independent replication (Glass et al., 1993; Wiley and Achim, 1994; Bell et al., 1998; Cherner et al., 2002). Thus, the neuro-AIDS field still emphasizes the connection between HIV-associated dementia and HIV encephalitis, as it has provided useful insight into the pathophysiology (Williams and Hickey, 2002). One reason why the correlation with HIV encephalitis is observed consistently might be that there are several changes that all can lead to the diagnosis. Also, the diagnosis of HIV encephalitis does not discriminate according to its severity; the category is broadly inclusive and "all or none". The lumping together of a constellation of changes instead of a single measurement stands a better chance of including many types of demented subjects. Nevertheless, some people with HIV-associated dementia do not fall into the broad HIV encephalitis category, and conversely, some people with HIV encephalitis were not demented. The lack of clear-cut cliniconeuropathological correlation remains the premier paradox in the neuro-AIDS field (Glass et al., 1993; Wiley and Achim, 1994). When the severity of neurocognitive change is compared specifically with the presence of multinucleated giant cells, a correlation is not obvious, especially in people with mild impairment (Cherner et al., 2002). Professor Budka's impression ("we are not dealing with a uniform disease") still seems apt (Budka, 1998). The challenge for the future is to sort out the fine variation in clinical and pathological appearance, including the influence of co-morbid factors, many of which do not have distinct neuropathologies. A hierarchical listing, grouped according to how much variation in neurocognitive function each anomaly might actually contribute, cannot be assembled from the data in hand.

Since brain cells that are productively infected with HIV-1 are mononuclear phagocytes, and infection of those cells is a key criterion used to diagnose HIV encephalitis, one would expect that the presence of HIV-1 antigenicity, mRNA and DNA to be strongly related to HIV-associated dementia. As already noted, these correlations are surprisingly weak (Glass et al., 1993, 1995). The most accessible compartment to assess HIV-1 replication in the CNS clinically is CSF. The concentration of HIV RNA in CSF shows some degree of correlation with dementia when virus replication is not suppressed (Brew 1997), but after effective virus suppression with HAART, the correlation does not persist (McArthur, 2004). Glass et al. (1995) undertook one of the only multiple comparisons in the HIV-associated dementia field, so as to examine relative histological contributions to dementia. When brain HIV-1 antigenicity and the prevalence of brain macrophages were compared, HIV-associated dementia was correlated with macrophage staining very strongly; HIV antigen staining correlated weakly. This suggested that brain inflammation was the key driving force behind dementia, and not the burden of replicating HIV-1 in the brain (Williams and Hickey, 2002). Several other abnormal findings pertaining to macrophage activation have been observed in demented patients, and support the suggestion that the process is driven primarily by mononuclear phagocyte activation. These include increased tumor necrosis factor alpha mRNA (Wesselingh et al., 1993, 1997), increased matrix metalloproteinase synthesis (Conant et al., 1999), increased monocyte chemoattractant protein-1 synthesis (Conant et al., 1998) and increased synthesis of inducible nitric oxide synthetase (Adamson et al., 1999; Vincent et al., 1999). Despite the substantial linkage of HIV-associated dementia with these signs of macrophage activation, it is not obvious how macrophage activation increases when the burden of replicating HIV-1 is not increased concomitantly. One explanation is that the mere contact of cells with neurotoxic components of HIV, such as HIV tat protein, or HIV gp120 envelopes, or assembled virions, can trigger lasting immune responses without productive virus infection. In that scenario, a "hit and run" type of macrophage activation occurs that is not produced by productive HIV-1 infection or virus accumulation (Nath et al., 1999). Another possible means of macrophage activation could be driven by HIV-1 infection of astrocytes (Thompson et al., 2004). Since astrocyte infection with HIV-1 produces only limited replication, astrocyte activation is not linked with the accumulation of virions. It is interesting that the now-classic brain macrophage correlation of Glass et al. (1995), and many other adjunctive macrophage-related changes mentioned above, have not been replicated in multiple cohorts independently (Table 18.1). That tends to limit cross-comparison between the various clinical correlations. This limitation is sure to be addressed more often in the future because brain specimens from the National NeuroAIDS Tissue Consortium have become available to the research community worldwide (Morgello et al., 2001).

Since many of the defining changes of HIV encephalitis often affect white matter primarily, and HIV encephalitis is correlated with dementia, white matter damage plays some role in producing dementia in some people. When structural damage and/or inflammation are present in the setting of HIV encephalitis, it is well known that subcortical structures are most vulnerable (Navia et al., 1986; Kure et al., 1991; Wiley and Achim, 1994; Glass et al., 1995; Everall et al., 1995; Bell, 2004). The concentration of replicating HIV-1 also is higher in subcortical structures, including basal ganglia (gray matter) and white matter both (Achim et al., 1994). Animal models of HIV encephalitis also suggest that white matter damage seems to predominate in the majority of cases (Raghavan et al., 1999; Xing et al., 2003). Thus, basal ganglia and subcortical white matter both are deemed to be critical regional targets of HIV-related brain pathology. In addition to myelin loss, axon injury occurs in white matter myelinated nerve fibers (Giometto et al., 1997). Prior to HAART it was shown that when the full spectrum of white matter injury was surveyed, there was a correlation with dementia in people with end-stage disease (Schmidbauer et al., 1992). However, when injured axons in white matter were examined, a significant correlation with dementia was not apparent (Giometto et al., 1997). Sensitive neuroimaging techniques also suggest that white matter changes and subcortical damage are clinically important (Jernigan et al., 1993). Brain spectroscopy (magnetic resonance spectroscopy, MRS) has shown that choline-containing compounds and myoinositol concentration are increased, which implies that membrane turnover is disturbed in white matter glial cells. All these neuropathological indicators are harmonious with the fact that HIV-associated dementia fits the phenotypic picture of a typical subcortical dementia (Lipton and Gendelman, 1995). As noted already, neocortical damage also has been observed (Wiley et al., 1991; Masliah et al., 1992b).

Several other more specific changes pertaining to white matter that might lead to HIV-associated dementia are listed in Table 18.1. In some of the cited work correlation with dementia was pursued. For example, association between white matter "pallor" and clinical dementia was established in a single cohort (Glass et al., 1993; Power et al., 1993). Paleness of myelin staining was caused by leakage of plasma protein due to a blood–brain barrier defect. Whether or not the blood–brain barrier change in white matter actually causes dementia, or is an epiphenomenon, is not clear. The mechanism for the change may be related to endothelial cell changes (Smith et al., 1990). As already noted, the loss of myelinated fibers in subcortical white matter also is likely to be correlated with dementia, although very few studies have specifically addressed that correlation (Schmidbauer et al., 1992; Bell, 2004). Other evidence that abnormal cells of white matter are related to dementia includes immunophilin synthesis (Achim et al., 2004), atypical gene expression in oligodendrocytes (Cosenza et al., 2002) and expansion of the lysosome system. The lysosomal expansion might be related to abnormal membrane turnover as visualized using MRS (Gelman et al., 2005).

HIV-1 does not infect neurons, but it is widely believed that HIV-associated dementia is a neurodegenerative disease. Neuropathological evidence supporting the presence of neurodegeneration rests on three main observations: (1) reports have described neurons and other brain cells undergoing apoptotic death (Petito and Roberts, 1995; Olano et al., 1996; Shi et al., 1996; Adle-Biassette et al., 1999); (2) cumulative dropout of neurons has been measured stereologically (Ketzler et al., 1990; Everall et al., 1991); (3) fine structural changes occur in the neuropil, which may represent subtle and/or early degeneration of neurons (Masliah et al., 1997; Everall et al., 1999). It seems likely that the latter effect is related in some form to the poliodystrophy that was recognized early on (Budka et al., 1991). The meaning of the presence of apoptotic neurons remains unclear because correlation with dementia is not significant (Adle-Biassette et al., 1999; Gray et al., 2000). However, the lack of correlation does not mean that apoptosis does not play a role, because functional changes may not occur until a threshold amount of neuron death has accumulated (Terry et al., 1991; Jankovic, 2005). According to that principle, one would expect that the accumulated loss of neurons would be correlated strongly with HIV-associated dementia. However, data comparing neuron density with HIV-associated dementia do not confirm that suggestion very consistently (Wiley et al., 1991; Everall et al., 1993b, 1995, 1999; Weis et al., 1993; Seilhean et al., 1993; Asare et al., 1996; Korbo and West, 2000). For example, one study showed that loss of neurons occurred in one area of brain cortex, but not an adjacent sector of cortex. When neuron density was compared with dementia, linkage was not significant (Weis et al., 1993). In another study, the spatial arrangement of large and small neurons in brain cortex seemed to be related to dementia, but neuronal dropout itself was not correlated (Asare et al., 1996). Another study illustrated that neuronal subpopulations are affected quite selectively. Substantial loss of a subpopulation of small interneurons that immunostain for parvalbumin occurs in one sector of hippocampus (Masliah et al., 1992b). Subpopulations of neurons such as this may not turn out to be critical clinically. As well, the loss of staining in a vulnerable neuronal subpopulation may not reflect dropout (i.e. it could represent a decrease in expression of the marker in viable neurons). There are substantial technical difficulties involved with the analysis of neuronal population densities that makes clear-cut interpretation difficult. As for the mechanism of neurodegeneration, many scenarios have been verified experimentally that illustrate ways that HIV-1 and/or inflammation can produce neuronal injury (Kaul et al., 2001). Most of these exciting concepts were worked out in vitro; the challenge that remains is to verify that these postulated events take place in the brain cells of demented people.

A notable neuronal change that has withstood a test for correlation with neurocognitive dysfunction is synaptodendritic simplification (Masliah et al., 1997). A study that compared these synaptodendritic changes with neuronal dropout suggested that subtle synaptic changes are the more important clinically, at least in the early stages of dysfunction. In contrast, full-blown neuronal loss was not an independent factor in people with mild impairment, which suggests that it is a late-stage event (Everall et al., 1999). It is interesting that synaptodendritic simplification may be present without evidence of an inflammatory change, which contradicts the putative dogma that inflammation is the driving force behind dementia. Another important change in neuropil of people with HIV encephalitis that might pertain to the synaptodendritic changes is that extracellular proteoglycan matrix is degraded (Belichenko et al., 1997; Medina-Flores et al., 2004), possibly by macrophage proteases that are released into the extracellular compartment (Gelman et al., 1997; Conant et al., 1999). The lack of neuropathological change, especially in gray matter, remains a numerically important category of neurocognitive impairment in HIV/AIDS.

As many as half of the demented patients with end-stage AIDS may fall into that category (Glass et al., 1993). This ephemeral topic attracts little attention, although interest has increased as more subjects have mild neurocognitive impairment (McArthur, 2004). It is suggested that a functional disturbance of neurons might occur in people with HIV-associated dementia who do not have clear-cut histopathological changes (Glass et al., 1993; Zheng and Gendelman, 1997). In support of that, results from a gene array showed that several genes involved in excitation of neurons in brain cortex are expressed abnormally in people with HIV-associated dementia (Gelman et al., 2004). Some of the demented people with "channelopathies" did not have HIV encephalitis. The concept of acquired neuronal channelopathies agrees with results of cortical electroencephalographic (EEG) recordings, which show abnormal postsynaptic potentials in demented people (Baldeweg and Gruzelier, 1997; Polich et al., 2000). The fact that neurocognitive impairment can improve after HAART (Brew, 2004; McArthur, 2004) also supports a reversible disturbance of neuronal function as opposed to irreversible dropout of neurons. In sum, the accumulated evidence that certain neuronal populations are prone to drop out in people with HIV/AIDS is compelling. However, the damage varies according to brain region, neuron size, circuit function, and histological behavior of the chosen neuronal marker. Since the correlation of neuronal death with cognitive impairment still remains unclear, it is premature at this point to classify HIV-associated dementia as a classic neurodegenerative disease. Extreme neuropathological examples of leukoencephalopathy demonstrate clearly that white matter injury can predominate and produce the dysfunction (Langford et al., 2002).

18.5. Peripheral nerve pathology associated with HIV infection

The pathological changes that underlie painful peripheral neuropathy in HIV-1-infected people have been outlined and illustrated (Griffin et al., 1998; Pardo et al., 2001). Peripheral neuropathy can be produced by a variety of insults in HIV/AIDS, but the most important type is known as HIV-associated distal sensory polyneuropathy (H-DSP). The process is a distal dying-back type of axonal neuropathy in which axons degenerate distally and the damage progresses centripetally. Small and large myelinated fibers and small unmyelinated fibers all undergo degeneration. Small fibers are selectively vulnerable, with preferential loss of the small nonmyelinated fibers. This pattern of nerve fiber loss is not specific to H-DSP and is observed in amyloid neuropathy and diabetes. Nociceptive impulses and sensory information are carried in small nonmyelinated C and Aδ fibers, and thus neuropathic pain is the dominant change in H-DSP (Cornblath and McArthur, 1988). Neuropathic pain is believed to arise because proximal axons of degenerating small nociceptive fibers increase firing rates and generate pain impulses (Sheen and Chung, 1993). Current theory suggests that the increased excitability of these axons in the neuropathic pain syndromes is produced by changes in ion channel expression in dorsal sensory neurons (Wood et al., 2004).

There are two main factors involved with the pathogenesis of H-DSP. It is clear that dideoxynucleoside reverse transcriptase inhibitors (NRTI) that are used in HAART regimens, including ddI, d4T and ddC, are neurotoxic. They contribute to the production of peripheral neuropathy (Brinley et al., 2001). However, peripheral neuropathy was prevalent in HIV/AIDS before those drugs were introduced, so HIV-1 infection itself also produces the lesion. Drug-induced and HIV-induced influences on H-DSP are virtually identical clinically and neuropathologically. In practice, H-DSP usually is a combination of the adverse changes induced by HIV infection and NRTI toxicity both, and the underlying pathophysiology is likely to be mixed. The mechanism of NRTI-induced peripheral neuropathy is probably inhibition of mitochondrial DNA replication. The phosphorylated NRTI is postulated to compete with gamma DNA polymerase required for mtDNA replication (Dalakis, 2001). The mechanism for HIV-related H-DSP is not known. A straightforward attack of HIV-1 in diseased nerve or ganglion tissue is unlikely because HIV-1 antigenicity within nerve or ganglia is not correlated (Gherardi et al., 1998). Neuropathologically, Wallerian-type degeneration of distal nerve fibers is observed. Characteristic changes are observed, including macrophage infiltration, but these changes are not specific as to the cause. A few pathological changes have been observed in the dorsal sensory ganglia, which contain the cell bodies of the affected sensory nerve fibers. One change noted in dorsal root ganglia of people with HIV/AIDS is the presence of nodules of Nageotte (NoN), which is a nonspecific reflection of ganglion cell dropout (Pardo et al., 2001; Keswani et al., 2003). These structures also are present in neurologically normal adults with age. A robust clinical correlation between neuropathy and NoN still is needed; it is not clear that dropout of the perikaryon of the sensory nerve is the cause of neuropathy (Keswani et al., 2003). Inflammatory changes consisting of HIV-1-infected macrophages within dorsal root ganglia have been observed in some people, which suggests that the driving force might be a subtle ganglionitis fueled by released cytokines (Rizzuto et al., 1995; Watkins et al., 1995; Pardo et al., 2001; Delakis, 2001; Zhu et al., 2005). Nevertheless, ganglionitis remains unseen in *routine* examinations of most people who exhibit distal sensory nerve fiber dropout at autopsy (observations of the author). Thus, it remains unknown whether the nonspecific anomalies such as NoN or macrophage infiltration of dorsal root ganglia are the cause of neuropathy. In terms of prevalence, H-DSP was present in at least 35% of the HIV/AIDS population prior to HAART (Brinley et al., 2001; Schifitto et al., 2002). A diminution of H-DSP after HAART, comparable to what was observed in HIV-associated dementia, was not observed in the HAART era. One reason was that the use of HAART drugs that were neurotoxic continued, and contributed to the problem. In recent times the use of potentially neurotoxic drugs such as ddI and d4T has been curtailed in cohorts in the USA, but the impact on neuropathy still remains unclear. From all the information at hand, peripheral neuropathy has become the most prevalent neurological complication of HIV/AIDS in the HAART era (Schifitto et al., 2002; Keswani et al., 2003; Morgello et al., 2004).

Sural nerve biopsy is an effective way to diagnose sensory nerve fiber loss and H-DSP neuropathologically. It is not practical to do in the clinic, however, because it is surgically invasive and nerve fiber analysis requires electron microscopy. Nerve conduction studies and electrophysiological technique also are not very useful in the clinic because the amplitude of conducted impulses is not very sensitive to small nerve fiber loss (Schifitto et al., 2002). Counting the number or intraepidermal axons in a skin biopsy from the foot is a promising new approach that has been used to measure distal small fiber loss in H-DSP. Skin biopsy is minimally invasive but is not practiced routinely because immunostaining intraepidermal axons and counting them is difficult to do. Preliminary data show that the technique might be more sensitive than sural nerve biopsy or electrophysiological testing (Herrmann et al., 2004). The histological changes in epidermal axons are not diagnostic for HIV-associated changes, but could prove to be practical for longitudinal study of neuropathy evolution and response to treatment.

18.6. Spinal cord pathology associated with HIV-1 infection

The spectrum of changes that can occur in the brains of HIV-1-infected people also can spread in the CNS to the spinal cord. Opportunistic infection in the spinal cord seems to have waned in the HAART era in parallel with changes elsewhere in the CNS. Since HIV encephalitis intensity may be declining as fewer people have end-stage AIDS, its prevalence in the spinal cord also appears to be less than before. The lesion known as AIDS-associated vacuolar myelopathy (VM) was the most common problem encountered prior to HAART, with an autopsy prevalence of at least 35%, although wide geographical differences were noted (Dal Pan et al., 1994). Often the nerve fiber degeneration in VM is not extensive enough to produce neurological symptoms or signs so the clinical prevalence is much lower. When the lesion is severe enough, it produces the expected neurological signs, i.e. spastic paraparesis, sensory ataxia and urinary incontinence. VM almost always occurred in end-stage AIDS, and nowadays, the lesion has become unusual in HAART-treated populations. However, in those who already had symptoms (and substantial pre-existing nerve fiber loss), HAART was not effective in reversing the disease (McArthur et al., 2005). The neuropathology of VM mimics subacute combined degeneration as observed in people with deficiency of vitamin B12 (Petito et al., 1985). Nerve fibers in ascending and descending myelinated fiber tracks undergo a vacuolar degeneration, especially in the thoracic segments. Electron microscopy has suggested that the vacuoles arise due to intralamellar splits, which suggests "primary demyelination" with relative sparing of axons (Maier et al., 1989). It has long been known that VM does not correlate with how much HIV-1 replication is present in the spinal cord. VM is not likely to be the direct effect of a direct attack by HIV-1 infection on spinal cord cells. Instead, a secondary metabolic disturbance akin to a functional cobalamin deficiency is suggested (Di Rocco et al., 2002). In recent times the most common anomaly in the spinal cord at autopsy is atrophy of the gracile tracts (i.e. loss of ascending myelinated long-fiber tracts in the dorsal columns). The increased dropout of these nerve fibers seldom is associated with myelopathic signs clinically. The prevalence of this lesion is due in part to the fact that sensory peripheral neuropathy is prevalent in this population, and gracile tracts undergo atrophy in association with sensory peripheral neuropathy and sensory neuron loss. As well, ascending long-fiber tract atrophy occurs during aging, and HIV-1-infected decedents increasingly live to older age nowadays.

18.7. Opportunistic infection and tumors in the CNS

Prior to the HAART era, generalized immunodeficiency in people with AIDS produced a high prevalence of opportunistic infection and primary lymphoma in the CNS. These conditions continue to produce substantial morbidity in the HAART era, albeit with probable reductions in their incidence, prevalence and mortality (Jellinger et al., 2000; Masliah et al., 2000; Morgello et al., 2002; Gray et al., 2003). There is no apparent change concerning the criteria for neuropathological diagnosis of these conditions, so the reader is referred to previous histological descriptions (Petito, 1993; Budka, 1998).

18.8. A neuro-AIDS research agenda for the future

Considering that HIV dementia is a completely new category of disease, much progress was made in the first quarter-century of the HIV/AIDS pandemic. The beneficial effect on CNS complications of suppressing HIV-1 replication with HAART is well established. But neurocognitive impairment, CNS pathology and peripheral neuropathy still persist and great challenges lie ahead. We need to elucidate the factors that make specific people vulnerable and invulnerable to HIV-associated dementia, HIV encephalitis and neuropathy. Differences in virus strain and host factors both might produce CNS response variation. The recent cataloguing of human gene polymorphisms and new technology to survey the human "haplotype map" (Diaz-Arrastia et al., 2004; International HapMap Consortium, 2005) provide a prime opportunity to elucidate these host genetic factors. The cliniconeuropathological correlation is only partly worked out because there are multiple disease categories and underlying pathophysiologies in play. Indeed, there is a highly diverse list of neuropathological events that can appear, and which tend to be lumped together for the purpose of a diagnosis. In the future, a more quantitative approach to the pathological characterization of these changes is needed for future experimentation and studies that pursue the cliniconeuropathological correlation. Specifically, a histological staging analogous to those established for research in the senile dementias is badly needed in the neuro-AIDS field (Braak et al., 1993). As well, future studies need to pursue aggressively the clinical impact of the various qualitative changes that are observed. For example, the mixture of leukoencephalopathy-like and neurodegeneration-like changes that occur in highly variable combinations begs for clarification. The field must determine eventually which neuropathological anomalies produce clinical effects, and determine precisely which neurocognitive functions are impacted (analogous to what has been accomplished in the senile dementias; see Grober et al., 1999). Since the putative neurodegeneration is fairly selective as to brain region, it will be important to define the specific anatomical "brain circuits" that become dysfunctional, or recover normal function when HIV-1 replication is suppressed. Increased effort also is needed to elucidate the effect of co-morbid factors that can influence the clinical and pathological course of HIV diseases of the nervous system, especially drug addiction. Ultimately, a hierarchical listing of pathological changes is needed in which relative clinical importance, or lack of, is assigned. In short, the neuro-AIDS research agenda needs to become both a pathologically defined and clinically relevant discipline. The drive to coordinate basic research with human pathology has become more feasible in the HAART era because brain specimens from the National NeuroAIDS Tissue Consortium are available worldwide.

18.9. Summary

HIV encephalitis remains a prevalent and important neuropathological correlate of dementia in the HAART era. It is probably the only neuropathological variable that has undergone substantial independent verification in multiple autopsy cohorts. HIV encephalitis has provided the field with a unifying concept to pursue the pathophysiology of dementia. Drawbacks remain that the diagnosis is an overly diverse constellation of anomalies in both gray and white matter, and correlation with dementia remains too vague.

The overall incidence of HIV-associated dementia in the HAART era declined by about half, but autopsy cohorts still exhibit a substantial prevalence of HIV encephalitis, probably because they are enriched with people with end-stage disease. The intensity of HIV encephalitis might have undergone a decrease in the HAART era, but HIV encephalitis remains essentially an "all or none" diagnosis.

The criteria for a diagnosis of HIV encephalitis remain essentially intact but variants might have appeared in the HAART era. Examples include an increase in the number of cases with predominantly white matter necrosis ("leukoencephalopathy variant"), cases with only minimal or pathological change ("functional variant"), cases that have scars that lack evidence of HIV-1 after treatment with HAART ("burnt out variant"), cases with age-associated degeneration ("elderly variant") and cases in which immune reconstitution inflammatory syndrome has played a role ("inflammatory variant").

The partial disassociation between clinical dementia and HIV encephalitis remains a longstanding paradox of the neuro-AIDS field. It suggests that more than one underlying pathophysiological mechanism is present, and that co-morbid factors may be important. Progress in understanding the cliniconeuropathological correlation should advance because brain specimens collected under standardized conditions are available from the National NeuroAIDS Tissue Consortium.

The incidences of opportunistic infection and tumors in the CNS underwent decreases after HAART. HAART prevents these processes from developing, but initiation of therapy when there is an ongoing pathological process can provoke immune reconstitution (immune reconstitution inflammatory syndrome) and an intense pathological response in the CNS that is potentially fatal.

HIV/AIDS might exacerbate diseases of brain aging, and aging could worsen HIV-1-associated neurocognitive disability. The neurobiological interaction between persistent HIV-1 infection and brain aging remains unclear. Brain aging is universal to all populations, so neuropathological study in people with HIV/AIDS must control for chronological age.

In the peripheral nervous system painful H-DSP is the most prevalent neurological problem of HIV/AIDS in the HAART era. The distal dying-back small fiber type of neuropathy still produces a disabling neuropathic pain. Measuring the loss of distal nerve fibers in the epidermis of a skin biopsy is a new diagnostic approach that is potentially useful to track disease progression and/or nerve regeneration experimentally. Even though the pathology occurs in the distal terminals, it remains likely that the driving force is a more proximal dorsal root ganglionopathy that is not characterized neuropathologically, or experimentally.

Signs of HIV encephalitis and other opportunistic infection in the spinal cord have declined in the HAART era. The prevalence of HIV-associated vacuolar myelopathy has waned primarily because the lesion was predominantly observed in end-stage of AIDS, which is less prevalent. Gracile tract atrophy remains prevalent, and probably results from both sensory peripheral neuropathy and aging.

References

Achim CL, Wang R, Miners DK, et al (1994). Brain viral burden in HIV infection. J Neuropathol Exp Neurol 53: 284–294.

Achim CL, Masliah E, Schindelar J, et al (2004). Immunophilin expression in the HIV-infected brain. J Neuroimmunol 157: 126–132.

Adamson DC, McArthur JC, Dawson TM, et al (1999). Rate and severity of HIV-associated dementia (HAD): correlations with Gp41 and iNOS. Molecular Med 5: 98–109.

Adle-Biassette H, Levy Y, Colombel M, et al (1999). Neuronal apoptosis does not correlate with dementia in HIV infection but is related to microglial activation and axonal damage. Neuropathol Appl Neurobiol 25: 125–133.

Asare E, Dunn G, Glass J, et al (1996). Neuronal pattern correlates with the severity of human immunodeficiency virus-associated dementia complex: usefulness of spatial pattern analysis in clinicopathological studies. Am J Pathol 148: 31–38.

Baldeweg T, Gruzelier JH (1997). Alpha EEG activity and subcortical pathology in HIV infection. Int J Psychophysiol 26: 431–442.

Belichenko PV, Miklossy J, Celio MR (1997). HIV-I induced destruction of neocortical extracellular matrix components in AIDS victims. Neurobiol Dis 4: 301–310.

Bell JE (2004). An update on the neuropathology of HIV in the HAART era. Histopathology 45: 549–559.

Bell JE, Brettle RP, Chiswick A, et al (1998). HIV encephalitis, proviral load and dementia in drug users and homosexuals with AIDS. Effect of neocortical involvement. Brain 121(Pt 11): 2043–2052.

Braak H, Braak E, Bohl J (1993). Staging of Alzheimer-related cortical destruction. Eur Neurol (Basel) 33: 403–408.

Brew BJ (2004). Evidence for a change in AIDS dementia complex in the era of highly active antiretroviral therapy and the possibility of new forms of AIDS dementia complex. AIDS 18(Suppl 1): S75–S78.

Brew BJ, Pemberton L, Cunningham P, et al (1997). Levels of human immunodeficiency type 1 RNA in cerebrospinal fluid correlate with AIDS dementia stage. J Infect Dis 175: 963–966.

Brinley FJ Jr, Pardo CA, Verma A (2001). Human immunodeficiency virus and the peripheral nervous system. Arch Neurol 58: 1561–1566.

Budka H (1986). Multinucleated giant cells in brain: a hallmark of the acquired immune deficiency syndrome (AIDS). Acta Neuropathol (Berl) 69: 253–258.

Budka H (1998). HIV-associated neuropathology. In: HE Gendelman, SE Lipton, L Epstein, S Swindells (Eds.), The Neurology of AIDS. Chapman & Hall, New York, pp. 241–260.

Budka H, Wiley CA, Kleihues P, et al (1991). HIV-associated disease of the nervous system: review of nomenclature and proposal for neuropathology-based terminology. Brain Pathol 1(3): 143–152.

Cherner M, Masliah E, Ellis RJ, et al (2002). Neurocognitive dysfunction predicts postmortem findings of HIV encephalitis. Neurology 59(10): 1563–1567.

Conant K, Garzino-Demo A, Nath A, et al (1998). Induction of monocyte chemoattractant protein-1 in HIV-1 Tat-stimulated astrocytes and elevation in AIDS dementia. Proc Natl Acad Sci USA 95(6): 3117–3121.

Conant K, McArthur JC, Griffin DE, et al (1999). Cerebrospinal fluid levels of MMP-2, 7, and 9 are elevated in association with human immunodeficiency virus dementia. Ann Neurology 46: 391–398.

Cornblath DR, McArthur JC (1988). Predominantly sensory neuropathy in patients with AIDS and AIDS-related complex. Neurology 38: 794–796.

Corral I, Quereda C, Garcia-Villanueva M, et al (2004). Focal monophasic demyelinating leukoencephalopathy in advanced HIV infection. Eur Neurol 52: 36–41.

Cosenza MA, Zhao M-L, Shankar SL, et al (2002). Up-regulation of MAP2e-expressing oligodendrocytes in the white matter of patients with HIV-1 encephalitis. Neuropathol Appl Neurobiol 28: 480–488.

Dalakis MC (2001). Peripheral neuropathy and antiretroviral drugs. J Periph Nerv Syst 6: 14–20.

Dal Pan GJ, Glass JD, McArthur JC (1994). Clinicopathological correlations of HIV-1-associated vacuolar myelopathy: an autopsy based case-control study. Neurology 44: 2159–2164.

Deayton JR, Wilson P, Sabin CA, et al (2000). Changes in the natural history of cytomegalovirus retinitis following the introduction of highly active antiretroviral therapy. AIDS 14: 1163–1170.

Diaz-Arrastia R, Gong Y, Kelly CJ, et al (2004). Host genetic polymorphisms in HIV-related neurological disease. J Neurovirol 10(Suppl 1): 67–73.

Di Rocco A, Bottiglieri T, Werner P, et al (2002). Abnormal cobalamin-dependent transmethylation in AIDS-associated myelopathy. Neurology 58: 730–735.

Du Pasquier RA, Koralnik IJ (2003). Inflammatory reaction in progressive multifocal leukoencephalopathy: harmful or beneficial? J Neurovirol 9(Suppl 1): 25–31.

Esiri MM, Biddolph SC, Morris CS (1998). Prevalence of Alzheimer plaques in AIDS. J Neurol Neurosurg Psych 65: 29–33.

Everall IP, Luthert PJ, Lantos PL (1991). Neuronal loss in the frontal cortex in HIV infection. Lancet 337(8750): 1119–1121.

Everall I, Luthert P, Lantos P (1993a). A review of neuronal damage in human immunodeficiency virus infection: its assessment, possible mechanism and relationship to dementia. J Neuropathol Exp Neurol 52: 561–566.

Everall IP, Luthert PJ, Lantos PL (1993b). Neuronal number and volume alterations in the neocortex of HIV infected individuals. J Neurol Neurosurg Psych 56: 481–486.

Everall I, Barnes H, Spargo E, et al (1995). Assessment of neuronal density in the putamen in human immunodeficiency virus (HIV) infection. Application of stereology and spatial analysis of quadrats. J Neurovirol 1: 126–129.

Everall IP, Heaton RK, Marcotte TD, et al (1999). Cortical synaptic density is reduced in mild to moderate human immunodeficiency virus neurocognitive disorder. HNRC Group. HIV Neurobehavioral Research Center. Brain Pathol 9: 209–217.

Fischer CP, Jorgen G, Gundersen H, et al (1999). Preferential loss of large neocortical neurons during HIV infection: a study of the size distribution of neocortical neurons in the human brain. Brain Res 828(1–2): 119–126.

Gelman BB (1993). Diffuse microgliosis associated with cerebral atrophy in the acquired immunodeficiency syndrome. Ann Neurol 34: 65–70.

Gelman BB, Guinto FC Jr (1992). Morphometry, histopathology, and tomography of cerebral atrophy in the acquired immunodeficiency syndrome. Ann Neurol 32: 32–40.

Gelman BB, Schuenke KW (2004). Brain aging in AIDS: increased ubiquitin-protein conjugate and correlation with decreased synaptic protein but not Aβ-stained diffuse plaque. J Neurovirol 10: 98–108.

Gelman BB, Dholakia S, Casper K, et al (1996). Expansion of the cerebral ventricles and correlation with acquired immunodeficiency syndrome: neuropathology in 232 patients. Arch Pathol Lab Med 120: 866–871.

Gelman BB, Wolf DA, Rodriguez-Wolf M, et al (1997). Mononuclear phagocyte hydrolytic enzyme activity associated with cerebral HIV infection. Am J Pathol 151: 1437–1446.

Gelman BB, Soukup VM, Holzer CE 3rd, et al (2005). Potential role for white matter lysosome expansion in HIV-associated dementia. J Acquir Immun Defic Syndr 39: 422–425.

Gherardi RK, Chretien F, Delfau-Larue MH, et al (1998). Neuropathy in diffuse infiltrative lymphocytosis syndrome: an HIV neuropathy, not a lymphoma. Neurology 50: 1041–1044.

Giometto B, An SF, Groves M, et al (1997). Accumulation of β-amyloid precursor protein in HIV encephalitis: relationship with neuropsychological abnormalities. Ann Neurol 42: 34–40.

Glass JD, Wesselingh SL, Selnes OA, et al (1993). Clinical–neuropathologic correlation in HIV-associated dementia. Neurology 43: 2230–2237.

Glass JD, Fedor H, Wesselingh SL, et al (1995). Immunocytochemical quantitation of human immunodeficiency virus in the brain: correlations with dementia. Ann Neurol 38: 755–762.

Gray F, Adle-Biassette H, Brion F, et al (2000). Neuronal apoptosis in human immunodeficiency virus infection. J Neurovirol 6(Suppl 1): S38–S43.

Gray F, Chretien F, Vallat-Decouvelaere AV, et al (2003). The changing pattern of HIV neuropathology in the HAART era. J Neuropathol Exp Neurol 62: 429–440.

Green DA, Masliah E, Vinters HV, et al (2005). Brain deposition of beta-amyloid is a common pathologic feature in HIV positive patients. AIDS 19: 407–411.

Griffin JW, Crawford TO, McArthur JC (1998). Peripheral neuropathies associated with HIV infection. In: HE Gendelman, SA Lipton, LE Epstein, S Swindells (Eds.), The Neurology of AIDS. Chapman & Hall, New York, pp. 275–290.

Grober E, Dickson D, Sliwinski MJ, et al (1999). Memory and mental status correlates of modified Braak staging. Neurobiol Aging 20: 573–579.

Heaton RK, Grant I, Butters N, et al (1995). The HNRC 500: neuropsychology of HIV infection at different disease stages. HIV Neurobehavioral Research Center. J Int Neuropsychol Soc 1: 231–251.

Herrmann DN, McDermott MP, Henderson D, et al (2004). Epidermal nerve fiber density, axonal swellings and QST as predictors of HIV distal sensory neuropathy. Muscle Nerve 29: 420–427.

International HapMap Consortium (2005). A haplotype map of the human genome. Nature 437(7063): 1299–1320.

Jacobson MA, Zegans M, Pavan PR, et al (1997). Cytomegalovirus retinitis after initiation of highly active antiretroviral therapy. Lancet 349: 1143–1145.

Jankovic J (2005). Progression of Parkinson disease: are we making progress in charting the course? Arch Neurol 62 (3): 351–352.

Jellinger KA, Setinek U, Drlicek M, et al (2000). Neuropathology and general autopsy findings in AIDS during the last 15 years. Acta Neuropathol (Berl) 100: 213–220.

Jernigan TL, Archibald S, Hesselink JR, et al (1993). MRI morphometric analysis of cerebral volume loss in HIV infection. Arch Neurol 50: 250–255.

Kaul M, Garden GA, Lipton SA (2001). Pathways to neuronal injury and apoptosis in HIV-associated dementia. Nature 410: 988–994.

Keswani SC, Hoke A, Schifitto G, et al (2003). Incidence of and risk factors for HIV-associated distal sensory neuropathy. Neurology 61: 279–280.

Ketzler S, Weis S, Haug H, et al (1990). Loss of neurons in the frontal cortex in AIDS brains. Acta Neuropathol (Berl) 80: 92–94.

King MD, Perlino CA, Cinnamon J, et al (2002). Paradoxical recurrent meningitis following therapy of cryptococcal meningitis: an immune reconstitution syndrome after initiation of highly active antiretroviral therapy. AIDS 13: 724–726.

Klatt EC (1988). Diagnostic findings in patients with acquired immune deficiency syndrome (AIDS). J Acquir Immun Defic Syndr 1(5): 459–465.

Korbo L, West M (2000). No loss of hippocampal neurons in AIDS patients. Acta Neuropathol 99: 529–533.

Kure K, Llena JF, Lyman WD, et al (1991). Human immunodeficiency virus type 1 infection of the nervous system: an autopsy study of 268 adult, pediatric, and fetal brains. Hum Pathol 22: 700–710.

Langford TD, Letendre SL, Marcotte TD, et al (2002). Severe, demyelinating leukoencephalopathy in AIDS patients on antiretroviral therapy. J Acquir Immun Defic Syndr 16: 1019–1029.

Lipton SA, Gendelman HL (1995). Dementia associated with the acquired immunodeficiency syndrome. N Engl J Med 332: 934–940.

Maier H, Budka H, Lassmann H, et al (1989). Vacuolar myelopathy with multinucleated giant cells in the acquired immune deficiency syndrome (AIDS). Light and electron microscopic distribution of human immunodeficiency virus (HIV) antigens. Acta Neuropathol (Berl) 78: 497–503.

Masliah E, Ge N, Achim CL, et al (1992a). Selective neuronal vulnerability in HIV encephalitis. J Neuropathol Exp Neurol 51: 585–593.

Masliah E, Achim CL, Ge N, et al (1992b). Spectrum of human immunodeficiency virus-associated neocortical damage. Ann Neurol 32: 321–329.

Masliah E, Heaton RK, Marcotte TD, et al (1997). Dendritic injury is a pathological substrate for human immunodeficiency virus-related cognitive disorders. HNRC Group. The HIV Neurobehavioral Research Center. Ann Neurol 42: 963–972.

Masliah E, DeTeresa RM, Mallory ME, et al (2000). Changes in pathological findings at autopsy in AIDS cases for the last 15 years. AIDS 14: 69–74.

McArthur JC (2004). HIV dementia: an evolving disease. J Neuroimmunol 157(1–2): 3–10.

McArthur JC, Brew B, Nath A (2005). Neurological complications of HIV infection. Lancet Neurol 4: 543–555.

Medina-Flores R, Wang G, Bissel SJ, et al (2004). Destruction of extracellular matrix proteoglycans is pervasive in simian retroviral neuroinfection. Neurobiol Dis 16: 604–616.

Miller RF, Isaacson PG, Hall-Craggs M, et al (2004). Cerebral CD^{8+} lymphocytosis in HIV-1 infected patients with immune restoration induced by HAART. Acta Neuropathol 108: 17–23.

Miralles P, Berenguer J, Lacruz C, et al (2001). Inflammatory reactions in progressive multifocal leukoencephalopathy after highly active antiretroviral therapy. AIDS 15: 1900–1902.

Morgello S, Gelman BB, Grant E, et al (2001). The National NeuroAIDS Tissue Consortium. Neuropathol Appl Neurobiol 27: 326–335.

Morgello S, Mahboob R, Yakoushina T, et al (2002). Autopsy findings in a human immunodeficiency virus-infected population over 2 decades: influences of gender, ethnicity, risk factors, and time. Arch Pathol Lab Med 126: 182–190.

Morgello S, Estanisllao L, Simpson D, et al (2004). HIV-associated distal sensory polyneuropathy in the era of highly active antiretroviral therapy: the Manhattan HIV Brain Bank. Arch Neurol 61: 546–551.

Nath A, Hersh LB (2005). Tat and amyloid: multiple interactions. AIDS 19: 203–204.

Nath A, Conant K, Chen P, et al (1999). Transient exposure to HIV-1 Tat protein results in cytokine production in macrophages and astrocytes. A hit and run phenomenon. J Biol Chem 274: 17098–17102.

Navia BA, Cho ES, Petito CK, et al (1986). The AIDS dementia complex: II. Neuropathology. Ann Neurol 19: 525–535.

Neuenburg JK, Brodt HR, Herndier BG, et al (2002). HIV-related neuropathology, 1985 to 1999: rising prevalence of HIV encephalopathy in the era of highly active antiretroviral therapy. J Acquir Immun Defic Syndr 31: 171–177.

Olano JP, Wolf DA, Keherly MJ, et al (1996). Quantifying apoptosis in banked human brains using flow cytometry. J Neuropathol Exp Neurol 55: 1164–1172.

Pardo CA, McArthur JC, Griffin JW (2001). HIV neuropathy: insights in the pathology of HIV peripheral nerve disease. J Periph Nerv Syst 6: 21–27.

Petito CK (1993). Neuropathology of acquired immunodeficiency syndrome. In: JS Nelson, JE Parisi, SS Schochet Jr (Eds.), Principles and Practice of Neuropathology. Mosby, St Louis, MO, pp. 88–108.

Petito CK, Cash KS (1992). Blood–brain barrier abnormalities in the acquired immunodeficiency syndrome: immunohistochemical localization of serum proteins in postmortem brain. Ann Neurol 32(5): 658–666.

Petito CK, Roberts B (1995). Evidence of apoptotic cell death in HIV encephalitis. Am J Pathol 146: 1121–1130.

Petito CK, Navia BA, Cho ES, et al (1985). Vacuolar myelopathy pathologically resembles subacute combined degeneration in patients with the acquired immunodeficiency syndrome. N Engl J Med 312: 874–879.

Petito CK, Adkins B, McCarthy M, et al (2003). CD^{4+} and CD^{8+} cells accumulate in the brains of acquired immunodeficiency syndrome patients with human immunodeficiency virus encephalitis. J Neurovirol 9: 36–44.

Polich J, Ilan A, Poceta JS, et al (2000). Neuroelectric assessment of HIV: EEG, ERP, and viral load. Int J Psychophysiol 38: 97–108.

Power C, Kong PA, Crawford TO, et al (1993). Cerebral white matter changes in acquired immunodeficiency syndrome dementia: alterations of the blood–brain barrier. Ann Neurol 34: 339–350.

Raghavan R, Cheney PD, Raymond LA, et al (1999). Morphological correlates of neurobiological dysfunction in macaques infected with neurovirulent simian immunodeficiency virus. Neuropathol Appl Neurobiol 25: 285–294.

Raja F, Sherriff FE, Morris CS, et al (1997). Cerebral white matter damage in HIV infection demonstrated using beta-amyloid precursor protein immunoreactivity. Acta Neuropathol (Berl) 93: 184–189.

Rempel HC, Pulliam L (2005). HIV-1 Tat inhibits neprilysin and elevates amyloid beta. AIDS 19: 127–135.
Rizzuto N, Cavallaro T, Monaco S, et al (1995). Role of HIV in the pathogenesis of distal symmetrical peripheral neuropathy. Acta Neuropathol 90: 244–250.
Sabri F, Titanji K, De Milito A, et al (2003). Astrocyte activation and apoptosis: their roles in the neuropathology of HIV infection. Brain Pathol 13: 84–94.
Sacktor NC, Lyles RH, Skolasky R, et al (1999). The Multicenter AIDS Cohort Study. HIV-1 related neurological disease incidence changes in the era of highly active antiretroviral therapy. Neurology 52: A252–A253.
Schifitto G, McDermott P, McArthur JC, et al (2002). Incidence of and risk factors for HIV-associated distal sensory polyneuropathy. Neurology 58: 1764–1768.
Schmidbauer M, Budka H, Okeda R, et al (1990). Multifocal vacuolar leucoencephalopathy: a distinct HIV-associated lesion of the brain. Neuropathol Appl Neurobiol 16: 437–443.
Schmidbauer M, Huemer M, Cristina S, et al (1992). Morphological spectrum, distribution and clinical correlation of white matter lesions in AIDS brains. Neuropathol Appl Neurobiol 18: 489–501.
Seilhean D, Duyckaerts C, Vazeux R, et al (1993). HIV-1-associated cognitive/motor complex: absence of neuronal loss in the cerebral neocortex [see comment]. Neurology 43 (8): 1492–1499.
Sheen KM, Chung JM (1993). Signs of neuropathic pain depend on signals from injured nerve fibers in a rat model. Brain Res 610: 62–68.
Shelburne SA, Visnegarwala F, Darcourt J, et al (2005). Incidence and risk factors for immune reconstitution inflammatory syndrome during highly active antiretroviral therapy. AIDS 19: 399–406.
Shi B, De Girolami U, He J, et al (1996). Apoptosis induced by HIV-1 infection of the central nervous system. J Clin Invest 98: 1979–1990.
Smith TW, DeGirolami U, Henin D, et al (1990). Human immunodeficiency virus (HIV) leukoencephalopathy and the microcirculation. J Neuropathol Exp Neurol 49: 357–370.
Streit WJ (2005). Microglia and neuroprotection: implications for Alzheimer's disease. Brain Res Brain Res Rev 48: 234–239.
Tantisiriwat W, Tebas P, Clifford DB, et al (1998). Progressive multifocal leukoencephalopathy in patients with AIDS receiving highly active antiretroviral therapy. Clin Infect Dis 28: 1152–1154.
Terry RD, Masliah E, Salmon DP, et al (1991). Physical basis of cognitive alterations in Alzheimer's disease: synapse loss is the major correlate of cognitive impairment. Ann Neurol 30: 572–580.
Thompson KA, Churchill MJ, Gorry PR, et al (2004). Astrocyte specific viral strains in HIV dementia. Ann Neurol 56: 873–877.
Vago L, Bonetto S, Nebuloni M, et al (2002). Pathological findings in the central nervous system of AIDS patients on assumed antiretroviral therapeutic regimens: retrospective study of 1597 autopsies. AIDS 16: 1925–1928.
Vincent VA, De Groot CJ, Lucassen PJ, et al (1999). Nitric oxide synthase expression and apoptotic cell death in brains of AIDS and AIDS dementia patients. AIDS 13: 317–326.
Vendrely A, Bienvenu B, Gasnault J, et al (2005). Fulminant inflammatory leukoencephalopathy associated with HAART-induced immune restoration in AIDS-related progressive multifocal leukoencephalopathy. Acta Neuropathol (Berl) 109: 449–455.
Watkins LR, Goehler LE, Relton J, et al (1995). Mechanisms of tumor necrosis factor-alpha (TNF-alpha) hyperalgesia. Brain Res 692: 244–250.
Weis S, Haug H, Budka H (1993). Neuronal damage in the cerebral cortex of AIDS brains: a morphometric study. Acta Neuropathol (Berl) 85: 185–189.
Wesselingh SL, Power C, Glass JD, et al (1993). Intracerebral cytokine messenger RNA expression in acquired immunodeficiency syndrome dementia. Ann Neurol 33: 576–582.
Wesselingh SL, Takahashi K, Glass JD, et al (1997). Cellular localization of tumor necrosis factor mRNA in neurological tissue from HIV-infected patients by combined reverse transcriptase polymerase chain reaction in situ hybridization and immunohistochemistry. J Neuroimmunol 74: 1–8.
Wiley CA, Achim CL (1994). Human immunodeficiency virus encephalitis is the pathological correlate of dementia in acquired immunodeficiency syndrome. Ann Neurol 36: 673–676.
Wiley CA, Masliah E, Morey M, et al (1991). Neocortical damage during HIV infection. Ann Neurol 29(6): 651–657.
Williams KC, Hickey WF (2002). Central nervous system damage, monocytes and macrophages, and neurological disorders in AIDS. Ann Rev Neurosci 25: 537–562.
Wood JN, Boorman JP, Okuse K, et al (2004). Voltage gated sodium channels and pain pathways. J Neurobiol 61: 55–71.
Wood ML, MacGinley R, Eisen DP, et al (1998). HIV combination therapy: partial immune reconstitution unmasking latent cryptococcal infection. AIDS 12: 1491–1494.
Xing HQ, Moritoyo T, Mori K, et al (2003). Simian immunodeficiency virus encephalitis in the white matter and degeneration of the cerebral cortex occur independently in simian immunodeficiency virus-infected monkey. J Neurovirol 9: 508–518.
Yamaguchi H, Sugihara S, Ogawa A, et al (2001). Alzheimer beta amyloid deposition enhanced by apoE epsilon4 gene precedes neurofibrillary pathology in the frontal association cortex of nondemented senior subjects. J Neuropathol Exp Neurol 60: 731–739.
Zheng J, Gendelman HE (1997). The HIV-1 associated dementia complex: a metabolic encephalopathy fueled by viral replication in mononuclear phagocytes. Curr Opin Neurol 10: 319–325.
Zhu Y, Jones S, Tsutsui S, et al (2005). Lentevirus infection causes neuroinflammation and neuronal injury in dorsal root ganglia: pathogenic effects of STAT-1 and inducible nitric oxide synthase. J Immunol 175: 1118–1126.

Further Reading

Brew BJ, Rosenblum M, Cronin K, et al (1995). AIDS dementia complex and HIV-1 brain infection: clinical–virological correlations. Ann Neurol 38: 563–570.

Budka H (1991). Neuropathology of human immunodeficiency virus infection. Brain Pathol 1(3): 163–175.

Everall I, Gray F, Barnes H, et al (1992). Neuronal loss in symptom-free HIV infection. Lancet 340(8832): 1413.

Everall IP, Glass JD, McArthur J, et al (1994). Neuronal density in the superior frontal and temporal gyri does not correlate with the degree of human immunodeficiency virus-associated dementia. Acta Neuropathol (Berl) 88: 538–544.

Garden GA, Budd SL, Tsai E, et al (2002). Caspase cascades in human immunodeficiency virus-associated neurodegeneration. J Neurosci 22: 4015–4024.

Glass JD, Wesselingh SL (1998). Viral load in HIV-associated dementia. Ann Neurol 44: 150–151.

Grassi MP, Clerici F, Vago L, et al (2002). Clinical aspects of the AIDS dementia complex in relation to histopathological and immunohistochemical variables. Eur Neurol 47: 141–147.

Keswani SC, Pardo CA, Cherry CL, et al (2002). HIV-associated sensory neuropathies. AIDS 16: 2105–2117.

McArthur JC, McClernon DR, Cronin MF, et al (1997). Relationship between human immunodeficiency virus-associated dementia and viral load in cerebrospinal fluid and brain. Ann Neurol 42: 689–698.

Sanders VJ, Pittman CA, White MG, et al (1998). Chemokines and receptors in HIV encephalitis. AIDS 12: 1021–1026.

Sevigny JJ, Albert SM, McDermott MP, et al (2004). Evaluation of HIV RNA and markers of immune activation as predictors of HIV-associated dementia. Neurology 63: 2084–2090.

Stoll M, Schmidt RE (2004). Adverse events of desirable gain in immunocompetence: the immune restoration inflammatory syndromes. Autoimmunity Rev 3(4): 243–249.

Subbiah P, Mouton P, Fedor H, et al (1996). Stereological analysis of cerebral atrophy in human immunodeficiency virus-associated dementia. J Neuropathol Exp Neurol 55: 1032–1037.

That fact implies that the onset of cognitive impairment might actually precede the appearance of neuropathological changes instead of following them.

Thompson KA, McArthur JC, Wesselingh SL (2001). Correlation between neurological progression and astrocyte apoptosis in HIV-associated dementia. Ann Neurol 49: 745–752.

Trillo-Pazos G, Everall IP (1996). Neuronal damage and its relation to dementia in acquired immunodeficiency syndrome (AIDS). Pathobiology 64(6): 295–307.

Handbook of Clinical Neurology, Vol. 85 (3rd series)
HIV/AIDS and the Nervous System
P. Portegies, J. R. Berger, Editors

Chapter 19

Neuropharmacology of HIV/AIDS

SIDNEY A. HOUFF* AND EUGENE O. MAJOR

Department of Neurology, University of Kentucky, Lexington, KY and Laboratory of Molecular Medicine and Neuroscience, NINDS, NIH, Bethesda, MD, USA

19.1. Introduction

Neurological diseases have been a prominent feature of HIV/AIDS since the pandemic began. The introduction of highly active antiretroviral therapy (HAART) has resulted in significant reductions in the opportunistic infections and cancers that were common early in the pandemic. With the reduction in cancers and opportunistic infections, survival has significantly increased with the average age of patients with HIV/AIDS reaching into the forties. Despite these improvements, HIV-related neurological diseases continue to have a significant impact on patients with HIV/AIDS (Neuenburg et al., 2002). HIV encephalitis, HIV-associated dementia and mild cognitive/motor disorders remain clinical and therapeutic challenges for physicians caring for patients with HIV/AIDS.

The pathogenesis of HIV-associated neurological disease is complex involving both the virology of retroviruses as well as the features of cell responses to injury, proinflammatory cytokines and other messenger molecules (Gonzalez-Scarano and Martin-Garcia, 2005; and see Chapter 4 of this volume). Alterations in nigrostriatal system functions by HIV, HIV with drugs of abuse and pharmacological agents used to treat various aspects of HIV-related neurological disease will provide further challenges to the successful management of HIV/AIDS (Berger and Arendt, 2000). The prolonged survival of patients with HIV/AIDS on HAART, while shifting the prevalence of HIV-related neurological diseases to older age groups, also is creating a population who are at risk for developing neurodegenerative diseases of later life (Neuenburg et al., 2002). The older HIV/AIDS patient will be approaching a time when they would otherwise be at risk for Alzheimer's disease, Parkinson's disease and other neurodegenerative disorders. It is conceivable that these patients, even without HIV-related neurological disease, will be at increased risk for developing neurodegenerative diseases which could be more severe and begin at an earlier age as a result of prior damage to basal ganglia and other neuropharmacological systems by HIV. Complicating an already complex situation, drugs used to treat neurodegenerative diseases may potentiate HIV-related effects on these disorders.

19.2. Neuropharmacology of antiretroviral agents

The development of antiretroviral therapies has had a significant impact on the neurological manifestations of HIV/AIDS. Zidovudine (ZDV), the first approved HIV drug, was quickly found to have significant effects on the cognitive symptoms of HIV-associated dementia (Schmitt et al., 1988; Hamilton et al., 1992; Nordic Medical Research Council, 1992; Tozzi et al., 1993). The development of additional antiretroviral agents including other nucleoside reverse transcriptase inhibitors (NRTIs), nonnucleoside reverse transcriptase inhibitors (NNRTIs) and protease inhibitors (PIs) have changed HIV/AIDS from an acute/subacute fatal infection to a chronic disease. The recent introduction of fusion inhibitors will likely further change the spectrum of HIV/AIDS but, because of absence of specific targets for fusion inhibitors in the brain, these drugs may have little direct effect on the neurological features of HIV/AIDS (Kilby and Eron, 2003).

19.3. Neurological disease in the HAART era

The introduction of multidrug therapy for HIV/AIDS has resulted in significant changes in the clinical and neuropathological features of neurological diseases in

*Correspondence to: Sydney A. Houff, MD, PhD, Department of Neurology, University of Kentucky, Lexington, KY 40536-0284, USA. E-mail: sahouf2@email.uky.edu.

Table 19.1

Neuropharmacological features of HIV-associated diseases

Basal ganglia and its neural circuits primary target of HIV infection
Neurotransmitters and receptors are involved in neurotoxic pathways leading to neurological disease
Dopamine, dopamine agonists and drugs of abuse that act through dopaminergic system facilitate HIV replication; accelerate HIV-associated neurological diseases
Pharmacological synergy between HIV-encoded proteins and neurotransmitters and drugs amplify neurotoxic effects in the brain
Cell death cascades encountered in neurodegenerative diseases are central in the pathogenesis of HIV-associated neurological disease
HIV alters neural progenitors affecting neural circuits
Lipid boats are disrupted by HIV-encoded proteins

HIV-infected patients. HAART has resulted in significant reductions in the incidence of opportunistic infections and cancers, including those of the nervous system (Langford et al., 2003). AIDS-defining events are no longer the major causes of death in HIV-infected patients who now die more frequently from hepatitis B- or C-associated cirrhosis, non-HIV-related malignancies, cardiovascular events, suicide, overdose or treatment-related fatalities (Bonnet et al., 2002).

It is useful to divide the AIDS epidemic into at least four stages based on the use of antiretroviral therapies (ART) (Langford et al., 2003). The first stage from 1982 to 1987 corresponds to the period before the introduction of ART and reflects the natural history of HIV/AIDS. The second stage, or monotherapy stage, corresponds to the period 1987 to 1992 when early developed NRTIs were used alone. The introduction of ZDV and ddI characterize ART in the second stage. The third stage from 1992 to 1995 saw the introduction of newer NRTIs and the practice of combining two ARVs in therapeutic regimens. The fourth stage from 1996 to the present encompasses the introduction of PIs and NNRTIs and the practice of combining at least three ARVs in treatment regimens. It is during this fourth stage that the most profound changes in the neurological features of HIV/AIDS have occurred.

During the pretreatment and early treatment eras, opportunistic infections of the brain and respiratory system were the most common causes of death. Some 63% of all autopsies in AIDS patients had evidence of nervous system involvement by opportunistic infections or HIV (Langford et al., 2003). The most common HIV-associated neuropathologies included HIV encephalitis, vacuolar myelopathy, lymphocytic meningitis, diffuse poliodystrophy, granulomatous angitis, cytomegalovirus (CMV) encephalitis, aspergillosis, progressive multifocal leukoencephalopathy (PML), bacterial meningitis, tuberculosis and *Mycobacterium avium* complex (Masliah et al., 1992; Jellinger et al., 2000). HIV encephalitis was associated with multinucleated giant cells in the parenchyma and perivascular spaces, myelin pallor, astrogliosis, microglial cell activation and microglial cell nodular formation. HIV encephalitis lesions were found predominantly in the basal ganglia, frontal cortex and white matter. White matter damage was characterized by focal demyelination, astrogliosis, mild infiltration of macrophages with rare instances of extensive white matter tract destruction (Gray et al., 2003; Langford et al., 2003). Lesions in the hippocampus, brainstem, cerebellum and thalamus were also occasionally seen.

The severity, neuropathological features and distribution of central nervous system (CNS) lesions were found to be associated with the degree of HIV viral load in the nervous system. Cases with abundant multinucleated giant cells and extensive white matter involvement were associated with high HIV viral loads (Masliah et al., 1992). Lower viral loads were associated with less severe disease characterized by lymphocytic meningitis, microglial proliferation and moderate astrogliosis without multinucleated giant cells. Langford has proposed that these findings suggest HIV-associated disease of the nervous system can be divided into several stages (Langford et al., 2003). First, meningeal inflammation occurs with perivascular mononuclear cell infiltration associated with altered blood–brain barrier permeability. Following the breaching of the blood–brain barrier, HIV-infected mononuclear cells and HIV reach the brain parenchyma resulting in microglial cell activation and proliferation with formation of multinucleated giant cells and microglial cell nodular formation. Activation of microglia and macrophages leads to inflammatory events and secondary injury to neural cells. Alterations in myelin with myelin pallor and astrogliosis also follow. This scenario is consistent with that seen with experimental models in monkeys infected with SIV and SCID mice grafted with HIV-infected macrophages (Lackner et al., 1991; Pesidsky et al., 1997).

The spectrum of neurological diseases has changed in stages 3 and 4 of the treatment era. Recent studies have reported an increase in HIV-related brain lesions in AIDS patients with long-term survival (Suarez et al., 2001). HIV encephalitis occurred in 40% of patients when survival was short at the beginning of the epidemic (Jellinger et al., 2000; Masliah et al., 2000; Suarez et al., 2001). The incidence of HIV encephalitis fell with the introduction of ZDT in 1987 and remained low in patients continuing on the drug until death (Langford et al., 2003). The rate of HIV encephalitis increased during the later years as new cases occurred mainly in patients who had discontinued ZDT for various reasons (Gray et al., 1994; Maehlen et al., 1995; Suarez et al., 2001).

The incidence of HIV encephalitis has fluctuated from year to year without a clear trend (Jellinger et al., 2000; Masliah et al., 2000; Langford et al., 2003). The frequency of CMV encephalitis and toxoplasmosis has shown a definitive downward trend. The frequency of PML, *Mycobacterium avium* complex and herpes simplex virus has remained constant over time while non-Hodgkin's lymphoma and fungal infections have varied widely. In an autopsy study from Vienna, the frequency of nonspecific CNS changes such as meningeal fibrosis increased (Jellinger et al., 2000). In an autopsy study of drug abusers with HIV/AIDS, HIV encephalitis, PML, bacterial infections, hepatic encephalopathy and negative CNS findings were more frequent than in non-drug users who showed increased incidence of CMV, toxoplasmosis or other opportunistic infections with nonspecific CNS findings (Davies et al., 1997). The frequency of focal white matter lesions of unknown etiology has also increased in the HAART era (Ammassari et al., 2000). In contrast to most autopsy studies, Sacktor (2001) reported a decrease in the frequency of HIV encephalitis during the HAART era. The consensus of these autopsy reports during the HAART era confirms the continued presence of neuropathological changes associated with HIV infection despite the use of multiple HAART regimens. Unfortunately, the makeup of HAART regimens for patients in these studies has not been reported in most instances. It remains possible that autopsies of patients taking neuroactive HAART regimens might have different neuropathological findings, i.e. a decrease in the frequency and severity of HIV-related pathological changes.

19.4. Antiretroviral treatment of HIV-related neurological diseases

A reduction in the incidence of HIV-related opportunistic infection and cancers of the nervous system was seen after the introduction of ZDT for the treatment of HIV infection. Along with these changes, improvement in the cognitive functioning of patients with HIV-associated dementia was also seen (Simpson, 1999). The addition of a second NNRTI has been followed by further improvement in cognition while the development of PIs and the practice of combining three or more ARTs (HAART) has been followed by a further reduction in HAD (Dore et al., 1999). Introduction of highly active antiretroviral therapy with at least three drugs including at least two NRTIs and one PI has led to a reduction in the incidence of HAD while the prevalence of HAD has increased with patients surviving for longer periods of time while on HAART (Dore et al., 2003).

ZDT is still the only ART approved for the treatment of HIV-related dementia (Simpson, 1999). In 1987, Yarchoan and colleagues reported the effects of ZDT on four patients, three of which had HAD (Yarchoan et al., 1987). ZDT was administered orally at 1500 mg/day in three patients. Another patient received intravenous ZDT. The three patients with HAD improved as measured by cognitive and motor skills. A randomized, double-blind trial of ZDT versus placebo was reported by Schmitt et al. (1988). The study included 150 patients with AIDS and 112 patients with AIDS-related complex who were treated with either ZDT 1500 mg/day or placebo. The trail was planned for 24 weeks but was stopped at 16 weeks because of significant reduction in mortality in the ZDT-treated arm. Cognition, including memory and attention, as well as motor skills were improved at both 8 and 16 weeks in patients receiving ZDT. Improvement was especially apparent in patients with AIDS. No significant difference was seen in affect between the ZDT-treated patients and those receiving placebo.

The Academic Medical Centre (Nordic Research Council) in Amsterdam reported the effects of ZDT in a case review of 536 HIV-infected patients (Nordic Medical Research Council, 1992). Forty patients met CDC criteria for HAD. Thirty-nine of these patients were not taking ZDT at the time of diagnosis. Twenty of the 40 were started on ZDT. Ten received ZDT after the diagnosis of HAD while the timing of initiation of ZDT therapy was not reported for the other 10 patients. ZDT was administered at doses ranging from 400 to 1200 mg/day in the patients who initiated treatment after the diagnosis of HAD. In this group, 3 patients showed marked improvement in cognition and motor skills. Improvement lasted from 16 to 32 months. Two other patients had slight improvement in cognition.

Although these trials had many limitations, they did demonstrate significant improvements in cognition and motor skills in some patients with HAD who received ZDT. Dosage recommendations could not be made from these studies.

The Nordic Medical Research Council's HIV Therapy Group conducted a randomized, double-blind, parallel-group, multicenter trial comparing the clinical outcomes in 474 patients including 126 patients with AIDS, 248 patients with HIV-related symptoms and 100 patients with low CD^4 T lymphocyte cell counts (Nordic Medical Research Council, 1992). Patients received ZDT at 400 mg, 800 mg or 1200 mg/day for a median of 19 months. Fewer cases of HAD were diagnosed in patients receiving the higher doses of ZDT (8% at 400 mg/day, 6% at 800 mg/day, 3% at 1200 mg/day). Quality of life did not differ between the treated groups. Anemia and leucopenia were most common at the highest dose of ZDT leading the authors to suggest that ZDT be used at doses between 400 and 600 mg/day.

The AIDS Clinical Trial Group (ACTG) reported the results of a randomized, double-blind, placebo-controlled, multicenter trial of ZDT treatment for HAD (Sidtis et al., 1993). Forty patients were randomly assigned to receive ZDT at 1000 or 2000 mg/day. Follow-up was on average for 64 weeks. Neuropsychological testing was performed every 4 weeks. Patients taking ZDT were found to have significant improvement in cognitive symptoms at 16 weeks compared to those taking placebo. Statistical significance was seen only in patients receiving 2000 mg/day. Patients who received placebo were re-randomized into one of the two ZDT doses. Significant improvement in neuropsychological testing at 16 and 32 weeks was seen in those treated with 2000 mg/day. This study was limited by small sample sizes, high dropout rate and failure to define a clear optimal dose of ZDT (Simpson, 1999). Tozzi and colleagues reported the long-term effects of ZDT treatment on severity of dementia in 30 consecutive patients with HAD (Tozzi et al., 1993). Patients received 500, 750 or 1000 mg/day. The Price and Brew staging system was used to evaluate the patients over a 12-month period. Some 73% of patients showed some improvement at 3 months and 83% after 6 months of ZDT treatment. The ZDT dose did not correlate with improvement. Subsequent deterioration occurred in 8 patients after 6 to 12 months of ZDT therapy. Five of the eight patients received 500 mg/day of ZDT, the lowest dose in the study. The authors concluded that ZDT was associated with a reversal of neurological dysfunction in patients with HAD.

The effect of treatment with alternating versus simultaneous regimens of ZDT and didanosine on cognitive functions and motor skills in symptomatic patients with HIV infection was reported by Brouwers et al. (1997). Thirty-eight patients were examined in a randomized, unblended trial where 21 patients had neurological symptoms, 12 of which were diagnosed with HAD. Patients in the simultaneous arm received ZDT 300 mg and didanosine 250 mg daily for 12 weeks. The remaining patients received ZDT 600 mg daily for 3 weeks alternating with 500 mg daily of didanosine for 3 weeks. Significant improvement in neuropsychological functioning, particularly in memory and attention batteries, was found in both treatment groups. Patients with neuropsychological symptoms at the beginning of the study showed the most improvement.

These studies suggest there is a dose-response effect with ZDT in HIV-infected patients with cognitive and motor symptoms. The optimum dose for patients with HAD or at risk of developing dementia was not clearly defined in these trials (Simpson, 1999).

The efficacy of prolonged ZDT treatment in the treatment of cognitive impairment with HIV has been evaluated in several studies (Portegies et al., 1993; Tozzi et al., 1993). Portegies et al. found that 5 of 10 patients who began ZDT treatment after HAD was diagnosed showed improvement in symptoms with 3 patients showing significant improvement at 16, 24 and 32 months. Tozzi and colleagues reported that 15 of 26 patients who showed improvement in neuropsychological functioning had clinical relapses after 6 to 12 months of therapy. Reinvang et al. (1991) conducted an open label study of 11 AIDS patients, 6 of whom had HAD, who were treated with ZDT in doses ranging from 800 to 2000 mg/day. Cognitive functioning improved during the first 4 to 6 months of the trial. The improvement seen at 6 months was not found at 9 to 12 months after initiating ZDT therapy. The authors suggested that ZDT was unlikely to have long-term benefits in the treatment of HAD. The significance of the Reinvang study is limited by its low power to detect differences given the small sample size and the wide range of doses of ZDT used. A continuation of the study by Reinvang found in 33 AIDS patients treated with ZDT significant improvement in neuropsychological tests given 6 months after baseline when compared to untreated AIDS and 16 asymptomatic HIV-infected patients (Karlsen et al., 1995). By 12 months, ZDT-treated AIDS patients who showed improvement at 6 months had declined in all areas of cognitive functions tested, especially in tests of verbal abilities. The conclusion of these two studies was that the effects of ZDT on cognitive functioning were transient and not sustained with prolonged therapy.

Several studies have examined the role of ZDT in prophylaxis for HAD. A retrospective study of a consecutive series of AIDS patients examined the incidence of HAD and cerebrospinal fluid (CSF) HIV-1 p24 antigen in patients treated with ZDT (Portegies et al., 1989). Of the 196 AIDS patients with neurological symptoms seen at the AIDS unit at the Academic Medical Centre in Amsterdam between 1982 and 1988, 20% were found to have the diagnosis of HAD and 45% had received ZDT. In this retrospective study, the incidence of HAD decreased with the increase in use of ZDT. Compared to patients who did not take ZDT, significantly fewer of the patients receiving ZDT developed HAD (2.3 versus 36%). The authors concluded that the introduction of ZDT had been accompanied by a significant decline in HAD. The ACTG conducted a randomized, double-blind, placebo-controlled trial of the prophylactic effect of ZDT in preventing HAD (Volberding et al., 1990).

The study included 1338 HIV-positive patients with CD^4 cell counts <500 cells/mm^3 and normal psychiatric function. Patients received either ZDT at 1500 mg/day, 500 mg/day or placebo and were followed for an average of 55 weeks. Two patients in the placebo group developed HAD. One patient receiving 1500 mg/day of ZDT developed HAD while none of those taking 500 mg/day developed HAD. The authors concluded that ZDT was safe and effective in this group even though the rates of HAD were too low to clearly show the protective effects of ZDT in preventing the development of HAD.

The effect of early versus late administration of ZDT has been examined in several studies. The Department of Veterans Affairs Cooperative Study Group on AIDS Treatment compared the outcomes of symptomatic HIV-infected patients without HAD who received daily doses of ZDT at 1500 mg/day either early or late after the patients' CD^4 T cell count had fallen to <200 cells/mm^3 (Hamilton et al., 1992). HAD developed in 6 of the group receiving ZDT late in their clinical course while none of those treated early developed HAD. The incidence of ZDT-associated side effects of leukopenia and anemia were the same in both groups. The authors concluded that the use of ZDT deserved consideration in both symptomatic and asymptomatic patients early and late in the disease course. The Concorde Coordinating Committee Study did not support the findings from the Veterans Affairs Cooperative Study. The Concorde Coordinating Committee examined the effects of early and deferred treatment in randomly assigned asymptomatic HIV-infected patients (Concorde Coordinating Committee, 1994). ZDT 1000 mg/day was started early in 877 patients and deferred to after the development of persistently low CD^4 T cell counts or the development of AIDS-related complex in 418 patients. Patients were followed for a median of 3.3 years during which HAD was the AIDS-defining illness in six patients in the early treatment group and seven in the deferred group. Anemia and neutropenia developed earlier in patients receiving ZDT early in the clinical course.

A prospective, open label study followed 500 homosexual men who had the diagnosis of AIDS for up to 7 years (McArthur, 1993). ZDT was given at variable doses but averaged less than 600 mg/day. Some 15% of the treated patients developed HAD. The use of ZDT either before or after the diagnosis of AIDS did not predict the development of HAD. The authors concluded that any benefit that ZDT might have provided may have been lost once AIDS develops. The Multicenter AIDS Cohort Study, a prospective, longitudinal study, described trends in the incidence of HIV-related neurological diseases from centers in Baltimore, Washington, Chicago, Pittsburgh and Los Angeles (Bacellar et al., 1994). HAD and sensory neuropathy were analyzed in 2641 men who were HIV-1 positive. The incidence of HAD did not change significantly between 1988 and 1992 and no protective benefit for ZDT could be seen. The time frame of the *Mycobacterium avium* complex study compared to that from the Netherlands may explain at least in part the differences in the outcome of the two studies. Baldeweg et al. (1995) examined the records of 141 HIV-infected men in a retrospective study of men receiving care in a HIV outpatient clinic. Patients received a median ZDT dose of 500 mg/day. Impaired neuropsychological functioning was associated with the clinical stage of HIV infection. Overall, there was no effect of current ZDT use on neuropsychological functioning. Among symptomatic HIV-infected and AIDS patients, significantly fewer of those currently taking ZDT were found to have abnormal electroencephalograms. Of the 98 men who had used ZDT for <1 month or >1 year, patients with long-term ZDT use with either symptomatic HIV infection or AIDS had significantly higher scores on tests of cognition compared to those who had been taking ZDT short term. The authors concluded that long-term ZDT therapy may reduce neurological, neuropsychological and electrophysiological abnormalities and thereby reduce the risk of HAD. This study is limited by the natural history design which does not support an assessment of cause and effect and the exclusion of women (Simpson, 1999). On the whole, these studies suggest that ZDT may forestall the development of HAD.

The effects of ZDT on HIV encephalitis have been reported in a number of studies (Glass et al., 1993; Vago et al., 1993; Gray et al., 1994; Amchlen et al., 1995; Bell et al., 1996). Vago and colleagues examined autopsy results at the Clinic of Infectious Diseases in Milan for patients seen between 1984 and 1990 (Vago et al., 1993). Eighty-two patients were taking ZDT at 600–1000 mg/day for 1 to 30 months. The incidence of multinucleated giant cells, HIV leukoencephalopathy and a history of severe dementia were significantly reduced in the ZDT-treated groups. HIV encephalitis and HIV encephalitis with HIV leukoencephalopathy were present less frequently in the ZDT-treated group. Effects of ZDT were time- and dose-related with the maximum effect of ZDT treatment seen in patients treated for 6 to 12 months with cumulative doses greater than 2 g. Similar results were reported by Glass et al. (1993) in an autopsy study of 56 patients dying of AIDS in the USA. In 40 patients with HAD, significantly fewer of the patients who had been treated with antiretroviral therapy for >12 months exhibited multinucleated giant cells and diffuse myelin pallor compared to the neuropathological findings in those who had been treated for shorter periods of time. Gray and colleagues examined the autopsy results of 192 patients dying of AIDS between 1982 and 1992 (Gray et al., 1994).

The prevalence of multinucleated giant cells and HIV encephalitis with leukoencephalopathy were significantly lower in patients who took ZDT for >3 months and until shortly before death than among patients not treated with ZDT. The prevalence of markers of HIV activity in the brain was intermediate in those patients who discontinued ZDT treatment at least 1 month prior to death. Amchlen et al. (1995) examined the pathological findings in 4 groups of patients including those that never took ZDT, those who took ZDT for >2 months and stopped treatment <1 month before death and those who stopped ZDT 2 to 6 months and those who ceased therapy greater than 6 months before death. The incidence of multinucleated giant cells was inversely related to the length of time since ZDT was discontinued but the time of cessation of treatment was less closely associated with the presence of diffuse white matter damage and microglial nodules. The authors concluded that ZDT reduced the risk of brain lesions only if treatment was continued until the patient was near death. Bell et al. (1996) evaluated the autopsy results of 66 patients who died of AIDS. ZDT use, particularly if used within 1 year of death, was significantly associated with lower risks of HIV encephalitis.

19.5. HIV-related neurological diseases in the HAART era

The introduction of HAART has resulted in significant changes in the neurological diseases seen in HIV/AIDS (Brew, 2001). HIV-associated dementia is an evolving disease in the HAART era (McArthur, 2004). Before the introduction of HAART, the annual incidence of HAD after AIDS was 7% and the cumulative risk for developing HAD during the lifetime of a HIV-infected patient was between 5 and 20% (McArthur et al., 1993). With HAART, the incidence of HAD dropped significantly. However, in 2003, the incidence rates began to rise again (McArthur, 2004). With the prolonged survival of HIV/AIDS patients, the prevalence of HAD is rising as well. In the NEAD cohort, the prevalence of HAD and minor cognitive/motor disorder remains high at 37% among those with advanced HIV/AIDS, even when treated with HAART (Sacktor, 2002; Schifitto et al., 2002). A comparison of the incidence of HAD in the Dana cohort prior to HAART was 25% at 1 year and 40% at 2 years while the NEAD cohort, at the time of HAART, showed a cumulative incidence of HAD to be 25% at 1 year and 38% at 2 years. The clinical predictors for HAD were also similar between those treated before HAART and those treated with HAART. The degree of neurological impairment, extrapyramidal signs and functional impairments predicted progression to HAD in both cohorts. In the Dana, but not the NEAD, cohort abnormalities on tests of executive function and psychomotor speed were predictive of HAD. Depression was highly predictive in both cohorts. The diagnosis of minor cognitive/motor disorder was highly predictive of the subsequent diagnosis of HAD in the NEAD cohort.

Most studies of antiretroviral therapies have used predominantly male cohorts. Recent evidence suggests that women may be at the same risk of progression to HAD as men when treated with HAART (Robertson et al., 2004). More subtle forms of cognitive/motor impairment defined as mild cognitive/motor disorders are present in at least 30% of symptomatic HIV-infected adults (Janssen et al., 1989; Sacktor et al., 2002). Mild cognitive/motor disorder has a significant functional impact on HIV/AIDS patients and can affect medication adherence, possibly increasing the risk of HIV progression (Albert et al., 1999). The markers associated with HAD and mild cognitive/motor disorder have also altered with the use of HAART regimens (McArthur, 2004). The measurement of CSF HIV RNA viral load and currently available immune activation markers may fail to delineate the milder degrees of HIV-associated neurocognitive disorders in HIV-infected patients on HAART. In the NEAD cohort, no significant associations were detected between neurological status and baseline markers including plasma and CSF HIV RNA levels, tumor necrosis factor (TNF)-α, monocyte chemotactic protein (MCP)-1 and M-CSF when adjusted for antiretroviral usage, site and CD^4 T cell counts (McArthur et al., 2004). These findings are in distinct contrast to studies in the pre-HAART era where associations between these markers and HAD/mild cognitive/motor disorder were robustly demonstrated. These recent findings suggest that the clinical phenotype and biological markers associated with HAD have changed with HAART (McArthur, 2004). While the use of HAART has undoubtedly altered these parameters, the reports of lower levels of HIV RNA in the CSF of untreated patients with HAD suggest actual attenuation of the neurological disease has occurred (Clifford et al., 2002; McArthur et al., 2004).

There has also been a change in the predictive markers associated with progression to HAD. Conflicting data have been reported on the predictive value of CSF HIV RNA levels and other biomarkers. Ellis and colleagues reported that CSF HIV RNA levels were predictive of the development of HAD and subsequent neurological deterioration (Ellis et al., 1997). No correlation was found between the CSF HIV RNA levels and progression to HAD in the NEAD cohort (Sevigny et al., 2004). In the NEAD cohort, 36% of patients reached the endpoint of HAD by 20.7 months. HAART was used by 73% of patients when entering the study. Plasma HIV RNA was undetectable in 23% of patients and CSF HIV RNA levels were undetectable in 48% of patients at baseline. In adjusted analysis, the type of antiretroviral agent used in HAART was not a predictor of progression to HAD. In addition, plasma and CSF HIV RNA levels were not associated with time to diagnosis of HAD. The levels of TNF-α and MCP-1 tended to be associated with time to diagnosis of dementia. The authors concluded that the lack of correlation between plasma and CSF HIV RNA levels suggests that HAART may be affecting the CNS viral dynamics leading to lower HIV RNA levels and weakening the usefulness of these measures for predicting progression to HAD. The elevation of TNF-α and MCP-1 are of considerable interest as markers of disease. TNF-α has been previously shown to be a marker for progression to AIDS and with virologic and immunologic failure (Aukrust et al., 1999). The elevation of TNF-α found in the NEAD cohort patients may suggest further immune dysregulation and possibly an incipient rise in HIV RNA levels. The association of MCP-1 levels and progression to HAD is also suggestive of dysregulation of the immune system. MCP-1 is produced by activated astrocytes, microglia and macrophages. MCP-1 has significant effects on monocyte and macrophage activation and monocyte trafficking between the periphery and the CNS. In the SIV model of HIV encephalitis, the CSF/plasma MCP-1 correlates with SIV encephalitis (Zink et al., 2001). In humans, CSF levels of MCP-1 have been associated with HIV encephalitis and HAD (Cinque et al., 1998; Conant et al., 1998; Kelder et al., 1998). The absence of a correlation between MCP-1 and CSF HIV RNA levels in patients using HAART suggests that other factors may have a role in elevating MCP-1 in these patients (Gisolf et al., 2000; McArthur, 2004). The authors suggest that upregulation of the CNS monocyte/macrophage population, direct toxic effects on components of the CNS, or recruitment of peripheral monocytes, and potentially more toxic cells, into the CNS may account for the association of MCP-1 and progression to HAD (Pulliam et al., 1997; Gartner, 2000).

19.6. The effect of HAART regimens on HIV dementia

The optimal treatment regimen for neurological disease has yet to be established. Several factors have made the determination of the most efficacious treatment regimen difficult to establish. The use of various combinations of NRTIs, NNRTIs and PIs in HAART regimens makes comparisons between individual studies difficult. Lack of accepted surrogate markers for HAD and HIV encephalitis complicate study design and evaluation. Measurement of CSF antiretroviral drug levels may not reflect tissue levels in the brain. Furthermore, most studies have examined drug concentrations in lumbar CSF which does not accurately reflect drug levels in ventricles and the subarachnoid space in the brain.

HAART regimens consist of three or more antiretroviral drugs in combination. Most regimens consist of two NRTIs and a PI. Recent studies have confirmed the efficacy in reversing cognitive deficits in HAD using HAART regimens including one PI (Ferrando et al., 1998; Sacktor et al., 2000). In the *Mycobacterium avium* complex study cohort, neurocognitive improvement was independent whether or not the HAART regimen included antiretroviral drugs with significant penetration into the CSF (Sacktor et al., 2001). An add-on study with high-dose abacavir and matched placebo added to a stable HAART regimen was performed to determine whether the addition of an antiretroviral with significant penetration into CSF would improve outcome in HAD (Brew et al., 2000). Abacavir 600 mg twice daily was given to 105 randomized HIV-1-infected patients with mild to moderate HAD. The primary outcome measure was a change in 8 neuropsychological tests over the 12-week study period. The median change from baseline for the neuropsychological tests was comparable in the two groups. Patients receiving abacavir had no additional decline in CSF HIV RNA levels. The addition of abacavir did not augment HAART. Neuropsychological improvements continued to accrue in most patients receiving HAART with or without abacavir beyond 8 weeks. These findings suggest reversal of neuropsychological deficits in HAD patients may be long lasting but may occur at a slower rate than expected (McArthur, 2004). Levels of β2-microglobulin at baseline were normal in the study suggesting HIV-related inflammation in the brain was insignificant. Additionally, 90% of the patients in this study had NRTI-resistant HIV variants in the CSF while 66% of isolates had three or more NRTI-resistant mutations. The presence of resistant HIV strains in the CSF, even to drugs employed in the HAART regimen, did not impact on the improved outcome seen in these patients.

From uncontrolled studies, a substantial proportion of patients with HAD and mild cognitive/motor disorder have partial reversal of their neuropsychological deficits. Cohen et al. (2001) reported that women taking HAART for 18 months had significant improvement in psychomotor and executive functions. Tozzi et al. (1999) reported sustained improvements in neurocognitive performance after 6 months of HAART. Others have reported similar results (Ferrando et al., 1998). Levels of CSF β2-microglobulin, when reported, are higher in those patients with HAD and mild cognitive/motor disorder that respond to HAART. These findings suggest that the present evidence for CNS inflammation in the CSF correlates with reversible neurological deficits with HAART (Dougherty et al., 2002).

In the era before HAART, the mean CD^4 T cell count at the time of diagnosis of HAD was between 50 to 100 cells/μl (Brew et al., 2001). The introduction of HAART has resulted in an increase in the CD^4 T cell count at the time of diagnosis of HAD. In an Australian cohort, the CD^4 count at the time of diagnosis of HAD averaged 160 cells/μl (Dore et al., 1999). HAART appears to have severed the link between CD^4 T cell count and severity of HAD that was reported in the pre-HAART era. While this appears correct, it may be that CD^4 T cell counts below 200 cells/μl allow continued HIV replication in the brain. In addition, the prolonged survival of HAART-treated patients also provides for longer periods of time for HAD to develop. Thus, HAART may have introduced two new risk factors for HAD; i.e. nadir CD^4 T cell counts and disease duration.

The natural history of HAD has also changed. In the era before HAART, the mean time to death was 6 months from the time of diagnosis (Navia, 1986). The time to death after diagnosis of HAD has increased to 44 months when patients are taking HAART (Dore et al., 2003). The potential conditions confounding development of HAD have changed as patients age. Hepatitis C, testosterone deficiency and the effects of aging are now the most important confounders in HAD (Dore et al., 2003).

Apart from the confounding factors, the actual cognitive deficits seen in HAD may be changing. The evidence is indirect but convincing. Most evidence comes from studies in asymptomatic patients. In one report, altered patterns of neuropsychological deficits were seen in HIV-infected asymptomatic patients that pointed to involvement of cortical areas instead of the basal ganglia-associated deficits seen in the pre-HAART era (Cysique et al., 2004). An increase in the frequency of verbal memory impairment along with classical distribution of impaired fine motor coordination suggests a change in the clinical picture of HAD.

19.7. Antiretroviral therapy of pediatric neuro-AIDS

A limited number of studies have examined antiretroviral therapy for HIV-related neurological diseases in the pediatric age group. The majority of these reports have examined the efficacy of ZDT in the treatment of HIV-related encephalopathy. A limited number of studies have been published examining the effects of HAART on neurological diseases in the pediatric age group.

The course of HIV encephalopathy in children generally falls into one of three categories: subacute progression, plateau progressive and static (Civitello, 1991, 1992). Neurological signs may develop before immunosuppression occurs (Tardieu et al., 2000). Subacute progressive HIV-1 encephalopathy is characterized initially by slow but normal development. After a period of normal development, social, language and motor skills begin to be lost. Symmetric motor deficits and pathological reflexes are associated with failure to attain age-appropriate milestones or loss of previously attained milestones (Bellman et al., 1985; Epstein et al., 1986). Neuroimaging studies reveal an acquired microcephaly and p24 antigen and HIV viral RNA are present in the CSF. Early in the HIV pandemic and before antiretroviral therapy was available, progressive HIV-1 encephalopathy was reported in as many as 50% of HIV-infected children.

Children with progressive HIV-1 encephalopathy also develop at a slow followed by a decline in the rate of developmental progress with little or no further acquisition of skills. Acquired microcephaly is present as in progressive HIV-1 encephalopathy but p24 antigen and HIV RNA are not found in the CSF. The static form of progressive HIV-1 encephalopathy is seen in a minority of children with HIV-related encephalopathy. With the static course, children are cognitively impaired and are late in acquiring motor and language skills. Microcephaly is not present and CSF studies are normal.

Early studies demonstrated the ability of ZDT to reduce HIV expression in the CSF of pediatric patients with HIV infection. McKinney et al. (1991) measured the levels of p24 antigen in CSF of children receiving ZDT orally at 180 mg/m^2 every 6 hours. Some 35% of patients expressed p24 antigen in the CSF at baseline. After 6 months of therapy, 8 of the 18 patients that tested positive for p24 antigen at baseline showed a decrease in p24 antigen concentrations and 6 tested negative. Sixteen of 33 patients whose CSF was negative for p24 antigen at baseline were also re-examined at 6 months. Only one patient had detectable p24 antigen in the CSF at 6 months. Laverda et al. (1994) examined the immunological and virological findings in the CSF of 15 HIV-infected children, four of whom had progressive encephalopathy. Patients were examined a baseline and at least 6 months after treatment with ZDT 600 mg/m^2/day. Substantial declines in the levels of interleukin (IL)-6, IL-1β and p24 antigen were found in the CSF of patients treated with ZDT.

Efficacy studies of ZDT in progressive HIV-1 encephalopathy were reported early after introduction of the drug. In a Phase I clinical trial, Pizzo et al. (1988) evaluated cognitive and adaptive functioning in 13 children receiving intravenous ZDT. Patients' ages ranged from 6 months to 12 years. Clinical evidence of progressive HIV-1 encephalopathy before therapy was present in 8 patients. ZDT was administered intravenously in dosages of 0.5, 0.9, 1.4 or 1.8 mg/kg/day. Cognitive and motor testing were assessed weekly. Significant improvement in the scores of cognitive tests was seen in children with and without progressive HIV-1 encephalopathy. Cognitive improvement was found to occur early in the treatment course and at all dosages used. Cognitive improvement was found to continue steadily and was maintained throughout the study period. An extension of the study re-evaluated the 13 children after 12 months of therapy (Brouwers et al., 1990). The gains in cognitive function and adaptive behavior seen at 6 months were mostly maintained in children with and without progressive HIV-1 encephalopathy. The efficacy of ZDT was also evaluated in an open label study of 88 children aged 3 months to 12 years with either AIDS or advanced HIV disease (McKinney et al., 1991). ZDT was administered orally at 180 mg/m^2 given every 6 hours. Cognitive function was tested at baseline and at 6 months after starting therapy in 55 children, 39 of whom had progressive HIV-1 encephalopathy. Increases in cognitive functioning were found in 42%. Significant improvement was seen in children under the age of 30 months. Older children did not show statistically significant improvement on cognitive scores at 6 months after starting ZDT therapy. Another open label, 6-month study examined the effects of ZDT on cognition, adaptive behavior and motor skills in 25 children aged 1 to 12 with symptomatic HIV infection of which 13 had progressive HIV-1 encephalopathy (Wolters et al., 1994). Twelve children received oral ZDT while another 13 received ZDT by continuous intravenous infusion. Significant improvements were seen in all neurodevelopmental areas measured at 6 months. Children with progressive HIV-1 encephalopathy had lower scores at baseline but no difference was seen between these patients and those without progressive HIV-1 encephalopathy at 6 months of therapy. The route of drug administration did not affect outcomes. Contrasting results were reported by Nozyce et al. (1994) in an open label 1-year study of the effects of oral ZDT in 54 children aged 2 months to 12 years with HIV infection, 4 of which had progressive HIV-1 encephalopathy. No significant changes in cognitive functioning were found after 12 months of ZDT therapy.

The optimal dose of ZDT in the treatment of pediatric HIV-related neurological disease has not been settled. In the study by Pizzo et al. (1988), the authors concluded that the optimum ZDT dosage ranged between 0.9 to 1.4 mg/kg/h based on the frequency of adverse reactions of anemia and neutropenia which were most common at doses of 1.4 and 1.8 mg/kg/h. There was no difference in cognitive functioning between any of the doses studied. Brady et al. (1996) reported a large, randomized, double-blind clinical trial with 424 patients randomized to receive either ZDT 90 mg/m^2 q6h or 180 mg/m^2 orally q6h. The HIV-infected patients ranged from 3 months to 12 years of age with mild to moderate symptoms. Progressive HIV encephalopathy was not present. Patients were followed for an average of 39 months. No significant differences were found in the treatment groups at any point in time when examined for the development of progressive HIV-1 encephalopathy, cognitive functioning or for survival. ZDT-induced anemia and neutropenia occurred at the same rate in both the 90 mg/m^2 q6h or 180 mg/m^2 treated groups.

The use of combination antiretroviral therapy using two ARTs has been reported for pediatric patients with HIV/AIDS. Several reports of multiple antiretroviral therapies have been reported in pediatric AIDS patients. A randomized, double-blind clinical trial examined the effects of a combination of ZDT and ddI therapy compared to either drug alone in antiretroviral-naive children aged 3 to 30 months (Raskino et al., 1999). Patients were stratified by age and randomly assigned to one of the three treatment groups. Serial neurological examinations, neurocognitive tests and brain growth assessments by neuroimaging were performed in 831 patients. During the initial 24 weeks of therapy, treatment with ZDT and ddI or ZDT alone resulted in statistically significant improvements from baseline for both cognitive scores and head circumference growth. The combination therapy arm also showed a significant reduction in motor dysfunction during the first 24 weeks. After the initial treatment period of 24 weeks, the effectiveness of ZDT monotherapy appeared to decline relative to combination therapy. Cognitive scores declined steeply after 24 weeks in the ZDT arm compared to combination therapy, becoming statistically significantly lower by week 96. Similar trends in favor of combination therapy was seen when head circumference and motor function were examined at 96 weeks. Patients treated with the combination of ZDT and ddI showed less cortical atrophy and better motor performance at 96 weeks compared to those treated with ZDT alone. When compared to ddI alone, combination therapy showed statistically significant benefits at the initial 24-week evaluation for neurocognitive function, head circumference growth and motor functioning. After 24 weeks, cognitive performance and head circumference were not significantly different between those receiving combination therapy or ddI alone. Cortical atrophy measured at 96 weeks appeared to be slightly increased in the ddI arm compared to those receiving combination therapy but results did not reach statistical significance. Motor function continued to be better in those receiving combination therapy compared to those treated with ddI alone. A comparable study using ZDT 60 to 180 mg/m^2 and ddI at 60 to

180 mg/m^2 did not find significant improvement in those receiving combination therapy when compared to those receiving either ZDT or ddI (Husson et al., 1994). Unlike the study by Raskino et al. (1999), patients in this trial were functioning in the normal range at entry which may have made it difficult to detect benefits with the use of combination therapy.

The use of HAART in the treatment of neurological aspects of HIV/AIDS in the pediatric population has been the subject of several reports. Shanbhag et al. (2005) reported a retrospective cohort study of 146 perinatally HIV-infected children born between June 1990 and May 2003 who had at least one neurocognitive test score while being seen in an outpatient clinic in an academic children's medical center. There was a statistically significant decrease in the prevalence of progressive HIV-1 encephalopathy, the combined prevalence of progressive HIV-1 encephalopathy and static HIV-1 encephalopathy, and the prevalence of static HIV-1 encephalopathy in children born after January 1996 compared to those born before that date. The mean neurocognitive scores were stable over time in all patients studied. A significant increase in the most recent neurocognitive scores examined during the study was seen in children born during the HAART era. The mean CD^4 T cell percentage during the 6 months prior to neurocognitive testing was associated with neurocognitive scores. HIV plasma viral load was marginally associated with neurocognitive scores while no association was detected between neurocognitive scores and the number of CSF-penetrating antiretroviral drugs used in individual HAART regimens.

A meta-analysis of 5 large studies of the Pediatric AIDS Clinical Trial Group was performed to examine the predictive value of antiretroviral treatment mediated changes in plasma HIV-1 RNA levels, CD^4 T cell counts and CD^4 percentage on weight growth failure, cognitive decline and survival in HIV-infected children (Lindsey, 2000). Patients were treated with NTRIs and NNRTIs but none were receiving PIs. Among children receiving nucleoside with or without NNRTIs, higher immunological and lower virological markers at 24 weeks were significant independent predictors of survival. Plasma HIV RNA load was a significant predictor of weight gain and cognitive functioning in children over 1 year of age. The findings in this study demonstrate the need to include age when assessing patient responses to antiretroviral drugs.

McCoig et al. (2002) reported the virological results in 25 perinatally HIV-infected, Panamanian children enrolled as part of a double-blind, randomized, multicenter, placebo-controlled trial conducted in the USA and Panama. ZDT- or ddI-experienced, HIV-infected children 7 months to 10 years were included. Patients were randomized to receive either abacavir/lamivudine/ZDT or lamivudine/ZDT with abacavir placebo. Initiation of a new antiretroviral regimen in children in this study using agents that cross the blood–brain barrier decreased CSF HIV RNA and improved both neurological and neuropsychological outcomes. The presence of different mutations associated with resistance in the CSF and plasma demonstrated a discordant viral evolution in CSF versus plasma compartments suggesting they evolved independently. The emergence of mutations associated with resistance at week 48 was thought to possibly account for the less than favorable virological response, especially in 12 children who had plasma HIV RNA loads greater than 4.00 $\log_{10}$ copies/ml. Eight of these 12 children had virus in plasma that developed new mutations associated with antiretroviral resistance. Only 4 children had paired data for CSF HIV RNA collected at baseline and week 48. Three of these patients had CSF HIV isolates that had developed new mutations associated with ART resistance. Despite the switch in therapy and the development of viral resistance to antiretroviral drugs in CSF isolates at week 48, all but one child showed improvement in neurological status at week 48.

Chriboga et al. (2005) reported the neurologic outcomes in HIV-infected children from the pediatric HIV program at the Harlem Hospital Center during the HAART era. Neurobehavioral and school placement was assessed prospectively in the year 2000 in 126 HIV-infected children. In 1992, 31% of patients from this cohort were diagnosed with progressive HIV-1 encephalopathy. The rate of active progressive HIV-1 encephalopathy in 2000 was 1.6% with a prevalence of 10%. Besides the introduction of HAART, other reasons for the low incidence include low HIV transmission rates since the advent of perinatal antiretroviral treatments that result in fewer susceptible children infected with HIV, the early identification of perinatal children infected with HIV and the implementation of well-defined treatment recommendations based on a child's age, CD^4 count and HIV RNA measures. There were also differences in the clinical features of progressive HIV-1 encephalopathy. Most children in the Harlem cohort presented with corticospinal tract signs and impaired developmental or cognitive abilities. Survivors in this cohort did not progress beyond a diagnosis of spastic diparesis. All children with the diagnosis of progressive HIV-1 encephalopathy improved with a change in antiretroviral therapy to HAART. Improvement in cognitive function and head size often preceded improvements in motor function. Residual neurological impairments among children with arrested progressive HIV-1 encephalopathy included high rates of nondisabling spastic diparesis and despite the significant improvement in cognitive and motor functioning required higher rates for placement in special education as compared to children who did not develop progressive HIV-1 encephalopathy.

19.8. ART drug distribution to the CNS

The free movement of drugs into the CNS is restricted by the blood–brain barrier located at the level of the cerebral endothelial cells and blood–CSF barrier formed by the choroids plexus and arachnoid membrane. Substances cross the capillary walls by a variety of routes including between the cells of the endothelium, through cell fenestrations or directly through the cell wall. In the CNS, these routes are restricted as a direct result of the differences in permeability directly linked to the ultrastructure of the endothelium (?). In the CNS, endothelial cells have a paucity of transcytotic vesicles and fenestrations, the presence of continuous belts of tight junctions between cells and a variety of intracellular enzyme systems (?). Molecular movement across the blood–CSF barrier is also restricted by the presence of tight junctions and enzyme systems (?). The passive movement of drugs across these barriers depends on the size, charge, hydrophobicity and hydrogen bonding potentials of the drug. Transporters present at the level of the blood–brain barrier and blood–CSF barrier influence the distribution of molecules into and out of the CNS and CSF (?).

The CNS distribution of ARTs is related, in part, to their ability to use transporters at the blood–brain barrier and blood–CSF barrier (Borst et al., 2000). Probencid sensitive transporters and ATP-binding cassette transporters are well established as responsible for the removal of certain NRTIs and PIs. The activity of these transporters is important in designing ART therapies since combinations of anti-HIV drugs can compete for the same transport system into or out of the brain (Enting et al., 1998). Competition between ARTs theoretically could result in subtherapeutic levels of ARTs in the brain thus allowing low-level HIV replication to continue with the possibility of development of resistant HIV strains within the brain (Lambotte et al., 2003). Of additional importance is the ability of HIV to alter the permeability of the CNS barriers and alter expression of drug transporters within the brain (?). Until the interactions of ARTs with drug transporters in the CNS are better understood, it may be best to select combinations of ARTs that do not interact for transport at the level of the blood–brain barrier and blood–CSF barrier.

19.9. Drug movement between the brain and CSF

The level of drug in the CSF does not necessarily correlate with drug concentrations in the brain or the ability of drugs to cross the blood–brain barrier. The blood–brain barrier capillaries have a much larger surface area than the blood–CSF barrier at the choroids plexus (5). A drug that crosses the choroids plexus and reaches the CSF may not be able to cross the blood–brain barrier. This is related to the higher paracellular permeability features of the choroids plexus compared to the endothelial cell barriers of the blood–brain barrier (6). Although tight junctions are present at both barriers, those that form the blood–CSF barrier do not form a continuous barrier like that seen at the blood–brain barrier (?). Therefore, some drugs may cross from the blood into the CSF through the intercellular route that cannot do the same at the blood–brain barrier. Once the drug has entered the CSF, it is free to move across the ependymal surface into the brain since there are no tight junctions at the ependymal surface. The movement of drug between the extracellular fluid and the CSF is extremely complex. Drug movement across the ependymal surface is by bulk diffusion, a relatively inefficient process. Furthermore, the bulk fluid movement of the extracellular fluid is predominantly from brain to CSF. Thus, the level of drug and the depth of penetration of drug from the CSF to the brain parenchyma limits the efficiency of this route for drug to enter the brain.

The three main classes of drugs used in the treatment of HIV infection are the NRTIs, the NNRTIs and the PIs. The more recent introduction of drugs that inhibit binding of HIV to cellular receptors provides another avenue of therapy.

19.10. CNS distribution of NRTIs

NRTIs remain the principal drugs used for the treatment of drug-naive patients (Arendt and von Giesen, 2002). AZT, D4T, ddI, ddC, 3TC and abacavir have been employed in the treatment of HIV/AIDS patients in the context of HAART therapy (?). The distribution of these drugs will be briefly discussed below.

19.10.1. AZT

3′-Azido 3″-deoxythymidine (ZDT, AZT) was the first drug to be found to be effective in the treatment of HIV infection. AZT was found to be effective in reducing the incidence of HAD (Greg et al., 1999; Hamilton et al., 1992; Pizzo et al., 1988; Schmitt et al., 1988). AZT also resulted in a decline in the CSF of markers of HIV infection in the CSF and improvement in the neuropathological findings in treated patients (Brady et al., 1996; Gadeneg et al., 1995; Vago et al., 1993). AZT can reach the CSF in HIV-infected patients with and without AIDS (Bell et al., 1995). Animal studies support the ability of AZT to reach the brain (?). Most studies support a non-facilitated transfer of AZT into the brain and CSF from the blood (?).

Considerable evidence is present for the active removal of AZT from the blood and CSF by a probenecid sensitive transport system (Takasawa et al., 1997; Takasawa et al., 1997b). Probenecid is a nonspecific inhibitor of organic anion transport and can inhibit multidrug resistance associated protein (MRP), organic anion transporting polypeptide (OATP) and organic anion transport (OAT) transporters. CNS removal of AZT has also been shown to be sensitive to ρ-aminohippurate (PAH) which is a substrate for the OAT system. Currently, OAT1 and OAT3 are strong candidates for the CNS removal of AZT. The possible involvement of the MRP transporter has been indicated at the brain endothelial cell barrier. The attempts to use probenecid to increase levels of AZT in the brain led to unwanted side effects that prevented its use. The possible efficacy of benzbromarone, that does inhibit AZT efflux, may provide an alternative adjunct therapy to potentiate the CNS penetration of AZT.

19.10.2. D4T

2′,3′-Didehydro-3′-deoxythymidine (D4T, stavudine) is a potent thymidine analogue against HIV in vitro (Borst et al., 2000). A small study investigating the effects of NRTIs on psychomotor slowing found that D4T and AZT were equally effective in improving CNS function. D4T has also demonstrated efficacy in the HIV-1/SF162 human aggregate brain model where there was a reduction in neuronal cell loss and astrocytosis.

D4T has been detected in the CSF of healthy and HIV-1-infected individuals at concentrations above the 40% effective concentrations (ED50) against clinical isolates of stavudine-sensitive HIV-1 (?). Combined with 3TC, D4T has been shown to reduce HIV-1 RNA levels in the CSF.

The distribution of D4T into the CSF and brain has been studied in a wide variety of animal models. Discrepancy between the results of various studies may likely reflect the models used to study drug distribution. The penetration of D4T into brain parenchyma is limited at best with several studies demonstrating no movement of D4T into the brain. CSF penetration of D4T appears to be non-saturable suggesting simple diffusion of the drug across the blood–CSF barrier (?). Active efflux of D4T has been proposed to occur at both blood–brain barrier and blood–CSF barrier (?). The structural similarity of AZT and D4T has been used to propose that a probenecid sensitive efflux process may alter D4T distribution (?). This hypothesis remains to be tested.

19.10.3. ddI

2′,3′-Dideoxyinosine (ddI, didanosine) was the second NRTI to be approved for treatment of HIV infection. ddI given alone improved CNS function measured by psychomotor tests in HIV-infected patients. ddI with AZT has been shown to be more effective than either drug alone (Borst et al., 2000). When ddI was used alone in patients sensitive to AZT, no reduction in the development of HAD was found (?). An additional study did not detect any effect of ddI on the incidence of brain disease.

CSF/plasma ratios in animal studies range from 1.5 to 2% in rats to 4–10% in monkeys. In humans, the CSF/plasma ratio has been found to be between 21 and 27% (?). The differences between animal studies and those in humans likely reflect the alterations of the blood–brain barrier seen in HIV/AIDS patients. The distribution of ddI in the animal brain has varied widely in many studies but most investigators believe that 4% over 2 hours is a likely distribution (Gibbs et al., 2003). These studies suggest that ddI does cross the blood–brain barrier and blood–CSF barrier but only to limited extent. ddI enters the CSF in a non-saturable process in monkeys and rats. The role of nucleoside transporters including an equilibrative transporter has been proposed. The distribution of ddI in the guinea pig brain is not affected by the presence of other NRTIs suggesting that the use of ddI with other NTRIs would not alter their levels in the brain. Considerable accumulation of ddI occurs in the CSF. AZT, 3TC and abacavir compete with ddI for uptake at level of the choroids plexus. The organic anion transporter has been implicated in the distribution of ddI across the blood–CSF barrier. An OAT (Oatp2)-like transporter at the blood-facing side of the choroids plexus has been implicated in facilitating ddI entry across the blood–CSF barrier. The removal of ddI across the blood–brain barrier and blood–CSF barrier has been shown to be sensitive to probenecid implicating these transporters in the removal of ddI from brain and CSF. Overall, the evidence suggests that multiple transporters are involved in the movement of ddI across blood–brain barrier and blood–CSF barrier.

19.10.4. ddC

2′,3′-Dideoxycytidine (ddC, zalcitabine) has a more favorable safety profile than other NRTIs and is unlikely to induce hematological and hepatic adverse effects, neuropathy or myopathy (?). ddC used as monotherapy has been shown to have a significantly lower mortality and longer AIDS-free survival than those receiving initial therapy with regimens limited to AZT, ddI or ddC (?). CSF penetration has been shown to be 6% of plasma levels at 2 hours after administration in adults and 0–46% in children. CSF concentrations were found to be higher in lumbar fluid compared to ventricular fluid in primates (?). Transport of 3TC across the blood–brain barrier in the guinea pig model was negligible (?). Transport inhibitors have been proposed but have yet to be identified (?).

19.10.5. Abacavir

Abacavir [(-)-(1S,4R)-4-[2-amino-6-(cyclopropylamino)-G-purin-9-y1]-2-cyclopentene-1-methanol; 1592-U89] is a carbocyclic nucleoside analogue that produces marked and sustained reductions of plasma HIV-1 RNA. Abacavir is a uniquely activated HIV reverse transcriptase inhibitor that is anabolized to carbovir triphosphate which is a potent inhibitor of HIV reverse transcriptase. Addition of abacavir to existing ATR regimens did not reveal any significant additional neuropsychological improvement compared to placebo (?). These individuals had advanced disease which may have influenced the results. Abacavir has been shown to have significant CNS side effects including headache and neuropsychiatric disorders (?). These studies suggest that abacavir penetrates the CNS in significant levels.

CSF/plasma ratios in patients treated with 200 mg orally TID resulted in 18% ratios (?). In the guinea pig model, abacavir penetration into the brain was significantly higher than that seen in the CSF. The distribution of abacavir into the CNS was non-saturable suggesting the drug enters by non-facilitated diffusion. The evidence for abacavir efflux has been shown in the guinea pig model. Abacavir interacts with the sodium-dependent nucleoside transporter of the N3 type. The N3 transporter has been demonstrated on rat microglia cell lines and the rabbit choroids plexus.

19.11. NNRTIs

The NNRTIs include nevirapine (NVP), efavirenz (EFV) and delaviridine (DLV). The NNRTIs are of special interest because of the ability of NVP and EFV to be detected in the CSF. Development of drug resistance is a major problem in the use of these drugs. Although these drugs are structurally diverse, they act through a common pathway to inhibit reverse transcription.

19.11.1. Nevirapine

NVP is a highly potent inhibitor of reverse transcriptase. NVP was the first NNRTI to be approved for clinical use. The use of NVP as monotherapy or added to failing drug regimens demonstrated the use of the drug results in rapid accumulation of drug-resistant mutants. The development of resistance to PIs has led to a renewed interest in NVP in HAART. They offer several advantages over PI regimens including convenient administration regimens, fewer drug interactions and CNS penetration. NVP has significant antiviral activity when used in combination with drugs of the nucleoside and protease inhibitor classes. NVP in combination with NRTIs significantly improves HIV-1-associated psychomotor slowing compared to NRTIs alone.

NVP CSF levels are higher than those seen for EFV and other NRTIs alone. Indirect evidence suggests that NVP reaches the CNS by the appearance of headaches and neuropsychiatric complications associated with NVP use. The CSF/plasma ratio for NVP reaches 40% in children. CSF concentrations for NVP reached IC95 levels for HIV-1 in vitro. Overall, studies suggest that NVP crosses the blood–brain barrier and reaches brain tissue. NVP rapidly crossed bovine endothelial cells where passage appears to be by passive diffusion (?). There is no change in the passage of NVP in the presence of NRTIs, PIs and the transport inhibitors probenecid and verapamil. NVP produces more promising CSF levels compared to EFV.

19.11.2. Efavirenz

EFV [(S)-(-)-6-chloro-4-(cyclopropylethynl)-4-(triflouromethyl)-2,4-dihydro-1H-3,1-benzoxazin-2-one] is effective in combination with NRTIs in reducing viral load and in preventing AIDS-defining events in therapy-naive patients (Borst et al., 2000). Combination of EFV with two nucleoside analogues significantly improves psychomotor slowing (?). CNS disturbances include headaches, concentration difficulties and neuropsychiatric complications.

CSF/plasma ratios range from 0.65 to 1.19%. Patients receiving 600 mg/day had an average CSF concentration of 35 nM. Although these levels are significantly lower than that seen with NVP, the concentrations of EFV in CSF effectively inhibit HIV replication in human microglia.

19.11.3. Delaviridine

Delaviridine meyslate [DLV; 1-[(5-methanesulfonamido-1-H-indol-2-yl)-carbonyl]-4-[3-[(1-methylethyl)amino]-pyridinyl]piperazine monomethanesulfonate] is a potent NNRTI with a 95% inhibitory concentration against wild-type HIV-1 (Borst et al., 2000). DLV is less used because of high pill burden and TID dosing. It is the only NNRTI that inhibits cytochrome P450 allowing a reduction in the dose of associated PIs. DLV does not significantly cross the blood–brain barrier in mice where brain concentrations are less than 3% of plasma concentrations. In a bovine blood–brain barrier, there is no movement of DLV across the blood–brain barrier in vitro.

19.11.4. Protease inhibitors

PIs are commonly prescribed with reverse transciptase inhibitors where their use has resulted in large reductions in plasma viral load (Arendt and von Giesen, 2002). Regression of periventricular white matter and basal ganglia abnormalities in HIV encephalopathy has been correlated with cognitive improvement in patients receiving PIs (?). These changes have occurred despite the limited distribution of PIs in the CNS (Borst et al., 2000).

Saquinavir and ritonavir were not detected in the CSF when they were administered together in HIV-infected patients (Gimenez et al., 2004). CSF HIV viral load was not suppressed below limits of detection after 12 weeks of therapy in contrast to patients receiving D4T. Nelfinavir did not penetrate the CSF in patients with or without HAD. Lopinavir was also undetectable in the CSF of HIV-infected patients. Indinavir has been detected in the CSF of HIV-infected patients where it reached concentrations above its in vitro 95% inhibitory concentration for HIV. An animal study showed that indinavir reached the brain with a low brain/plasma ratio of 0.18 at steady state. The differences between indinavir, lopinavir and nelfinavir appear to be related to the ability to bind plasma proteins with values of 61, 98 and 98%, respectively. Amprenavir does cross the blood–brain barrier in an in vitro bovine model (?). However, in a single oral dose study, amprenavir was undetectable in the CSF in 24 hours.

Overall, the limited distribution of PIs in the CNS is thought to be secondary to efflux pumps. The multidrug resistance transporter P-glycoprotein (P-gp) and possibly the MRP efflux pathway limit the levels of PIs in the brain (Haisman et al., 2002). It has been suggested that ritonavir may facilitate brain uptake of other PIs by being a potent inhibitor of P-gy. However, experimental support for this mechanism has been difficult to acquire. A cell line study and murine study using saquinavar or amprenavir failed to find differences in cell or CNS concentrations of PIs. Specific inhibitors of P-gy and MRP-1 have been shown to alter the distribution of PIs in the CNS.

19.12. Neurological complications of ART therapy

Adverse events related to ART therapy can be classified as those directly related to alterations in neurological function by ART drugs and those that appear as a result of reconstitution of the immune system.

19.12.1. Neurotoxic neuropathy

Several drugs used in the treatment of HIV-related disease can cause distal symmetric polyneuropathy (DSP). A majority of patients treated with chemotherapeutic regimens develop symptoms and signs of DSP. This is especially true of those treated with vincristine. Isonazid used to treat tuberculosis and thalidomide used in the treatment of aphthous ulcers have been implicated in the development of DSP.

The nucleoside analogue antiretroviral agents have several dose-limiting toxicities. AZT is limited by hematologic toxicity but is not associated with DSP (Arendt and von Giesen, 2002). Dideoxynucleoside analogues didanosine (ddI), zalcitabine (ddC) and stavudine (d4T) are well recognized to cause neurotoxicity. Peripheral neuropathy was first reported with ddC therapy (Schifitto et al., 2002). The neuropathy associated with ddC is clinically and electrophysiologically similar to HIV-related DSP. Burning, parathesias, or aching in the lower extremities are common early symptoms. Reduced pinprick, temperature, light touch and vibratory sensation in the lower extremities with absent ankle reflexes are typical findings. Electrophysiological studies are similar to those found in HIV-associated DSP indicating a distal axonopathy. The acute onset and rapid progression of nucleoside-related DSP helps distinguish drug-associated neuropathy from HIV-related DSP.

The occurrence of toxic neuropathy is dose dependent. In one series, all patients who received high-dose ddC (0.12–0.24 mg/kg/day), 80% of those receiving intermediate dose (0.04 mg/kg/day) and only 11% receiving low dose (0.02 mg/kg/day) developed DSP. Withdrawal of ddC resulted in improvement except in those receiving high-dose ddC that continued to experience increasing symptoms for several weeks after stopping the drug before improvement occurred. Toxic effects of ddC may be reduced by alternating therapy with AZT but the highest incidence of DSP in the ACTG Protocol 175 was in the AZT/ddC arm (Arendt and von Giesen, 2002).

Painful DSP is also common with ddI therapy where the symptoms may be dose-limiting (Schifitto et al., 2002). Variable incidence of DSP in ddI therapy may reflect different dosing schedules and doses. DSP was reported in 23% of patients followed for 10 months. Like ddC, patients taking higher daily doses (27.2 mg/kg) and with higher cumulative doses (2.6 g/kg) were more likely to develop symptoms and signs of DSP. Once the symptoms resolved, most patients tolerated rechallenge with ddI at half-dose. Patients with low CD^4 counts are at higher risk to develop DSP at lower cumulative doses of ddI. A history of prior neuropathy, older age and poor nutrition are associated with the development of ddI-associated DSP.

d4T may also cause DSP. Some 55% of patients taking the maximum d4T dose of 4 mg/kg/day developed dose-limiting peripheral neuropathy with clinical features similar to that seen with ddC and ddI. Dose-dependent neuropathy was seen in a randomized trial using d4T. A total of 31% of those taking 2.0 mg/kg/day developed DSP 8 to 16 weeks after beginning therapy. Doses of 0.5 and 1.0 mg/kg/day were associated with DSP in 6 and 15% of treated patients, respectively. Patients with a history of neuropathy, low CD^4 counts, Karnofsky scores less than 80 and hemoglobin levels less than 10 g/dl were at increased risk of developing DSP with d4T.

Hydroxyurea is neurotoxic. Toxicity is especially prominent when hydroxyurea is used with ddI and d4T, a common adjunct therapy in clinical practice. The combination of ddI, d4T and hydroxyurea substantially increases the risk of DSP over the incidence of neuropathy in patients receiving ddI monotherapy.

The pathogenesis of toxic neuropathies related to dideoxynucleoside derivatives is unknown. In vitro and animal studies suggest that nucleoside analogues inhibit mitochondrial DNA gamma polymerase. Numerous clinical toxicities seen with antiretroviral nucleoside analogues have been attributed to mitochondrial toxicity. Lactic acidosis and other features of mitochondrial toxicity have been reported with nucleoside antiretroviral therapy. A rapidly progressive neuromuscular weakness resembling Guillain–Barre syndrome associated with lactic acidosis has been recently reported in patients treated with d4T (Rosso et al., 2003). Although attractive, mitochondrial toxicity is yet to be established as the mechanism leading to DSP in patients treated with nucleoside analogues.

Deficiencies in carantine have been suggested as a cause for mitochondrial dysfunction. Carantine is a critical substrate in mitochondrial metabolism. Reduced levels of acetyl-carnitine in patients with DSP related to ddI, ddC and d4T have led to speculation that reduction in carantine availability leads to mitochondrial failure. However, in the ACTG 291 study, serum carantine levels did not correlate with severity of HIV infection or drug-related DSP.

3TC, NNRTIs and PIs have not been associated with neurotoxicity in the peripheral nervous system.

19.13. Neuropharmacology of antiretroviral drugs in the CNS

The concentration of antiretroviral drugs in the brain and CSF depends on a number of factors, some of which are still poorly understood. Therapeutic agents are known to cross the blood–brain barrier and blood–CSF barrier by a variety of mechanisms including passive diffusion, transcytosis, receptor-mediated absorptive endocytosis and/or facilitated/active transport systems (Thomas, 2004). Once across these barriers, drug accumulation in the brain can be further impaired by a number of mechanisms including passive efflux in the bulk flow of the CSF (sink effect), metabolic degradation and active efflux transport including those involving the P-glycoprotein (P-gp) and MRP systems (Thomas, 2004; Pardridge, 2003).

Most ARTs appear to cross the blood–brain barrier and blood–CSF barrier by passive diffusion (Enting et al., 1998). At the level of the blood–brain barrier, tight junctions between endothelial cells prevent drug passage between cells. Although the blood–CSF barrier also has tight junctions, some intercellular transport of drugs may occur. The nucleoside analogues used in treating HIV infection do not appear to be transported by nucleotide transporters present at the blood–brain barrier. The modification of the 3′-sugar moiety alters the substrate availability of the nucleoside analogues for the nucleoside transporters (Thomas and Segal, 1998). However, there is some evidence for the limited transport of a small number of ARTs by active transport at the blood–brain barrier and blood–CSF barrier (80).

Efflux mechanisms that transport antiretroviral drugs out of the brain and CSF have an important role in determining the level of drug present in the brain and CSF. Mutlidrug transport systems were first recognized as a mechanism of drug resistance in cancer treatment (Dano, 1973; Juliano and Ling, 1976). Many drug transporters have been well characterized in peripheral tissues and are known to be involved in the influx and efflux of drugs identified in the brain (Tamai and Tsuji, 2000; Borst et al., 2000). These include organic cation, organic anion, nucleoside, P-gp and MRP transporters (80). Drug transporters influence drug absorption, distribution and elimination outside the nervous system. Many of these same transporters identified in systemic organs have also been identified in the blood–brain barrier and blood–CSF barrier. The discussion here will be restricted to those transport systems in the blood–brain barrier and blood–CSF barrier that affect antiretroviral pharmacokinetics in the CNS.

19.14. Organic anion transport systems

The organic anion transport systems include several families of multispecific organic anion transporters that belong to two main families, the organic anion transporter polypeptide (oatp) and the organic anion transporter (OAT) (Sekine et al., 2000). Organic anion transporters are classified into three classes depending on their energy requirements: sodium-dependent OATs, sodium-independent facilitators or exchangers and active OATs that require ATP. The active and sodium-dependent OATs possess broad substrate specificity. OAT systems have been localized along the blood–brain barrier and blood–CSF barrier (Lee et al., 2001). These transporters are responsible for the active elimination of organic anions from the brain and CSF.

The role of these transporters in the pharmacentics of ARTs is just beginning to be recognized. Didanosine has been shown to be transported from the CSF into the blood by a probenecid-sensitive OAT system in the rat (Takasawa et al., 1997; 1997b). ZDT is recognized by this system but is not transported by OAT proteins. Recently, the multi-resistant protein homologues, MRP1 and MRP2, have been shown to be responsible for cellular exclusion of organic anions (Kusuhara et al., 1998). MRP1 transports the PIs saquinavir, ritonavir, indinavir and nelfinavir in the MRP1-overexpressing cell line VM1–5 (Srinivas et al., 1998). In an in vivo study using canine epithelial cells transfected with human MRP-expression vectors, MRP2 was found to efficiently transport saquinavir, ritonavir and indinavir (Huisman et al., 2002). Addition of probenecid enhanced PI transport. Sulfinpyrazone also enhanced MRP2-mediated saquinavir transport but not the other PIs examined. MRP1, MRP3, MRP5 and Bcrp1 did not efficiently transport the PIs studied.

Ritonavir and saquinavir inhibited MRP1 in a study examining the toxicity of doxorubicin in the MRP1 expressing T lymphocyte cell line CEM/VM-1–5 (Srinivas et al., 1998). Both saquinavir and ritonavir were able to inhibit MRP1 in isolated pig brain microvessels (Miller et al., 2000). Ritonavir, but not indinavir or amprenavir, was able to block MRP1 activity in the UMCC-1/VP cell line overexpressing MRP1 at a level comparable to probenecid (Olson et al., 2002).

19.15. Nucleoside transport systems

The brain is dependent on a continuous and balanced supply of purine and pyrimidine nucleosides from both synthesis in situ and the blood (Lee et al., 2001). A number of tissues including brain are deficient in de novo synthetic pathways for nucleosides and rely on the salvage of exogenous nucelosides to maintain nulcoside pools and to meet metabolic demands. Therefore the brain is dependent on a continuous and balanced supply of purine and pyrimidine nucleoside constituents from both synthesis in situ and the blood (Kraupp and Marz, 1995).

Several nucleoside transporters have been identified in the blood–brain barrier and blood–CSF barrier (Lee et al., 2001). Active nucleoside transporters are also present in brain parenchyma. Purine and pyrimidine nucleosides are transported by multiple distinct transporters. Eight functionally distinct nucleoside influx transporters have been identified. Equilibrative processes for nucleoside transport are widely distributed in mammalian cells. However, concentrative nucleoside transport has only been identified in macrophages, choroids plexus, microglia, leukemia cells, splenocytes, intestinal cells and renal brush-border membrane vesicles (Wu et al., 1994; Hong et al., 2000). There is also a saturable carrier system for the efflux of nucleosides out of the brain (Wu et al., 1993).

Active transport has been demonstrated for ZDT, stavudine and ddI in several cell systems (Lee et al., 2001). However, nucleoside transporters (N3) that have been identified at the blood–brain barrier and blood–CSF barrier are not involved in the transport of nucleoside analogue drugs (Lee et al., 2001). A rat microglia cell line has been shown to possess a sodium-dependent nucleoside transporter. ZDT is a potent noncompetitive inhibitor for thymidine uptake by microglia that use the sodium-dependent nucleoside transporter but is not a substrate for the system (Wang et al., 1997). A novel electrogenic AZT/hydrogen-dependent transporter has been identified in microglia (Hong et al., 2001). IAZT uptake was sensitive to organic cations but was unable to transport the standard OCT substrate, tetraethylammonium. The system was also sensitive to classic organic anions and probenecid. Other nucleoside analogues including lamivudine, abacavir, didanosine and zalcitabine did not inhibit ZDT uptake by microglia. Although the system has some features of an organic cation transporter, it involves a carrier that is distinct from other OCT carriers.

In vitro studies of nucleoside analogue transport in a choroids plexus model system have shown that AZT is transported by both concentrative, sodium-dependent nucleoside transporters, CNT1 and CNT3, and the equilibrative nucleoside transporter ENT2 (Straazielle et al., 2003). Human choroids plexus exhibits a broadly specific transport system similar to that seen in rabbit choroids plexus, of the CNT3 type. The affinity of ZDT for the three transporters was much lower than that seen with natural nucleosides. However, ZDT influx into the CSF is not altered by the presence of endogenous nucleosides suggesting this pathway is of minor importance in comparison to diffusion for the entry of ZDT and other nucleoside analogues into the CSF.

19.16. Efflux transport systems

19.16.1. P-glycoprotein

P-gp is a 170-kDa plasma membrane, energy-dependent efflux pump belonging to the ABC superfamily of membrane transporters (Kin, 2003). Originally described in Chinese hamster ovary cells selected for colchicine resistance, P-gp expressing cells expressed a wide range of cross-resistance to anti-neoplastic drugs. This multidrug resistance is the result of expression of the MDR1 isoform (Golden et al., 2000). The MDR2 isoform functions as a phosphatidyltranslocase in the liver. The primary sequence encoding the P-gp protein is organized in two tandem repeats joined by a linker region. Each repeat is composed of an amino-terminal hydrophobic domain containing six potential transmembrane segments followed by a hydrophilic intracytoplasmic domain encoding an ATP-binding site. P-gp substrates include a wide range of naturally occurring antineoplastic agents, immunosuppressive drugs, cardiac glycosides, antibiotics and pesticides. The protease inhibitors saquinavir, indinavir, ritonavir and nelfinavir are also substrates for P-gp (Gimenez et al., 2004). Inhibitors of P-gp include calcium channel blockers (verapamil), calmodulin antagonists (trifluoperazine), quinolines (quinidine), cyclosporine A and PIs (Kim, 2003; Gimenez et al., 2004). Many drugs have been found to function as both P-gp substrates resulting in efflux of the drug as well as P-gp inhibitors. In the instance of multiple effects of individual drugs on the P-gp transporter, most investigators believe that the drug interacts with more than one binding site on the P-gp protein (Golden et al., 2000).

The mechanism by which P-gp extrudes substrates from the cell cytoplasm is unknown. The current model is the P-gp acts as a "flippase" where membrane bound drug is relocated between inner and outer lipid leaflets of the plasma membrane (Golden et al., 2000). The location of P-gp expression in the brain remains controversial. Most studies have localized P-gp expression at the level of the cerebrovascular endothelial cell (Cordon-Cardo et al., 1989). Other studies have suggested that P-gp expression is located on the abluminal surface of astrocyte foot processes (Golden et al., 2000). The primary P-gp isoform detected on brain endothelial cells is MDR1a while MDR1b is the main isoform present in brain parenchyma. P-gp is located on the apical surface of choroids plexus epithelial cells (De Lange, 2004). The action of P-gp at the apical surface of epithelial cells is opposed by MRP expression at the basal epithelial membrane which pumps drugs into choroids plexus capillaries. These opposing actions of P-gp and MRP at the level of the choroid plexus epithelial cell add to the complexity of drug movement at the blood–CSF barrier.

P-gp is expressed by primary rat astrocyte cultures but at lower levels compared to the level of expression in primary cerebrovascular endothelial cell cultures (Ronaldson et al., 2004). Surprisingly most factors that are known to induce astroglial activation in culture fail to increase P-gp oexpression. Interferon-γ and IL-6 family of cytokines increase the level of P-gp in astrocyte cultures (Monville et al., 2002). Interestingly, HIV Tat also increases P-gp expression in astrocyte cell cultures (Hayashi et al., 2005). P-gp is also expressed by brain microglia (Dallas et al., 2003). The expression of functional P-gp by endothelial cells of the blood–brain barrier, epithelial cells in the choroids plexus and astrocytes and microglia within the brain parenchyma suggests antiretroviral drug levels in the CNS are regulated at multiple levels.

P-gp has a significant role in the efflux of PIs at the level of the blood–brain barrier (Bachmeier et al., 2005; Gimenez et al., 2004). Pretreatment with inhibitors of P-gp transport results in increased concentrations of PIs in brain parenchyma. Multiple drugs are known to inhibit P-gp function including cyclosporine A, erythromycin, ketoconazole, mibefradil, quinidine, verapamil and others. Experiments using the P-gp inhibitor LY335979 resulted in a 25-fold increase in brain nelfinavir levels with only a 1.8-fold increase in plasma concentrations (Anderson et al., 2006). Intravenous administration of indinavir, nelfinavir and saquinavir to knock-out mice without the mdr1a gene (mdra-/-) resulted in an 8- to 10-fold higher concentration of indinavir and saquinavir and 40-fold increase in nelfinavir when compared to wild-type mice. Brain penetration of saquinavir was 1.8-fold higher in mdr1a knockout mice when compared to wild-type mice. In general, ARTs appear to enter the CNS by passive P-gp-mediated cerebral transport of saquinavir. When given intravenously or orally, saquinavir brain concentrations were increased in P-gp deficient mice by a factor of 5 for IV and 10 for oral administration. IV administration of indinavir resulted in a 5.8 higher brain concentration of drug in mdr1a knockout mice compared to wild-type mice at 6 minutes after administration, twice as high at 2 hours, and unchanged at 4 and 24 hours following dosing. Plasma concentrations were unchanged in both mdr1a knockout mice and wild-type mice.

19.16.2. MRP family

MRP represents a second efflux transport protein subfamily that belongs to the ABC protein superfamily and can confer multidrug resistance (Lee et al., 2001). MRPs can export a variety of organic anions from cells including drugs used in HAART (Thomas, 2004). All MRPs contain two transmembrane domains of six α-helices each (the P-gp-like core) connected to a cytoplasmic linker region. The linker region is absolutely necessary for protein transport properties. All MRPs are energy-dependent efflux pumps. Conjugates of lipophilic compounds with glutathione, glucuronate, or sulfate are the preferred physiological substrates of MRP 1, 2 and 3 (Sekine et al., 2000). Substrates for MRP 4 and 5 include cyclic adenosine monophosphate, cyclic guanosine monophosphate and nucleoside analogues (?). MRP 6 does not transport prototypical MRP 1 like substrates but is reported to transport the anionic cyclopentapeptide BQ123 (?). The functions of MRP 7 are currently unknown but its location on chromosome 6p12 suggests a role in glutathione metabolism.

Localization of MRPs in the brain has only recently been studied. The expression profile of MRP mRNA and protein was studied in brain endothelial cells and astrocytes (Borst et al., 2000). MRP 1 is the dominant member of the MRP family expressed by both astrocytes and endothelial cells (Takasawa et al., 2000b; Kusuhara et al., 1998). MRP 1 mRNA expression was approximately 60% higher in astrocytes compared to brain endothelial cells. Based on constitutive expression, MRPs have significant roles in efflux mechanisms in the brain. MRP 3 mRNA is primarily expressed in astrocytes and MRP 4 mRNA is mainly found in brain endothelial cells. The relatively low levels of expression of MRP 2, 6 and 7 in astrocytes and endothelial cells suggest these MRPs are not important transporters in the blood–brain barrier. MRP 1 has also been detected in the choroids plexus where MRP 1 mRNA is 5-fold higher than that observed in lung, an organ where high MRP 1 expression has been previously demonstrated.

The protease inhibitors saquinavir, ritonavir, indinavir and nelfinavir are substrates for MRP 1 in the VM1–5 MRP 1 expressing cell line (Srinivas et al., 1998). Saquinavir has been shown to be a substrate for MRP 1 and 2 as well as P-gp. Transport ranking of these efflux proteins revealed an apparent rank order of P-gp > MRP 2 > MRP 1. MRP 1 transported ritonavir and indinavir but not amprenavir in a pig kidney epithelial cell system. The rank order for these PIs was P-gp > MRP 1. In human lymphocytes, saquinavir and ritonavir were substrates for MRP 1, including lymphocytes from HIV-infected individuals. Ritonavir and saquinavir efflux correlated with expression of MRP 1. In MDCKII cells expressing human MRP 1, 2, 3, 5 and Bcrp 1, saquinavir, ritonavir and indinavir were transported by MRP 2 but not MRP 1, 3, and 5.

Protease inhibitors also inhibit MRP function. Saquinavir and ritonavir but not indinavir and nelfinavir increased by 3-fold the cytotoxicity of doxorubicin indicating that saquinavir and ritonavir were MRP 1 inhibitors (Tamai and Tsuji, 2000). Ritonavir but not indinavir and amprenavir was able to block MRP 1 activity in UMCC-1/VP cell lines overexpressing MRP 1 in an etoposide cytotoxicity assay. The inhibitory effects of ritonavir and saquinavir have been studied in isolated pig brain microvessels expressing MRP 1 and MRP 2 (?). Both ritonavir and saquinavir inhibited the transport of sulforhodamine 101, a substrate for MRPs.

19.16.3. Alterations in P-glycoproteins and MRP expression in HIV infection

Altered expression of P-gp has been reported in HIV encephalitis (Lee et al., 2001). In patients without HIV encephalitis, P-gp was found primarily on endothelial cells. In those with HIV encephalitis, P-gp immunoreactivity was found to be extensively expressed in astrocytes and microglial cells. Neuronal expression of P-gp was not found. The most intense staining for P-gp was present in white matter. A subset of HIV encephalitis patients displayed intense staining of astrocytes associated with blood vessels. The expression of P-gp was significantly correlated with the level of HIV RNA present in brain. In HIV encephalitis patients, brain viral burden and P-gp levels were significantly higher than those found in HIV encephalitis negative patients.

Polymorphism of the MDR 1 gene in humans has been reported to predict immune recovery after initiation of antiretroviral therapy (Kin, 2003). In one study 123 HIV patients with either long-term viral suppression with HAART or started on HAART during the study were analyzed for an association of MDR 1 polymorphism and CD^4.

Recent studies have shown that HIV Tat upregulates the expression of P-gp and MRPs (Hayashi et al., 2005). Primary mouse brain endothelial cell cultures exposed to HIV Tat_{1-72} resulted in overexpression of P-gp at both the mRNA and protein levels. These changes were confirmed in brain vessels of mice injected with Tat_{1-72} into the hippocampus. Pretreatment with a NF-κB inhibitor prevented upregulation of P-gp expression. The expression of P-gp was shown to be functional by increased rhodamine 123 efflux. Exposure of brain endothelial cells and astrocytes to HIV Tat_{1-72} also alters expression of MRPs (Hayashi et al., 2006). HIV Tat_{1-72} specifically induced MRP 1 mRNA and protein expression by both astrocytes and brain endothelial cells. The alterations in MRP 1 expression was accompanied by enhanced MRP 1 mediated efflux functions by 30 to 40%. Tat mediated overexpression of MRP 1 by brain endothelial cells was mediated through the ERK1/2 pathway. Both the ERK1/2 pathway and the JNK pathway participated in Tat mediated overexpression of MRP 1 by astrocytes. These findings suggest HIV infection may contribute to drug resistance in the CNS by altering the expression of efflux proteins at the levels of brain endothelial cells and astrocytes.

19.17. Neuroactive antiretroviral therapy

The persistence of HIV in the CNS, the development of drug mutations in brain and CSF isolates and the failure of HAART to prevent neurological disease in some patients has encouraged the search for neuroactive antiretroviral therapies. To date, the most effective HAART regimen to prevent or treat neurological disease has not been defined. A number of problems are responsible for the failure to define an acceptable antiretroviral regimen for the treatment of neurological HIV infection and disease. The poor penetration of antiretroviral drugs into the CNS presents significant challenges. However, a number of antiretroviral drugs reach significant levels in CSF and possibly brain yet low-grade viral replication persists. The emergence of antiretroviral mutants in the CSF and brain may also contribute to the failure of HAART therapy to prevent neurological disease in some patients. Furthermore, differences in the events leading to HIV expression in the CNS at different stages of HIV infection may also alter the response to antiretroviral therapies.

Compartmentalization of HIV replication with the development of viral reservoirs in sites such as testes and brain has been well established. A viral reservoir is defined as a cell or anatomical site in association with which a replication-competent form of virus accumulates and persists with more stable kinetic properties than in the main pool of actively replicating virus (Blankson et al., 2002; Lambotte et al., 2004). There are several lines of evidence supporting the CNS as a reservoir for HIV replication. Comparison of HIV replication in CSF and blood suggests viral replication differs in the two compartments. There is no correlation between HIV viral load in plasma and CSF in the same drug-naive individual (Stingele et al., 2001). The presence of mutations in the HIV genome also varies between CSF and plasma. In a study examining paired HIV isolates from blood and CSF, 9% of a total of 118 mutations associated with drug resistance occurred in blood isolates only while 11% were detected in the CSF only. In another study, paired CSF and plasma samples from patients on multiple antiretroviral therapies were examined for mutations resulting in antiretroviral resistance (Venturi et al., 2000). The mutation patterns were different in the two compartments in most patients. Genotypic resistance to protease inhibitors was detected in both plasma and CSF in one patient while accessory protease inhibitor mutations at polymorphic sites were different in plasma and CSF in several other patients. Differences in HIV genome sequences are also found in the brain as well as CSF. In one study, mosaic structures suggesting frequent recombination events were found in brain isolates compared to lymphoid tissue isolates in the gp120, gp41 and p17gag regions (Morris et al., 1999). Independent evolution of drug-resistant mutations have been found in HIV directly cloned from various regions of the brain (Smit et al., 2004). Independent evolution of both primary and secondary drug-resistant mutations to both reverse transcriptase inhibitors and protease inhibitors were found in diverse regions of the CNS of HIV-infected patients with and without dementia who were on antiretroviral therapies, including HAART (DeLuca et al., 2002). Astrocyte specific viral strains have been found in the V3 region of the HIV envelope gene in specimens isolated by laser capture microdissection from two patients with HAD (?). Both V3 sequences were distinct from that found in adjacent macrophages or multinucleated giant cells and were characteristic of CCR5 using (R5) HIV isolates. Similar results have been found in the SIV-infected macaque monkey model (Ryzhova et al., 2002). Phylogenetic analyses of envelope clones obtained from the brain and spleen found that brain sequences were significantly divergent from corresponding splenic clones from the same animal. Intra- and inter-animal comparisons revealed six brain-specific amino acid substitutions not found in splenic clones. Many of the changes were associated with the acquisition of macrophage tropism while others were associated with fusion of micro/macroglia. Interestingly, cell-specific amplification of V3 from individual multinucleated giant cells found sequences similar or identical to the original inoculum. These clones had accumulated few or no changes during 2 years of infection.

The kinetics of CSF viral load with antiretroviral therapy as well as the above studies suggest that the HIV present in CSF is derived from two sources. CSF viral loads have consistently been found to decline rapidly in asymptomatic subjects commencing therapy for the first time (Harrington et al., 2005; Haas et al., 2003). These findings suggest that the vast majority of HIV present in CSF in these patients is produced by short-lived cells. Two possible explanations could account for these findings. First, the vast majority of HIV-1 present in the CSF during the asymptomatic stage of infection is produced by short-lived CD^4 T cells infected in the periphery with brain-derived HIV contributing little to the bulk viral decline in CSF. In this case, a rapid decline of periphery-derived HIV in the CSF obscures the slower decline of any CNS-derived variants. Alternately, short-lived cells infected in both the periphery and the CNS contribute to the HIV-1 population pool in the CSF of asymptomatic patients. To distinguish between these two possibilities, Harrington and colleagues examined the genetic variants seen in the CSF and plasma in asymptomatic patients undergoing antiretroviral therapy. Some 85% of the HIV-1 population in the CSF was enriched and/ or unique to the CSF when compared to HIV genetic variants in the peripheral blood plasma suggesting that a significant portion of the HIV-1 population in the

CSF of asymptomatic subjects was produced from locally infected cells in the CNS. These compartmentalized variants declined rapidly during the initial days of therapy, some more rapidly than CSF variants shared with the plasma, suggesting that the majority of CSF producing cells in asymptomatic patients was produced by short-lived cells. The cell type responsible for the rapidly declining viral load is as yet undefined. It is unlikely that monocyte/macrophage cells are responsible for the early decline in HIV viral load since these cells would be expected to decline with a half-life of several weeks or more compared to the 1 to 3 days seen in these studies. Peripheral blood activated CD^4 T cells rapidly turn over and their clearance corresponds with the rapid decline of HIV-1 RNA in plasma seen during the initial days of therapy. Small numbers of CD^4 T cells enter the CSF suggesting that they may be a source of HIV in CSF during this period. In fact, some studies have suggested there is a relationship between lymphocyte pleocytosis in HIV viral load in the CSF. If these CD^4 T cells are infected solely in the periphery, these findings cannot explain the presence of subpopulations of CD^4 T cells appearing to be productively infected with HIV-1 variants unique to CSF.

A more plausible model to explain these findings suggest that uninfected CD^4 T cells migrate to the CSF in asymptomatic subjects where they become infected with HIV-1 produced by long-lived macrophages and microglia residing in the CSF compartment (Haas, 2000). These newly infected CD^4 T cells amplify monocyte lineage-derived, CNS-compartmentalized variants to the concentrations found in CSF. CNS macrophages and resident microglia do not directly contribute substantial amounts of CNS variants since they cannot access the CSF from deep brain areas. Unique CSF variants are unlikely to arise from CD^4 T cells infected in the periphery since these would also be detectable in the HIV from the plasma. Although the site at which CD^4 T cells are infected with CNS variants has yet to be defined, studies of HIV variants present in cells of the choroids plexus suggest this may be a likely site for these events to occur. The choroids plexus has been shown to harbor replication-competent viral populations that contain viruses from peripheral origin and format in the CNS. Overall, viral variants in the choroids plexus are more genetically like those found in brain but still harbor sequences associated with virus variants found in the periphery, which was most similar to, although distinct from that, found in the brain. It also contained some viral variants with high similarity to those of peripheral origin. These findings suggest the choroids plexus may have a significant role in the bidirectional dissemination of HIV between the periphery and the CNS (Burkala et al., 2005).

Most studies have supported the role of a more autonomous CSF infection predominating in the late stages of HIV infection that is sustained by long-living cells infected by HIV in the CNS (Eggers et al., 2003). The importance of neuroactive antiretroviral drugs in this setting has yet to be established. Staprans et al. (1999) found that the long-term reduction in CSF HIV RNA load was comparable when neuroactive regimens were compared to those with less penetrating drugs, including nelfinavir. However, this study which examined the CSF RNA levels almost monthly over a period of 6 to 12 months was limited to 15 HIV-1-infected patients who were either undergoing a change in antiretroviral regimen or were initiating therapy. All patients were considered to be in late stage HIV infection. Most patients experienced a parallel reduction in CSF and plasma HIV RNA load. A smaller number had a slower decay in HIV RNA load in the CSF. The slower decline correlated with lower CD^4 T cell counts when initiating therapy. Despite the slow decay seen in some patients, long-term responses in CSF were generally substantial, comparable or better than that seen in plasma. Decline in CSF HIV RNA load did not correlate with the use of neuroactive antiretroviral drugs. Other investigators have supported the need for neuroactive antiretrovirals to control HIV replication. In a cross-sectional and longitudinal study of CSF HIV RNA levels in symptomatic HIV-infected individuals, HAART effectively reduced CSF HIV RNA levels in advanced patients. Patients with a history of HAART showed an eightfold higher probability of having undetectable HIV RNA in CSF at lumbar puncture. A significant correlation was found between plasma and CSF HIV RNA load in HAART-experienced but not drug-naive patients. In the longitudinal portion of the study, almost 50% of patients reached undetectable CSF HIV-1 RNA levels by 3 months of therapy. More than 75% had a CSF virological response as defined by either an undetectable CSF HIV RNA load or greater than $1\log_{10}$ reduction in HIV RNA load. There was a strong correlation between the greatest response in the CSF and indinavir exposure. CSF levels of indinavir are high, especially after low-dose ritonavir boosts. CSF levels of indinavir in vitro exceed the 95% HIV inhibitory concentration and persist longer than in plasma. Correlations between CSF RNA viral load reduction and antiretroviral therapy were also seen with ZDT, stavudine and lamivudine. NNRTs did not show benefit but the number of patients treated was small. The presence of neurological disease was also associated with failure to completely suppress HIV RNA levels in CSF.

Only in later stage HIV-1 infection does the CSF HIV RNA load correlate with the presence or severity of

Table 19.2

Pathogenesis of HIV-associated disease

HIV infection occurs predominantly in mononuclear cells in the brain; astrocytes are infected, neurons rarely infected
Immune response out of proportion to level of virus infection leading to high levels of proinflammatory molecules
Neuronal dysfunction and death follows from the neurotoxic effects of viral proteins, neuromodulatory molecules and dopaminergic transmitters
Alterations in astrocyte functions directed at maintaining neurons and extracellular milieu by HIV infection
Neurotoxic pathways converge on cell-death cascades including apoptosis, oxidative stress, mitochondrial dysfunction similar to that seen in Alzheimer's disease, Parkinson's disease and other neurodegenerative disorders

HAD (Antinori et al., 2002; Eggers et al., 2003). The factors influencing virological responses to HAART in the CSF of patients with advanced disease differ from those in asymptomatic patients (?). HAART effectively reduces HIV-1 replication in the CSF in these patients. However, the decline of CSF HIV RNA levels in patients with HIV encephalitis is significantly longer when compared to patients without neurological disease (Eggers et al., 2003). In the study by Eggers et al. (2003), the slope of the decline in HIV RNA in CSF did not correlate with HAART regimen, baseline CSF cell count, or genotype or phenotype resistance in plasma or CSF prior to institution of HAART. The only variable that correlated with slow declines of HIV RNA in CSF was the presence of HIV encephalitis.

19.18. Adjunctive therapies for HIV-associated neurologic disease

The continued development of neurological diseases associated with HIV infection despite HAART has led to attempts to address molecular pathways involved in the pathogenesis of HIV-related neurological disease. Since neurons themselves are not infected with HIV but rather succumb to soluble mediators released by HIV-infected cells and activated macrophages and glial cells, there is an excellent opportunity to develop agents that may block these pathways (Perry et al., 2005). Research has suggested that normal defense mechanisms against microbial agents such as oxidative burst by macrophages and cytokine/chemokine production are greatly activated in HIV-infected patients. These substances themselves have neurotoxic properties that can injure and kill neurons. Controlling the synthesis of these molecules and/or blocking their effects offers another avenue for treating HIV-related neurological disease.

Several placebo-controlled trials of neuroprotective agents for HIV-associated disease have been reported. These studies have been to assess safety and tolerability as primary endpoints with efficacy as a secondary outcome measure. The variable designs, endpoints, neuropsychological batteries and methods of reporting make comparison of these studies difficult at best. The absence of an accepted surrogate marker for HAD lends additional difficulties to assessing the significance of these reports. All trials to date have been phase II. Two have been late phase II trials. No phase III trial has yet been conducted using these agents.

Nimodipine is an L-type calcium channel antagonist that has been evaluated in a phase II trial using high and low dose arms (Turchan et al., 2003). Nimodipine was demonstrated early on to block the effects of HIV gp120 and Tat mediated increases in intracellular calcium and subsequent excitotoxic neurotoxicity. Subsequent studies demonstrated that nimodipine was capable of blocking gp120 effects on neurons but not on astrocytes. The phase II trial had three arms with two different dosages, 60 mg 5 times/day and 30 mg 5 times/day and placebo. The neuropsychological percentage change score was reported since subscores and individual Z scores showed similar changes. There were no significant differences between the three arms at 16 weeks. Analysis of data using observation carry forward (LOCF) approach also did not show any differences in the three groups. However, using LOCF, the study did show an improvement in neuropsychological tests at 4 and 8 weeks in the high-dose arm. Statistical analysis and *P* values were not reported.

Peptide T is an octapeptide with a pentapeptide sequence homologous to vasointestinal polypeptide. It was shown to block gp120 binding to brain cells and block gp120-mediated neurotoxicity in vitro (Perry et al., 2005). A large phase II trial using intranasal administration of peptide T did not demonstrate any effect on the primary end points that included neuropsychological tests. An adjustment of baseline CD^4 T cell counts resulted in a significant improvement in abstract thinking and speed of information processing in the peptide T-treated group. Additional post hoc analysis showed that patients with CD^4 counts greater than 200 cells/mm^3 had significant improvement in the global neuropsychological scores and in speed of information processing and working memory. The drug was well tolerated with increased mood disturbances and allergic reactions reported during the study.

OPC-14117 is a lipophilic compound structurally similar to vitamin E that acts as an antioxidant by scavenging superoxide radicals. Using LOCF analysis, no significant difference in neuropsychological performance was found between treated and untreated groups. Selegiline has been used in several studies. A phase I and phase II study of 36 patients with HIV-associated dementia or minor cognitive/motor disturbance were treated in the pre-HAART era with selegiline 2.5 mg three times/week. No statistical significant differences were found in neuropsychological tests or recall as well as psychomotor processing. However, there was a trend favoring selegiline in psychomotor tests and improvement in memory scores seen at 4 weeks continued to increase at 10 weeks. A small phase I and II study using transdermal administration of selegiline in 14 patients with HIV minor cognitive/motor disorder and Alzheimer's disease in the pre-HAART era showed significant improvement in delayed recall and psychomotor speed with the dominant hand. No differences in safety or tolerability were found. A larger phase II trial is currently underway in the USA.

Lexipafant is a platelet-activating factor (PAF) antagonist that blocks the neurotoxic effects of CSF from HIV-1-infected patients in neuronal cell cultures (Navia et al., 2005). A randomized, double-blind, placebo-controlled trial using lexipafant in HIV-associated minor cognitive/motor disorder and HAD showed the drug was safe and well tolerated. A trend to improvement was seen in only the verbal memory component of the neuropsychological battery examined. Memantine has been approved for use in patients with Alzheimer dementia. Memantine is an N-methyl-d-aspartate (NMDA) antagonist and would therefore block the excitoxicity seen with glutamate and quinolimate implicated in the pathogenesis of HAD. A large double-blind, placebo-controlled study of 140 patients with HAD on HAART was designed as a late phase II study to evaluate safety, efficacy and tolerability. At the end of the 16-week study, there was a trend towards improvement in a summary measure of neuropsychological testing in the memantine-treated group. The differences were not significantly different. Significant differences in MR spectroscopy were found between the two groups. At 20 weeks, four weeks after the wash out period, LOCF analysis identified a trend for improvement in neuropsychological scores in the memantine-treated group. Side effects included disorientation, agitation, insomnia, lightheadedness, photosensitivity and tremors in a small number of patients receiving memantine.

CPT-1189 is a lipophilic antioxidant that scavenges superoxide anion radicals and can block the neurotoxicity of gp120 and TNF-α (Navia et al., 2005). A double-blind, placebo-controlled study of patients with HIV-associated minor cognitive/motor disorder and dementia showed improvement in psychomotor speed in the Grooved Pegboard tests with the non-dominant hand. There was no other significant improvement in neuropsychological tests reported.

Several additional avenues for adjunctive therapy have been suggested by current experimental reports. Estrogen has been shown to have neuroprotective features (?). Women with HIV infection frequently develop menstrual abnormalities with amenorrhea (Turchan et al., 2003). Plasma estradiol levels are lower in HIV-infected women. Nath and colleagues have shown that estradiol in physiological and easily achievable pharamacological concentrations can protect against neurotoxic effects of HIV proteins and the combined effects of HIV proteins and drugs of abuse (?). Estradiol protects the mitochondria of neurons in a receptor-dependent manner. Here it interacts with a subunit of the mitochondrial F0F1-ATP syntase/ATPase required for the coupling of a proton gradient across the F0 sector of the enzyme in the mitochondrial membrane to ATP synthesis in the F1 sector of the enzyme. This property may account for the antioxidant properties of 17β-estradiol previously reported. Estradiol can also suppress proinflammatory effects of HIV proteins. The use of hormonal therapy may be limited by side effects including the unwanted feminization in males and the increased risk for development of cancers.

To supervene the unwanted side effects of estrogens, investigators have examined the use of plant estrogens in preventing neurotoxicity of HIV proteins. Favinoids, the phytoestrogens made by plants, have antioxidative and neuroprotective properties. Usually, this class of drugs has weak estrogen binding properties and thus fewer estrogen-related side effects, including feminization. Nath has shown that the plant estrogen present in fenugreek and yam, diosgenin, can prevent neurotoxicity by HIV proteins and CSF from patients with HAD (Turcham et al., 2003). Resveratrol, another plant estrogen, has the additional ability to protect dopaminergic neurons. Genistein, daidzein and quercetin are all estrogen-like compounds found in soy beans. Quercetin is a better antioxidant than either genistein or daidzein but all are well absorbed and have good bioavailability.

Selenium is an essential component of the antioxidant enzyme glutathione peroxidase. HIV-infected patients have low levels of selenium which has been associated with a high rate of mortality in these patients (Perry et al., 2005). Although selenium administration is safe and easy to accomplish, glutathione levels themselves may be low in HIV-infected cells suggesting there may not be a recognizable effect after selenium administration. Glutathione is an endogenous antioxidant found in cells. It normally is present in a reduced form and can scavenge free radicals by its sulfhydryl groups. There is considerable evidence that glutathione levels are decreased in HIV infection and the decreased levels may contribute to the progression of HAD. Tat and gp41 can cause decreases in glutathione levels in neurons (?). N-acetylcysteine (NAC) is an excellent source for sulfhydryl groups and is readily metabolized in the body into substances that stimulate glutathione synthesis. NAC is widely available as a mucolytic agent used to treat acetaminophen toxicity and prevent liver toxicity. A controlled trial of NAC in HIV-infected patients showed a reduction in TNF levels in serum and reduction in the decline of CD^4 T cell counts. L-2-oxothiazolidine 4-carboxylate is another pro-glutathione drug that has been evaluated in HIV-infected patients. NAC has the more potent free radical scavenging properties as well a much more potent antiviral activity in vitro.

Other free radical scavenging drugs include diethyldithiocarbamate, BG-104 (a Chinese herbal medicine), ferulic acid, S-nitrosoglutathione and vitamin E (Navia et al., 2005). Uric acid is a potent antioxidant but is insoluble and lacks the ability to cross the blood–brain barrier. New water-soluble analogues are being developed that may be of therapeutic benefit.

Anti-glutaminergic agents are another possible target for drug intervention in HIV-associated disease. Considerable evidence supports the ability of HIV proteins to induce neurotoxicity mediated by excitotoxicity by direct stimulation of excitatory amino acid receptors, release of quinolinate or blocking glutamate uptake by legal cells. Memantine discussed above is a wide-spectrum glutamate antagonist that shows neuroprotection and may benefit AHD patients. There may be a role for other drugs that block excitotoxicity including ramecimide, triluzole and lamotrigine. Pentamidine used for PCP in HIV-infected patients also has NMDA antagonist properties. Several pentamidine analogues have been developed that are more potent glutamate antagonists compared to pentamidine.

Strategies to block TNF-α have also been developed (Navia et al., 2005). Immunohistochemical analysis of macrophages in brains of HIV-infected patients with HAD and HIV encephalitis show increased expression of TNF-α as well as other cytokines and nitric oxide. TNR-α can be induced by HIV-Tat protein in glial cells. TNF-α levels are also an important indicator for HAD. The role of TNF-α in neurotoxicity is as yet unclear. TNF-α-mediated events that could contribute to neurotoxicity include reactive oxygen species formation and inhibition of glutamate uptake by astrocytes. However, under certain conditions, TNF-α can be neuroprotective. TNF-α also contributes to immune dysregulation by inducing IL-6, IL-8 and colony stimulating factors as well as upregulating the expression of adhesion molecules on endothelial cells. TNF-α upregulates HIV replication in microglia and in macrophages that are co-cultivated with astrocytes. This suggests that TNF-α requires a co-factor which has proven to be PDF. Several agents are available that block the activity of

Table 19.3

Neurotoxicity of HIV-encoded proteins

HIV protein	Location	Toxic conc.	Direct toxicity	Indirect toxicity
gp120	Membrane bound and extracellular	Picomolar	NMDA stimulation, EEA cytotoxicity, oxidative stress, apoptosis	↑NO, ↑tyrosine kinase, ↑cytokine release
gp41	Membrane bound and extracellular	Nanomolar	None known	Requires astrocytes, mitochondrial dysfunction, ↑glutamate release
Tat	Extracellular (taken up by neurons and astrocytes)	Nanomolar	Non-NMDA and NMDA stimulation, integrin binding, ↑intracellular calcium, EEA cytotoxicity, oxidative stress, apoptosis	Cytokine, chemokine and metalloproteinase release, impaired astrocyte functions
Vpr	↑CSF, not extracellular	Micromolar	Creates Na^+ channels in membranes with inward cation current, caspase 8 activation	?
Nef	Intracellular	Nanomolar	? nef gene sequences in neurons, activation of caspases, K^+ currents with membrane depolarization	Cytokine release, apoptosis of astrocytes, K^+ currents
Vpu	Intracellular	?	No known interactions with neurons, does form ion channels in COS cells	?
Rev	Intracellular	?	Interacts with neurons to alter cell membranes	?

TNF-α. Thalidomide, pentoxifylline, rolipram and pirfenidone have been used to block the activities of TNF-α. Pentoxifylline has been shown to block TNF neurotoxicity and HIV replication in tissue culture systems (?). It has also been effective in the treatment of multi-infarct dementia. In a small pilot study of patients with HAD, pentoxifylline was poorly tolerated. Thalidomide has unfortunate teratogenic potential and can cause peripheral neuropathy. A water-soluble analogue of thalidomide has been developed with better bioavailability. Rolipram is 500 times more potent than pentoxyfylline in inhibiting TNF-α production. Pirfenidone downregulates the synthesis of several proinflammatory cytokines and is another possible candidate for the treatment of HIV-associated neurologic disease. Care must be exercised given the worsening of patients with multiple sclerosis who received anti-TNF agents (?). An option might be to block TNF-related apoptosis-inducing ligand (TRAIL) that is necessary for mediating its toxic effects.

Other novel anti-inflammatory compounds that are available include curcumin which downregulates cytokine synthesis and inhibits HIV replication. Cyclooxygenase 2 inhibitors may also be helpful given the increased levels of cyclooxygenase in HIV-infected macrophages and induction in brain by HIV gp120.

Glycogen synthase kinase provides another target molecule for adjunctive therapy in HIV-associated neurologic disease (Perry et al., 2005). PAF mediates its biological effects through activation of glycogen synthase kinase beta (GSK-3b). Drugs that inhibit GSK-3b include lithium and valproate that may have therapeutic potential. HIV Tat also induces GSK-3b activity and hence this may represent a final common pathway in PAF- and Tat-induced neurotoxicity. RP55778, a PAF antagonist, inhibits HIV replication in T cells as well as TNF production independent of its anti-PAF activity. Other PAF antagonists are also available for trial in HIV-related neurologic disease.

Neuroprotective strategies directed at neuronal bioenergetics have been an important area of investigation. Profound bioenergetic changes in mitochondria occur after exposure to HIV-1 neurotoxins including Tat and PAF. Hyperpolarization of mitochondrial membrane potential in neurons is associated with recycling of synaptic vesicles prior to apoptosis. These findings are provocative since conventional wisdom suggests that excitotoxic stress induced by HIV-1 neurotoxins should be associated with loss of electrochemical gradients between mitochondrial membranes and neuronal cytosol prior to cell death by apoptosis. Hyperpolarization of mitochondrial membranes has also been seen in populations of peripheral blood mononuclear cells associated with an activated T-cell phenotype more susceptible to apoptotic stimuli. Protease inhibitors prevented the changes in mitochondrial membrane potential seen in PBMCs. The potassium channel antagonist tolbutamide also reversed this change in PBMC mitochondrial membrane hyperpolarization as well as Tat-induced neuronal apoptosis. At least one study has shown that tolbutamide protects against neuronal cell death induced by activation of neuronal potassium-ATP channels. Tolbutamide may also alter mitochondrial hyperpolarization by impacting glucose availability to or within neurons or by altering calcium regulation. Tolbutamide also has direct mitochondrial uncoupling properties which could protect neurons by limiting the production of reactive oxygen species, preventing mitochondrial release of pro-apoptotic products, or by reducing activation of cytotoxic adenosine receptors. Mitochondrial uncoupling has already been shown to protect against ischemic damage in brain and heart. These findings suggest that mitochondrial uncouplers may be one of the more promising adjunctive therapeutic approaches to HIV-associated neurologic disease. Perry et al. (2005) suggest that therapeutics that agonize or otherwise control activation or expression of endogenous mitochondrial uncoupling proteins such as the beta-adrenergic agonist CL-316,243 mya hold the most promise as viable treatment approaches to reduce neuronal damage from hyper-oxidative stress.

Neurotrophic agents offer another avenue for possible adjunct therapy in HIV-related neurologic disease. In patients with HIV encephalitis, nerve growth factor (NGF) and beta fibroblast growth factor (FGF) mRNA levels are significantly elevated. NGF immunoreactivity is present in macrophages in perivascular regions (?). FGF is expressed by astrocytes. This suggests that NGF and FGF induction is not sufficient to protect the brain in HIV encephalitis. Brain-derived nerve growth factor (BDNF) levels are not elevated in HIV encephalitis. Immunoreactivity for BDNF is found in infiltrating macrophages and neuritis in the striatum (Nosheny et al., 2005). BDNF is expressed by neuritis and cell bodies of neurons in the cortex. BDNF has been found to have potent mediator/modifier potential in synaptic plasticity in the adult mammalian brain and a unique neuroprotective agent. BDNF binds to both the p75 neurotrophin receptor (p75NTR) and TrkB. Binding to p75NTR is nonspecific whereas binding to TrkB is specific for BDNF. BDNF binding to p75NTR may result in cell survival or cell death depending on the physiological environment. While binding to p75NTR can activate tow cell death mediators, NF-κB and ceramide signal cascade, the receptor is also capable of neuroprotection since neurotrophin activation of p75NTR rescues neuroblastoma cells from Tat-mediated cell death. P75NTR is also crucial for axonal elongation and guidance. Most neuroprotective properties of BDNF occur via activation of the high affinity receptor TrkB. Some TrkB isoforms which lack the catalytic kinase domain may actually reduce neurotrophic effects of BDNF. It will be important to understand the contribution of each receptor type in designing BDNF therapeutic protocols for the treatment of HIV-related neurological disease. TrkB binding by BDNF is sufficient for pro-survival effects of BDNF. Some neurons do not express TrkB and are therefore unresponsive to BDNF. This may be a problem in late stages of HIV-related neurologic disease when several neuronal populations appear to die. Thus, BDNF may need to be administered early on alone or in conjunction with other growth factors. A more challenging problem is BDNF delivery into the brain. BDNF does not cross the blood–brain barrier. BDNF injection at peripheral sites could make the growth factor available to spinal cord and spinal ganglia neurons as BDNF is transported by axonal flow. The real challenge will be to get BDNF into the brain. BDNF has been injected into lateral ventricles and brain parenchyma by microinfusion techniques. This would present formidable problems in patients. Grafting of immortal cells' over-expressing growth factors is another approach to bringing BDNF into contact with central nervous system neurons (?). The human glial cell (SVG) that is immortalized using the SV40 T antigen synthesizes and releases BDNF. This cell type may offer a more physiological approach to growth factor delivery since the cell is constitutively expressing BDNF and may be able to recognize molecular signals from the environment that regulate growth factor expression. Other approaches include facilitating diffusion across the blood–brain barrier by conjugating growth factors to other molecules that facilitate diffusion across the blood–brain barrier or blood–CSF barrier. This strategy has been used for NGF using NGF joined to antibodies to the transferring receptor which can cross the blood–brain barrier. Finally, adenovirus-associated vectors may offer another means of transporting BDNF into the brain. Delivery of a rAAV incorporating a neuron-specific promoter driving expression of BDNF would offer a unique approach to targeting neurons. This approach has already proven successful in targeting BDNF for expression in the rat striatum where BDNF increased motor neuron revival and improved behavioral outcomes in a rat model of focal cerebral ischemia. This was reported early to block the effects of HIV gp120 on excitotoxicity through the NMDA receptor and was used with NRTI at doses of 60 mg 5 times/day or 30 mg 5 times/day. There was a trend towards improvement of neuropsychological performance at the highest dose only. Peptide T, a possible chemokine receptor blocker, was administered 2 mg TID by intranasal route. NRTIs were used as concomitant antiretroviral therapy. Patients were treated for 6 months without any recognizable effect. OPC-14117, an antioxidant, was administered at 240 mg/day along with MRTO for 12 weeks. Treated patients showed a trend for improvement in memory and timed gait tests. Selegiline, an antioxidant neuroprotectant, at 2.5 mg three times a week with Thoctate 1200 mg/day and NRTI was given for 10 weeks. Improvement in verbal learning with a trend for improvement in recall and psychomotor speed was noted. Selegiline transdermal administration along with NRTI was associated with a positive effect on neurocognition. Lexipafant, a PAF antagonist, was given at 500 mg/day along with HAART to 13 patients for 10 weeks. A trend for improvement of verbal learning and timed gait tests was seen in treated patients. Memantine, an NMDA antagonist, at 40 mg/day by week 4 along with HAART for 16 weeks resulted in a positive effect only after completion of the double-blind phase. Finally, CPI-1189, a TNF antagonist, at either 100 or 50 mg/day along with HAART for 10 weeks was associated with improvement of Peg Board test at the highest dose but with no effect on neurocognition.

Future development of adjunct therapies for HIV-associated neurologic diseases will be important as patients age and the appearance of neurodegenerative diseases complicate the course of HIV-infected individuals in late middle age and older. Continuing research further defining the molecular events involved in the pathogenesis of HIV-associated neurological disease will provide additional targets for potential adjunct drug therapies.

19.19. Complications and adverse reactions associated with antiretroviral therapy

The use of antiretroviral agents has been associated with a number of adverse reactions and unusual complications. Neurologists most often encounter adverse effects of NTRIs on mitochondria that lead to peripheral neuropathy and possibly Guillain–Barre syndrome (see below). The other adverse reactions seen with antiretroviral therapies involve systemic reactions that rarely involve the nervous system.

19.19.1. Nucleoside/nucleotide reverse transcriptase inhibitors

Current guidelines recommend the use of two NRTIs with either a PI or NNRTI as a complete HAART regimen. Because of their common use within HIV treatment regimens, side effects associated with NRTIs are likely to pose a significant obstacle to effective, long-term therapy.

Class-wide side effects include lactic acidosis, hepatic steatosis and lipoatrophy. More significant side effects include anemia, cardiomyopathy, gastrointestinal distress, drug-induced hypersensitivity, myopathy, peripheral neuropathy, nephrotoxicity, pancreatitis, ototoxicity and retinal lesions.

Mitochondrial toxicity is the foundation of many NRTI-associated adverse reactions. The enzyme responsible for ensuring functional mitochondrial DNA (mtDNA) is DNA polymerase-γ which closely resembles the HIV-encoded reverse transcriptase. NRTIs therefore will inhibit human DNA polymerase-γ leading to interference with mitochondrial DNA formation. Mitochondrial DNA replication does not have repair mechanisms so alterations by NRTIs cannot be repaired. Premature termination of mitochondrial mRNA and transfer RNA synthesis leads to faulty transcription and translation. Besides the inhibition of mtDNA polymerase-γ, mitochondrial toxicity also arises from impairment of normal oxidation of long chain fatty acids in the mitochondria which then leads to esterification of triglycerides and an increase in nonesterified fatty acids (?). Impaired energy production and possibly toxic effects of nonesterified fatty acids associated with dicarboxylic acids and free radicals are thought to contribute to the clinical manifestations of mitochondrial toxicities including peripheral neuropathy, myopathy, lactic acidosis and lipoatrophy. In vitro studies have revealed the following hierarchy of mitochondrial DNA polymerase-γ inhibition NRTIs: zalcitabine > didanosine > stavudine > ZDT > lamivudine = abacavir = tenofovir.

19.19.2. Peripheral neuropathy with NRTIs

To a varying degree, and in a dose-dependent manner, all NRTIs are neurotoxic. Peripheral neuropathy is most frequently seen in patients receiving zalcitabine, didanosine and stavudine. NRTI effects on peripheral nerve have not been extensively studied. In most cases, interference with oxidative metabolism, which can lead to reduction in acetyl-carnitine production, has been the most consistent finding. Low serum hydroxycobalamine levels and an inhibitory effect on nerve growth factor have also been reported. Pre-existing neuropathy and low CD^4 cell counts (nadir CD^4 count < 200 cells/mm^3) are predisposing factors for all NRTIs to cause peripheral neuropathy. Combination therapy may also be a risk factor. Zalcitabine is more neurotoxic than didanosine, stavudine and lamivudine but the combination of stavudine and didanosine is more toxic than either drug alone. Prior peripheral neuropathy and co-administration of other neurotoxins can also increase the risk of NRTI-induced peripheral neuropathy (?). Avoiding combination of didanosine and stavudine as well as caution in the use of individual NRTIs in patients with prior peripheral neuropathies appears prudent. Removal of the drug will often lead to resolution of symptoms. A "coasting" effect, with worsening of symptoms after stopping the drug before improvement occurs, may lead to confusion as to the role of NRTIs in causing the neuropathy. Pain and subjective symptoms respond to amytriptiline, gabapentin and possibly acetyl-L-carnitine. The role of B vitamins in the treatment of NRTI-associated neuropathy is unknown.

19.19.3. Immune reconstitution inflammatory syndrome in the nervous system

The recognition of immune reconstitution inflammatory syndrome (IRIS) did not become widespread until after the introduction of HAART in the mid 1990s. Prior to that time, there was ample evidence that reconstitution of the immune system with antiretroviral therapy was a possibility. Paradoxical responses were described to *Mycobacterium tuberculosis* and leprae had been described with antituberculous therapy. ZDT monotherapy was associated with atypical, localized presentations of *Mycobacterium avium intracellulare* (MAI) infections. Initially, the induction of immunity to hepatitis B in patients treated with HAART was documented by antibody titers that correlated with the clinical course. Subsequent studies using T cell techniques supported that IRIS occurred in the setting of a measurable increase in immune response to an underlying opportunistic infection. To date, IRIS has been associated with a number of opportunistic infections of the CNS as well as autoimmune diseases and syndromes with unknown initiating factors.

19.19.3.1. Definition of IRIS

It has been difficult to reach a consensus on the definition of IRIS (De Simone et al., 2000). However, most investigators support the general concept that cases of IRIS need to have an inflammatory component occurring in the setting of immune reconstitution that cannot be explained by either drug toxicity or a new opportunistic infection. Risk factors for the development of IRIS include young age at the time of HAART, antiretroviral naive patients, and initiation of HAART near the time of diagnosis of an opportunistic infection. Most patients with IRIS have rapidly declining HIV-1 RNA levels and increasing CD^4 T lymphocyte counts. A rapidly declining HIV-1 RNA level appears to be more sensitive than the increase in CD^4 T cells as a predictor of IRIS.

19.19.3.2. Clinical spectrum of IRIS

IRIS-related diseases occur in two settings. Most patients develop IRIS-related disease in the first 8 to 12 weeks of

HAART (Andersson et al., 1998; De Simone et al., 2000). IRIS-related disorders can also occur at later times, some occurring months to several years after initiation of HAART. The majority of IRIS cases have been associated with underlying viral infections and mycobacteria. Given the differences in opportunistic infections based on geographical locale, the disorders seen with IRIS may also vary as HAART therapy becomes available to underdeveloped countries.

The number of patients and timing of IRIS has been remarkably similar in all studies reported to date. Between 15 and 25% of patients developed IRIS with the majority of cases appearing within 3 months of beginning HAART (De Simone et al., 2000). The clinical spectrum of IRIS-related diseases includes atypical opportunistic infections, sarcoid-like disease and autoimmune disorders.

The pathogenesis of IRIS-related diseases is incompletely understood. The reconstitution of the immune system following HAART provides insight into the antigen-specific arm of IRIS. HAART is initially followed by a rapid fall in HIV RNA levels with a 90% decrease in the first 1 to 2 weeks of therapy. The viral load continues to decline for the first 8 to 12 weeks with levels often reaching <200 HIV RNA copies/ml. An increase in immune effector cells appears coincident with the reduction in HIV viral load. In some instances, the increase in CD^4 T cells may lag by weeks after the fall in HIV RNA levels. The initial expansion of T cells occurs in memory CD^4 T lymphocytes bearing the CD^45RO^+ phenotype. These cells have been previously activated by exposure to antigen. This initial increase in CD^4 memory T cells appears to be a redistribution rather than proliferation and generation of new T lymphocytes (?). Continuation of HAART leads to proliferation in different CD^4 T lymphocyte subsets. After 4 to 6 weeks of therapy, naive CD^4 T cell phenotypes CD^45RA^+ and $CD62L^+$ begin to appear. The increase in CD^4 naive T cells occurs in both the blood and lymph node suggesting the increase in naive CD^4 T cells results from proliferation and not redistribution. The long-term rise in CD^4 T cells is mainly accounted for by the persistent increase in naive CD^4 T cells. At the same time that a rise in CD^4 T cells occurs, the receptor repertoire increases in diversity and the cytokine profile switches from Th-2 to Th-1 with an increase in IL-2 and interferon-γ levels.

HAART also results in an increase in the number of memory CD^8 T lymphocytes in the first few weeks of therapy. With prolonged HAART, memory CD^8 T cells decline and are gradually replaced by naive CD^8 T cells. The CD^8 T cells show reduced activation markers with a broadened T cell receptor profile. The total number of CD^8 T cells remains steady leading to an increase in the CD^4/CD^8 ratio with the increase in CD^4 T cells.

The increase in T lymphocyte numbers is associated with improved functional activity. Within 4 weeks of beginning HAART, there is increase in delayed hypersec infectivity and in vitro lymphocyte proliferative responses to common antigens such as candida. Immune functions continue to improve with increased duration of therapy as defined by lymphocyte responses to antigens of common opportunistic organisms. Enhanced immune responses appear to be especially marked toward widely prevalent organisms such as CMV and mycobacteria.

IRIS has been implicated in several neurological diseases seen after the institution of HAART. JC virus, *Cryptococcus neoformans* and herpes zoster have been the predominant opportunistic infections encountered in IRIS-related neurological disease. Autoimmune diseases including Guillain–Barre syndrome have also been reported. Finally, a fulminant demyelinating disease has been encountered in patients receiving HAART whose pathogenesis is yet to be determined.

19.19.3.3. Opportunistic infections in IRIS

19.19.3.3.1. Cryptococcus neoformans

Case reports of IRIS associated with *C. neoformans* have mainly involved culture-negative meningitis (De Simone et al., 2000). The presentation and clinical course of IRIS-related *C. neoformans* infection are unusual in HIV/AIDS treated with HAART. Prior to HAART, *C. neoformans* meningitis presented with little or no meningeal signs, extremely high serum and CSF cryptococcal antigen titers and ease in culturing yeasts from the CSF. Patients with IRIS-related *C. neoformans* meningitis have significant meningeal signs including headache and stiff neck, CSF inflammatory responses, lower serum and CSF cryptococcal antigen titers and difficulty in isolating viable yeasts (?). The difficulty in isolating *C. neoformans* from the CSF has been attributed to the absence of viable yeasts in patients with IRIS-related *C. neoformans* meningitis. However, in some instances of IRIS-related *C. neoformans* meningitis yeasts have been isolated from the CSF suggesting that viable organisms are present. Unlike HIV-associated *C. neoformans* infections prior to HAART, unusual systemic presentations with granulomas, lymphadenopathy and cryptococcomas in the CNS have been encountered. Biopsy of these lesions has shown significant infiltration with macrophages with engulfed organisms. In most instances, isolation of *C. neoformans* from these lesions has been unsuccessful.

About 30% of all HIV-infected patients hospitalized for *C. neoformans* infection who receive HAART are subsequently readmitted with symptoms attributed to an inflammatory response. Patients most likely to develop IRIS-related *C. neoformans* disease include those with severe immunosuppression and high HIV viral loads before institution of HAART, drug naive, high cryptococcal antigen prior to HAART, higher intracranial pressure, disseminated *C. neoformans* and institution of HAART within 30 days of diagnosis of *C. neoformans* disease. At the time of diagnosis of IRIS-related *C. neoformans* infection, HIV viral load has fallen significantly and CD^4 T cell counts have increased in almost all patients. Treatment of IRIS-related *C. neoformans* meningitis has included the use of various anti-inflammatory treatments including high-dose steroids for severe CNS manifestations (?). The increase in intracranial pressure may require further therapy including shunting. Thalidomide, an anti-TNF-α agent, has been used to treat cyptococcomas at extraneural sites (?).

The association of *C. neoformans* meningitis and cryptococcomas with HAART has raised the issue of when to begin therapy in patients known to have *C. neoformans* infection. Most investigators recommend avoiding HAART introduction until at least cryptococcosis is microbiologically controlled (?). In those patients in whom HAART unmasks a previously unrecognized cryptococcal infection, withholding HAART and/or instituting steroid therapy must be made on an individual basis at the present time.

Other fungal and bacterial infections. Individual reports of bacterial and other fungal infections of the CNS have been reported as manifestations of IRIS. Miliary tuberculosis has been associated with intracranial tuberculomas (De Simone et al., 2000; Andersson et al., 1998). Disseminated histoplasmosis in HIV-infected patients treated with HAART has developed intracranial histoplasmomas. In both instances, the conversion from histocytic infiltrates before HAART to well-formed epithelioid and giant cell granulomas after HAART is compatible with the restoration of T-cell-dependent macrophage activation (Breton AIDS, 2006).

19.19.3.3.2. Progressive multifocal leukoencephalopathy (PML)

The institution of HAART has resulted in prolonged survival of patients with HIV-related PML (Albrecht et al., 1998; Berger et al., 1998; Clifford et al., 1999; Dworkin et al., 1999; Gasnault et al., 1999; Tassie et al., 1999). Most investigators believe that reconstitution of the immune system accounts for the improvement in outcome of patients with PML and HIV/AIDS. However, not all patients receive significant benefit from HAART (De Luca et al., 1998; Clifford et al., 1999; Dworkin et al., 1999; Gasnault et al., 1999; Tassie et al., 1999; Berenger et al., 2003). The reasons for the failure of HAART have yet to be clearly defined.

Many PML patients develop evidence of immune reconstitution after starting HAART. Clinical improvement is often associated with contrast enhancement of PML lesions on MRI. Brain biopsies in patients with contrast-enhancing lesions have shown a correlation between MRI enhancement and perivascular T cell infiltrates (Berenguer et al., 2003). The timing of clinical improvement and evidence of immune reconstitution varies with clinical improvement after preceding contrast enhancement in MRI studies. In some instances, immune reconstitution is followed by progression of PML and death. The reasons for the paradoxical response to HAART in these patients are unclear. Cinque et al. (2003) in a study of 43 patients with HIV-associated PML found that eight patients (19%) developed PML at a time consistent with IRIS. The interval between PML and HAART was uniformly between 13 and 59 days in patients whose disease progressed, within the time frame of IRIS. In patients who stabilized, the interval between PML and HAART was either less than 26 days or greater than 71 days (Cinque et al., 2003). These investigators also found increased percentage of PML patients that progressed in later calendar years suggesting that increased efficacy and compliance with HAART seen in the later years was associated with progressive disease.

Progression of PML has been associated with marked cellular infiltration in the brains of patients with progressive disease while on HAART. When examined, CD^8 T cells have been the predominant T cell subset found in the brain parenchyma and perivascular regions in brain biopsy and autopsy studies. Miralles (2001) reported the presence of inflammation in the brain biopsies in 5 of 28 brain biopsies. Four of the five patients were receiving HAART while the fifth patient was receiving two nucleosides. Inflammatory reactions were found in four of nine patients receiving HAART and in 1 of 19 patients not receiving HAART although the findings did not reach statistical significance. CD^8 suppressor T cells were the predominant T lymphocyte subset found in these biopsies. Kotecha et al. (1998) found CD^8 T cells to be the predominant T lymphocyte subset in perivascular infiltrates in a patient with PML progressing while on HAART. Vendrely et al. (2005) found a predominance of CD^8 T cells in the brain biopsy and brain at autopsy in a patient with PML treated with HAART. CD^8 T cells were found in both the perivascular infiltrates and in the brain parenchyma. CD^4 T cells were absent from both the perivascular infiltrates and brain parenchyma. Interestingly, pathological changes consistent with acute disseminated encephalomyelitis were also present. These lesions did not contain JC virus DNA. An imbalance of CD^8/CD^4 T cells has been observed in early stages of multiple sclerosis and acute disseminated encephalomyelitis supporting the possibility that IRIS is associated with a dysregulation of the immune system (Gay, 2003).

The findings reviewed above suggest that the infiltration of CD^8 T cells without an adequate CD^4 T cell response may lead to paradoxical progression of PML while on HAART. It is also possible that some patients fail to reconstitute JC virus-specific T cells known to be important in the control of JC virus infection. A shift from proinflammatory cytokine profiles to Th2 cytokine profiles may also alter the immune response or control of JC virus replication. Decreasing levels of interferon-α and IL-12 have been implicated in increasing hepatitis C virus viremia and cryptococcal meningitis, respectively (DeSimone et al., 2000).

19.19.3.3.3. Varicella zoster virus

Unlike other opportunistic infections in HIV-infected patients, herpes zoster develops at a relatively consistent rate regardless of degree of immunosuppression (Buchbinder et al., 1992). There was a two- to fivefold increase in herpes zoster in HIV-infected patients treated with HAART compared to those without HAART (Andersson et al., 1998; Martinez et al., 1998). In a study of 316 patients initiating HAART therapy, 8% developed herpes zoster within 17 weeks of starting therapy (Domingo et al., 2001). Some 30% of HIV-infected children in Thailand developed herpes zoster during a period consistent with IRIS (Puthanakit et al., 2006).

Clinically most patients develop classic symptoms and signs of herpes zoster radiculitis. Most cases develop by week 16 of HAART (Andersson et al., 1998). HAART appears not to result in reconstitution of varicella zoster virus cell-mediated immunity (Weinberg et al., 2004). Varicella zoster virus-specific CMI did not change over 3 years in children treated with HAART. Those who developed herpes zoster during the study developed virus-specific CMI if HIV viral load was controlled. CD^8 T cells appear to be important in the pathogenesis of varicella zoster virus reactivation in HAART. Domingo et al. (2001) found that the only factor associated with development of herpes zoster was an increase in CD^8 T lymphocytes from before the initiation of HAART to 1 month before the development of herpes zoster. Martinez et al. (1998) also found an increase in CD^8 T lymphocytes at baseline and 1 month as independent risk factors for herpes zoster.

19.19.3.3.4. Guillain–Barre syndrome (GBS)

GBS was recognized as a complication of HIV infection early in the epidemic. A precipitous decrease in HIV viral load with the development of anti-HIV immune response was associated with the development of a number of autoimmune disorders including GBS. Institution of HAART has been associated with GBS through two mechanisms. Development of lactic acidosis in patients receiving stavudine-containing HAART regimens has resulted in GBS, including the Miller Fisher variant (Marcus et al., 2002; Rosso et al., 2003; Shah et al., 2003). GBS has also been associated with IRIS (Makela et al., 2002; Piliero et al., 2003). Neither patient had lactic acidosis or was receiving stavudine. The patient reported by Makela and co-workers had a history of GBS 5 years prior to starting HAART. GBS was associated with increasing CD^4 T cell counts and a reduction in HIV viral load. Antibodies to GM1 and GQ1b gangliosides were not detected. The patient improved after IVIG and intravenous steroids. A patient receiving HAART, including the HIV fusion inhibitor enfuvirtide, developed GBS 12 days after starting therapy (Piliero et al., 2003). Lactic acidosis was not present. GBS developed in the context of a significant rise in CD^4 and CD^8 T cell counts and a reduction in HIV viral load. He was treated with plasmapharesis with stabilization of respiratory deterioration. He elected not to have further therapy and died of respiratory failure. Antibodies to *Campylobacter* species and GM1 ganglioside were not present. Foci of CD^8 T lymphocytes were present in the brachial plexus at autopsy.

19.19.4. Miscellaneous neurological disorders associated with antiretroviral therapy

Single case reports have attributed various neurological syndromes to immune reconstitution and inflammation. Cerebral vasculitis was seen in a 30-year-old HIV-infected individual treated with HAART (van der Ven et al., 2002). Neurological findings resolved with cessation of HAART only to return when the same regimen was restarted. Leptomeningeal and brain biopsy revealed a lymphocytic vasculitis mainly affecting the small blood vessels. Surrounding brain parenchyma showed activation of microglia and some infiltration of macrophages. Polymerase chain reaction was positive for *Mycobacterium chelonae* but special stains were negative. HIV p24 was not detected. *M. chelonae* has not been associated with CNS infection prior to this case and the significance of the presence of this organism awaits further reports.

Leber hereditary optic neuropathy has been reported in one patient with a family history of the condition (Shaikh et al., 2001). He had received ZDT without alterations in vision. However, after starting HAART, he noted a progressive decline in vision. DNA testing revealed a mutation in the mitochondrial DNA at nucleoside position 11,778 consistent with Leber hereditary optic atrophy. The authors postulated that mitochondrial damage by antiretroviral drugs had precipitated his visual deterioration. Opsoclonus-myoclonus has been reported in an infant 21 months old shortly after beginning HAART. The patient subsequently died of disseminated

CMV infection. The author proposed that IRIS was the most likely cause of the patient's neurological disease.

19.19.5. Leukoencephalopathy of undetermined origin

HIV-infected patients on HAART more frequently have extensive focal white matter lesions on neuroimaging studies compared to patients not taking antiretroviral drugs (Ammassari et al., 2000). Antinori et al. (2001) found that most patients with "not determined" leukoencephalopathy had CD^4 T cell counts above 200×10^6/l. The pathogenesis of these white matter lesions is still unclear but likely related to several pathogenic pathways.

IRIS and other mechanisms have been proposed to explain the development of white matter lesions in patients without opportunistic infections to account for demyelination. Miller et al. (2004) reported two patients with fatal leukoencephalopathy which appears to have resulted from IRIS while on HAART. In both patients, a progressive leukoencephalopathy followed institution of HAART with decreasing HIV viral load and increasing CD^4 T cell counts. One patient had HIV-associated dementia while the second had confusion and withdrawal. The first patient died 11 weeks after beginning HAART while the second died after 4 months of HAART. Neuropathological findings in both patients included the diffuse presence of activated microglia and infiltration of the white matter, leptomeninges and gray matter by $CD^3/CD^5/CD^8/CD^4$ lymphocytes and macrophages. Both cell types were present in perivascular regions as well. Upregulation of MHC class II antigens and polymerase chain reaction evidence of HIV DNA were present in both brains. Evidence for HIV replication was not found in either brain.

The morphological findings in these two patients are distinct from those previously reported in AIDS patients or those with asymptomatic HIV infection. A comparison of brains from HIV-infected patients with low CD^4 counts and HIV-associated dementia who had not received HAART and those with asymptomatic HIV infection showed distinct differences with only discrete localized cellular infiltration in the meninges and perivascular spaces with minimal extension into the brain parenchyma. T cell subtypes in HIV encephalitis not treated with HAART were a mixed population of CD^4 and CD^8 T cells localized to the perivascular spaces (Petito et al., 2003). Furthermore, the changes seen in these patients have not been normally associated with HAART. These findings bear some similarities to the patients described by Langford et al. (2002) discussed below with the presence of perivascular infiltrates. However, the patients of Langford and co-workers had failed HAART and lymphocytes did not infiltrate the white matter and their phenotypes are not described.

These patients appear to fulfill the criteria for IRIS. Both patients had evidence of HIV infection prior to beginning HAART. In one patient, HIV-associated dementia was present before HAART was initiated. The second patient had neurological symptoms suggestive of HIV-related neurological disease. It may be that patients with asymptomatic HIV infection of the CNS are also at risk for IRIS-related leukoencephalopathy with HAART. Both patients had infiltration of CD^8 T cells into the brain. The presence of CD^8 T cells with undetectable CD^4 T cells is also consistent with IRIS-related neurological disease. CD^8 T lymphocytes form the majority of cytotoxic cells and are present in small numbers in normal brain (Neumann et al., 2002). In some circumstances, a strong immune-mediated response may result in increases of CD^8 T cells where they outnumber CD^4 T cells (Poleuektova et al., 2002). The deleterious role of CD^8 T cells in IRIS is supported by the following: (1) CD^8 T lymphocytes are increased in the peripheral blood of patients treated with HAART, (2) CD^8 T cells are the only risk factor for herpes zoster in patients treated with HAART, (3) clinical hepatitis caused by hepatitis B and C virus appears to result from an increase in CD^8 T lymphocyte count, (4) suppression of HIV viremia correlates with a decrease in CD^8 T lymphocyte responses and (5) the association of CD^8 T lymphocyte infiltration in brains of PML patients with IRIS-related deterioration (Carr and Cooper, 1996, 1997; Kelleher et al., 1996; John et al., 1998; Martinez et al., 1998; Mirrlaes et al., 2001). In contrast, low circulating CD^8 T cells in children predict development of HIV encephalitis (Sanchez-Ramon et al., 2003).

19.20. The nigrostriatal system in HIV/AIDS

The nigrostriatal system has a central role in HIV-related neurological disease (Table 19.4). HIV-infected individuals exhibit a wide range of symptoms and signs directly attributable to dysfunction of dopaminergic systems. Basal ganglia involvement by HIV occurs early in the disease as evidenced by slowed cognition and motor reaction times even in asymptomatic HIV-infected patients. Impaired psychomotor speed is often the first manifestation of HAD (Arendt et al., 1990; Navia et al., 1986a, 1986b). With disease progression, patients develop a subcortical dementia clinically indistinguishable from that encountered in patients with Parkinson's disease, Huntington's disease and other disorders of the basal ganglia. Associated with the subcortical dementia are extrapyramidal motor signs including bradykinesia, postural instability, gait abnormalities, hypometric facies, slowed reaction times and decreased rapid alternating movements.

Table 19.4

The nigrostriatal system in HIV/AIDS

Clinical disorders have characteristics of basal ganglia disorders
Dopamine levels depleted early in the clinical course
Dopaminergic neurons appear to be specifically targeted by HIV-related neurotoxic pathways
Dopamine acts synergistically with HIV Tat and possibly other viral proteins in neurotoxic pathways
Dopamine and dopaminergic drugs accelerate SIV encephalitis in monkeys and possibly HIV encephalitis in humans
Dopamine upregulates HIV expression in peripheral blood mononuclear cells

Several lines of evidence support involvement of the dopaminergic system in HIV/AIDS. As noted above, patients who develop HAD had many features reminiscent of subcortical dementias seen in degenerative diseases of the basal ganglia. A smaller number of patients develop a Parkinsonian syndrome indistinguishable from idiopathic Parkinson's disease (Gonzalez-Scarano and Martin-Garcia, 2005). The predominant neuropathological features of HIV/AIDS are found in the putamen and caudate nuclei of the basal ganglia (Kure et al., 1990). HIV infection is most frequently found in the basal ganglia. The HIV envelope glycoprotein gp41 and the core protein p24 are found predominantly in the dopamine-rich regions of the basal ganglia (Itoh et al., 2000; Cartier et al., 2005). Neuronal degeneration is present in the substantia nigra in brains of patients dying in the late stages of HIV/AIDS (Reyes et al., 1991). Using cell counting techniques, significant neuronal cell loss has been noted in the pars compacta of the substantia nigra (Aylward et al., 1993). The remaining cell bodies are more heavily pigmented and shrunken in size. Gadolinium enhancement in the basal ganglia and reduced basal ganglia volume on MR imaging have been reported in HIV/AIDS patients with dementia but not those without cognitive changes (Gonzalez-Scarano and Martin-Garcia, 2005; Rottenberg et al., 1996). Altered striatal glucose metabolism is found using positron emission tomography scanning. Hypermetabolic changes are seen in the early stages of HIV disease while hypometabolic alterations are encountered during the late stages of HIV infection (Wang et al., 2004). Finally, HIV-infected individuals are extremely sensitive to dopaminergic blocking agents and dopamine agonists (Gonzalez-Scarano and Maratin-Garcia, 2005). Even extremely low doses of these medications can result in profound and permanent Parkinsonism.

Alterations in dopamine metabolism have been described in HIV-infected patients. Dopamine concentrations are reduced in brain structures associated with the nigrostriatal dopaminergic pathway. CSF dopamine and homovalinic levels are reduced in patients with HIV-1 infection, the lowest levels being found in those with HAD (Gonzalez-Scarano and Maratin-Garcia, 2005). Recently, abnormalities in dopaminergic transporters have been demonstrated in patients with HAD (Nath and Geiger, 1998). Using positron emission tomography to assess availability of dopamine transporters ([^{11}C] cocaine) (DAT) and dopamine D2 receptors ([^{11}C]raclopride), HIV patients with HAD were found to have significantly lower DAT availability in the putamen and ventral striatum when compared to normal controls and HIV-infected patients without dementia. A higher plasma HIV viral load in HAD patients correlated with lower DAT in the caudate and putamen. Dopamine D2 receptor availability was either mildly impaired or normal in HAD patients. The greater DAT decrease in the putamen than caudate as seen in HIV-1-infected patients with dementia parallels those changes observed in Parkinson's disease. The inverse relationship between HIV plasma viral load and DAT availability supports a direct relationship between HIV infection and reduced dopamine transport.

The HIV proteins gp120 and Tat are neurotoxic in selected populations of neurons (Berger and Nath, 1997). Although these subpopulations of neurons are not completely defined, several studies have shown that dopaminergic neurons are particularly susceptible to the neurotoxic effects of HIV proteins. Injection of Tat into the lateral ventricles of rats results in apoptosis of striatal neurons (Magnuson et al., 1995). Nuclear magnetic resonance spectroscopy shows loss of the N-acetylaspartate peak suggesting neuronal cell loss (Bennett et al., 1996). Pathological studies have shown loss of nigrostriatal fibers following injection of Tat peptides into the striatum. The excitatory amino acid receptor antagonist MK801 and blockers of nitric oxide synthase protected against loss of these cells (Jones et al., 1998). Treatment of mesencephalic neurons with gp120 blocked dopamine uptake by neurons and resulted in loss of dopaminergic neuronal processes without affecting the number of cells (Zauli et al., 2000). Tat inhibits tyrosine hydroxylase activity as well as synthesis and release of dopamine from cathecholaminergic cell lines (Koutsilieri et al., 2002). Taken together, these studies suggest Tat and gp120 may alter dopamine metabolism and availability by destruction of dopaminergic neurons through excitatory amino acid cytotoxicity and cell death cascades and by altering dopamine synthesis and dopamine uptake.

In the SIV-infected macaque, dopamine levels are reduced in the striatum within the first two months of infection. Reduction in dopamine content is associated with an increase in the dopamine metabolite

3,4-dihydroxyphenlacetic acid (DOPAC) late in the disease (Jenuwein et al., 2004). The low levels of dopamine seen early in the disease are likely to be secondary to increased dopamine turnover as suggested by increased CSF concentrations of DOPAC in asymptomatic monkeys. In addition, cAMP and the cAMP response element-binding protein (CREB), two factors involved in dopamine signaling pathways, are also altered in SIV-infected macaques within the first 4 to 19 weeks after, suggesting that dopamine signaling may be altered early in infection (Nath et al., 2000). The findings in SIV-infected monkeys suggest that alterations in dopamine metabolism and signaling may also occur early in HIV-infected patients.

19.20.1. Response to dopamine mediators

Several medications commonly administered to HIV-infected patients directly modulate dopaminergic neurotransmission (Jenuwein et al., 2004). HIV/AIDS patients, even when treated with mild dopamine blocking drugs, such as prochlorperazine, perpherazine, trifluperazine, low-dose haloperidol, thiothicine, chlorpromazine or metoclopramide, may develop severe and permanent Parkinsonism (Edelstein and Knight, 1987; Kieburtz et al., 1991; Mirsattari et al., 1998; Hriso et al, 1991). In one study, the likelihood of developing Parkinsonism was two- to four-fold greater in patients with AIDS when controlled for mean drug dose and body weight (Mintz et al., 1996). Parkinsonism developed in 50% of AIDS patients who received less than 4 mg/kg of chlorpromazine equivalents per day and 78% of those who received more than 4 mg/kg of chlorpromazine equivalents per day.

L-Dopa has been shown to reverse some Parkinsonian symptoms in a subset of HIV-infected patients (Mirsattari et al., 1998; Dana Consortium, 1998). Dopamine receptor agonists consistently improve motor dysfunction in pediatric patients with HIV and signs of Parkinsonism (Dana Consortium, 1998). However, the response to dopamine receptor agonists in adult HIV-infected patients with Parkinsonism has been variable (Kieburtz et al., 1991). Selegiline, a monoamine oxidase-B inhibitor, improves cognitive deficits in HIV-infected patients and SIV-infected monkeys (Koutsilieri et al., 2001; Koutsilieri et al., 2001b).

Psychostimulants are often given to HIV patients to treat depression and cognitive slowing. These drugs modulate dopaminergic neurotransmission through inhibition of dopamine reuptake and/or release of dopamine through calcium independent pathways. Psychostimulants increase cognitive performance and reduce fatigue and depression, common complaints in HIV-infected patients (Fernandez et al., 1995; Breitbart et al., 2001; Nath et al., 2002b).

Concerns have been raised of the possibility that dopaminergic agents may accelerate HIV-related disease. Dopamine is clearly toxic to neurons and may act synergistically with HIV-encoded proteins and other agents including drugs of abuse (see below). Several instances of acceleration of dementia in patients with HAD who were treated with dopaminergic agents have been reported. For instance, Nath et al. (2000) reported a patient with HAD whose dementia rapidly worsened after treatment with psychostimulants.

Both L-Dopa and selegiline accelerate SIV infection in macaque monkeys (Czub et al., 2001, 2004; Poli et al., 1990). Selegiline and L-Dopa increase dopamine release and availability in macaque monkeys. The incidence and severity of SIV encephalitis was significantly increased in drug-treated animals. Neuropathological changes of SIV infection appeared earlier in monkeys receiving dopaminergic agents when compared to untreated controls. Drug-treated SIV-infected monkeys had increased SIV viral load in the brain, increases in the severity of infection-related neuropathological changes, and ultrastructural alterations in dendrites of neurons in dopaminergic regions of the brain within 8 and 20 weeks after infection. A spongiform polioencephalopathy restricted to dopaminergic areas of the brain was frequently encountered. This is a unique neuropathological change not found in SIV-infected macaques that have not been treated with dopaminergic agents. L-Dopa and selegiline both increased the expression of TNF-α mRNA levels in microglia in SIV-infected monkeys treated with L-Dopa or selegiline suggesting TNF-α may contribute to the synergism between SIV and dopaminergic agents through its neurotoxic effects (Scheller et al., 2000). The rapid appearance of neuropathological changes and viral load in SIV-infected monkeys treated with dopaminergic agents suggests that synergistic interactions between SIV and dopaminergic agents can provoke the early appearance of SIV encephalitis with enhanced levels of microglia-derived SIV and TNF-α in the absence of immunodeficiency (Poli et al., 1990). Similar synergism between HIV and dopaminergic agents may explain, at least in part, the acceleration of HIV-related neurological changes in HIV-infected patients taking dopaminergic agents, including drugs of abuse (see below).

Dopaminergic agents may also alter HIV expression in peripheral immune cells. Dopaminergic agents act synergistically on lymphoid cells to alter immune responses. Lymphocytes produce, transport and bind dopamine present in the plasma. Circulating monocytes, T lymphocytes and B lymphocytes have dopamine receptors and mononuclear cells may synthesize dopamine. Dopamine upregulates HIV expression in a dose-dependent manner in T lymphoblasts chronically infected with HIV-1 (Rohr et al., 1999). Dopamine-induced activation of HIV infection was attenuated by glutathione and N-acetylcysteine suggesting dopamine acts by inducing oxidative stress through glutamate-dependent mechanisms. Support for this mechanism is found in the altered cellular redox states found in dopamine treated T lymphoblasts. HIV activation appears to be tightly linked to intracellular oxidant/antioxidant levels that are perturbed by dopamine. Dopamine effects on HIV transcription are mediated through the NF-κB element within the long terminal repeat of HIV (Droge et al., 1987). Dopaminergic effects on HIV replication in peripheral lymphoid cells may be augmented by proinflammatory cytokines. For instance, HIV replication is increased 15-fold in the presence of TNF-α.

19.21. Other neurotransmitter systems in HAD

19.21.1. Excitatory amino acid neurotransmitters in HAD

The effects of HIV gp120, gp41 and Tat on glutamate-mediated neurotoxicity have been reviewed above. Neuronal apoptosis following exposure to the viral proteins gp120 and Tat can be blocked by receptor antagonists for which excitatory neurotransmitters such as glutamate act as a substrate (Medina et al., 1999; Berger and Nath, 1997). Plasma glutamate levels are significantly elevated in HIV-infected patients, including those who are asymptomatic (Gurwitz and Kloog, 1997). CSF glutamate levels are increased in patients with HIV/AIDS with and without HAD (Koutsilieri et al., 1999). These findings are consistent with decreased astrocyte uptake of glutamate from the synaptic cleft, elevated plasma glutamate or both. NMDA receptor density is consistently reduced in postmortem studies of patients with HAD suggesting alterations of receptor synthesis following chronic overstimulation. Experiments using the SIV-infected macaque monkey model suggest alterations in glutamate metabolism occur early during the course of the disease. CSF levels of glutamate, but not aspartate, begin to increase 11 weeks after SIV infection (Heyes et al., 1989). The increase in glutamate appears to originate in microglia and correlates with high levels of viral antigen.

Other excitatory amino acid receptor agonists may possibly be involved in HAD. The most extensively studied of these is quinolinic acid. Quinolinic acid, an endogenous metabolite of L-tryptophan, is increased early and remains elevated in the CSF of HIV-infected patients (Heyes et al., 1989). Activated macrophages, both HIV-infected and un-infected, produce and secrete quinolinic acid and are the likely source of quinolinic acid in HIV-infected brains and CSF (Heyes et al., 1992). Elevated CSF levels of quinolinic acid correlate with dementia in the late stages of HIV/AIDS and in SIV-infected monkeys with neurological symptoms (Sarter and Podell, 2000; Lipton, 1998).

19.21.2. Cholinergic system in HAD

Essentially nothing is known about the cholinergic system in patients infected with HIV. Neuropsychological deficits are encountered frequently in HIV-infected patients before the development of HIV/AIDS (Brew, 2004). Early signs of cognitive decline in HIV-infected patients may be dominated by impaired attentional deficits and reductions in perceptual and psychomotor speed (Sarter et al., 1996). The cortical cholinergic system has an essential role in attentional functions and regulation of processing capacity suggesting this system may have a role in HIV-related abnormalities in these cognitive domains (Koutsilieri et al., 2000). Unfortunately, neuropathological studies of the brains of patients with HIV/AIDS have not addressed changes in the cholinergic system. However, the cholinergic system has been studied in SIV-infected monkeys. The activity of choline acetyltransferase (ChAT) is significantly reduced in the SIV-infected macaque (Koutsilieri et al., 2001). Reduced ChAT activity was most prominent in the putamen and hippocampus in asymptomatic monkeys. These changes did not correlate with brain SIV load, neuropathological changes or alterations in systemic immunity. ChAT activity could be completely restored by selegiline at doses that possess dopaminergic activity (Nath et al., 2002). The findings in the SIV-infected monkey and the cognitive changes seen early in HIV-infected individuals suggest that further attention to alterations in cholinergic transmission in HIV/AIDS may provide valuable insight into the pathogenesis of perceptual alterations and memory deficits in HIV-infected patients.

19.22. Drugs of abuse in HIV/AIDS

The pharmacological effects of drugs of abuse on HIV-related neurological disease is becoming of increasing importance. The role of drugs of abuse in facilitating risky behavior that leads to HIV infection was recognized early in the HIV/AIDS pandemic. During the ensuing years, drugs of abuse have been increasingly recognized as a prominent risk factor for HIV/AIDS. The most recent evidence suggests the HIV epidemic is, in part, being driven by drug abuse (Bell et al., 1998). The fastest growing populations with HIV infection in the USA and Western Europe acquire HIV through elicit drug use. Drugs of abuse appear to accelerate HIV infection and increase the

Table 19.5

Drugs of abuse and synergism with HIV proteins

Drug	Mode of toxicity	HIV protein synergism
Methamphetamine	Oxidative stress; mitochondrial damage	Tat: increased oxidative stress/ apoptosis
Opioid	Decreased dopamine levels; apoptosis	Tat: increased oxidative stress/apoptosis
Alcohol	Sensitized glutamate receptors	Gp120, Tat increased EEA toxicity

severity of HIV-related disease as illustrated by the increased prevalence of HIV encephalitis among drug abusers (Grassi et al., 1997). The pharmacological effects of drugs of abuse impact the same neuropharmacological systems altered by HIV infection. Drugs of abuse can increase HIV viral load by upregulating HIV replication in mononuclear cells.

Studies assessing the frequency of drug abuse in patients with HAD have produced variable results. Grassi et al. (1997) found a negative influence of drug abuse on cognitive function among Italian HIV-infected individuals. The *Mycobacterium avium* complex study also failed to find a difference in the incidence of HAD in HIV-infected drug users (Bouwman et al., 1998). A subsequent study reported that a history of injection drug use and psychomotor slowing at presentation heralded more rapid progression of neurological disease that was associated with more abundant macrophage activation in the brain (Koutsilieri et al., 1997). A cohort study in Scotland found that 56% of the brains of HIV-infected drug users had features of HIV encephalitis including the presence of p24 antigen and multinucleated giant cells compared to only 17% of homosexual HIV-infected men without a history of drug abuse having features of HIV encephalitis (Grassi et al., 1997).

The commonly abused drugs including amphetamines, cocaine and opiates have dopaminergic activation properties which could accelerate the loss of dopaminergic neurons in an already compromised dopaminergic system in HIV-infected patients (Table 19.5). Supporting this contention is the finding of more severe neuronal cell loss and shrunken neuronal cells in the substantia nigra in HIV-infected drug users when compared to HIV-infected patients without a history of drug abuse (Aylward et al., 1993).

19.22.1. HIV and methamphetamine/cocaine

Methamphetamine and cocaine act synergistically with HIV Tat and gp120 in neurotoxicity studies of dopaminergic systems. Neuronal cell lines exposed to dopamine, cocaine or morphine along with supernatants from HIV-infected cells show significant cell death and oxidative stress (Turchan et al., 2001). Acute exposure of human neuronal cell cultures to methamphetamine and cocaine in the presence of gp120 or Tat results in oxidative stress and cell death in a subpopulation of neurons (Chang et al., 2005). The mechanism of synergy in these studies appears to be, at least in part, through alterations in mitochondrial membrane potentials leading to oxidative stress. 17β-estradiol prevents the synergistic toxicities of methamphetamine and cocaine with dopamine in these studies (Maragos et al., 2002; Cass et al., 2003).

Oxidative stress is a possible mechanism for the synergism between cocaine and HIV proteins. Mitochondria are critical cellular targets for cocaine toxicity as evidenced by decreased mitochondrial respiration and increased synthesis of reactive oxygen species in animals exposed to cocaine. HIV Tat also produces oxidative stress in primary neuronal cell cultures with alterations in mitochondrial membrane potentials similar to those seen with cocaine. Oxidative changes in cellular proteins occur rapidly after injection of Tat into the rat striatum. Since both cocaine and Tat produce oxidative stress by targeting mitochondria, cocaine and HIV could have synergistic interactions in producing mitochondrial dysfunction, oxidative stress and cell death.

Clinical and experimental evidence suggests methamphetamine and HIV act synergistically to alter neural functions. Nath et al. (2002) reported the acceleration of HAD in a HIV-infected patient using methamphetamine. MR spectroscopy studies in HIV-infected patients with a history of chronic methamphetamine use suggest an additive effect of methamphetamine with HIV (Maragos et al., 2002). The neuronal marker N-acetylaspartate was reduced most prominently in the basal ganglia of HIV-infected patients who were chronic methamphetamine users compared to HIV-infected patients without a history of drug abuse. The glial cell marker myoinositol was significantly increased in the frontal lobes of these same patients. The authors suggest the reduction in N-acetylaspartate in the basal ganglia reflected neuronal cell loss while the increase in frontal lobe myoinositol could be explained by astrocytosis following up-regulation of HIV replication by methamphetamine.

Methamphetamine and HIV act synergistically in cell culture studies of neurotoxicity. Treatment of human neuronal cell cultures with methamphetamine or cocaine along with gp120 or Tat results in neuronal cell death that is prevented by treatment with 17β-estradiol (Maragos et al., 2002). These neurotoxic effects were associated with mitochondrial membrane potential changes and oxidative stress. Only a subset of cultured neurons was affected. Several studies have suggested that the subset of neurons affected in these studies is dopaminergic neurons.

HIV-1 Tat acts synergistically with methamphetamine to deplete striatal dopamine levels in an in vivo rat model (Cass et al., 2003). HIV-1 Tat potentiates methamphetamine-induced decreases in overflow of dopamine in the rat striatum (Nakayama et al., 1993). Injection of Tat into the rat striatum followed by intraperitoneal methamphetamine results in a 70 to 78% reduction in striatal dopamine overflow and content compared to 20% reduction in amphetamine-evoked overflow of dopamine and 16% decrease in dopamine content in rats treated with Tat alone. The interactions are specific for Tat and methamphetamine since neither heat-inactivated bovine serum nor HIV-1 gp120 were shown to synergize with methamphetamine. The synergism was also neurotransmitter specific since methamphetamine and Tat treatment did not alter serotonin levels. DOPAC levels are reduced in rats treated with Tat and methamphetamine. DOPAC levels reflect intraneural dopamine catabolism in dopamine synapses. The reduced DOPAC levels in rats treated with Tat and methamphetamine suggest either destruction of dopamine synapses has occurred or dopamine catabolism is shifted to another compartment. Tat alone increases the level of homovanilic acid, a product of extraneural dopamine metabolism. Homovanilic acid levels were unaffected in rats treated with Tat and methamphetamine. These findings suggest that Tat and methamphetamine may have different effects on the degradation pathways for dopamine metabolism and/or destruction of dopamine terminals by the combination of Tat and methamphetamine preventing synthesis of DOPAC. Loss of dopamine terminals may reflect repeated exposure to methamphetamine. Administration of methamphetamine to experimental animals, and possibly humans, over prolonged periods of time results in destruction of dopamine terminals and reduced striatal dopamine levels (Dykens, 1994).

Oxidative stress appears to be the primary mechanism leading to loss of dopaminergic neurons and dopamine in animals treated with Tat and methamphetamine. Mitochondrial membrane potentials are reduced in neurons exposed to HIV-1 Tat and methamphetamine (Chang et al., 2005; Cass et al., 2003). The combined toxicity of Tat and methamphetamines is blocked by antioxidants and 17β-estradiol. These studies suggest that, even though mitochondria may be a source of reactive oxygen species in other circumstances, the blockade of oxidative stress by antioxidants in Tat- and methamphetamine-treated neurons suggest oxidative stress precipitates rather than results from mitochondrial dysfunction when Tat and methamphetamine are present together. This finding is not unprecedented since several mitochondrial complexes can be inactivated by oxygen reactive species (Zhang et al., 1990; Peterson et al., 1993).

Methamphetamine also alters HIV expression outside the nervous system. Methamphetamine stimulates HIV replication in human peripheral blood mononuclear cells by increasing production of TNF-α (Ellis et al., 2003). These findings suggest that methamphetamine may increase viral loads by altering cytokine synthesis and release. The effects of methamphetamine on immune cell functions need to be more fully addressed. Increased HIV plasma loads have been found in patients who are using methamphetamines (Hauser et al., 1994). While upregulation of HIV replication would appear to be the most likely explanation for these findings, increased HIV plasma loads were only seen in patients taking HAART therapy. These paradoxical findings are most likely explained by poor compliance with HAART therapy by patients abusing methamphetamine. Poor compliance in the timing and dose of HAART therapy could allow for the escape of viral mutants that replicate despite suppressive therapy. Whether such changes in HIV replication occur in the central nervous system has not been studied. The acceleration of HAD in patients taking methamphetamine suggests this may be a possibility that deserves further study.

19.22.2. Opioid drugs

Opioids are typically proapoptotic when they cause cell death. Mu receptor drugs such as morphine and fentanyl induce toxicity in cerebellar Purkinje cell cultures and in the limbic system of rats when given in high doses (Kofke et al., 1996; Nair et al., 1997). Fentanyl exacerbates the effects of ischemia-induced damage to the basal ganglia (Nair et al., 1997). When combined with proapoptotic agents, opioids can exacerbate cell death (Singhal et al., 1998; Goswaami et al., 1998). For instance, mu agonists enhance staurosporine- and wortmannin-induced apoptosis in embryonic chick neurons or neuronal cell lines (Goswami et al., 2000; Nestler and Aghajanian, 1997).

Opioids may alter the susceptibility of dopaminergic neurons to viral damage by altering dopamine turnover. Endogenous opioids and opiate drugs interact with dopaminergic neurons through several pathways (Koob, 2000; Hauser et al., 1998; Rooney et al., 1991). Opioids decrease dopamine levels by reducing the activity of inhibitory interneurons that synapse with dopaminergic neurons in the ventral tegmentum. Opioids also influence dopamine cellular responses directly. Opioid receptor agonists increase D2 but not D1 dopamine receptor binding sites in rat striatum (De Vries et al., 1999; Gurwell et al., 2001). Repeated intermittent morphine exposure increases D2 receptor induced adenyl cyclase activity in the rat striatum.

Disruption of dopaminergic function with the development of tolerance and dependence or during opioid withdrawal may be of more significance when considering the effects of opioid drugs on HIV/AIDS (Koob, 2000). Chronic opioid drug exposure is accompanied by disruption of second messenger cascades, altered patterns of gene-activation, and increased oxidative stress (Koob, 2000; Rooney et al., 1991). These changes, especially those leading to oxidative stress, may increase neuronal susceptibility to HIV neurotoxic pathways.

Opioids and HIV Tat protein act synergistically to produce neuronal cell death. Human and mouse neurons undergo programmed cell death when exposed to Tat and morphine. Opioid antagonists prevent apoptosis in cells exposed to Tat and morphine (Stiene-Martin et al., 1998). The synergistic effects of morphine and HIV Tat are mediated by mitochondrial toxicity acting through Akt kinase, PI-3 kinase and caspases 1, 3 and 7 (Bell et al., 1998). As seen in dopamine-induced neurotoxicity, opioids are toxic to only a subset of neurons. Neuronal susceptibility to apoptosis following exposure to HIV-encoded proteins and opioids may depend, at least in part, on the distribution of opioid receptors in the brain. Subpopulations of striatal astrocytes and microglia express opioid receptors and may mediate the toxic effects of morphine and Tat on neurons (Chao et al., 1996; Donahue and Valhov, 1996). Finally, the administration of exogenous opioids may alter endogenous opioid peptide levels that could also participate in neurotoxic pathways.

Opiates may alter HIV expression outside the nervous system. Opioid drugs and HIV proteins act synergistically to destabilize immunity by altering monocyte and lymphocyte functions. The "opiate cofactor hypothesis" has been proposed as a mechanism in the pathogenesis of HIV/AIDS (Chao et al., 2001). The hypothesis is based on the findings that opioids promote HIV replication in immune cells and suppress immune functions. Subpopulations of leukocytes express mu, delta and kappa opioid receptors. Opioid drugs can modulate neuroimmune functions through direct and indirect actions that involve peripheral and central neural mechanisms. Opiates may have contradictory effects on immune functions depending on the receptor type involved. For instance, mu receptor stimulation is followed by increased HIV expression in monocytes while kappa receptor activation leads to inhibition of HIV expression in monocytes and lymphocytes (Nakayama et al., 1993; Donahue and Valhov, 1998; Rogers et al., 2000). Opiates also increase the expression of chemokine receptors that serve as co-receptors for HIV infection in susceptible cells thus possibly increasing the number of HIV susceptible immune cells (Baldwin et al., 2000).

19.22.3. Alcohol

Alcohol, like other drugs of abuse, is likely a significant risk factor for HIV infection. Individuals who abuse alcohol often engage in high-risk sexual behavior and have significant compromises of immune function both from the direct effects of alcohol as well as nutritional deficiencies (Dingle and Oei, 1997; Watzl and Watson, 1992; Tyor and Middaugh, 1999). Alcohol abuse can therefore alter both immune system function and nutritional status that could potentially increase the risk for development of HIV/AIDS-related diseases such as HAD (Pillai et al., 1991; Tabakoff, 1994; Johson-Greene et al., 1997).

Chronic alcohol intake stimulates the synthesis of reactive oxygen species, inhibits neuronal growth factor expression and reduces cerebral glucose utilization. These changes from chronic alcohol exposure could render neurons and glial cells more susceptible to the neurotoxic effects of HIV-related pathogenic pathways.

Chronic alcoholism alters neuronal function and sensitizes glutamate neurotransmitter receptors making them more susceptible to damage by HIV-1 neurotoxins. The neurotoxic effects of Tat and gp120 have been shown to be, at least in part, mediated by NMDA receptor overactivity leading to calcium-mediated excitatory cell death (Pendergast et al., 2000; Bell et al., 1998). Adaptive changes in NMDA receptors during long-term chronic alcohol abuse have been well documented. Chronic ethanol exposure in animals and primary neuronal cell cultures results in compensatory increases in the density and sensitivity of NMDA type glutamate receptors in cortical and hippocampal regions (Pendergast et al., 2000b; Hu and Ticku, 1995). NMDA receptor-mediated elevations in intracellular Ca^{2+} during alcohol withdrawal can result in neuronal cell death (190). The NMDA receptor mediated cell death in chronic alcoholism is blocked by NMDA receptor antagonists (Becker et al., 2004; Pendergast et al., 2000b; Chandler et al., 1993; Nebuloni et al., 2001). Alterations in NMDA activity seen with chronic alcohol abuse could potentiate the neurotoxic effects of Tat and/or gp120 acting through excitatory amino acid cell death cascades (Berger and Nath, 1997). The oxidative stress seen in chronic alcohol exposure may also enhance that seen as a result of HIV-encoded proteins. Thus, chronic alcohol exposure and alcohol withdrawal may act synergistically with HIV by increasing cell death through either excitatory amino acid or oxidative cell death cascades.

19.23. Neurodegenerative diseases in HIV/AIDS

HAART therapy has reduced mortality in patients infected with HIV (Berger and Arendt, 2000). While the incidence of HAD and other neurological complications of HIV has declined, the prevalence of HAD has increased as the HIV-infected population lives longer. The increased age of HIV-infected patients using HAART is likely to provide new challenges as neurodegenerative diseases of later life began to complicate the course of HIV/AIDS. Neuronal cell damage from HIV is likely to result in increased susceptibility to neurodegenerative diseases that may appear earlier than expected because of HIV-related compromise of neurotransmitter systems and other neural functions.

Cognitive decline appears to be more frequent and severe in older HIV-infected patients. In a community-based, sentinel survey of HIV-infected patients with cognitive impairment, the prevalence of cognitive disorders among HIV-infected individuals was significantly higher in those over 50 years of age compared to those less than 50 years old (Epstein et al., 1999). In the over 50 age group, dementia was the most common cognitive disorder seen where milder cognitive impairment was more common in the under 50 year old cohort. In these patients, alcohol was a significant risk factor for dementia whereas greater education was a protective factor. HIV viral load on entry into the study was significantly higher among those who developed cognitive impairment over the year of the study.

The demographic changes during the HAART era suggest patients may be at increased risk for developing Alzheimer's disease and possibly other neurodegenerative disorders. Although speculative, there are several reasons for this concern. Increasing age, high serum lipids, axonal injury and effects of Tat and quinolinic acid on the brain may lower the threshold for development of Alzheimer's disease. The increased mean age in a prevalence survey of neuropsychological abnormalities in an outpatient tertiary clinic was 49 ± 8.8 years compared to 38.5 ± 7 years (Kramer-Hammerle et al., 2005). The number of patients with elevated cholesterol and triglyceride levels is increasing as a result of age, HIV disease itself and HAART therapy, especially protease inhibitor drugs (Gabuzda and Wang, 2000). Elevated cholesterol levels have been consistently shown to be a risk factor for Alzheimer's disease. Patients with axonal injury are also at increased risk of developing Alzheimer's disease (Green et al., 2005). Axonal injury is a consistent finding in the brains of HIV-infected individuals, even early in the course of the disease. Deposition of β-amyloid in the brains of HIV-infected patients and alterations of β-amyloid metabolism in HIV infection may increase the risk of Alzheimer's disease in these patients.

While the above is speculative, studies have shown deposition of β-amyloid in the brains of HIV/AIDS patients. A strong association between HIV encephalitis and β-amyloid precursor protein (APP) deposition has been documented in several studies. The distribution of HIV p24 correlates with the distribution of APP aggregates (Mankowski et al., 2002). APP aggregates often take the form of intraxonal inclusions indicating that axonal injury has occurred. APP may also be found in extracellular aggregates in close proximity to HIV antigens. Similar APP depositions have also been found in SIV-infected macaques (Esiri et al., 1998).

Esiri et al. (1998) first described the presence of APP depositions in the brains of AIDS patients. A more comprehensive study found perivascular APP plaques that stained with 4G8 suggesting the deposits could be primarily vascular in origin. Green et al. (2005) found a statistically significant increase in parenchymal β-amyloid deposition in older patients on HAART. The low level replication of HIV in the brain in patients on HAART may lead to local inflammatory production and increased APP synthesis and susceptibility to amyloid deposition. The proinflammatory cytokine environment of HIV infection in the brain could lead to increases in cytokines IL-1α and S-100 which are known to cause overproduction of neuronal APP (Rempel et al., 2002). Alternatively, HAART therapy itself may contribute to the overall increase in amyloid deposition. It will be important to determine the functional significance of APP plaques in HIV-infected brains and whether other neuropathological changes associated with Alzheimer's disease are accelerated in HIV-infected patients.

Alterations in β-amyloid metabolism have been shown to be associated with HIV infection. HIV Tat inhibits neprilysin which degrades β-amyloid and elevates amyloid deposits in tissue culture systems (Liu et al., 2000). The absence of neprilysin on the cell surface of neurons exposed to Tat could help explain the increase in amyloid deposits in the extracellular spaces. Tat can also interact with the low-density lipoprotein receptor and thus inhibit uptake of its ligands including apolipoprotein E4 and β-amyloid (Landgford et al., 2003). Both β-amyloid and Tat are neurotoxic suggesting that synergistic interactions of Tat and β-amyloid may have a role in the pathogenesis of HAD and possibly Alzheimer's disease in HIV-infected patients.

Parkinson's disease represents another degenerative disorder that is likely to increase in frequency in the HIV-infected population as they age. It seems reasonable to expect the incidence of Parkinson's disease to increase as patients enter the age range in which idiopathic Parkinson's disease becomes more frequent. Impaired dopaminergic transmission is seen early and frequently in HIV/AIDS (Berger and Arendt, 2000). Many of the neuropathogenic pathways leading to HAD impact on the nigrostriatal system. HIV-1 Tat alters tyrosine presenlin activity in mixed neuronal glial cell cultures. Whether HIV alters γ-synthase activity is unknown. The ability of L-Dopa and selegiline to accelerate HIV and SIV related disease is likely to complicate the treatment of Parkinson's disease (Czub et al., 2004; Czub et al., 2001; Poli et al., 1990). The effects of dopamine agonists on the course of HIV/SIV disease will be important to examine. If dopamine agonists do not accelerate HIV-related disease, they may offer the safest approach to treating Parkinson's disease in the HIV-infected individual. These findings suggest that HIV-infected patients may develop Parkinson's disease earlier and may develop more severe disease that is more difficult to treat with conventional Parkinson agents.

19.24. Conclusion

Neuropharmacology offers an illuminating approach to understanding HIV-related neurological diseases. Unlike conventional viral diseases that destroy or alter cells they infect, HIV-related neurological disease is mediated through neuropharmacological pathways including neurotransmitter systems, oxidative stress and programmed cell death. The dopaminergic system has a central role in the pathogenesis of HIV-related neurological diseases. Neurotransmitter pathways mediate excitatory amino acid cellular cytotoxicity. The ability of several HIV regulatory proteins to form ion channels in plasma membranes and the recent demonstration of upregulation of genes encoding ion channels in the brains of HIV-infected patients with HAD illustrates the fundamental importance of neuropharmacology in the study of HIV-related diseases. The increasing importance of drugs of abuse as risk factors for HIV infection is likely to have a profound effect on the epidemiology of HIV as well as the neurological disorders encountered in HIV-infected patients. The increasing age of HIV-infected patients will likely be followed by the appearance of neurodegenerative diseases common in older age groups. The risk of accelerating HIV disease with the use of dopaminergic drugs is likely to be only one of many confounding issues that will challenge clinicians and researchers in the future. In this changing environment, the neuropharmacologist will have an increasingly important role in the study and treatment of HIV-related neurological diseases.

References

Albert SM, Weber C, Todak G (1999). An observed performance test of medication management ability in HIV: relation to neuropsychological status and adherence outcomes. AIDS Behav 3: 121–128.

Albrecht H, Hoffman C, Degen O, et al (1998). Highly active antiretroviral therapy significantly improves the prognosis of patients with HIV-associated progressive multifocal leukoencephalopathy. AIDS 12: 1149–1154.

Amchlen J, Dunlop O, Liestel K, et al (1995). Changing incidence of HIV-induced brain lesions in Oslo, 1983–1994: effects of zidovudine treatment. AIDS 9: 1165–1169.

Ammassari A, Cingolani A, Pezzoti P, et al (2000). AIDS-related focal brain lesions in the era of highly active antiretroviral therapy. Neurology 55: 1194–1200.

Andersson J, Fehniger TE, Patterson BK, et al (1998). Early reduction of immune activation in lymphoid tissue following highly active HIV therapy. AIDS 12: F123–F129.

Antinori A, Giancola ML, Grisetti S, et al (2002). Factors influencing virological response to antiretroviral drugs in cerebrospinal fluid of advanced HIV-1-infected patients. AIDS 16: 1867–1876.

Arendt G, Hefter H, Elsing C, et al (1990). Motor dysfunction in HIV-infected patients without clinically detectable central-nervous deficit. J Neurol 42: 891–896.

Aukrust P, Muller F, Lien E, et al (1999). Tumor necrosis factor (TNF) system levels in human immunodeficiency virus-infected patients during highly active antiretroviral therapy: persistent TNF activation is associated with virologic and immunologic failure. J Infect Dis 179: 74–82.

Aylward EH, Henderer JD, McArthur JC, et al (1993). Reduced basal ganglia volume in HIV-1-associated dementia: results from quantitative neuroimaging. Neurology 43: 2099–2104.

Bacellar H, Munoz A, Miller EN, et al (1994). Temporal trends in the incidence of HIV-1-related neurologic diseases: Multicenter AIDS Cohort Study, 1985–1992. Neurology 44: 1892–1900.

Baldeweg T, Catalan J, Lovett E, et al (1995). Long-term zidovudine reduces neurocognitive deficits in HIV-1 infection. AIDS 9: 589–596.

Baldwin JA, Maxwell CJ, Fenaughty AM, et al (2000). Alcohol as a risk factor for HIV transmission among American Indian and Alaska native drug. Am Indian Alask Native Ment Heath Res 9: 1–16.

Bell JE, Brettle RP, Chiswick A, et al (1998). HIV encephalitis, proviral load and dementia in drug users and homosexuals with AIDS. Effect of neocortical involvement. Brain 121: 2043–2052.

Bell JE, Donaldson YK, Lowrie S, et al (1996). Influence of risk group and zidovudine therapy on the development of HIV

encephalitis and cognitive impairment in AIDS patients. AIDS 10: 493–499.
Belman AL, Ultmann MH, Horoupian D, et al (1985). Neurological complications in infants and children with acquired immunodeficiency syndrome. Ann Neurol 18: 560–566.
Bennett BA, Rusyniak DE, Hollingsworth CK (1995). HIV-1 gp120-induced neurotoxicity to midbrain dopamine cultures. Brain Res 705: 168–176.
Berenguer J, Miralles P, Arrizabalaga J, et al (2003). Clinical course and prognostic factors of progressive multifocal leukoencephalopathy in patients treated with highly active antiretroviral therapy. Clin Infect Dis 36: 1047–1052.
Berger JR, Nath A (1997). HIV dementia and the basal ganglia. Intervirology 40: 122–131.
Berger JR, Arendt G (2000). HIV dementia: the role of the basal ganglia and dopaminergic systems. J Psychopharmacol 14: 214–221.
Berger JR, Pall L, Lanska D, et al (1998). Progressive multifocal leukoencephalopathy in patients with HIV infection. J NeuroVirol 4: 59–68.
Bonnet F, Morlat P, Chene G, et al (2002). Causes of death among HIV-infected patients in the era of highly active antiretroviral therapy, Bordeaux, France, 1998–1999. HIV Med 3: 195–199.
Borst P, Evers R, Kool M, et al (2000). A family of drug transporters: the multidrug resistance-associated proteins. J Natl Cancer Inst 92: 1295–1302.
Bouwman FH, Skolasky RL, Hess D, et al (1998). Variable progression of HIV-associated dementia. Neurology 50: 1814–1820.
Brady MT, McGrath N, Brouwers P, et al (1996). Randomized study of the tolerance and efficacy of high- versus low-dose zidovudine in human immunodeficiency virus-infected children with mild to moderate symptoms. J Infect Dis 173: 1097–1106.
Breitbart W, Rosenfeld B, Kaim M, et al (2001). A randomized, double-blind, placebo-controlled trial of psychostimulants for the treatment of fatigue in ambulatory patients with human immunodeficiency virus disease. Arch Intern Med 161: 411–420.
Brew BJ (2001). AIDS dementia complex. In: HIV Neurology, Oxford University Press, Oxford, pp. 53–90.
Brouwers P, Moss H, Wolters P, et al (1990). Effect of continuous-infusion zidovudine therapy on neuropyschologic functioning in children with symptomatic human immunodeficiency virus infection. J Pediatr 117: 980–985.
Brouwers P, Henricks M, Lietzau JA, et al (1997). Effect of combination therapy with zidovudine and didanosine on neuropsychological functioning in patients with symptomatic HIV disease: a comparison of simultaneous and alternating regimens. AIDS 11: 59–66.
Buchbinder SP, Katz MH, Hessol NA, et al (1992). Herpes zoster and human immunodeficiency virus infection. J Infect Dis 166: 1153–1156.
Carr A, Cooper DA (1997). Restoration of immunity to chronic hepatitis B infection in HIV-infected patients on protease inhibitor. Lancet 349: 995–996.
Carr A, Emery S, Kelleher A, et al (1996). CD8+ lymphocyte responses to antiretroviral therapy of HIV infection. J Acquir Immune Defic Syndr Hum Retrovirol 13: 320–326.
Cartier L, Hartley O, Dubois-Dauphin M, et al (2005). Chemokine receptors in the central nervous system: role in brain inflammation and neurodegenerative disorders. Brain Res Rev 48: 16–42.
Cass WA, Harned ME, Peters LE, et al (2003). HIV-1 protein Tat potentiation of methamphetamine-induced decreases in evoked overflow of dopamine in the striatum of the rat. Brain Res 984: 133–142.
Chang L, Ernst T, Speck O, et al (2005). Additive effects of HIV and chronic methamphetamine use on brain metabolite abnormalities. Am J Psychiat 162: 361–369.
Chao CC, Gekker G, Hu S, et al (1996). Kappa opioid receptors in human microglia down regulate human immunodeficiency virus 1 expression. Proc Natl Acad Sci USA 93: 8051–8056.
Chao CC, Gekker G, Sheng WS, et al (2001). U50488 inhibits HIV-1 expression in acutely infected monocyte-derived macrophages. Drug Alcohol Depend 62: 149–154.
Chriboga CA, Fleishman S, Champion S, et al (2005). Incidence and prevalence of HIV encephalopathy in children with HIV infection receiving highly active anti-retroviral therapy (HAART). J Pediatr 146: 402–407.
Cinque P, Vago L, Mengozzi M, et al (1998). Elevated cerebrospinal fluid levels of monocyte chemotactic protein-1 correlate with HIV-1 encephalitis and local viral replication. AIDS 12: 1327–1332.
Cinque P, Bossolasco S, Brambilla AN, et al (2003). The effect of highly active antiretroviral therapy-induced immune reconstitution on development and outcome of progressive multifocal leukoencephalopathy: study of 43 cases with review of the literature. J NeuroVirol 9(Suppl 1): 73–80.
Civitello LA (1991–1992). Neurologic complications of HIV infection in children. Pediatr Neurosurg 17: 104–112.
Clifford DB, Yiannoutsos C, Glickman M, et al (1999). HAART improves prognosis in HIV-associated progressive multifocal leukoencephalopathy. Neurology 52: 623–625.
Clifford DB, McArthur JC, Schifitto G, et al (2002). A randomized, placebo-controlled phase II clinical trial of CP1-1189 for HIV-associated motor-cognitive impairment. Neurology 59: 1568–1573.
Cohen RA, Boland R, Paul R, et al (2001). Neurocognitive performance enhanced by highly active antiretroviral therapy in HIV infected women. AIDS 15: 341–345.
Conant K, Garzino-Demo A, Nath A, et al (1998). Induction of monocyte chemoattractant protein-1 in HIV-1 Tat-stimulated astrocytes and elevation in AIDS dementia. Proc Natl Acad Sci USA 95: 3117–3121.
Concorde Coordinating Committee (1994). Concorde: MRC/ANRS randomize double-blind controlled trial of immediate and deferred zidovudine in symptom-free HIV infection. Lancet 343: 871–881.
Czub S, Koutsilieri E, Sopper S, et al (2001). Enhancement of central nervous system pathology in early simian immunodeficiency virus infection by dopaminergic drugs. Acta Neuropathol 104: 85–91.

Czub S, Czub M, Koutsilieri E, et al (2004). Modulation of simian immunodeficiency virus neuropathology by dopaminergic drugs. Acta Neuropathol 107: 216–226.

Dana Consortium (1998). A randomized, double-blind, placebo-controlled trial of deprenyl and thiotic acid in human immunodeficiency virus-associated cognitive impairment. Dana Consortium on the Therapy of HIV Dementia and Related Cognitive Disorders. Neurology 50: 645–651.

Davies J, Everall I, Weich S, et al (1997). HIV-associated brain pathology in the United Kingdom: an epidemiological study. AIDS 11: 1145–1150.

De Luca A, Ammmassari A, Cingolani A, et al (1998). Disease progression and poor survival of AIDS-associated progressive multifocal leukoencephalopathy despite highly active antiretroviral therapy. AIDS 12: 1937–1938.

DeSimone JA, Pomerantz RJ, Babinchak TJ (2000). Inflammatory reactions in HIV-1-infected persons after initiation of highly active antiretroviral therapy. Ann Intern Med 133: 447–454.

De Vries TJ, Shoffelmeer AN, Binnekade R, et al (1999). Dopaminergic mechanisms mediating the incentive to seek cocaine and heroin following long-term withdrawal of IV drug self-administration. Psychopharmacol (Berl) 143: 254–260.

Domingo P, Torres OH, Ris J, et al (2001). Herpes zoster as an immune reconstitution disease after initiation of combination antiretroviral therapy in patients with human immunodeficiency virus type-1 infection. Am J Med 110: 605–609.

Donahue RM, Valhov D (1998). Opiates as potential cofactors in progression of HIV-1 infections in AIDS. J Neuroimmunol 83: 77–87.

Dore G, Correll P, Kaldor J, et al (1999). Changes to the natural history of AIDS dementia complex in the era of HAART. AIDS 13: 1249–1253.

Dore BJ, McDonald A, Li Y, et al (2003). Marked improvement in survival following AIDS dementia complex in the era of highly active antiretroviral therapy. AIDS 17: 1539–1545.

Dougherty RH, Skolasky RL, McArthur JC (2002). Progression of HIV-associated dementia treated with HAART. AIDS Read 12: 69–74.

Droge W, Eck HP, Betzler M, et al (1987). Elevated plasma glutamate levels in colorectal carcinoma patients and in patients with acquired immunodeficiency syndrome (AIDS). Immunobiology 174: 473–479.

Dworkin MS, Wan PC, Hanson DL, et al (1999). Progressive multifocal leukoencephalopathy: improved survival of human immunodeficiency virus-infected patients in the protease inhibitor era. J Infect Dis 180: 621–625.

Dykens JA (1994). Isolated cerebral and cerebellar mitochondria produce free radicals when exposed to elevated Ca^{2+} and Na^{+}: implications for neurodegeneration. J Neurochem 63: 584–591.

Edelstein H, Knight RT (1987). Severe parkinsonism in two AIDS patients taking prochlorperazine. Lancet 2: 2312–2322.

Eggers C, Hertogs K, Sturenburg H-J, et al (2003). Delayed central nervous system virus suppression during highly active antiretroviral therapy is associated with HIV encephalopathy, but not with viral drug resistance or poor central nervous system drug penetration. AIDS 17: 1897–1906.

Ellis R, Hsia K, Spector S, et al (1997). CSF HIV-1 RNA levels are elevated in neurocognitively impaired individuals. Ann Neurol 42: 679–688.

Ellis RJ, Childers ME, Cherner M, et al (2003). Increased human immunodeficiency virus loads in active methamphetamine users are explained by reduced effectiveness of antiretroviral therapy. J Infect Dis 188: 1820–1826.

Ellis RJ, Childers ME, Cherner M, et al (2003). Increased human immunodeficiency virus loads in active methamphetamine users are explained by reduced effectiveness of antiretroviral therapy. J Infect Dis 188: 1820–1826.

Enting RH, Hoetelmans RM, Lange JM, et al (1998). Antiretroviral drugs and the central nervous system. AIDS 12: 1941–1955.

Epstein LG, Gelbard HA (1999). HIV-1-induced neuronal injury in the developing brain. J Leuko Biol 65: 453–457.

Esiri MM, Biddolph SC, Morris CS (1998). Prevalence of Alzheimer plaques in AIDS. J Neurol Neurosurg Psychiatry 65: 29–33.

Fernandez F, Levy JK, Samley HR, et al (1995). Effects of methylphenidate in HIV-related depression: a comparative trial with desipramine. Int J Psychiatry 25: 53–67.

Ferrando S, van Gorp W, McElhiney M, et al (1998). Highly active antiretroviral treatment in HIV infection: benefits for neuropsychological function. AIDS 12: F65–F70.

Gabuzda D, Wang J (2000). Chemokine receptors and mechanisms of cell death in HIV neuropathogenesis. J NeuroVirol 6 (Suppl 1): S24–S32.

Gartner S (2000). HIV infection and dementia. Science 287: 602–604.

Gasnault J, Taoufik Y, Goujard C, et al (1999). Prolonged survival without neurological improvement in patients with AIDS-related progressive multifocal leukoencephalopathy on potent combined antiretroviral therapy. J NeuroVirol 5: 442–449.

Gisolf EN, van Praag RM, Jurriaans S, et al (2000). Increasing cerebrospinal fluid chemokine concentrations despite undetectable cerebrospinal fluid HIV RNA in HIV-1-infected patients receiving antiretroviral therapy. J Acquir Immune Defic Syndr 25: 425–433.

Glass JD, Wesselingh SL, Selnes OA, et al (1993). Clinical–neuropathologic correlation of HIV-associated dementia. Neurology 43: 2230–2237.

Gonzalez-Scarano OA, Martin-Garcia J (2005). The neuropathogenesis of AIDS. Nature Rev Immunol 5: 69–81.

Goswami R, Dawson SA, Dawson G (1998). Cyclic AMP protects against staurosporine and wortmannin-induced apoptosis and opioid-enhanced apoptosis in both embryonic and immortalized (F-11 kappa7) neurons. J Neurochem 70: 1376–1382.

Grassi MP, Perin C, Clerici F, et al (1997). Effects of HIV seropositivity and drug use cognitive function. Eur Neurol 37: 48–52.

Gray F, Belec L, Keohane C, et al (1994). Zidovudine therapy and HIV encephalitis: a 10-year neuropathological survey. AIDS 8: 489–493.

Gray G, Chretien F, Vallat-Decouvelaere AV, et al (2003). The changing pattern of HIV neuropathology in the HAART era. J Neuropathol Exp Neurol 62: 429–440.

Green DA, Masliah E, Vinters HV, et al (2005). Brain deposition of beta-amyloid is a common pathologic feature in HIV positive patients. AIDS 19: 407–411.

Gurwell JA, Nath A, Sun Q, et al (2001). Synergistic neurotoxicity of opioids and human immunodeficiency virus-1 Tat protein in striatal neurons in vitro. Neuroscience 102: 555–563.

Gurwitz D, Kloog Y (1997). Elevated cerebrospinal fluid glutamate in patients with HIV-related dementia. JAMA 277: 1931.

Hamilton JD, Hartigan PM, Simberkoff MS, et al (1992). A controlled trial of early versus late treatment with zidovudine in symptomatic human immunodeficiency virus infection: results of the Veterans Affairs Cooperative Study. N Engl J Med 326: 437–443.

Hauser KF, Gurwell JA, Turbek CS (1994). Morphine inhibits Purkinje cell survival and dendritic differentiation in organotipic cultures of the mouse cerebellum. Exp Neurol 130: 95–105.

Hauser KF, Harris-White ME, Jackson JA, et al (1998). Opioids disrupt Ca^{2+} homeostasis and induce carbonyl oxyradical production in mouse astrocytes in vitro: transient increases and adaptation to sustained exposure. Exp Neurol 151: 70–76.

Heyes MP, Rubinow D, Lane C, et al (1989). Cerebrospinal fluid quinolinic acid concentrations are increased in acquired immune deficiency syndrome. Ann Neurol 26: 275–277.

Heyes MP, Brew BJ, Martin A, et al (1991). Quinolinic acid in cerebrospinal fluid and serum in HIV-1 infection; relationship to clinical and neurological status. Ann Neurol 29: 202–209.

Heyes MP, Saito K, Markey SP (1992). Human macrophages convert L-tryptophan into the neurotoxin quinolinic acid. Biochem J 283(Pt 3): 633–635.

Hriso E, Kuhn T, Masdeu JC, et al (1991). Extrapyramidal symptoms due to dopamine-blocking agents in patients with AIDS encephalopathy. Am J Psychiatry 148: 1558–1561.

Husson RN, Mueller BU, Farley M, et al (1994). Zidovudine and didanosine combination therapy in children with human immunodeficiency virus infection. Pediatrics 93: 316–322.

Itoh K, Mehraein P, Weis S (2000). Neuronal damage of the substantia nigra in HIV-1 infected brains. Acta Neuropathol 99: 376–384.

Janssen RS, Saykin AJ, Cannon L, et al (1989). Neurological and neuropsychological manifestations of HIV-1 infection: association with AIDS-related complex but not asymptomatic HIV-1 infection. Ann Neurol 26: 592–600.

Jellinger KA, Setinek U, Drlick M, et al (2000). Neuropathology and general autopsy findings in AIDS during the last 15 years. Acta Neuropathol (Berl) 100: 213–220.

Jenuwein M, Scheller C, Neuen-jacob E, et al (2004). Dopamine deficits and regulation of the cAMP second messenger system in brains of simian immunodeficiency virus-infected rhesus monkeys. J NeuroVirol 10: 163–170.

John M, Flexman J, French MA (1998). Hepatitis C virus-associated hepatitis following treatment of HIV-infected patients with HIV protease inhibitors: an immune restoration disease. AIDS 12: 2289–2293.

Jones M, Olafson K, Del Bigio MR, et al (1998). Intraventricular injection of human immunodeficiency virus type 1 (HIV-1) Tat protein causes inflammation, gliosis, apoptosis, and ventricular enlargement. J Neuropathol Exp Neurol 57: 563–570.

Karlsen NR, Reinvang I, Frasoland SS (1995). A follow-up study of neuropsychological functioning in AIDS-patients: prognostic significance and effect of zidovudine therapy. Acta Neurol Scand 91: 215–221.

Kelder W, McArthur JCF, Nance-Sproson T, et al (1998). Beta-chemokines MCP-1 and RANTES are selectively increased in cerebrospinal fluid of patients with human immunodeficiency virus-associated dementia. Ann Neurol 44: 831–835.

Kelleher AD, Carr A, Zaunders J, et al (1996). Alteration in the immune response of human immunodeficiency virus (HIV)-infected subjects treated with an HIV-specific protease inhibitor, ritonavir. J Infect Dis 173: 321–329.

Kieburtz KD, Eppstein LG, Gelhard HA, et al (1991). Excitotoxicity and dopaminergic dysfunction in the acquired immunodeficiency syndrome dementia complex. Therapeutic implications. Arch Neurol 48: 1281–1284.

Kofke WA, Garman RH, Stiller R, et al (1996). Opioid neurotoxicity: fentanyl dose-response effects in rats. Anesth Analg 83: 1298–1306.

Koob GF (2000). Neurobiology of addiction. Toward the development of new therapies. Ann NY Acad Sci 909: 170–185.

Kotecha N, George MJ, Smith TW, et al (1998). Enhancing progressive multifocal leukoencephalopathy: an indicator of improved status? Am J Med 105: 541–543.

Koutsilieri E, Gotz ME, Sopper S, et al (1997). Regulation of glutathione and cell toxicity following exposure to neurotropic substances and human immunodeficiency virus-1 in vitro. J Neurovirol 3: 342–349.

Koutsilieri E, Sopper S, Heinemann T, et al (1999). Involvement of microglia in cerebrospinal fluid glutamate increase in SIV-infected rhesus monkeys (Macacca mulatta). AIDS Res Hum Retrovir 15: 471–477.

Koutsilieri E, Czub S, Scheller C, et al (2000). Brain choline acetyltransferase activity deficits in SIV infection. An index of early dementia? Neuroreport 11: 2391–2393.

Koutsilieri E, ter Meulen V, Riederer P (2001a). Neurotransmission in HIV associated dementia: a short review. J Neural Transm 108: 767–775.

Koutsilieri E, Scheller C, Sopper S, et al (2001b). Selegiline completely restores choline acetyltransferase activity deficits in simian immunodeficiency infection. Eur J Pharmacol 411: R1–R2.

Koutsilieri E, Sopper S, Scheller C, et al (2002). Involvement of dopamine in the progression of AIDS dementia complex. J Neural Transm 109: 399–410.

Kramer-Hammerle S, Rothenaigner I, Wolff H, et al (2005). Cells of the central nervous system as targets and reservoirs of the human immunodeficiency virus. Virus Res 1111: 194–213.

Kure K, Lyman WD, Weidenheim KM, et al (1990). Cellular localization of an HIV-1 antigen in subacute AIDS encephalitis using an improved double-labeling immunohistochemical method. Am J Pathol 136: 1085–1092.

Lackner A, Smith M, Munn R, et al (1991). Localization of simian immunodeficiency virus in the nervous system of rhesus monkeys. Am J Pathol 139: 609–621.

Lambotte O, Deiva K, Tardieu M (2003). HIV-1 persistence, viral reservoir, and the central nervous system in the HAART era. Brain Pathol 13: 95–103.

Langford TD, Letendre SL, Marcotte TD, et al (2002). Severe, demyelinating leukoencephalopathy in AIDS patients on antiretroviral therapy. AIDS 16: 1019–1029.

Langford TD, Letendre SL, Larrea GJ, et al (2003). Changing patterns in the neuropathogenesis of HIV during the HAART era. Brain Pathol 13: 195–210.

Laverda AM, Gallo P, DeRossi A, et al (1994). Cerebrospinal fluid analysis in HIV-1-infected children: immunological and virological findings before and after AZT therapy. Acta Pediatr 83: 1038–1042.

Lindsey JC, Hughes MD, McKinney RE, et al (2000). Treatment-mediated changes in human immunodeficiency virus (HIV) type 1 RNA and CD^4 cell counts as predictors of weight growth failure, cognitive decline, and survival in HIV-infected children. J Infect Dis 182: 1385–1393.

Lipton SA (1998). Neuronal injury associated with HIV-1: approaches to treatment. Annu Rev Pharmacol Toxicol 38: 159–177.

Liu Y, Jones M, Hingten CM, et al (2000). Uptake of HIV-1 Tat protein mediated by low-density lipoprotein receptor-related protein disrupts the neuronal metabolic balance of the receptor ligands. Nat Med 6: 1380–1387.

Maehlen J, Dunlop O, Leistol K, et al (1995). Changing incidence of HIV-induced brain lesions in Oslo, 1983–1994: effects of zidovudine treatment. AIDS 9: 1165–1169.

Magnuson DSK, Knudsen BE, Geiger JD, et al (1995). Human immunodeficiency virus type 1 tat activates non-N-methyl-D-aspartate excitatory amino acid receptors and causes neurotoxicity. Ann Neurol 37: 373–380.

Makela P, Howe L, Glover S, et al (2002). Recurrent Guillain-Barre syndrome as a complication of immune reconstitution in HIV. J Infect 44: 186–193.

Mankowski JL, Queen SE, Tarwater PM, et al (2002). Accumulation of beta-amyloid precursor protein in axons correlates with expression of SIV gp41. J Neuropathol Exp Neurol 61: 85–90.

Marcus K, Truffa M, Boxwell D, et al (2002). Recently identified adverse events secondary to NRTI therapy in HIV-infected individuals: cases from the FDA's adverse event reporting system. In: Programs and Abstracts of the 9th Conference on Retroviruses and Opportunistic Infections (Seattle), Foundation for Retrovirology and Human Health, Alexandria, VA, abstract LB14.

Martinez E, Gatell J, Moran Y, et al (1998). High incidence of herpes zoster in patients with AIDS soon after therapy with protease inhibitors. Clin Infect Dis 27: 1510– 1513.

Masliah E, DeTeresa R, Mallory M, et al (2000). Changes in pathological findings at autopsy in AIDS cases for the last 15 years. AIDS 14: 69–74.

McArthur JC (2004). HIV dementia: an evolving disease. J Neuroimmunol 157: 3–10.

McArthur JC, Hoover DR, Bacellar H, et al (1993). Dementia in AIDS patients: incidence and risk factors. Neurology 43: 2245–2252.

McArthur JC, McDermott MP, McClernon D, et al (2004). Attenuated CNS infection in advanced HIV/AIDS with combination antiretroviral therapy. Arch Neurol 61: 1687–1696.

McCoig C, Castrejon M, Castano E, et al (2002). Effect of combination antiretroviral therapy on cerebrospinal fluid HIV RNA, HIV resistance, and clinical manifestations of encephalopathy. J Pediatr 14136–14144.

McKinney RE, Maha MA, Connor EM, et al (1991). A multicenter trial of oral zidovudine in children with advance human immunodeficiency virus disease. N Engl J Med 324: 1018–1025.

Medina I, Ghose S, Ben-Ari Y (1999). Mobilization of intracellular calcium stores participates in the rise of $[Ca^{2+}]_i$ and the toxic actions of the HIV coat protein gp120. Eur J Neurosci 11: 1167–1178.

Miller RF, Isaacson PG, Hall-Craggs M, et al (2004). Cerebral CD8+ lymphocytosis in HIV-1 infected patients with immune restoration induced by HAART. Acta Neuropathol 108: 17–23.

Mintz M, Tardieu M, Hoyt L, et al (1996). Levodopa therapy improves motor function in HIV-infected children with extrapyramidal syndromes. Neurology 47: 1583–1585.

Mirsattari SM, Power C, Nath A (1998). Parkinsonism with HIV infection. Mov Disord 13: 684–689.

Nair MP, Schwartz SA, Polasani R, et al (1997). Immunoregulatory effects of morphine on human lymphocytes. Clin Diagn Lab Immunol 4: 127–132.

Nakayama M, Koyama T, Yamashita I (1993). Long-lasting decrease in dopamine uptake sites following repeated administration of methamphetamine in the rat striatum. Brain Res 601: 209–212.

Nath A, Geiger J (1998). Neurobiological aspects of human immunodeficiency virus infection: neurotoxic mechanisms. Prog Neurobiol 54: 19–33.

Nath A, Anderson C, Jones M, et al (2000). Neurotoxicity and dysfunction of dopaminergic systems associated with AIDS dementia. J Psychopharmacol 14: 222–227.

Nath A, Hauser KF, Wojna V, et al (2002a). Molecular basis for interactions of HIV and drugs of abuse. J Acq Immun Def Syn 31: S62–S69.

Nath A, Maragos WF, Avison MJ, et al (2002b). Acceleration of HIV dementia with methamphetamine and cocaine. J NeuroVirol 7: 66–71.

Nestler EJ, Aghajanian GK (1997). Molecular and cellular basis of addiction. Science 278: 58–63.

Neumann H, Medana IM, Bauer J, et al (2002). Cytotoxic T lymphocytes in autoimmune and degenerative CNS diseases. Trends neuro sci 25: 313–319.

Nordic Medical Research Council's HIV Therapy Group (1992). Double blind dose-response study of zidovudine in AIDS and advanced HIV infection. Brit Med J 304: 13–17.

Nozyce M, Hoberman M, Apradi S, et al (1994). A 12-month study of the effects of oral zidovudine on neurodevelopmental functioning in a cohort of vertically HIV-infected inner-city children. AIDS 8: 635–659.

Perry SW, Norman JP, Gelbard HA (2005). Adjunctive therapies for HIV-1 associated neurologic disease. Neurotoxic Res 8: 161–166.

Persidsky Y, Stins M, Way D, et al (1997). A model of monocyte migration through the blood-brain barrier during HIV-1 encephalitis. J Immunol 158: 3499–3510.

Peterson PK, Gekker G, Schut R, et al (1993). Enhancement of HIV-1 replication by opiates and cocaine: the cytokine connection. Adv Exp Med Biol 335: 181–188.

Petito C, Adkins B, McCarthy M, et al (2003). CD^4+ and CD8+ cells accumulate in the brains of acquired immunodeficiency syndrome patients with human immunodeficiency virus encephalitis. J NeuroVirol 9: 36–44.

Piliero PJ, Fish DG, Preston S, et al (2003). Guillain-barre syndrome associated with immune reconstitution. Clin Infect Dis 36: e111–e114.

Pizzo PA, Eddy J, Fallon J, et al (1988). Effect of continuous intravenous infusion of zidovudine (AZT) in children with symptomatic HIV infection. N Engl J Med 319: 889–896.

Poleuektova LY, Munn DH, Persidsky Y, et al (2002). Generation of cytotoxic T cells against virus-infected human brain macrophages in a murine model of HIV-1 encephalitis. J Immunol 168: 3941–3949.

Poli G, Kinter A, Justement J, et al (1990). Tumor necrosis factor α functions in autocrine manner in the induction of human immunodeficiency virus expression. Proc Natl Acad Sci USA 87: 782–785.

Portegies P, de Gans J, Lange JM, et al (1989). Declining incidence of AIDS dementia complex after introduction of zidovudine treatment. Brit Med J 299: 819–821.

Portegies P, Enting RH, de Gans J, et al (1993). Presentation and course of AIDS dementia complex: 10 years of follow-up in Amsterdam, The Netherlands. AIDS 7: 669–675.

Pulliam L, Gascon R, Stubblebine M, et al (1997). Unique monocyte subset in patients with AIDS dementia. Lancet 349: 692–695.

Puthanakit T, Oberdorfer P, Akarathum N, et al (2006). Immune reconstitution syndrome after highly active antiretroviral therapy in human immunodeficiency virus-infected Thai children. Pediatr Infect Dis 25: 53–58.

Raskino C, Pearson DA, Baker CJ, et al (1999). Neurologic, neurocognitive, and brain growth in human immunodeficiency virus-infected children receiving different nucleoside antiretroviral regimens. Pediatrics 104: 32.

Reinvang I, Froland SS, Karlsen NR, et al (1991). Only temporary improvement in impaired neuropsychological function in AIDS patients treated with zidovudine. AIDS 5: 228–229.

Rempel H, Buffum D, Pulliam L (2002). HIV-1 tat inhibits the amyloid beta degrading enzyme, neprilysin. J Neurovirol 8(Suppl 1): 12.

Reyes MG, Faraldi F, Senseng CS, et al (1991). Nigral degeneration in acquired immune deficiency syndrome (AIDS). Acta Neuropathol (Berl) 82: 39–44.

Rogers TJ, Steele AD, Howard OM, et al (2000). Bidirectional heterologous desensitization of opioid and chemokine receptors. Ann NY Acad Sci 917: 19–28.

Rohr O, Sawaya BE, Lecestre D, et al (1999). Dopamine stimulates expression of the human immunodeficiency virus type 1 via NR-κB in cells of the immune system. Nucl Acids Res 27: 3291–3299.

Rooney KF, Armstrong RA, Sewell RD (1991). Increased dopamine receptor sensitivity in the rat following acute administration of sufentanil, U50, 488H and D-Ala2-D-Leu5-enkaphalin. Naunyn Schmiedebergs Arch Pharmacol 343: 458–462.

Rosso R, Di Biago A, Ferrdazin A, et al (2003). Fatal lactic acidosis and mimicking Guillain-Barre syndrome in an adolescent with human immunodeficiency virus infection. Pediatr Infect Dis 22: 668–670.

Rottenberg DA, Sidits JJ, Strother SC, et al (1996). Abnormal cerebral glucose metabolism in HIV-1 seropositive subjects with and without dementia. J Nucl Med 37: 1133–1141.

Sacktor N (2002). The epidemiology of human immunodeficiency virus-associated neurological disease in the era of highly active antiretroviral therapy. J NeuroVirol 8(Suppl 2): 115–121.

Sacktor N, Lyles RH, Skolasky R, et al (2001). HIV-associated neurologic disease incidence changes: Multicenter AIDS Cohort Study, 1990–1998. Neurology 56: 257–260.

Sanchez-Ramon S, Bellon JM, Resino S, et al (2003). Low blood CD8+ T-lymphocytes and high circulating monocytes are predictors of HIV-1-associated progressive encephalopathy in children. Pediatrics 111: E168–E175.

Sarter AM, Podell M (2000). Preclinical psychopharmacology of AIDS-associated dementia: lesson to be learned from the cognitive psychopharmacology of other dementias. J Psychopharmacol 14: 197–204.

Sarter AM, Bruno JP, Givens B, et al (1996). Neuronal mechanisms mediating drug-induced cognition enhancement: cognitive activity as a necessary intervening variable. Brain Res Cogn Brain Res 3: 329–343.

Scheller C, Sopper S, Jassoy C, et al (2000). Dopamine activates HIV in chronically infected T lymphoblasts. J Neural Transm 107: 1483–1489.

Schifitto G, McDermott MP, McArthur JC, et al (2002). Incidence of and risk factors for HIV-associated distal sensory polyneuropathy. Neurology 58: 1764–1768.

Schmitt FA, Bigley JW, McKinnis R, et al (1988). Neuropyschological outcome of zidovudine (AZT) treatment of patients with AIDS and AIDS-related complex. N Engl J Med 319: 1573–1578.

Selnes OA, Galai N, McArthur JC, et al (1997). HIV infection and cognition in intravenous drug users: long-term follow-up. Neurology 48: 223–230.

Sevigny JJ, Albert SM, McDermott MP, et al (2004). Evaluation of NIV RNA and markers of immune activation as predictors of HIV-associated dementia. Neurology 63: 2084–2090.

Shah SS, Rodriguez T, McGowan JP (2003). Miller Fisher variant of Guillain-barre syndrome associated with lactic acidosis and stavudine therapy. Clin Infect Dis 36: e131–e133.

Shaikh S, Ta C, Bashma AA, et al (2001). Leber hereditary optic neuropathy associated with antiretroviral therapy for human immunodeficiency virus infection. Am J Ophthalmol 131: 143–145.

Shanbhag MC, Rutstein RM, Zaoutis T, et al (2005). Neurocognitive functioning in pediatric human immunodeficiency virus infection. Arch Pediatr Adolsec Med 159: 651–656.

Sidtis JJ, Gatsonis C, Price R, et al (1993). Zidovudine treatment of AIDS dementia complex: results of a placebo controlled trial. Ann Neurol 33: 343–349.

Simpson DM (1999). Human immunodeficiency virus-associated dementia: review of pathogenesis, prophylaxis, and treatment studies of zidovudine therapy. Clin Infect Dis 29: 19–34.

Singhal PC, Sharma P, Kapasi AA, et al (1998). Morphine enhances macrophage apoptosis. J Immunol 160: 1886–1893.

Staprans S, Marlowe N, Glidden D, et al (1999). Time course of cerebrospinal fluid responses to antiretroviral therapy: evidence for variable compartmentalization of infection. AIDS 13: 1051–1061.

Stiene-Martin A, Zhou R, Hauser KF (1998). Regional, developmental, and cell cycle-dependent differences in mu, delta, and kappa opioid receptor expression among cultures mouse astrocytes. Glia 22: 249–259.

Suarez S, Baril L, Stankoff B, et al (2001). Outcome of patients with HIV-1-related cognitive impairment on highly active antiretroviral therapy. AIDS 15: 195–200.

Tardieu M, LeChenadee J, Persoz A, et al (2000). HIV-1-related encephalopathy in infants compared with children and adults. French Pediatric HIV Infection Study and the SEROCO Group. Neurology 54: 1089–1095.

Tassie JM, Gasnault J, Bentata M, et al (1999). Survival improvement of AIDS-related progressive multifocal leukoencephalopathy in the era of protease inhibitors. Clinical Epidemiology Group. French Hospital Database on HIV. AIDS 13: 1881–1887.

Thomas SA (2004). Anti-HIV drug distribution to the central nervous system. Curr Pharm Design 10: 1313–1324.

Tozzi V, Narciso P, Galgani S, et al (1993). Effects of zidovudine in 30 patients with mild to end-stage AIDS dementia complex. AIDS 7: 683–692.

Tozzi V, Balestra P, Galgani S (1999). Positive and sustained effects of highly active antiretroviral therapy on HIV-1 associated neurocognitive impairment. AIDS 13: 1889–1897.

Turchan J, Anderson C, Hauser KF, et al (2001). Estrogen protects against the synergistic toxicity by HIV proteins, methamphetamine and cocaine. BMC Neurosci 2: 3–13.

Vago L, Castagna A, Lazzarin A, et al (1993). Reduced frequency of HIV-induced brain lesions in AIDS treated with zidovudine. J AIDS 6: 42–45.

van der Ven AJ, van Ostenbrugge RJ, Kubat B, et al (2002). Cerebral vasculitis after initiation antiretroviral therapy. AIDS 16: 2362–2364.

Vendrely A, Bienvenu B, Gasnault J, et al (2005). Fulminant inflammatory leukoencephalopathy associated with HAART-induced immune restoration in AIDS-related progressive multifocal leukoencephalopathy. Acta Neuropathol 109: 449–455.

Volberding PA, Lagakos SW, Koch MA, et al (1990). Zidovudine in asymptomatic human immunodeficiency virus infection: a controlled trial in persons with fewer than 500 CD^4-positive cells per cubic millimeter. N Engl J Med 322: 941–949.

Wang G-J, Chang L, Volkow ND, et al (2004). Decreased brain dopaminergic transporters in HIV-associated dementia patients. Brain 127: 2452–2458.

Weinberg A, Wiznia AA, LeFleur BJ, et al (2004). Varicella-zoster virus-specific cell-mediated immunity in HIV-infected children receiving highly active antiretroviral therapy. J Infect Dis 190: 267–270.

Wolters PL, Brouwers P, Moss HA, et al (1994). Adaptive behavior of children with symptomatic HIV infection before and after zidovudine therapy. J Pediatr Psychol 19: 47–61.

Yarchoan R, Brouwers P, Spitzer AR, et al (1987). Response of human-immunodeficiency-virus-associated neurological disease to 3′-azido-3′-deoxythimidine. Lancet 1: 132–135.

Zauli G, Secchiero P, Rodella L, et al (2000). HIV-1 Tat-mediated inhibition of the tyrosine hydroxylase gene expression in dopaminergic neuronal cells. J Biol Chem 275: 4159–4165.

Zhang Y, Marcillat O, Giulivi C, et al (1990). The oxidative inactivation of mitochondrial electron transport chain components and ATPase. J Biol Chem 265: 16330–16336.

Zink MC, Coleman GD, Mankowski JL, et al (2001). Increased macrophage chemoattractant protein-1 in cerebrospinal fluid precedes and predicts simian immunodeficiency virus encephalitis. J Infect Dis 184: 1015–1021.

Further Reading

Adamson DC, Wildermann B, Sasaki MD, et al (1996). Immunologic NO synthase: elevation in severe AIDS dementia and induction by HIV-1 gp41. Science 274: 1917–1920.

Adamson DC, Kopinsky KL, Dawson TM, et al (1999a). Mechanisms and structural determinants of HIV-1 coat protein, gp41-induced neurotoxicity. J Neurosci 19: 64–71.

Ahern KB, Lustig HS, Greenberg DA (1994). Enhancement of NMDA toxicity and calcium responses by chronic exposure of cultured cortical neurons to ethanol. Neurosci Lett 165: 211–214.

Anderson BD, May MJ, Jordan S, et al (2006). Dependence of nelfinavir brain uptake on dose and tissue concentrations of the selective P-glycoprotein inhibitor zosuquidar in rats. Drug Metab Deposit 34: 653–659.

Anderson CE, Tomlinson GS, Pauly B, et al (2003). Relationship of Nef-positive and GFAP-reactive astrocytes to drug use in early and late HIV infection. Neuropath Appl Neurobiol 29: 378–388.

Arendt G, von Giesen HJ (2002). Antiretroviral therapy regimens for neuron-AIDS. Curr Drug Targets Infect Disord 2: 187–192.

Arendt G, Giesen H, Hefter H, et al (2001). Therapeutic effects of nucleoside analogues on psychomotor slowing in HIV infection. AIDS 15: 493–500.

Aukrust P, Liabakk NB, Muller F, et al (1994). Serum levels of tumor necrosis factor-alpha (TNF alpha) and soluble TNF receptors in human immunodeficiency virus type infection-correlations with clinical, immunologic, and virologic parameters. J Infect Dis 169: 420–424.

Bachmeier CJ, Spitzenberger TJ, Elmquist WF, et al (2005). Quantitative assessment of HIV-1 protease inhibitor interactions with drug efflux transporters in the blood-brain barrier. Pharmaceut Res 22: 1259–1268.

Barks JD, Sun R, Malinak C, et al (1995). gp120, an HIV-1 protein, increases susceptibility to hypoglycemic and ischemic brain injury in perinatal rats. Exp Neurol 132: 123–133.

Becker JT, Lopez OL, Dew MA, et al (2004). Prevalence of cognitive disorders differs as a function of age in HIV virus infection. AIDS 18(Suppl 1): S11–S18.

Blankson JN, Persaud D, Silicano RF (2002). The challenge of viral reservoirs in HIV-1 infection. Annu Rev Med 53: 557–593.

Brew BJ (2004). Evidence for a change in AIDS dementia complex in the era of highly active antiretroviral therapy and the possibility of new forms of AIDS dementia complex. AIDS 18(Suppl 1): S75–S78.

Brew BJ, Rosenblum M, Cronin K, et al (1995). AIDS dementia complex and HIV-1 brain infection; clinical–virological correlations. Ann Neurol 38: 563–570.

Brew B, Pemberton L, Cunningham P, et al (1997). Levels of human immunodeficiency virus type 1 RNA in cerebrospinal fluid correlate with AIDS dementia stage. J Infect Dis 175: 963–966.

Brooks PJ (1997). DNA damage, DNA repair, and alcohol toxicity — a review. Alcohol Clin Exp Res 21: 1073–1082.

Burkala EJ, He J, West JT, et al (2005). Compartmentalization of HIV-1 in the central nervous system: role of the choroid plexus. AIDS 19: 675–684.

Caffrey M, Braddock DT, Louis JM, et al (2000). Biophysical characterization of gp41 aggregates suggests a model for the molecular mechanism of HIV-associated neurological damage and dementia. J Biol Chem 275: 19877–19882.

Chandler LJ, Newsom H, Summers C, et al (1993). Chronic ethanol exposure potentates NMDA excitotoxicity in cerebral cortical neurons. J Neurochem 60: 1578–1581.

Chen P, Mayne M, Power C, et al (1997). The Tat protein of HIV-1 induces tumor necrosis factor-α production: implications for HIV associated neurological disease. J Biol Chem 272: 2235–2238.

Chen W, Sulcove J, Frank I, et al (2002). Development of a human cell model for human immunodeficiency virus (HIV)-infected macrophage-induced neurotoxicity: apoptosis induced by HIV type 1 primary isolates and evidence for involvement of the Bcl-e/Bcl-xL-sensitive intrinsic apoptosis pathway. J Virol 76: 9407–9419.

Cheng J, Nath A, Knudsen B, et al (1999). Neuronal excitatory properties of human immunodeficiency virus type 1 tat protein. Neuroscience 82: 97–106.

Corder EH, Robertson K, Lannfelt L, et al (1998). HIV-infected subjects with the E4 allele for APOE have excess dementia and peripheral neuropathy. Nat Med 4: 1182–1184.

Cutler RG, Haughey NJ, Tammara A, et al (2004). Dysregulation of sphingolipid and sterol metabolism by ApoE4 in HIV dementia. Neurology 63: 626–630.

Dingle GA, Oei TP (1997). Is alcohol a cofactor of HIV and AIDS? Evidence from immunological and behavioral studies. Psychol Bull 122: 56–71.

Drejer J, Meier E, Schousboe A (1983). Novel neuron-related regulatory mechanisms for astrocytic glutamate and GABA high affinity uptake. Neurosci Lett 37: 301–306.

Dreyer EB, Kaiser PK, Offermann JT, et al (1990). HIV-1 coat protein neurotoxicity prevented by calcium channel antagonists. Science 20: 364–367.

Du Pasquier RA, Clark KW, Smith PS, et al (2001). JCV-specific cellular immune response correlates with a favorable clinical outcome in HIV-infected individuals with progressive multifocal leukoencephalopathy. J NeuroVirol 7: 318–322.

Du Pasquier RA, Koralnik IJ (2003). Inflammatory reaction in progressive multifocal leukoencephalopathy: harmful or beneficial? J NeuroVirol 9(Suppl 1): 25–31.

Du Pasquier et al (2001).

Ekdahl CT, Claasen JH, Bonde S, et al (2003). Inflammation is detrimental for neurogenesis in adult brain. Proc Natl Acad Sci USA 100: 13632–13637.

Ellis RJ, Gamst AC, Cappraelli E, et al (2000). Cerebrospinal fluid HIV RNA originates from both local CNS and systemic sources. Neurology 54: 927–946.

Ewart GD, Sutherland T, Gage PW, et al (1996). The vpu protein of human immunodeficiency virus type 1 forms cation-selective ion channels. J Virol 70: 7108–7115.

Fantini J, Garmy N, Mahfoud R, et al (2002). Lipid rafts: structure, function and role in HIV, Alzheimer and prion diseases. Expert Rev Mol Med 2002: 1–22.

Foga IO, Nath A, Hasinoff BB, et al (1997). Antioxidants and dipyrimadole inhibit HIV-1 gp120-induced free radical-based oxidative damage to human monocytoid cells. J Acquir Immune Defic Syndr Hum Retrovirol 16: 223–229.

Garden GA, Guo W, Jayadev S, et al (2004). HIV associated degeneration requires p53 in neurons and microglia. FASEB J 18: 1141–1143.

Gasnault J, Kaharaman M, de Herve G, et al (2003). Critical role of JC virus-specific CD^4 T-cell responses in preventing progressive multifocal leukoencephalopathy. AIDS 17: 1443–1449.

Gay GW, Drye TJ, Dick GW, et al (1997). The application of multifactorial cluster analysis in the staging of plaques in early multiple sclerosis. Identification and characterization of the primary demyelinating lesion. Brain 120: 1461–1483.

Gisslen M, Norkrans G, Svennerholm B, et al (1997). The effect on cerebrospinal fluid HIV RNA levels after initiation of zidovudine or didanosine. J Infect Dis 175: 434–437.

Gonzalez ME, Carrasco L (1998). The human immunodeficiency virus type 1 vpu protein enhances membrane permeability. Biochemistry 37: 13710–13719.

Goswami R, Dawson SA, Dawson G (2000). Multiple polyphosphoinositide pathways regulate apoptotic signaling in a dorsal root gangli-derived cell line. J Neurosci Res 59: 136–144.

Groothuis DR, Levy RM (1997). The entry of antiviral and antiretroviral drugs into the central nervous system. J NeuroVirol 3: 387–400.

Gulick RM, Mellors J, Havlir D, et al (1997). Treatment with indinavir, zidovudine and lamivudine in adults with human immunodeficiency virus infection and prior antiretroviral therapy. N Engl J Med 337: 734–739.

Haughey NJ, Nath A, Chan SL, et al (2002). Disruption of neurogenesis by amyloid beta-peptide, and perturbed neural progenitor cell homeostasis in models of Alzheimer's disease. J Neurochem 83: 1509–1524.

Haughey NJ, Cutler RGM, Tanara A, et al (2004). Perturbation of sphingolipid metabolism and ceramide production in HIV-dementia. Ann Neurol 25: 257–267.

Hayman M, Arbuthnott G, Harkiss G, et al (1993). Neurotoxicity of peptide analogues of the transactivating protein tat from Maedi-Visna virus and human immunodeficiency virus. Neuroscience 53: 1–6.

Hinkin CH, Castellon SA, Hardy DJ, Farinpour DJ, et al (2001). Methylphenidate improves HIV-1 associated cognitive slowing. J Neuropsychiatry Clin Neurosci 13: 248–254.

Hong M, Schlichter L, Bendayan R (2000). A Na^+-dependent nucleoside transporter in microglia. J Pharmacol Exp Ther 292: 366–374.

Hu XJ, Ticku MK (1995). Chronic ethanol treatment upregulates the NMDA receptor function and binding in mammalian cortical neurons. Brain Res Mol Brain Res 30: 347–356.

Ilyin SE, Plata-Salaman CR (1997). HIV-1 envelope glycoprotein 120 regulates brain IL-1 beta system and TNF-alpha mRNAs in vivo. Brain Res Bull 44: 67–73.

Jana A, Pahan K (2004). Human immunodeficiency virus type 1 gp120 induces apoptosis in human primary neurons through redox-regulated activation of neutral sphingomyelinase. J Neurosci 24: 9531–9540.

Johnson-Greene D, Adams KM, Gilman S, et al (1997). Effects of abstinence and relapse upon neuropsychological function and cerebral glucose metabolism in severe chronic alcoholism. J Clin Exp Neuropathol 19: 378–385.

Johnston JB, Zhang JB, Silva C, et al (2001). HIV-1 Tat neurotoxicity is prevented by matrix metalloproteinase inhibitors. Ann Neurol 49: 230–241.

Kao AW, Price RW (2004). Chemokine receptors, neural progenitor cells, and the AIDS dementia complex. J Infect Dis 190: 211–215.

Kaul M, Lipton SA (1999). Chemokines and activated macrophages in HIV gp120-induced neuronal apoptosis. Proc Natl Acad Sci USA 96: 8212–8216.

Kaul M, Lipton SA (2004). Signaling pathways to neuronal damage and apoptosis in human immunodeficiency virus type 1-associated dementia: chemokine receptors, excitotoxicity, and beyond. J NeuroVirol 10: S97–S101.

Kepler TB, Perelson AS (1998). Drug concentration heterogeneity facilitates the evolution of drug resistance. Proc Natl Acad Sci USA 95: 11514–11519.

Khurdayan VK, Buch S, El-Hage N, et al (2004). Preferential vulnerability of astroglia and glial precursors to combined opioid and HIV-1 Tat exposure in vitro. Eur J Neurosci 19: 3171–3182.

Koller H, von Giesen HJ, Schaal H, et al (2001). Soluble cerebrospinal fluid factors induce Ca^{2+} dysregulation in rat cultured cortical astrocytes in HIV-1 associated dementia complex. AIDS 15: 1789–1792.

Kort JJ (1998). Impairment of excitatory amino acid transport in astroglial cells infected with the human immunodeficiency virus type 1. AIDS Res Hum Retrovir 14: 1329–1339.

Koutsilieri E, Scheller C, ter Meulen V, et al (2004). Monoamine oxidase inhibition and CNS immunodeficiency infection. NeuroTox 25: 267–270.

Krathwohl MD, Kaiser JL (2004). HIV-1 promotes quiescence in human neural progenitor cells. J Infect Dis 190: 216–226.

Kruman II, Nath A, Maragos WF, et al (1999). Evidence that Par-4 participates in the pathogenesis of AIDS dementia. Am J Pathol 155: 39–46.

Kuhn HG, Dickinson-Anson H, Gage FH (1996). Neurogenesis in the dentate gyrus of the adult rat: age-related decrease of neuronal progenitor proliferation. J Neurosci 16: 2027–2033.

Kutsch O, Oh J, Nath A, et al (2000). Induction of the chemokines interleukin-8 and IP-10 by human immunodeficiency virus type 1 tat in astrocytes. J Virol 74: 9214–9221.

Lannuzel A, Lledo PM, Lamghitnia HO, et al (1995). HIV-1 envelope proteins gp120 and gp160 potentiate NMDA-induced $[Ca^{2+}]_i$ increase, alter $[Ca^{2+}]_i$ homeostasis and induce neurotoxicity in human embryonic neurons. Eur J Neurosci 7: 2285–2293.

Levi G, Patrizio M, Bernado A, et al (1993). Human immunodeficiency virus coat protein gp120 inhibits beta-adrenergic regulation of astroglial and microglial functions. Proc Natl Acad Sci USA 90: 1541–1545.

Levy DN, Refaeli Y, MacGregor RR, et al (1994). Serum vpr regulates productive infection and latency of human immunodeficiency virus type 1. Proc Natl Acad Sci USA 91: 10873–10877.

Lipton SA, Sucher NJ, Kaiser PK, et al (1991). Synergistic effects of HIV coat protein and NMDA receptor-mediated neurotoxicity. Neuron 7: 111–118.

Lipton SA, Yeh M, Dryer EB (1994). Update on current models of HIV-related neuronal injury: platelet-activating factor, arachidonic acid and nitric oxide. Adv Neuroimmunol 4: 181–188.

Ma M, Nath A (1997). Molecular determinants for cellular uptake of tat protein of human immunodeficiency virus type 1 in brain cells. J Virol 71: 2495–2499.

Mabrouk K, Van Rietschoten J, Vives E, et al (1991). Lethal neurotoxicity in mice of the basic domains of HIV and SIV rev proteins. Study of these regions by circular dichroism. FEBS Lett 289: 13–17.

Maggirwar SB, Tong N, Ramirez S, et al (1999). HIV-1 tat-mediated activation of glycogen synthase kinase-3β contributes to tat-mediated neurotoxicity. J Neurochem 73: 578–586.
Maragos WF, Young KL, Turchan JT, et al (2002). Human immunodeficiency virus-1 Tat protein and methamphetamine interact synergistically to impair striatal dopaminergic function. J Neurochem 83: 955–963.
Maragos WF, Tillman P, Jones M, et al (2003). Neuronal injury in hippocampus with human immunodeficiency virus transactivating protein, Tat. Neuroscience 117: 43–53.
Mattson MP, Haughey NJ, Nath A (2005). Cell death in HIV dementia. Cell Death Different 12: 893–904.
McArthur J, McClernon D, Cronin M, et al (1997). Relationship between HIV-associated dementia and viral load in CSF and brain. Ann Neurol 42: 689–698.
Merrill JE, Koyanagi Y, Zack J, et al (1992). Induction of interleukin-1 and tumor necrosis factor alpha in brain cultures by human immunodeficiency virus type 1. J Virol 66: 2217–2225.
Meucci O, Miller RJ (1996). gp120-induced neurotoxicity in hippocampal pyramidal neuron cultures: protective action of TGF-1. J Neurosci 16: 4080–4088.
Milani D, Mazzoni M, Borgatti P, et al (1996). Extracellular human immunodeficiency virus type-1 tat protein activates phosphatidylinositol 3-kinase in PC12 neuronal cells. J Biol Chem 271: 22961–22964.
Minagar A, Shpashak P, Duran EM, et al (2004). HIV-associated dementia, Alzheimer's disease, multiple sclerosis, and schizophrenia: gene expression review. J Neurol Sci 224: 3–17.
Mollace V, Nottet HS, Clayette P, et al (2001). Oxidative stress and neuroAIDS: triggers, modulators and novel antioxidants. Trends Neurosci 24: 411–416.
Monje ML, Toda H, Palmer TD (2003). Inflammatory blockade restores adult hippocampal neurogenesis. Science 302: 1760–1765.
Nath A, Conant K, Chen P, et al (1999). Transient exposure to HIV-1 Tat protein results in cytokine production in macrophages and astrocytes. A hit and run phenomenon. J Biol Chem 11: 17098–17102.
Navia B, Jordan BD, Price RW (1986a). The AIDS dementia complex: I. Clinical features. Ann Neurol 19: 517–524.
Navia BA, Cho ES, Petito CK, et al (1986b). The AIDS dementia complex: II. Neuropathology. Ann Neurol 19: 525–535.
Nebuloni M, Pellegrinelli A, Ferri A, et al (2001). Beta amyloid precursor protein and patterns of HIV p24 immunohistochemistry in different brain areas of AIDS patients. AIDS 15: 571–575.
Nishino H, Hida H, Takei N, et al (2000). Mesencephalic neural stem (progenitor) cells develop to dopaminergic neurons more strongly in dopamine-depleted striatum than in intact striatum. Exp Neurol 164: 209–214.
Nuttall JJC, Wilmshurst JM, Ndondo AP, et al (2004). Progressive multifocal leukoencephalopathy after initiation of highly active antiretroviral therapy in an immunodeficiency virus infection: a case of immune reconstitution inflammatory syndrome. Pediatr Infect Dis 23: 683–685.
Orsini MJ, Debouck CJ, Webb CL, et al (1996). Extracellular human immunodeficiency virus type 1 tat proteins promotes aggregation and adhesion of cerebellar neurons. J Neurosci 16: 2546–2552.
Overholser ED, Coleman GD, Bennett JL, et al (2003). Expression of simian immunodeficiency virus (SIV) nef in astrocytes during acute and terminal infection and requirement of nef for optimal replication of neurovirulent SIV in vitro. J Virol 77: 6855–6866.
Patel CA, Mukhtar M, Ponerantz RJ (2000). Human immunodeficiency virus type 1 Vpr induces apoptosis in human neuronal cells. J Virol 74: 9717–9726.
Patel CA, Mukhtar M, Harley S, et al (2002). Lentiviral expression of HIV-1 Vpr induces apoptosis in human neurons. J NeuroVirol 8: 86–99.
Patrick MK, Johnston JB, Poser C (2002). Lentiviral neuropathogenesis: comparative neuroinvasion, neurotropism, neurovirulence and host neurosusceptibility. J Virol 76: 7923–7931.
Patton HK, Zhou ZH, Bubien JK, et al (2000). Gp120-induced alterations of human astrocyte function: Na(+)H(+) exchange, K(+) conductance, and glutamate flux. Am J Physiol 279: C700–C708.
Pendergast MA, Harris BR, Blanchard JA II, et al (2000a). In vitro effects of ethanol withdrawal and spermidine on viability of hippocampus from male and female rat. Alcohol Clin Exp Res 24: 1855–1861.
Pendergast MA, Harris BR, Mayer S, et al (2000b). Chronic, but not acute, nicotine exposure attenuates ethanol withdrawal-induced hippocampal damage in vitro. Alcohol Clin Exp Res 24: 1583–1592.
Persaud-Sawin DA, McNamara JO, Ryolva S, et al (2004). A galactosylceramide binding domain is involved in trafficking of CLN3 from golgi to rafts via recycling endosomes. Pediatr Res 54: 449–463.
Petito CK, Roberts B (1995). Evidence of apoptotic cell death in HIV encephalitis [see comments]. Am J Pathol 146: 1121–1130.
Phillipon V, Vellutini C, Gambarelli D, et al (1994). The basic domain of the lentiviral Tat protein is responsible for damages in mouse brain: involvement of cytokines. Virology 205: 519–529.
Pillai R, Nair BS, Watson RR (1991). AIDS, drugs of abuse and the immune system: a complex immunotoxicological network. Arch Toxicol 65: 609–617.
Piller SC, Ewart GD, Jans DA, et al (1999). The amino-terminal region of vpr from human immunodeficiency virus type 1 forms ion channels and kills neurons. J Virol 73: 4230–4238.
Ramirez SH, Sanchez JF, Dimitri CA, et al (2001). Neurotrophins prevent HIV Tat-induced neuronal apoptosis via a nuclear factor-kappaB (NF-kappaB)-dependent mechanism. J Neurochem 78: 874–889.
Rausch DM, Heyes MP, Murray EA, et al (1994). Cytopathologic and neurochemical correlates of progression to

motor/cognitive impairment in SIV-infected rhesus monkeys. J Neuropathol Exp Neurol 53: 165–175.

Rice QC, Bullock MR, Shelton KL (2004). Chronic ethanol consumption transiently reduces adult progenitor cell proliferation. Brain Res 11: 94–98.

Robertson K, Fiscus S, Kapoor C, et al (1998). CSF, plasma viral load and HIV associated dementia. J Neurovirol 4: 90–94.

Sabatier JM, Vives E, Marbourk E, et al (1991). Evidence for neurotoxicity of Tat from HIV. J Virol 65: 961–967.

Sacktor N, Tartwater PM, Skolasky RL, et al (2001). CSF antiretroviral drug penetrance and the treatment of HIV-associated psychomotor slowing. Neurology 57: 542–544.

Sadar A, Rubocki RJ, Horvath JA, et al (2002). Fatal immune restoration disease in human immunodeficiency virus type1-infected patient with progressive multifocal leukoencephalopathy: impact of antiretroviral therapy-associated immune reconstitution. Clin Infect Dis 35: 1250–1257.

Seth P, Major EO (2005). Human brain derived cell culture models of HIV-1 infection. Neurotox Res 8: 1–8.

Shors TJ, Miesegaes G, Beylin A, et al (2002). Neurogenesis in the adult is involved in the formation of trace memories. Nature 28: 1030–1104.

Singhal PC, Reddy K, Frankl N, et al (1997). Morphine induces splenocyte apoptosis and enhanced mRNA expression of catharsis-B. Inflammation 21: 609–617.

Solas C, Lafeuillade A, Halfon P, et al (2003). Discrepancies between protease inhibitor concentrations and viral load in reservoirs and sanctuary sites in human immunodeficiency virus-infected patients. Antimicrob Agents Chemother 47: 238–243.

Stanley LC, Mark RE, Woody RC, et al (1994). Glial cytokines as neuropathogenic factors in HIV infection: pathogenic similarities to Alzheimer's disease. J Neuropathol Exp Neurol 53: 231–238.

Sung JH, Shin SA, Park HK, et al (2001). Protective effect of glutathione in HIV-1 lytic peptide 1-induced cell death in human neuronal cells. J NeuroVirol 7: 454–465.

Sweetman PM, Saab OH, Wroblewski JT, et al (1993). The envelope glycoprotein of HIV-1 alters NMDA receptor function. Eur J Neurosci 5: 276–283.

Tabakoff B (1994). Alcohol and AIDS: is the relationship all in our heads? Alcohol Clin Exp Res 18: 415–416.

Tashima KT, Caliendo AM, Ahmad M, et al (1999). Cerebrospinal fluid human immunodeficiency virus type 1 (HIV-1) suppression and efavirenz drug concentrations in HIV-1 infected patients receiving combination therapy. J Infect Dis 180: 862–864.

Thomas SA, Segal MB (1997). The passage of azidodeoxythymidine into and within the central nervous system: does it follow the parent compound, thymidine? J Pharmacol Exp Ther 281: 1211–1218.

Thompson KA, Churchill MJ, Gorry PR, et al (2004). Astrocyte specific viral strains in HIV dementia. Ann Neurol 56: 873–877.

Toggas SM, Masliah E, Rockenstein EM, et al (1994). Central nervous system damage produced by expression of the HIV-1 coat protein gp120 in transgenic mice. Nature 367: 188–193.

Torres-Munoz J, Stockton P, Tacoronte N, et al (2001). Detection of HIV-1 gene sequences in hippocampal neurons isolated from postmortem AIDS brains by laser capture microdissection. J Neuropath Exp Neuro 60: 885–892.

Trillo-Pazos G, McFarlane-Abdulla E, Campbell IC, et al (2000). Recombinant nef HIV-IIIB protein is toxic to human neurons in culture. Brain Res 864: 315–326.

Trono D (1995). HIV accessory proteins: leading role for the supporting cast. Cell 82: 189–192.

Turchan J, Pocernich CB, Gairola C, et al (2003a). Oxidative stress in HIV demented patients and protection *ex vivo* with novel antioxidants. Neurology 28: 307–314.

Tyor WR, Middaugh LD (1999). Do alcohol and cocaine abuse alter the course of HIV-associated dementia complex? J Leukoc Biol 65: 475–481.

van de Bovenkamp M, Nottet HS, Pereira CF (2002). Interactions of human immunodeficiency virus-1 proteins with neurons: possible role in the development of human immunodeficiency virus-1-associated dementia. Eur J Clin Invest 32: 619–627.

van Marle G, Henry S, Todoruk T, et al (2004). Human immunodeficiency virus type 1 Nef protein mediates neural cell death: a neurotoxic role for IP-10. Virology 329: 302–318.

van Prang RME, Weverling GJ, Portegies P, et al (2000). Enhanced penetration of indinavir in cerebrospinal fluid and semen after the addition of low-dose ritonavir. AIDS 14: 1187–1194.

van Toorn R, Rabie H, Warwick JM (2005). Opsoclonus-myoclonus in an HIV-infected child on antiretroviral therapy-possible immune reconstitution syndrome. Eur J Paediat Neurol 9: 423–426.

Wahl LM, Corcoran ML, Pyle SW, et al (1989). Human immunodeficiency virus glycoprotein (gp120) induction of monocyte arachidonic acid metabolites and interleukin 1. Proc Natl Acad Sci USA 86: 621–625.

Walker DW, Heaton MG, Lee N, et al (1993). Effect of chronic ethanol on the septohippocampal system: a role for neurotrophic factors? Alcohol Clin Exp Res 17: 12–18.

Wang J, Schaner ME, Thomassen S, et al (1997). Functional and molecular characterisitics of na+-dependent nucleoside transporters. Pharm Res 14: 1524–1531.

Watzl B, Watson RR (1992). Role of alcohol abuse in nutritional immunosuppression. J Nutr 122: 733–737.

Werner T, Ferroni S, Saermark T, et al (1991). HIV-1 nef protein exhibits structural and functional similarity to scorpion peptides interacting with K^+ channels. AIDS 5: 301–308.

Williams KC, Hickey WF (2002). Central nervous system damage, monocytes and macrophages, and neurological disorders in AIDS. Annu Rev Neurosci 25: 537–562.

Wu X, Hui AC, Giacomini KM (1993). Formycin B elimination from the cerebrospinal fluid of the rat. Pharm Res 10: 611–615.

Wu X, Gutierrez NM, Giacomini KM (1994). Further characterization of the sodium-dependent nucleoside transporter (N3) in choroid plexus from rabbit. Biochim Biophys Acta 1191: 190–196.

Wynn HE, Brundage RC, Glecher CV (2002). Clinical implications of CNS penetration of antiretroviral drugs. CNS Drugs 16: 595–609.

Yao R, Cooper GM (1995). Requirement for phosphatidylinositol-3 kinase in the prevention of apoptosis by nerve growth factor. Science 267: 2003–2006.

Zachary I, Rozengurt E (1992). Focal adhesion kinase ($p125^{FAK}$): a point of convergence in the action of neuropeptides, integrins, and oncogenes. Cell 71: 891–894.

Zauli G, Gibellini D, Milani D, et al (1993). Human immunodeficiency virus type 1 tat protein protects lymphoid, epithelial and neuronal cell lines from death by apoptosis. Cancer Res 53: 4481–4485.

Zauli G, La Placa M, Vignoli M, et al (1995). An autocrine loop of HIV type-1 tat protein responsible for the improved survival/proliferation capacity of permanently tat-transfected cells and required for optimal HIV-1 LTR transactivating activity. J AIDS 10: 306–316.

Zauli G, Milani D, Mirandola P, et al (2001). HIV-1 Tat protein down-regulates CREB transcription factor expression in PC 12 neuronal cells through a phosphatidylinositol 3-kinase AKT/cyclic nucleoside phosphodiesterase pathway. FASEB J 15: 483–491.

Zhou X-J, Havlir DV, Richman DD, et al (2000). Plasma population pharmacokinetics and penetration into cerebrospinal fluid of indinavir in combination with zidovudine and lamivudine in HIV-1-infected patients. AIDS 14: 2869–2976.

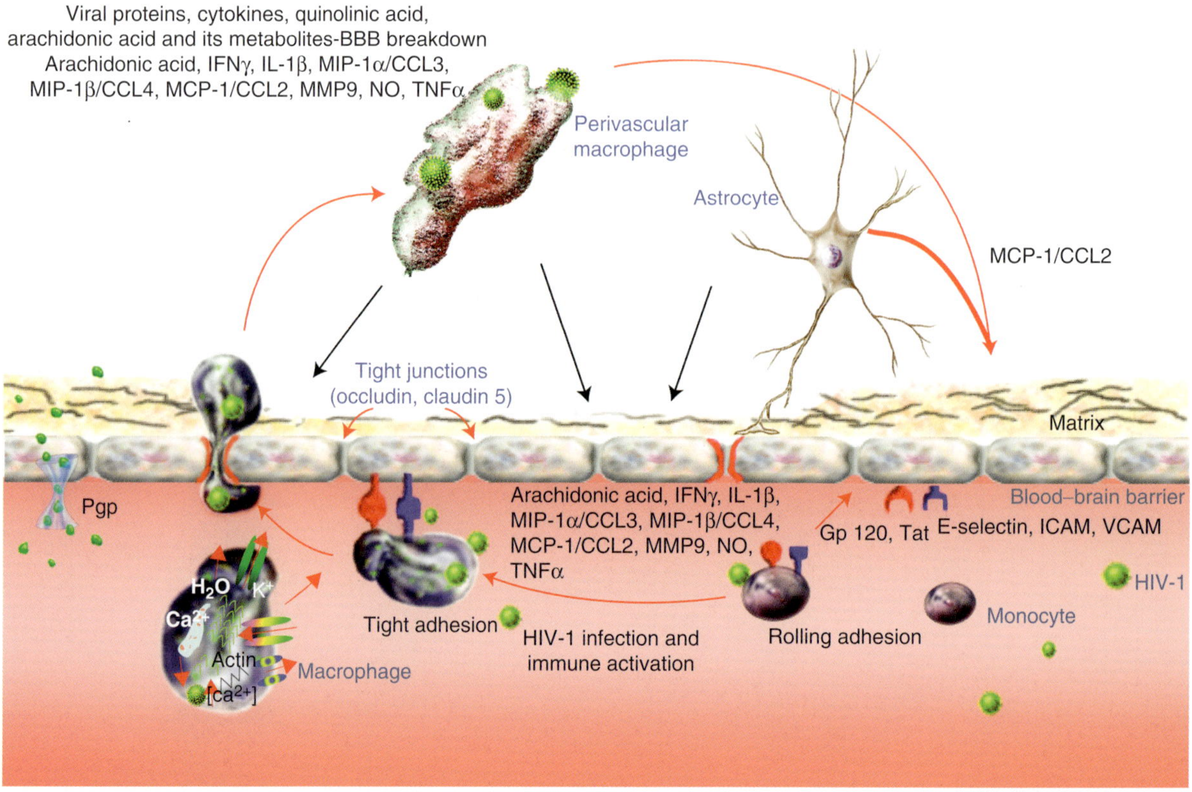

Fig. 4.1. BBB and HIV-1 entry into an inflamed CNS. The structure of the BBB consists of matrix and BMVEC connected by tight junctions consisting of claudin-5 and occludin proteins among others. Astrocyte end-feet are in close proximity to the matrix and the BMVEC and contribute to the integrity of the BBB. Pgp is an ATP-dependent transporter located in the BMVEC that negatively impacts drug delivery into the CNS through its removal of antiretroviral drugs from the CNS. Entry of virus into the brain and across the BBB involves monocyte–macrophage maturation, ion channel expression, viral infection and immune activation leading to the production and release of viral and cellular toxins that affect the integrity and function of the BBB. The factors released by MPs that damage the BBB include, but are not limited to, viral proteins (Tat and gp120), pro-inflammatory cytokines (including TNF-α and IL-1β), MMP9 and free radicals such as NO. Chemokines such as MCP-1/CCL2 and MIP-1α/β (CCL3/CCL4) are released from astrocytes, microglia, endothelial cells and neurons and attract more macrophages into the brain. Virus-infected and immune-activated macrophages and microglia MPs alter the BBB by increasing the number of cell adhesion molecules present on the BMVEC, promote rolling and tight adhesion of macrophages with BMVEC through upregulation of cell adhesion molecules and changes in cell shape and volume mediated by alterations in ion channels in macrophages. This leads to the transendothelial migration of macrophages and serving as an increased nidus for continued inflammatory activities and perpetuating the entry of virus and macrophages into the brain. (See page 48.)

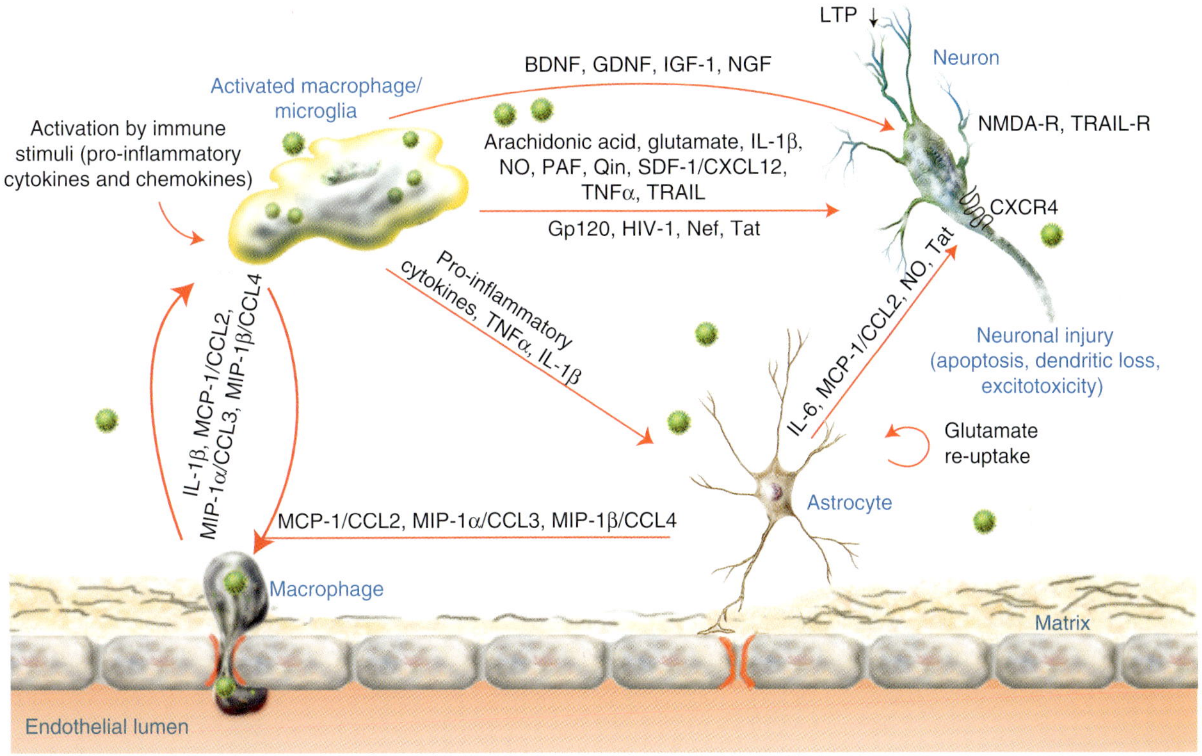

Fig. 4.2. Mechanisms for HIV-1 neuropathogenesis. MP secretory products affect a cascade of immunomodulatory activities that engage neurons and astrocyte effector functions. Activated perivascular macrophages and microglia secrete viral and cellular neurotoxins which include, but are not limited to, viral proteins (Nef, Tat and gp120), pro-inflammatory cytokines (for example, TNF-α and IL-1β), PAF, free radicals such as NO, glutamate-like agonists, and quinolate, arachidonic acid and its metabolites and amines. Chemokines such as MCP-1/CCL2 and MIP-1α/β (CCL3/CCL4) are also produced from glial cells attracting more macrophages into an inflamed brain. The cumulative effects of MP neurotoxic factors include LTP inhibition and neuronal injury through excitotoxicity, apoptosis and dendritic process loss. Astrocyte activation also plays a critical regulatory role in neuronal injury. Innate astrocyte immunity includes the secretion of chemokines, cytokines and ROS (such as MCP-1/CCL2, IL-6 and NO). (See page 51.)

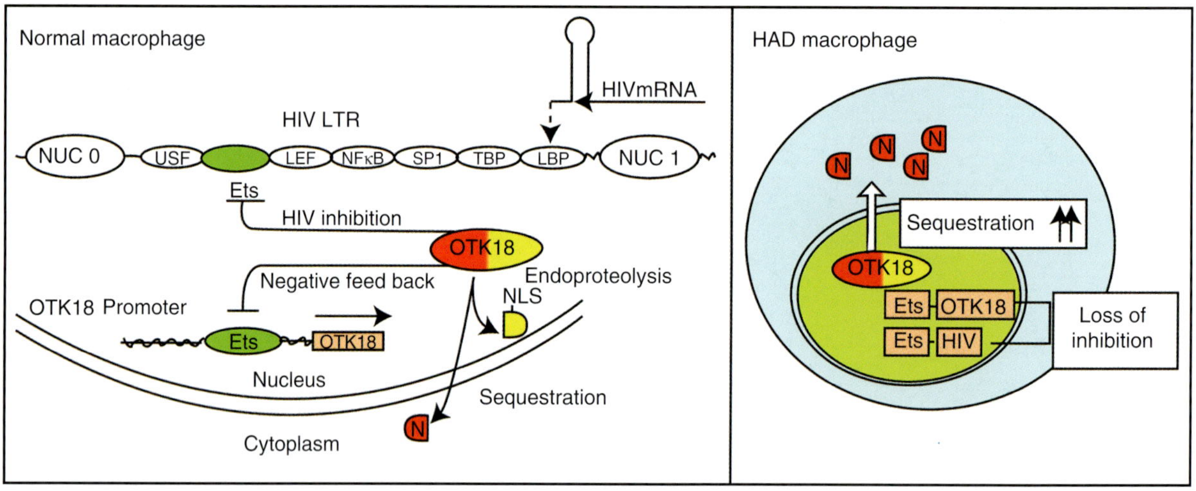

Fig. 4.3. Regulation and antiretroviral activities of OTK18). In macrophages, HIV-1 infection induces OTK18 expression mediated through a proximal promoter region (Ets). Viral suppression results when OTK18 binds to the HIV-1 LTR Ets. Transient expression of OTK18 occurs as the result of a negative feedback where OTK18 binds to an Ets proximal to its own transcriptional start site. In HIVE, endoproteolysis of OTK18 is enhanced due to chronic viral infection and inflammation, leading to increased endoproteolysis and accumulation of N-terminal fragments in the cytosol. This results in failure of OTK18 suppression of the HIV-1 LTR and OTK18 promoter. As a result the cytoplasmic accumulation of OTK18 in macrophages may serve as a predictor of advanced HIVE and HAD. (See page 57.)

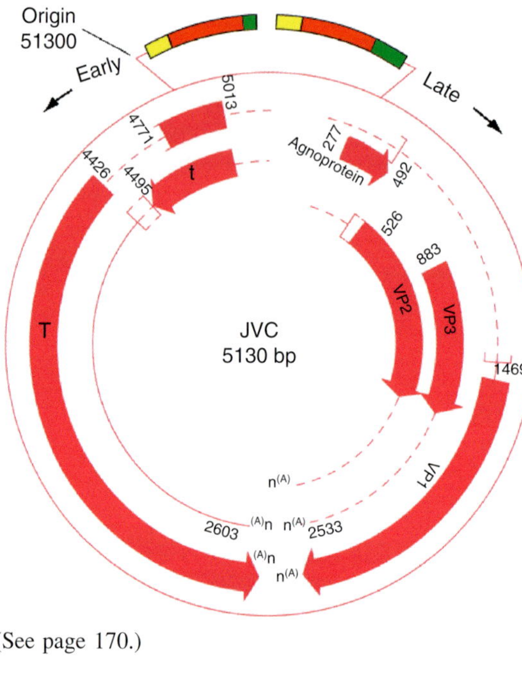

Fig. 13.1. The JC virus genome. (See page 170.)

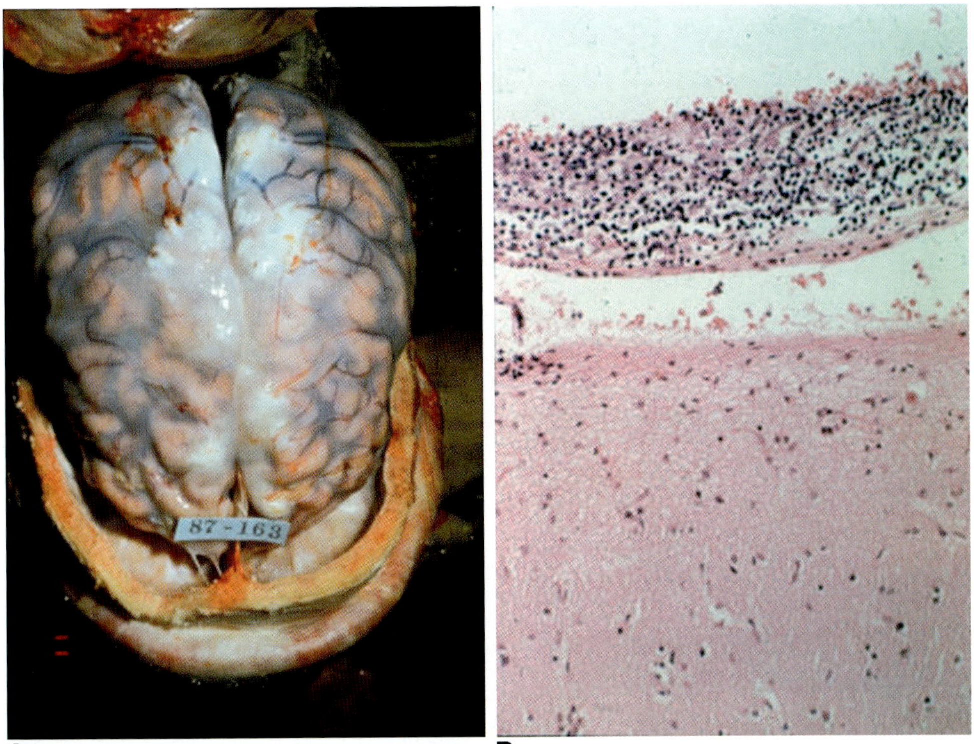

A B

Fig. 14.1. Syphilitic meningitis. A. Gross pathology of syphilitic meningitis with opacification of the meninges. B. Histopathological examination showing lymphocytic infiltration of the meninges. (See page 190.)

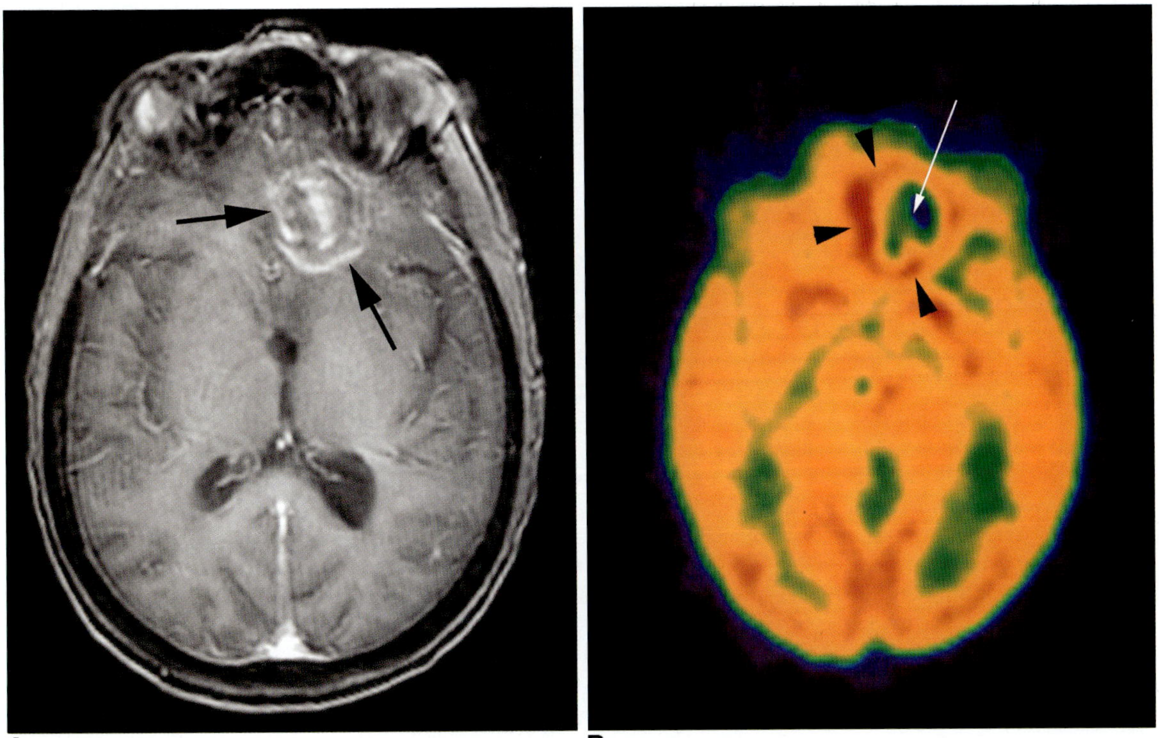

A B

Fig. 16.22. Primary CNS lymphoma, utility of PET imaging. A. Contrast-enhanced axial T1-weighted MR image reveals a ring-enhancing lesion (black arrows) within the inferior portions of the left frontal lobe with central areas of necrosis. B. Axial image from a FDG-18 PET scan reveals hypermetabolic activity associated with the rim of the lesion (arrowheads). There is hypometabolic activity within center of the lesion corresponding to areas of necrosis (white arrow). (See page 251.)

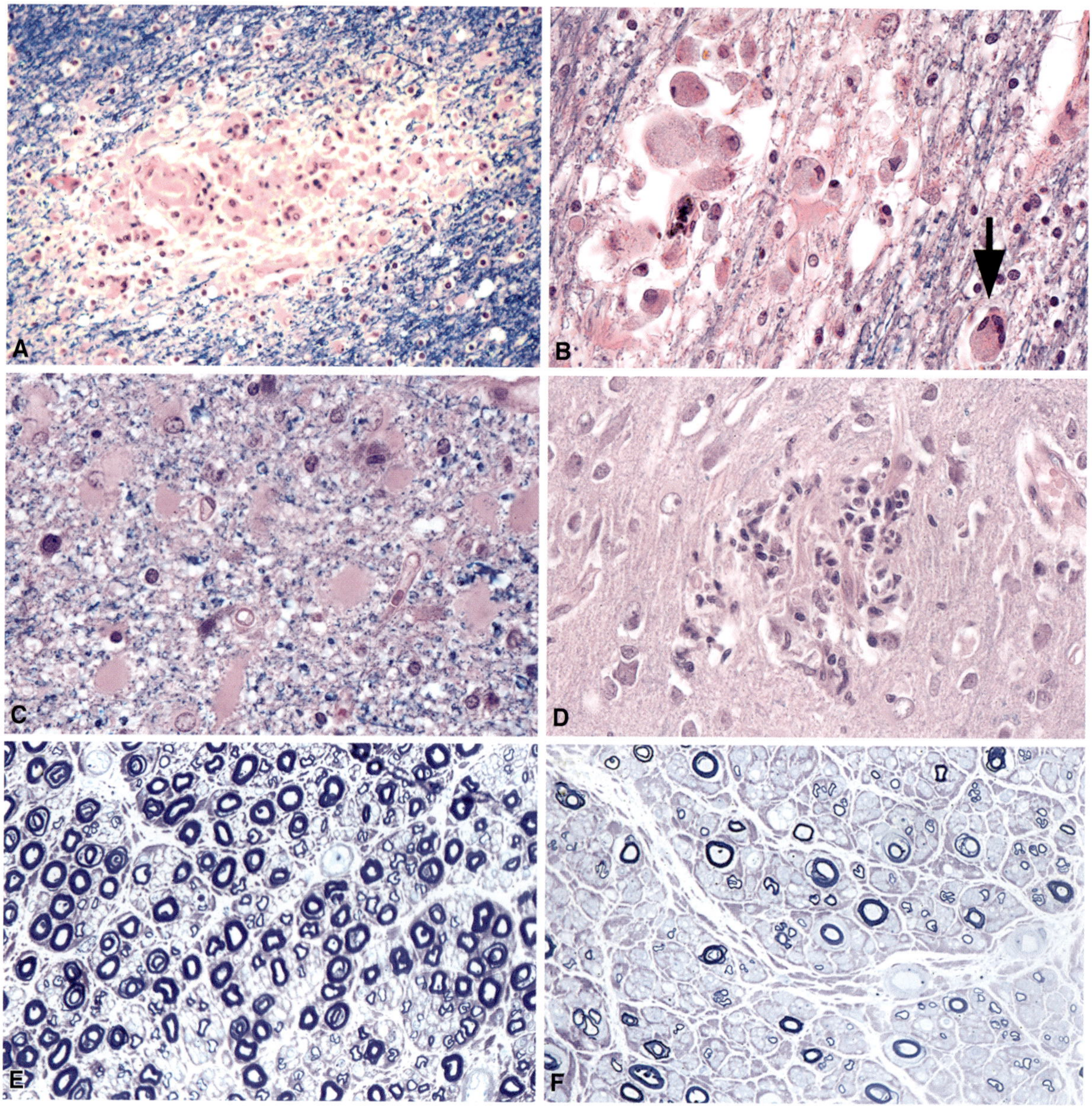

Fig. 18.1. Neuropathological changes in HIV-associated encephalitis and peripheral neuropathy. A. A white matter lesion in HIV encephalitis contains numerous macrophages and microglial cells. Note the loss of blue-stained myelin in this lesion. B. A lesion in HIV encephalitis contains abundant plump macrophages; some contain more than one nucleus. A typical HIV-infected multinucleated cell is indicated by the arrow. C. Blue-stained myelin contains many abundant plump reactive astrocytes (pink areas). Astrocytes can support a nonproductive HIV infection and might participate in the pathophysiology of HIV encephalitis. D. A large microglial nodule is shown in brain gray matter. These clusters of microglial cells are usually present in HIV encephalitis, and often are immunostained with antibodies against HIV antigens. E, F. Cross-sections of the sural nerve from two HIV-infected subjects. E shows a subject that did not have neuropathy; F is from a subject with distal sensory peripheral neuropathy (DSP) and reveals a lack of myelinated nerve fibers. DSP is not in general correlated with the amount of HIV replication in the nerve tissue. (A–D: Luxol Fast Blue/hematoxylin/eosin stain; E, F: Semithin sections stained with toluidine blue.) (See page 295.)

Index

Note: Page numbers in bold relate to figures or tables